MW00835476

Scale Relativity and Fractal Space-Time

A New Approach to Unifying Relativity and Quantum Mechanics

Scale Relativity and Fractal Space-Time

A New Approach to Unifying Relativity and Quantum Mechanics

Laurent Nottale

Paris-Meudon Observatory, France

Imperial College Press

Published by

Imperial College Press
57 Shelton Street
Covent Garden
London WC2H 9HE

Distributed by

World Scientific Publishing Co. Pte. Ltd.

5 Toh Tuck Link, Singapore 596224

USA office: 27 Warren Street, Suite 401-402, Hackensack, NJ 07601

UK office: 57 Shelton Street, Covent Garden, London WC2H 9HE

Library of Congress Cataloging-in-Publication Data
Nottale, Laurent.
    Scale relativity and fractal space-time : a new approach to unifying relativity and quantum
mechanics / Laurent Nottale.
        p. cm.
    ISBN 978-1-84816-650-9 (hardcover)
    1. Space and time. 2. Fractals. 3. Relativity (Physics) I. Title.
    QC173.59.S65N675 2011
    530.11--dc22
                                                                    2010053905

British Library Cataloguing-in-Publication Data
A catalogue record for this book is available from the British Library.

Typeset by Stallion Press
Email: enquiries@stallionpress.com

Printed in Singapore by World Scientific Printers.

PREFACE

This book aims at giving the state of the art in the development of the theory of the relativity of scales. The basis and principles of this new theoretical approach were first presented in 1993 in the book entitled *Fractal Space-Time and Microphysics: Towards a Theory of Scale Relativity* [353]. Since then, the theory has been developed and extended, and applications have been proposed in various sciences. In many cases, the theoretical predictions that have been made at that time have now been put to the test with positive results on new experimental and observational data. It seemed, therefore, necessary to collect these developments and results in a monography, so that the interested readers could more easily find the information, which is presently scattered in many different papers and books.

I am, therefore, particularly grateful to Laurent Chaminade, Publisher at Imperial College Press, for his invitation to publish such a project. I also warmly thank Charles Alunni, Pierre Anglès, Charles Antoine, Charles Auffray, Fayçal Ben Adda, Vincent Bontems, Eric Brian, Jean-Eric Campagne, Roland Cash, David Ceccolini, Marie-Noëlle Célérier, Jean Chaline, Jacky Cresson, Daniel da Rocha, Laurent di Menza, Hans Diebner, Bérengère Dubrulle, Eric Eveno, Cédric Foellmi, Maxime Forriez, Patrick Galopeau, Jean Gay, Pierre Grou, Frédéric Héliodore, Raphaël Hermann, Felix Herrmann, Cédric Jacob, Eric Lefèvre, Serge Lefranc, Thierry Lehner, Philippe Martin, Michel Mendès-France, Marc Pocard, Etienne Rouleau, Gérard Schumacher, Pierre Timar, Nguyet Tran Minh and Matthieu Tricottet, who have participated at a level or another (sometimes high) either to the applications of the theory or to the study of its consequences in various sciences during these seventeen years, and many other colleagues whom I unfortunately cannot all cite, for their encouraging questions, remarks and comments. I acknowledge helpful support and interest in this work from Patrice Abry, Georges Alecian, Salvatore Capozziello, Jacques

Colin, Jean-Patrick Connerade, Jacques Dubois, Daniel Dubois, Mohamed El Naschie, Bernard Fort, Uriel Frisch, Ana Gomez, Peter Hunter, Benoît Mandelbrot, Denis Noble, Garnet Ord, Otto Rössler, Evry Schatzman, Gautam Sidharth, Didier Sornette and Peter Weibel, and stimulating discussions with them.

Additional special thanks go to Charles Auffray and Denis Noble for their kind support to the publication of this book, to Marie-Noëlle Célérier, Thierry Lehner and Louis de Montera for their reading of the manuscript and for helpful suggestions, to Stephan LeBohec for his detailed editing of the manuscript and his very useful suggestions of corrections and improvements, and to my wife Béatrice, for her patience and loving support during the writing of this book.

CONTENTS

LIST OF FIGURES

Part I: General Introduction

GENERAL INTRODUCTION

One of the main concerns of this book is the foundation of quantum mechanics. As it is now well known, the principle of relativity (of motion) underlies the foundation of most of classical physics. Up to now, quantum mechanics, though it is harmoniously combined with special relativity in the framework of relativistic quantum mechanics and quantum field theories, is founded on different grounds. Its present foundation, actually, is mainly axiomatic, i.e. it is based on postulates and rules, which are not derived from underlying more fundamental principles.

The theory of scale relativity offers an original solution to this fundamental problem. In its framework, quantum mechanics may indeed be founded on the principle of relativity itself, provided this principle, up to now only applied to position, orientation and motion, be extended to scales. One generalizes the definition of reference systems by explicitly including variables characterizing their scale. Then, one can generalize the possible transformations of these reference systems by adding, to the relative transformations already accounted for (translation, velocity and acceleration of the origin, rotation of the axes), the transformations of the scale variables, namely, their relative dilations and contractions. In the framework of such a further generalized relativity theory, the laws of physics may be given a general form that transcends and includes both the classical and the quantum laws, allowing, in particular, to study in a renewed way the poorly understood nature of the classical to quantum transition.

A related important concern of this book is the question of the geometry of space-time at all scales. In the new framework of scale relativity, and in analogy with Einstein's construction of general relativity of motion, which is based on the generalization of flat space-times to curved Riemannian geometry, it is suggested that a new generalization of the description of space-time is now needed toward a nondifferentiable and fractal geometry. New mathematical and physical tools are therefore developed in order

3

to implement such a generalized description, which goes far beyond the standard view of differentiable manifolds.

The principle of relativity

Let us first make some brief historical remarks about the principle of relativity and the evolution of relativity theories. The relativity of inertial motion was discovered in Western science by Galileo, who expressed it by writing that "for all things that take part equally in it, it does not act, it is as if it were not; [...] the motion is as nothing", and erected it to a principle: "Let us therefore set as a principle that, whatever be the motion that one attributes to the Earth, it is necessary that, for us who [...] partake of it, it remains perfectly imperceptible and as not being" [190].[a]

Einstein's first insight about the relativity of gravitation, which he has qualified "the most fortunate of all his life" is of a similar nature. Namely, he realized in 1907 that if a man moves in free fall in a gravitational field, he no longer feels his own weight (see, e.g. [425]). Einstein has then translated this first statement into the principle of equivalence, according to which a gravitational field and a uniform acceleration field are locally equivalent. Indeed, a gravitational field can be locally canceled or conversely, an apparent gravity field can be created by the choice of an accelerating reference system.

In both cases, the key point raised by Galileo and Einstein is, both for motion and gravitation, their lack of absolute existence. The velocity of a body, its acceleration, and the gravitational force felt by a body are not intrinsic or local physical properties of the body, but relative properties that depend on the reference system. They are not individual properties of objects, but inter-properties between objects, which have therefore the status of relations. By changing the reference system, the property changes (this is relativity), and, in the limit, there is one reference system (namely, the proper reference system), in which the property vanishes (this is "emptiness").

A final general remark about the nature of the principle of relativity concerns its connection with space-time theories. It is also remarkable that

---

[a]It is remarkable that this Galileo statement has been anticipated by the principle of relativity-emptiness of Buddhist philosophy, according to which the various phenomena are empty of proper (absolute) existence and appear only in a relative and interdependent way. For example, one finds in Nagarjuna's 2000 years old writings [327] an analysis of the relativity of motion (among other phenomena), which is almost word by word identical to Galileo's analysis: "The agent of motion does not move", "Motion, its beginning and its cessation are analogous to motion", "The agent of motion, motion, and the place of motion do not exist (according to their proper nature)".

all successive theories of relativity (Galileo's relativity, Einstein–Poincaré special relativity and Einstein general relativity) are naturally achieved in terms of progress in the description of the geometry of space-time (respectively Euclidean space, then Minkowskian and Riemannian space-times). This is easily understandable if one realizes that space-time is defined as an inter-relational level of description between objects, as is manifest in the metric of general relativity, which is a generalization of Pythagoras' relation. The account of more and more general transformations between coordinate systems, therefore, naturally leads to more and more general geometries of space-time.

Now coming back to today's physics, this general view of the principle of relativity and of its link to the evolution of space-time geometric theories reveals to be extremely powerful. Up to now, relative properties of physical objects, such as position, orientation, velocity and acceleration, are incorporated in relativity theories. The inclusion of gravitation has marked another step for physics, since gravitation was considered since Newton as a universal field, while it becomes a relative manifestation of the curved geometry of space-time in Einstein's theory.

But there are nowadays yet many properties of physical objects (masses, spin, charges), fields (electromagnetic, weak and strong fields) and apparently universal phenomena (quantum laws), which are considered not to come under a relativity theory. This does not mean that these properties and phenomena do not satisfy to the present theory of relativity (all of them do have a "relativistic" description in the sense of motion special relativity), but that their very existence is still not founded on the principle of relativity-emptiness, but is derived most of the time from experiment only.

Premise of scale relativity theory

Following such a line of thought, I was led in the years 1979–1980 to suggest that the quantum properties could be the result of a new manifestation of the principle of relativity, generalized to a scale-dependent, i.e. fractal, geometry of space-time [387]. The first paper on this proposal was submitted for publication in 1981 under the title *Fractals and Nonstandard Analysis: a model for the microstructure of space*, but its content was finally published later [344, 349], then extended in *Fractal Space-Time and Microphysics* [353].

This suggestion was based on the remark that the failure of a large number of attempts to understand the quantum behavior in terms of standard differentiable geometry indicated that a possible "quantum geometry" should be of a completely new nature. Moreover, following the lessons of Einstein's construction of a geometric theory of gravitation, it was clear that any geometric property to be attributed to space-time itself, and not only to particular objects or systems, was necessarily universal.

Fortunately, the founders of quantum theory had already brought to light a universal and fundamental behavior of the quantum realm in contrast with the classical world: the explicit dependence of the measurement results on the apparatus resolution, as described by the Heisenberg uncertainty relations.

This motivated me to ask the following two questions: Was it possible to describe intrinsically a space-time whose geometry would be explicitly dependent on the scale of observation? Could such a geometry be able to give rise to the quantum behavior of the objects embedded into it and participating to it?

Now the concept of a scale-dependent geometry (at the level of objects and media) had already been introduced and developed by Benoit Mandelbrot, who coined the word "fractal" to describe it [305, 306]. The new program amounted to use fractal geometry, not only for describing "objects" (that remain embedded in an Euclidean space), but also for describing in an intrinsic way the geometry of space-time itself.

A preliminary work toward such a goal may consist in introducing the fractal geometry in Einstein's equations of general relativity at the level of the source terms. This amounts to give a better description of the density of matter in the Universe accounting for its hierarchical organization and fractality, then to solve Einstein's field equations for such a scale dependent momentum-energy tensor. Such a program remains a challenge to cosmology. A complementary approach consists of considering the consequences of the inhomogeneous distribution of matter through the effects of gravitational lensing by planetary size objects, stars, galaxies, clusters of galaxies and large scale structures of the Universe (see, e.g. the review paper [348] and references therein). Several of these predicted effects have subsequently been discovered observationally, such as the Sachs–Wolfe effect, the global effect of clusters of galaxies as gravitational lenses [247, 339, 499], etc.

But a more direct connection of the fractal geometry with fundamental physics comes from its use in describing not only the distribution of matter in space, but also the geometry of space-time itself. Such a program may be considered as the continuation of Einstein's program of generalization of the geometric description of space-time. In general relativity, the essence of universal gravitation is understood as being the various manifestations of space-time curvature. In the new fractal space-time theory [344, 420, 349, 353, 165], the essence of quantum physics is a manifestation of the nondifferentiable and fractal geometry of space-time.

Another line of thought leading to the same suggestion comes not from relativity and space-time theories but from quantum mechanics itself. Indeed, it has been discovered by Feynman [176] that the typical quantum mechanical paths, i.e. those that contribute in a main way to the path integral, are nondifferentiable and fractal. Namely, Feynman has shown that, although a mean velocity can be defined for these quantum paths, the mean-square velocity, which is given by $\langle v^2 \rangle \propto \delta t^{-1}$, does not exist at any point. One recognizes in this expression the behavior of a curve of fractal dimension $D_F = 2$ [1]. Based on these premises, the reverse proposal, according to which the laws of quantum physics find their very origin in the fractal geometry of space-time, has been developed along three different and complementary approaches.

(i) Ord and co-workers [420, 316, 423, 424], extending the Feynman chessboard model, have worked in terms of probabilistic models, in the framework of the statistical mechanics of binary random walks.

(ii) El Naschie has suggested to give up not only the differentiability, but also the continuity of space-time. This leads him to work in terms of a "Cantorian" space-time [165, 166, 167], and therefore to use in a preferential way the mathematical tool of number theory.

(iii) The scale relativity approach [344, 349, 353, 360], which is studied and developed in the present book, is on the contrary founded on a fundamentally continuous geometry of space-time, which therefore includes the differentiable and nondifferentiable cases, constrained by the principles of motion and scale relativity.

Summary of fractal space-time and microphysics

Let us give a brief summary, adapted from Chapter 1 of the content of *Fractal Space-Time and Microphysics* [353], of which the present book is a continuation and a development.

Many phenomena in *physical, biological and other sciences* are characterized by a fundamental and explicit scale dependence. Hence in quantum mechanics the results of measurements explicitly depend on the resolution of the measurement apparatus, as described by the Heisenberg relations;

in cosmology, it is the whole set of interdistances between the objects of the Universe that depends on a time varying universal scale factor (this is the expansion of the Universe). Moreover, scale laws and scaling behavior are encountered in many situations, at small scales (microphysics), large scales (extragalactic astrophysics and cosmology) and intermediate scales (complex self-organized systems), but most of the time such laws are found in an empirical way, since we still lack a theory allowing us to understand them from fundamental principles.

The proposal that was developed at that time, and which is still at the root of the present work, is that such a fundamental principle upon which a theory of scale laws may be founded is the principle of relativity itself [349, 352]. But, by "principle of relativity" we mean something more general than its application to particular laws: we actually mean a universal method of thought. Following Einstein, it may be expressed by postulating that "the laws of Nature must be such that they apply to reference systems whatever their state" (of position, orientation, motion and now scale).

The new proposal has consisted of applying the principle of relativity to laws of scale. Taking advantage of the relative character of every length and time scales in Nature, one defines the resolution of measurements (more generally, the characteristic scale of a given phenomenon) as the state of scale of the reference system. This allows one to set a principle of scale relativity, according to which the laws of physics must be such that they apply to coordinate systems whatever their state of scale. The mathematical translation of this principle is the requirement of scale covariance of the equations of physics, which means that they should keep their form under scale transformations of the reference system.

In order to describe physical laws complying to these principles of scale relativity and covariance, one needs some mathematical tools able to achieve such a fundamental and explicit dependence of physics on scale in their very definition. Two tools naturally come to mind in this regard: Mandelbrot's fractal geometry, and Wilson's many scales of length approach using the renormalization group. Both tools have been considered in *Fractal Space-Time and Microphysics*, and connected together by the fact that the standard measures on fractals (based on their topological dimension, such as length, area, volume...) can be derived as solutions of *renormalization* group-like equations.

To this end, we have described and developed fractal geometry, including a reminder about fractals and to a first attempt at going beyond fractal objects seen as embedded in Euclidean space, in order to reach an intrinsic definition of the concept of fractal space (more generally, of fractal space-time). New mathematical tools for physics have been introduced, such as non-standard analysis and scale-dependent fractal functions, that allowed to deal with the nondifferentiability and the infinities characterizing fractal geometry.

Then we considered the behavior of quantum mechanical paths in the light of the fractal tool. We have shown that the Heisenberg relations can be translated in terms of a fractal dimension of all four space-time coordinates jumping from $D_F = 2$ in the quantum and quantum relativistic domain to an effective value $D_F = 1$ in the classical domain (which corresponds in this case to nonfractality), the transition being identified as the de Broglie scale $\lambda_{\mathrm{dB}} = \hbar/p$, where p is the momentum of the particle.

A reversed demonstration has also been given, according to which generalized Heisenberg relations may be deduced from the conjectured fractal structure. Then the geometric effects of nondifferentiability and fractality have been combined in the construction of a covariant total derivative, in terms of which the equation of motion takes the form of a geodesic equation, i.e. of Galileo's free motion law, $\widehat{d}\mathcal{V}/dt = 0$. Then a wave function is constructed as a reformulation of the action, and, in its terms, the Schrödinger equation can be derived from this geodesic equation. This fundamental result, which is confirmed and extended in Chapters 5 and 6 of the present book has therefore implemented for the first time the relativity-emptiness principle in the case of quantum properties, since they vanish in the proper (geodesic) reference system and reappear in their standard form after expansion of the covariant derivative (as in the case of the geodesic equation of general relativity).

Another fundamental result of the theory was obtained, according to which the general solution to the special relativity problem (i.e. the problem of finding the general linear transformations that come under the principle of relativity) can be proved to have the mathematical structure of the Lorentz group, for motion but also for scales (in logarithmic form) [352]. This new "log-Lorentz" form of scale transformations, in conjunction with the breaking of the scale relativity symmetry at the de Broglie scale, led to a demonstration of the existence of a universal, unreachable lower scale in Nature, that is invariant under dilations and plays the same role for scale as that played by the velocity of light in vacuum c for motion. We have identified it with the Planck scale and then generalized the de Broglie and Heisenberg relations. Then we have shown that this new suggested structure of high energy physics, when accounted for in renormalization group equations, has many consequences for elementary particle physics (convergence of masses and charges, solution to the hierarchy problem and to the contradictions encoutered by the minimal SU(5) "grand unification", etc.), which are reconsidered and confirmed in the present book (Chapter 11).

Finally, the scale relativity tools and methods have been applied to cosmology and gravitational structuring in the case of planetary systems. By generalizing dilation laws to log-Lorentz scale laws toward large scales, we came to the conclusion that there should exist a universal unreachable upper scale $\mathbb{L}$ in Nature, which is invariant under dilations, replaces the infinite (as an horizon) and is symmetrical to the Planck scale. By identifying this scale with the cosmological constant scale $\Lambda^{-1/2}$, we have implemented Mach's principle and Dirac's large numbers hypothesis in an exact way and therefore obtained a numerical prediction of the cosmological constant, which is now supported by its observational value, precisely known since 1998 (Chapter 12 of the present book). Finally, an application of the scale relativity methods to chaotic systems on time-scales large with respect to the horizon of predictability implied by chaos, allowed to apply a generalized Schrödinger type approach to the formation of planetary systems. The predicted peaks of probability were in remarkable agreement with the observed positions of planets and asteroids in our Solar System, and, moreover, new peaks were predicted in the intramercurial and trans-Neptunian regions. Since 1995, i.e. after these theoretical predictions, a large number of extra-solar planetary systems have been discovered and many Kuiper belt objects which support them, in addition to many other gravitational structures in the Universe (Chapter 13 of the present book).

Content of the book

After the present general introductory Part I, the Part II of the book is devoted to the theoretical foundation of the theory of scale relativity. We recall the basic first principles that underlie both current theoretical physics and the new approach (Chapters 1 and 2), then we describe the new physical and mathematical tools, which have been developed in a specific way to deal with the nondifferentiability and its associated infinities and divergences. We recall and prove again the fundamental theorem according to which a nondifferentiable continuum is fractal, i.e. explicitly dependent on internal scale ("resolution") variables and even scale divergent (Chapter 3), then the equations for this dependence are written under the form of partial differential equations acting in the "scale space" of these variables. These scale differential equations are physically constrained by the requirement that they come under the principle of scale relativity (Chapter 4). Their solutions allow us to derive the standard scale invariant fractal laws, but also to suggest several scale covariant generalizations to them.

Then we apply these methods to the question of the geometric foundation of quantum mechanics (Chapter 5). We derive the quantum postulates, and therefore the quantum laws, as manifestations of a nondifferentiable and fractal continuous space-time coming under the principle of relativity. This is done first in the non-relativistic case, by constructing complex wave functions and by deriving the Schrödinger equation as an integral of the geodesic equation of a fractal space (Chapter 5), then in the relativistic case, by deriving the Klein–Gordon equation, then by constructing spinor wave functions, which are solutions of the Dirac equation (Chapter 6), itself derived from the geodesic equation of a fractal and nondifferentiable full space-time.

In Chapter 7 we show that gauge fields, in the Abelian and non-Abelian cases, can also be understood in such a framework as manifestations of the fractal geometry of space-time, in analogy with gravitation manifesting its curvature. The theoretical Part II ends with Chapter 8, in which we tentatively suggest the development of a quantum-like mechanics in scale space.

The Part III of the book is devoted to applications of the theory to various sciences. After an introduction in Chapter 9, we consider applications to laboratory and Earth scale physics in Chapter 10, to elementary particle and high energy physics in Chapter 11, to cosmology in Chapter 12, to gravitational structuring in astrophysics in Chapter 13, and we finally conclude in Chapter 14 with prospective applications to other non-physical sciences, in particular to biology.

Audience

This book is a review and re-organization of original works in this new domain of research. These works have been published in various and separate contexts (physics, astrophysics, fractal studies, biology), so that it is a unique occasion to collect them in a unified and organized way, which should give the reader a clearer understanding of the wide ability of the new theory to treat and solve open problems in various sciences.

The audience that may be interested in this book or in parts of it includes mathematical physicists, physicists, astrophysicists, mathematical biologists, and graduate students in these sciences, in the areas of foundation of physics, relativity, foundation of quantum mechanics and its possible generalizations, self-organization processes, fractal and scale-dependent phenomena, and applications to laboratory physics, particle and high energy physics, astrophysics, and to life and human sciences.

The prerequisites are a basic knowledge in mathematical physics, in particular in classical and quantum mechanics, special and general relativity, and fluid and statistical mechanics.

We have given, as much as possible, in addition to the main text, a list of Exercises, which are already solved problems, with most of the time their solutions and hints about their derivation, and of Open Problems, which may not be yet fully solved problems (for lack of time), with hint of possible way to solve them, and may be sometimes still fully unsolved (to our knowledge) open questions. These Open Problems also allow us to give a hint of what will be the future continuation of the "scale relativity program".

Let us end this general introduction with a remark about the vocabulary. The development of a new theory of relativity poses a problem in this respect, since, up to now, the words "relativity" and "relativistic" referred to the theory of the relativity of motion (including actually also position in space and time, as described by the four translations of the Poincaré group, and orientation, as described by the three rotations in space in addition to the three Lorentz boosts). Moreover, this theory of the relativity of motion has itself three levels, Galilean, special and general (see, e.g. [265, 266]), a situation, which has already provoked a confusion of vocabulary since physicists are now accustomed to call "nonrelativistic" a description coming under Galilean relativity, and "relativistic" in the two other cases, possibly specified as "special relativistic" or "general relativistic". Another confusion comes from the expression "general relativity" instead of Einstein's own formulation "generalized relativity" [158], which may suggest that the current relativity theory is fully general, which it is clearly far to be. With the addition of scale relativity to motion relativity, the risk of confusion becomes still higher. We have, therefore, specified "scale relativity", "scale relativistic", "special scale relativistic", etc., in the new scale case (now including the Galilean case in the acceptance of "scale relativity"), but we have kept the standard expressions in the motion case, except in situations of possible confusion between them ("motion relativistic" instead of "relativistic", etc.).

Part II: Theory

Chapter 1

INTRODUCTION

The theory of scale-relativity is an attempt to extend today's theories of relativity, by applying the principle of relativity not only to motion transformations, but also to scale transformations of the reference system. Recall that, in the formulation of Einstein [158], the principle of relativity consists in requiring that "the laws of nature be valid in every system of coordinates, whatever their state". Since Galileo, this principle had been applied to the states of position (origin and orientation of axes) and of motion of the system of coordinates (measured by velocity and acceleration). These states are characterized by their relativity, namely, they are never definable in an absolute way, but only in a relative way. This means that the state of any system (including reference systems) can be defined only with respect to another system.

We have suggested that, in addition to position, orientation and motion, the observation scale, i.e. the resolution at which a system is observed or experimented, should also be considered as characterizing the state of reference systems. It is an experimental fact long known that the scale of a system can only be defined in a relative way: namely, only scale ratios do have a physical meaning, never absolute scales.

This led us to propose that the principle of relativity should be generalized in order to apply also to relative scale transformations of the reference system, i.e. dilations and contractions of space-time resolutions. Note that, in this approach, one reinterprets the resolutions, not only as a property of the measuring device and/or of the measured system, but more generally as a property that is intrinsic to the geometry of space-time itself. In other words, space-time is considered to be fractal, not as an hypothesis, but as a very consequence of a generalization of the geometric description to a nondifferentiable continuum. Moreover, we connect the fractal geometry with relativity, so that the resolutions are assumed to characterize the state of scale of the reference system in the same way as velocity characterizes its state of motion. The principle of relativity of scale then consists in requiring that "the fundamental laws of nature apply whatever the state of scale of the coordinate system".

It is clear that the present state of fundamental physics is far from coming under such a principle. In particular, in today's prevailing view there are two physics, a quantum one toward small scales and a classical one toward large scales. The principle of scale-relativity amounts to requiring a re-unification of these laws, by writing them under a unique more fundamental form, which becomes respectively the usual quantum laws and the classical laws depending on the relative state of scale of the reference system.

There are other motivations for adding such principle to fundamental physics. It allows one to generalize the current description of the geometry of space-time. The standard description (curved geometry) is usually reduced to at least two-time differentiable manifolds (even though singularities are possible at certain particular points). So a way of generalization of current physics consists in trying to abandon the hypothesis of differentiability of space-time coordinates. This means to consider general continuous manifolds, which may therefore be differentiable or nondifferentiable. These manifolds include as a sub-set the usual differentiable ones, and therefore all the Riemannian geometries that subtend Einstein's generalized relativity of motion. Then in such an approach, the standard classical physics is naturally recovered, in limits which will be studied throughout the present book.

However, it should be clear from now on that the partition of the set of general continuous spaces in two subsets, the differentiable ones and the nondifferentiable ones, is far from being symmetric. In order to get an idea of this point, consider, for example, all the continuous curves that relate two given points in an Euclidean plane. Nondifferentiable curves are for sure infinitely more numerous than differentiable ones. In other words, differentiable curves are actually a very particular case of continuous curve, most of them being nondifferentiable. The same is true for continuous manifolds, namely, differentiable (Riemannian) manifolds actually constitute a very particular and small subset of all possible continuous manifolds, most of them being nondifferentiable.

New physics is expected to emerge as a manifestation of the new non-differentiable geometry. One can indeed prove that a continuous and non-differentiable space is fractal, under an acceptation close to Mandelbrot's general definition of this concept [305, 306], i.e. the coordinates acquire an explicit dependence on resolutions and diverge when the resolution interval tends to zero [353, 42]. We shall recall the proof of this fundamental theorem of the scale relativity theory in Chapter 3 and develop its consequences for the new differential calculus (in position space and scale space), which is one of the main tools of the theory.

This leads one to apply the concept of fractality not only to objects in a given space, but to the geometry of space-time itself. Such generalization is possible provided one becomes able to define a fractal space-time in an intrinsic way, in analogy with Gauss's intrinsic characterization of the surface of a sphere by its metric, which opened the road to non-Euclidean geometries. An internal characterization of a fractal geometry is indeed possible [349, 353], based on the explicit scale dependence and divergence of a fractal metric.

Hence the tool of fractals, whose universality has now been recognized in almost every sciences (see e.g. the volumes [416, 295], other volumes of these series and references therein), is also expected to play a central role in fundamental physics. Note, however, that (as will be recalled in this book), even though continuity and nondifferentiability implies fractality, nondifferentiable space-times do also have new properties that are irreducible to the mere fractality.

One of the main strengths of space-time theories, pioneered by Einstein's general relativity [158], is their ability to identify the trajectories of "free" particles with the space-time geodesics. One does not need in such theories to add a force equation to the field equations (as, e.g. in electrodynamics, where the Lorentz force must be postulated in addition to Maxwell's equations), since the equation of motion is fully known as a geodesic equation once the space-time geometry is known. This is precisely one of the main specificities of the scale relativity theory to write the equations of motion as geodesic equations in a nondifferentiable and fractal space-time. As we shall see, these geodesic equations, written in terms of covariant derivatives built from the geometric effects on elementary displacements, are able to generate both the quantum laws and the gauge fields.

One of the first fundamental consequences of the nondifferentiability and fractality of space-time will be the nondifferentiability and fractality of its geodesics. The introduction of nondifferentiable paths in physics dates back to Feynman's pioneering works on the path integral formulation of quantum mechanics [176]. Specifically, Feynman has demonstrated that the typical quantum mechanical paths that contribute in a dominant way to the path integral are fractal nondifferentiable curves of fractal dimension $D_F = 2$ [1, 344].

In the fractal space-time approach, one is naturally led to consider the reverse question: could quantum mechanics itself find its origin in the fractality and nondifferentiability of space-time? Such a suggestion, first made at the beginning of the eighties [420, 344], has been subsequently developed [349, 421, 316, 352, 165, 353].

Note that the introduction of nondifferentiable trajectories was also underlying the various attempts of construction of a stochastic mechanics [331, 197]. But stochastic mechanics is now known to have self-consistency problems [205, 362], and, moreover, to be in contradiction with quantum mechanics [526] for multitime measurements. The scale relativity approach, even if it shares some common features with stochastic mechanics, due to the necessity to use stochastic variables as a consequence of the nondifferentiability, is fundamentally different and is not subjected to the same difficulties [362].

Note also that we do not have to assume that the paths are fractal and nondifferentiable, since this becomes a consequence of the fractality of space-time, which is itself a consequence of its continuity and nondifferentiability. As we shall see, the Schrödinger equation (and more generally the Klein–Gordon, Pauli and Dirac equations) are, in the scale-relativity approach, integrals of the equation of geodesics [353, 360, 93, 96]. Note that we consider here only geodesical curves (of topological dimension 1), but it is also quite possible to be more general and to consider subspaces of larger topological dimensions (fractal strings of topological dimension 1, whose trajectories are therefore of dimension 2 [275], fractal membranes, etc.).

The present Part II of this book is mainly devoted to the theoretical aspects of the scale-relativity approach. We shall first develop at various levels the description of the laws of scale that come under the principle of scale relativity (Chapter 4).

Then we shall recall how the description of the effects on motion of the internal fractal and nondifferentiable structures of the paths (with which we identify the wave-particles) lead to write a geodesic equation that is integrated in terms of quantum mechanical equations (Chapter 5). The Schrödinger equation is derived in the (motion) non-relativistic case, that corresponds to a space-time of which only the spatial part is fractal [353]. Looking for the motion-relativistic case (Chapter 6) amounts to work in a full fractal space-time, in which the Klein–Gordon equation is derived [356]. Finally the Pauli and Dirac equations are derived as integrals of geodesic equations when accounting for the breaking of the reflection symmetry of space differential elements that is expected from nondifferentiability [93, 96].

Now, the three minimal conditions under which these results are obtained (infinity of paths, fractality of the paths, breaking of differential time reflection invariance) may be achieved in more general systems than only the microscopic realm. As a consequence, we expect the existence of macroscopic systems which may share some properties with the standard quantum mechanical systems of microphysics (but not all quantum properties). This generalization will play an important role in the next parts devoted to the applications of the theory.

Chapter 7 is devoted to the interpretation of the nature of gauge transformation and of gauge fields in the scale relativity framework, as regards first Abelian (U(1)-like) gauge fields [356, 360], then their non-Abelian generalization [394]. We attribute the emergence of such field to the effects of coupling between scale and motion, in the framework of a general scale relativity theory. In other words, the internal resolutions become themselves "fields", which are functions of the coordinates.

We finally consider hints of a new tentative extension of the theory (Chapter 8), in which quantum mechanical laws are written in the scale space itself, i.e. in the space of the new internal scale variables ("resolutions"). In this case the internal relative fractal structures become described by probability amplitudes, from which a probability density of their "position" in scale space can be deduced. A new quantized conservative quantity, that we have called "complexergy", is defined (it plays for scale laws the same role as played by energy for motion laws), whose increase corresponds to an increase of the number of hierarchically imbricated levels of organisations in the system under consideration.

Chapter 2

STRUCTURE OF THE THEORY

2.1. Development Levels of the Theory

The theory of scale relativity is constructed by supplementing the standard laws of classical physics (laws of motion in space, i.e. of displacement in space-time) by new scale laws, in which the space-time resolutions are considered as variables intrinsic to the description. We hope such a stage of the theory to be only provisional, and the motion and scale laws to be treated on the same footing in the final theory. However, before reaching such a goal one must realize that the various possible combinations of scale laws and motion laws already lead to a large number of sub-sets of the theory to be developed. Indeed, three principal domains are to be considered in order to implement the scale relativity program [371]:

(i) *Scale laws*: description of the internal fractal structures of paths in a non-differential space-time at a given point/event.

(ii) *Induced effects of scale laws on the equations of motion*: generation of quantum mechanics as mechanics in a nondifferentiable space-time.

(iii) *Scale-motion coupling*: effects of dilations induced by displacements, that we interpret as gauge fields.

Now, concerning the first step (i) alone, several levels of description of scale laws can be considered. These levels are quite parallel to that of the historical development of the theory of motion laws:

(i.a) *Galilean scale relativity*: standard laws of dilation, that have the structure of a Galileo group (fractal power law with constant fractal dimension). When the fractal dimension of trajectories is $D_F = 2$,

21

the induced motion laws are that of standard quantum mechanics [353, 360].

(i.b) *Special scale relativity*: generalization of the laws of dilation to a Lorentzian form [352, 353]. The fractal dimension itself becomes a variable and plays the role of a fifth dimension (similar for scale laws of what is time for motion laws) that we have called "djinn". An impassable length-time scale, invariant under dilations, appears in the theory; it replaces the zero point, owns all its physical properties (e.g. an infinite energy-momentum would be needed to reach it) and plays for scale laws the same role as played by the velocity of light for motion.

(i.c) *Scale dynamics*: while the first two cases correspond to "scale freedom", one can also consider distortion from strict self-similarity that would come from the effect of a "scale-force" or "scale-field" [362, 363, 381, 409, 410].

(i.d) *General scale relativity*: in analogy with the gravitational field being ultimately attributed to the geometry of space-time, a future more profound description of the scale-field could be done in terms of the geometry of a five-dimensional "space-time-djinn" and its couplings with the standard classical space-time. This case also involves scale-motion couplings (level iii) and leads to a new interpretation of gauge fields as a manifestation of the fractal geometry of space-time.

(i.e) *Quantum scale relativity*: the above cases assume differentiability of the scale transformations. If one assumes them to be continuous, but, as we have assumed for space-time, nondifferentiable, one is confronted for scale laws to the same conditions that lead to quantum mechanics in space-time. One may therefore attempt to construct a new quantum mechanics in scale-space, thus achieving a kind of "third quantization".

The possible complication of the theory becomes apparent when one realizes that these various levels of the description of scale laws lead to different levels of induced dynamics (ii) and scale-motion coupling (iii). Moreover, other sublevels are also to be considered, depending on the status of motion laws (non-relativistic, special-relativistic, general-relativistic).

The present Part II of this book is devoted to the theoretical developments of these various levels in the progressive construction of the theory. This construction itself relies on first principles, the main one being the principle of relativity. Let us make a brief historical reminder of the evolution of ideas in physics, of their link to relativity and of the fundamental first principles on which today's physics is founded [16, 33]. As we shall see, the scale relativity theory is based on the same principles, extended to more complete physical laws that include an explicit dependence on scale variables.

2.2. Evolution of Concepts in Physics

Let us first briefly recall some features of the evolution of ideas in physics. We propose as an example the history of gravitational theories and the motion of astronomical bodies. The physical approach has shifted from mainly descriptive modelling to genuine predictive theories founded on first principles, and from a way of investigation based on setting *a priori* hypotheses and/or postulates (such as Newton's theory of gravitation and today's quantum mechanics) to a new way of thinking pioneered by Einstein, that consists of actively abandoning fundamental axioms underlying previous physical theories. For example, the construction of Einstein's theory of general relativity was rendered possible by the giving up of the general belief that the geometry of space can be only Euclidean. In this theory flat space becomes a very particular case of all the possible curved geometries.

Such an approach proceeds through important generalizations, which involve the creation of new concepts aimed at describing the multiple structures, which are expected to emerge. It is an extension of the frame of work and of thought, in which the apparently unavoidable hypothesis in the previous paradigm are given up and revealed for what they actually were: unnecessary blocking constraints. For example, before the discovery of curved geometry by Gauss, Bolyai and Lobachevsky, one could not even imagine that geometry could be anything else than Euclidean. This extended framework leads to a deeper understanding of nature. The structures which are explained by this method are understood as the manifestation of principles of a greater generality: hence, in general relativity, the various aspects of gravitation are understood as the multiple manifestations of space-time curvature. In the theory of scale relativity, the implicit axiom of differentiability, which underlies the classical theoretical description of all natural sciences, is abandoned.

Ptolemy's model

The first step of our knowledge of the motion of astronomical bodies was merely descriptive. Ptolemy's epicyclic model of planetary motions was intended to account for the direct positions of planets in the sky. As there are as many parameters as degrees of freedom in such a model, no genuine understanding can be gained, and it only describes in a direct way the observational data.

Kepler's model

The second step was Kepler's model of planetary motion that relies on Copernicus' discovery that the planets follow orbits around the Sun. Its ability to synthesize a huge amount of observational data is already extraordinary. Indeed, it allows a profound understanding of many features of planetary motion, but it nevertheless remains a model since the three Kepler laws (elliptical orbits, law of area, relation between periods and semi-major axes) are inferred from an analysis of observations but have no theoretical basis, although Kepler himself did look for a central force.

Newton's theory

The third step was Newton's universal theory of gravitation. The conceptual framework becomes in this case profound enough that it deserves the name of theory, although, as remarked by Newton himself, the basis of the theory remains axiomatic and badly founded. In this framework the three Kepler laws are now proved, understood and generalized to parabolic and hyperbolic motion thanks to the introduction of the gravitational force, and the theory acquires a large predictive power (e.g. by the prediction of the planet Neptune). But the expression for the force, proportional to mass and inversely proportional to the square of the distance, remains a postulate of unknown origin.

Field theory

The fourth step was field theory (Poisson, Laplace, Maxwell, Lorentz). The concept of force, which was defined locally between two bodies, is extended to that of field, which is now assumed to be created by a body (namely, by its active "charge"), fills the whole space, and is felt by another body provided it has a passive charge. The description of this force field now involves two concepts, the field itself and the potential, from which the field derives. The theory includes field equations (Poisson equation, generalized to Maxwell equations for electromagnetism) and motion equations, which amount to the fundamental equation of dynamics under the force F. However, at this level of physical theories, the motion equations (i.e. the expression of the force) must still be postulated and cannot yet be derived from the knowledge of the field.

Relativistic theories

The fifth step was Einstein's general relativity of motion, which is also a relativistic theory of gravitation. In order to better understand its meaning and its profound consequences, let us briefly summarize the evolution of ideas concerning the theories of relativity.

Galilean relativity

The first level of motion relativity theories is Galileo's relativity of inertial motion. Galileo has set as a principle that "for all things that participate to it, motion remains perfectly imperceptible and as if it were not" [190]. This means that motion is not a property of individual bodies, but a relative property between two bodies. It cannot be defined in the absence of a reference system, and it therefore disappears in the proper reference system (which participates in it).

Special relativity

The second level is Poincaré–Einstein's special theory of relativity of motion. The problem is the same as in Galileo's relativity, namely, to find the laws of transformation of inertial coordinate systems under changes of their relative motion. However, the general solution found by Poincaré [450] and Einstein [155] to this problem, the transformation called "Lorentz transformation" by Poincaré, generalizes the Galileo transformation and includes it as a particular degenerate case (namely, when the limit velocity $c \to \infty$). The physical meaning of this transformation is that space and time are no longer separated as in Galilean relativity, but become subspaces of a four-dimensional space-time. As a consequence, motion is nothing but rotations in space-time, which explains the relativistic contraction of length and the dilation of time as projection effects (under such rotations) and relates the relativity of motion to that of orientation.

Einstein's general relativity

The third level is Einstein's general relativity of motion. In this theory, the principle of relativity is applied not only to inertial motion, but also to accelerated motion. This has been made possible by Einstein's discovery of the principle of equivalence, according to which a gravitational field is locally equivalent to an acceleration field.

Accelerated motion, which seemed to be definable in an absolute way under Newton's view (through the appearance of inertial forces), revealed to be once again only relative to the choice of the reference system. Namely, a body in free fall accelerates toward the Earth in the Earth reference frame, and this acceleration is described as resulting from a gravitational force in Newton's approach, then from the effect of the gravitational potential of the Earth in a field theory approach. But Einstein realized that another body in free fall was as well habilitated to be taken as reference system, and that two close bodies in free fall are either at rest or in inertial motion one relatively to the other.

Therefore, neither the acceleration nor the gravitational force or field constructed from it can be considered any longer to exist in an absolute way. As a consequence, Einstein's theory of general relativity of motion is also a theory of the relativity of gravitation. The very existence of gravitation becomes relative, dependent on the choice of the reference system.

In this theory, the concepts of gravitational force, potential, and field, are replaced by characteristics of the space-time geometry. While the Minkowskian space-time of special relativity was still absolute (even though separated space and time became relative) and non structured (flat, i.e. non curved), the space-time of general relativity is curved and dependent on its energy-momentum content. What is called gravitation is nothing but the various manifestations of the curved geometry, with free particles following the geodesics of the Riemannian space-time constrained by Einstein's field equations. Therefore, the equation of motion (i.e. the equation of dynamics) becomes an equation of geodesics, and, as such, is no longer postulated in a separate manner. It is instead deduced as a direct consequence of the geometry.

Quantum theory

A sixth step in the evolution of ideas in physics is the quantum theory. Unfortunately, we can no longer use the example of gravitation to illustrate it, since there is not yet any fully self-consistent quantum theory of gravitation. But quantum theories of the other fundamental fields (strong, weak and electromagnetic) do exist and have yielded extraordinary results. The quantum approach is based on a representation, which is completely different from the classical one. The description tool is a wave function (or probability amplitude) whose square modulus gives the probability density of observables. Through this description, elementary particles have a triple aspect of particle, wave and field. This also becomes the case for what was classically considered as the source (fermions, of half-integer spin) or the field (bosons, of integer spin). Therefore, in its framework, matter (such as quarks and electrons) has wave properties, as discovered by de Broglie in 1923 [130], while force fields have particle properties, described in the electromagnetic case by Einstein in 1905 [156] in terms of light quanta, later called photons. The quantum theory has achieved an impressive unification of previously separated concepts. The trouble is that its foundation remains axiomatic: indeed, the mathematical tools and the equations themselves are postulates instead of being derived from first principles.

Toward a unified view

From the viewpoint of this history (strongly summarized above), the theory of scale relativity aims at preparing the emergence of a seventh step by unifying the quantum and relativity concepts on the basis of first principles [353, 93, 394], in a framework where the quantum fields become themselves manifestations of the geometry of space-time.

It is worth stressing from now on a fundamental difference in such a program with the current view about the future evolution of physics. It is widely considered that the main remaining challenge for theoretical physics amounts to unifying general relativity and quantum mechanics under the form of a theory of quantum gravity. Such a goal is founded on the view that the quantum laws are fundamental, and that the theories of relativity have reached their completion with Einstein's general theory. Since there does exist a well-behaved special relativistic quantum theory, the only remaining missing link is, in this frame of thought, the construction of a quantum theory of gravitation.

The diagnostic and proposed remedies for the present crisis of theoretical physics are different in the scale relativity theory. In its framework, one distinguishes between the principle of relativity and the theories that are constructed in order to implement it [289]. The main problem for today's physics is considered to be the absence of a foundation of the quantum laws themselves on the principle of relativity, while the question of building a quantum gravity theory is considered to be a secondary problem (although certainly very interesting, but also very difficult since it involves Planck scale physics, which corresponds to energies 10^{16} larger than the largest energies reached by today's accelerators).

2.3. Founding First Principles in Relativity

2.3.1. *Optimisation principle: least action and conservation laws*

The whole of theoretical physics (classical and quantum) relies on a fundamental optimization principle [266], from which the basic equations of physics can be constructed under the form of Euler–Lagrange equations, relating fundamental quantities, such as energy, momentum and angular momentum. This principle of least action becomes, as we shall recall hereafter, a geodesic principle in the framework of relativity theories (i.e. the action is identified with the proper time).

Recall that one can describe a physical system by generalized coordinates x_i and generalized velocities $v_i = dx_i/dt$ (where the index i runs on the degrees of freedom of the system; it is omitted in what follows in order to simplify the writing) and that there exists some function $L(x, v, t)$, called Lagrange function, the integral of which is the action,

$$S = \int_{t_1}^{t_2} L(x, v, t) dt. \tag{2.1}$$

The principle of least action states that the motion of the system between $x(t_1)$ and $x(t_2)$ is such that it optimizes the value of this action to a constant, minimum value. This is expressed by the fact that the first variation of the action vanishes, i.e.

$$\delta S = 0. \tag{2.2}$$

From this principle, a general form of the equations of motion can be derived as Euler–Lagrange equations, that read

$$\frac{d}{dt}\left(\frac{\partial L}{\partial v}\right) - \frac{\partial L}{\partial x} = 0. \tag{2.3}$$

For a system of material points, these equations become Newton's fundamental equation of dynamics. Their Lagrange function reads $L = \sum \frac{1}{2}mv^2 - \phi(x)$, where $T = \sum \frac{1}{2}mv^2$ is their kinetic energy, whose form derives from relativity [265, 266], and where ϕ is the potential energy of the system. In this case the Euler–Lagrange equations read

$$m \frac{dv}{dt} = -\nabla \phi, \qquad (2.4)$$

for each particle of the system. In the case of free particles, it takes the form of a geodesic equation in an Euclidean space, i.e. of the Galileo equation for inertial motion,

$$\frac{dv}{dt} = 0. \qquad (2.5)$$

This equation is the archetype for strongly covariant/geodesic equations of motion in physics. One of the main goals of relativity theories amounts, by the construction of covariant derivatives, to writing the equations of motion in general situations under this simplest possible form, which states that, whatever the complexity of the motion and of the apparent forces, which create it, from the viewpoint of the proper reference system swept along with the body, one recovers freedom and there is actually no motion.

Linked to this principle, one proves the existence of fundamental physical quantities, which are subjected to conservation laws. These quantities and their conserved (invariant) character result from symmetries of the underlying basic variables of description (this is Noether's theorem). For example, energy is the conservative quantity that results from the uniformity of time, momentum from the uniformity of space and angular momentum from the isotropy of space. In quantum field theories, the charges are the conserved quantities that result from the symmetries of the phases of the wave function.

2.3.2. *The principle of relativity*

Let us now briefly recall the fundamental principles that underlie, since the work of Poincaré (1905) [450] and Einstein (1905, 1916) [155, 158], the foundations of relativity theories. We shall express them here under a general form that transcends particular theories of relativity, namely, they can be applied to any state of the reference system (origin, orientation, motion, scale, etc.).

The basic principle is the principle of relativity, which requires that the laws of physics should be of such a nature that they apply for any state of the reference system (in [158], Einstein specified "state of motion"). In other words, it means that physical quantities are not defined in an absolute way, but are instead *relative to the state* of the reference system (which can be static or dynamic). This principle, which is still of philosophical nature at this stage, is subsequently implemented in physics by three related and interconnected principles and is equipped with their corresponding mathematical tools.

2.3.2.1. *Principle of covariance*

The principle of covariance requires that the equations of physics keep their form under changes of the state of the reference systems. As remarked by Weinberg [528], it should not be interpreted in terms of simply providing the most general (arbitrarily complicated) form to the equations, which would be meaningless. It rather means that, knowing that the fundamental equations of physics have a simple form in some particular coordinate systems, they will keep this simple form when considering more general coordinate systems. With this meaning in mind, two levels of covariance can be defined:

Strong covariance, according to which one recovers the simplest possible form of the equations, which is the Galilean form they have in the vacuum devoid of any force. In general relativity, under this principle, the equations of motion take the free inertial form

$$\frac{Du_\mu}{ds} = 0. \tag{2.6}$$

Here D denotes Einstein's covariant derivative (we briefly recall the way it is built in the next Sec. 2.3.2.4), which is the main mathematical tool of the theory. The vector $u_\mu = dx_\mu/ds$ is the four-dimensional velocity, defined in the four-dimensional space-time of general relativity by the index μ running on time and space coordinates, i.e. $(0, 1, 2, 3)$ for (t, x, y, z). The variable s represents the proper time, i.e. the time measured by a clock that is taken along with the system considered, and ds is the proper time differential element, which is the fundamental invariant of the theory. One of the main tools for implementing strong covariance is the tensor calculus, which is a natural generalization of vectors to several indices, and which allows to compactify the writing of the equations [14], namely, tensors have a particularly simple way to transform under changes of coordinate systems.

Weak covariance, according to which the equations keep the same simple form under any coordinate transformation, but not as simple as the free Galilean-like equation. A large part of the general relativity theory is only weakly covariant in this context. For example, Einstein's field equations have a source term, the stress-energy tensor of matter, while Einstein's initial hope was to construct a purely geometric theory in which the sources themselves would be of geometric origin. Another case of weak covariance in general relativity concerns the gravitational fields (the Christoffel symbols), which are not tensors, but transform themselves under changes of coordinates in a more complicated, nonlinear way.

2.3.2.2. *Principle of equivalence*

The principle of equivalence is a more specific statement of the principle of relativity, when it is applied to a given physical domain. In general relativity, it states that a gravitational field is locally equivalent to an acceleration field, i.e. it expresses that the very existence of gravitation is relative to the choice of the reference systems and it specifies the nature of the coordinate systems in which gravitation locally disappears. Therefore, in such an accelerating coordinate system the field of gravitation vanishes, so that the motion equation naturally takes the strongly covariant form of free motion devoid of any force, i.e. once again $Du_\mu/ds = 0$.

In scale relativity, one may, as we shall see, make a similar proposal and set down generalized equivalence principles according to which the quantum behavior is locally equivalent to a fractal and nondifferentiable motion, while the gauge fields are locally equivalent to expansions or contractions of the internal resolution variables needed to describe a nondifferentiable manifold.

2.3.2.3. *Geodesic principle*

The geodesic principle states that the free trajectories are the geodesics of space-time. It plays a very important role in a geometric relativity theory, since it means that the fundamental equation of dynamics is completely determined by the geometry of space-time, and therefore has not to be set down as an independent equation. Moreover, in such a framework the action dS can be identified (modulo a constant) with the fundamental metric invariant, which is nothing but the proper time ds itself:

$$dS = -mc\,ds. \tag{2.7}$$

The action principle becomes nothing else than the geodesic principle. As a consequence, its meaning becomes very clear and simple: the physical trajectories are those, which minimize the proper time itself, i.e. it becomes a general Fermat–Maupertuis principle. Moreover, in general relativity, once the geometry is known, i.e. the field, the trajectories are also known since the geodesics are completely defined by the geometry. This is quite different from what happens in other fundamental physical theories, which need to specify both the field equations and the motion equations, such as in Newton's and Maxwell's theories.

As we shall see, one may even go one step further in the scale relativity framework. While in general relativity one still considers that the geodesics describe the trajectories followed by "particles" (of matter or radiation), in scale relativity one may identify the particles with the geodesics themselves. They are, therefore, no longer viewed as trajectories of "something", but only as purely geometrical paths (in continuity with Feynman's path integral view), from which the various properties of the wave-particle emerge.

There is another formulation of the geodesic principle, which is a direct consequence of the relativity principle under its "emptiness" formulation (see the General Introduction). It amounts to generalizing the initial Einstein statement that led him to the principle of equivalence for gravitation, according to which, for someone who is in free fall in a gravitational field, the acceleration of gravity has vanished. In other words, in a reference system, which is driven with the trajectory (the geodesic reference system), the various forces disappear and the local expression of the motion on the geodesical line takes the Galilean form of rectilinear inertial uniform motion, $Du/ds = 0$. One recovers the result obtained from the strong covariance and equivalence principles. But here, the meaning is that of a geodesic equation, since the geodesics of an Euclidean space are indeed straight lines travelled through with uniform velocity.

2.3.2.4. *Covariant derivative*

The concept of covariant derivative is the main tool designed by Einstein in order to implement these principles. This tool includes the effects of geometry through a new definition of the derivative. This is contrary to the standard field approach, in which the effects of the field are considered to be externally applied to the system.

In general relativity, it amounts to substracting the geometric effects to the total increase of a vector [266], leaving only the inertial part. Einstein's covariant derivative is therefore defined (for a covariant vector) as

$$DA_\mu = dA_\mu - \Gamma^\rho_{\mu\nu} A_\rho dx^\nu, \tag{2.8}$$

and for a contravariant one,

$$DA^\mu = dA^\mu - \Gamma^\mu_{\nu\rho} A^\nu dx^\rho \tag{2.9}$$

so that the geodesic equation reads

$$\frac{Du^\mu}{ds} = \frac{du^\mu}{ds} - \Gamma^\mu_{\nu\rho} u^\nu u^\rho = 0. \tag{2.10}$$

The three-index Christoffel symbols $\Gamma^\rho_{\mu\nu}$ naturally enter this expression thanks to Einstein's convention about the summation of up and down indices. They represent the general relativistic expression of the gravitational field.

One of the most remarkable results of general relativity is that the three principles (strong covariance, equivalence and geodesic/least action principle) lead to this same form for the motion equation, $Du^\mu/ds = 0$ (see e.g. [266]) which is the simple Galileo form for the inertial motion equation of a free body submitted to no force.

2.4. Methods of Scale Relativity Theory

The basic method of scale relativity amounts to explicitly introducing scale variables in the physical description. Therefore it can be decomposed into three steps.

(i) Finding the laws of scale at a given point and instant. These laws are obtained as solutions of differential equations acting in scale space (i.e. that describe the effect on physical quantities of an infinitesimal zoom), constrained by the principle of scale relativity.

(ii) For each of these underlying scale laws, finding the laws of motion in standard space(-time), i.e. the fundamental equations of dynamics. They are written in terms of a geodesic equation using a covariant derivative tool that includes the effects of nondifferentiability and fractality in the differentiation process itself. As we shall see, the laws of motion constructed in this way acquire a quantum-type form.

Hence, the theory of scale relativity follows a line of thought similar to general motion relativity, by the construction of new covariant tools enabling the description of the geodesics of a nondifferentiable and fractal space-time geometry founded on the principle of the relativity of scales.

(iii) Finding the laws of coupling of scale and motion. In this case the scale variables become themselves functions of coordinates. These resulting scale fields, which are manifestations of the fractal geometry, can actually be identified to gauge fields of the Abelian and non-Abelian types, and are finally reinserted in the motion equations.

Chapter 3

NONDIFFERENTIABLE GEOMETRY
AND FRACTAL SPACE-TIME

3.1. Analysis of the Differentiability Hypothesis

Consider a differentiable function $f(x)$. Its derivative is defined since the work of Newton and Leibniz during the 17th century and its formalization during the 19th century as

$$f'(x) = \lim_{dx \to 0} \frac{f(x + dx) - f(x)}{dx}. \tag{3.1}$$

For example, for $f(x) = x^2$, it is easy to compute this limit and to find that $f'(x) = 2x$. Critical analyses of the differential calculus goes back to its origin. Hence Berkeley wrote in 1734 ([49], Sect. XXXV):

> "It must, indeed, be acknowledged that he [Newton] used Fluxions, like the Scaffold of a building, as things to be laid aside or got rid of, as soon as finite Lines were found proportional to them. But then these finite Exponents are found by the help of Fluxions. Whatever therefore is got by such Exponents and Proportions is to be ascribed to Fluxions, which must, therefore, be previously understood. And what are these Fluxions? The Velocities of evanescent Increments? And what are these same evanescent Increments? They are neither finite Quantities nor Quantities infinitely small, nor yet nothing. May we not call them the Ghosts of departed Quantities?"

One may reconsider this criticism under the light of modern physics. From the mathematical viewpoint, the modern definition in terms of limit could be considered as an answer to Berkeley's criticism. But then another question arises: does the mathematical definition of derivative agree with its physical meaning?

It is easy to show that this is not the case and that the mathematical derivative is an ideal concept, which is very far from what is actually done in physics. Consider for example the motion of a point-like body. Its velocity is theoretically defined as

$$v(t) = \lim_{dt \to 0} \frac{x(t + dt) - x(t)}{dt}. \tag{3.2}$$

In order to evaluate the validity of such a definition, let us attempt to perform a thought experiment (Gedanken experiment) in the Galileo–Einstein way that would implement it. If really the physical velocity was to be defined in this way, this would involve our ability to make measurements of the position of the body on time intervals, say $1\,\text{s}$, then $10^{-1}\,\text{s}, 10^{-2}\,\text{s}, \ldots, 10^{-10}\,\text{s}, 10^{-20}\,\text{s}, \ldots, 10^{-100}\,\text{s}$, etc., without any limit. We recover here one of Kant's antinomies, according to which the zero point is actually unreachable and is, therefore, fundamentally an infinite.

Such an infinite series of measurements is indeed impossible for two main reasons.

(i) We are unable to make measurements of position and of time below some minimal resolution scale, however, this minimal resolution scale depends on the quality of our measurement apparatus and it is indeed one of the main progress of experimental physics since the discovery of the microscope to continuously improve their resolution, so this is not a fundamental limitation in the framework of a thought experiment.

(ii) Their number is infinite. In other words, even if one may hope that whatever small a resolution length-scale δx or time-scale δt may be, one will be able in the future to make measurements at these scales, anyway a measurement at $\delta x = 0$ or at $\delta t = 0$ can be considered as definitively impossible.

Now the view point of standard physics is that one does not need to go effectively to the zero limit, since, assuming that the position function $x(t)$ is differentiable, the value obtained for smaller and smaller dt's differs in a smaller and smaller way from the theoretical limit (see Fig. 3.1). Namely, performing a Taylor expansion yields $\{x(t + dt) - x(t)\}/dt = x'(t) + (1/2)x''(t)\,dt + \ldots$, which indeed tends to $x'(t)$ when $dt \to 0$.

However the 20th century experimental physics has brought a fundamental denial to this view. If one really performs the experiment of measuring positions and instants with smaller and smaller space and time

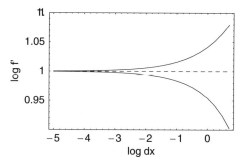

Fig. 3.1: **Scale dependence of a derivative.** Variation in terms of scale of the finite element derivatives of a standard differentiable function, $f'_+(x, dx) = [f(x+dx)-f(x)]/dx$ and $f'_-(x, dx) = [f(x) - f(x - dx)]/dx$. When $dx \to 0$, i.e. when $\log|dx| \to -\infty$, these functions of two variables tends to the standard derivative $f'(x)$. But, instead of taking only the limit $dx \to 0$, we consider here the way dx tends to zero from its maximal possible value ($dx \lesssim x$). The figure corresponds to $f(x) = x^2$, i.e. $f'_\pm = 2x \pm dx$, for $x = 5$.

resolution intervals, one rather rapidly reaches scales where the quantum laws manifest themselves. Instead of the expected vanishing difference $\delta v \approx (x''/2)\delta t \approx (x''/2x')\delta x$, the velocity becomes more and more badly defined with decreasing scale according to Heisenberg's relation, $\delta v \approx (\hbar/m)\delta x^{-1}$.

This fundamental contradiction has led Heisenberg to construct the quantum theory on the basis of the giving up of the classical concepts of position and velocity. However, from the above analysis, the key hypothesis, which seems to be falsified, is the assumption of differentiability of the position variables, i.e. of space (and more generally of space-time in a motion-relativistic framework), not their very existence.

Let us finally give another equivalent way to exhibit the fundamental contradiction that lies at the heart of today's physics. The Heisenberg's relations also tell us that, if one wants to perform an experiment, e.g. a measurement of position, at decreasing time resolution δt, one needs an increasing energy $E \approx \hbar/\delta t$. The construction of experimental devices aiming at scanning the microscopic world have definitely confirmed this fundamental fact, with magnifying glasses and microscopes using visible light energy, then electronic microscopes, then field effects microscopes, then particle accelerators which now reach several TeV energy scales.

Therefore, if one wanted to really implement by a true experiment the very definition of the derivative, $\psi'(t) = \lim_{dt \to 0}\{\psi(t + dt) - \psi(t)\}/dt$, one would need an infinite energy. Now, this definition is currently used in quantum mechanics, e.g. the derivative of the wave function in the

Schrödinger equation, while it is quantum mechanics itself, which shows that taking this limit is impossible.[a]

A possible way to escape this difficulty, which anticipates on the scale relativity method, amounts to being interested, not only in the limit $dx \to 0$ itself, but also in the way the function evolves when dx tends to this limit. To this purpose, let us now compute the finite difference expressions without going to the limit. We find in that case that f' becomes an explicit function of x and dx, $f'(x, dx)$. Now, since there are two expressions for the derivative, we find two scale dependent functions, $f'_+(x, dx)$ and $f'_-(x, dx)$. We shall see in Chapter 5 that this fundamental two-valuedness of derivatives is considered in this framework to be at the origin of the complex number nature of the wave function in quantum mechanics. For example, when $f(x) = x^2$, one finds

$$f'_+(x, dx) = \frac{(x + dx)^2 - x^2}{dx} = 2x + dx, \qquad (3.3)$$

$$f'_-(x, dx) = \frac{x^2 - (x - dx)^2}{dx} = 2x - dx. \qquad (3.4)$$

We recover the usual derivative as $f'(x) = f'(x, 0) = 2x$. More generally, one can perform a Taylor expansion of the function $f(x)$ and one obtains, provided it is differentiable to higher orders,

$$f'_\pm(x, dx) = \frac{f(x \pm dx) - f(x)}{\pm dx} = f'(x) \pm \frac{1}{2}f''(x)dx + \frac{1}{6}f'''(x)dx^2 + \cdots. \qquad (3.5)$$

This behavior is illustrated in Fig. 3.1. It is clear that $f'_\pm(x, dx)$ still contains the information about the standard definition of the derivative $f'(x)$, i.e. the limit $dx \to 0$), but it also contains more, namely it includes information about the specific approach to this limit. It is also remarkable that its first order expression, $f'_\pm(x, dx) = f'(x) \pm (1/2)f''(x)dx$, looks quite like the expression for a physical quantity with its uncertainty.

The next fundamental question is now: what about a nondifferentiable continuous function $f(x)$? As we shall see in what follows, it is possible

---

[a]It is remarkable that after the first publications of this kind of analysis in the eighties, [344, 420, 349, 353], it has indeed been found by Berry [47] and Hall [222] in the nineties that fractal and nondifferentiable wave functions can be obtained in standard quantum mechanics. This is clearly one of the predictions of the scale relativity approach [371, 403], in which, as we shall see in Chapter 5, the existence of wave functions and the equations they satisfy can be understood as manifestations of the nondifferentiable geometry of space-time.

to define a derivative in an extended and more physical way even in the nondifferentiable case. Indeed, by "nondifferentiability" we mean here that a function has no derivative in the usual sense (because it is either infinite or undefined due to infinite fluctuations at the limit $dx \to 0$). But since we keep the continuum hypothesis, such a function can still be differentiated.

This is a very important and crucial point in all the scale relativity approach. It, therefore, differs in a fundamental way from other later attempts to develop a fractal space-time theory, such as El Naschie's Cantorian space-time theory [165, 166], in which continuity is also abandoned, leading him to use number theory as a mathematical tool to implement the theory.

Here our viewpoint is fundamentally different, since the giving up of the continuity of space-time has no physical meaning in a theory based on the principle of relativity. Indeed, this would mean that some regions of space and time would be totally disconnected from other parts, so that they would, therefore, have no physical existence. In general, authors who claim to introduce a discontinuity of space, for example by describing space as a dust of points, also introduce communication between these points. If they did not, these points would be totally disconnected and would then behave as totally separate universes. But, by introducing these connections, they have actually reintroduced the continuity of space. Indeed, in a relativistic view of space-time, the individual points have no existence by themselves, only their relations have a physical meaning.

Let us come back to the question of differentiation. We stress once again that, because of continuity, there is no obstacle to define arbitrarily small differential elements of the coordinates, dx and dt, and differentials of continuous functions of these coordinates, $df = f(x + dx) - f(x)$. Nondifferentiability only means that the limit of their ratio df/dx when $dx \to 0$ is undefined. We shall now see that it is nevertheless possible to define and to use a differential calculus even in this nondifferentiable situation.

3.2. Beyond Differentiability

Since the time of Newton and Leibniz, the founders of the integro-differentiation calculus, the hypothesis of differentiability has been put forward in our description of physical phenomena. The strength of this hypothesis has been to allow physicists to write the equations of physics in terms of differential equations. However, there exists neither a prime principle nor any definite experiment that would impose the fundamental laws of physics to be differentiable. On the contrary, it has been shown by Feynman that typical quantum mechanical paths are nondifferentiable [176].

The basic idea that underlines the theory of scale relativity is, therefore, to give up the hypothesis of differentiability of space-time, while nevertheless keeping the mathematical tool of partial differential equations. Namely, one uses in this approach a generalized description keeping continuity, but including differentiable and nondifferentiable systems. In such a framework, the successes of present day standard differentiable physics could be understood as applying to domains where the approximation of differentiability (or integrability) is good enough, i.e. at scales such that the effects of nondifferentiability are smoothed out. But conversely, we expect the usual differential method, which assumes the existence of derivatives in the standard meaning, to fail when it is confronted with truly nondifferentiable or nonintegrable phenomena. This concerns in the first place very small and very large length scales, i.e. microphysical quantum physics and cosmology, but also possibly a large class of mesoscopic complex systems.

3.3. Fractal Geometry

As we shall see in the next section, the fractality of a continuous and nondifferentiable manifold is the key to its mathematical description. Before establishing this founding theorem of the scale relativity theory, we need therefore to give a short reminder about fractals. The concept of fractals and its applications to all sciences have now become so wide that an exhaustive account of what they are now seems to be impossible. For a more complete view, we refer the reader to the list of books compiled by Mandelbrot [308].

Here we shall only summarize Chapter 3 of *Fractal Space-Time and Microphysics* [353] and recall some elementary definitions and properties of continuous fractals, in the present context of an intrinsic description of a fractal space-time.

The term "fractal" was coined in 1975 by B. Mandelbrot [305, 306], who has defined it as follows:

Definition (Mandelbrot) Objects, curves, functions or sets are fractal when "their form is extremely irregular and/or fragmented at all scales".

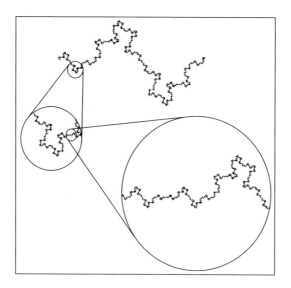

Fig. 3.2: **Successive zooms on a self-similar fractal curve.**

Therefore, fractals are characterized by the fact that they show structures at all scales. This means that fractals can be described only through scale transformations, i.e. successive zooms (Fig. 3.2). Among the most general fractals, a special class is made of the self-similar ones, which show the same structure under scale dilatations or contractions (Fig. 3.3). But, as stressed by Mandelbrot himself, one should not identify fractality with self-similarity, which characterizes only some particular fractals and also apply to several non-fractal geometric shapes (in particular the most simple ones).

Such a fragmentation at all scales implies an explicit scale dependence of the various properties of a fractal object, in particular of its topological measure (length of a curve, area of a surface, etc.). This scale dependence constitutes, as we shall see, the main tool of the scale relativity method.

An important point to remember about fractals is that this concept does not affect the topological dimension. Recall that the definition of topological dimensions remains relative:

Definition Two sets have the same topological dimension if and only if one may define a continuous and one-to-one transformation between them.

Topological dimensions, therefore, remain integer, even in the case of fractal objects and sets. One is then led to consider fractal dusts of points

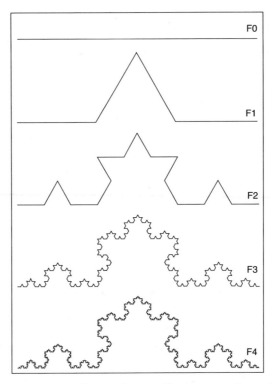

Fig. 3.3: Iterative construction of the von Koch curve. Example of an iterative construction of a self-similar fractal curve [250]. Its generator is made of $p = 4$ segments of length $1/q = 1/3$, so that its fractal dimension is $D_F = \log p / \log q = \log 4 / \log 3$. The resolution interval at step n of its construction is $(\varepsilon/\lambda) = q^{-n}$, while its length is $L_n = L_0 (p/q)^n = L_0 (q^{\log p / \log q}/q)^n = L_0 (q^n)^{D_F - 1} = L_0 (\lambda/\varepsilon)^{D_F - 1}$. One, therefore, recovers the standard scale variation of the length of a fractal curve as a power-law, in terms of a scale exponent $\tau = D_F - D_T$, where $D_T = 1$ is the topological dimension of the curve. Note that, strictly, it is characterized by a discrete self-similarity (see Chapter 4), since one recovers exactly the same shape only for zooms by factor 3^n.

of topological dimension $D_T = 0$, fractal curves of topological dimension $D_T = 1$, fractal surfaces of topological dimension $D_T = 2$, etc.

Now many fractal objects may also be characterized by alternative definitions of dimension, which do not take the place of the topological definition but are complementary. These definitions have been given by Mandelbrot the generic name of "fractal dimensions" (Fig. 3.3). There are several different fractal dimensions, such as the Hausdorff–Besicovitch dimension, the box-counting dimension, the covering dimension, the similarity dimension, etc.

Owing to the existence of this concept, there has been attempts to define fractals as sets of topological dimension D_T and fractal dimension D_F, such that $\tau = D_F - D_T > 0$.

However, this definition revealed itself to be too restrictive, since some objects or sets that are clearly fractal under Mandelbrot's definition because they show structures at all scales, may nevertheless have a fractal dimension equal to their topological dimension, $D_F = D_T$, or may have no constant or well-defined dimension. Some examples of such fractals will be given in the next sections.

Indeed the existence of a fractal dimension (different from its topological dimension) corresponds to a power-law divergence of, e.g. the length of a fractal curve. Namely, this length is given by [344]

$$\mathcal{L}(s, \varepsilon) = s \left(\frac{\lambda}{\varepsilon}\right)^{\tau}, \tag{3.6}$$

where $s = \mathcal{L}(s, \lambda)$ is a renormalized curvilinear coordinate along the fractal curve and where the scale exponent τ is given by

$$\tau = D_F - 1 = D_F - D_T. \tag{3.7}$$

This expression can be easily generalized to curves depending on several scale variables (see Sec. 3.6). For example, the length of a curve of fractal dimension D_F in a plane measured with a resolution ε_x in one direction and ε_y in the other reads

$$\mathcal{L}(s, \varepsilon_x, \varepsilon_y) = s \left\{ \left(\frac{\lambda_x}{\varepsilon_x}\right)^{\tau} + \left(\frac{\lambda_y}{\varepsilon_y}\right)^{\tau} \right\}. \tag{3.8}$$

This relation may be still generalized by accounting for a possible correlation between the two resolutions (see Exercise 7).

When writing such relations, one should be cautious from now on about the meaning of the scale variable ε. Under this form, ε is a space resolution. It corresponds, e.g. to a measurement of the length of the curve by covering it with balls of size ε, so that it can be identified with the length resolution itself, namely, $\varepsilon = \delta\mathcal{L}$.

But one may consider other ways to measure the length of a fractal curve. In particular, one may move on it and then measure its length in terms of a time resolution $\varepsilon = \delta t$. Now, as shown in [353], one can define a naturally renormalized curvilinear coordinate along a fractal curve. Indeed, though the distance is infinite between any couple of points (in the limit $\varepsilon \to 0$), the ratios of distances do remain finite. Therefore, the variable

$s = \mathcal{L}_\infty(s)/\mathcal{L}_\infty(1)$ in Eq. (3.6) remains finite on a curve of constant fractal dimension (this is in agreement with the principle of scale relativity, according to which only scale ratios have a physical meaning, not the scales themselves). If the fractal curve is covered at constant speed, this variable is proportional to time. Such a result will prove to play an essential role in what follows, since it allows one to define fractal geodesics, despite the infinite length of fractal paths.

By considering small differences on the fractal curve whose length is described by Eq. (3.6), one obtains

$$\delta\mathcal{L} = \delta s \left(\frac{\lambda}{\delta\mathcal{L}}\right)^\tau, \tag{3.9}$$

so that the two resolutions intervals $\delta\mathcal{L}$ and δs are linked by the fundamental relation

$$\delta\mathcal{L}^{D_F} \propto \delta s. \tag{3.10}$$

When skipping to differential calculus, since ds is a standard differential element, this means that $d\mathcal{L} \propto ds^{1/D_F}$ is a differential element of non integer order. By replacing $\varepsilon = \delta\mathcal{L}$ by $\delta s^{1/D_F}$ in Eq. (3.6), one obtains the expression of the length divergence in terms of the s resolution $\varepsilon_s = \delta s$,

$$\mathcal{L}(s,\varepsilon_s) = s \left(\frac{\lambda_s}{\varepsilon_s}\right)^{\tau'}, \tag{3.11}$$

where the scale exponent τ' is now given by

$$\tau' = 1 - \frac{1}{D_F}, \tag{3.12}$$

and more generally $\tau' = 1 - D_T/D_F$ for a topological dimensions D_T.

This result applies in particular to a fractal function (see examples of construction herebelow). A fractal function $y(x)$ may be described as an explicit function of the x resolution, i.e. one may write it as $y = y(x, \delta x)$ or of the y resolution, $y = y(x, \delta y)$. Its length is given by $\mathcal{L} = \sum \sqrt{\delta x^2 + \delta y^2} \approx \sum |\delta y|$, and the relation between the x and y resolutions is again $\delta y^{D_F} \propto \delta x$. This point should not be forgotten when estimating fractal dimensions from real data.

Another important result as concerns the physical applications considered in this book is the fact that the coordinates of a fractal curve (and more generally, of a surface, volume, etc.) can be themselves defined as fractal functions of the intrinsic curvilinear coordinate s defined along the fractal curve (see Fig. 3.4), and that they have the same fractal dimension as the curve itself [353] (except for infinitely anisotropic curves, see Exercise 9 in Chapter 4).

This description can also be generalized to fractal surfaces and to fractals of higher topological dimensions. We have given in [349, 353] several ways to build fractal surfaces from the iterated application of a two-dimensional generator. We refer the reader to these references for more detail. An example of continuous fractal surface that one can build from the folding of a lacunar fractal is given in Fig. 3.5.

Note that the power law case is only a very particular case of fractal law. For slower divergences, for example logarithmic divergences, no fractal dimension different from the topological one is defined although new fragmented structures (of decreasing size) may appear at all scales, thus still deserving the name of fractal. Still more generally, one may define variable fractal dimensions, depending on the position and time variables, or on scale itself [353], as we shall see in Chapter 4.

Let us now specify what we mean by fractality, when it is applied to a space (more generally to a space-time). Here, as in [353] and in subsequent works, we shall translate Mandelbrot's definition of fractals ("irregular and/or fragmented at all scales") by defining fractality by the explicit scale dependence and divergence, namely:

Definition Let $\mathcal{M}$ be the D_T-measure of an object, curve, function or set, i.e. the measure built from its topological dimension D_T, number of points for a dust, length for a curve, area for a surface, volume for a three-dimensional manifold, etc. Such a set is fractal provided $\mathcal{M}$ is an explicit function of a scale variable (resolution interval) ε, and such that $\mathcal{M} = \mathcal{M}(\varepsilon) \to \infty$ when $\varepsilon \to 0$.

For example, a fractal curve will be defined by having a length which is no longer a constant number between two given points, but a function of a scale variable ε, such that $\mathcal{L} = \mathcal{L}(\varepsilon) \to \infty$ when $\varepsilon \to 0$. In other words, we identify here fractality with the property of scale-divergence. We are conscious that this definition may be considered as more restrictive than Mandelbrot's, since one may construct curves that show smaller and smaller structures at all scales, in such a way that their length, though very large, nevertheless remains finite. Such curves may indeed genuinely appear as being fractal, though they are of finite length and therefore differentiable (see Lebesgue's theorem herebelow).

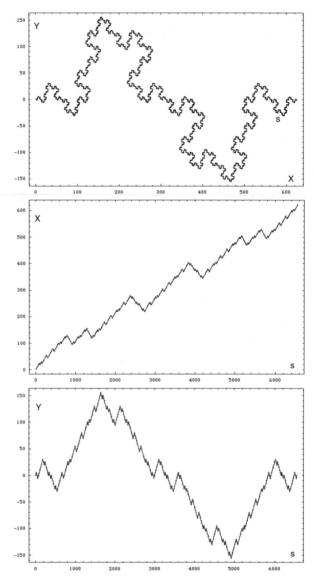

Fig. 3.4: **Fractal curve and its coordinates as fractal functions.** The X and Y coordinates of a fractal curve (top figure) are plotted as functions of a continuous curvilinear coordinate s defined along the curve (see [353]). They are, therefore, fractal functions of this variable, $X = X(s, \delta s)$ and $Y = Y(s, \delta s)$. In the example, the generator is made of $p = 9$ segments of length $1/q = 1/5$, so that the fractal dimension of the curve is $D_F = 3 \log 3 / \log 5$. The fractal functions which describe the coordinates have the same fractal dimension.

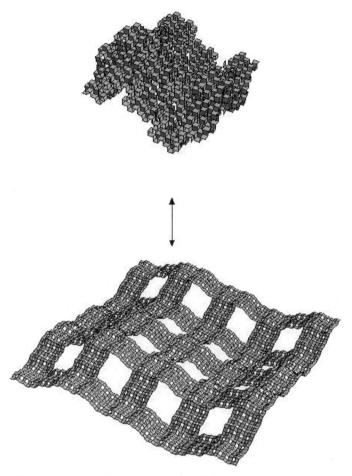

Fig. 3.5: **Fractal surface.** Example of the folding and unfolding of a fractal surface built from the iterated application of a scaled generator (see Chapter 3.6 of [353]). A continuous fractal surface can be constructed in this way by the folding of a lacunar one. Here the partly unfolded fractal surface is plotted at order 2 of the iteration process.

But here we do not use fractals in the same spirit: contrary to the definition of fractal objects, fractality is not the fundamental characteristics of our object of study, but, as we shall see, a secondary property owned by continuous and nondifferentiable manifolds.

Two remarks are relevant about this definition:

(i) The scale dependence by itself is insufficient to characterize the fractality. Indeed, if one considers successive approximations of a smooth, differentiable function by segments of decreasing sizes, it is clear that the length of the approximations does continuously increase, and is therefore an explicit function $\mathcal{L}(\varepsilon)$ of the scale ε. But it converges to the finite length of the curve, $\mathcal{L}(\varepsilon) \to L$ when $\varepsilon \to 0$. Therefore, the scale divergence in the limit $\varepsilon \to 0$ is an essential ingredient of fractality, in the acceptation we adopt here for a space or space-time.

(ii) The various definitions of fractal dimensions correspond to various ways to define the variable ε that characterizes the scale at which the object is observed, measured or defined. We shall not attempt to be exhaustive here about these definitions (see [306]), but we shall consider two main classes which are relevant in the scale relativity context, namely:

(a) The size of the object or of a window which selects a part of the object (for example, a window on the galaxy distribution in the Universe): applying a dilation ρ to this size yields a change of measure (of length, surface, volume, etc.) by ρ^{D_s}, where D_s is in this case a similarity dimension. For example, changing the size of a square by ρ yields a surface of the square increased by ρ^2 (in this non-fractal case the similarity and topological dimension coincide), while the same operation on the von Koch curve (Fig. 3.3) yields a length increase by $\rho^{\ln 4/\ln 3}$.

(b) The resolution interval of observation or of measurement: this general class contains many possible subdefinitions of scale variables, which themselves belong to two main subclasses:

- Definitions in an experimental or observational context: uncertainty, error bars (pixel, smoothing ball, filter, scanning interval).
- Definitions in a theoretical context: dissection interval, infinitesimals, differential element, finite differences.

The fractal dimensions that are relevant in this context are therefore mainly boxcounting or covering dimensions [306], that we may include into a general definition of "resolution dimension", which characterizes the way a given physical quantity varies when one changes the measurement resolution.

Problems and Exercises

Open Problem 1: Is it possible to define a non-integer topological dimension?

Hint: The definition of the topological dimension being relative, one characterizes the topological dimension of a given object or set by comparing it to a template, e.g. the real line for topological dimension 1, a plane for topological dimension 2, etc. Therefore this problem amounts to finding templates in the non-integer case. ∎

Exercise 1 Define the generator of a fractal curve of the von Koch type by "structural constants". Give its fractal dimension.

Answer: (See [344, 349] and [353] Chapter 3). Let such a generator $F1$ be made of p segments of length $1/q$, in terms of an initial segment F_0 of length $L_0 = 1$ (see Figs. 3.4, 3.3 and many examples in [349, 353].)

It can be defined by the complex coordinates (either Cartesian or polar) of the origins of the p segments Z_j, for $j = 0$ to $p - 1$, with $Z_0 = 0$ (while the end of the last segment is at $Z_p = 1$).

It can also be defined by the polar angles of the segments, ω_j, which are such that $Z_{j+1} - Z_j = q^{-1}e^{i\omega_j}$, or by the relative angles between segments $j - 1$ and j, $\alpha_j = \omega_j - \omega_{j-1}$.

If one applies the same generator at each step of the construction, the fractal dimension of the final curve is $D_F = \ln p / \ln q$. ∎

Exercise 2 Define a curvilinear coordinate on a fractal curve of the von Koch type in function of the structural constants that define its generator.

Answer: (See [344, 349] and [353] Chapter 3). One may number each segment of the generator by an integer $s = 0$ to $p - 1$. Then any position on the fractal curve can be marked by identifying the successive rank of the segments to which it belongs at each step of the iterative construction. This leads to define a curvilinear coordinate in number base p as

$$s = 0.s_1 s_2 \cdots s_k \cdots = \sum_{k=1}^{\infty} s_k p^{-k}. \tag{3.13}$$

When the segments of the generator are equal and when the iterative construction is strictly self-similar, this curvilinear coordinate is but the renormalized length on the fractal curve from the origin to the point considered, $\chi(s) = \mathcal{L}(s)/\mathcal{L}(1)$. ∎

Exercise 3 Give the equation of a fractal curve in a plane in function of the structural constants that define its generator.

Answer: (See [344, 349] and [353] Chapter 3). The complex coordinate in the plane of a point of curvilinear coordinate $s = 0.s_1 s_2 \cdots s_k \cdots$ written in base p (see previous exercises) is given by the equation

$$Z(s) = Z_{s_1} + q^{-1}e^{i\omega_{s_1}}[Z_{s_2} + q^{-1}e^{i\omega_{s_2}}[\cdots]] = \sum_{k=1}^{\infty} Z_{s_k} q^{1-k} e^{i\sum_{j=0}^{k-1} \omega_{s_j}},$$

$$(3.14)$$

where we have formally set $\omega_{s_0} = 0$ in order to obtain a compact self-consistent formula. It is noticeable that this formula simply reproduces the hierarchical structure of the fractal curve and in this aim it involves a combination of the fundamental transformations of physics, namely translations Z_{s_k} and rotations $e^{i\omega_{s_k}}$, to which one now adds scale transformations (contractions) q^{-k}.   ∎

Exercise 4 Give the equation of a fractal curve in a plane in a differential way, i.e. give the equation of the relative angle between one segment to the following at a given finite resolution.

Answer: (See [344, 349] and [353] Chapter 3). Consider the fractal curve at resolution $\varepsilon_n = q^{-n}$. The curvilinear coordinate written in base p of a given point on the fractal curve reads at this resolution $s(\varepsilon_n) = 0.s_1 s_2 \cdots s_n$. Now the values of s_n, s_{n-1}, etc. can be null if the point considered is on the first segment of the generator at the corresponding scales. Let us, therefore, define the rank h of the last non-zero digit of s. This means that s reads in base p:

$$s(\varepsilon_n) = 0.s_1 s_2 \cdots s_h 0 \cdots 0. \tag{3.15}$$

The relative angle of rotation is given by the much simple formula [344]

$$\alpha(s) = \alpha_{s_h}. \tag{3.16}$$

As noticed in [353], this formula reveals itself to be very efficient in the computer drawing of fractal curves.   ∎

Exercise 5 Generalize the above exercises to a fractal curve in a three dimensional space.

Answer: (See [353] Chapter 3). The generalization is straightforward, by replacing the two-dimensional rotation complex operators $e^{i\omega_k}$ by the three-dimensional rotation matrices R_k. One may also use quaternions in this purpose, see [259]. In the same way, the elementary relative rotation from a segment to the following at a given resolution is $A(s) = A_{s_h}$, where the A_k's are the relative rotation matrices on the generator.   ∎

Exercise 6 Construct fractal functions by iterative cutting up of base 2 (bisection). Deduce the properties of this function (scale variation of the length, fractal dimension, etc.) from the geometry of the elementary doubling.

Generalize to other bases and to more complicated cutting up.

Hint: (See Fig. 3.6). When going from F_n to F_{n+1}, the curve is subjected to a relative length increase given by

$$\rho = \frac{AC + CB}{AB} = \frac{\sqrt{a^2 + (h+b)^2} + \sqrt{a^2 + (h-b)^2}}{2\sqrt{a^2 + b^2}}. \tag{3.17}$$

Reciprocally, this means that, in order to obtain a given length ratio ρ from step n to step $n+1$, one needs to apply a difference h given by

$$h = \rho(a^2 + b^2)\left[\frac{\rho^2 - 1}{\rho^2 a^2 + (\rho^2 - 1)b^2}\right]^{1/2}. \tag{3.18}$$

When $\rho > 1$ (needed for length increase) and $b \gg a$ (which becomes true when n becomes large, see Fig. 3.7), one obtains $h \approx \rho b$. By iteration, for a constant value of ρ, the length at step n is a power law, $\mathcal{L} \propto \rho^n$. For a given fractal dimension D_F, one gets $2b = \delta y \approx \delta x^{1/D_F}$, so that $\mathcal{L} \approx \sum |\delta y| \approx \sum \delta x^{1/D_F} \approx \delta x^{(1/D_F)-1}$, in agreement with Eqs. (3.11) and

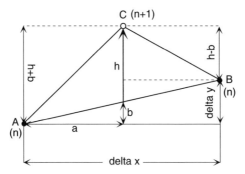

Fig. 3.6: **Elementary construction of a fractal function by bisections.** A fractal function is built through iterated scale doubling (successive bisections). At step n of the construction, one considers a given segment AB of F_n, of x and y coordinate differences respectively given by $\delta x = 2^{-n}$ and $\delta y = 2b$. At step $n+1$, one constructs F_{n+1} at scale $\delta x = 2^{-(n+1)}$ by adding a new point C at x coordinate $(x_A + x_B)/2$, which differs by a distance h from the y coordinate of the middle of segment AB, $(y_A + y_B)/2$. The properties of the final fractal function can be reduced to the law of the h values (see text).

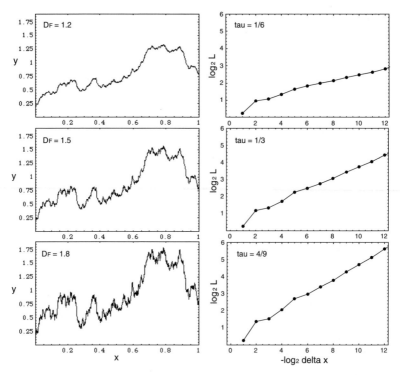

Fig. 3.7: **Construction of fractal functions by bisections.** Examples of construction of fractal functions by successive bisections of the x coordinate (see Fig. 3.6 for an explanation of the method), for three values of the fractal dimension (left figures). The h differences applied at each step of this construction are given by $h = \eta \times \delta x^{1/D_F}$, where η is a pseudo random variable such that $\langle \eta \rangle = 0$ and $\langle \eta^2 \rangle = 1$. We have used the same seed for the random variable in the three cases, which explains the similar (sef-affine) shapes of the three curves. The variation of the length in terms of the step number n (such that $\delta x = 2^{-n}$) is expected to be given by $L \propto \delta x^\tau$, with $\tau = 1 - 1/D_F$. An explicit measurement of the slope of the $(\log \delta x, \log L)$ curve (figures on the right) supports this expectation: one finds, up to a resolution of 2^{-12}, $\tau = 0.167, 0.331, 0.439$, which fits rather well the expected values $1/6$, $1/3$ and $4/9$.

(3.12). Examples of construction of fractal functions following this bisection method are given in Fig. 3.7.   ■

3.4. Fractality of a Nondifferentiable Continuum

3.4.1. *Position of the problem*

The construction of a continuous but nondifferentiable space-time may be viewed as a new "frontier" for mathematical physics. Indeed, the presently most developed theory of relativity is Einstein's "generalized relativity" of motion [158]. This theory is usually called "general" relativity. This denomination is in contradiction with Einstein's own attempts to generalize it in order to find an origin to quantum properties and other fields than gravitation. It also contradicts the fact that it relies on Riemannian geometry, which is (at least two times) differentiable. Indeed, Riemannian differentiable manifolds generalize Gauss's curved geometry, which is by definition locally flat.

The theory of scale relativity and fractal space-time precisely amounts to find a new generalization of the possible geometries that describe the physical space-time. To this purpose, one should relax some of the hypotheses that underlie Einstein's construction. Since our aim is to apply such a generalized geometric description in particular to the microscopic scales, it seems natural to relax the local flatness hypothesis, and, therefore, the differentiability hypothesis.

Set in such terms, the project may seem extraordinarily difficult. According to Einstein, who considered the possibility to give up differentiability [161, 179], it amounts to "try to breathe in empty space". In a letter to Pauli of 1948 [163], Einstein wrote (author's translation):

> "[...] this complete description could not be content with the fundamental concepts used in point mechanics. I have told you more than once that I am an inveterate supporter, not of differential equations, but quite of the principle of general relativity whose heuristic force is indispensable to us. However, despite much research, I have not succeeded in satisfying the principle of general relativity in another way than thanks to differential equations; maybe someone will find out another possibility, provided he searches with enough perseverance."

This letter points out the fundamental problem that seems to be posed by nondifferentiability. For Einstein at that time, it seemed evident that giving up differentiability implies to give up the differential calculus. This would mean to give up differential equations, while since the works of Leibniz and Newton, the equations of physics are differential equations. Many other physicists, mathematicians and philosophers have also considered this possibility, in connection with the quantum mechanical behavior, and have arrived at the same conclusion (see, e.g. Alunni [14] and references therein about attempts by Buhl, Bachelard and other authors). Feynman, in the framework of his path integral formulation of quantum mechanics and of his attempts to come back to a space-time representation, writes ([176], p. 177):

> "Typical paths of a quantum-mechanical particle are highly irregular on a fine scale. Thus, although a mean velocity can be defined, no

mean-square velocity exists at any point. In other words, the paths are nondifferentiable."

Under this form of Feynman's statement about the nondifferentiability of quantum mechanical paths, the infinity of the mean-square velocity is translated in terms of its non existence, i.e. as a loss of the concept of derivative. Yet, in the same page of *Quantum mechanics and path integrals*, the same behavior is expressed in a fundamentally different way:

> "If some average velocity is defined for a short time interval Δt, as, for example, $[x(t + \Delta t) - x(t)]/\Delta t$, the "mean" square value of this is $-\hbar/(im\Delta t)$. That is, the "mean" square value of the velocity averaged over a short time interval is finite, but its value becomes larger as the interval becomes shorter."

Under this form, one recovers a tool for dealing with the infinities of nondifferentiable curves. Namely, the velocity does exist, not as a number but as a function of Δt, and it is only at the unphysical limit $\Delta t \to 0$ that it becomes undefined. This is just the view point of the scale relativity approach. Moreover, anticipating on the following, one may remark from now that the Feynman result $\langle v^2 \rangle \propto \Delta t^{-1}$ may be compared with the expected fractal law $\langle v^2 \rangle = \langle (\Delta x/\Delta t)^2 \rangle \propto \Delta t^{(2/D_F)-2}$, i.e. $\Delta x^{D_F} \propto \Delta t$, which is a signature of the fractal dimension $D_F = 2$ of typical quantum mechanical paths [1, 344, 420], [353, Chapter 4].

Unfortunately this avenue has not been followed any longer at that time and it has not been realized that the Feynman analysis was actually containing the solution for a rigorous treatment of infinities in physics. For example, Finkelstein (also quoting Einstein's attempt) also reached, more recently, a negative conclusion about the use of the differential calculus in nondifferentiable physics [179]:

> "The vector/space-time compound emerges only in the limit $\Delta t \to 0$ of the continuum, where the chord joining two points becomes a tangent vector asymmetrically assigned to one of the points. This is the limit where the differential calculus works. The small physical constant that is neglected in the old physics and which will be the insignia of the new, if this prediction comes true, is the cut-off value for the limit $\Delta t \to 0$ and is therefore a small time. I infer that the physics of differential equations is a transient phase, and that it will evolve into a purely algebraic physics. Einstein (1936) considered this possibility without committing himself to it."[b]

---

[b]Einstein's letter of 1948 to Pauli quoted hereabove, in which he says that he has devoted "much research" to this problem, contradicts this last statement.

One of the main features of the theory of scale relativity is precisely to reconsider this question and to resolve it, but from a completely different perspective. Specifically, it is not necessary to give up the differential calculus in order to construct a nondifferentiable physics. A description of nondifferentiable functions and nondifferentiable physical quantities is actually possible in terms of differential equations, more precisely, in terms of a double differential calculus that works both in position space and in scale space [353].

The fundamental key which allows such a result is the very concept of fractals and of scale transformations. Indeed, as we are now going to see, continuity and nondifferentiability implies explicit scale-dependence (leading to divergence), i.e. fractality. A nondifferentiable function $f(t)$ can therefore be described in terms of an explicitly scale dependent function, $f(t, \delta t)$, which may be differentiated with respect to the variable t (for all values of $\delta t \neq 0$) and with respect to the variable δt itself.

3.4.2. *Continuity, nondifferentiability and fractality*

A proof of this essential property, according to which a continuous and nondifferentiable function is fractal, has been given in [353, p. 82], in the framework of non-standard analysis. Here we shall give a more detailed and intuitive proof relying on Lebesgue's theorem, from [360, 362] (see also [42, 115, 116]). Lebesgue's theorem does not apply to nondifferentiable curves, but instead to finite and differentiable curves. However, as we shall see it has important consequences for nondifferentiable curves, functions and spaces.

3.4.2.1. *Lebesgue's theorem*

Lebesgue's theorem states that

Theorem 3.4.1 (Lebesgue's theorem) *A continuous curve of finite length is differentiable or almost everywhere differentiable.*

A detailed proof can be found in [514].

3.4.2.2. *From nondifferentiability to scale divergence*

Now one can prove the following theorem [353], which is the key stone of the scale relativity theory:

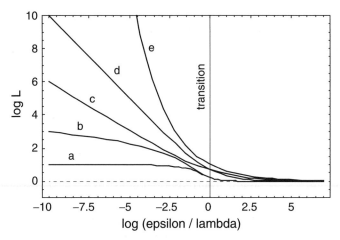

Fig. 3.8: **Scale dependence of the length of standard versus fractal curves.**
Various modes of scale dependence of a curve length toward small scales (including
a transition to scale independence toward large scales). (a) A standard curve of
finite length, which is therefore differentiable, according to Lebesgue's theorem. (b)
Logarithmic divergence, such as $L/L_0 = \ln(1 + (\lambda/\varepsilon)^\tau)$. (c) and (d) Power law
divergences, of the self-similar fractal type, such as $L/L_0 = 1 + (\lambda/\varepsilon)^\tau$, for two different
values of the exponent τ. (e) Exponential divergence, such as $L/L_0 = e^{\lambda/\varepsilon}$.

Theorem 3.4.2 *The length of a continuous and nowhere or almost
nowhere differentiable curve is explicitly dependent on the resolution ε, and,
further, it diverges when the resolution interval tends to zero: $\mathcal{L}(\varepsilon) \to \infty$
when $\varepsilon \to 0$. In other words, a continous and nondifferentiable curve is
fractal (under the above acceptation of scale-divergence).*

Proof The proof involves three steps:

(i) Consider a continuous but nondifferentiable function $f(x)$ between two
points $A_0[x_0, f(x_0)]$ and $A_N[x_N, f(x_N)]$ (see Fig. 3.9).

Since f is non-differentiable, there exists a point A_1 of coordinates
$[x_1, f(x_1)]$ with $x_0 < x_1 < x_N$, such that A_1 is not on the segment
A_0A_N (see Fig. 3.9). Then the total length is such that $\mathcal{L}_1 =
\mathcal{L}(A_0A_1) + \mathcal{L}(A_1A_N) > \mathcal{L}_0 = \mathcal{L}(A_0A_N)$, while for a differentiable
function, one would have $\mathcal{L}_1 \geq \mathcal{L}_0$.

We can now iterate the argument and find two coordinates x_{01}
and x_{11} with $x_0 < x_{01} < x_1$ and $x_1 < x_{11} < x_N$, such that
$\mathcal{L}_2 = \mathcal{L}(A_0A_{01}) + \mathcal{L}(A_{01}A_1) + \mathcal{L}(A_1A_{11}) + \mathcal{L}(A_{11}A_N) > \mathcal{L}_1 > \mathcal{L}_0$
(the indices are written in base 2).

By iteration we finally construct successive approximations $f_0, f_1, \ldots, f_n$ of $f(x)$ whose lengths $\mathcal{L}_0, \mathcal{L}_1, \ldots, \mathcal{L}_n$ increase monotonically when the resolution $\varepsilon_x \approx (x_N - x_0) \times 2^{-n}$ tends to zero. In other words, continuity and nondifferentiability implies a monotonous dependence and increase of the length of f as a function of the scale, i.e. $\mathcal{L} = \mathcal{L}(\varepsilon)$.

(ii) From the above Lebesgue theorem, a curve of finite length is differentiable or almost everywhere differentiable. Therefore, since f is continuous and almost everywhere nondifferentiable, its length cannot be finite, since if it were finite it would be differentiable. Then:

Lemma 3.4.3 *The length of a continuous and everywhere or almost everywhere nondifferentiable curve is infinite.*

(iii) One can finally combine the scale dependence and the infinity of the length of f by stating that $\mathcal{L} = \mathcal{L}(\varepsilon_x) \to \infty$ when the resolution $\varepsilon_x \to 0$, i.e. f is *scale dependent* and *fractal*. ∎

Note that the various points A_1, A_{01}, etc. can be chosen as close as wanted from the bisecting lines of the various segments, so that the construction of the nondifferentiable fractal function may proceed by bisection or quasi-bisection. Let us prove this statement.

Proof If the point of abscissa $(x_0 + x_N)/2$ on the nondifferentiable function is not on the segment $A_0 A_N$, it can be taken as the new point A_1.

If this point is on the segment $A_0 A_N$ (namely, it is in this case its middle point I), consider a point B on the segment $A_0 A_N$ arbitrary close to I (see Fig. 3.9). We may now apply the above argument to the segment BI, and we are, therefore, certain that there is a point of the nondifferentiable function which is not on BI itself and which is even closer than B from the bisecting line. ∎

It is noticeable that this theorem establishes the scale divergence of the length of continuous nondifferentiable functions, but lets open the question about the way it diverges. Therefore, the power law divergence of self-similar fractals is only a very particular case, and several other modes of divergence are possible, including slower divergences of the logarithmic type or faster divergences of the exponential type (see Fig. 3.8) and even superexponential divergences toward a finite small scale different from zero, such as the log-Lorentz law encountered in the framework of special scale relativity (see Sec. 4.4).

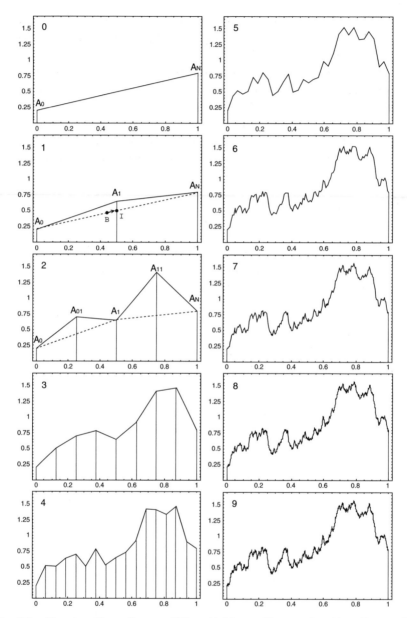

Fig. 3.9: **Construction of a nondifferentiable continuous function.** Due to the nondifferentiability, more precise versions of the function have a tendency to always depart from a linear approximation. As a consequence, the length is explicitly increasing when the resolution scale decreases and it diverges when the scale interval tends to zero.

3.4.2.3. *Inverse problem: from scale divergence to differentiability or nondifferentiability*

Let us now consider the inverse problem: is a continuous function whose length is infinite between any couple of points (such that $x_A - x_B$ finite), i.e. $\mathcal{L}_{AB} = \mathcal{L}_{AB}(\varepsilon) \to \infty$ when $\varepsilon \to 0$, differentiable or nondifferentiable?

The answer depends on the nature of the divergence of the function length, in particular on whether it is homogeneous (when the type of divergence is the same for all points) or inhomogeneous (when different modes of divergences co-exist on the same curve).

Let us prepare the proof by establishing a simple connection between length and slope for a continuous function.

Consider a continuous function $y = f(x)$ in the Euclidean plane between $x = 0$ and $x = 1$. For a given value of the resolution scale δx, the curve can be approximated by the segments that relate the points of the curve of coordinates $\{x_k = k\delta x, y_k = f(x_k)\}$, with $k = 0$ to $N = 1/\delta x$ (see Fig. 3.9). Let us set $\delta y_k = y_k - y_{k-1}$. The length of this approximation, i.e. of this version of the scale-dependent curve at resolution scale δx, is, therefore,

$$\mathcal{L}(\delta x) = \sum_{k=1}^{N} \sqrt{\delta x^2 + \delta y_k^2} \,. \tag{3.19}$$

Introducing the *absolute value of the slope* of each segment, $v_k = |\delta y_k/\delta x|$, one therefore obtains the relation

$$\mathcal{L}(\delta x) = \sum_{k=1}^{N} \delta x \sqrt{1 + v_k^2} = \frac{1}{N} \sum_{k=1}^{N} \sqrt{1 + v_k^2} = \left\langle \sqrt{1 + v_k^2} \right\rangle. \tag{3.20}$$

As a consequence we have:

Theorem 3.4.4 *A necessary and sufficient condition for the length to be infinite (scale divergent) is that the average absolute value v of the slope be infinite, namely $\mathcal{L}$ infinite $\Leftrightarrow \langle v \rangle$ infinite. Moreover, when $\mathcal{L}(\delta x)$ reaches large values,*

$$\mathcal{L} \simeq \langle v \rangle, \tag{3.21}$$

i.e. the length and the mean slope share the same mode of divergence when $\delta x \to 0$.

This result leads us to consider two different modes of divergence of the length, namely:

(i) *Inhomogeneous divergence.* In this case there may exist curves such that only a subset of null measure of their points have divergent slopes, in such a way that the length is nevertheless infinite in the limit $\delta x \to 0$.

As an example, consider a function such that the N segments at resolution δx are distributed as:

(a) N_1 segments such that for them $\langle v_1 \rangle = (\lambda_1/\delta x)^\kappa$,
(b) $N_2 = N - N_1$ segments such that $\langle v_2 \rangle$ remains finite, for example $\langle v_2 \rangle = V[1 - (\delta x/\lambda_2)^{\kappa_1}]$, where V is a finite constant and where $\kappa_1 > 0$, while the ratio N_1/N_2 is itself given by a power law that is a function of δx, e.g. $N_1/N_2 = (\delta x/\lambda_3)^\chi$, with $\chi > 0$. Therefore, $N_1/N_2 \to 0$ when $\delta x \to 0$, i.e. the subset of points where the curve has divergent slopes is indeed of null measure, as required.

In this case the mean value of the absolute value of the slope is

$$\langle v \rangle = \frac{N_1 \langle v_1 \rangle + N_2 \langle v_2 \rangle}{N} \simeq \left(\frac{\lambda_4}{\delta x} \right)^{\kappa - \chi}, \tag{3.22}$$

and the length of the function therefore diverges according to a standard self-similar fractal law

$$\mathcal{L} = \mathcal{L}_0 \left(\frac{\lambda_4}{\delta x} \right)^\tau \tag{3.23}$$

with $\tau = \kappa - \chi = 1 - 1/D_F$. Such a function may, therefore, be almost everywhere differentiable and in the same time be characterized by a genuine fractal law of divergence of its length, of fractal dimension D_F. The same reasoning may be applied to other types of divergences, such as logarithmic, exponential, etc. In this case an infinite curve may be either differentiable or nondifferentiable whatever its divergence mode, i.e. there is no inverse theorem. This conclusion corrects [360, 362], in which we erroneously claimed that the nondifferentiability was ensured for divergences as fast as or faster than power laws.

When it is applied to physics, this result means that a fractal behavior may result from the action of singularities (in infinite number even though forming a subset of null measure) in a space or space-time that nevertheless remains almost everywhere differentiable, such as, for example, Riemannian manifolds in Einstein's general relativity. This comes in support of Mandelbrot's view about the origin of fractals, which are discovered to be extremely frequent in many natural phenomena that yet seem to be well described by standard differential equations: this could come from the existence of singularities in differentiable physics (see e.g. [306], Chapter 11).

But the viewpoint of the scale relativity theory is more radical, since the main problem we aim at solving in its framework is not the (yet very interesting) question of the origin of fractals, but the issue of the foundation of the quantum theory and of gauge fields from geometric first principles. As we shall show, a fractal space-time is not sufficient to reach this goal (in particular as concerns the emergence of complex numbers). One needs to work in the framework of nondifferentiable manifolds, which are indeed fractal (i.e. scale-divergent) as now proved. But the fractality is not central in this context, it mainly appears as a derived (and very useful) property.

(ii) *Homogeneous divergence.* In this case the various slopes are assumed to diverge in the same way for all points of the curve. In other words, we assume that, for any couple of points the absolute values v_1 and v_2 of their scale-dependent slopes verify: $\exists K_1$ and K_2 finite, such that, $\forall \delta x$, $K_1 < v_2(\delta x)/v_1(\delta x) < K_2$. Then the mode of divergence of the mean is the same as the divergence of the slope on the various points and it is also the mode of divergence of the length. In this case the inverse theorem is true, namely,

Theorem 3.4.5 *In the case of homogeneous divergence, the length $\mathcal{L}(f)$ of a fractal curve or function f satisfies:*
$\mathcal{L}(f)$ infinite (i.e. $\mathcal{L} = \mathcal{L}(\delta x) \to \infty$ when $\delta x \to 0$) $\Leftrightarrow$ f nondifferentiable.

3.5. Explicit Scale Dependence on Resolution

The above theorem (3.4.2) is the key for a description of nondifferentiable processes in terms of differential equations. It leads to explicitly introducing the resolutions in the expressions of the main physical quantities, and, as a consequence, in the fundamental equations of physics.

This means that a physical quantity f, usually expressed in terms of space-time variables x, i.e. $f = f(x)$, must be now described as also depending on resolutions, $f = f(x, \varepsilon)$. In other words, rather than considering only the strictly nondifferentiable mathematical object $f(x)$ defined at the limit $\varepsilon \to 0$, we consider its various realizations at all possible resolutions.

Let us now be more specific about the meaning of this proposal, by recalling the physical definition we have given to fractal function in [353]. Indeed, the meaning of ε in a fractal function $f(x, \varepsilon)$ is not that of an usual variable. It instead owns a status similar to that of uncertainties and error bars and this should be included in its very definition.

The goal here is to have a definition of theoretical functions that would match the experimental and observational reality of measured quantities. Namely, it is a well-known fact that various physical quantities may be measured, thanks to the experimental progress, with better and better resolution. The result of such successive measurements usually yields, for a classical variable and provided there has been no large unidentified systematic errors, newly measured values that agree within error bars with all the preceding ones (see e.g. [435] and Fig. 3.10). It is definitely this particular process that we want to include in the very definition of a scale-dependent function. The difference is that it should now apply to all of its points, instead of applying it to only one point.

From a purely mathematical point of view, one could define a scale-dependent fractal function as

$$f(x, \varepsilon) = \int_{-\infty}^{+\infty} \Phi(x, y, \varepsilon) f(x + y) dy \qquad (3.24)$$

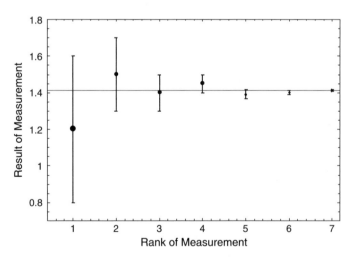

Fig. 3.10: **Measurements of a physical quantity at improving resolutions.** Example of the successive measurements of a given physical quantity at various resolutions. If the error bars have been correctly estimated at each step, the results of new measurements at improved resolutions should agree within uncertainties with the previous ones.

where $\Phi(x, y, \varepsilon)$ is a smoothing function centered on x, for example a step function of width $\approx 2\varepsilon$, or a Gaussian of standard deviation $\approx \varepsilon$ (which may, in general, be a function of position). In other words, this is a kind of wavelet transformation, but using a filter that is not necessarily conservative. Wavelets are now widely used for data analysis of fractal, scale dependent and hierarchically organized phenomena, and one needs to construct a theoretical equivalent.

However the above definition is not satisfactory for our physical purpose, since it assumes that the limiting function $f(x) = f(x, 0)$ is already known. This is in contradiction with the non-reductionist view that underlies the present work: new information is expected to appear when one changes the scale. We shall show in Chapter 4 that the true physical nature of the scale variables leads one to naturally represent them in terms of $\ln \varepsilon$ — more precisely, accounting for the dimensionality of the scale variable, of $\ln(\varepsilon/\lambda)$ — rather than ϵ. Under this form the fractal function reads $f = f(x, \ln \varepsilon)$, so that the limit function $f(x)$ now reads $f(x) = f(x, -\infty)$. This exhibits the true nature of the zero point as actually being an infinity. We definitely consider that $f(x) = f(x, -\infty)$ is devoid of physical meaning, the limit $\varepsilon \to 0$, i.e. $\ln \varepsilon \to -\infty$ being an unreachable horizon. Only the transformation from one finite scale to another finite scale does have physical meaning (i.e. a local and finite scale dilatation or contraction).

In Sec. 3.8 of [353], we have proposed a definition of fractal functions based on two steps. The first step defines the way by which one relates the function seen at a given finite scale to another finite scale. It simply ensures that $f(x, \varepsilon')$ can be obtained from a smoothing out of $f(x, \varepsilon)$ for $\varepsilon' > \varepsilon$ both finite, but it no longer assumes that $f(x, \varepsilon)$ is known for $\varepsilon = 0$. The second step of the definition consists of ensuring that ε does have a status similar to uncertainties, i.e. to define the equality of fractal functions within a given resolution, for all values of this resolution. Note that, as remarked by Cresson [115], one should reverse the order of the two steps for them to be fully self-consistent.

Definition A fractal function $f(x, \varepsilon)$ is a function of a variable x and of a resolution scale ε that satisfies the relations:

(i) Change of scale:

$$\forall x, \forall \varepsilon' > \varepsilon, \quad f(x, \varepsilon') = \int_{-\infty}^{+\infty} \Phi(x, y, \varepsilon') f(x + y, \varepsilon) dy, \qquad (3.25)$$

where $\Phi(x, y, \varepsilon')$ is a smoothing function of resolution ε'.

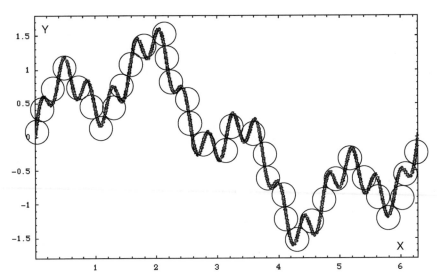

Fig. 3.11: **Fractal function plotted at two resolutions.** Example of a fractal function plotted at two different resolutions by covering it with balls of varying radii. Its equation is $\sum_{k=0}^{\infty} \sin(4^k x)/2^k$.

(ii) Equality (equivalence class within resolution):

$$f \equiv g \Leftrightarrow \forall \varepsilon, \forall x, \exists x', |x - x'| < k\varepsilon, f(x, \varepsilon) = g(x', \varepsilon), \qquad (3.26)$$

where k defines the accepted statistical level of agreement. For example, if the filter is a Gaussian of dispersion ε, the choice $k = 3$ corresponds to a 3σ statistical agreement.

Other equivalent definitions have been given in [353]. Indeed, the agreement between two curves in the (x, y) plane can be checked by using either the x resolution, as in the above definition or the y resolution, or an uncertainty ellipse combining both of them (see next section). For example, one may more generally define a fractal function as an explicit function of x, δx and δy, $y = f[x(\delta x), \delta y]$, and define the equality (within $k\sigma$) of two fractal functions of this type, $f[x(\delta x), \delta y]$ and $g[x(\delta x), \delta y]$, as (see Fig. 3.12):

$$f \equiv g \Leftrightarrow \forall x, \delta x, \delta y, \delta x', \delta y', \exists x', |x' - x| < k\sigma_x, |y' - y| < k\sigma_y, \qquad (3.27)$$

with $\sigma_x = \sqrt{\delta x^2 + \delta x'^2}$ and $\sigma_y = \sqrt{\delta y^2 + \delta y'^2}$. In any case, it amounts to applying to all points of a full curve (see Figs. 3.11, 3.12) the various statistical methods, which are now well established for data analysis, for example in particle physics (see e.g. [435], Statistics).

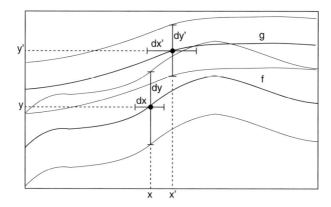

Fig. 3.12: **Equality of fractal functions.** Two explicitly resolution dependent functions are equal up to (within) their resolutions. Their equality is ensured provided the statistical agreement between the two curves remains true for all resolutions and for all points of the curve.

Such a definition is expected to be particularly well adapted to applications in physics. Indeed, as we have already stressed, any real measurement is always performed with a finite resolution (see [353, 349] for additional comments on this point).

In this framework, $f(x)$ becomes the $\varepsilon \rightarrow 0$ limit of the family of functions $f(x, \varepsilon)$. But while $f(x, 0)$ is nondifferentiable, the fractal function $f(x, \varepsilon)$ is differentiable for all $\varepsilon \neq 0$.

The problem of the physical description of the processes where the function f intervenes is now laid down in a new way. In standard differentiable physics, it amounts to finding differential equations involving the derivatives of f, namely $\partial f/\partial x, \partial^2 f/\partial x^2$ and describing the laws of displacement and of motion. The integro-differentiation method amounts to performing such a local description, then integrating to get the global properties of the system under consideration. Such a method has often been called "reductionist", and it was indeed adapted to most classical problems where no new information appears at different scales.

But the situation is completely different for systems involving nondifferentiability at a fundamental level, like the space-time of microphysics itself as suggested in [344, 349, 353]. At high energies, the properties of quarks, of nucleons, of the nucleus, of atoms are interconnected but not reducible ones to the others. In living systems, the scales of DNA bases, chromosomes, nuclei, cells, tissues, organs, organisms, then social scales, do coexist, are related one with another, but are certainly not reducible to one particular scale, even the smaller one. In such cases, new, original information may exist at different scales, and the project to reduce the behavior of a system at one scale (in general, the large one) from its description at another scale (in general, the smallest one, described by the limit $\delta x \to 0$ in the framework of the standard differentiable tool) seems to lose its meaning and to be hopeless. Our suggestion consists precisely to give up such a hope and to introduce a new frame of thought where all scales coexist simultaneously in a scale space, but are connected together via scale differential equations. As we shall see, the solutions of the scale equations that come under the principle of scale relativity are able to describe not only continuous scaling behavior on some ranges of scales, but also the existence of sudden transitions at some particular scale.

Indeed, in non-differentiable physics, $\partial f(x)/\partial x = \partial f(x,0)/\partial x$ does not exist any longer. But the physics of the given process will be completely described if we succeed knowing $f(x,\varepsilon)$ for all values of ε, which is differentiable when $\varepsilon \neq 0$, and can be the solution of differential equations involving $\partial f(x,\varepsilon)/\partial x$ but also $\partial f(x,\varepsilon)/\partial \ln \varepsilon$. More generally, if one looks for nonlinear laws, one expects the equations of physics to take the form of second order differential equations, which then contain, in addition to the previous first derivatives, operators like $\partial^2/\partial x^2$ (laws of motion), $\partial^2/(\partial \ln \varepsilon)^2$ (laws of scale), but also $\partial^2/\partial x \partial \ln \varepsilon$, which corresponds to a coupling between motion laws and scale laws.

3.6. Nature of Resolutions: Covariance Matrix

Before skipping to the step by step description of the implementation of our program, let us give a hint of what will be a more complete description of the resolution variables.

The genuine nature of the scale variables, as they are defined in experimental situations, is actually tensorial. Indeed, the uncertainties are described, in the case of multivariable measurements, by a covariance matrix. In two dimensions it defines an error ellipse, and more generally an error ellipsoid in N dimensions. We shall rarely use this tensorial representation in the present book, except for applications to non-Abelian gauge theories (Chapter 7.5). Due to the high complexity of the program, whose present state of the art is described here, we have voluntarily chosen to proceed step by step and to construct the various structures originating from nondifferentiability by beginning with the simplest ones and then by generalizing to a more complete description. However, we also need to prepare future developments, which is the purpose of this section.

We shall consider here in detail only the two-dimensional case. The generalization to a larger number of dimensions, i.e. to a resolution ellipsoid, is straightforward.

It is clear that a scalar representation of the resolution intervals (in terms of a single scale variable ε, or equivalently of a global dilation ρ) is only a simplified description of their true behavior.

A next step would consist of using a vectorial representation. However, a vectorial description such as given by finite differences or differentials $(\delta x, \delta y)$ can be only an incomplete approximation of their true nature. To be convinced of this fact, one may simply remark that if $(\delta x, \delta y)$ was to be a vector, then one could rotate the reference system in such a way that, in the new coordinate system, the resolution vector becomes $(0, \delta y')$. Our whole analysis falsifies such a possibility, since we have shown that a vanishing resolution scale has actually no physical meaning.

This leads us to introducing a tensorial (or matrix) description of resolutions. Such a description agrees with the nature of these variables, which have a status similar to "uncertainties" or "errors". Such multidimensional uncertainties are fully described by error ellipses (see e.g. [435]). Recall that we have made the choice of the word "resolution" to designate them, in order to definitively give up the classical view according to which physical variables could be defined or known with an infinite precision, which implies that a measurement with a finite precision would be "uncertain" or "with error". The scale relativistic view [349, 353, 368] is instead that the finite character of the resolution at which a measurement is performed and more generally through which any physical quantity can be defined is consubstantial of the physical description.

We are, therefore, led to introduce a resolution tensor similar to a variance-covariance matrix (index 1 is for x and 2 for y)

$$\varepsilon_{jk} = \begin{pmatrix} \varepsilon_x^2 & \rho_{xy}\varepsilon_x\varepsilon_y \\ \rho_{xy}\varepsilon_x\varepsilon_y & \varepsilon_y^2 \end{pmatrix}.$$

Accounting for the fact that the resolutions can be defined only relatively to a given reference scale, one more correctly defines the dimensionless matrix

$$\eta_{jk} = \frac{\varepsilon_{jk}}{\lambda^2} = \begin{bmatrix} (\varepsilon_x/\lambda)^2 & \rho_{xy}(\varepsilon_x/\lambda)(\varepsilon_y/\lambda) \\ \rho_{xy}(\varepsilon_x/\lambda)(\varepsilon_y/\lambda) & (\varepsilon_y/\lambda)^2 \end{bmatrix}.$$

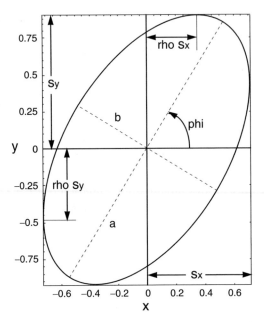

Fig. 3.13: **Geometric elements of a resolution ellipse.** Note that s_x in the figure stands for ε_x in the text and s_y for ε_y, a for ε_M, b for ε_m, phi for φ and rho for ρ.

The number of degrees of freedom is, therefore, three in two dimensions instead of two, since one should also account for the correlation ρ_{xy}, and more generally, since the resolution tensor is symmetric, $N(N+1)/2$ degrees of freedom in N dimensions.

Then one may define a resolution ellipse (in two dimensions), in similarity with error ellipses. Its equation is (omitting the xy index of ρ for simplifying the writing):

$$\frac{1}{1-\rho^2}\left(\frac{x^2}{\varepsilon_x^2} - 2\rho\frac{xy}{\varepsilon_x\varepsilon_y} + \frac{y^2}{\varepsilon_y^2}\right) = 1. \tag{3.28}$$

Let $a = \varepsilon_M$ be its semi-major axis, $b = \varepsilon_m$ its semi-minor axis, and φ its orientation angle (see Fig. 3.13). The relations of these elements to ρ, ε_x and ε_y read:

$$\varepsilon_x^2 = \varepsilon_M^2 \cos^2\varphi + \varepsilon_m^2 \sin^2\varphi, \tag{3.29}$$

$$\varepsilon_y^2 = \varepsilon_M^2 \sin^2\varphi + \varepsilon_m^2 \cos^2\varphi, \tag{3.30}$$

$$\rho\varepsilon_x\varepsilon_y = (\varepsilon_M^2 - \varepsilon_m^2)\sin\varphi\cos\varphi. \tag{3.31}$$

Reversely, the angle of the major axis of the ellipse and the maximal and minimal resolutions (i.e. the semi-major and semi-minor axes) are given by

$$\tan 2\varphi = \frac{2\rho\varepsilon_x\varepsilon_y}{\varepsilon_x^2 - \varepsilon_y^2}, \tag{3.32}$$

$$\varepsilon_M^2 = \frac{1}{2}\left(\varepsilon_x^2 + \varepsilon_y^2 + \sqrt{(\varepsilon_x^2 - \varepsilon_y^2)^2 + 4\rho^2\varepsilon_x^2\varepsilon_y^2}\right), \tag{3.33}$$

$$\varepsilon_m^2 = \frac{1}{2}\left(\varepsilon_x^2 + \varepsilon_y^2 - \sqrt{(\varepsilon_x^2 - \varepsilon_y^2)^2 + 4\rho^2\varepsilon_x^2\varepsilon_y^2}\right). \tag{3.34}$$

The parametric equation of the resolution ellipse (Fig. 3.13) is finally given by

$$x = \varepsilon_M \cos\varphi \cos\theta - \varepsilon_m \sin\varphi \sin\theta, \tag{3.35}$$

$$y = \varepsilon_M \sin\varphi \cos\theta + \varepsilon_m \cos\varphi \sin\theta, \tag{3.36}$$

for $\theta = 0$ to 2π, where φ, ε_M and ε_m are given in terms of the data $(\rho, \varepsilon_x, \varepsilon_y)$ by the three previous relations (3.32)–(3.34).

Now one may consider the resolution tensor as a contravariant metric-like tensor, $G^{jk} = \rho^{jk}\varepsilon^j\varepsilon^k$, with $\rho^{kk} = 1$, by adopting also a contravariant notation for the ε's. Its covariant form is given by the relation (adopting Einstein's notation about the summation over covariant and contravariant indices)

$$G^{ji}G_{ik} = \delta_k^j, \tag{3.37}$$

namely,

$$G_{jk} = \frac{1}{1 - \rho^2}\begin{pmatrix} 1/\varepsilon_x^2 & -\rho_{xy}/\varepsilon_x\varepsilon_y \\ -\rho_{xy}/\varepsilon_x\varepsilon_y & 1/\varepsilon_y^2 \end{pmatrix}.$$

We recognize its expression in the ellipse equation, which, therefore, may be written under a metric-like form

$$G_{jk} x^j x^k = 1, \tag{3.38}$$

where the resolutions and the correlation coefficient define the "resolution metric potentials". More generally, one may define concentration ellipses

$$\frac{1}{1 - \rho^2}\left(\frac{x^2}{\varepsilon_x^2} - 2\rho\frac{xy}{\varepsilon_x\varepsilon_y} + \frac{y^2}{\varepsilon_y^2}\right) = \chi^2, \tag{3.39}$$

which are homothetic to the indicatrix ellipse and allow one to vary the statistical level of agreement that defines the identity (within resolution) between two curves.

This description of the resolution variables in terms of a resolution ellipse solves the hereabove considered problem of the resolution obtained after a rotation of the axes. Indeed, a rotation can no longer yield a zero resolution scale, as in the vectorial representation. The minimal possible resolution is now obtained along the minor axis of the ellipse ε_m, after a rotation that suppresses the correlation and reduces the resolution matrix to a diagonal form.

This remark leads one to ask the question of the value ε_β of the resolution that is expected along any direction with position angle β. This problem is now easily solved, since the answer is given by intersecting the ellipse with the line of equation $y = x \tan \beta$. We obtain (setting $x = x^1$ and $y = x^2$)

$$\varepsilon_\beta^2 = (G_{jk}\, x_\beta^j x_\beta^k)^{-1}, \tag{3.40}$$

with $x_\beta = x_\beta^1 = \cos \beta$ and $y_\beta = x_\beta^2 = \sin \beta$. This relation may be used to cover a curve with resolution ellipses of arbitrary orientation, as seen in Fig. 3.14 for a standard curve and in Fig. 3.15 for a fractal one. Let $y = f(x)$ be such a curve, then we cover it with a resolution ellipse characterized by ρ, ε_x and ε_y. Provided the size of the ellipse matches the curve resolution, the distance between two ellipses is given by $2\varepsilon_\beta$, with $\tan \beta = f' = dy/dx$,

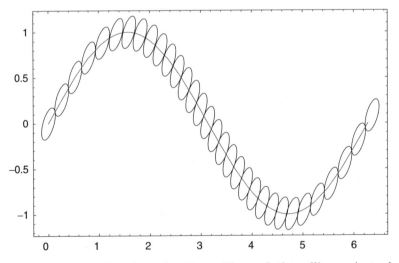

Fig. 3.14: **Covering of a sinus function with resolution ellipses.** A standard differentiable function (here, a sinus) is covered with non orthogonal resolution ellipses.

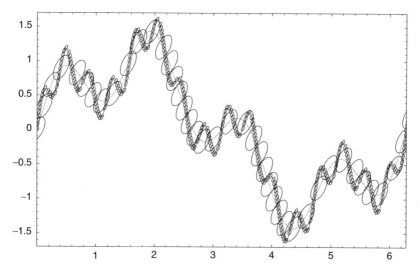

Fig. 3.15: Covering of a fractal function with resolution ellipses. Covering of the fractal function $\sum_{k=0}^{\infty} \sin(4^k x)/2^k$ with non orthogonal resolution ellipses, at two different resolutions.

with

$$\varepsilon_\beta^2 = \frac{dx^2 + dy^2}{G_{jk} dx^j dx^k}. \tag{3.41}$$

It is remarkable that this is the ratio of the standard Euclidean metric over the resolution metric. This formula, therefore, combines the theoretical view about the curve, described by the ideal differential elements (dx, dy) and the experimental view, described by the resolution matrix depending on $(\rho, \varepsilon_x, \varepsilon_y)$.

In three dimensions, the resolution tensor reads:

$$\varepsilon^{ij} = \begin{pmatrix} \varepsilon_1^2 & \rho_{12}\varepsilon_1\varepsilon_2 & \rho_{13}\varepsilon_1\varepsilon_3 \\ \rho_{12}\varepsilon_1\varepsilon_2 & \varepsilon_2^2 & \rho_{23}\varepsilon_2\varepsilon_3 \\ \rho_{13}\varepsilon_1\varepsilon_3 & \rho_{23}\varepsilon_2\varepsilon_3 & \varepsilon_3^2 \end{pmatrix}.$$

The inverse matrix, such that $\varepsilon_{ij}\varepsilon^{jk} = \delta_i^k$, reads:

$$\varepsilon_{ij} = \frac{1}{1-\rho^2} \begin{pmatrix} \frac{1-\rho_{23}^2}{\varepsilon_1^2} & \frac{\rho_{13}\rho_{2,3}-\rho_{12}}{\varepsilon_1\varepsilon_2} & \frac{\rho_{12}\rho_{23}-\rho_{13}}{\varepsilon_1\varepsilon_3} \\ \frac{\rho_{13}\rho_{23}-\rho_{12}}{\varepsilon_1\varepsilon_2} & \frac{1-\rho_{13}^2}{\varepsilon_2^2} & \frac{\rho_{12}\rho_{13}-\rho_{23}}{\varepsilon_2\varepsilon_3} \\ \frac{\rho_{12}\rho_{23}-\rho_{13}}{\varepsilon_1\varepsilon_3} & \frac{\rho_{12}\rho_{13}-\rho_{23}}{\varepsilon_2\varepsilon_3} & \frac{1-\rho_{12}^2}{\varepsilon_3^2} \end{pmatrix}$$

where $\rho^2 = \rho_{12}^2 + \rho_{23}^2 + \rho_{13}^2 - 2\rho_{12}\rho_{23}\rho_{13}$. It allows to define an indicatrix resolution ellipsoid according to the metric-like equation

$$\varepsilon_{ij}\xi^i\xi^j = 1. \tag{3.42}$$

Applied to differential elements, it takes a metric form $d\sigma^2 = \varepsilon_{ij}d\xi^i d\xi^j$.

Problems and Exercises

Exercise 7 Generalize the expression of the scale dependence of the length of a fractal curve to a tensorial description of the resolution including a correlation between the x and y resolutions, i.e. find the general law $\mathcal{L} = \mathcal{L}(\varepsilon_x, \varepsilon_y, \rho_{xy})$ for a curve of constant fractal dimension D_F.

Hint: By performing a rotation of the reference system, one may skip to uncorrelated resolutions ε_m and ε_M. In terms of these resolutions, the length is given by

$$\mathcal{L} = \mathcal{L}_0 \left\{ \left(\frac{\lambda_m}{\varepsilon_m} \right)^\tau + \left(\frac{\lambda_M}{\varepsilon_M} \right)^\tau \right\}. \tag{3.43}$$

Then replace ε_m and ε_M by their expressions (Eqs. (3.33) and (3.34)). ■

Exercise 8 Generalize the above matrix definition of resolutions as resolution tensors and resolution ellipsoids to four space-time dimensions.

Hint: Concerning resolutions (which are a re-interpretation of what is usually considered as uncertainties or "errors"), the signature of the resolution metric is $(+,+,+,+)$, i.e. the space and time resolutions (uncertainties) combine themselves quadratically, despite the minus sign on coordinates in the Minkowski metric. ■

Chapter 4

LAWS OF SCALE TRANSFORMATIONS

4.1. Introduction

As we have seen in the previous chapters, the first step in the construction of the theory consists in finding the laws of explicit scale dependence, which arise as a manifestation of the principle of scale relativity. In analogy with the physics of motion, we assume that these laws are solutions of differential equations, but now acting both in scale-space and in classical space-time. In other words, we consider an infinitesimal dilation of the scale variable, i.e. an infinitesimal "zoom", and we attempt to establish its effect on a physical quantity, which becomes an explicit function of the resolution, in particular the fractal coordinates and the associated velocities themselves. We shall therefore be led to consider first order partial differential equations of scale, which allow us to recover the usual self-similar fractals of constant fractal dimensions, but also to generalize them. In particular, this more general behavior includes scale symmetry breaking and variable fractal dimensions. Another generalization consists in writing second order differential equations of scale transformation, in analogy with the laws of dynamics in motion physics.

In what follows the variable, to which we apply these methods, is the length of a fractal curve that may also represent a coordinate in a fractal reference system. But all the results obtained may be easily generalized to fractal surfaces, fractal volumes, etc. The only difference between these cases is their topological dimension D_T. Since it is not the fractal dimension D_F, which directly appears in these equations, but a scale exponent $\tau = D_F - D_T$ (which will be identified with a "scale-time"), the fractal dimension is simply $D_F = 1 + \tau$ for lengths, $2 + \tau$ for surfaces, $3 + \tau$ for volumes, etc.[a]

Note also that in order to simplify the development we consider only in this chapter the dependence on scale. In other words, we describe internal scale structures at a given "point" of standard space-time. However, it must be clear from the very beginning that it is only a first step of the approach, and that the functions we consider ultimately depend on scale variables, on positions (space) and on instants (time). This more complete description will be the subject of the subsequent chapters.

The results contained in this chapter are mainly based on the works published in [353, 355, 363, 381, 409, 410] and they extend them.

4.2. Scale Invariance and Galilean Scale-relativity

Consider a non-differentiable (fractal) curvilinear coordinate $\mathcal{L}(x, \varepsilon)$ that depends on some parameter x and on the resolution ε. Such a coordinate generalizes to non-differentiable and fractal space-times the concept of curvilinear coordinates introduced for curved Riemannian space-times in Einstein's general relativity [353]. $\mathcal{L}(x, \varepsilon)$ being differentiable when $\varepsilon \neq 0$, it can be the solution of partial differential equations involving the derivatives of $\mathcal{L}$ with respect to both x and ε.

4.2.1. *Differential dilation operator*

Let us apply an infinitesimal dilation $\varepsilon \to \varepsilon' = \varepsilon(1 + d\rho)$ to the resolution. Being, at this stage, interested in pure scale laws, we omit the x dependence in order to simplify the notation and we obtain, to first order,

$$\mathcal{L}(\varepsilon') = \mathcal{L}(\varepsilon + \varepsilon \, d\rho) = \mathcal{L}(\varepsilon) + \frac{\partial \mathcal{L}(\varepsilon)}{\partial \varepsilon} \, \varepsilon \, d\rho = (1 + \tilde{D} \, d\rho) \, \mathcal{L}(\varepsilon), \qquad (4.1)$$

where $\tilde{D}$ is, by definition, the dilation operator. The identification of the two last members of this equation yields

$$\tilde{D} = \varepsilon \frac{\partial}{\partial \varepsilon} = \frac{\partial}{\partial \ln \varepsilon}. \qquad (4.2)$$

This is the well-known form of the infinitesimal dilation operator, obtained above by the "Gell-Mann–Levy method", which allows to find

---

[a]The case of fractal "volumes" constructed by iteratively removing smaller and smaller parts from an initial three-dimensional object should be taken with caution, since their true topological dimension is $D_T = 2$.

the currents corresponding to a given symmetry [7]. It clearly shows that the natural variable for the resolution is $\ln \varepsilon$, and that the expected new differential equations involve quantities like $\partial \mathcal{L}(x, \varepsilon)/\partial \ln \varepsilon$.

The renormalization group equations, in the multi-scale-of-length approach proposed by Wilson [538, 539], already describe a similar kind of scale dependence. The scale relativity approach allows to suggest more general forms for these scale groups (and for their symmetry breaking).

4.2.2. Simplest differential scale law

The simplest differential equation of explicit scale dependence which one can write is first order and states that the variation of $\mathcal{L}$ under an infinitesimal scale transformation $d \ln \varepsilon$ depends only on $\mathcal{L}$ itself. Using the previous derivation of the form of the dilation operator, we thus write

$$\frac{\partial \mathcal{L}(s, \varepsilon)}{\partial \ln \varepsilon} = \beta(\mathcal{L}). \tag{4.3}$$

This is reminiscent of some renormalization group equations, for example that of coupling constants in quantum field theories (see Chapter 11).

The function β is *a priori* unknown. However, still looking for the simplest form of such an equation, we expand $\beta(\mathcal{L})$ in powers of $\mathcal{L}$: $\beta(\mathcal{L}) = a + b\mathcal{L} + \cdots$. Disregarding the s dependence for the moment, we obtain, to first order, the following linear equation, in which a and b are constants:

$$\frac{d\mathcal{L}}{d \ln \varepsilon} = a + b\mathcal{L}. \tag{4.4}$$

In order to simplify the writing of the solution of this equation, let us change the names of the constants as $\tau = -b$ and $\mathcal{L}_0 = a/\tau$, so that $a + b\mathcal{L} = -\tau(\mathcal{L} - \mathcal{L}_0)$. We obtain the equation

$$\frac{d\mathcal{L}}{\mathcal{L} - \mathcal{L}_0} = -\tau \, d \ln \varepsilon, \tag{4.5}$$

i.e.

$$\frac{d \ln(\mathcal{L} - \mathcal{L}_0)}{d \ln \varepsilon} = -\tau. \tag{4.6}$$

Its solution reads

$$\mathcal{L} = \mathcal{L}_0 + \mathcal{L}_1 \, \varepsilon^{-\tau}, \tag{4.7}$$

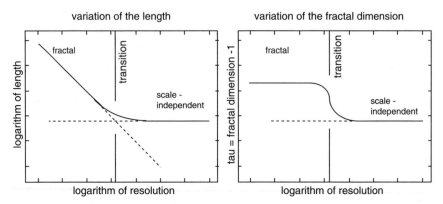

Fig. 4.1: **Fractal length and fractal dimension: self-similar fractal.** The two figures show the scale dependence of the length of a fractal curve (Eq. 4.8) and of its effective fractal dimension (Eq. 4.19) in the case of "inertial" or "Galilean-like" scale laws, which are solutions of a simple first order scale-differential equation. Toward the small scale one obtains a scale-invariant law with constant fractal dimension, while the explicit scale-dependence is lost at scales larger than some transition scale λ.

where $\mathcal{L}_1$ is an integration constant. After a new redefinition of the constants, it can be put under the form (see Fig. 4.1)

$$\mathcal{L}(\varepsilon) = \mathcal{L}_0 \left\{ 1 + \left(\frac{\lambda}{\varepsilon} \right)^\tau \right\}. \tag{4.8}$$

This solution corresponds to a length measured on a fractal curve up to a given point. One can now generalize it to a variable length that also depends on the position characterized by the parameter s. One obtains

$$\mathcal{L}(s, \varepsilon) = \mathcal{L}_0(s) \left\{ 1 + \zeta(s) \left(\frac{\lambda}{\varepsilon} \right)^\tau \right\}, \tag{4.9}$$

in which, in the most general case, the fractal dimension $D_F = 1 + \tau$ may itself be a variable depending on the position.

The same kind of result is obtained for the projections on the various axes of such a fractal length. As recalled in [353], such projections have usually the same fractal dimension as the original curve (see Exercise 9).

Let $X(s, \varepsilon)$ be one of these projections, it reads

$$X(s, \varepsilon) = x(s) \left\{ 1 + \zeta_x(s) \left(\frac{\lambda}{\varepsilon} \right)^\tau \right\}. \tag{4.10}$$

In this case $\zeta_x(s)$ becomes a highly fluctuating function, even possibly stochastic, as can be seen in Fig. 4.2.

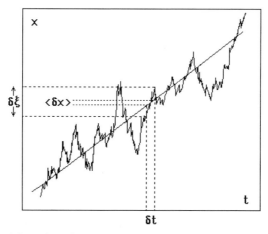

Fig. 4.2: **Fractal function.** An example of fractal function is given by the projections of a fractal curve on Cartesian coordinates, in function of a continuous and monotonous parameter (here the time t) which marks the position on the curve. The figure also exhibits the relation between space and time differential elements for such a fractal function, and compares the differentiable and non-differentiable parts of the space elementary displacement. While the "classical" coordinate variation $\delta x = \langle \delta X \rangle$ is of the same order as the time differential δt, the fractal fluctuation becomes much larger than δt when $\delta t \ll T$, where T is a transition time scale, and it depends on the fractal dimension D_F as: $\delta \xi \propto \delta t^{1/D_F}$. Therefore the two contributions to the full differential displacement are related by the fractal law $\delta \xi^{D_F} \propto \delta x$, since δx and δt are differential elements of the same order.

The important point here is that the solution obtained is the sum of two terms, a classical (differentiable) part that depends only on the position and a fractal (nondifferentiable) part that depends on the position and on ε in a divergent way. By differentiating these two parts in the above projection, we obtain the differential formulation of this essential result,

$$dX = dx + d\xi, \tag{4.11}$$

where dx is a classical differential element, while $d\xi$ is a differential element of fractional order, namely $d\xi \propto dx^{1/D_F}$ (see Fig. 4.2, in which the parameter s that characterizes the position on the fractal curve has been taken to be the time t). This relation plays a fundamental role in what follows.

In Sec. 3.3, in the context of the description of empirical fractal laws, we pointed out that the resolution variable ε may have various physical meanings. The same caution is relevant in the present context of scale laws obtained as solutions of scale differential equations. Let us recall again that there are two main ways to characterize a fractal curve (such as, e.g. the length of a coast in geography [306]).

(i) One may measure it at various length scales, for example by varying the map scale. In this case, which is the one considered up to now, the resolution ε is a length interval, $\varepsilon = \delta X$, and one obtains the scale dependent length given, by definition, by the law

$$X(s, \delta X) = X_0(s) \times \left(\frac{\lambda}{\delta X} \right)^{D_F - 1} . \tag{4.12}$$

The exponent in this generic solution is then identified in the scale-dependent part of Eq. (4.10) as $\tau = D_F - 1$.

Now one may use the time t instead of a space variable as the position parameter s, and if one travels on the curve at constant velocity, one obtains $X_0(t) = a\, t$. Then a differential version of Eq. (4.12) reads

$$\delta X = a\, \delta t \left(\frac{\lambda}{\delta X} \right)^{D_F - 1} , \tag{4.13}$$

so that we recover the usual fundamental relation between space differential elements and time differential elements on a fractal curve or function,

$$\delta X^{D_F} \propto \delta t. \tag{4.14}$$

Namely, they are differential elements of different orders, as illustrated by Fig. 4.2.

(ii) One may travel on the curve and measure its length at constant time intervals, then change the time scale. In this case the resolution ε is a time interval, $\varepsilon = \delta t$. One may therefore replace, in Eq. (4.12), the space resolution δX by its expression in function of the time resolution, $\delta X \propto \delta t^{1/D_F}$. Then the fractal length is now found to depend on the time resolution as

$$X(s, \delta t) = X_0(s) \times \left(\frac{T}{\delta t} \right)^{1 - 1/D_F} . \tag{4.15}$$

Therefore, in that case one may identify the exponent τ in the generic solution (4.9) with $\tau = 1 - 1/D_F$. This alternative relation between the scale exponent and the fractal dimension will be very useful in the subsequent construction of the theory.

Let us come back to the simplest definition of the resolution as a space interval, namely, $\varepsilon = \delta X$. In the asymptotic regime $\varepsilon \ll \lambda$, $\tau = -b$ is constant, and one obtains a power-law dependence on resolution that reads

$$\mathcal{L}(s, \varepsilon) = \mathcal{L}_0(s) \left(\frac{\lambda}{\varepsilon}\right)^\tau. \tag{4.16}$$

This pure scaling formula is, more generally, valid also for $(\mathcal{L} - \mathcal{L}_0)$ in the formula Eq. (4.9) that contains a constant term.

When τ is constant, it is therefore given in this case by $\tau = \ln(\mathcal{L}/\mathcal{L}_0)/\ln(\lambda/\varepsilon)$. This relation naturally leads to define the scale exponent $\tau = D_F - D_T$ in a more general way. Following Mandelbrot [305, 306], it can be defined differentially as[b]

$$\tau = \frac{d\ln\mathcal{L}}{d\ln(\lambda/\varepsilon)}. \tag{4.17}$$

In the case, mainly considered here, in which we apply these methods to the length of a curve, the topological dimension is $D_T = 1$ and the scale exponent is related to the fractal dimension D_F by

$$\tau = D_F - 1. \tag{4.18}$$

Anticipating on what follows, the definition (4.17) can also be applied on any fractal curve, not only the strictly self similar ones. In this case τ becomes itself a scale-dependent variable. In particular, one can define an "effective" or "local" scale exponent from the derivative of the complete solution (4.9), that jumps from zero to its constant asymptotic value at the transition scale λ (see the right part of Fig. 4.1 and the following figures). Indeed, derivating the logarithm of Eq. (4.9) with respect to $\ln(\lambda/\varepsilon)$ yields an effective exponent given by

$$\tau_{\text{eff}} = \frac{\tau}{1 + (\varepsilon/\lambda)^\tau}. \tag{4.19}$$

---

[b]Note the change of notation with respect to previous publications, in which this scale exponent $D_F - D_T$ was most of the time denoted δ. As we shall see in the following, it can be generalized from a constant to a varying exponent, and it becomes, as such, a fundamental variable that we have called "djinn". This variable plays for scale laws a role similar to that plays for motion laws by time. It is therefore a kind of "scale-time". This is the basis for the newly adopted notation τ, which seems preferable to δ since we shall be led to differentiate it. In this regard, the case of a constant τ studied in these first sections is the analog of statics for motion laws.

Note also that we have considered here (in Eq. (4.8) and Fig. 4.1) the case of a fractal divergence toward the small scales and of a transition to scale independence toward the large scales. But the reverse situations may also occur in nature (for example, in geography [310]). They are easily obtained from Eq. (4.8) by a simple inversion of the variables, $\varepsilon \to 1/\varepsilon$ and $\mathcal{L} \to 1/\mathcal{L}$, which makes in all four possible types of transitional configurations (respectively, power law increase of the length toward the small or large scale, scale independence at large or small scales). In terms of the relevant variables, which are logarithmic, this inversion leads to the symmetric variables, $\ln(\varepsilon/\lambda) \to -\ln(\varepsilon/\lambda)$ and $\ln(\mathcal{L}/\mathcal{L}_0) \to -\ln(\mathcal{L}/\mathcal{L}_0)$. One obtains for example, when the transition to scale independence is toward the small scales, a law of transition

$$\mathcal{L} = \frac{\mathcal{L}_0}{1 + (\varepsilon/\lambda)^\tau}, \tag{4.20}$$

which is relevant in many real situations [413].

Problems and Exercises

Exercise 9 Show that under reasonable isotropy conditions, the fractal dimensions of the projections of a fractal curve on the coordinate axes, which are fractal functions, are the same as the fractal dimension of the original curve.

Inversely, show that if one constructs a fractal curve, e.g. in two dimensions, from projections $X(s, \delta s)$ and $Y(s, \delta s)$, which are two fractal functions of different fractal dimensions, it would be infinitely anisotropic.

Find possible applications in theoretical physics.

Hint: writing in the scaling asymptotic domain $X = X_0(\lambda_x/\varepsilon)^{\tau_x}$ and $Y = Y_0(\lambda_y/\varepsilon)^{\tau_y}$, with e.g. $\tau_y > \tau_x$, one finds that $Y/X \propto \varepsilon^{\tau_x - \tau_y} \to \infty$ when $\varepsilon \to 0$. The coordinate X is therefore vanishing at the ultraviolet limit relatively to coordinate Y when both are seen at the same resolution. This could be a new way, different from small scale compactification, to introduce extradimensions at some scales while they would spontaneously vanish at others.   ∎

4.2.3. Galilean relativity of scales

Let us now check that the fractal part of this solution, described by a self-similar scaling law $\mathcal{L} = \mathcal{L}_0(\lambda/\varepsilon)^\tau$, does come under the principle of relativity extended to scale transformations of the resolutions. This simple verification is an essential point of the scale relativity approach, since in its framework the laws of scale to be derived must be constrained (and at some level, constructed) by the principle of relativity itself. More generally, all the following statements remain true for the complete scale law including the transition to scale-independence, by replacing $\mathcal{L}$ by $(\mathcal{L} - \mathcal{L}_0)$.

The above quantities transform, under a scale transformation $\varepsilon \to \varepsilon'$, as

$$\ln \frac{\mathcal{L}(\varepsilon')}{\mathcal{L}_0} = \ln \frac{\mathcal{L}(\varepsilon)}{\mathcal{L}_0} + \tau(\varepsilon) \ln \frac{\varepsilon}{\varepsilon'}, \tag{4.21}$$

$$\tau(\varepsilon') = \tau(\varepsilon). \tag{4.22}$$

We have considered here for full generality, that the two variables of this transformation, $\ln \mathcal{L}$ and τ, may be functions of the resolution (which characterizes in a relative way the state of scale of the reference system). But, clearly, the case under study here is particular since $\tau = $ cst. Note also that we neglect for simplicity the position and time dependence, which means that we consider a given point on the fractal curve instead of a running point.

We also remark from now on that, in accordance with the principle of scale relativity according to which no absolute scale does exist in nature, but only scale ratios, the resolutions ε and ε' do not appear directly in this transformation, but only their ratio $\rho = \varepsilon'/\varepsilon$.

These transformations have exactly the mathematical structure of the Galileo group (applied here to scale instead of motion), as confirmed by the dilation composition law, $\varepsilon \to \varepsilon' \to \varepsilon''$, which reads

$$\ln \frac{\varepsilon''}{\varepsilon} = \ln \frac{\varepsilon'}{\varepsilon} + \ln \frac{\varepsilon''}{\varepsilon'}. \tag{4.23}$$

Let us indeed compare this scale transformation law with the Galileo group of motion transformation, that reads

$$X' = X + TV, \tag{4.24}$$

$$T' = T, \tag{4.25}$$

where V is the velocity of the coordinate system K with respect to the coordinate system K'. The law of velocity composition between three reference systems K, K' and K'' reads

$$V'' = V + V'. \tag{4.26}$$

The two groups have the same structure, accounting for the correspondence $X \leftrightarrow \ln \mathcal{L}$, $T \leftrightarrow \tau$ and $V \leftrightarrow \ln \rho$. Since the Galileo group of motion transformations is known to be the simplest group that implements the principle of relativity (it is nothing but a degenerate case of the Lorentz group under the limit $c \to \infty$), the same is true for scale transformations. We shall come back in more detail on this point in Sec. 4.4.

It is important to realize that this is more than a mere analogy: the physical problem is the same in both cases, and it is, therefore, solved in terms of similar mathematical structures (since the use of logarithms transforms what would have been a multiplicative group into an additive group). Indeed, in both cases, it amounts to finding the transformation law of a position variable (X for motion in a Cartesian system of coordinates, $\ln \mathcal{L}$ for scales in a fractal system of coordinates) under a change of the state of the coordinate system (change of velocity V for motion and of resolution $\ln \rho$ for scale), knowing that these state variables are defined only in a relative way: V is the relative velocity between the reference systems K' and K, and ρ is the relative scale ratio between the scales ε and ε'. Note that ε and ε' have indeed disappeared in the transformation law, only their ratio remains. This remark founds the status of resolutions (in logarithm form) as (relative) "scale velocities" and of the scale exponent τ as a "scale time".

It is also worth noticing that the situation here is actually a degeneracy respectively to the case of motion laws. Indeed, time T flows in general while here the scale-time τ is constant. This situation is therefore an analog of statics, or equivalently of a description of the Galilean laws of motion at a given time. However, we shall see in forthcoming sections that we are naturally led to generalize this constant scale exponent to a variable one.

4.2.4. *Scale relativity versus scale invariance*

Let us briefly be more specific about the way the scale-relativity viewpoint differs from "scaling" or simple "scale invariance". In the standard concept of scale invariance, one considers scale transformations of the coordinate,

$$X \to X' = q \times X, \tag{4.27}$$

then one looks for the effect of such a transformation on some function $f(X)$. It is scaling when

$$f(qX) = q^{\alpha} \times f(X). \tag{4.28}$$

The scale relativity approach involves a more profound level of description, since the coordinate X is now explicitly resolution-dependent, i.e. $X = X(\varepsilon)$. Therefore we now look for a scale transformation of the resolution,

$$\varepsilon \to \varepsilon' = \rho\varepsilon, \tag{4.29}$$

which implies a scale transformation of the position variable, that reads in the self-similar case

$$X(\rho\varepsilon) = \rho^{-\tau} X(\varepsilon). \tag{4.30}$$

But now the scale factor applied on the variable gets a physical meaning, which goes beyond a trivial change of units. It corresponds to a coordinate measured on a fractal curve of fractal dimension $D_F = 1+\tau$ at two different resolutions. Finally, one can also consider again a scaling function of a fractal coordinate like the one in Eq. (4.28). Under a change of resolution, then of the position variable, it now transforms as

$$f[X(\rho\varepsilon)] = f(\rho^{-\tau} X) = \rho^{-\alpha\tau} \times f[X(\varepsilon)]. \tag{4.31}$$

This transformation now involves two combined exponents instead of one.

In the framework of the analogy with the laws of motion and displacement, the dilation (4.27) is the equivalent of a static translation $x' = x + a$. Indeed, it reads in logarithmic form

$$\ln \frac{X'}{\lambda} = \ln \frac{X}{\lambda} + \ln q. \tag{4.32}$$

Note that it can also be generalized to four different dilations on the four coordinates, $\ln(X'_\mu/\lambda) = \ln(X_\mu/\lambda) + \ln q_\mu$. One jumps from static translation $x' = x + a$ to motion by introducing a time dependent translation $a = -vt$, so that one obtains the Galileo law of coordinate transformation, $x' = x - vt$. The transition from a simple dilation law $\ln X' = \ln X + \ln q$ to the law of scale transformation of a fractal self-similar curve, $\ln X' = \ln X - \tau \times \ln \rho$ is therefore of the same nature. In other words, fractals are to scale invariance what motion is to static translations.

These "scale-translations" should not be forgotten when constructing the full scale-relativistic group of transformations, in similarity with the Poincaré group, that adds four space-time translations to the Lorentz group of rotation and motion in space (i.e. rotation in space-time).

It is also noticeable here that such a scale-relativity group will be different and larger than a conformal group, for the two reasons outlined in this chapter:

(i) The conformal group adds to the Poincaré group a global dilatation and an inversion (that leads to four special conformal transformations when combined with translations), yielding a 15 parameter group (four translations, three rotations, three Lorentz boosts, one dilation, four special conformal transformations). But these transformations are applied to the coordinates without specification of their physical cause. In scale relativity, the cause is the fractality, i.e. the resolution dependence of the coordinates. For example, the symmetric element in a resolution transformation is $\ln(\lambda/\varepsilon') = -\ln(\lambda/\varepsilon)$, which is nothing but a resolution inversion $\varepsilon' = \lambda^2/\varepsilon$. A fractal coordinate, which is resolution-dependent as a power law, $L(\varepsilon) = (\lambda_0/\varepsilon)^\tau$, is therefore itself transformed by an inversion, namely $L(\varepsilon') = L_1/L(\varepsilon)$, where $L_1 = (\lambda_0/\lambda)^\tau$.

(ii) More generally we need to define four independent resolution transformations on the four coordinates. Such a transformation does not preserve the angles and it therefore goes beyond the conformal group.

(iii) Ultimately, as we have shown in Sec. 3.6, the true nature of the resolutions is tensorial, since they have a status similar to errors and uncertainties, which are described by a covariance matrix, namely, $\varepsilon_{ij} = \rho_{ij}\varepsilon_i\varepsilon_j$, where ρ_{ij} are correlation coefficients.

4.2.5. *Scale transition*

An important result apparent in Eq. (4.9) is that the scale differential equation we have derived from simple and fundamental arguments have provided us not only with the expected standard fractal behavior, which is typical of many natural scaling systems, but also with a transition from a fractal (i.e. explicitly scale-dependent) behavior to a non-fractal (i.e. scale-independent) behavior at scales larger than some transition scale λ. The existence of such a breaking of scale invariance is also a fundamental feature of most natural systems, which in most cases was misunderstood.

The advantage of the way it is derived here is that it appears as a natural, spontaneous, but only effective symmetry breaking, since it does not affect the underlying scale symmetry. Indeed, the obtained solution is the sum of two terms, the scale-independent contribution that we have called "classical part" (i.e. differentiable part), and the explicitly scale-dependent and divergent contribution that we have called "fractal part" (i.e. nondifferentiable part) [93]. At large scales the scaling part becomes dominated by the classical part, but it is still underlying even if it is hidden. There is therefore an apparent symmetry breaking (see Fig. 4.1), though the underlying scale symmetry actually remains unbroken.

This is to be compared with a renormalization group-like approach, under which the constant a would vanish in Eq. (4.4). Two arguments at least lead us to retain this constant:

(i) A research of generality: we look for the most general among the simplest laws, so that there is no *a priori* reason to suppress the first term of the Taylor expansion.

(ii) The fact that the new scale laws do not take the place of motion-displacement laws, but instead must be combined with them. What we want to describe here is general coordinates, that go beyond Cartesian coordinates, and also beyond curvilinear coordinates in order to include fractality. But all these previous descriptions should also be recovered in the new approach, which proceeds by generalization. One cannot therefore exclude the case when a classical (non fractal) change of coordinate system is applied to a fractal coordinate.

This means that starting from a strictly scale-invariant law, $a = 0$, $\mathcal{L} = \mathcal{L}_0(\lambda/\varepsilon)^\tau$, and adding a translation in standard position space ($\mathcal{L} \to \mathcal{L} + \mathcal{L}_1$), we obtain

$$\mathcal{L}' = \mathcal{L}_1 + \mathcal{L}_0 \left(\frac{\lambda}{\varepsilon}\right)^\tau = \mathcal{L}_1 \left\{ 1 + \left(\frac{\lambda_1}{\varepsilon}\right)^\tau \right\}. \tag{4.33}$$

Therefore, we indeed recover the broken solution ($a \neq 0$). This solution is now asymptotically scale-dependent (in a scale-invariant way) only at small scales, and becomes independent of scale at large scales, beyond some transition λ_1, which is partly determined by the translation itself. This shows the relative character of the transition, that explicitly depends on the state of the reference system: position, orientation and motion.

The scale symmetry is, therefore, spontaneously broken by the very existence of the standard space-time symmetries. Here, the translations, that are part of the full Poincaré group of space-time transformations including also the rotations and the Lorentz boosts. The symmetry breaking is not achieved here by a suppression of one law to the benefit of the other, but instead by a domination of each law (scale versus motion) over the other, respectively toward the small and large scales. Since the transition is itself relative to the state of motion of the reference system, this implies that one can jump from one behavior to the other by a change of the reference system.

As we shall see in what follows, this transition plays an important role in the fractal space-time approach to quantum mechanics, in which it can be identified with the Einstein–de Broglie scale. Therefore the quantum-classical transition acquires a geometric interpretation as being nothing but the fractal-non fractal transition in scale-space [353].

4.2.6. *Multiple scale transitions*

Let us now go one step further and consider the simple generalization in which one keeps the quadratic term in the Taylor expansion of the function $\beta(\mathcal{L})$ in the scale differential equation, namely,

$$\frac{d\mathcal{L}}{d\ln\varepsilon} = a + b\mathcal{L} + c\mathcal{L}^2 + \cdots \tag{4.34}$$

The symmetry between the microscopic and the macroscopic cases can be seen again in this case, directly from the properties of this equation. Let us indeed transform the two variables $\mathcal{L}$ and ε by inversion, i.e. $\mathcal{L} \to \mathcal{L}' = 1/\mathcal{L}$ and $\varepsilon \to \varepsilon' = 1/\varepsilon$. We find that Eq. (4.34) becomes

$$\frac{d\mathcal{L}'}{d\ln\varepsilon'} = c + b\mathcal{L}' + a\mathcal{L}'^2 + \cdots \tag{4.35}$$

This is exactly the same equation up to the exchange of the constants a and c. In other words, Eq. (4.34) is covariant (i.e. form invariant) under the inversion transformation, which transforms the small scales into the large ones and reciprocally, but also the upper symmetry breaking scale into a lower one. Hence the inversion symmetry, which is clearly not achieved in nature at the level of the observed structures, may nevertheless be an exact symmetry at the level of the fundamental laws.

This is confirmed by directly looking at the solutions of Eq. (4.34) while keeping now the quadratic term.

One of the solutions of this equation describes a scaling behavior which is broken toward both the small scales and the large scales, in accordance with most real fractal systems, such as the Britain coast, etc. [305, 306].

Indeed, let us start from the following form of equation (4.34)

$$\frac{d\mathcal{L}}{d\ln(\lambda_0/\varepsilon)} = a + b\mathcal{L} + c\mathcal{L}^2 \tag{4.36}$$

and consider it in the case when the three coefficients may be written under the form

$$a = -\frac{\tau L_0 L_1}{L_0 - L_1}, \quad b = \tau \frac{L_0 + L_1}{L_0 - L_1}, \quad c = -\frac{\tau}{L_0 - L_1}. \tag{4.37}$$

The equation reads

$$\frac{d\mathcal{L}}{d\ln(\lambda_0/\varepsilon)} = -\tau \frac{(\mathcal{L} - L_0)(\mathcal{L} - L_1)}{L_0 - L_1}. \tag{4.38}$$

By introducing a new length-scale λ_1 defined by the relation

$$\frac{L_0}{L_1} = \left(\frac{\lambda_1}{\lambda_0}\right)^\tau, \tag{4.39}$$

the solution can finally be written as

$$\mathcal{L} = L_0 \frac{1 + (\lambda_0/\varepsilon)^\tau}{1 + (\lambda_1/\varepsilon)^\tau}. \tag{4.40}$$

One therefore recovers toward the large scales, i.e. when $\varepsilon > \lambda_1$, the previous generic solution $\mathcal{L} = L_0(1+(\lambda_0/\varepsilon)^\tau)$. This new solution describes a behavior of constant fractal dimension $D_F = 1 + \tau$, with two transitions to scale independence (non fractal behavior) at scales λ_1 and λ_0 toward respectively the small and the large scales. Indeed, as can be seen in Fig. 4.3,

(i) when $\varepsilon < \lambda_1 < \lambda_0$, one has $(\lambda_0/\varepsilon) \gg 1$ and $(\lambda_1/\varepsilon) \gg 1$, so that $\mathcal{L} = L_0(\lambda_0/\lambda_1)^\tau \approx$ cst, independent of scale;

(ii) when $\lambda_1 < \varepsilon < \lambda_0$, one has $(\lambda_0/\varepsilon) \gg 1$ but $(\lambda_1/\varepsilon) \ll 1$, so that the denominator disappears, and one recovers the previous pure scaling law $\mathcal{L} = L_0(\lambda_0/\varepsilon)^\tau$;

(iii) when $\lambda_1 < \lambda_0 < \varepsilon$, one has $(\lambda_0/\varepsilon) \ll 1$ and $(\lambda_1/\varepsilon) \ll 1$, so that $\mathcal{L} = L_0 =$ cst, again independent of scale.

This behavior will play an important role in the subsequent construction of a new macroscopic quantum-type mechanics (see Chapter 10), since it will help us to characterize the scale range conditions under which such a construction is possible.

4.2.7. *Generalization to several resolution variables*

In the above scale laws, only one variable of scale was considered. However, there are many situations where a multidimensional system is measured with resolutions, which are different for the various coordinates. The

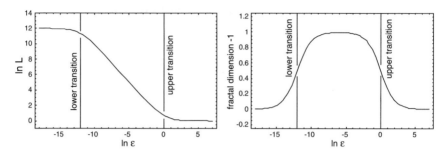

Fig. 4.3: **Fractal length and fractal dimension: two transitions.** Scale dependence of the length $\mathcal{L}$ (left) and of the effective scale exponent $\tau = D_F - 1$, defined
as the fractal dimension minus the topological dimension (right) in the case of a self-
similar behavior (constant fractal dimension) involving two (upper and lower) scales of
transition from scale-dependence to scale-independence (see text). This behavior is a
solution of a first order scale differential equation. Though the ratio of the upper and
lower cut-offs is, in this example, as large as $e^{12} \approx 1.6 \times 10^5$, one finds that, due to the
fact that the transitions occur on a finite range of scales (each of them cover a factor of
about $e^4 \approx 50$), the range of scales over which the effective fractal dimension is constant
remains small (≈ 50).

question that we now address is the following: are the fractal laws obtained
in such a case (see Sec. 3.6) still derivable from a differential equation?

As we shall now see, the first order scale differential equation whose
solutions are the standard self-similar fractals may easily be generalized
in that case. Indeed, let us first consider the two-dimensional case. As
recalled in Sec. 3.3, the length of a fractal curve in a plane, measured with a
resolution interval ε_x along the x coordinate and ε_y along the y coordinate,
reads (in the absence of correlation, see also Sec. 3.6)

$$\mathcal{L}s = \mathcal{L}_0 \left\{ \left(\frac{\lambda_x}{\varepsilon_x} \right)^{\tau} + \left(\frac{\lambda_y}{\varepsilon_y} \right)^{\tau} \right\}, \tag{4.41}$$

i.e. in a more compact writing, $\mathcal{L} = a\,\varepsilon_x^{-\tau} + b\,\varepsilon_y^{-\tau}$. One recovers the same
relations as in the one-dimensional resolution case, but now in terms of
partial derivatives, namely,

$$\varepsilon_x \frac{\partial \mathcal{L}}{\partial \varepsilon_x} = -a\tau\varepsilon_x^{-\tau}, \quad \varepsilon_y \frac{\partial \mathcal{L}}{\partial \varepsilon_y} = -b\tau\varepsilon_y^{-\tau}. \tag{4.42}$$

Therefore we obtain

$$\frac{\partial \mathcal{L}}{\partial \ln \varepsilon_x} + \frac{\partial \mathcal{L}}{\partial \ln \varepsilon_y} = -\tau(a\,\varepsilon_x^{-\tau} + b\,\varepsilon_y^{-\tau}). \tag{4.43}$$

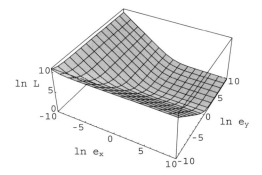

Fig. 4.4: **Fractal length: two scale variables.** Dependence on the x and y resolutions, ε_x and ε_y (noted e_x and e_y in the figure), of the length $\mathcal{L}$ of a fractal curve in a plane, including a transition to scale independence.

This means that the fractal law Eq. (4.41) is a solution of the first order partial differential equation of scale

$$\mathrm{div}_{\mathrm{V}}\mathcal{L} + \tau\mathcal{L} = 0, \tag{4.44}$$

where the $\mathrm{div}_{\mathrm{V}}$ operator acts here in the scale space $(\ln\varepsilon_x, \ln\varepsilon_y)$, namely, $\mathrm{div}_{\mathrm{V}} = \partial/\partial\ln\varepsilon_x + \partial/\partial\ln\varepsilon_y$.

This differential equation can easily be generalized to any number of resolution variables and to account for a transition to scale independence through the replacement of $\mathcal{L}$ by $(\mathcal{L} - \mathcal{L}_0)$, which is equivalent to adding a constant term in the equation. One obtains

$$\sum_k \frac{\partial\mathcal{L}}{\partial\ln\varepsilon_k} + \tau\mathcal{L} + \alpha = 0, \tag{4.45}$$

which is indeed a natural generalization of Eq. (4.4) to several resolution variables (see Fig. 4.4).

4.3. Generalized Scale Laws

4.3.1. *Discrete scale invariance and log-periodic behavior*

Among the possible generalizations of pure scale invariance one of them is potentially important, namely, the log-periodic correction to power laws that is provided, e.g. by complex exponents or complex fractal dimensions [275]. Sornette *et al.* (see [497] and references therein) have shown that such a behavior provides a very satisfactory and possibly statistically predictive model of the time evolution of many critical systems, including earthquakes and market crashes [498]. More recently, it has been applied to the analysis of major event chronology of the evolutionary tree of life [102, 373], of human development [86] and of the main economic crisis of western and Precolumbian civilizations [373, 241].

Let us show how one can recover log-periodic corrections from requiring scale covariance of the scale differential equations [363]. Consider a scale-dependent function $\mathcal{L}(\varepsilon)$, (for example, it may be the length measured along a fractal curve). In the applications to temporal evolution quoted above, the scale variable is identified with the time interval $|t - t_c|$, where t_c is the date of the crisis.

Assume as a first step that $\mathcal{L}$ satisfies a first order "scale-inertial" differential equation,

$$\frac{d\mathcal{L}}{d \ln \varepsilon} - \nu \mathcal{L} = 0, \tag{4.46}$$

whose solution is a power law $\mathcal{L}(\varepsilon) \propto \varepsilon^{\nu}$. Now looking for corrections to this law, we remark that simply jumping to a complex value of the exponent ν would lead to large log-periodic fluctuations rather than to a controlable correction to the power-law. So let us assume that the right-hand side of Eq. (4.46) actually differs from zero, i.e.

$$\frac{d\mathcal{L}}{d \ln \varepsilon} - \nu \mathcal{L} = \chi. \tag{4.47}$$

We can now apply the scale covariance principle and require that the new function χ be a solution of an equation, which keeps the same form as the initial equation,

$$\frac{d\chi}{d \ln \varepsilon} - \nu' \chi = 0. \tag{4.48}$$

Setting $\nu' = \nu + \eta$, we find that $\mathcal{L}$ must be a solution of a second-order equation,

$$\frac{d^2 \mathcal{L}}{(d \ln \varepsilon)^2} - (2\nu + \eta) \frac{d\mathcal{L}}{d \ln \varepsilon} + \nu(\nu + \eta)\mathcal{L} = 0. \tag{4.49}$$

The solution writes $\mathcal{L}(\varepsilon) = a\varepsilon^{\nu}(1 + b\varepsilon^{\eta})$, and finally, the choice of an imaginary exponent $\eta = i\omega$ yields a solution whose real part includes a log-periodic correction,

$$\mathcal{L}(\varepsilon) = a\,\varepsilon^{\nu}\,[1 + b\cos(\omega \ln \varepsilon)]. \tag{4.50}$$

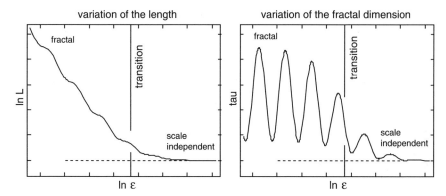

Fig. 4.5: **Fractal length and fractal dimension: log-periodicity.** Scale dependence of the length and of the fractal dimension in the case of a log-periodic correction to constant fractal dimension, including a fractal to nonfractal transition (see text).

As for the previously obtained scale laws, it is easy to generalize this result to the more realistic behavior that includes a symmetry breaking of scale invariance at some given scale. We simply replace $\mathcal{L}$ by $(\mathcal{L} - \mathcal{L}_0)$ in all the above equations. This yields a scaling solution with log-periodic fluctuations, which presents a transition to scale independence toward the large scales (see Fig. 4.5), which may be written as

$$\mathcal{L}(\varepsilon) = \mathcal{L}_0 \left\{ 1 + \left(\frac{\lambda}{\varepsilon}\right)^\tau \left[1 + b\cos\left(\omega\ln\frac{\varepsilon}{\lambda}\right)\right] \right\}, \qquad (4.51)$$

where we have replaced $-\nu$ by τ in order to match with the previous notations.

Let us now give another physically meaningful way to obtain an equivalent behavior without making use of imaginary exponents. Define a log-periodic local scale exponent:

$$\tau = \frac{\partial \ln \mathcal{L}}{\partial \ln \varepsilon} = \nu - b\omega\sin(\omega\ln\varepsilon). \qquad (4.52)$$

It leads after integration to a scale-divergence that reads

$$\mathcal{L}(\varepsilon) = a\varepsilon^\nu e^{b\cos(\omega\ln\varepsilon)}, \qquad (4.53)$$

whose first order expansion is Eq. (4.50). Such a law is a solution of a scale stationary wave equation:

$$\frac{\partial^2}{(\partial \ln \varepsilon)^2}\left(\ln\frac{\mathcal{L}}{\mathcal{L}_0}\right) + \omega^2\ln\frac{\mathcal{L}}{\mathcal{L}_0} = 0, \qquad (4.54)$$

where $\mathcal{L}_0 = a\varepsilon^\nu$ is the strictly self-similar solution. Hence the log-periodic behavior can be viewed as a stationary wave in the scale-space (this prepares Chapter 8, in which we tentatively introduce a quantum-type wave scale equation). Note that these solutions can apply to fractal lengths only for $b\omega < \nu$, since the local scale exponent should remain positive. This behavior is typical of what is observed when measuring the resolution-dependent length of fractal curves of the von Koch type, which are built by iteration and, strictly, have only discrete scale invariance instead of a full continuous scale invariance.[c] But such laws also apply to other kinds of variables (for example market indices or ion concentrations near earthquake zones, see [497]) for which local decreases are relevant.

Examples of application of such log-periodic laws to the analysis of the chronology of species evolution, of society evolution and of human development are given in Chapter 14.

4.3.2. *Multifractal behavior: multiple fractal dimensions*

Let us consider another example of second order differential equation of scale, which also yields solutions whose behavior is typical of real natural systems. Setting again $\mathbb{V} = \ln(\lambda_0/\varepsilon)$, it reads

$$\frac{d^2\mathcal{L}}{d\mathbb{V}^2} - (\tau_0 + \tau_1)\frac{d\mathcal{L}}{d\mathbb{V}} + \tau_0\tau_1\mathcal{L} = 0. \tag{4.55}$$

One of its solutions is the sum of two power laws, $\mathcal{L} = \gamma_0\varepsilon^{-\tau_0} + \gamma_1\varepsilon^{-\tau_1}$, and can be written as

$$\mathcal{L} = \mathcal{L}_0\left\{\left(\frac{\lambda_0}{\varepsilon}\right)^{\tau_0} + \left(\frac{\lambda_1}{\varepsilon}\right)^{\tau_1}\right\}. \tag{4.56}$$

One may easily introduce in this solution a large scale transition to nonfractality by the same method as before, i.e. by replacing $\mathcal{L}$ by $(\mathcal{L} - \mathcal{L}_0)$ in the equation, which hence acquires a constant term,

$$\frac{d^2\mathcal{L}}{d\mathbb{V}^2} - (\tau_0 + \tau_1)\frac{d\mathcal{L}}{d\mathbb{V}} + \tau_0\tau_1\mathcal{L} + A = 0. \tag{4.57}$$

Taking $A = -\tau_0\tau_1\mathcal{L}_0$, the solution reads

$$\mathcal{L} = \mathcal{L}_0\left\{1 + \left(\frac{\lambda_0}{\varepsilon}\right)^{\tau_0} + \left(\frac{\lambda_1}{\varepsilon}\right)^{\tau_1}\right\}. \tag{4.58}$$

---

[c]However it is possible to generalize them to a continuous construction, see [353] p. 54.

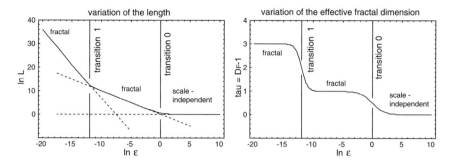

Fig. 4.6: **Fractal length and fractal dimension: multifractal behavior.** Scale dependence of the length and of the fractal dimension in the case of a fractal curve having a multifractal behavior. The "multifractality" means here that the system is described by different values of the fractal dimension on different scale ranges. Such a behavior is a solution of a scale differential equation of second order (see text).

Finally, the effective fractal dimension $D_{eff} = 1 + \tau_{eff}$ can be computed from the derivative $\tau_{eff} = d \ln \mathcal{L}/d\mathbb{V}$. One finds

$$\tau_{eff} = \frac{\tau_0 \left(\frac{\lambda_0}{\varepsilon}\right)^{\tau_0} + \tau_1 \left(\frac{\lambda_1}{\varepsilon}\right)^{\tau_1}}{1 + \left(\frac{\lambda_0}{\varepsilon}\right)^{\tau_0} + \left(\frac{\lambda_1}{\varepsilon}\right)^{\tau_1}}. \tag{4.59}$$

These solutions show different fractal dimensions on different scale intervals, as can be seen in Fig. 4.6. This is one of the two meanings, which have been given to the word "multifractal", the other describing statistical systems whose various moments are characterized by different fractal dimensions [307, 207, 427, 187].

Such a behavior is actually observed in several real fractal systems, such as the Britain coast [305]. Indeed, in such a case the upper cut-off is given by the size of Britain itself, while toward the small scales a drastic change of regime is usually found, depending on the region considered. For example, if one reaches a sand beach one first jumps to scale independence, then at the level of sand grains to a new fractal dimension; if one reaches rocks at small scales, one may skip from the fractal dimension of the coast to that of the rocks, etc.

4.3.3. *Lagrangian approach to scale laws*

The Lagrangian approach can be used in the scale space in order to obtain physically relevant generalizations of the above scale laws. This proposal has been first made in [352, 353] in the framework of special scale relativity, then it has been generalized to other cases in [363, 381, 410]. With this aim in view, we are led to reverse the definition and meaning of the variables.

This reversal is analogous to what was achieved by Galileo as concerns motion laws. Indeed, from the Aristoteles viewpoint, "time is the measure of motion". In the same way, the fractal dimension $D_F = D_T + \tau$, in its standard (Mandelbrot's) acception, is defined from the topological measure of the fractal object (length of a curve, area of a surface, etc.) and from resolution, namely (see Eq. 4.17)

$$\text{``}t = \frac{x}{v}\text{''} \quad analogous\ with \quad \tau = \frac{d \ln \mathcal{L}}{d \ln(\lambda/\varepsilon)}. \tag{4.60}$$

In the case where $\mathcal{L}$ represents a length (i.e. more generally, a fractal coordinate), the topological dimension is $D_T = 1$ so that $\tau = D_F - 1$, but, as recalled in the introduction of this section, it can be easily generalized to surfaces or volumes. With Galileo, time becomes a primary variable, and the velocity is deduced from space and time, which are, therefore, treated on the same footing, even though the Galilean space-time remains degenerate because of the implicitly assumed infinite velocity of light.

In analogy, the scale exponent $\tau = D_F - 1$ becomes a primary variable that plays, for scale laws, the same role as played by time in motion laws. We have suggested to call this variable scale-exponent "djinn".[d]

Carrying on the analogy in the same way as the velocity is the derivative of position with respect to time, $v = dx/dt$, we expect the derivative of the logarithm of position with respect to scale time τ to be a "scale velocity". Consider as reference the self-similar-like case that reads $\ln \mathcal{L} = \tau \ln(\lambda/\varepsilon)$. Derivating with respect to τ, now considered as a variable, yields $d \ln \mathcal{L}/d\tau = \ln(\lambda/\varepsilon)$, i.e. the logarithm of the relative resolution. By extension, we assume that this scale velocity provides a new general definition of the resolution even in more general situations, namely,

$$\mathbb{V} = \ln \left(\frac{\lambda}{\varepsilon} \right) = \frac{d \ln \mathcal{L}}{d\tau}. \tag{4.61}$$

It is noticeable that, because the resolution ε is dimensioned, we are obliged to take its ratio with some reference scale λ in order to skip to the logarithm. This is but a manifestation of the principle of scale relativity. Moreover,

---

[d]It means "gift" in Tibetan, under the sense of "natural quality", "talent", "marvelous potential". Note that it has been denoted as δ in previous papers, but we prefer here the less misleading notation τ, for "scale-time".

this new definition comes in support of the relativistic view of scale laws. Namely, in the same way as the velocity defines the relative state of motion of a system, the log of resolution, identified to a scale velocity, defines its relative state of scale.

We omit here the reference scale $\mathcal{L}_0$ for the sake of writing simplicity, since it disappears in the derivative of the logarithm, but one must keep in mind that it is implicitly present for reason of physical dimensionality.

Recall that in classical physics of motion (see, e.g. [265]), one introduces a Lagrange function $L(x, v, t)$, where $v = \dot{x}$, then an action

$$S = \int_{t_1}^{t_2} L(x, v, t)\, dt. \tag{4.62}$$

The least-action principle $\delta S = 0$, applied on this action, allows one to obtain the equations of motion under the form of Euler–Lagrange equations,

$$\frac{d}{dt}\frac{\partial L}{\partial v} = \frac{\partial L}{\partial x}. \tag{4.63}$$

In analogy, a scale Lagrange function $\widetilde{L}(\ln \mathcal{L}, \mathbb{V}, \tau)$ is introduced, and a scale action is constructed

$$\widetilde{S} = \int_{\tau_1}^{\tau_2} \widetilde{L}(\ln \mathcal{L}, \mathbb{V}, \tau)\, d\tau. \tag{4.64}$$

The application of the action principle yields a scale Euler–Lagrange equation that writes

$$\frac{d}{d\tau}\frac{\partial \widetilde{L}}{\partial \mathbb{V}} = \frac{\partial \widetilde{L}}{\partial \ln \mathcal{L}}. \tag{4.65}$$

Continuing the analogy with the physics of motion, in the absence of any "scale-force" (i.e. $\partial \widetilde{L}/\partial \ln \mathcal{L} = 0$), and when one takes the simplest possible form for the Lagrange function, i.e. $\widetilde{L} \propto \mathbb{V}^2$, the Euler–Lagrange equation becomes

$$\partial \widetilde{L}/\partial \mathbb{V} = \text{const} \Rightarrow \mathbb{V} = \text{const.}, \tag{4.66}$$

which is the equivalent for scale of what inertia is for motion. The constancy of $\mathbb{V} = \ln(\lambda/\varepsilon)$ means here that it is independent of the "djinn" τ. Eq. (4.61) can therefore be integrated to give the usual power law behavior, $\mathcal{L} = \mathcal{L}_0(\lambda/\varepsilon)^\tau$.

This reversed viewpoint has several advantages, which allow a full implementation of the principle of scale relativity:

(i) The "djinn" τ is given its actual status of additional dimension or scale time and the logarithm of the resolution, $\mathbb{V}$, its status of scale velocity (see Eq. 4.61). This is in accordance with its scale-relativistic definition, in which it characterizes the state of scale of the reference system, in the same way as the velocity $v = dx/dt$ characterizes its state of motion.

(ii) This leaves open the possibility of generalizing this formalism to the case of four-independent space-time resolutions, $\mathbb{V}^\mu = \ln(\lambda^\mu/\varepsilon^\mu) = d\ln\mathcal{L}^\mu/d\tau$. This amount to jump to a five-dimensional geometric description in terms of a space-time-djinn.

To be more specific it must be clear from now on that the four-dimensional nature of the classical space-time is not put in question here. Recall indeed that we have found that, accounting for the nondifferentiable nature of the coordinates, their differentials dX^μ are naturally decomposed as the sum of two contributions, a classical differentiable one dx^μ and a fractal nondifferentiable one $d\xi^\mu \propto (dx^\mu)^{1/D_F}$. The classical parts are not scale dependent and therefore do not combine with τ. The five-dimensional fractal space-time-djinn concerns the four new fractal contributions $d\xi^\mu$ to which the "djinn" τ may be added as a fifth dimension.

However, as stated in Sec 3.6, a vectorial description of resolutions is not the last word. Indeed, the genuine nature of resolutions is ultimately tensorial, $\varepsilon^\nu_\mu = \varepsilon_\mu\varepsilon^\nu = \rho_{\mu\lambda}\varepsilon^\nu\varepsilon^\lambda$ and it, therefore, involves correlation coefficients $\rho_{\mu\lambda}$, in analogy with variance-covariance error matrices.

Let us consider, in particular, the Hölder exponent version of the scale time τ, when it is applied to the fractal differential calculus. As we have seen, the elementary displacement $dX = dx + d\xi$ on a fractal curve can be described in terms of the sum of a classical part dx and of a fractal fluctuation $d\xi$. The fundamental relation between the two differential elements reads

$$d\xi^{D_F} = \eta^{D_F}\lambda D_F - 1\,dx. \tag{4.67}$$

Here η is a dimensionless variable that describes in a normalized way the fluctuating behavior of $d\xi$, i.e. $\langle\eta\rangle = 0$ and $\langle\eta^2\rangle = 1$. In the case when this describes displacements in a nondifferentiable and fractal space(-time), it is taken to be a stochastic variable (see the justification of this choice in Sec. 4.5). The fractal character is expressed by the fact that dx and $d\xi$ are differential elements of different orders. Let us set

$$\widetilde{d\xi} = \langle d\xi^2 \rangle^{1/2}. \tag{4.68}$$

In the scale relativity framework, according to which the differential elements are now considered as full variables, which tend to zero but without explicitly reaching it, the above relation may be written under the form

$$\ln \frac{\widetilde{d\xi}}{\lambda} = \frac{1}{D_F} \ln \frac{dx}{\lambda}, \tag{4.69}$$

where the Hölder exponent $H = 1/D_F$ appears.

In this formula one can interpret $\ln(dx/\lambda)$ as being the resolution variable, i.e. identify it with a scale velocity,

$$\mathbb{V} = \ln \frac{dx}{\lambda}. \tag{4.70}$$

In the asymptotic fractal domain where $d\xi \gg dx$, we have $dX \approx d\xi$, and therefore the quantity

$$\mathbb{X} = \ln \frac{\widetilde{d\xi}}{\lambda} \tag{4.71}$$

can be identified with the main length variable. Under this interpretation, the above relation (4.69) becomes a Galilean inertial-like scale relativity relation, which reads $\mathbb{X} = \mathbb{V}\tau$ in terms of a scale time $\tau = H = 1/D_F$.

This identification can easily be generalized to a variable scale time τ, which leads to the natural definition

$$\mathbb{V} = \ln \frac{dx}{\lambda} = \frac{d\ln(\widetilde{d\xi}/\lambda)}{d\tau}. \tag{4.72}$$

The generalization to a vectorial version of resolutions and fractal fluctuations is now straighforward

$$\mathbb{V}^k = \ln \frac{dx^k}{\lambda^k} = \frac{d\ln(\widetilde{d\xi}^k/\lambda^k)}{d\tau}. \tag{4.73}$$

The same is true for a final generalization to tensorial resolutions. Indeed, the fractal differential elements being essentially fluctuations, which can be represented by stochastic variables, they are more completely described by their correlation $d\xi^j d\xi^k$, whose derivative will give a tensorial result, namely,

$$\sigma^{jk} = \frac{d\ln\langle d\xi^j d\xi^k \rangle}{d\tau}. \tag{4.74}$$

(iii) A final advantage of this new representation is that scale laws more general than the simplest self-similar ones can be derived from more general scale Lagrangians [363], as we shall see in what follows.

Note, however, that there is also a shortcoming in this approach. Contrary to the case of motion laws, in which time is always flowing toward the future, except possibly in elementary particle physics at very small time scales [353], the variation of the scale-time may be non-monotonous, as exemplified by the previous case of log-periodicity. Therefore, this Lagrangian approach is restricted to monotonous variations of the fractal dimension, or, more generally, it applies to scale intervals on which it varies in a monotonous way.

4.3.4. Scale dynamics

Our previous discussion indicates that the scale invariant behavior corresponds, in the framework of a scale physics, to freedom, i.e. to a behavior which is free of a scale force. However, in the same way as there exists forces in nature that imply departure from inertial, rectilinear uniform motion, we expect most natural fractal systems to present also distortions in their scale behavior with respect to the pure scale invariance. This means taking nonlinearity in scale space into account. Such distortions may be, as a first step, attributed to the effect of a scale "dynamics", i.e. of a "scale-field". Clearly, at this level of description, this is only an analog of dynamics, which acts on the scale axis, i.e. on the internal structures of the system under consideration, not in space-time. The effects of coupling with space-time displacements will be studied in forthcoming chapters.

Let us be more specific about the meaning of this approach to generalizing scale laws. Once again, the analogy with the laws of motion and their history will be enlightening.

Recall that field theories have known five essential steps; (1) Galileo's free inertial motion; (2) Newton's conception of a force acting at distance between two bodies; (3) Classical field theory, in which a body produces a field that propagates and derives from a potential; the force is a function of the field and of the velocity; (4) Einstein's relativity theory, in which the concepts of potential, field and force become secondary aspects, understood as relative manifestations of the geometry of space-time and (5) Quantum field theory, in which the field results from the exchange of quantum boson particles.

Note that one of the frontiers of fundamental physics amounts to unifying the space-time description (4) and the quantum one (5). This is one of the goals of the scale relativity theory which tackles this question in a reversed way as compared with most other approaches. Namely, instead of basing itself on the quantum laws considered as more fundamental in other attempts, it starts from a space-time geometric representation and derives the quantum laws and the fields from (nondifferentiable) geometry.

Let us come back to the analogy between motion laws and scale laws. We have used it again here in order to point out that, since Einstein's work, the concept of force can be used only as a practical, simplified and approximative tool that represents nothing else but the manifestations of space-time geometry. It is precisely in this spirit that we introduce here a scale force. It is intended to be a simple first step representation of the effects of the geometry of the scale space. In other words, this means that the introduction of a scale dynamics is a simplified way to introduce a generalized relativity in/of the scale space. In such a more profound description, the scale potential is expected to become derived from a scale metric potential of pure geometric nature. However, several intermediate steps are needed before such a scale general relativity could be constructed, the first being to generalize the Galilean-like scale laws to Lorentzian-like special scale relativistic laws (see Sec. 4.4).

When a scale force is present, the Lagrange scale-equation takes the form of Newton's equation of dynamics,

$$F = \mu \frac{d^2 \ln \mathcal{L}}{d\tau^2}, \tag{4.75}$$

where μ is a "scale mass", which measures how the system resists to the scale force, and where $\Gamma = d^2 \ln \mathcal{L}/d\tau^2 = d \ln(\lambda/\varepsilon)/d\tau$ is the scale acceleration.

We shall now attempt to define physical, generic, scale-dynamical behaviors, which could be common to very different systems. For various systems the scale force may have very different origins, but in all the cases where it has the same form (constant force, harmonic oscillator, etc.), the same kind of scale behavior is expected to be obtained. As stressed hereabove, such a Newtonian-type approach is itself considered to be only an intermediate step while waiting for a fully developed general scale relativity. Thus the scale forces are expected to be finally recovered as approximations of the manifestations of the geometry of the scale space.

4.3.5. *Constant scale force*

Let us first consider the case of a constant scale-force. The potential is $\varphi = F \ln \mathcal{L}$, and Eq. (4.75) writes

$$\frac{d^2 \ln \mathcal{L}}{d\tau^2} = G, \tag{4.76}$$

where $G = F/\mu = cst$. It is easily integrated in terms of a parabolic solution, which is the equivalent for scale laws of parabolic motion in a constant field:

$$\mathbb{V} = \mathbb{V}_0 + G\tau; \quad \ln \mathcal{L} = \ln \mathcal{L}_0 + \mathbb{V}_0 \tau + \frac{1}{2} G\tau^2, \tag{4.77}$$

where $\mathbb{V} = d\ln \mathcal{L}/d\tau = \ln(\lambda/\varepsilon)$.

However the physical meaning of this result is not clear under this form. This is due to the fact that, while in the case of motion laws we search for the evolution of the system with time, in the case of scale laws we search for the dependence of the system on resolution, which is the directly measured observable. Since the reference scale λ is arbitrary, we can re-define the variables in such a way that $\mathbb{V}_0 = 0$, i.e. $\lambda = \lambda_0$. Indeed, from Eq. (4.77) we get $\tau = (\mathbb{V} - \mathbb{V}_0)/G = [\ln(\lambda/\varepsilon) - \ln(\lambda/\lambda_0)]/G = \ln(\lambda_0/\varepsilon)/G$. Then we obtain

$$\tau = \frac{1}{G} \ln \left(\frac{\lambda_0}{\varepsilon} \right), \quad \ln \left(\frac{\mathcal{L}}{\mathcal{L}_0} \right) = \frac{1}{2G} \ln^2 \left(\frac{\lambda_0}{\varepsilon} \right). \tag{4.78}$$

The scale exponent τ, and therefore also the fractal dimension $D_F = 1+\tau$, become linear functions of the resolution, and the $(\log \mathcal{L}, \log \varepsilon)$ relation is now parabolic instead of linear (see Fig. 4.7). Note that, as in previous cases, we have considered here only the small scale asymptotic behavior, and that we can once again easily generalize this result by including a transition to scale-independence at large scale (see Fig. 4.7). This cannot be achieved in this more general case by simply replacing $\mathcal{L}$ by $(\mathcal{L} - \mathcal{L}_0)$ in all the equations as in the linear case (Fig. 4.1), because $\ln^2(\lambda_0/\varepsilon)$ increases again when $\varepsilon \to 0$. A possible solution, shown in Fig. 4.7, consists of including the transition in the "djinn" τ by replacing $e^\tau = (\lambda_0/\varepsilon)^{1/G}$ by $e^{\tau_{\text{eff}}} = 1 + (\lambda_0/\varepsilon)^{1/G}$, i.e.

$$\tau_{\text{eff}} = \ln(1 + e^{\mathbb{V}/G}), \quad \ln \left(\frac{\mathcal{L}}{\mathcal{L}_0} \right) = \frac{1}{2} G\tau_{\text{eff}}^2. \tag{4.79}$$

There are several physical situations where, after careful examination of the data, the power-law models were rejected since no constant slope could be defined in the $(\log \mathcal{L}, \log \varepsilon)$ plane. In the several cases where a

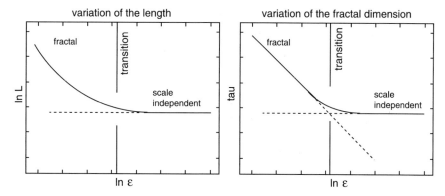

Fig. 4.7: **Fractal length and fractal dimension: constant scale force.** Scale dependence of the length of a fractal curve $\ln \mathcal{L}$ and of its effective fractal dimension, $D_F = D_T + \tau_{\mathrm{eff}}$, where D_T is the topological dimension, in the case of a constant "scale force", with an additional fractal to nonfractal transition (see text). Other solutions are possible when the curvature is reversed and when the transition to scale independence is toward the small scales.

clear curvature appears in this plane, e.g. turbulence, sand piles, physical geography, etc., the physics could come under such a "scale-dynamical" description. In these cases it might be of interest to identify and study the scale-force responsible for the scale distorsion, i.e. for the deviation to standard scaling.

4.3.6. *Scale harmonic oscillator*

Another interesting case is that of a repulsive harmonic oscillator potential, $\varphi = -(k/2)\ln^2 \mathcal{L}$, in the scale space. The scale differential equation reads in this case (we omit the reference scale $\mathcal{L}_0$ in order to simplify the exposition):

$$\frac{d^2 \ln \mathcal{L}}{d\tau^2} = k \ln \mathcal{L}. \tag{4.80}$$

Setting $k = 1/\tau_0^2$, where τ_0 is constant, one of its solutions reads

$$\ln \mathcal{L} = a \sinh\left(\frac{\tau}{\tau_0} + \alpha\right), \tag{4.81}$$

so that the scale velocity, which is its derivative with respect to τ, reads

$$\mathbb{V} = \ln\frac{\lambda}{\varepsilon} = \frac{a}{\tau_0}\cosh\left(\frac{\tau}{\tau_0} + \alpha\right). \tag{4.82}$$

As in the previous section, we may now re-express $\ln \mathcal{L}$ in function of the resolution thanks to the relation $\cosh^2 x - \sinh^2 x = 1$. We obtain

$$\frac{1}{\tau_0^2} \ln^2 \mathcal{L} - \ln^2 \frac{\lambda}{\varepsilon} = -\frac{a^2}{\tau_0^2}. \tag{4.83}$$

Finally, reintroducing a reference scale for $\mathcal{L}$ and changing the name of the constants, the solution can be put under the form

$$\ln \frac{\mathcal{L}}{\mathcal{L}_0} = \tau_0 \sqrt{\ln^2 \frac{\lambda}{\varepsilon} - \ln^2 \frac{\lambda}{\lambda_1}}. \tag{4.84}$$

Note the correction to previous publications [363, 377], in which we considered only a restricted solution, this leading to the erroneous claim that the constants λ, λ_1 and τ_0 should be always related. Actually the new scale of transition λ_1 can be defined independently of the scale force-free transition λ. One should also be aware not to confuse the asymptotic constant exponent τ_0 with the now variable exponent τ.

For $\varepsilon \ll \lambda$ it gives the standard Galilean-type case $\mathcal{L} = \mathcal{L}_0 (\lambda/\varepsilon)^{\tau_0}$, i.e. constant fractal dimension $D_F = 1 + \tau_0$. But its intermediate-scale behavior is particularly interesting, since, owing to the form of the mathematical solution, resolutions larger than a scale λ_1 are no longer possible. This new kind of transition therefore separates small scales from large scales, i.e. an "interior" (scales smaller than λ_1) from an "exterior" (scales larger than λ_1). It is characterized by an effective fractal dimension that becomes formally infinite. This behavior may prove to be particularly interesting for applications to biology (see Chapter 14).

Here λ is the fractal to nonfractal transition scale for the asymptotic domain, i.e. it is the transition scale, which would have been observed in the absence of the additional scale force (see Fig. 4.8).

Another possible interpretation of this scale harmonic oscillator model consists of considering the variable ε in the above equations as a distance r from a centre (i.e. a scaling coordinate instead of a scale-resolution), as for example in the renormalization group approach to radiative corrections, where the variable r in the Callan–Symanzik [79, 503] equations for the electric charge is the distance to the "centre" of the bare electron. Then it describes a system in which the effective fractal dimension of trajectories diverges at some distance $r = \lambda_{\max}$ from the center, is larger than 1 in the inner region and becomes 1 (i.e. non-fractal) in the outer region. Since the increase of the fractal dimension of a curve corresponds to the increase of its "thickness" [305, 306] (see [353] p. 80 for an example of a fractal

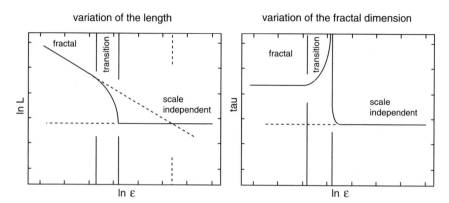

Fig. 4.8: **Fractal length and fractal dimension: linear scale force.** Scale dependence of the length of a fractal curve and of its effective fractal dimension (minus topological dimension) in the case of the application of a harmonic oscillator-like scale-potential (see text).

curve whose fractal dimension varies with position), such a model can be interpreted as describing a system in which the inner and outer domains are separated by a wall (Fig. 4.8).

We hope such a behavior to provide a model for confinement in QCD. Indeed, the gauge symmetry group of QCD, SU(3), is the dynamical symmetry group of a three dimensional isotropic harmonic oscillator, while gauge invariance can be re-interpreted in scale relativity as scale invariance on space-time resolutions (see Chapters 7 and 11).

Another solution of the same equation is with a reverse sign under the square root. It reads

$$\ln \frac{\mathcal{L}}{\mathcal{L}_0} = \tau_0 \sqrt{\ln^2 \frac{\lambda}{\varepsilon} + \frac{1}{\tau_0^2}}, \qquad (4.85)$$

and it therefore provides another example of smooth transition from the self-similar regime $\mathcal{L} = \mathcal{L}_0(\lambda/\varepsilon)^{\tau_0}$, to the scale-independent regime, $\mathcal{L} = \mathcal{L}_0$. This solution is equivalent to the scale-Galilean simplest solution, except during the transition.

4.4. Special Relativity of Scales

4.4.1. *Theory*

The question that we shall now address is that of finding the general laws of scale transformations that come under the principle of scale relativity.

Up to now, we have characterized typical scale laws as the simplest possible laws, namely, those, which are solutions of linear scale differential equations. This reasoning has provided us with the standard, power-law behavior with a constant fractal dimension, including a spontaneous effective scale symmetry breaking. We have subsequently verified that these laws do come under the principle of scale relativity, under its simplified, Galilean-like, form. Then they have been naturally generalized to scale dynamical laws, in analogy with the history of the evolution of ideas concerning motion.

But are the simplest possible laws those chosen by nature? Experience in the construction of the former physical theories suggests that the correct and general laws are the simplest among those, which satisfy some fundamental principle, rather than those, which are written in the simplest way. Anyway, the simplest laws are often recovered as approximations in the framework of more general theories: good examples of such relations between theories are given by Einstein's motion special relativity that includes the Galilean laws of inertial motion as a particular degenerate case ($c \to \infty$), and by Einstein's general relativity, which includes Newton's theory of gravitation as a limit case. In both situations the more general laws are constructed from the requirement of covariance rather than from the too simple requirement of invariance.

The theory of special scale relativity [352, 353] proceeds along a similar reasoning. The principle of scale relativity may be implemented by requiring that the equations of physics be written in a covariant way under scale transformations of resolutions. We have already shown in Sec. 4.2.3 that the standard scale laws, those described by first order scale differential equations whose solution is a fractal power-law behavior, are scale-covariant, since they form a Galilean-type group of transformation.

In such a frame of thought, the problem of finding the laws of linear transformation of a fractal (scale-dependent) length $\mathcal{L} = \mathcal{L}_0(\lambda/\varepsilon)^\tau$ in a scale transformation ($\varepsilon \to \varepsilon'$) characterized by a scale velocity $\mathbb{V} = \ln \rho = \ln(\varepsilon/\varepsilon')$ amounts to finding four functions, $a(\mathbb{V}), b(\mathbb{V}), c(\mathbb{V})$, and $d(\mathbb{V})$, such that the transformation

$$\ln \frac{\mathcal{L}'}{\mathcal{L}_0} = a(\mathbb{V}) \ln \frac{\mathcal{L}}{\mathcal{L}_0} + b(\mathbb{V})\tau,$$

$$\tau' = c(\mathbb{V}) \ln \frac{\mathcal{L}}{\mathcal{L}_0} + d(\mathbb{V})\tau$$

(4.86)

comes under the principle of relativity. Note that linearity is but an expression of the "special relativity" case, which is addressed here. This does not mean that we exclude nonlinear transformations, but that they will be considered in an enlarged, "general scale-relativistic" framework, to which the scale dynamical laws (Sec. 4.3) are a first approach.

Once the question is formulated in this way, it immediately appears that the current scale invariant transformation law of the standard form (Eqs. 4.21, 4.22), given by $a = 1, b = \mathbb{V}, c = 0$ and $d = 1$, is indeed a solution of the problem that corresponds to a Galileo-like scale group.

However, the general solution to the "special relativity problem" (namely, find the four functions a, b, c and d from the principle of relativity itself) is the Lorentz group [289, 352], of which the Galileo group is only a particular case. We have proved [352] that, for two variables, only two axioms were needed in addition to the linearity axiom (which characterizes the special relativity problem), namely, internal composition law and reflection invariance. Then we have suggested to generalize the standard law of dilatation, $\varepsilon \to \varepsilon' = \varrho \times \varepsilon$ in terms of a new log-Lorentzian relation [352]. Because of the importance of this result in the context of the present book, we shall now recall this proof in detail, following the derivation of [352, 353] (with corrections of some minor misprints in the original version).

4.4.2. *Derivation of the Lorentz transformation from the mere relativity principle*

As remarked by Levy-Leblond [289], very little freedom is allowed for the choice of a relativity group, so that the Poincaré group is an almost unique solution to the problem [34]. In his original paper, Einstein derived the Lorentz transformation from the (sometimes implicit) successive assumptions of (i) linearity; (ii) invariance of c, the light velocity in vacuum; (iii) existence of a composition law; (iv) existence of a neutral element and (v) reflection invariance.

But one can demonstrate that the postulate of the invariance of some absolute velocity is not necessary for the construction of the special theory of relativity. Indeed it was shown by Levy-Leblond [289] that the Lorentz transformation can be obtained through six successive constraints: {1} homogeneity of space-time (translated by the linearity of the transformation of coordinates), {2} isotropy of space-time (translated by the reflection invariance), group structure (i.e. {3} existence of a neutral element, {4} existence of an inverse transformation and {5} composition law yielding a new transformation, which is a member of the group, viz. which is internal) and {6} the causality condition. The last group axiom, associativity, is in fact straightforward in this case and leads to no new constraint.

Actually this set of hypotheses is still overdetermined to derive the Lorentz transformation. We have indeed demonstrated that the Lorentz transformation may be obtained from the only assumptions of {a} linearity; {b} internal composition law and {c} reflection invariance. All the other assumptions, in particular the postulate of the existence of an inverse transformation, which is a member of the group, may be deduced as consequences of these purely mathematical constraints. The importance of this result, especially concerning scale relativity, is that we do not have to postulate a full group law in order to get the Lorentz behavior, since the hypothesis of a (reflexive) semi-group structure is sufficient. The full group structure therefore emerges as a consequence instead of being set as an axiom.

Since the linearity axiom characterizes the special relativity problem that is addressed here, this means that the Lorentz transformation can be deduced from the only two postulates of internality and reflection invariance, which are themselves direct consequences of the principle of relativity (see the end of this section).

The proof that follows is valid both for motion transformations and for scale transformations. In the case of motion transformation, x is a space variable, t a time variable and v a velocity.

In the case of scale transformations,

(i) x denotes the logarithm of a length (more generally, a surface, a three-dimensional manifold, etc.), $x = \ln(\mathcal{L}/\mathcal{L}_0)$ or, including a scale symmetry breaking around the length $\mathcal{L}_0$,

$$x = \ln\left(\frac{\mathcal{L} - \mathcal{L}_0}{\mathcal{L}_0}\right),$$

(ii) t represents the scale exponent or scale time,

$$t = \tau,$$

(iii) and v represents the logarithm of resolution ratios,

$$v = \mathbb{V} = \ln\frac{\varepsilon'}{\varepsilon}.$$

As recalled in Sec. 4.2.2, depending on the choice of the resolution variable, the relation that links the "djinn" τ to the fractal dimension may differ. In particular, for a space resolution $\varepsilon = \delta\mathcal{L}$ it reads $\tau = D_F - 1$, while for a time resolution $\varepsilon = \delta t$, it reads $\tau = 1 - 1/D_F$. While in previous publications only the first case has been considered, we shall in the next sections develop applications to both cases and study their interrelations.

Note finally that, more generally, the coordinates x and t and the state variable v in the following derivation may refer to any kind of variables coming under a principle of relativity, in the special relativity case.

Let us start from a linear transformation of the coordinates:

$$x' = a(v)x - b(v)t, \tag{4.87}$$

$$t' = \alpha(v)t - \beta(v)x. \tag{4.88}$$

Eq. (4.87) may be written as $x' = a(v)[x - (b/a)t]$. But we may *define* the "velocity" v as $v = b/a$, so that, without any loss of generality, linearity alone leads to the general form

$$x' = \gamma(v)[x - vt], \tag{4.89}$$

$$t' = \gamma(v)[A(v)t - B(v)x], \tag{4.90}$$

where $\gamma(v) = a(v)$, and A and B are new functions of v. Let us now perform two successive transformations of the form (4.89, 4.90):

$$x' = \gamma(u)[x - ut], \tag{4.91}$$

$$t' = \gamma(u)[A(u)t - B(u)x], \tag{4.92}$$

$$x'' = \gamma(v)[x' - vt'], \tag{4.93}$$

$$t'' = \gamma(v)[A(v)t' - B(v)x']. \tag{4.94}$$

This results in the transformation

$$x'' = \gamma(u)\gamma(v)[1 + B(u)v]\left[x - \frac{u + A(u)v}{1 + B(u)v}t\right], \tag{4.95}$$

$$t'' = \gamma(u)\gamma(v)[A(u)A(v) + B(v)u]\left[t - \frac{A(v)B(u) + B(v)}{A(u)A(v) + B(v)u}x\right]. \tag{4.96}$$

Then the principle of relativity tells us that the composed transformation (4.95, 4.96) keeps the same form as the initial one (4.89, 4.90), in terms of a composed velocity w given by the factor of t in (4.95), i.e. the law of composition is internal. We obtain four conditions:

$$w = \frac{u + A(u)v}{1 + B(u)v},$$ (4.97)

$$\gamma(w) = \gamma(u)\gamma(v)[1 + B(u)v],$$ (4.98)

$$\gamma(w)A(w) = \gamma(u)\gamma(v)[A(u)A(v) + B(v)u],$$ (4.99)

$$\frac{B(w)}{A(w)} = \frac{A(v)B(u) + B(v)}{A(u)A(v) + B(v)u}.$$ (4.100)

The third postulate is reflection invariance. It reflects the fact that the choice of the orientation of the x (and x') axis is completely arbitrary and should be indistinguishable from the alternative choice $(-x, -x')$. With this new choice, the transformation (4.91, 4.92) becomes $\{-x' = \gamma(u')(-x - u't), t' = \gamma(u')[A(u')t + B(u')x]\}$ in terms of the value u' taken by the relative velocity in the new orientation. The requirement that the two orientations be indistinguishable yields $u' = -u$. This leads to parity relations for the three unknown functions γ, A and B [289]:

$$\gamma(-v) = \gamma(v), \quad A(-v) = A(v), \quad B(-v) = -B(v).$$ (4.101)

Combining Eqs. (4.97), (4.98) and (4.99) yields the relation

$$A\left[\frac{u + A(u)v}{1 + B(u)v}\right] = \frac{A(u)A(v) + B(v)u}{1 + B(u)v}.$$ (4.102)

Making $v = 0$ in this equation gives

$$A(u)[1 - A(0)] = uB(0).$$ (4.103)

Making $u = 0$ yields only two solutions, $A(0) = 0$ or 1. The first case gives $A(u) = uB(0)$. Since $B(0) \neq 0$ is excluded by reflection invariance (4.101), we obtain $A(u) = 0$. Then Eq. (4.100) becomes $A(w) = B(w)u$ so that $B(w) = 0$. Since $t' = 0 \ \forall u$, this is a case of complete degeneracy to only one effective variable, which can thus be excluded. We are left with $A(0) = 1$, which implies $B(0) = 0$, and the existence of a neutral element is therefore demonstrated.

Let us now make $v = -u$ in (4.102) after accounting for (4.101), and introduce a new even function $F(u) = A(u) - 1$, which verifies $F(0) = 0$.

We obtain

$$2F(u)\frac{1 + F(u)/2}{1 - uB(u)} = F\left[\frac{uF(u)}{1 - uB(u)}\right]. \qquad (4.104)$$

We shall now use the fact that B and F are continuous functions and that $B(0) = 0$. This implies that:

$\exists \eta_0 > 0$ such that in the interval $-\eta_0 < u < \eta_0$, $1 - uB$ and $1 + F/2$ become bounded to $k_1 < 1 - uB(u) < k_2$ and $k_3 < 1 + F(u)/2 < k_4$ with k_1, k_2, k_3 and $k_4 > 0$.

The bounds on $1 + F/2$ and $1 - uB$ allow us to bring the problem back to the equivalent equation

$$2F(u) = F[uF(u)]. \qquad (4.105)$$

The continuity of F at $u = 0$ reads, owing to the fact that $F(0) = 0$:

$$\forall \varepsilon, \exists \eta \text{ such that } |u| < \eta \Rightarrow |(F(u)| < \varepsilon.$$

Start with some $u_0 < \eta$, yielding $F(u_0) = F_0 = 2^{-n} < \varepsilon$. Then, from Eq. (4.105), $F(u_0 F_0) = 2F_0$. Set $u_1 = u_0 F_0$ and iterate. After p iterations, one obtains

$$F(u_p) = F[2^{p[(p-1)/2-n]}u_0] = 2^{p-n}.$$

In particular one gets after n iterations: $F[2^{-n(n+1)/2}u_0] = 1$ if n is an integer. In the general case where n is not integer, one gets after $\text{Int}[n]$ iterations a value of F larger than $1/2$. This is in contradiction with the continuity of F, since $u_n < u_0 < \eta$ while $F(u_n) > \varepsilon$. Then the only solution is $F = 0$ in a finite non null interval around the origin, and from step to step, this is true whatever the value of u, so that

$$A(u) = 1. \qquad (4.106)$$

As a consequence (4.102) becomes $B(u)v = B(v)u$, a relation which finally constrains the B function to be

$$B(v) = \kappa v, \qquad (4.107)$$

where κ is a constant. At this stage of our demonstration, the law of transformation of velocities is already fixed to the Poincaré–Einstein–Lorentz form,

$$w = \frac{u + v}{1 + \kappa uv}, \qquad (4.108)$$

and it is easy to verify that a full group law is achieved. Namely, the existence of an identity transformation and of an inverse transformation are ensured without having been presupposed. Consider now the γ factor. It verifies the condition

$$\gamma\left(\frac{u+v}{1+\kappa uv}\right) = \gamma(u)\gamma(v)(1+\kappa uv). \tag{4.109}$$

Let us consider the case $u = -v$. Eq. (4.109) reads $\gamma(0) = \gamma(v)\gamma(-v)(1 - \kappa v^2)$. For $v = 0$ it becomes $\gamma(0) = [\gamma(0)]^2$, implying $\gamma(0) = 1$, and we obtain

$$\gamma(v)\gamma(-v) = \frac{1}{1-\kappa v^2}. \tag{4.110}$$

The final step to the Lorentz transformation is straighforward from reflection invariance: this is a key point of Einstein's derivation in his 1905 paper [155]. The invariance by reflection implies $\gamma(v) = \gamma(-v)$ (Eq. 101) and fixes the γ factor under its Lorentz form,

$$\gamma(v) = \frac{1}{\sqrt{1-\kappa v^2}}. \tag{4.111}$$

The case $\kappa < 0$, which actually corresponds to an Euclidean space (x,t), is clearly excluded (for these variables) since it yields a non-ordered group. Indeed, being unlimited by a "light cone", the combination of two successive positive "velocities", which is given by a spherical rotation in this case, may yield a negative one. We are left with the only two physical solutions, the Lorentz group ($\kappa = c^{-2} > 0$) describing a Minkowski space-time, and the Galileo group ($\kappa = 0$), which is a particular case of the Lorentz group. Three of their properties: existence of a neutral element, of an inverse element and commutativity (for one space dimension), have not been postulated, but deduced from the initial axioms.

This does not mean that we need a fourth axiom in order to derive the correct special relativity group. Indeed, both solutions, Euclidean and Minkowskian, are allowable relativity groups, which are actually achieved in Nature. They correspond to different types of variables to be transformed. If the two variables to be transformed are both space variables x and y, the group of transformation is just the Euclidean group of rotations, which is included in the full Lorentz group and more generally in the Poincaré group, which also includes the four space-time translations. But, as we have seen, for space and time variables (motion laws) and log-space and "djinn" variables (scale laws), only the Minkowskian solution is admissible.

Note finally that the last Galilean solution, $\kappa = 0$, is also obtained as the limit $\kappa \to 0$ of both Euclidan and Minkowskian solutions.

Therefore, the searched transformation finally takes the well-known Lorentz form

$$x' = \frac{x - vt}{\sqrt{1 - v^2/c^2}}, \qquad (4.112)$$

$$t' = \frac{t - vx/c^2}{\sqrt{1 - v^2/c^2}}, \qquad (4.113)$$

and for the composition of velocities,

$$w = \frac{u + v}{1 + uv/c^2}, \qquad (4.114)$$

but this result is now more general, since it applies to linear transformations, which come under the principle of relativity but may be different from usual motion transformations.

Let us make some historical remarks on this occasion about the Lorentz transformation of motion. Recall that this special relativistic transformation has been established in its correct form by Poincaré [450], then by Einstein [155]. It has been given by Poincaré the name of Lorentz in his honour, though the transformation published by Lorentz [294] was incorrect, as well concerning the transformation of velocities (this have been pointed out by Pais [425]), as concerning the space and time transformations themselves (see [368]).

It is noticeable that Lorentz's mistake came from his view that the length contraction was a real mechanical contraction of the electron in the direction of motion instead of a simple projection effect due to relativity, which led him not to take into account the $-vt$ term in the position transformation, and, therefore, to obtain a totally wrong time transformation (see [294] p. 14, Eqs. (4) and (5)). On the contrary, the correct result was obtained by Poincaré in his June 1905 paper [450] (three months before Einstein's paper of 26 September [155]) precisely because he had identified this problem as a question of relativity several years ago and he was, therefore, looking for a generalization of the Galileo transformation $x' = x - vt$. Now, Poincaré missed in his demonstration the key point of reflection invariance that plays a leading role in the proof, (see the passage from Eq. (4.110) to Eq. (4.111) hereabove), which led him to finally fix the value of the γ factor by mere hypothesis, while Einstein used it in a very natural way, therefore, obtaining the final result in a more physical way.

Let us end this section by a comment that enlightens the power of the principle of relativity [352, 353].

We have shown that, once we have retained the hypothesis of linearity, the Lorentz transformation may be obtained through the only postulates of internal composition law and reflection invariance. Linearity is not a constraint by itself, since it characterizes the special relativity problem that is addressed here. Relaxing this hypothesis would correspond to addressing a general relativity problem, which will be considered in the following.

With regard to the other two postulates, they are nothing but a direct translation of the Galilean principle of relativity, which would not be the case of the axiom of the existence of a symmetric element, which cannot be imposed by relativity alone. Indeed the hypothesis that the composed coordinate transformation $(K \to K' \to K'')$ and the transformation in the reversed frame $(-K \to -K')$ must keep the same form as the initial one $(K \to K')$ is nothing but an application to the laws of coordinates transformation themselves, of Einstein's formulation of the Galilean principle of relativity, according to which "the laws of nature must keep the same form in different inertial reference systems". Indeed the laws of transformations between coordinate systems are clearly an essential part of these laws to which the principle of relativity should apply. In other words, one applies the principle of relativity to itself. So the general solution to the problem of inertial motion, without adding any postulate to the way it might have been stated at the Galileo and Descartes epoch, is actually Poincaré–Einstein's special relativity, whose Galilean relativity is a special case $(c = \infty)$.

But the main point that we want to stress here is that the foundation of the special theory of relativity can be based on the physical principle of relativity alone, and that the mathematical axioms from which the proof is obtained are derived from this principle instead of being prime. In other words, they are mathematical axioms but no longer physical axioms. Moreover, the principle of relativity, when it is translated in mathematical axioms, does not lead to a full group structure (at the level of the axioms): actually, only one group axiom is needed, the internality of the composition law. The identification that is often made between relativity theories and group theories is therefore not valid (once again, to be clear, at the foundation level). Relativity rather leads to a reflexive semi-group structure. Now, once this structure is stated at the axiomatic level, the final result is indeed that the set of transformations does form a group. But this is a result instead of a founding hypothesis.

Because of the generality of the way the problem has been posed and of the initial definition of the variables, the same conclusion holds for scale laws. However, in that case, one should also take into account the effective scale symmetry breaking, which is typical of the natural fractal behavior and which lead to a more complicated form of relativistic law.

4.4.3. *Lorentzian scale transformation*

Let us now recall the results obtained by applying this general derivation of the Lorentz transformation to scale transformations of the resolution variables, as they are given in [352, 353, 360].

We have argued in Sec. 4.2.3 that the scale transformations also come under a relativity theory, and that the group of transformations of standard fractal laws (of constant fractal dimension) could be given the mathematical form of a Galileo group. The generality of the above demonstration of the Lorentz transformation as implementing the principle of special relativity naturally suggests that a generalization of scale laws to a Lorentz form could be relevant.

This claim is supported by reconsidering from its origin the question of deriving the scale laws, which come under the principle of relativity. Start from the standard writing of the length of a fractal curve,

$$\mathcal{L} = \mathcal{L}_0 \left(\frac{\lambda}{\varepsilon}\right)^\tau,$$
(4.115)

where $\tau = D_F - 1$ when $\varepsilon = \delta\mathcal{L}$ and $\tau = 1 - 1/D_F$ when $\varepsilon = \delta t$. Under a scale transformation $\varepsilon \to \varepsilon'$, this law transforms as

$$\ln \frac{\mathcal{L}(\varepsilon')}{\mathcal{L}_0} = \ln \frac{\mathcal{L}(\varepsilon)}{\mathcal{L}_0} + \tau(\varepsilon) \ln \frac{\varepsilon}{\varepsilon'},$$
(4.116)

$$\tau(\varepsilon') = \tau(\varepsilon),$$
(4.117)

which is, as we have seen, a simplified form of Galilean-type scale transformation.

Set in a general way, the problem of scale transformation now consists of looking for a two-variable transformation $\ln \mathcal{L}' = f_1(\ln \mathcal{L}, \tau)$, $\tau' = f_2(\ln \mathcal{L}, \tau)$, depending on one parameter, the relative state of scale $\mathbb{V} = \ln(\varepsilon/\varepsilon')$.

In order to solve this problem let us analyse how the mathematical axioms on which was founded the above derivation of the Lorentz transformation are physically translated in the case of scale transformations.

In scale relativity, the quantities which play the role of lengths and times are now respectively the logarithm of the length (relative to some reference length) $\ln(\mathcal{L}/\mathcal{L}_0)$, and the "djinn" (or scale time) τ.

(i) The first axiom, *linearity*, could be, as for motion relativity, deduced from the uniformity of these variables. However, their uniformity is not *a priori* straighforward, even though it is already ensured in the scale laws of present physics. But linearity, as already specified, may be inferred from a hypothesis of simplicity. More precisely, linearity is the simplest choice to make, and so comes as a provisional specialization of the present theory. Therefore, as we have already stressed, the hypothesis of linearity of the transformation identifies with the special relativity problem itself. This also means that it is only a provisional hypothesis that must be relaxed in the general case. Such a "general scale relativity" will be considered in the following of this book: note that some of its aspects have already been considered (under a Newtonian version) when we have studied nonlinear scale laws issued from the action of scale forces. Another aspect is the emergence of gauge fields (see Chapter 7).

(ii) The second axiom, *internality of the composition law*, is a direct application of the principle of relativity: there is no difference here between motion and scale. It simply means that the product of two scale transformations, i.e. of two scale ratios, is also a scale transformation (a scale ratio), which is a straightforward fact.

(iii) The third axiom, *reflection invariance*, means that one may equally work with either $\ln(\mathcal{L}/\mathcal{L}_0)$ or $\ln(\mathcal{L}_0/\mathcal{L})$, to which would respectively correspond scale states $\ln(\varepsilon/\varepsilon_0)$ and $\ln(\varepsilon_0/\varepsilon)$; this is indeed straightforward.

The last question to be solved is the choice between $\kappa < 0$ (that corresponds to a Euclidian space-djinn) and $\kappa > 0$ (Minkowskian space-djinn) in Eqs. (4.108) and (4.111). The case $\kappa < 0$ is excluded in motion relativity since it would break down causality. As we have seen, it is also clearly excluded for scale laws, since when applying two successive dilatations we could obtain in this case a resulting contraction.

So, from the above result according to which the general solution to the linear relativity problem is the Lorentz transformation, we have concluded [352, 353] that the laws of scale transformation must also generally take a Lorentzian form, instead of being restricted to the Galilean form, which is a particular case of this more general transformation.

Let us now make explicit this Lorentz scale transformation, by starting with the Einstein–Poincaré law of composition of "velocities", now expressed in terms of scale dilatation and contractions, namely,

$$\mathbb{W} = \frac{\mathbb{U} + \mathbb{V}}{1 + \mathbb{U} \times \mathbb{V}/\mathbb{C}^2}, \tag{4.118}$$

where $\mathbb{U}, \mathbb{V}, \mathbb{W}$ are now scale velocities, i.e. logarithms of scale ratios, namely $U = \ln \nu$, $V = \ln \varrho$, $W = \ln \mu$. Here $\mathbb{C}$ is a fundamental dimensionless scale constant that may also be written under the same form, i.e. as the logarithm of a fundamental scale ratio [352]

$$\mathbb{C} = \ln \mathbb{K}. \tag{4.119}$$

(See Part III for the physical identification and interpretation of these constants.) We, therefore, obtain the generalized law of composition of scale transformations,

$$\ln \mu = \frac{\ln \varrho + \ln \nu}{1 + \ln \varrho \times \ln \nu / \ln^2 \mathbb{K}}. \tag{4.120}$$

Note that, under this form, this transformation is, as physically expected, invariant under the change of the logarithm basis [353, 353]. The constant $\mathbb{K}$ also appears as a natural basis, since, in this basis, the constant is absorbed in the variables, namely, the law of dilation composition reads

$$\log_{\mathbb{K}} \mu = \frac{\log_{\mathbb{K}} \varrho + \log_{\mathbb{K}} \nu}{1 + \log_{\mathbb{K}} \varrho \times \log_{\mathbb{K}} \nu}. \tag{4.121}$$

When the new scale constant $\mathbb{K}$ is infinite, one recovers the usual law of composition of dilatations, $\mu = \varrho \times \nu$. This is analogous to the recovering of the Galilean laws of composition of velocities, $w = u + v$, as the limit when $c \to \infty$ of the Lorentz laws. In this scale-relativistic generalization of the usual law of dilatation, the successive application of two dilatations ϱ and ν now yields a dilatation μ, which is smaller than the product $\varrho \times \nu$, namely,

$$\mu = (\varrho \times \nu)^{1/(1 + \ln \varrho \times \ln \nu / \ln^2 \mathbb{K})}. \tag{4.122}$$

Let us now write the scale-Lorentz law for the transformation of a fractal length $\mathcal{L}$ under a dilation ϱ. One obtains an analog of the law of transformation of space coordinate, according to the correspondence

$x \to \ln(\mathcal{L}/\mathcal{L}_0)$, $v \to \mathbb{V} = \ln \varrho$ and $t \to \tau$,

$$\ln \frac{\mathcal{L}'}{\mathcal{L}_0} = \frac{\ln(\mathcal{L}/\mathcal{L}_0) + \tau \times \mathbb{V}}{(1 - \mathbb{V}^2/\mathbb{C}^2)^{1/2}}. \tag{4.123}$$

Let us show, in order to be complete, that one evidently recovers the usual fractal law of transformation characterized by a constant fractal dimension in the limit $\mathbb{C} \to \infty$. Start with a length that depends on resolution as

$$\frac{\mathcal{L}}{\mathcal{L}_0} = \left(\frac{\lambda}{\varepsilon}\right)^\tau, \tag{4.124}$$

then skip to another resolution ε' such that $\varepsilon/\varepsilon' = \varrho$, one obtains $\mathcal{L}' = \mathcal{L} \times \varrho^\tau$, i.e.

$$\ln \frac{\mathcal{L}'}{\mathcal{L}_0} = \ln \frac{\mathcal{L}}{\mathcal{L}_0} + \tau \times \ln \varrho, \tag{4.125}$$

as expected.

The "djinn", which was previously invariant, is now transformed under the same scale transformation $\varepsilon \to \varepsilon' = \varepsilon/\varrho$ according to the relation

$$\tau' = \frac{\tau + \mathbb{V} \times \ln(\mathcal{L}/\mathcal{L}_0)/\mathbb{C}^2}{(1 - \mathbb{V}^2/\mathbb{C}^2)^{1/2}}. \tag{4.126}$$

The fractal dimension is therefore variable and equal to $D_F(\varrho) = 1 + \tau(\varrho)$.

In particular, if one starts from the reference scale λ and dilate it to ε, thus applying a transformation $\mathbb{V} = \ln \varrho = \ln(\lambda/\varepsilon)$, the "djinn" jumps from τ_0 at scale λ to τ at scale ε, while the length jumps from $\mathcal{L}_0$ to $\mathcal{L}$, following a simplified form of the transformation (see Fig. 4.9),

$$\tau = \frac{\tau_0}{\sqrt{1 - \ln^2(\lambda/\varepsilon)/\mathbb{C}^2}}, \quad \ln \frac{\mathcal{L}}{\mathcal{L}_0} = \frac{\tau_0 \times \ln(\lambda/\varepsilon)}{\sqrt{1 - \ln^2(\lambda/\varepsilon)/\mathbb{C}^2}}. \tag{4.127}$$

The covariance of these expressions is explicit, since it means that we have conserved the form of the standard fractal law $\mathcal{L} = \mathcal{L}_0(\lambda/\varepsilon)^\tau$, but now with a variable "djinn" $\tau(\varepsilon)$ as given by Eq. (4.127). In the new expression, the range of possible dilatations is now limited by the constant $\mathbb{K} = e^\mathbb{C}$, in analogy with velocities that cannot become larger than the limit velocity c in special motion relativity. But the length continues to diverge, now when tending to the scale ratio $\mathbb{K}$ instead of to an infinite scale ratio.

As expected, the standard fractal law $\mathcal{L} = \mathcal{L}_0(\lambda/\varepsilon)^{\tau_0}$ is also recovered as an approximation for small dilations and contractions, namely, when

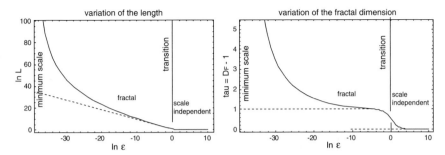

Fig. 4.9: **Fractal length and fractal dimension: log-Lorentz scale law.** Scale dependence of the logarithm of the length and of the effective fractal dimension, $D_F = 1 + \tau$, in the case of scale-relativistic Lorentzian scale laws including a transition to scale independence toward large scales. The resolution is defined here as $\varepsilon = \delta \mathcal{L}$. The constant $\mathbb{C}$ has been taken here to be $\mathbb{C} = 4\pi^2 \approx 39.478$, which is a fundamental value for scale ratios in elementary particle physics in the scale relativity framework (see Chapter 11), while the effective fractal dimension jumps from $D_F = 1$ to $D_F = 2$ at the transition, then increases without any limit toward small scales. We have identified here the transition to scale independence with the scale-Galilean to scale-Lorentzian transition, but they may be different. In this case, the exponent τ jumps to $\tau = 1$ and remains constant on a given range of scales, then it begins to increase only after a new (now smooth) transition.

$\mathbb{V} \ll \mathbb{C}$. Let us establish this result and also give the "scale-relativistic" correction to this law up to next-to-leading order. One may write the fractal length as

$$\mathcal{L} = \mathcal{L}_0 \left(\frac{\lambda}{\varepsilon}\right)^{\tau_0/\sqrt{1-\ln^2(\lambda/\varepsilon)/\mathbb{C}^2}}, \qquad (4.128)$$

and it becomes to leading and next-to-leading orders, for $\mathbb{V} \ll \mathbb{C}$, i.e. $(\lambda/\varepsilon) \ll \mathbb{K}$,

$$\mathcal{L} \approx \mathcal{L}_0 \left(\frac{\lambda}{\varepsilon}\right)^{\tau_0\left[1+\frac{1}{2\mathbb{C}^2}\ln^2\frac{\lambda}{\varepsilon}\right]} \approx \mathcal{L}_0 \left(\frac{\lambda}{\varepsilon}\right)^{\tau_0} \left\{1 + \frac{\tau_0}{2\mathbb{C}^2} \ln^3\left(\frac{\lambda}{\varepsilon}\right)\right\}. \qquad (4.129)$$

Note the initially slow, then rapidly increasing variation of the scale-relativistic correction (in $\mathbb{V}^3/\mathbb{C}^2$), which may play an important role, e.g. in elementary particle and high energy physics (see Chapters 11 and 12).

However, the above laws are not yet the complete laws of special scale relativity transformations. Indeed, they are a generalization of pure scaling laws, while we have seen that physically relevant scale laws incorporate a transition to effective scale independence. Let us now recall how the account of this spontaneous scale symmetry breaking leads to the emergence of a dimensionned fundamental scale constant.

4.4.4. *Scale symmetry breaking and invariant length-scale*

A scale symmetry breaking can be easily included in the new law by the same method as previously. Namely, we replace $\mathcal{L}/\mathcal{L}_0$ by $(\mathcal{L} - \mathcal{L}_0)/\mathcal{L}_0$ from the very beginning of the derivation and we obtain a spontaneous transition to scale independence following the law

$$\mathcal{L} = \mathcal{L}_0 \left\{ 1 + \left(\frac{\lambda}{\varepsilon} \right)^{\tau_0/\sqrt{1 - \ln^2(\lambda/\varepsilon)/\mathbb{C}^2}} \right\}, \qquad (4.130)$$

as plotted in Fig. 4.9.

However, this transition has also a still more important consequence. Consider indeed the log-Lorentz law of dilation composition. Let us start from the transition scale λ_0, taken as a reference scale, and let us apply a scale transformation, which leads to a new scale $\lambda_1 = \nu \lambda_0$. Then we start again from λ_0 (in a real physical experiment, this amounts to preparing the system again under the same conditions), and we apply another transformation with a factor μ leading to a new scale $\lambda_2 = \mu \lambda_0$. The nature of the new transformation is such that the dilation from λ_1 to λ_2 is $\varrho \neq \lambda_2/\lambda_1$. Now one can also take λ_0 as a reference for the fundamental dilation $\mathbb{K}$, which leads to formally introduce a length scale $\mathbf{\Lambda}$ defined as

$$\mathbf{\Lambda} = \mathbb{K} \times \lambda_0. \qquad (4.131)$$

At this level of the analysis $\mathbf{\Lambda}$ seems to be far from invariant, since it seems to depend on λ_0, which can be any scale and is also expected to depend on a change of the reference system, in particular under a motion Lorentz transformation.

However the composition law now takes the form

$$\ln \frac{\lambda_2}{\lambda_0} = \frac{\ln(\lambda_1/\lambda_0) + \ln \varrho}{1 + \ln \varrho \ln(\lambda_1/\lambda_0)/\ln^2(\mathbf{\Lambda}/\lambda_0)}. \qquad (4.132)$$

Let us consider the behavior of the length scale $\mathbf{\Lambda}$ in this expression. Assume that we start with this length, i.e. $\lambda_1 = \mathbf{\Lambda}$ and that we apply to it the dilatation or contraction ϱ. From Eq. (4.132), we find that this results into a length given by $\ln(\lambda_2/\lambda_0) = \ln(\mathbf{\Lambda}/\lambda_0)$, i.e. finally, $\lambda_2 = \mathbf{\Lambda}$, *whatever the value of the reference scale* λ_0. Starting from any scale larger than Λ (assuming that Λ is toward the very small scales), and applying any finite contraction, we get a scale always larger than Λ. The scale Λ can be the result only of infinite contraction or of an infinite product of contractions, i.e. it plays the same role as the zero point of the previous theory.

Hence the principle of relativity, once applied to scales, combined with the existence of a fractal to nonfractal transition, leads to the existence of a universal length in nature, which is *invariant under dilations and contractions*. Motion relativity immediately ensures that this will be also true for time and that an invariant time scale $\mathbb{T} = \Lambda/c$ does exist in nature. Note that a particular case of scale transformations is the motion Lorentz length contraction and time dilatation: as a consequence it is straighforward that Λ and Λ/c will be also invariant under a Lorentz transformation, i.e. independent of the relative velocity of the reference system in which they are observed.

In the above derivation, we have identified the reference scale λ_0 with the nonfractal to fractal transition λ_F. We indeed know that for scales larger than λ_F, which we shall prove in Chapter 5 to be the de Broglie scale or its generalization, one jumps to scale independent classical laws, which are the analog for scales of what is statics for motion. This regime is not contradictory with scale relativity, in the same way as statics do not contradict motion relativity, but is a degenerate case. There is no longer any explicit resolution dependence of physical laws in this regime, but the laws of dilation and contraction between scale intervals still apply, and macroscopic measurements at laboratory scales prove that if one measures a length l_3 with a unit l_2 and finds $l_3 = \rho_2 l_2$, then l_2 with a unit l_1 and finds $l_2 = \rho_1 l_1$, the measurement of l_3 with the unit l_1 will yield $l_3 = \rho_2 \rho_1 l_1$ (but note that this is no longer the case in cosmology at very large scales). Therefore, the standard dilation composition law takes a Galilean form in the scale-independent regime, and it is only beyond λ_F that a Lorentzian law may become possible. However, λ_0 is, strictly, a scale-Galilean to scale-Lorentzian transition and it can therefore be different from the nonfractal to fractal transition λ_F, as remarked by Célérier (private communication). In this case, one obtains an effective fractal dimension $D_F = 1$ at scales larger than λ_F, then $D_F = 2$ (the quantum mechanical value, see Sec. 5) between λ_F and λ_0, then an increase of the fractal dimension below λ_0. As we shall see in the following of this book, we have suggested [352, 353], in the case of the application of the theory to standard quantum mechanics, to identify the fractal-nonfractal transition to the quantum-classical de

Broglie transition scale $\hbar/p$, and the Galilean–Lorentzian transition to the Einstein–Compton scale $\hbar/mc^2$, which is also the quantum-classical time transition $\hbar/E$ in rest frame.

One might be disturbed by the fact that $\mathbb{K}$ is not an universal constant, contrary to the structure expected from a pure special relativity theory. However, once λ_0 is fixed, λ_0/Λ is a constant for the system under consideration and one is brought back to the usual structure of special relativity. Namely, as a consequence of the constancy and universality of a dimensionned length-scale, the parameter $\mathbb{C} = \ln \mathbb{K}$ is a constant only for a given system. This makes a fundamental difference with motion special relativity, where the constant c is universal. But the difference is that scale relativity relies on motion relativity since, as we shall see, the transition scale can be identified with Einstein–de Broglie scales, which depend on velocity. Conversely it is rather satisfactory that in the same way as motion relativity led to the existence of an universal unexceedable velocity scale relativity leads to the existence of an universal invariant limit for all lengths and times (having the status of an horizon, not of a cut-off or barrier).

The final point to be elucidated is the nature of Λ. Actually, since the above scale laws are reversible i.e. they apply to scale contractions as well a dilatations, one is naturally led to consider the possibility of existence of two invariant scales, a minimal one toward small scales and a maximal one toward large scales. We have suggested that the minimal scale can be identified with the Planck length-scale, $l_{\mathbb{P}} = \sqrt{\hbar G/c^3}$ [352, 353], and the maximal scale with the cosmic scale $\mathbb{L} = \Lambda^{-1/2}$ defined by the cosmological constant Λ. These proposals and their consequences will be considered in more detail in Part III of this book.

4.4.5. *Laws of special scale relativity*

For simplicity, we shall consider in what follows only the one-dimensional case.

(i) We first define the resolution as $\varepsilon = \delta\mathcal{L}$. The new law of composition of dilatations reads, for $\varepsilon < \lambda_0$ and $\varepsilon' < \lambda_0$

$$\ln \frac{\varepsilon'}{\lambda_0} = \frac{\ln(\varepsilon/\lambda_0) + \ln \varrho}{1 + \ln \varrho \ln(\varepsilon/\lambda_0)/\ln^2(\Lambda/\lambda_0)}. \tag{4.133}$$

In the simplified case of a transformation from $\mathcal{L}_0$ to $\mathcal{L}$, the length measured along a fractal coordinate, which was previously scale-dependent (as $\ln(\mathcal{L}/\mathcal{L}_0) = \tau_0 \ln(\lambda_0/\varepsilon)$ for $\varepsilon < \lambda_0$) becomes (see Fig. 4.9)

$$\ln \frac{\mathcal{L}}{\mathcal{L}_0} = \frac{\tau_0 \ln(\lambda_0/\varepsilon)}{\sqrt{1 - \ln^2(\lambda_0/\varepsilon)/\ln^2(\lambda_0/\Lambda)}}. \tag{4.134}$$

Respectively to the scale invariant approaches, the main new feature of scale relativity is that the scale exponent τ and the fractal dimension $D_F = 1 + \tau$, which were previously constant ($D_F = 2, \tau = 1$), are now explicitly varying with respect to the scale (Fig. 4.9), following the law (starting again from $\mathcal{L}_0$)

$$\tau(\varepsilon) = \frac{\tau_0}{\sqrt{1 - \ln^2(\lambda_0/\varepsilon)/\ln^2(\lambda_0/\Lambda)}}, \tag{4.135}$$

so that the form of the scale dependence of the length, $\mathcal{L} = \mathcal{L}_0(\lambda_0/\varepsilon)^\tau$ is conserved, but now in terms of a scale dependent exponent. This means that scale invariance has been generalized to scale covariance [349, 352, 353, 145, 449, 146].

(ii) One may also consider the case in which the resolution is a standard space or time interval, namely, $\varepsilon = \delta x$ or $\varepsilon = \delta t$. This case has not been studied in previously published works about special scale relativity, though it may also be relevant in some situations.

Recall indeed that we have shown that a displacement on a fractal curve may be generally written as $dX = dx + d\xi$, where $dx = v\,dt$ is a classical displacement while $d\xi$ is a fractal fluctuation. These are differential elements of different orders, since they are related through the fractal dimension D_F by $d\xi^{D_F} \propto dt$. When defining the resolution or scale variable, one may therefore define it as being of the same nature as $d\xi$, which yields a scale variation of the length given by $\mathcal{L} \propto d\xi^{1-D_F}$, or of the same nature as dt or dx, which yields the other relation $\mathcal{L} \propto dt^{(1/D_F)-1}$.

In the case where the fractal dimension is constant, using one of the other relations, which correspond to two different ways of making length measurements, one obtains the same fractal dimension. For example, the Abbott and Wise approach [1] to the fractal dimension of typical paths of quantum mechanical particles was based on the first method, and therefore yielded from Heisenberg's relation $\mathcal{L} \propto \delta\mathcal{L}^{-1}$, while Feynman found $\mathcal{L} \propto \delta t^{-1/2}$. Both relations are quite compatible and lead to the same fractal dimension $D_F = 2$.

Now, in the special scale-relativistic framework described here, this unity is broken, since the two definitions may lead to different varying fractal dimensions. This situation is reminiscent of a similar one that already occurred for fractal objects, when different definitions of fractal dimensions (Hausdorff, Besicovitch, similarity, box counting, covering, etc.) which agree for the simplest fractal objects, may give different results for more complicated fractals [306].

Namely, in the special scale-relativistic generalization of the second case, $\mathcal{L} \propto dt^{(1/D_F)-1}$, the three above relations (4.133, 4.134, 4.135) remain true but now in terms of a scale exponent τ which is linked to the fractal dimension by the relation $\tau = (1/D_F) - 1$, so that

$$D_F = \frac{1}{1-\tau}. \tag{4.136}$$

The variation of the fractal dimension is therefore different in this case.

Problems and Exercises

Exercise 10 Compare the special scale relativistic variation of the fractal dimension in the two cases: (1) $\tau = D_F - 1$ and (2) $\tau = (1/D_F) - 1$ at scales smaller than the reference scale λ_0 but which remain close to it.

Hint: In this case one may perform a power series expansion to first order in $\mathbb{V}^2/\mathbb{C}^2$, since $\varepsilon \approx \lambda_0$ implies $\mathbb{V}^2/\mathbb{C}^2 \ll 1$. When $\tau = D_F - 1$, one finds

$$D_F = D_0 \left\{ 1 + \frac{1}{2} \left(1 - \frac{1}{D_0} \right) \frac{\mathbb{V}^2}{\mathbb{C}^2} + \cdots \right\}, \tag{4.137}$$

with $\mathbb{V} = \ln(\lambda_0/\delta\mathcal{L})$ in this case, while when $\tau = (1/D_F) - 1$, one obtains

$$D_F = D_0 \left\{ 1 + \frac{1}{2}(D_0 - 1) \frac{\mathbb{V}^2}{\mathbb{C}^2} + \cdots \right\}, \tag{4.138}$$

where the resolution variable is now $\mathbb{V} = \ln(T_0/\delta t)$. ∎

4.4.6. *Scale Lorentz laws from a Lagrangian approach*

Let us come back to the case $\varepsilon = \delta\mathcal{L}$. The new scale relativistic laws corresponds to a Minkowskian-like scale metric invariant that reads

$$d\sigma^2 = d\tau^2 - \frac{(d\ln \mathcal{L})^2}{\mathbb{C}^2}. \tag{4.139}$$

When it is written under this form, the meaning of σ is that of a "proper djinn" or "proper scale-time". This means that, although the apparent fractal dimension $D_F(\varepsilon)$ (which is equal to $D_F(\varepsilon) = 1 + \tau(\varepsilon)$ when $\varepsilon = \delta \mathcal{L}$) is variable, it is actually a projection of an invariant proper fractal dimension $D_{F0} = 1 + \sigma$.

From this metric approach, one can easily recover the Lorentzian scale laws as solutions of a differential equation in the scale space. Actually, since the mathematical structure of the scale equations is the same as that of the motion equations, by using Lorentz covariance, we can give these differential equation a very simple form similar to the strictly self-similar case of constant fractal dimension.

To this purpose we may use one of the main results of relativity theories, according to which the action is proportional to the invariant [266]. Recall that this fundamental result allows one to transform the action principle in a geodesic principle. Here, this means that the infinitesimal scale action reads (taking a scale mass $\mu = 1$)

$$dS = -\mathbb{C}\,d\sigma = -\mathbb{C}\sqrt{1 - \frac{(d\ln\mathcal{L}/d\tau)^2}{\mathbb{C}^2}}\,d\tau, \qquad (4.140)$$

i.e. since by definition $\mathbb{V} = \ln(\lambda/\varepsilon) = d\ln\mathcal{L}/d\tau$,

$$dS = -\mathbb{C}\sqrt{1 - \frac{\mathbb{V}^2}{\mathbb{C}^2}}\,d\tau, \qquad (4.141)$$

so that the scale Lagrange function reads, in analogy with the motion-relativistic case,

$$\widetilde{L} = -\mathbb{C}^2\sqrt{1 - \frac{\mathbb{V}^2}{\mathbb{C}^2}}. \qquad (4.142)$$

In the absence of an external force, the scale Euler–Lagrange equation takes the form

$$\frac{d}{d\tau}\frac{\partial\widetilde{L}}{\partial\mathbb{V}} = 0. \qquad (4.143)$$

This introduces the conservative quantity that we have called "scale momentum" in [352, 353]

$$\mathcal{P} = \frac{\partial\widetilde{L}}{\partial\mathbb{V}}, \qquad (4.144)$$

whose value, from the expression of the Lagrange function, is

$$\mathcal{P} = \frac{\mathbb{V}}{\sqrt{1 - \mathbb{V}^2/\mathbb{C}^2}}. \tag{4.145}$$

One recognizes here the analog for scale laws of the relativistic momentum of motion laws. In [352, 353], a full scale relativistic mechanics has been developed. We shall account for these results in Chapter 11 (see in particular Sec. 11.1.1.3).

Here our only aim amounts to showing that the differential equation now obtained in terms of a special relativistic Euler–Lagrange equation allows us to recover as a solution of this equation the Lorentz-like dependence of the fractal length as a function of the resolution variable. Indeed, the scale differential equation now reads

$$\frac{d\mathcal{P}}{d\tau} = 0, \tag{4.146}$$

whose solution is $\mathcal{P} = \text{cst}$, so that $\mathbb{V} = \text{cst}$ (meaning that it is not a function of τ). Therefore, we recover the form of the Galilean result, namely, since $d\ln\mathcal{L}/d\tau = \mathbb{V}$, we find

$$\ln\frac{\mathcal{L}}{\mathcal{L}_0} = \mathbb{V}\tau \Rightarrow \mathcal{L} = \mathcal{L}_0\left(\frac{\lambda}{\varepsilon}\right)^\tau. \tag{4.147}$$

But now, contrary to this Galilean case, τ is no longer a constant but it has become a function of the resolution, $\tau = \tau(\varepsilon)$. As already remarked in the previous derivation of the same result by other means, this is precisely an example of covariance (i.e. invariance of the form of equations) as generalizing simple invariance (here of the numerical value of the exponent). From the scale metric expression we find

$$d\tau = \frac{d\sigma}{\sqrt{1 - \mathbb{V}^2/\mathbb{C}^2}}, \tag{4.148}$$

so that we recover the previous expression Eq. (4.135) of the scale variation of the "djinn", namely

$$\tau = \frac{\tau_0}{\sqrt{1 - \mathbb{V}^2/\mathbb{C}^2}}, \tag{4.149}$$

where τ_0 is the value of the "djinn" just after the transition, i.e. the value it would have kept in the scale Galilean case.

From this expression we finally derive the form of the special-scale relativistic variation of the length,

$$\ln \frac{\mathcal{L}}{\mathcal{L}_0} = \frac{\tau_0 \mathbb{V}}{\sqrt{1 - \mathbb{V}^2/\mathbb{C}^2}}, \qquad (4.150)$$

where $\mathbb{V} = \ln(\lambda_0/\varepsilon)$. This expression, obtained from a scale differential equation written as a Euler–Lagrange equation, is, as expected, the same as Eq. (4.134), which was obtained from the axiomatic derivation of the scale Lorentz transformation.

Problems and Exercises

Open Problem 2: Generalize the log-Lorentz scale relativistic transformation to more than two variables. ■

4.4.7. *Toward a general theory of scale relativity*

The above approach in terms of "scale dynamics" is actually intended to be a provisional description. Indeed, in analogy with Einstein's general relativity of motion, in which Newton's gravitational force becomes a mere manifestation of space-time curvature, it should be clear from the very beginning that the scale dynamical forces introduced hereabove are only intermediate and practical concepts, which should ultimately be recovered as mere manifestations of the fractality of space-time.

While the Galilean and special scale relativity transformations remain linear, such a general geometric description will also imply to skip to nonlinear scale laws, for example by writing differential equations involving second order derivatives, such as $\partial^2/\partial \ln \varepsilon^2$. We shall therefore be led in future developments of the theory to look for the general nonlinear scale laws that satisfy the principle of scale relativity. This amounts to looking for the equivalent in scale space of what the Einstein group of general motion relativity is in standard space-time, knowing that the genuine nature of the resolutions is tensorial. This program of research remains an open problem that will not be pursued in the present book.

Indeed, as recalled in the introduction, the question of the working out of pure scale laws (at a given point) is only a first step of the scale relativity construction. It is followed by a second step, which amounts to setting up the motion equations under the form of geodesic equations. This will be the subject of Chapters 5 and 6: as we shall see, these geodesic equations can be given the form of the equations of quantum mechanics. Now the geodesics of a nondifferentiable and fractal space-time have internal structures, which are described by the scale laws established in the present chapter as coming under the scale relativity principle. The various solutions found, from the simplest solution of constant fractal dimension to special scale relativistic and scale dynamical solutions will then lead to various more and more complicated forms of geodesic equations. However, the standard quantum mechanics is recovered in the simplest case of a constant fractal dimension $D_F = 2$ of the geodesics. This means that a huge set of possible generalizations of the quantum mechanical equations is generated by the more general scale laws. In particular, the special scale relativistic laws give rise to high energy departures, which may be relevant in particle physics (see Chapter 11). There is therefore much work to be done for implementing the scale laws described in this chapter (most of the present book is devoted to this subject, in a yet non exhaustive way). This is the reason why we let open to future works the full description of general scale relativistic laws.

Now this is not the last word, since there still is a third step in the construction of the theory. Indeed, at a more profound level of the description of a fractal space-time, one is naturally led to introduce resolutions, which may now vary from place to place, i.e. become functions of the standard coordinates, $\varepsilon(x, y, z, t)$. This means to treat scale laws and motion laws on the same footing, and, for example, to also include in the partial differential equations second order derivative, such as $\partial^2/\partial x \partial \ln \varepsilon$. As we shall see, such an approach leads to found gauge field theories in a geometric way (Chapter 7). The Abelian U(1) case corresponds to one varying resolution variable, $\varepsilon(x, y, z, t)$. But, as we have noticed in Sec. 3.6, the true nature of the resolution variables is tensorial: in the case of non-Abelian gauge theories, it is a full ten component resolution tensor $\varepsilon_{\mu\nu}(x, y, z, t)$, which must be introduced, and which allows to suggest generalizations of these theories. Finally, this third step may also be considered to be part of the program of constructing a general scale relativity theory.

Problems and Exercises

Open Problem 3: Find the general scale-relativistic laws of transformations in scale space, accounting for the tensorial nature of the scale variables.   ∎

4.5. Metric of a Fractal Space-time

4.5.1. *Introduction*

Let us conclude this chapter by preparing the continuation of the scale relativity construction, which aims at writing the geodesic equations of a nondifferentiable space-time. As we have proven, such a space-time continuum is characterized, among other properties, by its fractality, i.e. following the definition accepted here, by the fact that it is everywhere or almost everywhere scale-divergent.

In Sec. 3.10 of [353], we reached the conclusion that one of the ways to describe a fractal space-time was to generalize the general relativity tools (metric potentials, Christoffel symbols, Ricci and Riemann tensors, etc.) to explicitly scale dependent functions. We shall develop further this proposal in the present section, according to the results of [378, 394].

4.5.2. *Reminder: general relativity tools*

Let us first briefly recall the nature of the main physical and mathematical tools of general relativity (see e.g. [266, 528, 323] for more detail), and give a short analysis of their properties and of the way they are constructed.

(i) The fundamental invariant is the proper time, which is given in terms of the coordinates and of the metric potentials by the metric relation

$$ds^2 = g_{\mu\nu} \, dx^\mu dx^\nu, \tag{4.151}$$

where we use Einstein's convention of summation on covariant and contravariant indices. The metric potentials have a geometric origin (curvature and/or state of the reference system) and generalize the Newton potential. They are used to lower and raise the indices.

(ii) Under a translation dx^ρ, a vector A^μ is changed by a total amount that account for the geometric and/or curvilinear effects in terms of a covariant derivative

$$DA^\mu = dA^\mu + \Gamma^\mu_{\nu\rho} A^\nu dx^\rho. \tag{4.152}$$

The Christoffel symbols $\Gamma^\mu_{\nu\rho}$ represent a geometric field or "space-time field" that generalizes the Newtonian field. They are given, as in every field theories, by derivatives of the potentials, namely,

$$\Gamma^\mu_{\nu\rho} = \frac{1}{2} g^{\mu\lambda} \left(\partial_\rho g_{\lambda\nu} + \partial_\nu g_{\lambda\rho} - \partial_\lambda g_{\nu\rho} \right). \tag{4.153}$$

The covariant derivative DA^μ represents the inertial part of the full derivative dA^μ remaining after having taken out the geometric part $-\Gamma^\mu_{\nu\rho} A^\nu dx^\rho$ [266, 16].

(iii) The principle of relativity (through its equivalence, covariance and geodesic versions) leads to writing the equation of motion for a particle in a gravitational field in terms of a geodesic equation that keeps the form of the equation of free Galilean (inertial) motion, namely

$$\frac{Du^\mu}{ds} = 0 \Rightarrow \frac{du^\mu}{ds} + \Gamma^\mu_{\nu\rho} u^\nu u^\rho = 0. \tag{4.154}$$

This equation can be obtained following three equivalent methods [266]: (1) From the principle of equivalence, the gravitational field is expected to locally vanish in the free fall curvilinear system of coordinates, so that the motion equation reads under the free form $Du^\mu = 0$; (2) From the strong principle of covariance, the equation of motion, when written in terms of the covariant derivative, takes the simplest possible form of the Galilean equation for inertial motion in the absence of any force and (3) From the geodesic principle, which is the relativistic version of the least action principle, one writes $\delta(ds^2) = 0$, which leads to Eq. (4.154).

(iv) The field theories that are anterior to general relativity are two-level theories, namely, the tool of description includes two concepts, the potential and the field that derives from this potential. For example, in Newton's theory, the potential energy ϕ is scalar and the force, identifiable to the field, reads $F_k = -\partial_k \phi$ and is therefore vectorial. In Maxwell's theory, the potential A^μ is vectorial and the field $F_{\mu\nu} = \partial_\mu A_\nu - \partial_\nu A_\mu$ is tensorial, while the force reads $F^\mu = eF^{\mu\nu} u_\nu$. In Einstein's theory, the tensorial potential $g_{\mu\nu}$, the three-index field $\Gamma^\mu_{\nu\rho}$, and the generalized four-force $-m\Gamma^\mu_{\nu\rho} u^\nu u^\rho$ (see Eq. (4.154)), are no longer sufficient to describe the various effects of a gravitational field (i.e. of Riemann curvature). Indeed, $D_\mu D_\nu$ and $D_\nu D_\mu$ no longer commute in general relativity, so that one is led to introduce an additional tensor from the commutator

$$(D_\mu D_\nu - D_\nu D_\mu)A_\rho = R^\lambda_{\rho\nu\mu}A_\lambda. \tag{4.155}$$

This is the Riemann tensor, which is obtained from the fields components $\Gamma^\mu_{\nu\rho}$ by a new derivation, namely

$$R^\lambda_{\rho\nu\mu} = \partial_\nu \Gamma^\lambda_{\rho\mu} - \partial_\mu \Gamma^\lambda_{\rho\nu} + \Gamma^\lambda_{\alpha\nu}\Gamma^\alpha_{\rho\mu} - \Gamma^\lambda_{\alpha\mu}\Gamma^\alpha_{\rho\nu}. \tag{4.156}$$

There has been long discussions to determine whether the gravitational field was to be identified with the Christoffel symbols (which are not tensors) or to the Riemann tensor. Arguments may be found in favor of the two choices: e.g. while the $\Gamma^{\mu}_{\nu\rho}$ do appear in the motion equation in a way which is a natural generalization of the Newton and Lorentz forces, the expression of the Riemann tensor in function of the Christoffel symbols seems to be the natural generalization of the relation betwen field and potential in Maxwell's theory (it is actually a non-Abelian generalization of the Yang–Mills type). We think that a solution to the dilemma would simply be to admit that Einstein's theory is a genuine three-level theory that needs three concepts, potential, field and say, superfield, instead of only two, in order to define in an exhaustive way the properties of the gravitational field.

(v) Contrary to the motion equations, the field equations of general relativity come only under the weak covariance principle, since they still contain a non-free source term. They read, in terms of the Ricci tensor $R_{\mu\nu} = g^{\alpha\beta}R_{\alpha\mu\beta\nu}$, of the scalar curvature $R = g^{\alpha\beta}R_{\alpha\beta}$ and of the momentum-energy tensor $T_{\mu\nu}$,

$$R_{\mu\nu} - \frac{1}{2}R\,g_{\mu\nu} - \Lambda\,g_{\mu\nu} = \chi\,T_{\mu\nu}, \qquad (4.157)$$

where Λ is the cosmological constant and where $\chi = -8\pi G/c^2$. It has been proven by Cartan [84] that the above Einstein tensor with a cosmological constant is the most general tensor of null divergence, and that it can therefore be made equal, up to a multiplicative constant, to the energy-momentum tensor (which owns the same property).

4.5.3. *Metric and curvature of a fractal surface*

Let us examplify the behavior of a fractal metric, i.e. of an explicitly scale-dependent and everywhere divergent metric, by the explicit example of a fractal surface. Such a surface, viewed in an intrinsic way, i.e. from the viewpoint of internal two-dimensional fractal coordinates (the two dimensions here refer to the topological dimensions) is equivalent to a fractal two-dimensional space. As we shall see, its metric elements and its curvature are everywhere explicitly scale dependent and divergent when the resolution scale tends to zero.

Let us consider the nondifferentiable surface defined as

$$z(x,y) = \sum_{k=0}^{\infty} \frac{\sin(p^k x) + \sin(p^k y)}{q^k}. \tag{4.158}$$

Its nondifferentiability when $p > q$ is easy to establish since

$$\frac{\partial z}{\partial x} = \sum_{k=0}^{\infty} \left(\frac{p}{q}\right)^k \cos(p^k x), \quad \frac{\partial z}{\partial y} = \sum_{k=0}^{\infty} \left(\frac{p}{q}\right)^k \cos(p^k y), \tag{4.159}$$

which are everywhere divergent. Its metric is given by

$$ds^2 = \left[1 + \left(\frac{\partial z}{\partial x}\right)^2\right] dx^2 + 2\left(\frac{\partial z}{\partial x}\right)\left(\frac{\partial z}{\partial y}\right) dx\, dy + \left[1 + \left(\frac{\partial z}{\partial y}\right)^2\right] dy^2, \tag{4.160}$$

and it is therefore undefined in the usual meaning. The scale relativity method simply amounts to replacing the limit surface by the scale dependent family of surfaces

$$z(x,y,n) = \sum_{k=0}^{n} \frac{\sin(p^k x) + \sin(p^k y)}{q^k}, \tag{4.161}$$

which are equivalent to viewing the limit surface at a variable resolutions $\varepsilon = \lambda \times p^{-n}$. Therefore the number n is the scale variable itself in its natural logarithmic form, $n = \log_p(\lambda/\varepsilon)$. We therefore obtain an explicitly scale dependent metric,

$$ds^2 = \left[1 + \left(\sum_{k=0}^{n} \left(\frac{p}{q}\right)^k \cos(p^k x)\right)^2\right] dx^2 + 2\sum_{k=0}^{n} \left(\frac{p}{q}\right)^k \cos(p^k x)$$

$$\times \sum_{k=0}^{n} \left(\frac{p}{q}\right)^k \cos(p^k y)\, dx\, dy + \left[1 + \left(\sum_{k=0}^{n} \left(\frac{p}{q}\right)^k \cos(p^k y)\right)^2\right] dy^2, \tag{4.162}$$

which is now defined for all finite values of n (see Fig. 4.10).

In the two-dimensional case, the space curvature is reduced to the Gaussian curvature, which is defined as a function of the derivatives of the metric elements g_{jk} and of the determinant g of the metric (see e.g. [528]). It is therefore everywhere infinite for the above surface under the standard acception. But here, since the metric elements are now scale dependent

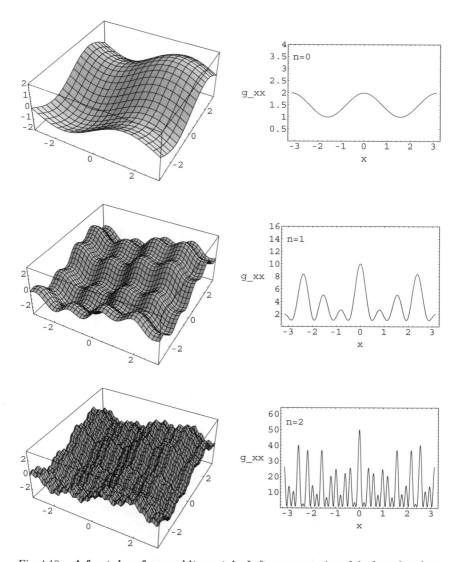

Fig. 4.10: **A fractal surface and its metric.** Left: representation of the fractal surface $z(x, y, n) = \sum_{k=0}^{n} 2^{-k}[\sin(4^k x) + \sin(4^k y)]$, at three resolution values ($n = 0$ to 2). Right: the scale divergent metric element g_{xx} of the same surface, plotted at increasing resolutions.

functions, this is the same for the Gaussian curvature, which is now defined as a fractal function, which diverges only in the limit $n \to \infty$, namely,

$$
K(x_1, x_2, n) = \frac{1}{2g} \left[2 \frac{\partial^2 g_{12}}{\partial x_1 \partial x_2} - \frac{\partial^2 g_{11}}{\partial x_2^2} - \frac{\partial^2 g_{22}}{\partial x_1^2} \right]
$$

$$
- \frac{g_{22}}{4g^2} \left[\left(\frac{\partial g_{11}}{\partial x_1} \right) \left(2 \frac{\partial g_{12}}{\partial x_2} - \frac{\partial g_{22}}{\partial x_1} \right) - \left(\frac{\partial g_{11}}{\partial x_2} \right)^2 \right]
$$

$$
+ \frac{g_{12}}{4g^2} \left[\left(\frac{\partial g_{11}}{\partial x_1} \right) \left(\frac{\partial g_{22}}{\partial x_2} \right) - 2 \left(\frac{\partial g_{11}}{\partial x_2} \right) \left(\frac{\partial g_{22}}{\partial x_1} \right) \right]
$$

$$
+ \frac{g_{12}}{4g^2} \left[\left(2 \frac{\partial g_{12}}{\partial x_1} - \frac{\partial g_{11}}{\partial x_2} \right) \left(2 \frac{\partial g_{12}}{\partial x_2} - \frac{\partial g_{22}}{\partial x_1} \right) \right]
$$

$$
- \frac{g_{11}}{4g^2} \left[\left(\frac{\partial g_{22}}{\partial x_2} \right) \left(2 \frac{\partial g_{12}}{\partial x_1} - \frac{\partial g_{11}}{\partial x_2} \right) - \left(\frac{\partial g_{22}}{\partial x_1} \right)^2 \right], \quad (4.163)
$$

(where $x_1 = x$ and $x_2 = y$) in terms of the $g_{jk}(x, y, n)$. Examples of such a scale divergent Gaussian curvature are given in Figs. 4.11 and 4.14. As concluded in [353], it corresponds to a fractal distribution of positive and negative curvatures tending to $+\infty$ and $-\infty$ when $n \to \infty$. Since the Gaussian curvature is a curvature invariant that does not depend on the

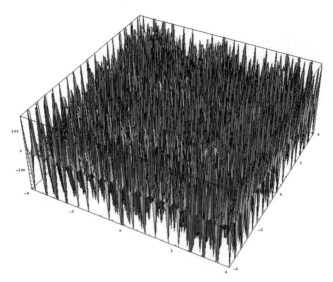

Fig. 4.11: **Gaussian curvature of a two-dimensional fractal space.** The Gaussian curvature of the fractal surface of Fig. 4.10, plotted at resolution level $n = 3$.

choice of the coordinate system (contrary to the metric elements), its scale divergence is a genuine inner property of a fractal two-dimensional space.

Although the metric is completely defined in the above example, it is highly fluctuating and becomes extremely complicated for high values of n (see Fig. 4.10). It can therefore be approximated by a stochastic function. We give in Fig. 4.13 the histogram of the distribution of the g_{xx} values for this example. We find in this case a probability distribution well fitted by an exponential, $P \propto \exp(-a g_{xx})$.

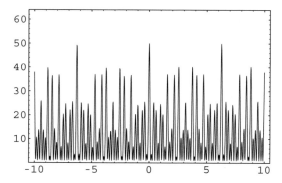

Fig. 4.12: **Metric of a fractal surface.** The scale divergent metric potential $g_{xx}(x, n) = 1 + \left(\sum_{k=0}^{n} 2^k \cos(4^k x)\right)^2$ of the fractal surface of Fig. 4.10 is plotted at resolution $n = 2$.

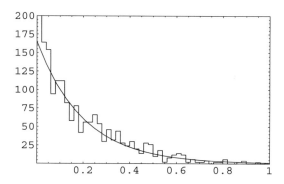

Fig. 4.13: **Probability distribution of the metric of a fractal surface.** The figure gives the histogram of the values of the renormalized scale divergent metric potential $g_{xx}/4^{n+1}$ (Fig. 4.12) of the fractal surface of Fig. 4.10, where $g_{xx}(x, n) = 1 + \left(\sum_{k=0}^{n} 2^k \cos(4^k x)\right)^2$, computed for $n = 10$. It is fitted to a probability law $P \propto \exp(-5 g_{xx})$.

Problems and Exercises

Exercise 11 Generalize the discrete fractal surface construction of Eq. (4.161) to a continuous construction, i.e. to a continuous definition of the scale variable.

Hint: Replace the sum by an integral, namely,

$$z(x, y, \mathbb{V}) = \int_0^{\mathbb{V}} \frac{\sin(p^k x) + \sin(p^k y)}{q^k} \, dk. \tag{4.164}$$

■

4.5.4. *Fractal metric*

Let us now generalize the concept of fractal (scale divergent) metric to a more general fractal space (or space-time). We have seen in the previous sections that the solution of scale differential equations may be generally written under the form of the sum of two terms, a classical differentiable part and a fractal, nondifferentiable fluctuation. In differential form it reads

$$dX^\mu = dx^\mu + d\xi^\mu. \tag{4.165}$$

Recall again that, since we have kept continuity, the nondifferentiability does not mean that we cannot differentiate, but that derivatives do not exist in their usual acception. Hence $d\xi/dx$ is infinite in the limit $dx \to 0$, but it can nevertheless be defined as an explicit function of dx, now considered as a variable in its own.

Anticipating the next chapters, let us first consider the simplified case of a constant fractal dimension $D_F = 2$ (as we shall see, this is the fractal dimension of geodesic paths in standard quantum mechanics, in agreement with Feynman's path integral approach [176]). In the one-body case, the fractal fluctuations (here in four dimensions) may be written in this case under the form

$$d\xi^\mu = \eta^\mu \sqrt{\lambda \, ds}. \tag{4.166}$$

for $\mu = 0$ to 3. The scale λ must be introduced in this relation for dimensional reason. We shall see in what follows that, when this approach is applied to standard quantum mechanics, it can be identified with the Compton length $\hbar/mc$. The η are stochastic variables such that $\langle \eta^\mu \rangle = 0$, $\langle (\eta^0)^2 \rangle = -1$ and $\langle (\eta^k)^2 \rangle = 1$ ($k = 1$ to 3). These stochastic variables describe the geometric effect of the fractal space-time on the elementary displacements. We shall now show that this description of the fractal

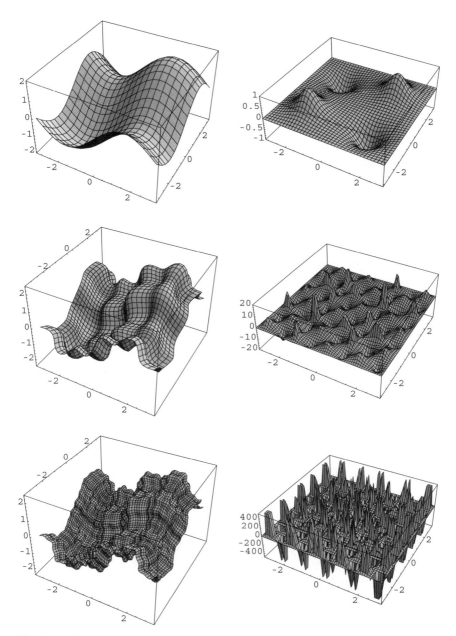

Fig. 4.14: **Fractal surface and curvature.** Left: a fractal surface drawn at three successive levels of resolution. Right: the corresponding Gaussian curvature of this surface seen at these three resolutions. It is itself fractal, i.e. an explicit and divergent function of the resolution scale (note the increase of the vertical units).

geometry can be expressed in terms of a fractal metric [349, 353, 378, 394] (valid at least in this one-body case and on the infinite set of geodesics).

The first step consists in defining the fractal fluctuations in terms of the classical coordinate differentials dx^μ instead of the invariant length differential ds. Since dx^μ and ds are differentials of the same order, we may write

$$d\xi^\mu = \eta^\mu \sqrt{\lambda^\mu \, dx^\mu}. \qquad (4.167)$$

The identification of equations (4.166) and (4.167) allows one to relate the new length and time scales λ^μ (which will be later identified with generalized Einstein–de Broglie relations) to the scale λ, namely,

$$\lambda_x = \frac{\lambda}{dx/ds} = \frac{\lambda}{u_x}, \quad \lambda_y = \frac{\lambda}{u_y}, \quad \lambda_z = \frac{\lambda}{u_z}. \qquad (4.168)$$

We are fully aware of the notation problem with the indices (in particular as concerns their covariant and contravariant status) here and in the developments below, which comes from the fact that the true nature of the resolutions is tensorial, while we consider only an incomplete vectorial notation for the simplicity. This section is only aimed at giving a hint of the nature of the dependence on scale of a fractal metric, knowing that a full study will be considered in forthcoming works.

Let us now assume for generality that space-time is curved at large scales, and therefore described by a standard Riemannian metric $g_{\mu\nu}$, but that it is also fractal toward the small scales, so that the elementary displacements involve fractal fluctuations in addition to the classical differentials. Then the invariant proper time dS reads in such a curved and fractal space-time

$$dS^2 = g_{\mu\nu} \, dX^\mu dX^\nu = g_{\mu\nu}(dx^\mu + d\xi^\mu)(dx^\nu + d\xi^\nu). \qquad (4.169)$$

Now replacing the $d\xi$'s by their expression (4.167), one obtains (on the fluid of geodesics of fractal dimension $D_F = 2$ which identifies with one particle)

$$dS^2 = g_{\mu\nu} \left(1 + \eta^\mu \sqrt{\frac{\lambda^\mu}{dx^\mu}} \right) \left(1 + \eta^\nu \sqrt{\frac{\lambda^\nu}{dx^\nu}} \right) dx^\mu dx^\nu, \qquad (4.170)$$

(where Einstein's convention about indices does not apply inside the square root). This proper time is now everywhere infinite at the limit $dx^\mu \to 0$ and would therefore be undefined in the standard description (this is the fundamental expression of the fractality). But in the scale relativity framework, the metric potentials are definitely defined in terms of explicitly dependent functions of the coordinate differential elements, which are now interpreted themselves as resolution variables, in agreement with the program of [349, 353]. Namely, the metric can now be written as

$$dS^2 = \tilde{g}_{\mu\nu}(x, dx)dx^\mu dx^\nu, \tag{4.171}$$

with

$$\tilde{g}_{\mu\nu}(x, dx) = g_{\mu\nu}(x)\left(1 + \eta^\mu\sqrt{\frac{\lambda^\mu}{dx^\mu}}\right)\left(1 + \eta^\nu\sqrt{\frac{\lambda^\nu}{dx^\nu}}\right), \tag{4.172}$$

(where Einstein's convention does not apply). We recover here our result [353] according to which, in the limit (dx^μ, $dt \to 0$), the metric is everywhere divergent (singular) at each of its points and instants, which is another expression of the fractality of space-time. Note that the concept of a stochastic space-time, manifested by a stochastic metric, has already been introduced by Frederick in 1976 [184] to attempt to comprehend quantum mechanics. However, in this approach the stochasticity of space-time is assumed to disappear at the position of a mass point, while, as we shall see in the next chapter, in the scale relativity approach we construct particles and their mass (and other conservative quantities) from the fractal geometry of space-time and of its geodesics itself.

In the simplified case where the underlying classical metric is Minkowskian, of assumed large scale signature $(-, +, +, +)$, which is a valid approximation in microphysics for scales larger than the Planck scale, the fractal metric (on a one-particle geodesic) reads

$$dS^2 = -c^2\left(1 + \eta_t\sqrt{\frac{\tau_E}{dt}}\right)^2 dt^2 + \left(1 + \eta_x\sqrt{\frac{\lambda_x}{dx}}\right)^2 dx^2$$
$$+ \left(1 + \eta_y\sqrt{\frac{\lambda_y}{dy}}\right)^2 dy^2 + \left(1 + \eta_z\sqrt{\frac{\lambda_z}{dz}}\right)^2 dz^2. \tag{4.173}$$

Its stochastic average reads

$$\langle dS^2 \rangle = -c^2\left(1 - \frac{\tau_E}{dt}\right)dt^2 + \left(1 + \frac{\lambda_x}{dx}\right)dx^2 + \left(1 + \frac{\lambda_y}{dy}\right)dy^2$$
$$+ \left(1 + \frac{\lambda_z}{dz}\right)dz^2. \tag{4.174}$$

As discussed in [349, 353], the metric signature changes to an Euclidean one $(+, +, +, +)$ for time scale $dt < \tau_E = \hbar/E$, the Einstein–de Broglie transition scale. This is a manifestation of the fact that the geodesics become able to run backward in time at these small time scales. This property is indeed observed in our laboratory coordinates in terms of the appearance of virtual particle-antiparticle pairs [353].

Let us give two comments about the form of this scale-dependent metric. We first note that its various terms can also be written as the sum of two differentials of different orders, for example $dx^2 + \lambda_x \, dx$. Viewed in this way, this means that the metric is no longer quadratic, but contains also a linear contribution (in this case of fractal dimension 2). The form of Eq. (4.174) has the advantage to make explicit the scale divergence and the transition from scale independance to fractality.

As a second comment, we may compare the new fractal terms in this metric form with those coming from curvature in a Schwarzschild metric, e.g. for the time component, $g_{00}(dt) = 1 - \tau_E/dt$ (where $\tau_E = \hbar/mc^2$ in the standard quantum case and in rest frame), with $g_{00}(r) = 1 - r_s/r$, where $r_s = 2GM/c^2$. This analogy identifies the Compton relation, which connects a space-time scale to the inertial mass, as the analog of the Schwarzschild relation, which connects a space-time scale to the active gravitational mass. But the Schwarzschild relation is but a simple form taken by Einstein field equations $S_{\mu\nu} = \chi T\mu\nu$ (which connects geometry to energy-momentum) in the (one-body + test particle) case. This remark anticipates on the derivation of the Compton relation (to come in Chapter 5), which establishes its fondamental status of (progenitor) of the field equation of a (one-particle) fractal space-time.

Before going on with the scale relativity program, we stress the fact that one cannot reduce the description of a nondifferentiable spacetime to the construction of such a fractal metric. Indeed, this metric form is valid only for one particle or, as an approximation, close to a particle in the multiparticle case, and only on the geodesics themselves, which nevertheless fill space-time like a fluid in the fractal geometry case. It has been given to provide a hint of the form, which a scale-divergent metric [353] may take, but the question of the general equations of a fractal space-time metric (i.e. the equivalent of Einstein's field equation for curved space-time) and of their solutions remains largely open. Moreover, as we shall show in the next chapters, the nondifferentiability has many other consequences (infinity of geodesics leading to a probabilistic description, multivaluedness of the fractal velocity field leading to complex and spinorial representations, coordinate dependence of the resolutions leading to gauge fields, etc.). In other words, this means that the general relativity tools, once they are generalized in order to include the scale divergence, although they participate in the description of the new geometry, are nevertheless insufficient to complete such a description. We shall see that a full description can be reached only by constructing new geometric tools, which can subsequently be shown to be equivalent to the physical and mathematical tools of quantum mechanics.

Problems and Exercises

Exercise 12 Write the metric of a fractal space or space-time in the cases of the various more general scale laws described in the present chapter (constant fractal dimension different from $D_F = 2$, variable fractal dimension in special scale relativity, in the case of the existence of a scale force, etc.

Hint: Express the elementary displacements dX as the sum $dX = dx + d\xi$ of a classical part dx and of a fractal part $d\xi$ in these various cases, then express the $d\xi$'s in terms of the dx's, and include these expressions in the metric element $ds^2 = g_{\mu\nu} \, dX^\mu dX^\nu$.   ■

Exercise 13 Derive the scale divergent metric elements of a fractal space-time from scale differential equations.

Comment: This can easily be done by the same method as that used for fractal lengths (i.e. for fractal coordinates). But this is only a first step of a more general theory that would treat scales and positions on the same footing in terms of a double differential calculus.   ■

4.5.5. *Intrinsic characterization of a fractal space*

The property of everywhere scale divergence establishes a nondifferentiable continuum space-time as fundamentally not Riemannian. Indeed, the fundamental Gauss hypothesis of local flatness, which underlies Riemannian geometry is clearly broken in this case. Moreover, it allows one to characterize a fractal space in an intrinsic way [353], i.e. from the only viewpoint of its inner properties. This is, since Gauss's work about the metric of a sphere generalized to hyperbolic surfaces (by Gauss, Bolyai and Lobachevsky, see e.g. [528]) then generalized by Riemann to any number of dimensions, a key point for a genuine definition of a new geometry.

Hence, we know that it is impossible to characterize in an intrinsic way whether a curve is straight or curved. Therefore, the concept of curved space is defined only for two and more dimensions, as can be seen from the two-dimensional nature of the Gaussian curvature, $K = 1/(R_1 R_2)$. The same is partially true for fractal spaces.

Indeed, one could think that one could characterize a fractal curve in an intrinsic way by the divergence (infinity at the limit) of the distance between any couple of points on the curve. However, this divergence is manifest only when comparing the distance to a fixed unit defined in an outer way. If one uses as unit the distance between two given points of the fractal curve defined along the curve, then a measure of distance is given by a ratio of distances, and it therefore remains finite when the fractal dimension is constant. This is still an expression of the principle of the relativity of scales. Now, if the fractal dimension is variable along the curve, distance measurements become scale dependent and an intrinsic characterization becomes possible even in a one-dimensional space.

It, however, remains that it is for two and more dimensions that the full subtleties of a geometric approach to physics can be used, in particular by the fundamental definition of geodesics and the identification of the free particle paths to these geodesics, and, more profoundly, by the identification of the "particles" themselves to the geodesics and their geometric properties (in the case of the application of the theory to elementary particles in standard quantum mechanics). This is the subject of the next chapters. In this case an intrinsic characterization of nondifferentiable fractal spaces is straightforward, by using the fact that their metric elements, which are potential energies, and their curvature invariants become at every point explicit functions of the scale variables, which diverge when these variables tend to zero [353].

Therefore, an experiment can easily be conceived in order to put to the test in an intrinsic way the nondifferentiability of space-time and its fractality toward small scales. It amounts to measuring energy and momentum for space and time intervals which become smaller and smaller, and to verifying that these measurements show ever increasing fluctuations. From the basic expression of fractal fluctuations, $m(\delta \xi^\mu)^{D_F} \propto \delta s$ for $\mu = 0$ (time coordinate) and $\mu = 1$, 2 and 3 (space coordinates), one obtains $m(\delta \xi^\mu/ds)(\delta \xi^\mu)^{D_F - 1} \approx \text{cst}$, so that one expects, in the fractal small scale regime where $\delta x \approx \delta \xi$, the energy-momentum fluctuations to diverge as

$$\delta p \sim \frac{1}{\delta x^{D_F - 1}}. \tag{4.175}$$

i.e. $\delta E \times \delta t \approx$ cst and $\delta p \times \delta x \approx$ cst when the fractal dimension is $D_F = 2$. This kind of experiment has been achieved many times, since it is nothing else but a standard experiment in microphysics, and it has definitely given the expected results, which are but the Heisenberg relations. They may therefore be viewed, in such a framework, as a direct and intrinsic experimental proof that the geometry of the quantum space-time is fractal [349].

Chapter 5

FROM FRACTAL SPACE TO NON-RELATIVISTIC QUANTUM MECHANICS

5.1. Introduction

The previous chapters were devoted to the study of pure scale laws, i.e. to the description of the scale dependence of fractal paths at a given point of space-time. The next step, which we consider now, consists of studying the effects on motion in standard space that are induced by these fractal structures.

Indeed, as we shall see, it is not sufficient to write a standard equation of dynamics for an "object" that would own fractal structures (one cannot actually separate what is "internal" from what is "external" to the fractal paths). Since these structures are those of the paths themselves, the equation of dynamics is itself fundamentally affected by the fractality. We are therefore led to ask again from the very beginning the question of the foundation of a fundamental equation of dynamics, now in a fractal and nondifferentiable space-time.

This is a vast question that must be solved step by step. We therefore proceed by first studying the induced effects on motion of the simplest scale laws (self-similar laws of fractal dimension 2 for the particle paths) under restricted conditions (only fractal space, then fractal space and time, and progressive account of new discrete symmetry breakings due to nondifferentiability). In this construction our main clue is once again the principle of relativity, under two of its most efficient consequences, the geodesic and covariance principles. The various effects of the nondifferentiability and of the fractality of space-time are included into the construction of a covariant derivative, which is actually a generalization of a total derivative. Then, in terms of this covariant derivative, the equation of motion is written under the form of a geodesic equation, i.e. of a free-like, Galileo-type inertial law of motion.

In this way, we successively recover more and more profound levels of quantum mechanical laws. In the present chapter, one derives non-relativistic quantum mechanics in terms of complex wave functions that are solutions of a Schrödinger equation, as first obtained in [353], then confirmed by many subsequent physical [360, 362, 144, 93, 409] and mathematical works, in particular by Cresson and Ben Adda [115, 117, 43, 44] and Jumarie [242, 243, 244, 245], including attempts of generalizations using the tool of the fractional integro-differential calculus [44, 119, 245].

In the next chapter, one derives relativistic quantum mechanics, at first without spin (Klein–Gordon equation [356, 360]), then including spin (Dirac equation [93]). The Pauli equation is obtained as a nonrelativistic limit of the Dirac equation [96], as in standard quantum mechanics. In all these cases, the theory allows one to construct the quantum tools (complex wave functions, then Pauli spinors and Dirac bispinors) in a geometric way and then to derive the equations they satisfy as integrals of geodesic equations.

More complicated situations (constant fractal dimension different from 2, variable fractal dimension, scale dynamical laws, special scale-relativistic log-Lorentzian behavior, etc), which lead to scale-relativistic corrections to standard quantum mechanics [358, 360], are subsequently briefly considered at the end of this chapter.

One of the main new results of this approach is that we no longer need to introduce a "particle" which would follow a "trajectory". Actually, the various properties of a quantum particle, which have been incompletely described using classical concepts as "wave", "particle" or "field" behaviors, can be recovered as properties of the geodesics themselves. This allows one to also derive from first principles other main axioms of quantum mechanics, such as the Born, von Neumann and projection postulates.

Under such a viewpoint, one does not even need to attribute to the particle internal characteristics, such as a mass, a spin or a charge. Indeed, these physical quantities can also be constructed as geometric characteristics of the geodesics, provided one considers geometry not only in space-time but also in the scale-space.

Hence, the mass can be identified with the fractal-nonfractal transition in scale-space (in rest frame), the spin with an internal angular momentum, which is specific of fractal curves of fractal dimension 2 (see Chapter 6 and [349, 353]), and the charges with the conservative quantities that originate from the symmetries of the scale-space (see Chapter 7). Though one cannot yet claim that this is achieved for all internal properties of particles, since many quantum numbers remain of unknown origin, this is nevertheless a way of approach, which is already partly implemented, so that the scale relativity theory allows one to contemplate the possibility of a future purely geometric description of quantum "objects" and of their motion.

5.2. General Link between Heisenberg–de Broglie Relations and Fractal Behavior

As a preliminary to this chapter, let us briefly provide a general argument [371] that allows one to directly connect Heisenberg's and de Broglie's relations to fractality. This question was one of the main issue of *Fractal Space-Time and Microphysics* [353], so that we refer the reader to this book and references therein, in particular [176, 1, 344, 420] for other arguments supporting this connection.

Once they are established, note that they play no longer any direct role in the theory. Indeed, if one refers, e.g. to Feynman's pioneering work on this subject, as recalled in Sec. 3.4, he established from quantum mechanical laws that the typical paths of quantum particles are continuous, nondifferentiable and of fractal dimension $D_F = 2$. In such an approach, the geometric characterization is therefore a property, which is deduced from the founding quantum properties. Now, in the theory of scale relativity and fractal space-time, what we attempt to do is the reverse, namely, to found quantum mechanical laws on nondifferentiability, i.e. on relaxing two hypotheses from the foundation of physics instead of adding new ones. This is the subject of the next sections.

Finkel [178] has proposed to use the information entropy for constructing generalized Heisenberg inequalities. His method consists of building the extremum state that achieves the wanted minimum. It is an essential result, since it allows one to construct exact Heisenberg inequalities for any couple of variables, while, except for standard deviations, only order of size relations are most of the time known. This method allows to establish the correct inequality in dependence to the precise definition of the uncertainties (which may be intervals, dispersions, standard deviations, etc), and to identify the probability distributions that achieve the minimum.

It can therefore be used to give a general solution to the problem, which dates back to the Abbott and Wise work [1], of translating quantum mechanical relations in terms of fractal geometric behavior [371].

As emphasized in [353], the key to this problem amounts to understanding the relation, for a given variable X, between $\langle |X| \rangle^2$, $\langle X \rangle^2$ and $\langle X^2 \rangle$ (where $\langle \rangle$ denotes averaging on a given probability density distribution). This question remains essential even nowadays. For example, in his proposal for an experiment to measure the Hausdorff dimension of quantum mechanical trajectories, Kröger [257] assumes that

$$\langle |\Delta x| \rangle^2 = \langle (\Delta x)^2 \rangle. \tag{5.1}$$

As already shown in [353], this relation is not strictly valid, but it is an extremum relation. Let us establish again this result using Finkel's method. Such a relation is important to be known, since it yields not only the small scale fractal behavior and the identification of the fractal dimension $D_F = 2$, but also the value of the transition length toward nonfractal behavior at large scale (i.e. the quantum to classical effective transition).

Consider two observables A and B and their averages, $\langle A \rangle = \sum_a \omega_a a$ and $\langle B \rangle = \sum_b \omega_b b$. For example, for obtaining the standard position-momentum Heisenberg relation, one will take $A = (x - \langle x \rangle)^2$ and $B = (p - \langle p \rangle)^2$.

The information entropy is given by

$$S(\varphi) = -\sum_a \omega_a \ln \omega_a, \tag{5.2}$$

with

$$\omega_a = |\langle a|\varphi \rangle|^2; \langle \varphi|\varphi \rangle = 1. \tag{5.3}$$

We shall now require $\langle A \rangle$ to be extremum, i.e.

$$\delta \langle A \rangle = 0. \tag{5.4}$$

Finally, Finkel's method amounts to maximizing the entropy (5.2) subject to the constraints (5.3) and (5.4). The solution of such an optimization is:

$$\langle a|\varphi \rangle = \exp\left(\frac{\Omega - \lambda a}{2}\right), \quad \Omega = -\ln\left(\sum_a e^{-\lambda a}\right), \quad \langle A \rangle_{ex} = \frac{\partial \Omega}{\partial \lambda}. \tag{5.5}$$

For example, when it is applied to $\langle A \rangle = \delta x$ and $\langle B \rangle = \delta p$, this method yields the position-momentum uncertainty relation for intervals,

$$\delta x \times \delta p \geq \frac{\hbar}{\pi}. \tag{5.6}$$

This minimum is achieved, e.g. by the ground state of the harmonic oscillator. One sees that, when intervals are used to define the uncertainty, the Heisenberg limit is no longer $\hbar/2$ as for standard deviations, but $\hbar/\pi$. This result will play an important role in subsequent works on the nature of charges and on the expectation of their bare value (see Sec. 11.1.2.2).

Let us now apply this method to typical quantum paths in the non-relativistic case. The length measured along a typical (nondeterministic) quantum trajectory is, in the non-relativistic case,

$$\mathcal{L} \propto \langle |v| \rangle \propto \langle |p| \rangle. \tag{5.7}$$

The problem posed here is therefore of finding a Heisenberg relation for the absolute value of a variable, $A = |z|$, rather than for the variable z itself. The extremum is given by

$$\langle a|\varphi \rangle = \exp \left(\frac{\Omega - \lambda|z|}{2} \right) \tag{5.8}$$

with

$$\Omega = -\ln \left(\int e^{-\lambda|z|} d|z| \right) = \ln \left(\frac{\lambda}{2} \right). \tag{5.9}$$

The maximized mean value is therefore

$$\langle A \rangle_{ex} = \frac{\partial \Omega}{\partial \lambda} = \frac{1}{\lambda}. \tag{5.10}$$

The probability distribution that achieves the extremum is

$$\omega_{ex}(|z|) = e^{\Omega - \lambda|z|} = \frac{1}{2\langle |z| \rangle} e^{-|z|/\langle |z| \rangle}. \tag{5.11}$$

Therefore we finally find the extremal relation

$$\langle |z| \rangle^2 = \langle z^2 \rangle. \tag{5.12}$$

When this method is applied to $z = \Delta x$, it yields the general inequality

$$\langle |\Delta x| \rangle^2 \leq \langle \Delta x^2 \rangle. \tag{5.13}$$

Let us now consider again the problem of typical quantum paths. Choosing $z = p$ and knowing that $\langle (\Delta p)^2 \rangle = \langle p^2 \rangle - \langle p \rangle^2$, we obtain the extremum relation

$$\langle |p| \rangle^2 = \langle p \rangle^2 + \langle (\Delta p)^2 \rangle. \tag{5.14}$$

Now $\langle p \rangle$ yields the de Broglie length, $\lambda = \hbar/\langle p \rangle$, while $\langle (\Delta p)^2 \rangle$ is linked through the Heisenberg relation to the position standard error, which is one of the way to characterize a resolution, $\Delta x \approx \hbar/\Delta p$. Since the length during a total time interval T is $\mathcal{L} = \langle |p| \rangle T/m$, we obtain

$$\mathcal{L} = \langle v \rangle T \left\{ 1 + \left(\frac{\lambda_x}{\Delta x} \right)^2 \right\}^{1/2}. \tag{5.15}$$

This is easily generalized to three coordinates, namely,

$$V = \langle v \rangle \left(1 + \frac{\lambda^2}{\sigma^2} \right)^{1/2}, \tag{5.16}$$

with

$$\lambda^{-2} = \lambda_x^{-2} + \lambda_y^{-2} + \lambda_z^{-2}, \quad \sigma^{-2} = \sigma_x^{-2} + \sigma_y^{-2} + \sigma_z^{-2}. \tag{5.17}$$

This result can now be compared with the prediction from a fractal model of trajectory. The length of a fractal trajectory is given by

$$\mathcal{L} = \int \left\{ 1 + \left(\frac{dX}{dt} \right)^2 \right\}^{1/2} dt. \tag{5.18}$$

Thanks to the general relation between length resolution and time resolution on a fractal curve (see Chapter 4), $\delta X^{D_F} \approx \lambda^{D_F - 1} \delta t$, we can integrate Eq. (5.18), and we obtain the final scale dependence

$$\mathcal{L} \approx \langle v \rangle T \left\{ 1 + \left(\frac{\lambda}{\delta X} \right)^{2(D_F - 1)} \right\}^{1/2}. \tag{5.19}$$

We recognize here exactly the quantum result, which allows us to identify the transition scale λ with the de Broglie length and to fix the fractal dimension to $D_F = 2$, in accordance with the Abbott–Wise and subsequent results (see e.g. [353] and references therein).

5.3. Main Consequences of Nondifferentiability

5.3.1. *Analysis of the problem*

We shall now implement one of the main goals of the initial scale relativity program: to recover quantum mechanical laws as arising from the behavior of the geodesics of a nondifferentiable space-time. In this chapter, we consider only a nondifferentiable and fractal space which corresponds to nonrelativistic quantum mechanics. The generalization to a fractal space-time, i.e. relativistic quantum mechanics, will be considered in the next chapter.

As we shall now see, the nondifferentiability of space involves three main consequences:

(i) Infinite number of geodesics.
(ii) Fractality of geodesics (with fractal dimension 2 playing a critical role).
(iii) Symmetry breaking under the reflection $(dt \leftrightarrow -dt)$.

These three (minimal) effects will be combined into the construction of a covariant derivative. Then we shall write a geodesic equation (coming under strong covariance) in terms of this covariant derivative. This free-form geodesic equation can finally be integrated, after a change of variables, in terms of a Schrödinger equation [353]. In this equation, the wave function is but a re-expression of the action, which is now complex because of the fundamental two-valuedness of derivatives that emerges from condition (iii) as a very consequence of nondifferentiability.

5.3.2. *Infinite number of geodesics*

Let us proceed step by step and consider first the restricted problem of the motion along a fractal geodesic, but by doing this we shall keep in mind that this is just an intermediate step in the description on the motion in a nondifferentiable space-time, since such a single geodesic has actually no physical meaning. As we shall see in what follows, the geodesics of a fractal space-(time) are in infinite number, so that "particles" are identified in the scale relativity approach with a fluid of geodesics that fills space. Let us consider a time dependent coordinates $X(t)$, where X is the position vector in space (we omit the indices for simplicity of the writing). Strictly, the nondifferentiability of the coordinates means that the velocity

$$V = \frac{dX}{dt} = \lim_{dt \to 0} \frac{X(t + dt) - X(t)}{dt} \tag{5.20}$$

is undefined. When dt tends to zero, either the ratio dX/dt tends to infinity or it fluctuates without reaching any limit.

However, as recalled in the previous chapters, continuity and nondifferentiability imply an explicit dependence on scale of the various physical quantities. As a consequence, the position becomes an explicit fractal function of the scale $X = X(t, dt)$ and the velocity V now becomes also defined as a fractal function $V = V(t, dt)$. The usual definition of velocity corresponds to $V(t, 0)$.

Nondifferentiability means that $V(t, 0)$ is undefined for any value of t, while $V(t, dt)$ is defined for all $dt \neq 0$. The small but significant change of the scale relativity approach with respect to the standard differential calculus therefore simply amounts to still considering what happens when $dt \to 0$, but without effectively taking the limit. Since the zero limit is actually an infinite, this changes nothing to the essence of the differential calculus, but it has an enormous advantage: dt is now a nonzero number, i.e. $(|dt|/T) \in \mathbb{R}_+$ (T being a time unit) whatever its smallness, so that the standard use of physicists to multiply, divide, etc. differential elements is now fully supported. Morever, it now holds even in the nondifferentiable case, namely one is now entitled to write expressions like (λ/dx), which would be forbidden in the standard use of the differential calculus because of its infinity.

However, this change of description is not yet sufficient. Under this form, it concerns a particular fractal path which may lie in an Euclidean or curved space, while we are now interested in defining the geodesics of a fractal space. This means to skip from an outside view to an inside view, as we began to do in Sec. 4.5.

One of the direct geometric consequences of nondifferentiability and of the subsequent fractal character of space itself (not only of the paths) is that there is an infinity of fractal geodesics relating any couple of points of this fractal space [353]. This theorem has been proved by Cresson [115] by using the fractional differential calculus (see [44, 119] for a development of this method and its use in scale relativity theory). With this tool, he writes the geodesic equations of a fractal manifold in terms of a system of equations of the form

$$\frac{d^\alpha}{dt^\alpha}\left(\frac{d^\alpha \gamma_j}{dt^\alpha}\right) + \Gamma_{ij}^k \left(\frac{d^\alpha \gamma_k}{dt^\alpha}\right) = X_i, \qquad (5.21)$$

which can be transformed into a system of linear fractional differential equations. Such a system is known to have an infinity of solutions.

Another simple way to derive this result consists just in remarking that a fractal space-time can be described in terms of a generalization to explicitly scale-dependent functions of the general relativity tools (see Sec. 4.5). In particular the metric elements and their derivatives the Christoffel symbols, which enter in the geodesic equation, become explicit and divergent functions of scale variables (in addition to being functions of the space-time coordinates). This implies an infinity of solutions to such a scale dependent equation.

This result can be easily understood and explained in a direct way. Consider a fractal surface as described in Sec. 4.5. It can be represented in terms of a fractal distribution of conic points of positive and negative infinite curvature, i.e. in terms of the new tool, of scale-divergent curvature (see [353], Secs. 3.6 and 3.10). We know that already in general relativity there may be a demultiplication of geodesics around a mass due to curvature, as exemplified by gravitational lens effects. The number of lensed images is two around a point mass, but it becomes infinite in case of perfect alignment, creating an Einstein ring [162]. It is $2n + 1$ for a galaxy potential and it increases due to the observed hierarchical distribution of matter in the universe composed of stars in galaxies, galaxies in clusters, etc. (see e.g. the review paper [348]). Here it is an infinite hierarchy of structures that is introduced, so that the number of geodesics is multiplied to infinity.

The number of geodesics is also infinite at the differential level, since each "point" of the fractal space has an infinite diffusive effect on the paths. The geodesics are then expected to fill the space or part of the space like a fluid.

As a consequence, we are led to replace the velocity $V(t, dt)$ on a particular geodesic by a fractal velocity field

$$V = V[x(t, dt), t, dt] \qquad (5.22)$$

of the flow of geodesics.

Such a fundamental and unavoidable loss of information of purely geometric origin already has an important physical consequence. It means the giving-up of determinism at the level of the particle paths and it therefore leads one to jump to a statistical and probabilistic description. But here, contrary to the view of standard quantum mechanics, the statistical nature of the physical tool is not set as a foundation of physics, but it is derived from geometric properties.

5.3.3. *Fractality of geodesics*

The fractality of the velocity field allows us to use the various tools that we have constructed in Chapter 4 in order to derive scale laws from first principles. We shall once again proceed by steps and first consider the simplest possible case, then once the laws of motion found in that case, generalize to more complicated scale laws.

In the simplest case we expect the geodesic velocity field to be solution of a first order scale differential equation of the type

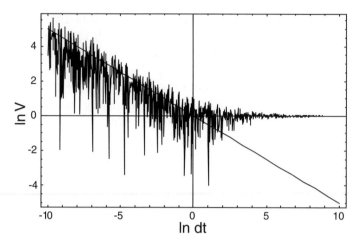

Fig. 5.1: Dependence on time-scale of velocity on a fractal geodesic. The figure shows the dependence on the time-scale, $\ln(dt/\tau_1)$, of the logarithm of the velocity $\ln(V/V_0)$ on a fractal geodesic, including a classical (differentiable) part, which is dominant at large scale (it tends to a scale-independent velocity toward the right in the figure), and a fractal (nondifferentiable) fluctuating stochastic part, which is dominant at small scales (to the left). The fluctuation has been taken in this figure to be Gaussian, but the theory applies to any probability distribution of this stochastic variable. The fluctuating fractal part $d\xi$ is of order $dt^{1/2}$ for fractal dimension 2, so that the velocity diverges toward small scales as $dt^{-1/2}$, which expresses the nondifferentiability of the fractal coordinate.

$\partial V/\partial \ln dt = a + bV + \cdots$ (see Sec. 4.2.2), which reads, when the fractal dimension of the geodesics is D_F (see Fig. 5.1),

$$V[x, t, dt] = v[x, t] + w[x, t, dt] = v[x, t]\left\{1 + \zeta \left(\frac{\tau_o}{dt}\right)^{1-1/D_F}\right\}, \quad (5.23)$$

where τ_o is a transition time-scale. We recover our result according to which the velocity is now the sum of two independent terms of different orders of differentiation, since their ratio v/w is, from the standard viewpoint, infinitesimal. In analogy with the real and imaginary parts of a complex number, we have called $v = v[x(t), t]$ the "classical part" or "differentiable part" of the velocity field and $w = w[x(t, dt), t, dt]$ its "fractal part" or "nondifferentiable part" [91, 93, 387, 400].

By "classical", we do not mean that the classical velocity field is necessarily a variable of classical physics (for example, as we shall see hereafter, it will become two-valued due to nondifferentiability, which is clearly not a classical property). We mean that it remains differentiable and, therefore, comes under classical differentiable equations and that it becomes a classical variable in the classical infrared limit.

The new component w is an explicitly scale-dependent fractal fluctuation, which would be infinite from the standard point of view where one makes $dt \to 0$. The nondifferentiability of the space implies a fundamental loss of determinism of the paths in this space, in particular of the geodesics, so that we are led to describe this fractal fluctuation by a stochastic variable ζ. It is normalized by choosing the reference time-scale τ_t in such a way that $\langle \zeta \rangle = 0$ and $\langle \zeta^2 \rangle = 1$, where ζ is now a purely mathematical dimensionless stochastic variable. The mean $\langle \rangle$ is taken upon the probability distribution of this variable. But, as we shall see, the final result does not depend on this distribution so that we do not have to specify it. This means that this description includes far more general processes than Markov, Brownian-like or Wiener processes.

The expression obtained for the velocity field means that, although its scale-dependent fractal part may be existing for all values of the scale, which means that space would be fractal at all scales, it is dominated by the differentiable part at large scales (see Figs. 4.1 and 5.1). This means that we shall recover the differentiable classical theory at large scales: this was a fundamental requirement for such an approach. Moreover, the reverse is also true, namely, the differentiable part is dominated by the fractal part toward the small scales. This fractal-nonfractal transition, which is here a geometric transition from scale-dependence to effective scale-independence, will subsequently be identified with the de Broglie scale of quantum to classical transition.

The continuation of the fractal part inside the classical differentiable domain without any limit (Fig. 5.1), even though it is masked by the classical contribution, is also a necessary ingredient of the theory. Indeed, the relative contributions of the classical and fractal parts depend on the value of the transition, which is itself relative, depending in particular on the state of motion of the reference system. Namely, the de Broglie length-scale is $\lambda_{\text{deB}} = \hbar/p = \hbar/mv$ for a free particle, while the de Broglie thermal scale is $\lambda_{\text{th}} = \hbar/(2mkT)^{1/2} = \hbar/(m\langle v^2 \rangle^{1/2})$ and the de Broglie time $\tau_{\text{deB}} = \hbar/E = (\hbar/\frac{1}{2}mv^2)$. This is a key point in order to allow for macroscopic quantum effects, such as superconductivity of Bose–Einstein, condensates to manifest themselves at large scales when the temperature decreases.

The maintenance of the classical behavior inside the quantum realm is a well-known result of the quantum theory. Indeed, the Ehrenfest theorem

[168] states that the average value of a given quantity, calculated from the probability density given by quantum mechanics, i.e. $P = |\psi|^2$, yields the classically expected result. As we shall see, this result of quantum mechanics is a manifestation of the fact that the differentiable parts of the variables are their average and do not vanish at small scales (i.e. in the quantum domain) even though they are dominated by the fractal fluctuation, while the fractal stochastic fluctuations, although dominant at small scales, are of zero average (see Fig. 5.1).

On the contrary, the maintenance of the quantum behavior inside the classical realm is, as far as we know, a new view of scale relativity, which may possibly lead to a test of the theory. Indeed, the standard quantum mechanics view about the passage from quantum behavior to classical behavior is in terms of wave packet reduction and decoherence, and does not usually consider that some quantum properties may be hidden inside classical phenomena. We shall come back to this point in Chapter 10, which is devoted to proposals of laboratory experiments aiming at applying the theory and putting it to the test.

Let us go on with the derivation of the motion equations by deriving a differential version of the expression for the velocity field. Eq. (5.23) multiplied by dt gives the elementary displacement in a fractal space, in particular along its geodesics, dX, as a sum of two infinitesimal terms of different differentiation orders

$$dX = dx + d\xi. \tag{5.24}$$

The variable dx is defined as the classical (or differentiable) part of the full displacement dX while $d\xi$ represents its fractal (or nondifferentiable) part.

We stress again the fact that, as previously recalled, the nondifferentiability here does not mean that we cannot differentiate, since we have kept the continuity of space-time (so that we can define $d\xi$), but that we can no longer calculate a derivative in the standard meaning (namely, $d\xi/dt$ is infinite when $dt \to 0$).

Due to the definitive loss of information implied by the nondifferentiability, we are lead to represent it in terms of a stochastic variable (see the justification in Sec. 4.5). We therefore write the classical and fractal parts under the form

$$dx = v\, dt, \tag{5.25}$$

$$d\xi = \zeta\lambda \left(\frac{|dt|}{\tau_o} \right)^{1/D_F}, \tag{5.26}$$

where $v = v[x(t), t]$ is the classical velocity field introduced hereabove and where λ and τ_o are respectively space and time scales introduced for dimensional reasons.

Arrived at that point, we are faced with the fundamental question of the fractal dimension of the geodesics of a fractal space. From the theoretical point of view, we do not want to cut it short and to give a unique and definitive answer to this question. We shall on the contrary use our step by step method and consider the various levels of possible complexity for the answer that have been given in Chapter 4, namely,

(i) constant fractal dimension $D_F = 2$;
(ii) constant fractal dimension different from 2;
(iii) variable fractal dimension.

This includes fractal dimensions that vary with scale, as in scale-dynamical laws, log-periodic laws and log-Lorentzian laws of the special scale-relativistic theory, but also those which vary with space-time variables or with both kinds of variables.

The main part of the present chapter is devoted to the study of the first case, constant fractal dimension $D_F = 2$ of the geodesics. A general method to treat the other situations will be given in Sec. 5.14, but their full development goes beyond the goal of the present book and will be let open to future works (except for the special scale-relativistic case, which may already have applications to high energy and elementary particle physics). However, it is not by lack of interest that we make such a choice, but by lack of place and time. Indeed, as we shall see, the fractal dimension $D_F = 2$ alone already provides a foundation for the whole of standard quantum mechanical laws and also for generalizations to a macroscopic quantum-type mechanics, which is no longer based on the Planck constant. Therefore, the differences with respect to this constant value $D_F = 2$ are expected to give rise to still new very interesting generalizations of the quantum laws.

The reasons for making conspicuous the particular value $D_F = 2$ of the fractal dimension are the following:

(i) It has been shown by Feynman [176] that the typical quantum mechanical paths, i.e. those that contribute most to the path integral, are of fractal dimension 2 (see Sec. 3.4). This result has been subsequently confirmed by many works [256, 1, 81, 10, 420, 344].

In the relativistic case ($v \rightarrow c$ and scales smaller than the Compton scale), although some authors have concluded that the fractal dimension fell down again from 2 to 1, i.e. to a nonfractal behavior, we have shown that this result was actually due to the failure to account for particle-antiparticle virtual pairs, which appear at relativistic energy [353]. By accounting for them, one finds that the fractal dimension 2 remains valid in both cases (nonrelativistic and relativistic) [420, 349, 353] and for all dimensions, the three spatial ones and the time dimension.

This divergence of both space (below the de Broglie length-scale $\lambda_{dB} = \hbar/p$) and time coordinates (below the Einstein time-scale $\tau_E = \hbar/E$) allows the observed velocity, which is a ratio of space intervals over time intervals, to remain limited by c [353], despite the divergence of each of them taken separately, which manifests itself in terms of an increasing number of virtual pairs for increasing energy, i.e. decreasing scale.

(ii) The fractal dimension 2 is that of Brownian motion and more generally of Markov processes, which correspond to cases when the various elementary displacements are independent of the previous and next ones. In other words, this fractal dimension is typical of uncorrelated motion, while $D_F \leq 2$ corresponds to correlated motion and $D_F \geq 2$ to anticorrelated motion (see, e.g. [514]). Now, as suggested in ([353], Chapter 5), the effect of a fractal and nondifferentiable space-time on these elementary displacements is equivalent to an infinite chaos, i.e. to a chaos with an infinite Lyapunov exponent λ_l, therefore with a vanishing chaos time $\tau_c = 1/\lambda_L$ (that defines the horizon of predictability $\tau_H \approx 10\tau_c$). It therefore leads to a full loss of predictability, on position, time and angles, which involves neither correlation nor anticorrelation and finally a fractal dimension 2.

(iii) We shall see by considering the more general case $D_F \neq 2$ that $D_F = 2$ plays actually a critical role in the theory (see [358, 360] and Sec. 5.14). The geodesic equation of motion one obtains for $D_F \neq 2$, which may still be given the form of a generalized Schrödinger equation, is explicitly scale-dependent in this case (i.e. it keeps the dependence on dt as a separated variable). It is only in the special case $D_F = 2$ that this explicit scale dependence disappears and that one gets the standard form of the Schrödinger equation [358].

(iv) One may define a finite internal angular momentum for a fractal spiral path of fractal dimension 2, while for $D_F < 2$ it is null and for $D_F > 2$ it becomes infinite [349, 353]. This result was obtained prior to the full construction of a quantum spin in the scale relativity theory [93] and has been supported by this construction [96] (see Sec. 6.4.3).

Therefore, in what follows, we consider a fractal dimension

$$D_F = 2. \tag{5.27}$$

See [353] for more detail about the physical significance of this dimension and about an explicit construction of fractal curves in space and space-time manifesting it. Such curves are very particular, since they are of topological dimension 1 and of fractal dimension 2, while they live in a space of topological dimension 3 (or 4 in the case of space-time). Though being curves from the viewpoint of their topological dimension, their measure is of Hausdorff dimension 2, namely, they are characterized by an area instead of a length, although they are not surfaces, i.e. their content should be given in m^2.

In this case, the fractal fluctuation may now be written as

$$d\xi = \zeta \sqrt{2\mathcal{D}|dt|}, \tag{5.28}$$

where ζ is a stochastic variable such that

$$\langle \zeta \rangle = 0, \quad \langle \zeta^2 \rangle = 1, \tag{5.29}$$

and where $\mathcal{D}$ is a fundamental parameter, which is introduced for dimensional reason. Indeed, since $d\xi$ is a length and dt a time, it is given by the relation

$$\mathcal{D} = \frac{1}{2} \frac{\langle d\xi^2 \rangle}{dt} \tag{5.30}$$

and its dimensionality is, therefore, $[L^2 T^{-1}]$. This choice was motivated by the analogy of this parameter with a diffusion coefficient, since Eq. (5.28) is similar to the equation for fluctuations in a Brownian motion. However, the analogy does not hold further, since the interpretation of this relation is here fundamentally different from a classical diffusion. This is an essential difference with Nelson's stochastic mechanics [331, 332], in which this relation is considered to be the result of a classical diffusion process (by, e.g. a sub-quantum medium). In our first attempts to derive a Schrödinger equation from a fractal space-time description [353, 357], we considered the new approach to be a "new formulation of stochastic mechanics". However, a more thorough comparison between the two theories proved that most of the equations and hypotheses of stochastic mechanics were unnecessary in the scale relativity approach and that the interpretation, including that of the above common fluctuation equation, was very different. A more detailed comparison of the two theories will be the subject of Sec. 10.5.1.1.

As we shall see further on, $2\mathcal{D}$ is a scalar quantity, which can be identified in standard quantum mechanics with $\hbar/m$, which means that it identifies up to a fundamental constant with the Compton scale $\lambda_c = \hbar/mc$, so that its physical meaning yields the mass of the particle itself, which so acquires a geometric meaning.

5.3.4. *Discrete symmetry breaking and two-valuedness of derivatives*

One of the most fundamental consequences of the nondifferentiable nature of space (more generally, of space-time) is the breaking of a new discrete symmetry, namely, of the reflection invariance on the differential element of (proper) time. Indeed, this implies a two-valuedness of the velocity, which can be subsequently shown to be the origin of the complex nature of the quantum tool.

The derivative with respect to the time t of a differentiable function f can be actually written twofold,

$$\frac{df}{dt} = \lim_{dt \to 0^+} \frac{f(t+dt) - f(t)}{dt} = \lim_{dt \to 0^+} \frac{f(t) - f(t-dt)}{dt}. \tag{5.31}$$

The two definitions are equivalent in the differentiable case. In the non-differentiable situation, both definitions fail, since the limits are no longer defined. In the new framework of scale relativity, the physics is related to the behavior of the function during the "zoom" operation on the time resolution δt, identified, in the theoretical description considered here, with the differential element dt. The nondifferentiable function $f(t)$ is replaced by an explicitly scale-dependent fractal function $f(t, dt)$, which, therefore, becomes a function of two variables, t (in space-time) and dt (in scale space). Two functions f'_+ and f'_- are, therefore, defined as explicit functions of the two variables t and dt,

$$f'_+(t, dt) = \frac{f(t+dt, dt) - f(t, dt)}{dt}, \quad f'_-(t, dt) = \frac{f(t, dt) - f(t-dt, dt)}{dt}. \tag{5.32}$$

One passes from one definition to the other by the transformation $dt \leftrightarrow -dt$. This is a differential time reflection, which is actually an implicit discrete symmetry of differentiable physics, but which is now broken in the nondifferentiable case.

Another way to describe this behavior could consist in using a single function $f'(t, dt)$ for positive and negative values of dt. The symmetry breaking is expressed in this case by the fact that this is not a symmetric function of dt, i.e. $f'(t, dt) \neq f'(t, -dt)$ and the nondifferentiability by the fact that it is not defined at $dt = 0$. However, we know that the scale relativity approach to such an explicitly scale dependent function amounts to finding it as a solution of a scale differential equation written in terms of the relevant scale variable, which is $d \ln |dt|$ according to the Gell-Mann–Levy method (see Chapter 4). This necessary and natural jump to a logarithmic representation implies to introduce two different functions for positive and negative values of dt, while it sends to infinity the two limits $dt \to 0^{\pm}$.

When they are applied to fractal space coordinates $x(t, dt)$, these definitions yield two velocity fields instead of one, that are fractal functions of the resolution, $V_+[x(t, dt), t, dt]$ and $V_-[x(t, dt), t, dt]$.

These two fractal velocity fields may in turn be decomposed in terms of their classical and fractal parts, namely,

$$V_+[x(t, dt), t, dt] = v_+[x(t), t] + w_+[x(t, dt), t, dt], \qquad (5.33)$$

$$V_-[x(t, dt), t, dt] = v_-[x(t), t] + w_-[x(t, dt), t, dt]. \qquad (5.34)$$

The V_+ and V_- fractal functions are *a priori* different functions, and the same is therefore true of their differentiable parts v_+ and v_-, which are scale-independent standard fluid mechanics-like velocity fields. While, in standard classical mechanics, the concept of velocity was one-valued, we must therefore introduce, for the case of a nondifferentiable space, two velocity fields instead of one, even when going back to the classical domain. But we stress that, contrary to the case of stochastic quantum mechanics [331], these velocities are not here backward and forward velocities. Indeed, the two-valuedness comes in scale relativity from the reversal of the scale variable dt, not of time itself, so that the two velocities are actually forward velocities, defined from $t - dt$ to t for V_- and from t to $t + dt$ for V_+ (with $dt > 0$). In recent papers, Ord [424] also insists on the importance of introducing "entwined paths" for understanding quantum mechanics (but without giving a mechanism for their emergence).

A simple and natural way to account for this doubling consists in using complex numbers and the complex product. As we recall hereafter, this is the origin of the complex nature of the wave function of quantum mechanics, since this wave function can be identified with the exponential of the complex action that is naturally introduced in this framework. We shall demonstrate in Sec. 5.4.1 that the choice of complex numbers to represent the two-valuedness of the velocity is a simplifying and "covariant" choice (in the sense of the principle of covariance, according to which the simplest possible form of the equations of physics should be conserved under all coordinate transformations). It is remarkable in this respect that a doubled Hilbert space has recently been proposed to solve the problem of defining a time operator in quantum mechanics and to represent the quantum-mechanical time evolution [260].

5.4. Covariant Total Derivative Operator

We are now led to describe the elementary displacements for both processes, $dX_\pm$, as the sum of a differentiable part, $dx_\pm = v_\pm\, dt$, and of a stochastic fluctuation about this differentiable part, $d\xi_\pm$, which is, by definition, of zero mean $\langle d\xi_\pm \rangle = 0$, namely

$$dX_+ = v_+\, dt + d\xi_+,$$

$$dX_- = v_-\, dt + d\xi_-. \tag{5.35}$$

More generally, one may define two classical derivative operators, d_+/dt and d_-/dt, which yield the twin classical velocities when they are applied to the position vector x,

$$\frac{d_+}{dt}x = v_+, \quad \frac{d_-}{dt}x = v_-. \tag{5.36}$$

As regards the fluctuations, the generalization to three dimensions of the fractal behavior of Eq. (5.26) reads

$$\langle d\xi_{\pm i}\, d\xi_{\pm j} \rangle = \pm 2\mathcal{D}\delta_{ij}\, dt \quad i,j = x,y,z, \tag{5.37}$$

since the $d\xi$'s are of null differentiable part and assumed to be mutually independent. The Krönecker symbol δ_{ij} in Eq. (5.37) means indeed that the mean crossed product $\langle d\xi_{\pm i}\, d\xi_{\pm j} \rangle$, with $i \neq j$, is null. In this expression, we consider that $dt > 0$ for the $(+)$ process and $dt < 0$ for the $(-)$ process, so that $\pm 2\mathcal{D}\, dt$ is always positive, in agreement with $\langle d\xi_\pm^2 \rangle$.

5.4.1. *Origin of complex numbers in quantum mechanics*

We now know that each component of the velocity field takes two values instead of one. This means that each component of the velocity becomes a vector in a two-dimensional space, or, in other words, that the velocity becomes a two-index tensor. The generalization of the sum of these quantities is straightforward, but one also needs to define a generalized product.

The problem can be put in a general way: it amounts to finding a generalization of the standard product that keeps its fundamental physical properties [84].

From the mathematical point of view, we are here exactly confronted with the well-known problem of the doubling of algebra (see e.g. [453]). Indeed, the effect of the symmetry breaking $dt \leftrightarrow -dt$ (or $ds \leftrightarrow -ds$ for the proper time in the relativistic case) is to replace the algebra $\mathcal{A}$ in which the classical physical quantities are defined, by a direct sum of two exemplaries of $\mathcal{A}$, i.e. the space of the pairs (a, b) where a and b belong to $\mathcal{A}$. The new vectorial space $\mathcal{A}^2$ must be supplied with a product in order to become itself an algebra (of doubled dimension).

The same problem is asked again when one takes also into account the symmetry breakings $dx^\mu \leftrightarrow -dx^\mu$ and $x^\mu \leftrightarrow -x^\mu$ [93]: this leads to new algebra doublings. The mathematical solution to this problem is well-known: the standard algebra doubling amounts to supplying $\mathcal{A}^2$ with the complex product. Then the doubling $\mathbb{R}^2$ of $\mathbb{R}$ is the algebra $\mathbb{C}$ of complex numbers, the doubling $\mathbb{C}^2$ of $\mathbb{C}$ is the algebra $\mathbb{H}$ of quaternions, the doubling $\mathbb{H}^2$ of quaternions is the algebra of Graves–Cayley octonions.

This mathematical solution, obtained by Cartan [84], fully justifies the use of complex numbers, then of quaternions, which are equivalent to spinors, (see Sec. 6.3), in order to describe the successive doublings due to discrete symmetry breakings at the infinitesimal level, which are themselves more and more profound consequences of space-time nondifferentiability [93, 403].

The problem with algebra doubling is that the iterative doubling leads to a progressive deterioration of the algebraic properties. Namely, one loses the order relation of reals in the complex plane, then the quaternion algebra is non-commutative, and the octonion algebra is also non-associative. But an important positive result for physical applications is that the doubling of a metric algebra is a metric algebra [453].

In what follows, we give complementary arguments of a physical nature, which show that the use of the complex product in the first algebra doubling $\mathbb{R} \rightarrow \mathbb{C}$ have a simplifying and covariant effect (the same is true of the passage to quaternions and biquaternions, see Chapter 6). We use here once again the word "covariant" in the original meaning given to it by

Einstein [158], namely, the requirement of form invariance of fundamental equations.

In order to simplify the argument, let us consider the generalization of scalar quantities, for which the product law is the standard product in $\mathbb{R}$. The result obtained can then be easily generalized to vectorial tensorial and other physical quantities.

The first constraint is that the new product must remain an internal composition law. We also make the simplifying assumption that it remains linear in terms of each of the components of the two quantities to be multiplied. A general bilinear product $c = a \otimes b$ reads

$$c^k = a^i \omega_{ij}^k b^j, \tag{5.38}$$

and it is therefore defined by eight numbers ω_{ij}^k in the case (considered here) of a two-valuedness of the quantities a, b, c.

The second physical constraint is the requirement to recover the classical variables and the classical product at the classical limit. The mathematical equivalent of this constraint is the requirement that $\mathcal{A}$ still be a sub-algebra of $\mathcal{A}^2$. Therefore, we identify $a_0 \in \mathcal{A}$ with $(a_0, 0)$ and we set $(0, 1) = \alpha$. This allows us to write the new two-dimensional vectors in the simplified form $a = a_0 + a_1 \alpha$, so that the product now writes

$$c = (a_0 + a_1\alpha)(b_0 + b_1\alpha) = a_0 b_0 + a_1 b_1 \alpha^2 + (a_0 b_1 + a_1 b_0)\alpha. \tag{5.39}$$

The problem is now reduced to find α^2, which is now defined by only two coefficients

$$\alpha^2 = \omega_0 + \omega_1 \alpha. \tag{5.40}$$

Let us now come back to the beginning of our construction. We have introduced two elementary displacements, each of them made of two terms, a differentiable part and a fractal part (see Eq. 5.35)

$$dX_+ = v_+ \, dt + d\xi_+,$$

$$dX_- = v_- \, dt + d\xi_-. \tag{5.41}$$

Let us first consider the two values of the differentiable part of the velocity. Instead of considering them as a vector of a new plane, (v_+, v_-), we shall use the above construction for defining them as an element of the doubled algebra [371]. Explicitely, we first replace (v_+, v_-) by the equivalent twin velocity field $[(v_+ + v_-)/2, (v_+ - v_-)/2]$, then we define

$$\mathcal{V} = \left(\frac{v_+ + v_-}{2} - \alpha \frac{v_+ - v_-}{2} \right). \tag{5.42}$$

This choice is motivated by the requirement that, at the classical limit when $v = v_+ = v_-$, the real part identifies with the classical velocity v while the new "imaginary" part vanishes.

The same operation can be made for the fractal parts. One can define velocity fluctuations $w_+ = d\xi_+/dt$ and $w_- = d\xi_-/dt$, so that we define a new element of the doubled algebra,

$$\mathcal{W} = \left(\frac{w_+ + w_-}{2} - \alpha \frac{w_+ - w_-}{2} \right). \tag{5.43}$$

Finally the total velocity field $\tilde{\mathcal{V}} = \mathcal{V} + \mathcal{W}$, including the differentiable part and the fractal divergent part, reads

$$\tilde{\mathcal{V}} = \left(\frac{v_+ + v_-}{2} - \alpha \frac{v_+ - v_-}{2} \right) + \left(\frac{w_+ + w_-}{2} - \alpha \frac{w_+ - w_-}{2} \right). \tag{5.44}$$

This quantity includes the complete information on the velocity field, as described by the two successive doublings issued from nondifferentiability, namely, the separation into a classical differentiable part and a fractal nondifferentiable part, and the separation into a $(+)$ and a $(-)$ process.

We shall see in what follows that a Lagrange function can be introduced in terms of the new two-valued tool that leads to a conserved form for the Euler–Lagrange equations. In the end, as we shall see, the Schrödinger equation is obtained as their integral. In [353] and subsequent publications, we have used the covariance principle to write the complex Lagrange function of the free particle as $\mathcal{L} = \frac{1}{2}m\mathcal{V}^2$, i.e. we have constructed it from only the differentiable part of the velocity. Now, to be more complete, the stochastic mean of the Lagrange function for the free particle should strictly be written as

$$\mathcal{L} = \frac{1}{2}m\langle\tilde{\mathcal{V}}^2\rangle = \frac{1}{2}m\langle(\mathcal{V} + \mathcal{W})^2\rangle, \tag{5.45}$$

namely, it should also include the fractal fluctuation. Since $\langle\mathcal{W}\rangle = 0$ by definition, and $\langle\mathcal{V}\cdot\mathcal{W}\rangle = 0$ because they are mutually independent, it reads

$$\mathcal{L} = \frac{1}{2}m(\langle\mathcal{V}^2\rangle + \langle\mathcal{W}^2\rangle) = \frac{1}{2}m(\mathcal{V}^2 + \langle\mathcal{W}^2\rangle). \tag{5.46}$$

We recover the previous expression, but with an additional term $\frac{1}{2}m\langle\mathcal{W}^2\rangle$. The presence of this term would greatly complicate all the subsequent developments toward the Schrödinger equation, since it would imply a fundamental divergence of quantum mechanics. Let us expand it. One finds

$$\langle \mathcal{W}^2 \rangle = \frac{1}{4} \langle [(w_+ + w_-) - \alpha(w_+ - w_-)]^2 \rangle, \tag{5.47}$$

i.e.

$$\langle \mathcal{W}^2 \rangle = \frac{1}{4} \langle (w_+^2 + w_-^2)(1 + \alpha^2) - 2\alpha(w_+^2 - w_-^2) + 2w_+ w_-(1 - \alpha^2) \rangle. \tag{5.48}$$

Since $\langle w_+^2 \rangle = \langle w_-^2 \rangle$ and $\langle w_+ \cdot w_- \rangle = 0$ (they are mutually independent), we finally find that $\langle \mathcal{W}^2 \rangle$ can vanish only provided

$$\alpha^2 = -1, \tag{5.49}$$

whose solution is $\alpha = \pm i$, the imaginary unit. In summary, we have found that

$$\alpha = \pm i \Leftrightarrow \langle \mathcal{W}^2 \rangle = 0, \tag{5.50}$$

and that in this case the stochastic mean Lagrange function reduces to $\mathcal{L} = \frac{1}{2} m \mathcal{V}^2$.

Therefore, the choice of the complex product in the algebra doubling plays an essential physical role, since it allows to suppress what would be additional infinite terms in the final equations of motion [93, 403]. One may think that these infinities, which have not been seen at first by the founders of the quantum theory since they have naturally taken a representation in terms of complex numbers (one may say that they are hidden in the choice of complexes), are the same, which re-appeared some years later in the framework of the quantum field theories and which led to the renormalization and renormalization group theories.

The two solutions $+i$ and $-i$ have equal physical meaning, since the final equation of Schrödinger (demonstrated in the following) and the wave function are physically invariant under the transformation $i \rightarrow -i$ provided it is applied to both of them, as is well-known in standard quantum mechanics.

Now this result yields an explanation of one of the badly explained features of quantum mechanics, according to which, if one replaces the wave function by its complex conjugate, $\psi \rightarrow \psi^\dagger$, one should also replace in the Schrödinger equation the time term $i\hbar \partial \psi / \partial t$ by its opposite, $-i\hbar \partial \psi / \partial t$. This is fully understood if one considers that the transformation considered is not a simple change of the sign of the phase, but instead a general mathematical transformation from i to $-i$, under which the whole quantum physics is actually invariant, in agreement with our present result.

It has also another consequence as regards our understanding of quantum mechanics. In its standard form, the complex character of the wave function (or equivalently the state function, or the probability amplitude in Feynman's approach) is set as one of the fundamental postulates of quantum mechanics. Here the choice of complex numbers becomes a possible representation of quantum mechanics, but other representations are now possible. The main physical effect is the two-valuedness of the derivative originating from nondifferentiability, while complex numbers are now only a way to represent this two-valuedness. Moreover, we shall see in Chapter 6 that in the scale relativity framework the spin has a similar origin, which leads one to obtain Pauli spinors in terms of quaternions, i.e. of "complex complexes", and Dirac bispinors in terms of biquaternions, i.e. complex quaternions.

Problems and Exercises

Open Problem 4: Is it possible to treat on the same footing the doublings of variables that lead to complex wave functions and to spinors, and therefore to define new representations in which what leads to complexity and what leads to spin could be rotated? Would this correspond to a new physical effect? ∎

5.4.2. *Complex velocity field*

We now combine the two derivatives to define a complex derivative operator from the classical (differentiable) parts that allows us to recover local differential time reversibility in terms of the new complex process [353],[a]

$$\frac{\hat{d}}{dt} = \frac{1}{2}\left(\frac{d_+}{dt} + \frac{d_-}{dt}\right) - \frac{i}{2}\left(\frac{d_+}{dt} - \frac{d_-}{dt}\right). \qquad (5.51)$$

---

[a]Note about the notation: in previous publications, this operator, which plays the central role of a covariant derivative, has been denoted by various notations different from that adopted here, in particular by d'/dt. However this notation was not clearly different from a standard derivative in some publications. For this reason, we choose here to define a clearly identifiable notation for this "quantum-covariant derivative", $\hat{d}/dt$, with the hope that it does not correspond to any already defined concept and that it will not be misleading.

Applying this operator to the position vector yields the differentiable part of the complex velocity field

$$\mathcal{V} = \frac{\hat{d}}{dt}x = V - iU = \frac{v_+ + v_-}{2} - i\frac{v_+ - v_-}{2}. \tag{5.52}$$

The full complex velocity field, $\widetilde{\mathcal{V}} = \mathcal{V} + \mathcal{W}$, introduced in Eq. (5.44) in which one can now make $\alpha = i$, also includes the divergent fractal part of zero mean $\mathcal{W}$. It will be taken fully into account in Sec. 5.8. This complex velocity field is one of the main tools of the scale relativity theory. In the fluid-type approach adopted here, the wave function plays for this velocity field the role of a potential of velocity. This means that it is not defined as the velocity of an assumed point-like object, but as that of the fluid of geodesics itself, and that the various properties usually attributed to a quantum mechanical "particle" are derived from it.

The minus sign in front of the imaginary term is chosen here in order to obtain the standard Schrödinger equation in terms of the wave function ψ. The reverse choice would give the Schrödinger equation for the complex conjugate of the wave function $\psi^\dagger$ and would be, therefore, physically equivalent.

The classical limit corresponds to situations where the two-valuedness has disappeared, so that there is no longer any separated U and V velocity fields. As we shall see in Sec. 10.2.2, there are classical limits of the theory where $v_+ = v_- = v$, so that the real part, V, of the complex velocity $\mathcal{V}$, identifies with the standard classical velocity, while its imaginary part, $U = 0$, but also other ones where $v_+ = -v_-$ so that V vanishes and only U remains.

5.4.3. *Complex time-derivative operator*

Contrary to what happens in the differentiable case, the total derivative with respect to time of a function $f[X(t, dt), t]$ of a fractal coordinate $X(t, dt) = x(t) + \xi(t, dt)$ contains finite terms up to higher orders [155], depending on the value of the fractal dimension. Let us consider the Taylor expansion

$$\frac{df}{dt} = \frac{\partial f}{\partial t} + \frac{\partial f}{\partial x_i}\frac{dX_i}{dt} + \frac{1}{2}\frac{\partial^2 f}{\partial x_i \partial x_j}\frac{dX_i dX_j}{dt}$$

$$+ \frac{1}{6}\frac{\partial^3 f}{\partial x_i \partial x_j \partial x_k}\frac{dX_i dX_j dX_k}{dt} + \cdots \tag{5.53}$$

where one makes the sum on indices, which appear twice (Einstein's convention). In this expression and in what follows we keep the notation $\partial f/\partial x$ for the derivative with respect to the coordinate, whether it is fractal or not (i.e. $\partial f/\partial x$ or $\partial f/\partial X$ are considered to be the same). But in the expression $df = (\partial f/\partial x)dX$, the fact that $dX = dx + d\xi$ becomes relevant and plays an essential role.

Note that it has been shown by Kolwankar and Gangal [251] that, even if the fractal dimension is not an integer, a fractional Taylor expansion can still be defined using the local fractional derivative (see [115] about the physical relevance of this tool and [243] for another more efficient proposal).

In our case, a finite contribution only proceeds from terms of D_F-order, while lesser-order terms yield an infinite contribution and higher-order ones yield infinitesimal terms which vanishes when $dt \to 0$. Therefore, in the special case of a fractal dimension $D_F = 2$, the total derivative writes

$$\frac{df}{dt} = \frac{\partial f}{\partial t} + \frac{\partial f}{\partial x_i}\frac{dX_i}{dt} + \frac{1}{2}\frac{\partial^2 f}{\partial x_i \partial x_j}\frac{dX_i dX_j}{dt} + \mathcal{O}(dt^{1/2}). \tag{5.54}$$

Since $d\xi$ is an infinitesimal of order $1/2$, the dX^3 term is an infinitesimal which varies as $dt^{1/2}$, the dX^4 term as dt, etc. We can therefore now come back to the standard differential calculus by taking the limit $dt \to 0$, and all these infinitesimal terms vanishes.

Let us now consider the differentiable ("classical") part of this expression (the complete calculus which also takes account of the divergent, nondifferentiable part will be given in Sec. 5.8). By definition, $\langle dX \rangle = dx$, so that the second term is reduced to $v \cdot \nabla f$. Now concerning the term $dX_i dX_j/dt$, it is usually infinitesimal, but here it writes:

$$\frac{dX_i dX_j}{dt} = \frac{(dx_i + d\xi_i)(dx_j + d\xi_j)}{dt}$$
$$= \frac{dx_i dx_j}{dt} + \frac{dx_i d\xi_j + dx_j d\xi_i}{dt} + \frac{d\xi_i d\xi_j}{dt}. \tag{5.55}$$

Since dx_i is a standard infinitesimal of order one while $d\xi_i$ is an infinitesimal of order $1/2$, the first term $dx_i dx_j/dt$ is an infinitesimal of order one and the second term an infinitesimal of order $1/2$. Both of them vanish in the limit $dt \to 0$. But the third term is finite. Therefore the classical (differentiable) part of this expression reduces to $\langle d\xi_i d\xi_j \rangle/dt$, which is equal to $2\mathcal{D}\delta_{ij}$ according to Eq. (5.37). The last term of the differentiable part of Eq. (5.54) amounts to a Laplacian and we obtain

$$\frac{d_\pm f}{dt} = \left(\frac{\partial}{\partial t} + v_\pm \cdot \nabla \pm \mathcal{D}\Delta \right) f. \tag{5.56}$$

Substituting Eqs. (5.56) into Eq. (5.51), we finally obtain the expression for the complex time derivative operator [353],[b]

$$\frac{\hat{d}}{dt} = \frac{\partial}{\partial t} + \mathcal{V} \cdot \nabla - i\mathcal{D}\Delta. \tag{5.57}$$

This is one of the main results and tool of the theory of scale relativity. Indeed, the passage from differentiable (or almost everywhere differentiable) geometry to the new nondifferentiable geometry can now be implemented by replacing the standard time derivative d/dt by the new complex operator $\hat{d}/dt$. In other words, this means that $\hat{d}/dt$ plays the role of a "covariant derivative operator", with which we shall write the fundamental equations of physics under the same form they had in the differentiable case.

When doing such a replacement, in particular when using the Leibniz rule for products or composed functions [446], one should of course be cautious with the fact that it is not a first order derivative. As can be seen in its expression, it involves a linear combination of first order and second order derivatives, so that this is true also of its Leibniz rules [387]. We shall come back to this point in Sec. 6.2.2.

It should be remarked, before going on with this construction, that we use here the concept of covariant derivative in analogy with the covariant derivative $D_j A^k = \partial_j A^k + \Gamma^k_{jl} A^l$ replacing $\partial_j A^k$ in Einstein's general relativity. But one shoud be cautious with this analogy, since the two situations are different. Indeed, the problem posed in the construction of general relativity was that of a new geometry, in a framework where the differential calculus was not affected. Therefore the Einstein covariant derivative amounts to substracting the new geometric effect $-\Gamma^k_{jl} A^l$ in order to recover the mere inertial motion, for which the Galilean law of motion $Du^k/ds = 0$ naturally holds [266]. Here there is an additional

---

[b]In this expression there is no two-valuedness on the partial differential $\partial/\partial t$, since we have considered a function f, which is not fractal by itself, but fractal only as a consequence of the fractality of the coordinates, i.e. $f = f[x(t, dt), t]$. In a more general situation, f could be by itself a fractal function, $f = f[x(t, dt), t, dt]$, and, as suggested by L. de Montera (private communication), the operator $\partial/\partial t$ could also become two-valued, i.e. $\partial/\partial t \to \hat{\partial}/\partial t$. In this case the covariant derivative operator becomes $\hat{d}/dt = \hat{\partial}/\partial t + \mathcal{V} \cdot \nabla - i\mathcal{D}\Delta$. This leads to a generalization of standard quantum mechanics, since an additional terms would appear in the Schrödinger equation.

difficulty: the new effects come not only from the geometry (see Chapter 7 for a scale covariant derivative acting in the same way as that of general relativity and giving rise to gauge fields) but also from the nondifferentiability and its consequences on the differential calculus.

The true status of the new derivative is actually an extension of the concept of total derivative. Already in standard physics, the passage from the free Galileo–Newton equation to its Euler form was a case of conservation of the form of equations in a more complicated situation, namely, $d/dt \rightarrow \partial/\partial t + v \cdot \nabla$ when $v(t) \rightarrow v[x(t), t]$. In the fractal and nondifferentiable situation considered here, the three consequences of nondifferentiability (infinity of geodesics, fractality and two-valuedness of derivative) lead to three new terms in the total derivative operator: the Eulerian term $V \cdot \nabla$, but also two additional imaginary contributions, $-iU \cdot \nabla$ and $-i\mathcal{D}\Delta$.

5.5. Complex Covariant Mechanics

5.5.1. *Lagrangian approach*

Let us now summarize the main steps by which one may generalize the standard classical mechanics using this covariance. We are now looking for the equations of motion in a fractal space. In what follows, we consider only the differentiable parts of the variables, which are independent of resolutions. However, we shall subsequently show that the same result i.e. construction of a complex wave function that is solution of a Schrödinger equation, can also be obtained by taking the full velocity field, including the fractal nondifferentiable part.

The effects of the internal nondifferentiable structures are now contained in the covariant derivative. Following the standard construction of the laws of mechanics, we assume that the differentiable part of the mechanical system under consideration can be characterized by a Lagrange function that keeps its usual form, but now in terms of the complex velocity $\mathcal{V}$, $\mathcal{L} = \mathcal{L}(x, \mathcal{V}, t)$, from which an action $\mathcal{S}$ is defined

$$\mathcal{S} = \int_{t_1}^{t_2} \mathcal{L}(x, \mathcal{V}, t)dt, \qquad (5.58)$$

which is therefore now complex since $\mathcal{V}$ is itself complex. As a consequence, the action principle is no longer a least action principle, since there is no order relation in the complex plane. However, as already remarked by de Broglie [131], the least action principle is, from the viewpoint of its mathematical implementation, actually a stationary action principle. Both are equivalent for real functions, but no longer in generalized cases. This finally poses no problem, since its stationary form $\delta S = 0$ still holds in the complex plane. It yields as a consequence two optimized real least actions, its real part and its imaginary part.

We have assumed here that the $(+)$ and $(-)$ velocity fields can be combined in the expression of the Lagrange function in terms of the complex velocity. In the previous Sec. 5.4.1, we have already given arguments for the choice made in the construction of the complex velocity to be simplifying and covariant. We shall now prove that the action can indeed be put under the above form and that it allows us to conserve the standard form of the Euler–Lagrange equations, i.e. to write generalized covariant Euler–Lagrange equations.

In a general way, the Lagrange function is expected to be a function of the variables x and of their time derivatives $\dot{x}$. We have found that the number of velocity components $\dot{x}$ is doubled, so that we are led to write

$$\pounds = \pounds\left(x, \dot{x}_+, \dot{x}_-, t\right). \tag{5.59}$$

In terms of this Lagrange function, the action principle reads

$$\delta S = \delta \int_{t_1}^{t_2} \pounds\left(x, \dot{x}_+, \dot{x}_-, t\right) dt = 0. \tag{5.60}$$

It becomes

$$\int_{t_1}^{t_2} \left(\frac{\partial \pounds}{\partial x}\delta x + \frac{\partial \pounds}{\partial \dot{x}_+}\delta \dot{x}_+ + \frac{\partial \pounds}{\partial \dot{x}_-}\delta \dot{x}_-\right) dt = 0. \tag{5.61}$$

The Lagrange function of Eq. (5.58), re-expressed in terms of $\dot{x}_+$ and $\dot{x}_-$, reads

$$\pounds = \pounds\left(x, \frac{1-i}{2}\dot{x}_+ + \frac{1+i}{2}\dot{x}_-, t\right), \tag{5.62}$$

and it is therefore equivalent to a function of $\dot{x}_+$ and $\dot{x}_-$ considered as independent variables. Therefore we obtain

$$\frac{\partial \pounds}{\partial \dot{x}_+} = \frac{1-i}{2}\frac{\partial \pounds}{\partial \mathcal{V}}; \quad \frac{\partial \pounds}{\partial \dot{x}_-} = \frac{1+i}{2}\frac{\partial \pounds}{\partial \mathcal{V}}, \tag{5.63}$$

while the new covariant time derivative operator reads

$$\frac{\hat{d}}{dt} = \frac{1-i}{2}\frac{d_+}{dt} + \frac{1+i}{2}\frac{d_-}{dt}. \tag{5.64}$$

Since $\delta \dot{x}_+ = d_+(\delta x)/dt$ and $\delta \dot{x}_- = d_-(\delta x)/dt$, the action principle takes the form

$$\int_{t_1}^{t_2} \left(\frac{\partial \mathcal{L}}{\partial x} \delta x + \frac{\partial \mathcal{L}}{\partial \mathcal{V}} \left[\frac{1-i}{2} \frac{d_+}{dt} + \frac{1+i}{2} \frac{d_-}{dt} \right] \delta x \right) dt = 0, \tag{5.65}$$

i.e.

$$\int_{t_1}^{t_2} \left(\frac{\partial \mathcal{L}}{\partial x} \delta x + \frac{\partial \mathcal{L}}{\partial \mathcal{V}} \frac{\hat{d}}{dt} \delta x \right) dt = 0. \tag{5.66}$$

This is the form one would have obtained directly from Eq. (5.58), which proves our statement according to which one may replace in a general way the velocity doublet (v_+, v_-) by its compact complex form $\mathcal{V}$, even when taking the derivative with respect to this variable.

The subsequent demonstration of the Lagrange equations from the stationary action principle relies on an integration by part. This integration by part cannot be performed in the usual way without a specific analysis, because it involves the new covariant derivative.

The first point to be considered is that such an operation involves the Leibniz rule for the covariant derivative operator $\hat{d}/dt$. Since $\hat{d}/dt = \partial/dt + \mathcal{V} \cdot \nabla - i\mathcal{D}\Delta$, it is, as already remarked, a linear combination of first and second order derivatives, so that the same is true of its Leibniz rule, namely, it is a linear combination of the first order and second order Leibniz rules. This implies the appearance of an additional term in the expression for the derivative of a product (see [446, 387] and Sec. 6.2.2), namely,

$$\frac{\hat{d}}{dt}\left(\frac{\partial \mathcal{L}}{\partial \mathcal{V}} \cdot \delta x \right) = \frac{\hat{d}}{dt}\left(\frac{\partial \mathcal{L}}{\partial \mathcal{V}} \right) \cdot \delta x + \frac{\partial \mathcal{L}}{\partial \mathcal{V}} \cdot \frac{\hat{d}}{dt} \delta x - 2i\mathcal{D}\nabla\left(\frac{\partial \mathcal{L}}{\partial \mathcal{V}} \right) \cdot \nabla \delta x. \tag{5.67}$$

Since $\delta x(t)$ is not a function of x, the additional term vanishes. Therefore the above integral becomes

$$\int_{t_1}^{t_2} \left[\left(\frac{\partial \mathcal{L}}{\partial x} - \frac{\hat{d}}{dt}\frac{\partial \mathcal{L}}{\partial \mathcal{V}} \right) \delta x + \frac{\hat{d}}{dt}\left(\frac{\partial \mathcal{L}}{\partial \mathcal{V}} \cdot \delta x \right) \right] dt = 0. \tag{5.68}$$

The second point to consider in this derivation of the covariant Euler–Lagrange equations is concerned with the integration of the covariant derivative itself. We define a new integral as being the inverse operation of the covariant derivation, i.e.

$$\oint \hat{d}f = f, \qquad (5.69)$$

in terms of which one obtains

$$\int_{t_1}^{t_2} \hat{d}\left(\frac{\partial \mathcal{L}}{\partial \mathcal{V}} \cdot \delta x\right) = \left[\frac{\partial \mathcal{L}}{\partial \mathcal{V}} \cdot \delta x\right]_{t_1}^{t_2} = 0, \qquad (5.70)$$

since $\delta x(t_1) = \delta x(t_2) = 0$ by definition of the variation principle. Therefore the action integral becomes

$$\delta S = \int_{t_1}^{t_2} \left(\frac{\partial \mathcal{L}}{\partial x} - \frac{\hat{d}}{dt}\frac{\partial \mathcal{L}}{\partial \mathcal{V}}\right) \delta x \, dt = 0. \qquad (5.71)$$

5.5.2. *Complex Euler–Lagrange equations*

Finally we obtain generalized complex Euler–Lagrange equations that read

$$\frac{\hat{d}}{dt}\frac{\partial \mathcal{L}}{\partial \mathcal{V}} = \frac{\partial \mathcal{L}}{\partial x}, \qquad (5.72)$$

where $x = \{x_k\}$ and $\mathcal{V} = \{\mathcal{V}_k\}$ are vectors, the index k running on all the degrees of freedom of the system. Therefore, thanks to the transformation $d/dt \to \hat{d}/dt$, they take exactly their standard classical form.

Let us consider in more detail the case of the free particle. The form of the free particle Lagrange function is a direct consequence of relativity [266]. In classical relativistic mechanics, the action is identified, up to a multiplicative constant, with the invariant proper time, $dS = -mc\,ds$. Now we have seen (see [352, 353] and Sec. 4.4) that the Poincaré–Lorentz transformation is the general solution to the special relativity problem, without adding any postulate to the principle of relativity. Therefore the Lagrange function of a free particle reads $L = -mc^2(1 - v^2/c^2)^{1/2}$ and its nonrelativistic approximation is $L = \frac{1}{2}mv^2$.

In the scale relativity theory, we therefore expect from covariance that the Lagrange function of a nonrelativistic free particle takes the form of a complex generalization of this expression, namely,

$$\mathcal{L}_{\text{free}} = \frac{1}{2}m\mathcal{V}^2, \qquad (5.73)$$

where $\mathcal{V}$ is a three-dimensional complex vector. This is easily generalized to an ensemble of n free particles as $\mathcal{L}_{\text{free}} = \sum \frac{1}{2}m_a\mathcal{V}_a^2$, where $a = 1$ to n runs on all the particles. The motion equation one deduces from it reads

$\hat{d}(\partial \pounds / \partial \mathcal{V})/dt = 0$, i.e. $\hat{d}\mathcal{V}/dt = 0$. It takes, as expected, the form of a free, strongly covariant, geodesic equation.

More generally, the Lagrange function of a particle in a potential is expected, from the covariance principle, to take the form,

$$\pounds(x, \mathcal{V}, t) = \frac{1}{2}m\mathcal{V}^2 - \phi. \tag{5.74}$$

Therefore the Euler–Lagrange equations read in this case

$$m\frac{\hat{d}}{dt}\mathcal{V} = -\nabla\phi, \tag{5.75}$$

namely, they keep the same form as Newton's fundamental equation of dynamics. This can be easily generalized to the many particle case (see Sec. 5.10).

5.5.3. *Conservative quantities*

Recall that in classical mechanics after having derived the Euler–Lagrange equations from the action principle, in which both ends of the action integral are fixed, one may construct the various conservative quantities, such as energy, momentum, and angular momentum from the symmetries of space and time, respectively, uniformity of time, uniformity of space and isotropy of space (Noether theorem). Then, as a second step, one may consider one of the ends of the action integral to be variable. This introduces the action as a function of the coordinates, $S = S(x, t)$. The conservative quantities can then be derived from this function following the fundamental relations $p_k = \partial S/\partial x_k$, $E = -\partial S/\partial t$ (see, e.g. [265]).

All these various steps can be recovered in the new complex formalism, thanks to covariance. Therefore, one can define from the homogeneity of position space a generalized complex momentum given by

$$\mathcal{P} = \frac{\partial \pounds}{\partial \mathcal{V}}. \tag{5.76}$$

If we now consider the action as a functional of the upper limit of integration in Eq. (5.58), the variation of the action from a trajectory to another nearby trajectory yields a generalization of the above relation of standard mechanics, namely,

$$\mathcal{P} = \nabla \mathcal{S}. \tag{5.77}$$

Concerning the energy, i.e. in terms of coordinates and momenta, the Hamilton function of the system, it reads

$$\mathcal{H} = -\frac{\partial \mathcal{S}}{\partial t}. \tag{5.78}$$

These relations will soon appear as the progenitors of the correspondence principle (see Sec. 5.6.3).

In the case where the Lagrange function reads $\mathcal{L} = \frac{1}{2}m\mathcal{V}^2 - \phi$ (particle in a scalar potential), the complex momentum $\mathcal{P} = \partial \mathcal{L}/\partial \mathcal{V}$ keeps its standard classical form, as a function of the complex velocity field,

$$\mathcal{P} = m\mathcal{V}. \tag{5.79}$$

As regards the generalized energy, its expression involves an additional term [362, 367, 446, 387]. Namely, the Hamilton function reads

$$\mathcal{H} = \frac{1}{2}m(\mathcal{V}^2 - 2i\mathcal{D}\nabla \cdot \mathcal{V}). \tag{5.80}$$

The additional term will be derived in Sec. 5.6.7 from a fully covariant form of the Hamilton function (see also Sec. 5.9). It is actually a direct consequence of the second order term in the covariant derivative. It is also closely linked to the existence of a quantum potential, which is, in the present framework, a manifestation of the fractal geometry of space. Its four-dimensional generalization will be discussed in Sec. 6.2.3, as well as methods to restore the strong covariance in the relativistic case [446, 387].

5.6. From Newton to Schrödinger Equation

5.6.1. *Geodesic equation*

As we have seen hereabove, the Euler–Lagrange equations of a closed system keeps the form of Newton's fundamental equation of dynamics

$$m\frac{\hat{d}}{dt}\mathcal{V} = -\nabla\phi, \tag{5.81}$$

which is now written in terms of complex variables and of the complex time derivative operator.

In the case when there is no external field, and also when the field can itself be constructed from a covariant geometric process, like gravitation in general relativity [158] and now gauge fields in scale relativity [356, 394, 405], the covariance is explicit and complete, so that Eq. (5.81) takes the simple form of Galileo's equation of inertial motion, i.e. of a geodesic equation

$$\frac{\hat{d}}{dt} \mathcal{V} = 0. \tag{5.82}$$

This equation can be obtained both from the action principle, which acquires the status of a geodesic principle in relativity theories, and from the strong covariance principle.

This is analogous to Einstein's general relativity, where the equivalence principle of gravitation and inertia leads to a strong covariance principle, expressed by the fact that one may always find a coordinate system in which the metric is locally Minkowskian. In this coordinate system, the covariant equation of motion of a free particle is that of inertial motion $Du_\mu = 0$ in terms of the general-relativistic covariant derivative D and four-vector u_μ. The expansion of the covariant derivative subsequently transforms this free-motion equation into a geodesic equation in a curved space-time, whose manifestation is the gravitational field. The same form can be obtained from the equivalence principle, from the strong covariance principle, and from the least-action principle that becomes a geodesic principle in a relativity theory thanks to the identification of the action with the invariant proper time (see e.g. [266]).

The same is true here, since the motion equation (5.82) could have also been written directly from the principle of strong covariance, or from a "quantum principle of equivalence" according to which a quantum motion is locally equivalent to a fractal nondifferentiable motion. In all cases, it amounts to implementing the most radical consequence of the principle of relativity, namely, the emptiness principle [327], which was the very initial statement of the principle of relativity by Galileo [190]: "for all things that participate to it, motion is as if it were not". Applied to motion, it means that whatever complicated and intricated be a path, in the proper reference system, which is swept along with the motion, there is no motion. The geodesic equation $\hat{d}\mathcal{V}/dt = 0$, which has the form of the equation of a free particle in vacuum, is a direct mathematical expression of this statement.

Now, after expansion of the covariant derivative, the free-form motion equations of general relativity can be transformed into a Newtonian equation in which a generalized force emerges, of which the Newton gravitational force is an approximation. In an analogous way, the covariance induced by scale effects leads to a transformation of the equation of motions, which, as we shall now show, becomes after integration the Schrödinger equation.

With or without an external field, the complex momentum $\mathcal{P}$ reads in both cases

$$\mathcal{P} = m\mathcal{V}, \tag{5.83}$$

so that, from Eq. (5.77), the complex velocity field $\mathcal{V}$ is potential (irrotational), namely it is given by the gradient of the complex action,

$$\mathcal{V} = \frac{\nabla\mathcal{S}}{m}. \tag{5.84}$$

5.6.2. *Complex wave function*

We now introduce a complex function ψ that will be subsequently identified with a wave function or state function, which is nothing but another expression for the complex action $\mathcal{S}$,

$$\psi = e^{i\mathcal{S}/S_0}. \tag{5.85}$$

One should not be misled by the form of this expression: since $\mathcal{S} = S_R - iS_I$ is complex, despite its apparent form, this ψ function has a phase $\theta = S_R/S_0$ and a modulus, $|\psi| = \exp(S_I/S_0)$.

The factor S_0 has the dimension of an action, i.e. of an angular momentum, and must be introduced for dimensional reasons. We show in what follows that when this formalism is applied to microphysics, i.e. to the standard quantum mechanics of, e.g. atomic and molecular physics, S_0 is nothing but the fundamental constant $\hbar$. But all the physical and mathematical structure of the theory is preserved whatever be the value of this constant: this fact will play an essential role for macroscopic applications of the approach (see Chapter 13).

5.6.3. *Correspondence principle*

From Eq. (5.84), we find that the function ψ is related to the complex velocity field as follows:

$$\mathcal{V} = -i\frac{S_0}{m}\nabla(\ln\psi). \tag{5.86}$$

This means that the complex velocity field is potential, and that $\ln\psi$ plays the role of a velocity potential (see e.g. [269]).

Momentum

Since the complex momentum is $\mathcal{P} = m\mathcal{V}$, this relation may be written under the form

$$\mathcal{P} = -iS_0 \nabla (\ln \psi), \tag{5.87}$$

i.e.

$$\mathcal{P}\psi = -iS_0 \nabla \psi. \tag{5.88}$$

In the case of standard quantum mechanics, for which $S_0 = \hbar$, this relation reads $\mathcal{P}\psi = -i\hbar\nabla\psi$. This is therefore a derivation of the principle of correspondence for momentum, $p \rightarrow -i\hbar\nabla$. Indeed, the real part of the complex momentum $\mathcal{P}$ is, in the classical limit, the classical momentum p. The "correspondence" is then understood here as between the real part of a complex quantity and an operator acting on the function ψ. But now, thanks to the introduction of the complex momentum of the geodesic fluid, it is no longer a mere correspondence, but it has become a genuine equality.

Energy

A similar result may be easily obtained for energy. From the relations $\mathcal{E} = \partial S / \partial t$ and $S = -iS_0 \ln \psi$, one obtains

$$\mathcal{E}\psi = iS_0 \frac{\partial \psi}{\partial t}, \tag{5.89}$$

which establishes a correspondence between the real part E of the complex energy, which is the classical energy in the classical limit, and the operator $iS_0\partial/\partial t$. In the case of standard mechanics, we recover the principle of correspondence for energy, $E \rightarrow i\hbar\partial/\partial t$.

Angular momentum

The classical angular momentum is defined as $M = r \times p$, so that it is easily generalized to the complex representation of scale relativity as

$$\mathcal{M} = r \times \mathcal{P}. \tag{5.90}$$

Therefore, using the expression (5.87) for $\mathcal{P}$, one obtains

$$\mathcal{M}\psi = -iS_0 r \times \nabla \psi, \tag{5.91}$$

so that we recover, in the standard quantum case $S_0 = \hbar$, the correspondence principle for angular momentum, $M \to -i\hbar r \times \nabla$. It now emerges, once again, as an equality instead of a mere correspondence.

5.6.4. *Remarkable identity*

We have now at our disposal all the mathematical tools needed to write the fundamental equation of dynamics Eq. (5.81) in terms of the new quantity ψ. It takes the form

$$iS_0 \frac{\hat{d}}{dt}(\nabla \ln \psi) = \nabla \phi. \tag{5.92}$$

Now one should be aware that $\hat{d}$ and ∇ do not commute. However, as we shall see in the following, $\hat{d}(\nabla \ln \psi)/dt$ is nevertheless a gradient in the general case.

Replacing $\hat{d}/dt$ by its expression, given by Eq. (5.57), yields

$$\nabla \phi = iS_0 \left(\frac{\partial}{\partial t} + \mathcal{V} \cdot \nabla - i\mathcal{D}\Delta \right)(\nabla \ln \psi), \tag{5.93}$$

and replacing once again $\mathcal{V}$ by its expression in Eq. (5.86), we obtain

$$\nabla \phi = iS_0 \left[\frac{\partial}{\partial t}\nabla \ln \psi - i\left\{ \frac{S_0}{m}(\nabla \ln \psi \cdot \nabla)(\nabla \ln \psi) + \mathcal{D}\Delta(\nabla \ln \psi) \right\} \right]. \tag{5.94}$$

Consider now the identity [353]

$$(\nabla \ln f)^2 + \Delta \ln f = \frac{\Delta f}{f}, \tag{5.95}$$

which proceeds from the following tensorial derivation

$$\begin{aligned}
\partial_\mu \partial^\mu \ln f + \partial_\mu \ln f \partial^\mu \ln f &= \partial_\mu \frac{\partial^\mu f}{f} + \frac{\partial_\mu f}{f}\frac{\partial^\mu f}{f} \\
&= \frac{f\partial_\mu \partial^\mu f - \partial_\mu f \partial^\mu f}{f^2} + \frac{\partial_\mu f \partial^\mu f}{f^2} \\
&= \frac{\partial_\mu \partial^\mu f}{f}.
\end{aligned} \tag{5.96}$$

When we apply this identity to ψ and take its gradient, we obtain

$$\nabla\left(\frac{\Delta\psi}{\psi}\right) = \nabla[(\nabla \ln \psi)^2 + \Delta \ln \psi]. \tag{5.97}$$

The second term in the right-hand side of this expression can be transformed, using the fact that ∇ and Δ commute, i.e.

$$\nabla \Delta = \Delta \nabla. \tag{5.98}$$

The first term can also be transformed thanks to another identity

$$\nabla (\nabla f)^2 = 2(\nabla f \cdot \nabla)(\nabla f), \tag{5.99}$$

that we apply to $f = \ln \psi$. We finally obtain [353]

$$\nabla \left(\frac{\Delta \psi}{\psi} \right) = 2(\nabla \ln \psi \cdot \nabla)(\nabla \ln \psi) + \Delta(\nabla \ln \psi). \tag{5.100}$$

This identity can be still generalized thanks to the fact that ψ appears only through its logarithm in the right-hand side of the above equation. By replacing in it ψ with ψ^α, we obtain the general remarkable identity [403]

$$\frac{1}{\alpha} \nabla \left(\frac{\Delta \psi^\alpha}{\psi^\alpha} \right) = 2\alpha (\nabla \ln \psi \cdot \nabla)(\nabla \ln \psi) + \Delta(\nabla \ln \psi). \tag{5.101}$$

5.6.5. *Schrödinger equation*

We recognize in the right-hand side of this equation the two terms of Eq. (5.94), which were respectively in factor of S_0/m and $\mathcal{D}$. Therefore, by writing the above remarkable identity in the case

$$\alpha = \frac{S_0}{2m\mathcal{D}}, \tag{5.102}$$

the whole motion equation becomes a gradient,

$$\nabla \phi = 2m\mathcal{D} \left\{ i \frac{\partial}{\partial t} \nabla \ln \psi^\alpha + \mathcal{D} \nabla \left(\frac{\Delta \psi^\alpha}{\psi^\alpha} \right) \right\}, \tag{5.103}$$

and it can therefore be generally integrated, in terms of the new function

$$\psi^\alpha = \left(e^{iS/S_0} \right)^\alpha = e^{iS/2m\mathcal{D}}. \tag{5.104}$$

One would have obtained the same result by directly setting $S_0 = 2m\mathcal{D}$ in the initial definition of ψ [353], but in the above proof it is obtained without making any particular choice. This relation is more general than in standard quantum mechanics, for which $S_0 = \hbar = 2m\mathcal{D}$. The Eq. (5.102) is actually a generalization of the Compton relation (see next section). This means that the function ψ becomes a wave function only provided it comes with a Compton–de Broglie relation. Without this relation, the equation of motion would remain of third order, with no general prime integral.

Indeed, the simplification brought by this relation is twofold: (i) several complicated terms are compacted into a simple one; (ii) the final remaining term is a gradient, which means that the fundamental equation of dynamics can now be integrated in a universal way. The function ψ in Eq. (5.85) is therefore finally defined as

$$\psi = e^{i\mathcal{S}/2m\mathcal{D}}, \tag{5.105}$$

and it is a solution of the fundamental equation of dynamics, Eq. (5.81), which now takes the form

$$\frac{\hat{d}}{dt}\mathcal{V} = -2\mathcal{D}\nabla\left\{i\frac{\partial}{\partial t}\ln\psi + \mathcal{D}\frac{\Delta\psi}{\psi}\right\} = -\nabla\phi/m. \tag{5.106}$$

Integrating this equation finally yields a generalized Schrödinger equation,

$$\mathcal{D}^2\Delta\psi + i\mathcal{D}\frac{\partial}{\partial t}\psi - \frac{\phi}{2m}\psi = 0, \tag{5.107}$$

up to an arbitrary phase factor, which may be set to zero by a suitable choice of the ψ phase. The standard Schrödinger equation of quantum mechanics is recovered in the special case $\hbar = 2m\mathcal{D}$. Therefore the Schrödinger equation is the new form taken by the Hamilton–Jacobi/energy equation (see [367] on this point) after a change of variable from the complex action to the function ψ.

Arrived at that point, several steps have already been made, toward the final identification of the function ψ with a wave function. It is complex, solution of a Schrödinger equation, so that its linearity is also ensured. Namely, if ψ_1 and ψ_2 are solutions, $a_1\psi_1 + a_2\psi_2$ is also a solution. Let us complete the proof by giving a derivation of other basic postulates of quantum mechanics.

5.6.6. *Generalized Compton and de Broglie relations*

The relation obtained above,

$$S_0 = 2m\mathcal{D}, \tag{5.108}$$

means that there is a natural link between the Compton relation and the Schrödinger equation. Indeed, in the case of standard quantum mechanics, S_0 is nothing but the fundamental action constant $\hbar$, while $\mathcal{D}$, which characterizes the amplitude of fractal fluctuations, also defines the fractal-nonfractal transition, i.e. the transition from explicit scale-dependence to scale-independence in the rest frame. This can be directly seen by writing the fractal fluctuation in a scale invariant way. This may be achieved by introducing a reference length scale λ_r and a reference time scales τ_r, namely,

$$\left\langle \left(\frac{d\xi}{\lambda_r} \right)^2 \right\rangle = \frac{dt}{\tau_r}. \qquad (5.109)$$

The comparison with the definition of $\mathcal{D}$, $\langle d\xi^2 \rangle = 2\mathcal{D}dt$, yields

$$2\mathcal{D} = \frac{\lambda_r^2}{\tau_r}. \qquad (5.110)$$

A first natural choice for the reference scales amounts to taking the same reference for length and time, namely, $\lambda_r = c\tau_r$, and one obtains the identification, respectively in the general case of any constant $\mathcal{D}$ (for the system under consideration) and of the particular standard QM case $\mathcal{D} = \hbar/2m$,

$$\lambda_c = \frac{2\mathcal{D}}{c} \text{ (gen.)}, \quad \lambda_c = \frac{\hbar}{mc} \text{ (QM)}. \qquad (5.111)$$

One indeed recognizes the definition of the Compton scale in the standard QM case, of which the left relation is therefore a generalization.

Another natural possibility amounts to relating the length and time natural scales through the speed of the particle and to chose $\lambda_r = v\tau_r$. In this case one obtains

$$\lambda_{\text{dB}} = \frac{2\mathcal{D}}{v} \text{ (gen.)}, \quad \lambda_{\text{dB}} = \frac{\hbar}{mv} \text{ (QM)}, \qquad (5.112)$$

in which one recognizes the standard nonrelativistic de Broglie length and its generalization (left relation).

There is another simple way to recover the de Broglie length from the Compton one directly through the basic scale relativity description of elementary displacements in a fractal space-time. Indeed, we have chosen to write the fractal fluctuation in terms of the time differential element dt. Its square reads $\langle d\xi^2 \rangle = \hbar\, dt/m$ and it is therefore a differential elements of order $1/2$. Now one could equivalently make the choice to express it in function of the classical space differential elements dx. Since dx is of the same order as dt, we are led to write

$$\langle d\xi_x^2 \rangle = \lambda_x \, dx, \qquad (5.113)$$

where the length scale λ_x must be introduced for dimensional reasons. The identification of these two relations implies $\lambda_x = \hbar/(m \, dx/dt)$, i.e.

$$\lambda_x = \lambda_{\mathrm{dB}} = \frac{\hbar}{mv_x}, \qquad (5.114)$$

(and similar relations for the other coordinates), which is the nonrelativistic expression of the de Broglie length.

We have actually recovered here the two main ways of descriptions of fractal curve and coordinates already emphasized in Chapter 4: a static description involving only space coordinates and resolutions, and a kinematic description involving time, speed and a time resolution. The two corresponding expressions of the full elementary displacements, now including the classical and the fractal parts, read for one variable, which is easily generalizable to several coordinates

$$dX = dx + \zeta\sqrt{\lambda_{\mathrm{dB}} dx} = dx \left(1 + \zeta\sqrt{\frac{\lambda_{\mathrm{dB}}}{dx}} \right), \qquad (5.115)$$

$$dX = v \, dt + \zeta\sqrt{\lambda_c c \, dt} = v \, dt \left(1 + \zeta\sqrt{\frac{c\lambda_c/v^2}{dt}} \right), \qquad (5.116)$$

where ζ is the normalized dimensionless stochastic fluctuation. Under their right-hand side forms, these equations clearly express the new fundamental role played by the de Broglie and Compton scales in the scale relativity theory. Namely, their meaning is that of a geometric transition from scale dependence to effective scale independence. We stress the fact that this is not a geometric property of the standard space-time, but of the scale space, since, as can be viewed in the above equations and in Fig. 4.1, it applies to the scale variables dx and dt.

The space transition is given by the de Broglie length $\hbar/p$, both in the motion nonrelativistic and relativistic cases. In the nonrelativistic case, the time transition scale is given by $c\lambda_c/v^2 = \hbar/mv^2$, i.e. up to a factor of 2, it is the non relativistic de Broglie time scale $\hbar/E = \hbar/(p^2/2m)$. In the motion relativistic case (see Chapter 6), the time transition is $\hbar/E$, with $E^2 = p^2c^2 + m^2c^4$, and it therefore becomes $\hbar/mc^2 = c\lambda_c$ in rest frame. Therefore, in this case the Compton length (and then the parameter $\mathcal{D}$)

not only helps defining the transition scale but it becomes itself such a transition.

Now the profound meaning of the Compton length amounts to defining the inertial mass (up to the fundamental constants $\hbar$ and c, which can both be identified to one in the special scale and motion relativity theory, see Chapter 2). Therefore, in the scale relativity framework, a new geometric meaning can be given to mass, in terms of the effective transition scale from fractality to scale-independence (at large scales) in rest frame. We note that this length-scale is to be understood as a structure of scale-space, not of standard space. This has an important consequence for the foundation of the theory, since it means that one does not need to consider a particle as "having" a mass as some inner degree of freedom. Here the mass is just one among the geometric properties of fractal geodesics (provided one considers geometry both in position space and in scale space), so that it has become a "large" scale property emerging from the geometry itself. Under such a view, the scale variation of the mass of particles, i.e. of their self-energy [530, 238] now given by the renormalization group equations [79, 503] becomes quite expected (in a framework where the radiative corrections due to particle-antiparticle pairs are no longer viewed as exterior to the particle, but as a part of its own scale-dependent nature [353]).

In the case $\hbar = 2m\mathcal{D}$ we finally recover the standard form of the Schrödinger equation

$$\frac{\hbar^2}{2m}\Delta\psi + i\hbar\frac{\partial}{\partial t}\psi = \phi\psi. \tag{5.117}$$

The Planck constant $\hbar$ becomes a geometric property of the fractal space itself. It is defined through the fractal fluctuations as $\hbar = m\langle d\xi^2\rangle/dt$. This means that the fractal fluctuations of the space(-time) plays the role of an infinite energy thermostat, and that the Planck constant can be viewed as the amplitude of an explictly scale dependent and divergent temperature, according to the relation

$$E = \frac{1}{2}m\left\langle\left(\frac{d\xi}{dt}\right)^2\right\rangle = kT(dt) = \frac{1}{2}\frac{\hbar}{dt}. \tag{5.118}$$

This relation holds in the nonrelativistic domain, i.e. for $dt > \hbar/mc^2$, and is generalized to $E = \hbar c/ds$ in the relativistic case (see the next Chapter).

5.6.7. *Hamiltonian approach and improved covariance*

Let us now show how one can obtain the Schrödinger equation in the framework of a Hamilton-like mechanics. We shall first show how one can directly integrate the Euler–Lagrange equations as an energy equation expressed in terms of the complex velocity field.

Start again from the Euler–Lagrange equations of a particle in a scalar potential

$$m\frac{\hat{d}}{dt}\mathcal{V} = -\nabla\phi. \tag{5.119}$$

By expanding the expression of the covariant derivative, it reads

$$m\left(\frac{\partial \mathcal{V}}{\partial t} + \mathcal{V} \cdot \nabla\mathcal{V} - i\mathcal{D}\Delta\mathcal{V}\right) = -\nabla\phi. \tag{5.120}$$

Now, we know that the velocity field $\mathcal{V}$ is potential, since it reads, in terms of the complex action, $\mathcal{V} = \nabla\mathcal{S}/m$. Thanks to this simplifying property, we have

$$m\frac{\partial \mathcal{V}}{\partial t} = \frac{\partial}{\partial t}(\nabla\mathcal{S}) = \nabla\left(\frac{\partial \mathcal{S}}{\partial t}\right), \tag{5.121}$$

$$m\mathcal{V} \cdot \nabla\mathcal{V} = \nabla\left(\frac{1}{2}m\mathcal{V}^2\right), \tag{5.122}$$

$$-i\mathcal{D}\Delta\mathcal{V} = -i\mathcal{D}\nabla(\nabla \cdot \mathcal{V}). \tag{5.123}$$

Therefore all the terms of the motion equation are gradients, so that one obtains after integration a prime integral of the motion equations, up to a constant which can be absorbed in a redefinition of the potential energy ϕ. One obtains a generalized Hamilton–Jacobi equation,

$$\frac{\partial \mathcal{S}}{\partial t} + \mathcal{H} = 0, \tag{5.124}$$

where $\mathcal{H}$ is the Hamilton (energy) function, which was searched for, and which reads

$$\mathcal{H} = \frac{1}{2}m\mathcal{V}^2 - im\mathcal{D}\nabla \cdot \mathcal{V} + \phi, \tag{5.125}$$

i.e. in terms of the complex momentum,

$$\mathcal{H} = \frac{\mathcal{P}^2}{2m} - i\mathcal{D}\nabla \cdot \mathcal{P} + \phi. \tag{5.126}$$

There appears an additional term of potential energy $-i\mathcal{D}\nabla \cdot \mathcal{P}$ in the energy expression. While the standard $\mathcal{V}^2$ term comes from the $\mathcal{V} \cdot \nabla$

contribution in the covariant derivative, the additional potential energy $-im\mathcal{D}\mathrm{div}\mathcal{V}$ comes from the second order derivative contribution, $i\mathcal{D}\Delta$. We shall give in Sec. 5.9 another equivalent method to obtain this result.

The above expression for the Hamilton function shows that the standard relation of classical mechanics, $H = v \cdot p - L$ (see e.g. [265]), is no longer true in the scale relativity case. This means that the energy is affected by the nondifferentiable and fractal geometry at the very fundamental level of its initial conceptual definition. This was not unexpected, owing to the fact that the scale relativity approach is aimed at founding quantum mechanics, while the quantum realm is known since its discovery to have brought radical new features of the energy concept, such as the vacuum energy and its divergence in quantum field theories.

The reason for this apparent breaking of covariance can be traced back to the initial construction of prime integrals. Since energy is the very conservative quantity constructed from the uniformity of time, it is obtained as a quantity such that $dE/dt = 0$, with $E = \dot{q} \cdot \partial L/\partial \dot{q} - L$, i.e. a quantity that remains constant during the time evolution of the system. But now the main new characteristics of the scale relativity approach is to change the total time derivative by adding in particular second order terms in its definition. These second order terms imply that the Leibniz rule for a product is no longer the first order Leibniz rule, and therefore that [446] $\mathcal{H} \neq \mathcal{P} \cdot \mathcal{V} - \mathcal{L}$. In particular, when $\mathcal{L} = \frac{1}{2}m\mathcal{V}^2 - \phi$ the complex Hamilton function is no longer given by $\mathcal{H} = \frac{1}{2}m\mathcal{V}^2 + \phi$.

However, this breaking of covariance is only apparent. Indeed, it has been shown by Pissondes [446, 447] that the strong covariance can be fully implemented by introducing new tools allowing to keep the form of the first order Leibniz rule, despite the presence of the second order derivatives.

We have suggested, to this purpose, to define a covariant velocity operator [387, 94]

$$\widehat{\mathcal{V}} = \mathcal{V} - i\mathcal{D}\nabla. \tag{5.127}$$

When it is written in terms of this operator, the covariant derivative recovers a strongly covariant form. Indeed, it then recovers the standard first order form of the expression of a total derivative in terms of partial derivatives:

$$\frac{\hat{d}}{dt} = \frac{\partial}{\partial t} + \widehat{\mathcal{V}} \cdot \nabla. \tag{5.128}$$

Then the equation of dynamics recovers the form of a Euler equation for the geodesic fluid, namely,

$$\left(\frac{\partial}{\partial t} + \hat{\mathcal{V}} \cdot \nabla\right)\mathcal{V} = -\frac{\nabla\phi}{m}. \tag{5.129}$$

More generally, one may define, for any function f, the operator

$$\frac{\widehat{df}}{dt} = \frac{\hat{d}f}{\partial t} - i\mathcal{D}\nabla f \cdot \nabla. \tag{5.130}$$

The covariant derivative of a product,

$$\frac{\hat{d}(fg)}{dt} = f\frac{\hat{d}g}{dt} + g\frac{\hat{d}f}{dt} - 2i\mathcal{D}\nabla f \cdot \nabla g, \tag{5.131}$$

now keeps its first order form in terms of this operator, namely,

$$\frac{\hat{d}(fg)}{dt} = f\frac{\widehat{dg}}{dt} + g\frac{\widehat{df}}{dt}. \tag{5.132}$$

Let us now consider again from its origin the general question of the construction of an Hamilton function in the new covariant mechanics. In classical mechanics one expands the total derivative of the action as $dS/dt = \partial S/\partial t + \sum \dot{q}(\partial S/\partial q)$, so that, since $L = dS/dt$, $H = -\partial S/\partial t$, $p = \partial S/\partial q$ and $v = \dot{q}$, one obtains $H = v.p - L$. In scale relativity, we look for an invariant under the full total derivative $\hat{d}/dt$. Therefore, the fully covariant equivalent of this proof reads, in terms of the complex action $\mathcal{S}$

$$\pounds = \frac{\hat{d}\mathcal{S}}{dt} = \frac{\partial \mathcal{S}}{\partial t} + \hat{\mathcal{V}} \cdot \nabla\mathcal{S}. \tag{5.133}$$

Since $\mathcal{P} = \nabla\mathcal{S}$ and $\mathcal{H} = -\partial\mathcal{S}/\partial t$ we finally obtain a strongly covariant form for the Hamilton function

$$\mathcal{H} = \hat{\mathcal{V}} \cdot \mathcal{P} - \pounds. \tag{5.134}$$

We have indeed recovered here in terms of the new complex quantities, the same form as the classical Hamilton function, $H = v \cdot p - L$ [265].

After expansion of the velocity operator, we obtain

$$\mathcal{H} = \mathcal{V} \cdot \mathcal{P} - i\mathcal{D}\nabla \cdot \mathcal{P} - \pounds. \tag{5.135}$$

We therefore find that the general form of the additional potential energy reads

$$\phi_F = -i\mathcal{D}\nabla \cdot \mathcal{P} \tag{5.136}$$

from which we recover its expression $-im\mathcal{D}\nabla \cdot \mathcal{V}$ in the particular case $\mathcal{P} = m\mathcal{V}$.

Let us finally conclude with the derivation of the motion equation as a Schrödinger equation. If one now replaces the action in Eq. (5.124) by its expression in terms of the wave function, one obtains a Hamilton form for the motion equation,

$$\mathcal{H}\psi = \left(\frac{1}{2}m\mathcal{V}^2 - im\mathcal{D}\nabla \cdot \mathcal{V} + \phi\right)\psi = 2im\mathcal{D}\frac{\partial \psi}{\partial t}. \tag{5.137}$$

However, this is not yet the quantum mechanical equation, since here $\mathcal{H}\psi$ is a product, not an operator product. In order to complete the proof, we therefore need to show that $\mathcal{H}\psi = \hat{H}\psi$, where $\hat{H}$ is the standard Hamiltonian operator. To this purpose, we replace the complex velocity field by its expression $\mathcal{V} = -2i\mathcal{D}\nabla \ln \psi$, and we finally obtain again a Schrödinger equation under the general form

$$\hat{H}\psi = 2im\mathcal{D}\frac{\partial \psi}{\partial t}, \tag{5.138}$$

where we find the Hamiltonian to be given by

$$\hat{H} = -2m\mathcal{D}^2\Delta + \phi, \tag{5.139}$$

as expected. By considering the standard quantum mechanical case $\hbar = 2m\mathcal{D}$, we finally demonstrate the correspondence principle for the kinetic energy, $T = p^2/2m \rightarrow -(\hbar^2/2m)\Delta$, and more generally for the total energy, $E = T + \phi \rightarrow -(\hbar^2/2m)\Delta + \phi$.

5.7. Born and von Neumann Postulates

5.7.1. *Fluid representation of geodesic equations*

We have given above two representations of the Euler–Lagrange fundamental equations of dynamics in a fractal and locally irreversible context. The first representation is the equation of geodesics, $d\hat{\mathcal{V}}/dt = 0$, that is written in terms of the complex velocity field, $\mathcal{V} = V - iU$ and of the covariant derivative operator, $\hat{d}/dt = \partial/dt + \mathcal{V} \cdot \nabla - i\mathcal{D}\Delta$.

The second representation is the Schrödinger equation, whose solution is a wave function ψ. Both representations are related by the transformation

$$\mathcal{V} = -2i\mathcal{D}\nabla \ln \psi. \tag{5.140}$$

Let us now write the wave function under the form $\psi = \sqrt{P} \times e^{i\theta}$, decomposing it in terms of a modulus $|\psi| = \sqrt{P}$ and of a phase θ. We shall now build a mixed representation, in terms of the real part of the complex velocity field, V, and of the square of the modulus of the wave function, $P = |\psi|^2$. This fluid-like representation allows one to come back to the initial view of the motion as following a fluid of fractal geodesics in a nondifferentiable space. We shall indeed obtain a fluid mechanics-type description of the differentiable part of the velocity field V, but with an added quantum potential, which profoundly changes the meaning and behavior of this description.

By separating the real and imaginary parts of the Schrödinger equation and by making the change of variables from ψ, i.e. (P, θ), to (P, V), we obtain respectively a generalized Euler-like equation and a continuity-like equation [374, 406]

$$\left(\frac{\partial}{\partial t} + V \cdot \nabla \right) V = -\nabla \left(\frac{\phi}{m} - 2\mathcal{D}^2 \frac{\Delta\sqrt{P}}{\sqrt{P}} \right), \tag{5.141}$$

$$\frac{\partial P}{\partial t} + \mathrm{div}(PV) = 0. \tag{5.142}$$

See Sec. 10.6 for a detailed derivation of these equations and for a full study of this transformation and of its reverse version. This system of equations is equivalent to the classical system of equations of fluid mechanics (with no pressure and no vorticity), except for the change from a density of matter to a density of probability, and for the appearance of an extra potential energy term Q that reads

$$Q = -2m\mathcal{D}^2 \frac{\Delta\sqrt{P}}{\sqrt{P}}. \tag{5.143}$$

This potential is sometimes interpreted as the effect of a "quantum pressure" p_Q [506]. However, this interpretation implies that $\nabla p_Q = -2\mathcal{D}^2 P \nabla(\Delta\sqrt{P}/\sqrt{P})$, which is not possible in general.

The existence of this potential energy is, in the scale relativity approach, a very manifestation of the geometry of space, namely, of its nondifferentiability and fractality, in similarity with Newton's potential being a manifestation of curvature in Einstein's general relativity. It is a generalization of the quantum potential of standard quantum mechanics [299, 58], which relies on the Planck constant $\hbar$. However, its nature was misunderstood in this framework, since the variables V and P were constructed from the wave function, which is set as one of the axiom of quantum mechanics, in the same way as the Schrödinger equation itself. On the contrary, in the scale relativity theory, we know from the very beginning of the construction that V represents the velocity field of the fractal geodesics, and the Schrödinger equation is derived from the very equation of these geodesics.

The von Neumann postulate, according to which just after the measurement, the system is in the state given by the measurement result, and Born's postulate, according to which the square of the modulus of the wave function $P = |\psi|^2$ gives the probability of presence of the particle, can now be inferred from the scale relativity construction.

5.7.2. *Derivation of the von Neumann postulate*

Indeed, we have identified the wave-particle with the various geometric properties of a subset of the fractal geodesics of a nondifferentiable space (more generally, space-time, see the next chapter). Under such an interpretation, a measurement, and more generally any knowledge acquired about the system, even when not linked to an actual measurement, amounts to a "selection" of the sub-sample of the geodesic family in which are kept only the geodesics having the geometric properties corresponding to the measurement result. Therefore, just after the measurement, the system is in the state given by the measurement result, in accordance with von Neumann's postulate of quantum mechanics.

Such a state is described by the wave function or state function ψ, which, in the present approach, is a manifestation of the various geometric properties of the geodesic fluid, whose velocity field is $\mathcal{V} = -2i\mathcal{D}\nabla \ln \psi$. Now, one should be cautious about the meaning of this selection process. It means that a measurement or a knowledge of a given state is understood as corresponding to a set of geodesics, which are characterized by common and definite geometric properties. Among all the possible virtual sets of geodesics of a nondifferentiable space-time, these geodesics are therefore "selected". But this does not mean that these geodesics, with their geometric properties were already existing before the measurement process. Indeed, it is quite possible that the interaction involved in the measurement process be at the origin of the geometric characteristics of the geodesics, as identified by the measurement result. There is no given, static space with given geodesics, but instead a dynamic and changing space whose geodesics are themselves dynamic and changing. In particular, any interaction and therefore any measurement participates in the definition of the space and of its geodesics, and in their change.

The situation here is even more radical than in general relativity. Indeed, in Einstein's theory, the concept of test-particle can still be used. For example, one may consider a given static space, such as given by the Schwarzschild metric around an active gravitational mass M.[c] Then the equation of motion of a test-particle of inertial mass $m \ll M$ does not depend on this particle, but only on the active mass M, which enters the Christoffel symbols and therefore the covariant derivative. This is expressed by saying that the active mass M has curved space-time, and that the test-particle then follows the geodesics of this curved space-time. Now, when m can no longer be considered as small with respect to M, one falls into a two-body problem, which becomes very intricated. Indeed, the motion of the bodies enters the stress-energy tensor, so that the problem is looped, the general exact solutions of Einstein's equations in this case become extremely complicated and are therefore unknown.

In scale relativity even the one-body problem is looped. Indeed, it is the inertial mass of the very particle whose motion equation is searched for that enters the covariant derivative. This is indeed expected of a microscopic description of space-time, which is at the level of its own objects, and in which, finally, one cannot separate what is space (the container) from what is the "object" (contained). This inseparability is fully achieved here, since the "objects" are identified with the geodesics of the fractal space, whose equation reduces to the Compton relation, thanks to elementarity.

---

[c] Recall that in general relativity one defines an active gravitational mass, which appears in the energy-momentum tensor (right-hand side of Einstein's field equations: it is the mass, which curves space-time) and a passive gravitational mass, which is the mass of a particle subjected to this curvature. The weak equivalence principle requires that the passive mass be equal to the inertial mass, and the strong principle that the three masses be equal.

5.7.3. *Derivation of the Born postulate*

As a consequence, the probability for the "particle" to be found at a given position must be proportional to the density of the geodesic fluid. We already know its velocity field, whose real part is given by V, identified, at the classical limit, with a classical velocity field. The density of geodesics ϱ has not yet been introduced at this level of the construction. This is in contradistinction with most stochastic approaches, in particular with Nelson's stochastic quantum mechanics [331, 332], where the probability density is introduced from the very beginning under a diffusion-like description. Moreover, this probability density is used in these approaches to define averages. This is one of their main difficulties, leading to contradictions with standard quantum mechanics, in particular concerning multitime measurements [205, 526].

One of the advantages of the present scale relativity approach is that it does not come under such problems. The reason is that we are, concerning the order of the derivation of the various "postulates", exactly in the same situation as during the historical construction of quantum mechanics. Indeed, the Born interpretation of the wave function as being complex instead of real and such that $P = |\psi|^2$ gives the probability of presence of the particle [63, 64] has been definitively fixed after the setting of the other postulates. Here, as we have seen, we have been able to derive the existence of a wave function, the correspondence principle for momentum and energy and the Schrödinger equation itself without using a probability density.

Now we expect the fluid of geodesics to be more concentrated at some places and less at others to fill some regions and to be nearly vanishing in others, as does a real fluid. This behavior should be described by a probability density of presence of the paths. However, these paths are not trajectories. They do exist as geometrical "objects", but not as material objects. For example, one may define the geodesical line between two points on the terrestrial sphere: we know that it is given by a great circle. This geodesical line does exist as a virtual path, let it be followed by a boat or a plane or not.

The idea is the same here. The fluid of geodesics is defined as an ensemble of virtual paths in a purely geometric way, not as an ensemble of real trajectories. Now, in a real experiment, one may emit zero, one, two, or a very large number of particles, under the conditions virtually described by the geometric characteristics of the space and its geodesics, namely, fractal-nonfractal transition that yields the mass, initial and possibly final conditions (in a probability amplitude-like description à la Feynman), limiting conditions, etc.

When the number of particles is small, the fluid density will manifest itself in terms of a probability density, as in particle by particle two-slit experiments. When the number of particles is very large, it will manifest itself as a continuous intensity, (e.g. of light in a two-slit experiment). But, since the interferences are those of the complex fluid of geodesics themselves (in a two-slit experiment when both slits are open), we expect them to exist even in the zero particle case. This will be studied in more detail in Chapter 10, in particular with the help of numerical simulations.

In order to calculate the probability density, we remark that it is expected to be a solution of a fluid-like Euler and continuity system of equations, namely,

$$\left(\frac{\partial}{\partial t} + V \cdot \nabla\right) V = -\nabla\left(\frac{\phi}{m} + Q\right), \tag{5.144}$$

$$\frac{\partial \rho}{\partial t} + \text{div}(\varrho V) = 0, \tag{5.145}$$

where ϕ describes an external scalar potential possibly acting on the fluid, and where Q is the potential that is expected to appear as a manifestation of the fractal geometry of space. This is a system of four equations (since Eq. (5.144) is vectorial and is therefore made of three equations) for four unknowns, (ϱ, V_x, V_y, V_z). The properties of the fluid would be therefore completely determined by such a system.

Now these equations are exactly the same as Eqs. (5.141, 5.142), except for the replacement of the square of the modulus of the wave function P by the fluid density ϱ. Therefore this result allows one to univoquely identify $P = |\psi|^2$ with the probability density of the geodesics, i.e. with the probability of presence of the particle [93, 403]. Moreover, one identifies the non-classical term Q with the new potential, which is expected to emerge from the fractal geometry. Numerical simulations in which the expected probability density can be obtained directly from the distribution of geodesics without writing the Schrödinger equation, confirm this result (see [227] and Sec. 10.5).

Note that this proof is founded on the general conditions under which one may transform the Schrödinger equation into a continuity and Euler fluid mechanics system. By examining these conditions in more detail (see Sec. 5.7.1), one finds that it heavily relies on the form of the free particle energy term in the Lagrange function, $p^2/2m$. In quantum mechanical terms this means that it relies on the free part of the Hamiltonian, $\widehat{T} = -(\hbar^2/2m)\Delta$. In particular, one finds that the quantum potential reads in terms of this operator

$$Q = \frac{\widehat{T}\sqrt{P}}{\sqrt{P}}. \qquad (5.146)$$

As we have argued, this form of the work operator is very general since it comes from motion relativity itself (in the nonrelativistic limit $v \ll c$). However, if there exists some (possibly effective) situations in which this form would no longer be true, the derivation of Eqs. (5.141) and (5.142) would no longer be ensured, and the Born postulate would therefore no longer hold in its current form. This could yield a difference of theoretical prediction between the scale relativity theory and standard quantum mechanics, for which the Born postulate $P = |\psi|^2$ is always true, and therefore this could provide us with a possibility to falsify one of the theories, as required of genuine physical theories.

5.8. Generalization to Nondifferentiable Wave Functions

5.8.1. *Complete geodesic equation*

In previous sections, we have obtained the Schrödinger equation by keeping only the differentiable parts, thanks to a stochastic average, which has suppressed the divergent fractal parts, which are by definition of zero mean. Let us now show how one can make a complete calculation where all terms are kept, including the fractal stochastic fluctuations [371].

One starts again with elementary displacements dX that are described in terms of an average dx and a stochastic fluctuation $d\xi$,

$$dX = dx + d\xi, \qquad (5.147)$$

with $dx = v\,dt$. To be more general, we relax the condition $D_F = 2$ and we consider a general value of the fractal dimension $D_F \geq D_T = 1$. The fractal fluctuations can be written as

$$d\xi = a\sqrt{2\mathcal{D}}(dt^2)^{\frac{1}{2D_F}}, \qquad (5.148)$$

where D_F is the fractal dimension of the paths, $\langle a \rangle = 0$, $\langle a^2 \rangle = 1$. In [371], we also added a correction to this expression by going to larger order of differentiation, but these terms lead to no additional final corrections in the Schrödinger equation, so that we omit them here.

Recall that one of the basic methods of the scale relativity approach consists of treating the differential elements as full variables. Non-standard analysis is one possible mathematical framework allowing one to develop such a method in a rigorous way [344, 353], but, as we have stressed in previous chapters, it is sufficient to consider the differential elements as real non-null variables of $\mathbb{R} - \{0\}$ to obtain a rigorous physical and mathematical tool.

It is remarkable that the $d\xi$'s exist only at the differential level, but do not contribute to the macroscopic space-time coordinates, since they are of zero mean. Hence, at the level of the space(-time) coordinates, X and x are indistinguishable, while at the level of their derivatives the dX's no longer reduce to the classical differential dx's. The new variables $d\xi$ are those that carry the underlying information about the fractal inner structure of space-time (here, in the nonrelativistic case of space) and are coupled to the fractal dimension. The scale laws to be constructed are those that describe their behavior. Strictly, they go beyond space-time. Indeed, the classical macroscopic space-time (x, y, z, t) emerges from the far more complicated structure $(dx, dy, dz, dt; d\xi_x, d\xi_y, d\xi_z, d\xi_t, \tau)$, in the general case including a scale-time or "djinn" (see Chapter 4).

Note also that the parameter $\mathcal{D}$, which reads $\mathcal{D} = \hbar/2m$, and is, therefore, in the simple case that leads to standard quantum mechanics, another expression for the Compton length, can be generalized to a tensorial form, which can become a full field in the general case where the fractal fluctuation reads $\langle d\xi_j d\xi_k \rangle = 2\mathcal{D}_{jk}(x, y, z, t)(dt^2)^{1/D_F}$ (see Secs. 5.11.5 and 5.14). It plays in this case the role of a metric form and it can be incorporated in a redefinition of the derivatives, in particular of the Laplacian [332].

Let us go on with this more complete calculation. Consider a function $f[X(t), t]$, and write its Taylor expansion

$$\frac{df}{dt} = \frac{\partial f}{\partial t} + \frac{\partial f}{\partial X}\frac{dX}{dt} + \frac{1}{2}\frac{\partial^2 f}{\partial X^2}\frac{dX^2}{dt} + \frac{1}{6}\frac{\partial^3 f}{\partial X^3}\frac{dX^3}{dt} + \cdots . \tag{5.149}$$

The velocity is given by

$$\frac{dX}{dt} = v + a\sqrt{2\mathcal{D}}(dt^2)^{\frac{1}{2D_F} - \frac{1}{2}}. \tag{5.150}$$

This means that it contains a divergent term of zero-mean, $w = a\sqrt{2\mathcal{D}}$ $(dt^2)^{\frac{1}{2D_F} - \frac{1}{2}}$ which manifests the nondifferentiability. Concerning the fluctuations, one obtains

$$\frac{dX^2}{dt} = 2a^2 \mathcal{D}(dt^2)^{\frac{1}{D_F}-\frac{1}{2}} + 2a\sqrt{2\mathcal{D}}v(dt^2)^{\frac{1}{2D_F}} + v^2 dt + \cdots, \quad (5.151)$$

$$\frac{dX^3}{dt} = a^3(2\mathcal{D})^{3/2}(dt^2)^{\frac{3}{2D_F}-\frac{1}{2}} + \cdots. \quad (5.152)$$

We verify that a finite contribution remains up to terms in dX^{D_F}/dt, while the preceding ones contain infinite contributions.

Let us come back to the special case $D_F = 2$. The derivative of a function f writes

$$\frac{df}{dt} = \frac{\partial f}{\partial t} + (v + w)\frac{\partial f}{\partial x} + \mathcal{D}\frac{\partial^2 f}{\partial x^2} + \cdots. \quad (5.153)$$

The new step amounts to introducing the two-valuedness of the derivative that comes from the fundamental irreversibility induced by the giving up of differentiability. The two velocities v and w are actually each of them two-valued, as a consequence of the breaking of the infinitesimal time symmetry ($dt \leftrightarrow -dt$). We therefore introduce, as previously done (see Sec. 5.4.1), a complete complex velocity field, which now includes the differentiable and fractal parts,

$$\tilde{\mathcal{V}} = \mathcal{V} + \mathcal{W} = \left(\frac{v_+ + v_-}{2} - i\frac{v_+ - v_-}{2}\right) + \left(\frac{w_+ + w_-}{2} - i\frac{w_+ - w_-}{2}\right). \quad (5.154)$$

The new divergent complex velocity is of zero mean $\langle \mathcal{W} \rangle = 0$, and, thanks to the use of complex numbers, its square is also of zero mean, $\langle \mathcal{W}^2 \rangle = \langle w_+ w_- \rangle - i(\langle w_+^2 \rangle - \langle w_-^2 \rangle)/2 = 0$. As proved in Sec. (5.4.1), this is an important result for the equation of motion. Indeed, the complete Lagrange function for a free particle can now be written

$$\mathcal{L} = \frac{1}{2}m\tilde{\mathcal{V}}^2 = \frac{1}{2}m(\mathcal{V} + \mathcal{W})^2 \quad (5.155)$$

and it reduces in the mean to the form that keeps only the differentiable parts, $\langle \mathcal{L} \rangle = \frac{1}{2}m\mathcal{V}^2$ [353].

An additional term must be now taken into account in the complete "quantum-covariant" derivative, which reads for $D_F = 2$

$$\frac{\widehat{\widetilde{d}}}{dt} = \frac{\partial}{\partial t} + \widetilde{\mathcal{V}} \cdot \nabla - i\mathcal{D}\Delta = \frac{\hat{d}}{dt} + \mathcal{W} \cdot \nabla, \qquad (5.156)$$

plus terms that vanish when $dt \to 0$. The equation of motion of a free "particle" (i.e. in our framework, the second order differential geodesic equation) therefore reads

$$\frac{\widehat{\widetilde{d}}}{dt}\widetilde{\mathcal{V}} = 0, \qquad (5.157)$$

where we recall that $\widetilde{\mathcal{V}} = \mathcal{V} + \mathcal{W}$. Expanding this equation yields

$$\frac{\hat{d}\mathcal{V}}{dt} + \left(\frac{\partial \mathcal{W}}{\partial t} + \nabla(\mathcal{V} \cdot \mathcal{W}) + \mathcal{W} \cdot \nabla\mathcal{W} - i\mathcal{D}\Delta\mathcal{W} \right) = 0, \qquad (5.158)$$

where new divergent terms of zero mean have therefore been added to the previous equation $\hat{d}\mathcal{V}/dt = 0$.

One may now consider various ways to derive a Schrödinger equation from this generalized geodesic equation. The first method, proposed in [371], consists of still defining the wave function from the only differentiable part, i.e. through the expression $\mathcal{V} = -2i\mathcal{D}\nabla \ln \psi$. In this case the additional terms appear as an exterior field. Indeed, Eq. (5.158) can be written under the general form

$$\frac{\hat{d}\mathcal{V}}{dt} + F\sqrt{2\mathcal{D}}\,dt^{-\frac{1}{2}} = 0, \qquad (5.159)$$

where $\langle F \rangle = 0$. Assuming that $F = \nabla G$, with $\langle G \rangle = 0$, it can be integrated under the form of a generalized Schrödinger equation (see [353, 360]),

$$\mathcal{D}^2\Delta\psi + i\mathcal{D}\frac{\partial\psi}{\partial t} + G\,dt^{-\frac{1}{2}}\psi = 0. \qquad (5.160)$$

The last term is infinite from the standard viewpoint, but of vanishing mean. Therefore, taking the stochastic average of this equation yields the free particle Schrödinger equation. An interesting consequence of this result is also that one could introduce fluctuations, such that $\langle G \rangle = \phi dt^{1/2}$. Such scale dependent fluctuations would be classically indistinguishable from $\langle G \rangle = 0$. However, their vanishing dependence on the scale variable dt combines in such a way with the divergent $dt^{-\frac{1}{2}}$ term that their product (classically given by 0) yields a finite scalar potential term in the Schrödinger equation, whose origin is purely geometric. Ord [424] has more recently obtained an equivalent result using a random walk model. These results anticipate on the construction of gauge fields from the fractal geometry, which will be the subject of Chapter 7.

5.8.2. *Nondifferentiable velocity field*

The second method [403] consists of defining the wave function from the full velocity field, including its fractal divergent part. The full velocity field of the fractal space-time geodesics actually contains the formally infinite term $\mathcal{W}$, which comes from the nondifferentiability of space(-time). We shall now show that this space nondifferentiability is expected to manifest itself in terms of a possible nondifferentiability of wave functions [403].

Let us start again from the full complex velocity field,

$$\widetilde{\mathcal{V}} = \mathcal{V} + \mathcal{W} = \left(\frac{v_+ + v_-}{2} - i \frac{v_+ - v_-}{2} \right) + \left(\frac{w_+ + w_-}{2} - i \frac{w_+ - w_-}{2} \right).$$

$$(5.161)$$

Although the fractal part is infinite and, therefore, undefined from the viewpoint of standard methods, in the scale relativity framework it can be defined as an explicit function of the scale variable dt, namely, $\mathcal{W} = \mathcal{W}[x(t, dt), t, dt]$, which becomes infinite only in the limit $dt \to 0$, i.e. $\ln dt \to -\infty$.

The full complex action, now defined in terms of the differentiable part and of the fractal (divergent) part of the velocity, reads

$$\nabla \widetilde{\mathcal{S}} = m\widetilde{\mathcal{V}}. \qquad (5.162)$$

It is equal to the previous mean action $\nabla \mathcal{S} = m\mathcal{V}$, plus terms of zero stochastic average (see previous section).

We now define a complete wave function $\widetilde{\psi}$ from this full action $\widetilde{\mathcal{S}}$,

$$\widetilde{\psi} = e^{i\widetilde{\mathcal{S}}/2m\mathcal{D}}, \qquad (5.163)$$

and the relation of the complete complex velocity to the complete wavefunction therefore reads

$$\widetilde{\mathcal{V}} = \mathcal{V} + \mathcal{W} = \nabla \widetilde{\mathcal{S}}/m = -2i\mathcal{D}\nabla \ln \widetilde{\psi}. \qquad (5.164)$$

In the scale relativity approach this equation keeps a mathematical and physical meaning in terms of fractal functions, which are explicitly dependent on the scale interval dt and divergent only when $dt \to 0$. In other words, the wavefunction $\widetilde{\psi}$ defined hereabove can now be nondifferentiable.

We shall now prove that this nondifferentiable wavefunction nevertheless remains solution of a Schrödinger equation.

Let us write the fractal parts of the velocities under the form:

$$w_+ = \zeta_+ \sqrt{\frac{2\mathcal{D}}{dt}}, \quad w_- = \zeta_- \sqrt{\frac{2\mathcal{D}}{dt}}, \tag{5.165}$$

where ζ_+ and ζ_- are stochastic variables such that $\langle \zeta_+ \rangle = \langle \zeta_- \rangle = 0$ and $\langle \zeta_+^2 \rangle = \langle \zeta_-^2 \rangle = 1$.

The two $(+)$ and $(-)$ derivatives read

$$\frac{d_+ f}{dt} = \frac{\partial f}{\partial t} + (v_+ + w_+)\nabla f + \mathcal{D}\zeta_+^2 \Delta f + \cdots, \tag{5.166}$$

$$\frac{d_- f}{dt} = \frac{\partial f}{\partial t} + (v_- + w_-)\nabla f - \mathcal{D}\zeta_-^2 \Delta f + \cdots, \tag{5.167}$$

where the next terms are infinitesimals. Let us now define the following complex stochastic variables:

$$\tilde{\zeta} = \frac{\zeta_+ + \zeta_-}{2} - i\frac{\zeta_+ - \zeta_-}{2}, \tag{5.168}$$

$$1 + \tilde{\kappa} = \frac{\zeta_+^2 + \zeta_-^2}{2} + i\frac{\zeta_+^2 - \zeta_-^2}{2}, \tag{5.169}$$

which are such that $\langle \tilde{\zeta} \rangle = 0$ and $\langle \tilde{\kappa} \rangle = 0$. The variable $\tilde{\zeta}$ is actually the stochastic variable appearing in the complex velocity fluctuation $\mathcal{W}$,

$$\mathcal{W} = \sqrt{\frac{2\mathcal{D}}{dt}}\,\tilde{\zeta}. \tag{5.170}$$

We can now combine the two derivatives in terms of the complete complex covariant derivative (plus the vanishing term $-i\mathcal{D}\tilde{\kappa}\Delta$),

$$\frac{\widehat{\tilde{d}}}{dt} = \frac{\partial}{\partial t} + (\mathcal{V} + \mathcal{W}) \cdot \nabla - i\mathcal{D}(1 + \tilde{\kappa})\Delta. \tag{5.171}$$

The next terms are infinitesimals, which vanish when $dt \to 0$ and are therefore neglected. We can write this covariant derivative operator under the form

$$\frac{\widehat{\tilde{d}}}{dt} = \left[\frac{\partial}{\partial t} + \mathcal{V} \cdot \nabla - i\mathcal{D}\Delta\right] + \sqrt{\frac{2\mathcal{D}}{dt}}\,\tilde{\zeta} \cdot \nabla - i\mathcal{D}\tilde{\kappa}\Delta. \tag{5.172}$$

It is therefore equal to the mean covariant derivative plus two additional stochastic terms of zero mean, the first being $\mathcal{W} \cdot \nabla$, which is infinite at the limit $dt \to 0$, and the second $-i\mathcal{D}\tilde{\kappa}\Delta$, in which $\tilde{\kappa}$ remains finite. The first of these terms was already introduced in [371] (see previous section), while the second was neglected since their ratio is an infinitesimal of order $dt^{1/2}$.

The motion equation, $\widehat{d\tilde{\mathcal{V}}}/dt = 0$, becomes, in the presence of an exterior potential ϕ, a covariant equation, which keeps the form of Newton's fundamental equation of dynamics

$$\frac{\widehat{\tilde{d}}}{dt}\tilde{\mathcal{V}} = -\frac{\nabla\phi}{m}. \tag{5.173}$$

We now expand the covariant derivative and we find

$$\left(\frac{\partial}{\partial t} + \tilde{\mathcal{V}} \cdot \nabla - i\mathcal{D}(1 + \tilde{\kappa})\Delta\right)\tilde{\mathcal{V}} = -\frac{\nabla\phi}{m}. \tag{5.174}$$

As we have seen, the stochastic term $\tilde{\kappa}\Delta\tilde{\mathcal{V}}$ is infinitesimal with respect to the other stochastic term $\mathcal{W}\cdot\nabla\tilde{\mathcal{V}}$, so that we can neglect it as we did in [371].

5.8.3. *Nondifferentiable solutions of Schrödinger equation*

Now introducing the full (fractal and nondifferentiable) wavefunction $\tilde{\psi}$ in this equation thanks to equation (5.164)

$$\tilde{\mathcal{V}} = -2i\mathcal{D}\nabla\ln\tilde{\psi}, \tag{5.175}$$

we obtain

$$\left(\frac{\partial}{\partial t} + (-2i\mathcal{D}\nabla\ln\tilde{\psi}) \cdot \nabla - i\mathcal{D}\Delta\right)(-2i\mathcal{D}\nabla\ln\tilde{\psi}) = -\frac{\nabla\phi}{m}. \tag{5.176}$$

In the standard framework, as remarked by Berry [47] and Hall [222], the very writing of this equation would be forbidden since $\tilde{\psi}$ is nondifferentiable and therefore its derivatives are formally infinite. But, as recalled above, the fundamental tool used in the scale-relativity approach, which was definitely constructed to solve this kind of problem (at the level of fractal space-time coordinates), can now be used in a similar manner at the level of the wavefunction. Namely, in terms of a fractal (explicitly scale dependent) wavefunction $\tilde{\psi}(x, t, dt)$, the various terms of equation (5.176) remain finite for all values of $dt \neq 0$. We are therefore in the same conditions as in previous calculations involving a differentiable wave function [353, 93], so that it can finally be integrated in terms of a generalized Schrödinger equation that keeps the same form as in the differentiable wave function case, namely,

$$\mathcal{D}^2 \Delta \tilde{\psi} + i\mathcal{D}\frac{\partial \tilde{\psi}}{\partial t} - \frac{\phi}{2m}\tilde{\psi} = 0. \tag{5.177}$$

This generalized Schrödinger equation now has nondifferentiable solutions, which come, in our framework, as a direct manifestation of the nondifferentiability of space. Such a result agrees with Berry's [47] and Hall's [222] similar findings obtained in the framework of standard quantum mechanics (see explicit examples of such nondifferentiable and fractal wavefunctions in Chapter 14). The research in laboratory experiments of such a behavior constitutes an interesting new challenge for quantum physics (see Chapter 10). It could also be interesting to put the scale relativity approach to the test by searching for a possible very small contribution (vanishing toward small scales) of the differential terms, which have been neglected in the final equation because they are differential elements of positive order.

Note, to be complete, that this equation may be still generalized by accounting for the fact that, $\tilde{\psi} = \tilde{\psi}(x, t, dt)$ being nondifferentiable, there could be an additional doubling of the partial time derivative, $\partial/\partial t \rightarrow (\partial_+/\partial t, \partial_-/\partial t)$ (de Montera, private communication). This leads to define

$$\frac{\widehat{\partial}}{\partial t} = \frac{1}{2}\left(\frac{\partial_+}{\partial t} + \frac{\partial_-}{\partial t}\right) - \frac{1}{2}i\left(\frac{\partial_+}{\partial t} - \frac{\partial_-}{\partial t}\right) \tag{5.178}$$

and to write a generalized Schrödinger equation

$$\mathcal{D}^2 \Delta \tilde{\psi} + i\mathcal{D}\frac{\widehat{\partial}}{\partial t}\tilde{\psi} - \frac{\phi}{2m}\tilde{\psi} = 0. \tag{5.179}$$

Though it keeps its standard form, the imaginary part of the new complex operator involves a new real term in this Schrödinger equation. It could be interesting to look for its possible effects in laboratory experiments. However, we doubt that such a new term could have physical meaning. Indeed, the Schrödinger equation is actually not well founded in this nonrelativistic way. One needs a full motion-relativistic treatment in order to obtain it in a fully consistent way: this is true in both standard quantum mechanics and scale relativity approach, in which a full fractal space-time should be taken into account. This will be the subject of Chapter 6. We shall see that the Schrödinger equation is then obtained as a spinless Pauli equation, itself being the nonrelativistic version of the

Dirac equation. In this derivation, the time coordinate becomes one among the four space-time coordinates, while there is no explicit dependence on the proper time s and, therefore, no intervention of the partial derivative $\partial/\partial s$. The possible new two-valuedness is therefore inefficient in this relativistic situation, and the new term is absent in the final Schrödinger equation, even in the case of fractal and nondifferentiable wavefunctions.

Let us conclude this section by emphasizing an important consequence of the results obtained. Since we now use the full velocity field in the construction of the covariant derivative and in the final derivation of the Schrödinger equation, we have reached a high level of generality. Namely, it finally appears that the separation of the elementary displacements into two parts, the classical one and the fractal one of zero mean, can be considered as a practical worktool, which gives a general description of these displacements without involving any special hypothesis, as can be seen from the fact that we assume no special form for the probability distribution of the stochastic variables that describe the fractal fluctuations. The only remaining constraint, the fractal dimension 2 of the fractal part, will be also relaxed in Sec. 5.14.

5.9. Potential Energy from Fractal Geometry

5.9.1. *Fractal potential*

In order to get a more precise understanding of the effects of nondifferentiability and of fractality, we shall now re-express the effect of the fractal fluctuation in terms of an effective "force" and of the potential energy from which it derives [362, 367]. In what follows, we shall separate the two effects of nondifferentiability: (i) doubling of time derivative expressed in terms of complex numbers and (ii) fractality (of fractal dimension $D_F = 2$), expressed by the occurence of second order terms in the total time derivative, then treat them in a different way. This will allow us to get a more detailed understanding of the origin of the additional term in the energy equation.

This separation will be made in analogy with general relativity, in which one may recover a Newtonian-like interpretation of the equation of geodesics in terms of the action of a generalized force. For this purpose, one starts with the covariant form of the geodesic equations, $D^2 x^\mu/ds^2 = 0$, one develops the covariant derivative and one obtains $d^2 x^\mu/ds^2 + \Gamma^\mu_{\nu\rho} u^\nu u^\rho = 0$, which generalizes Newton's equation, $m d^2 x^i/dt^2 = F^i$ in terms of a "force" $-m\Gamma^\mu_{\nu\rho} u^\nu u^\rho$.

Once the velocity two-valuedness is introduced, leading to the replacement $(V \to \mathcal{V})$, we may write the time derivative under the form of a standard total derivative expressed in terms of partial derivatives,

$$\frac{d}{dt} = \frac{\partial}{\partial t} + \mathcal{V} \cdot \nabla, \tag{5.180}$$

which is the Euler-like part of the total complex derivative. Under this form, the second-order term in the complete covariant derivative, $-i\mathcal{D}\Delta$, which finds its origin in the fractal geometry, is missing. In terms of this partially covariant derivative, the equation of a free particle still takes the form of Newton's equation of dynamics, but the missing term can now be treated as a right-hand member,

$$\frac{d}{dt}\mathcal{V} = i\mathcal{D}\Delta\mathcal{V}. \tag{5.181}$$

This right-hand member can therefore be identified with an external complex "fractal force" divided by m:

$$\mathcal{F} = im\mathcal{D}\Delta\mathcal{V}. \tag{5.182}$$

In the scale-relativistic, fractal-space(-time) approach, this 'force' is a manifestation of the very structure of space. When such a reasoning is applied to the standard quantum mechanical case of microphysics, we can require such a force to be universal, independent of the mass of the particle. This may give a justification for the fact that, in this case, $2m\mathcal{D}$ is a universal constant, namely, the Planck constant,

$$S_0 = 2m\mathcal{D} = \hbar. \tag{5.183}$$

Moreover, since

$$\Delta\mathcal{V} = -2i\mathcal{D}\Delta\nabla \ln \psi = -2i\mathcal{D}\nabla\Delta \ln \psi, \tag{5.184}$$

the force given by Eq. (5.182) derives from a complex "fractal potential energy" [362, 367],

$$\phi_F = -im\mathcal{D}\nabla \cdot \mathcal{V} = -2m\mathcal{D}^2\Delta \ln \psi. \tag{5.185}$$

This is but the additional term in the energy balance. As already explained in Sec. 5.6.7, the introduction of this potential energy allows us to derive the Schrödinger equation in a very fast way by the Hamilton–Jacobi approach [367]. Such a derivation gives a foundation to its standard quantum mechanical construction via the correspondence principle. We simply write the expression for the total energy, including the fractal potential plus a possible external potential ϕ,

$$\mathcal{E} = \frac{\mathcal{P}^2}{2m} + \phi_F + \phi, \tag{5.186}$$

then we replace $\mathcal{E}, \mathcal{P}$ and ϕ_F by their exact expressions, thus using equalities instead of a mere correspondence. This yields (with $2m\mathcal{D} = \hbar$)

$$i\hbar \frac{\partial}{\partial t} \ln \psi = \frac{(-i\hbar \nabla \ln \psi)^2}{2m} - \frac{\hbar^2}{2m} \Delta \ln \psi + \phi. \tag{5.187}$$

Thanks to the remarkable identity $(\nabla \ln \psi)^2 + \Delta \ln \psi = \Delta\psi/\psi$, this equation can be finally transformed into the standard Schrödinger equation, now obtained in a direct way instead of being integrated from the Lagrange equation, i.e.

$$\frac{\hbar^2}{2m} \Delta\psi + i\hbar \frac{\partial}{\partial t} \psi - \phi\psi = 0. \tag{5.188}$$

This approach yields another explanation of the origin of the new term in the free energy expression [367, 446],

$$\mathcal{E}_{\text{free}} = \frac{\mathcal{P}^2}{2m} + \phi_F = \frac{1}{2}m(\mathcal{V}^2 - 2i\mathcal{D}\nabla \cdot \mathcal{V}). \tag{5.189}$$

Such a term has been wrongly considered as breaking the strong covariance [447]. However, we have shown in Sec. 5.6.7 that a strongly covariant form for the Hamiltonian can be obtained by using the improved tool of the covariant velocity operator $\widehat{\mathcal{V}} = \mathcal{V} - i\mathcal{D}\nabla$. The covariance, i.e. the invariance of the form of the equations, is not expected to hold at all levels of the representation but only at the most fundamental ones. In particular, it has no reason to apply to the individual contributions of energy. On the contrary, one may remark that one of the most important results of relativity theories is precisely to introduce new contributions to the energy. Hence, when special relativity has yielded a new form for the energy with respect to the Galilean theory, $E = mc^2/\sqrt{1 - v^2/c^2} = mc^2 + \frac{1}{2}mv^2 + \cdots$, one has considered this to be a success of the theory instead of a breaking of covariance.

The same is true here. The theory of scale relativity yields a new contribution to the energy that comes from the very geometry of space-time. As can be seen in Eq. (5.187), this term contributes in an essential way to the final purely quantum term proportional to $\hbar^2$ in the Schrödinger equation. Moreover, as we shall now see, its real counterpart is the quantum potential, so that it is fundamentally linked to the vacuum energy in quantum mechanics.

5.9.2. *Quantum potential*

The fluid representation of the equations in terms of continuity and Euler equation with quantum potential plays a key role in the theory, since, as we have seen in Sec. 5.7, it conditions the derivation of the Born postulate of quantum mechanics.

We have obtained the Euler and continuity equation from the real and imaginary parts of the Schrödinger equation, in a way quite similar to the Madelung transformation [299]. However, since the Schrödinger equation is itself obtained as a reformulation of the geodesic equation, it should be possible to go directly from the covariant equation of dynamics $m\hat{d}\mathcal{V}/dt = -\nabla\phi$ to the fluid mechanics equations without neither defining the wave function nor passing through the Schrödinger equation.

This will enlight the meaning and origin of the quantum potential. In the end, the effect of the nondifferentiable and fractal geometry leads to the emergence of a real scalar "fractality field" [392] yielding a quantum force and a quantum potential. Our goal here is to trace in more detail the origin of this quantum potential in the fractal space framework.

For this purpose, let us explicitly introduce the real and imaginary parts of the complex velocity $\mathcal{V} = V - iU$ in the geodesic equation $\hat{d}\mathcal{V}/dt = 0$ (the generalization to the case when an exterior potential is present is straighforward). One obtains

$$\frac{\hat{d}\mathcal{V}}{dt} = \left(\left\{ \frac{\partial}{\partial t} + V \cdot \nabla \right\} - i \left\{ U \cdot \nabla + \mathcal{D}\Delta \right\} \right) (V - iU) = 0. \qquad (5.190)$$

In this expression, we see that the real part of the covariant derivative, $\hat{d}_R/dt = \partial/\partial t + V \cdot \nabla$, is the standard total derivative expressed in terms of partial derivatives, while the new terms are included in the imaginary part, $\hat{d}_I/dt = -(U \cdot \nabla + \mathcal{D}\Delta)$.

By separating the real and imaginary parts, Eq. (5.190) reads

$$\left(\frac{\partial}{\partial t} + V \cdot \nabla \right) V - (U \cdot \nabla + \mathcal{D}\Delta)U$$

$$- i \left\{ (U \cdot \nabla + \mathcal{D}\Delta)V + \left(\frac{\partial}{\partial t} + V \cdot \nabla \right) U \right\} = 0. \qquad (5.191)$$

Therefore the real part of this equation takes the form of a Euler–Newton equation of dynamics,

$$\left(\frac{\partial}{\partial t} + V \cdot \nabla\right) V = (U \cdot \nabla + \mathcal{D}\Delta)U, \tag{5.192}$$

where the total derivative of the velocity field V takes its standard form $dV/dt = (\partial/\partial t + V \cdot \nabla)V$. Now the right-hand side of this equation can be identified with a force field, since it reads

$$\frac{dV}{dt} = \frac{F}{m}, \tag{5.193}$$

where the force F is given by

$$F = m(U \cdot \nabla U + \mathcal{D}\Delta U). \tag{5.194}$$

The new field therefore finds its origin in the additional terms $\hat{d}_I/dt$ in the derivative operator, acting only on the imaginary part U of the complex velocity.

Let us now show that this additional force derives from a potential. Since the complex velocity field $\mathcal{V}$ is a gradient, the same is true of its imaginary part U. Let us therefore define a new quantity P by the relation

$$U = \mathcal{D}\nabla \ln P. \tag{5.195}$$

This means that $\mathcal{D}\nabla \ln P$ is a velocity potential for U. Note that here P is directly introduced in this way without any reference to a wave function. The force becomes

$$F = m\mathcal{D}^2[(\nabla \ln P \cdot \nabla)(\nabla \ln P) + \Delta(\nabla \ln P)]. \tag{5.196}$$

Now, by introducing $\sqrt{P}$ in this expression, one makes explicitly appear the remarkable identity that is already at the heart of the proof of the Schrödinger equation ([353], p. 151), namely,

$$\frac{F}{2m\mathcal{D}^2} = 2(\nabla \ln \sqrt{P} \cdot \nabla)(\nabla \ln \sqrt{P}) + \Delta(\nabla \ln \sqrt{P}) = \nabla\left(\frac{\Delta\sqrt{P}}{\sqrt{P}}\right). \tag{5.197}$$

Therefore the force F derives from a potential energy $-2m\mathcal{D}^2\Delta\sqrt{P}/\sqrt{P}$, which is nothing but the standard "quantum potential", but here directly established from the geodesic equation as a mere manifestation of the nondifferentiable and fractal geometry.

The real part of the geodesic equation finally takes the standard form of the Euler equation of dynamics in presence of a scalar potential,

$$\frac{dV}{dt} = \left(\frac{\partial}{\partial t} + V \cdot \nabla\right) V = -\frac{\nabla Q}{m}, \tag{5.198}$$

where the expression for the potential,

$$Q = -2m\mathcal{D}^2 \frac{\Delta\sqrt{P}}{\sqrt{P}} \tag{5.199}$$

stands out as a "fractality field" equation, e.g., the analog of the Poisson equation for a gravitational field.

Consider now the imaginary part of the geodesic equation (5.191). By replacing U by its expression $\mathcal{D}\nabla \ln P$, it becomes

$$\left(\frac{\partial}{\partial t} + V \cdot \nabla\right) \nabla \ln P + (\nabla \ln P \cdot \nabla + \Delta)V = 0, \tag{5.200}$$

i.e.

$$\nabla \left\{ \frac{1}{P} \left(\frac{\partial P}{\partial t} + V \cdot \nabla P + P\nabla \cdot V \right) \right\} = 0, \tag{5.201}$$

which can finally be integrated in terms of the continuity equation

$$\frac{\partial P}{\partial t} + \operatorname{div}(PV) = 0. \tag{5.202}$$

The integration constant can be set to zero as we have seen in the derivation of the Schrödinger equation, since it actually corresponds to the indetermination in the definition of a potential ϕ, that can be changed to $\phi + A(t)$, because it appears only through its gradient in the fundamental equation of dynamics.

The situation is therefore quite comparable with the Newtonian limit of general relativity. Indeed, in this case the motion equation of a fluid of matter density ρ is given by

$$\frac{dV}{dt} = \left(\frac{\partial}{\partial t} + V \cdot \nabla\right) V = -\frac{\nabla \phi}{m} \tag{5.203}$$

and the field equation is the Poisson equation $\Delta\phi = 4\pi G\rho$, that can be written under the form

$$\phi = 4\pi G\Delta^{-1}\rho. \tag{5.204}$$

In both cases the potential is given as a function of the density of matter in the gravitational motion relativistic case, which describes a material fluid, and of probability in the quantum scale relativistic case, which describes a purely geometric geodesic fluid. Both potential energies are now understood as genuine geometric potential energies of curvature in the gravitational case and of fractality in the quantum case.

It is remarkable that at the level of the derivation of the equations the situation seems to be reversed in the scale relativity case compared with the motion relativity case. Indeed, in general relativity one needs to solve first the field equations to know the fields (the Christoffel symbols), and then to be able to write the motion equation as a geodesic equation. Here, the field equation has been obtained from the motion equation.

The reason is that, as already remarked, a real situation in general relativity is actually far more complicated and looped than described above. It is only in the test-particle case that one may proceed in this way, first determining the curvature and then the geodesics in the curved space-time. Already in the two body problem one must solve the field and geodesic equations together, since the trajectories participate in the energy-momentum tensor, which is the source term of Einstein's field equations.

In the case of a quantum mechanical particle considered in scale relativity, the loop between the motion (geodesic) equation and the field equation is even more tight. Indeed, here the concept of test-particle loses its meaning. Even in the case of only one "particle", the space(-time) geometry is determined by the particle itself and by its motion, so that the field equation and the geodesic equation now participate of the same level of description. This explains why the motion/geodesic equation, in its Hamilton–Jacobi form, which is the Schrödinger equation, is obtained without us having first written the field equation in an explicit way. Actually, it is the Compton relation which plays here the role of the field equations, and we have seen that it is derived at the same time as the Schrdinger equation.

The potential Q is implicitly contained in the Schrödinger form of the equations, and it is made explicit only when coming back to a fluid-like Euler–Newton representation. In the end, the particle is described by a wave function (constructed from the geodesics), of which only the square of the modulus P is observable. Therefore one expects the "field" to be given by a function of P, which is exactly what is found. One may, nevertheless, remark that such a potential exhibits a more profound level of "potentiality" than standard fields, since it is constructed from a probability density, that may itself be considered as describing a potentiality in physics.

Note finally that although here we have explicitly used only the differentiable part $(\mathcal{V} = V - iU)$ of the velocity field, all the calculations can be done in the same way with the full velocity field $\widetilde{\mathcal{V}} = \mathcal{V} + \mathcal{W}$.

5.9.3. *Invariants and energy balance*

Let us now make explicit the energy balance by accounting for this additional potential energy. This question has already been discussed in Sec. 5.6.7 and the previous sections, but we propose here a different presentation. We shall express the energy equation in terms of the various equivalent variables, which we use in scale relativity, namely, the wave function ψ, the complex velocity $\mathcal{V}$, or its real and imaginary parts V and $-U$. All the results presented here can be easily generalized to the complete velocity field $\widetilde{\mathcal{V}}$ and to nondifferentiable wave functions $\widetilde{\psi}$, for which the form of the equations is conserved.

The first and main form of the energy equation is the Schrödinger equation itself, that we have derived as a prime integral of the geodesic equation. The Schrödinger equation is therefore the quantum equivalent of the metric form, i.e. of the equation of conservation of the energy. It may be written in the free case under the form

$$\mathcal{D}^2 \frac{\Delta\psi}{\psi} = -i\mathcal{D}\frac{\partial \ln\psi}{\partial t}. \tag{5.205}$$

In the stationary case with given energy E, it becomes

$$E = -2m\mathcal{D}^2\frac{\Delta\psi}{\psi}, \tag{5.206}$$

where E is now real. We can use the fundamental remarkable identity $\Delta\psi/\psi = (\nabla \ln\psi)^2 + \Delta \ln\psi$. Re-introducing the complex velocity field $\mathcal{V} = -2i\mathcal{D}\nabla \ln\psi$ in this expression we finally recover in another way the energy form

$$E = -2m\mathcal{D}^2\frac{\Delta\psi}{\psi} = \frac{1}{2}m(\mathcal{V}^2 - 2i\mathcal{D}\nabla \cdot \mathcal{V}). \tag{5.207}$$

When an exterior potential term is present, all these relations remain true by replacing E by $E - \phi$.

This is the non-relativistic equivalent of the relativistic relation, $\mathcal{V}^\mu\mathcal{V}_\mu + i\lambda\partial^\mu\mathcal{V}_\mu = 1$ [446] (see also Chapter 6). As already remarked, it has an additional term with respect to the expected strongly covariant form $\frac{1}{2}m\mathcal{V}^2$. We shall now connect this additional term to the quantum potential.

From equation (5.207) we know that the imaginary part of $(\mathcal{V}^2 - 2i\mathcal{D}\nabla \cdot \mathcal{V})$ is zero. By writing its real part in terms of the real velocities U and V, we find

$$E = \frac{1}{2}m(\mathcal{V}^2 - 2i\mathcal{D}\nabla \cdot \mathcal{V}) = \frac{1}{2}m(V^2 - U^2 - 2\mathcal{D}\nabla \cdot U). \tag{5.208}$$

The potential energy Q has been expressed in terms of the velocity field U, namely,

$$Q = -\frac{1}{2}m(U^2 + 2\mathcal{D}\nabla \cdot U). \tag{5.209}$$

We recover from this result the expression of Eq. (5.194) for the force $F = -\nabla Q$. Then we may finally write the energy balance under the three equivalent forms:

$$E = -2m\mathcal{D}^2\frac{\Delta\psi}{\psi} = \frac{1}{2}m(\mathcal{V}^2 - 2i\mathcal{D}\nabla \cdot \mathcal{V}) = \frac{1}{2}mV^2 + Q. \tag{5.210}$$

More generally, in presence of an external potential energy ϕ and in the non-stationary case, it reads

$$-\frac{\partial S_R}{\partial t} = \frac{1}{2}mV^2 + Q + \phi, \tag{5.211}$$

where S_R is the real part of the complex action (namely, $S_R/2m\mathcal{D}$ is the phase of the wave function). This equation is but the real part of the Schrödinger equation itself.

5.10. Quantum Mechanics of Many Particles

Let us now show that the scale relativity theory is also able to tackle the problem of the many particle quantum mechanics, including the many identical particle case [358, 360, 406].

The many-particle quantum theory plays a particularly important role in the understanding of quantum phenomena, since it is often considered as a definite proof that one cannot understand quantum mechanics in terms of a pure space-time description. Indeed the wave function must be defined in a configuration space having $3n+1$ dimensions, $\psi = \psi(x_1, x_2, \ldots, x_{3n}; t)$, in which the variables are *a priori* not separated, even though they are separated in the Hamiltonian, $H = \sum_i H_i(x_i, p_i)$. In the case of identical particles in the same state, they have, in addition, the original quantum property of indistinguishability.

This case is also important since it is first step towards the second quantization (see e.g. Bjorken and Drell [53]) and underlies the demonstration of the Pauli principle. Let us show now that the fractal/ nondifferentiable space-time approach of scale relativity allows one to easily recover the many identical particle Schrödinger equation and to understand the meaning of the non-separability of particles in the wave function.

The key to the solution of this problem clearly lies in the fact that the wave function is nothing but another expression for the action, now complex as a consequence of nondifferentiability. We define a generalized, complex Lagrange function that depends on all position and complex velocity components of the n particles,

$$\mathcal{L} = \mathcal{L}(x_1, x_2, \ldots, x_{3n}; \mathcal{V}_1, \mathcal{V}_2, \ldots, \mathcal{V}_{3n}; t), \qquad (5.212)$$

from which the complex action integral keeps its form,

$$\mathcal{S} = \int_{t_1}^{t_2} \mathcal{L}(x_1, x_2, \ldots, x_{3n}; \mathcal{V}_1, \mathcal{V}_2, \ldots, \mathcal{V}_{3n}; t) dt. \qquad (5.213)$$

The stationary action principle, applied to this integral, yields Euler–Lagrange equations, which read

$$\frac{\hat{d}}{dt} \frac{\partial \mathcal{L}}{\partial \mathcal{V}_k} - \frac{\partial \mathcal{L}}{\partial x_k} = 0, \qquad (5.214)$$

for $k = 1$ to $3n$.

Now considering the action as a function of the coordinates and of time at the upper limit of integration, one is led to define a unique wave function for the whole ensemble of particles

$$\psi = e^{i\mathcal{S}/\hbar}. \qquad (5.215)$$

5.10.1. *Many identical particles*

Let us first consider the case where the particles are identical. One may replace in Eq. (5.215) $\hbar$ by $2m\mathcal{D}$, i.e. in an equivalent way, by $\hbar = mc\lambda_c$, because the particles, being identical, have the same mass and the same Compton length. Then ψ is a function of the whole set of $3n$ coordinates of the n particles, $\psi = \psi(x_1, x_2, \ldots, x_{3n}; t)$. Therefore, in terms of the wave function the complex action still reads

$$S = -i\hbar \ln \psi. \tag{5.216}$$

Let us use the Hamilton approach for deriving the Schrödinger equation (see Sec. 5.6.7). It involves two relations. The first one is the expression for the momenta in terms of partial derivatives of the action, easily generalized to n particles,

$$\mathcal{P}_k = m\mathcal{V}_k = -im\lambda_c \partial_k \ln \psi, \tag{5.217}$$

where $k = 1$ to $3n$ runs on each of the three coordinates of the n particles. For the particular solutions of the many-particle equation, which are a product of solutions of the single-particle equation, they become a sum in terms of $\ln \psi$, so that the momentum of a given particle, being defined as the gradient of $\ln \psi$ relative to its own coordinates, depend only on them and not on the coordinates of the other particles. In this case, the particles are separated. However this is no longer true of the general solution, which is a linear combination of these particular ones.

The second relation is the expression for the energy, written in terms of the complex Hamilton function

$$\mathcal{H} = -\frac{\partial S}{\partial t}. \tag{5.218}$$

Inserting into this equation the expression of the complex action (Eq. 5.216), we finally obtain a Schrödinger equation

$$i\hbar \frac{\partial}{\partial t} \psi = \mathcal{H}\psi. \tag{5.219}$$

Recall that here $\mathcal{H}$ does not represent the Hamiltonian $\hat{H}$, which is an operator acting on ψ, but a Hamilton complex function, which is in product of ψ. In order to obtain the standard quantum mechanical equation, we therefore now need to prove that $\mathcal{H}\psi = \hat{H}\psi$.

The Hamilton function is classically given, by definition, by the relation $H = v \cdot p - L$, and it is generalized, in a covariant way, to complex quantities as (see Sec. 5.6.7)

$$\mathcal{H} = \widehat{\mathcal{V}}_k \mathcal{P}^k - \mathcal{L}, \tag{5.220}$$

where we use Einstein's convention about indices. The Lagrange function of n particles in a potential ϕ reads

$$\mathcal{L} = \frac{1}{2} m\mathcal{V}_k \mathcal{V}^k - \phi. \tag{5.221}$$

Finally, the Hamilton function reads

$$\mathcal{H} = \frac{1}{2}m\mathcal{V}_k\mathcal{V}^k - im\mathcal{D}\partial_k\mathcal{V}^k + \phi. \tag{5.222}$$

It is remarkable that the sum would have been exactly the same if it were made over the $3n$ coordinates of a unique particle in a $3n$-dimensional space. Namely, there is no way to distinguish different coordinates of a same particle from coordinates of different particles, since all coordinates are on the same footing in this expression. We are therefore brought back to the one-particle proof, with the 3-Laplacian Δ replaced by a $3n$-Laplacian, $\Delta_{3n} = \partial_k\partial^k$, which changes nothing to the formalism.

We therefore obtain a Schrödinger equation for the n identical particles that reads

$$\mathcal{D}^2\Delta_{3n}\psi + i\mathcal{D}\frac{\partial}{\partial t}\psi - \frac{\phi}{2m}\psi = 0. \tag{5.223}$$

The fact that the motion of n particles in quantum mechanics is irreducible to their classical motion, i.e. cannot be understood as the sum of n individual motions but instead must be taken as a whole, agrees with our identification of particles with the geodesics of a fractal space-time [353]. Being reduced to the geometric properties of virtual geodesics (here "geometry" is meant in terms of both displacements and scale transformations of resolutions, i.e. in position space and scale space), identical particles become totally indistinguishable. Since, in this framework, they are nothing but an ensemble of purely geometrical fractal curves, there is absolutely no physical property that could allow one to distinguish them.

Recall, moreover, that these lines themselves should not be considered in the same way as standard nonfractal curves. They are defined in a scale relative way, i.e. only the ratio from a finite scale to another finite scale does have meaning, while the zero limits, $dt \to 0$ and $dx \to 0$ are undefined. As a consequence, the concept of a unique geodesic itself loses its physical meaning. Whatever be the manner to unfold them or to select them and whatever the scale, the definition of a particle always involves an infinity of geodesics. Now, it is also clear, in this view, that the ensemble of geodesics that describes, e.g. two particles, is globally different from a simple sum of the geodesic ensemble of the individual particles. This is properly described by the wave function $\psi(x_1, x_2)$ (where x_1 and x_2 are here the position vectors of the two particles), which is solution of the above Schrödinger equation, from which one can recover the velocity fields of the particles, $\mathcal{V}_1 = -i(\hbar/m)\nabla_1 \ln\psi$ and $\mathcal{V}_2 = -i(\hbar/m)\nabla_2 \ln\psi$.

One of the consequences of these results is that, once the concept of spin constructed from first principles (see Sec. 6), we shall be able to derive the Pauli "principle" [358]. Indeed it is well-known that this is not an axiom of quantum mechanics. As shown, e.g. by Landau [267], it can be proved from the knowledge of the complex nature of the wave function, of the existence of spin, and of the symmetry properties of the many identical particle wave function.

5.10.2. *Many different particles*

In this case the parameter $\mathcal{D}$ is different for the various particles when they have different masses, $\mathcal{D}_k = \hbar/2m_k$. An archetype of this situation is the two-particle case, for example the hydrogen atom problem, in which a Schrödinger equation must be written for the proton and the electron together.

One may think that this would prevent one from having a unique space-(time) for the two particles. But actually this is not the case, since we have identified the particles, not with the space itself, but with its geodesics. A unique space, described by a unique wave function ψ, can contain different subsets of geodesics, of different velocity fields $\mathcal{V}_k = -2i\mathcal{D}_k\nabla_k \ln \psi$ and different geometric properties corresponding to different particles.

This is reminiscent of Einstein's general relativity where various bodies with different active gravitational masses M_k and therefore different Schwarzschild radii, each of them following their own geodesics, contribute to the same unique space-time. For example, in the two compact body problem, which has no exact solution, an approximate solution very near from each of the bodies is given by $g_{00} = 1 - r_{s1}/r$ and $g_{00} = 1 - r_{s2}/r$ respectively, with $r_{s1} = (2G/c^2)M_1$ and $r_{s2} = (2G/c^2)M_2$, and very far from the bodies by $g_{00} = 1 - r_s/r$ with $r_s = (2G/c^2)(M_1 + M_2)$. The Schwarzschild relation, which relies a length (space-time geometry) to a mass (energy-momentum), is nothing but a very simplified version of Einstein's field (space-time) equations.

Now the Compton relations and more generally the Einstein–de Broglie relations are the equivalent in scale relativity and quantum mechanics, of what are the Schwarzschild relations in motion general relativity. This is apparent in the fractal metric written in Sec. 4.5.4 and has been justified by the development of a special scale-relativistic mechanics [352, 353], in which both relations are two (symmetric) solutions of the same problem (see Sec. 11.1.1.3). They are therefore the progenitors of general field equations for the fractal space-time, which remain to be written. Like for the two-body problem in general relativity, for the two particle problem in quantum mechanics one may also define two Compton relations $\lambda_1 = \hbar/m_1 c$ and $\lambda_2 = \hbar/m_2 c$ near the particles, and a Compton relation $\lambda = \hbar/(m_1 + m_2)c$ at distances large with respect to the interdistance between the particles. This is clear from the transformation of the Hamiltonian [267]

$$\hat{H} = -\frac{\hbar^2}{2m_1}\Delta_1 - \frac{\hbar^2}{2m_2}\Delta_2 \qquad (5.224)$$

to its expression in the center of mass

$$\hat{H} = -\frac{\hbar^2}{2(m_1 + m_2)}\Delta_R - \frac{\hbar^2}{2m}\Delta, \qquad (5.225)$$

where $m = m_1 m_2/(m_1 + m_2)$ is the reduced mass. This well-known fact manifests itself experimentally in terms of interferences of atoms or molecules, in which the interference pattern is linked to the total mass of the system, which appears as a unique particle.

In this two-particle case, two different velocity fields of geodesics, $\mathcal{V}_1 = -2i\mathcal{D}_1\nabla_1\ln\psi$ and $\mathcal{V}_2 = -2i\mathcal{D}_2\nabla_2\ln\psi$ can be defined for a single wave function $\psi(x_1, y_1, z_1, x_2, y_2, z_2, t)$. The Lagrange function keeps the form of the classical Lagrange function for two particles, but now in terms of these complex velocity fields, i.e. $\mathcal{L} = (1/2)m_1\mathcal{V}_1^2 + (1/2)m_2\mathcal{V}_2^2 - \phi$. Then the complex Hamilton function, which was found to read $\mathcal{H} = \hat{\mathcal{V}} \cdot \mathcal{P} - \mathcal{L}$ in the one-particle case, becomes for two particles $\mathcal{H} = \hat{\mathcal{V}}_1 \cdot \mathcal{P}_1 + \hat{\mathcal{V}}_2 \cdot \mathcal{P}_2 - \mathcal{L}$, and the Hamilton–Jacobi energy equation $\mathcal{H} + \partial\tilde{\mathcal{S}}/\partial t = 0$ becomes $\mathcal{H}\psi = iS_0\partial\psi/\partial t$. The calculation is the same as for the one-particle case, except for the existence of two terms instead of one, so that, after replacing the complex velocity field by its expression in terms of ψ, one finally derives a two-particle Schrödinger equation

$$S_0\left[(\mathcal{D}_1\Delta_1 + \mathcal{D}_2\Delta_2)\psi + i\frac{\partial\psi}{\partial t}\right] = \phi\psi, \qquad (5.226)$$

where $S_0 = 2m_1\mathcal{D}_1 = 2m_2\mathcal{D}_2$. One recovers the two-particle Schrödinger equation of standard quantum mechanics for $\mathcal{D}_1 = \hbar/2m_1$, $\mathcal{D}_2 = \hbar/2m_2$ and $S_0 = \hbar$. This is easily generalizable to any number of particles.

Problems and Exercises

Exercise 14 Write the hydrodynamical form (Euler + continuity equations) of the multiparticle Schrödinger equation (5.226) and find the form of the quantum potential in this case.

Hints: Make a change of variables from ψ to the modulus of the wavefunction $\sqrt{P}$ and the real parts $V_1 = \nabla_1 S/m_1$ and $V_2 = \nabla_2 S/m_2$ of the complex velocity fields. One may bring back Eq. (5.226) to the form of the many identical particle case by making the change of variable $x_1 \rightarrow x_1/\sqrt{\mathcal{D}_1}$ (and a similar change on the other position variables), and then to the one-particle case in $3n$ dimensions.

Solution: The real part of the Schrödinger equation becomes the energy equation

$$E = -\frac{\partial S}{\partial t} = \frac{1}{2}m_1 V_1^2 + \frac{1}{2}m_2 V_2^2 + \phi + Q, \qquad (5.227)$$

where the quantum potential is generalized to the expression

$$Q = -S_0 \frac{(\mathcal{D}_1 \Delta_1 + \mathcal{D}_2 \Delta_2)\sqrt{P}}{\sqrt{P}}. \qquad (5.228)$$

Since $S_0 = 2m_1\mathcal{D}_1 = 2m_2\mathcal{D}_2$, it is formally the sum of each particle quantum potentials (but with a unique probability P common to both of them),

$$Q = -\frac{2m_1\mathcal{D}_1^2 \Delta_1 \sqrt{P} + 2m_2\mathcal{D}_2^2 \Delta_2 \sqrt{P}}{\sqrt{P}}. \qquad (5.229)$$

Then the two vectorial Euler equations are obtained for the two velocity fields by taking the gradients of the energy equation,

$$m_k \left(\frac{\partial}{\partial t} + V_k \cdot \nabla_k\right) V_k = F_k, \qquad (5.230)$$

with forces $F_k = -\nabla_k(\phi + Q)$, the index k running on the particle ranks.

Finally, the continuity equation is obtained as the imaginary part of the Schrödinger equation. It reads

$$\frac{\partial P}{\partial t} + \nabla_1 \cdot (PV_1) + \nabla_2 \cdot (PV_2) = 0. \qquad (5.231)$$

∎

5.11. Schrödinger form of other Fundamental Equations of Physics

The general method described above can be applied to any physical situation where the three basic conditions, namely, (i) infinity of geodesics, (ii) each geodesic is a fractal curve of fractal dimension 2, (iii) breaking of differential time reflection invariance, are achieved in an exact or in an approximate way.

These conditions find their origin in the full nondifferentiability, which implies a scale dependence without any limit toward the small scales, i.e. a scale divergence. We consider this case to correspond to standard quantum mechanics. In this situation there is no sub-quantum medium, i.e. no breaking of the scale dependence toward the small scales. Indeed, such a breaking would turn the theory into a hidden parameter theory, which is now excluded by EPR-type experiments.

The three above conditions may also be achieved, in an approximate way, in other situations than microphysics, so that a Schrödinger form can be obtained for the equations of motion of some macroscopic systems [353, 362]. This is made possible, in particular, by the fact that the whole mathematical structure of the derivation of the Schrödinger equation and of its solutions does not rely on the Planck constant $\hbar$, but is instead preserved for any value of the parameter $\mathcal{D}$. In this framework $\hbar = 2m\mathcal{D}$ is only a particular case and the theory opens the possibility of a quantum-type regime for some macroscopic systems [353], but which would share only some of the properties of the standard quantum theory. These new applications of a quantum-like theory, along with an analysis of the conditions under which it could emerge, will be considered in more detail in Chapters 10, 13 and 14. Here we shall only prepare these applications by showing that the scale relativity method can be applied to many equations of physics under the three fractal conditions.

Indeed, several fundamental equations of classical physics can be transformed to take a generalized Schrödinger form under these conditions [362]: namely, the equation of motion in the presence of an electromagnetic field (see also Chapter 7), the Euler and Navier–Stokes equations in the case of potential motion and for incompressible and isentropic fluids; the equations of the rotational motion of solids, the motion equation of dissipative systems; field equations (scalar field for one space variable).

5.11.1. *Particle in a vectorial field*

The scale relativity approach, which has been hereabove applied to the case of a particle (more generally, of many particles) in a scalar potential, can be easily generalized to the vectorial field case. An example of such a situation is a particle in an electromagnetic field. We consider here the field as externally applied on the particle. The question of the origin of the field itself, which can ultimately be described as a manifestation of the very fractal geometry, will be considered later (in Chapter 7).

Actually, we may use the same method as in standard quantum mechnics and introduce the field from a covariant derivative, which simply amounts to including into the full momentum and energy the momentum and the energy of the field, identified with its potential. The generalized momentum and energy of a particle in a vectorial potential A reads

$$\tilde{\mathcal{P}} = \mathcal{P} + qA, \quad \tilde{\mathcal{E}} = \mathcal{E} + q\phi, \tag{5.232}$$

where A is a three-dimensional vector. This leads to introduce a QED-covariant derivative [356]

$$2mi\mathcal{D}\tilde{\nabla} = 2mi\mathcal{D}\nabla + qA. \tag{5.233}$$

The resulting equations have the form of the Schrödinger equation in the presence of an electromagnetic field (of vector potential A and scalar potential ϕ), namely,

$$\mathcal{D}^2 \left(\nabla - i\frac{q}{2m\mathcal{D}}A \right)^2 \psi + i\mathcal{D}\frac{\partial}{\partial t}\psi - \frac{q\phi}{2m}\psi = 0. \tag{5.234}$$

In Chapter 7 we shall show how one can derive this equation from a free-like geodesic equation, in which both the quantum effects and the field are constructed from nondifferentiability and fractality.

5.11.2. *Particle in a tensorial field*

In its present acceptance gravitation is understood as the various manifestations of the geometry of space-time at large scales. Up to now, in the framework of Einstein's theory, this geometry is considered to be Riemannian.

Let us consider the motion of a free particle in a curved and fractal space-time. As a first step, one could define a motion and scale covariant derivative that combines the general-relativistic covariant derivative, which describes the effects of curvature and the scale-relativistic covariant derivative, which describes the effects of fractality. Namely, Einstein's covariant derivative reads

$$\frac{D}{ds}A^\mu = \frac{d}{ds}A^\mu + \Gamma^\mu_{\nu\rho}v^\nu A^\rho. \tag{5.235}$$

Using this covariant derivative, Einstein's geodesic equations are written in terms of the free particle equation of motion

$$\frac{D}{ds}v^\mu = 0 \Rightarrow \frac{d}{ds}v^\mu + \Gamma^\mu_{\nu\rho}v^\nu v^\rho = 0. \tag{5.236}$$

This equation could now be made scale-covariant, by replacing d/ds by $\hat{d}/ds$ at all the levels of the construction. We define a scale relativistic and motion relativistic covariant derivative

$$\frac{\bar{D}}{ds}A^\mu = \frac{\hat{d}}{ds}A^\mu + \Gamma^\mu_{\nu\rho}\mathcal{V}^\nu A^\rho, \tag{5.237}$$

i.e. using the relativistic expression of the quantum covariant derivative in terms of the complex four-velocity $\mathcal{V}^\mu = \hat{d}x^\mu/ds$,

$$\frac{\hat{d}}{ds} = (\mathcal{V}^\mu + i\mathcal{D}\partial^\mu)\partial_\mu, \tag{5.238}$$

one obtains

$$\frac{\bar{D}A^\mu}{ds} = (\mathcal{V}^\nu\partial_\nu + i\mathcal{D}\partial^\nu\partial_\nu)A^\mu + \Gamma^\mu_{\rho\nu}\mathcal{V}^\rho A^\nu. \tag{5.239}$$

The equation of motion of a free particle can now be written as a geodesic equation by using this covariant derivative. However, one should take care that the combination of the two covariant derivatives imply the appearance of a new term in the geodesic equation [142, 362, 447]. This is easily established by starting from the quadratic invariant, [446] $\mathcal{V}_\mu\mathcal{V}^\mu + 2i\mathcal{D}\partial_\mu\mathcal{V}^\mu = 1$, which is the relativistic equivalent (see Chapter 6) of the energy equation $\mathcal{E} = \frac{1}{2}m\mathcal{V}^2 - im\mathcal{D}\nabla\cdot\mathcal{V}$ (5.125). This equation becomes in the general-relativistic case

$$\mathcal{V}_\mu\mathcal{V}^\mu + 2i\mathcal{D}D_\mu\mathcal{V}^\mu = 1, \tag{5.240}$$

where we have $\mathcal{V}_\mu\mathcal{V}^\mu = g_{\mu\nu}\mathcal{V}^\mu\mathcal{V}^\nu$. The equations of motion can now be directly obtained by differentiating this relation, namely,

$$\frac{\hat{d}}{ds}\mathcal{V}^\mu + \Gamma^\mu_{\nu\rho}\mathcal{V}^\nu\mathcal{V}^\rho - i\mathcal{D}R^\mu_\nu\mathcal{V}^\nu = 0. \tag{5.241}$$

This equation now also contains an additional Ricci contribution. It can finally be integrated in terms of a generalized "Einstein–Klein—Gordon" equation of motion that reads

$$\frac{4\mathcal{D}^2}{c^2}\big(g_{\mu\nu}\partial^\mu\partial^\nu\psi + \partial_\nu\big(\ln\sqrt{-g}\big)\partial^\nu\psi\big) = -1, \qquad (5.242)$$

where g is the metric determinant.

Now, one must be cautious with the interpretation of this equation. It is obtained by assuming that the two (motion and scale) covariances do not interact with each other. This can only be a rough approximation. Indeed, in order to solve the problem of motion in a general, non flat fractal space-time, which is nothing but the problem of finding a theory of quantum gravity in our framework, one should strictly examine the geometrical effects of curvature and fractality at the level of the construction of the covariant derivatives, not only once they have been constructed. This problem, which is specific of the Planck scale physics, at which the quantum, gauge field and gravitational effects all become of the same order, reveals to be extraordinarily complicated and will not be considered further in the present book.

A second problem with this equation concerns the interpretation of the scale-covariant derivative in the motion-relativistic case. It is obtained by assuming that not only space but space-time is fractal, which implies that the trajectories of particles can go backward in time. This is not a problem in microphysics: on the contrary, it is even needed by the existence of virtual pairs of particle-antiparticles, through Feynman's interpretation of antiparticles as particles going backward in time (see [420, 349, 353] for a development of the fractal approach to this question). It is more difficult to make a similar interpretation in the case of macroscopic applications, so that, in this case, only the non(motion)-relativistic limit will be considered, in which only space becomes fractal. This leads to a generalized Newton's equation of dynamics,

$$\frac{\hat{d}}{dt}\mathcal{V} + \frac{\nabla\phi}{m} = 0, \qquad (5.243)$$

in which ϕ is the gravitational potential energy, that can be integrated in terms of the generalized gravitational Schrödinger equation. Many applications of this equation will be considered in Chapter 13.

5.11.3. *Euler–Schrödinger equation (potential motion)*

The scale-relativity approach can be generalized to fluid mechanics in a straightforward way. Actually we have already partly adopted a fluid description when introducing a velocity field $v = v[x(t), t]$, even if it remained, at this stage, a virtual fluid (of geodesics). The Euler equation for a fluid in a scalar potential ϕ reads

$$\frac{d}{dt}v = \left(\frac{\partial}{\partial t} + v \cdot \nabla\right)v = -\frac{\nabla p}{\rho} - \nabla\phi. \tag{5.244}$$

Let us now assume that the motion of the fluid is subjected to the three conditions that allow to construct a complex velocity $\mathcal{V} = V - iU$ and a quantum covariant derivative $\hat{d}/dt = \partial_t + \mathcal{V} \cdot \nabla - i\mathcal{D}\Delta$. The Euler equation is then transformed into the complex equation:

$$\frac{\hat{d}}{dt}\mathcal{V} = -\frac{\nabla p}{\rho} - \nabla\Phi. \tag{5.245}$$

In the general case $\nabla p/\rho$ is not a gradient and we cannot transform this equation into a Schrödinger-like equation. However, in the case of an incompressible fluid ($\rho = $ cst), where one can use the sound approximation and more generally in the case of an isentropic fluid (including perfect fluids), $\nabla p/\rho$ is the gradient of the enthalpy by unit of mass w (see e.g. [269])

$$\frac{\nabla p}{\rho} = \nabla w. \tag{5.246}$$

Under this approximation Eq. (5.245) becomes the Euler–Lagrange equation constructed from the Lagrange function $\mathcal{L}(x, \mathcal{V}, t) = \frac{1}{2}m\mathcal{V}^2 - \phi - w$. Therefore it derives from a stationary action principle working with the complex action $\mathcal{S} = \int \mathcal{L}dt$, so that the previous formalism is now recovered. We introduce the probability amplitude ψ (now defined for a unit mass):

$$\mathcal{S} = -2i\mathcal{D}\ln\psi, \tag{5.247}$$

so that the complex velocity field of the (now real) fluid reads

$$\mathcal{V} = -2i\mathcal{D}\nabla\ln\psi. \tag{5.248}$$

This means that this description is valid only for potential motion and that $\mathcal{S}$ plays the role of a velocity potential.

In terms of ψ, $\hat{d}\mathcal{V}/dt$ is a gradient

$$\frac{\hat{d}}{dt}\mathcal{V} = -2\nabla\left[\frac{\mathcal{D}^2\Delta\psi + i\mathcal{D}\partial\psi/\partial t}{\psi}\right], \tag{5.249}$$

so that the complex Euler equation can now be integrated, leading to a generalized Schrödinger-like equation,

$$\mathcal{D}^2\Delta\psi + i\mathcal{D}\frac{\partial}{\partial t}\psi - \frac{w+\phi}{2}\psi = 0. \tag{5.250}$$

Now we know that by taking the real and imaginary parts of this equation, one respectively recovers a Euler equation including a quantum potential and a continuity equation, in terms of the real velocity field V and of the fluid density $\rho = |\psi|^2$ (see Sec. 5.7.3). This means that, in this approach, the continuity equation was included into our initial complex Euler equation.

We shall see in Chapter 10 that a Schrödinger equation (of the magnetic type) can also be written for the equation of motion of a rotational fluid in many situations.

5.11.4. *Navier–Schrödinger equation (potential motion)*

A similar work can be performed with the Navier–Stokes equations, at least formally. The quantum-covariant generalized Navier–Stokes equations read

$$\left(\frac{\partial}{\partial t} + \mathcal{V}\cdot\nabla - i\mathcal{D}\Delta\right)\mathcal{V} = -\frac{\nabla p}{\rho} + \nu\Delta\mathcal{V}. \tag{5.251}$$

It is quite remarkable that the viscosity term in the Navier–Stokes equation plays a role similar to the parameter $\mathcal{D}$. This has suggested to us to combine them into a new complex parameter [362]

$$\widetilde{\mathcal{D}} = \mathcal{D} - i\nu. \tag{5.252}$$

In terms of $\widetilde{\mathcal{D}}$, the complex Navier–Stokes equation recovers the form of the complex Euler equation

$$\left(\frac{\partial}{\partial t} + \mathcal{V}\cdot\nabla - i\widetilde{\mathcal{D}}\Delta\right)\mathcal{V} = -\frac{\nabla p}{\rho}. \tag{5.253}$$

Once again, in the incompressible or isentropic cases, this equation can be integrated to yield a Schrödinger-like equation

$$\widetilde{\mathcal{D}}^2\Delta\psi + i\widetilde{\mathcal{D}}\frac{\partial}{\partial t}\psi - \frac{w}{2}\psi = 0. \tag{5.254}$$

This equation is also valid in the presence of a gravitational field or in the presence of any field that is the gradient of a potential energy ϕ. It becomes

$$\widetilde{\mathcal{D}}^2 \Delta\psi + i\widetilde{\mathcal{D}}\frac{\partial}{\partial t}\psi - \frac{w+\phi}{2}\psi = 0. \tag{5.255}$$

However, its interpretation is more difficult than in previous calculations. Indeed the complex nature of $\widetilde{\mathcal{D}}$ prevents the imaginary part of this equation to be an equation of continuity, so that the Born postulate is no longer directly ensured in terms of $|\psi|^2$ in this case.

Problems and Exercises

Open Problem 5: Separate the real and imaginary parts of the above Navier–Schrödinger equation (with a complex diffusion coefficient) Eq. (5.255) and construct a statistical interpretation of its variables.   ∎

5.11.5. *Rotational motion of solids*

The equation of the rotational motion of a solid body can be given the form of Euler–Lagrange equations and, therefore, easily comes under the scale relativity approach [362, 365]. We shall give here first hints about this case that we shall treat in more detail in Chapter 10 (in a different context). The role of the variables (x, v, t) is now played by the rotational coordinates, (ϕ, Ω, t) where ϕ denotes three rotational position angles, for example the Euler angles and Ω is the corresponding rotational velocity. The Euler–Lagrange equation reads in this case [265]

$$\frac{d}{dt}\frac{\partial L}{\partial \Omega} = \frac{\partial L}{\partial \phi}, \tag{5.256}$$

in terms of the Lagrange function L of the solid, that writes

$$L = \frac{1}{2}\mu V^2 + \frac{1}{2}I_{ik}\Omega_i\Omega_k - \phi. \tag{5.257}$$

Here I_{ik} is the tensor of inertia of the body and ϕ a potential term. We use throughout this section the tensorial notation where a sum is meant on two repeated indices. The right-hand member of Eq. (5.256) reads

$$K = \frac{\partial L}{\partial \phi} = -\frac{\partial U}{\partial \phi} = \sum r \times F, \tag{5.258}$$

which identifies with the total torque, i.e. the sum of the moments of all forces acting on the body. In the left-hand member one recognizes the angular momentum about the center of mass,

$$M_i = \frac{\partial L}{\partial \Omega_i} = I_{ik}\Omega_k, \qquad (5.259)$$

and we finally recover a rotational equation of dynamics,

$$\frac{d}{dt}M = K. \qquad (5.260)$$

Let us now apply the quantum covariance to these equations. We are in similar conditions as in the case of translational motion, but now the position angles have replaced the coordinates. We assume that, instead of deterministic values of the rotational velocities, there is an infinity of possible paths in the position angle space, that these paths are fractal of dimension 2, and that there is a discrete reflection invariance breaking on dt, so that one is led to introduce a two-valued fractal rotational velocity field, $\Omega_\pm[\varphi(t,dt),t,dt]$.

This angular velocity can be decomposed in terms of a differentiable part $\widetilde{\Omega}$ and of a fractal part W, such as

$$\langle W_j W_k \rangle = \frac{2\mathcal{D}_{jk}}{dt}, \qquad (5.261)$$

where $\mathcal{D}_{jk}$ is now a tensor. We then build a quantum-covariant derivative,

$$\frac{\hat{d}}{dt} = \frac{\partial}{\partial t} + \widetilde{\Omega}_k\partial_k - i\mathcal{D}_{jk}\partial_j\partial_k. \qquad (5.262)$$

The quantization of Eq. (5.260) is straighforward using this scale-covariant derivative. One obtains the complex equation

$$I_{jk}\frac{\hat{d}\widetilde{\Omega}_k}{dt} = -\partial_j\phi. \qquad (5.263)$$

We now introduce the wave function as another expression for the action $\mathcal{S}$, $\psi = \exp(i\mathcal{S}/\mathcal{S}_0)$, where $\mathcal{S}_0$ is still a constant having the dimension of an angular momentum. In the case of translational motion, we have seen that this constant was given by $\mathcal{S}_0 = 2m\mathcal{D}$, i.e. in other words, that the fractal fluctuation parameter was given by $\mathcal{D} = \frac{1}{2}\mathcal{S}_0 m^{-1}$. In the rotational motion case now considered, we need a tensorial generalization of this relation. Introducing the inverse inertia tensor, I^{-1}, one finds the generalized matrix relation

$$\mathcal{D}_{jk} = \frac{1}{2}\mathcal{S}_0 I_{jk}^{-1}. \qquad (5.264)$$

See more detail about the derivation of this expression in Chapter 10. Note also the correction to [362, 365], in which the final Schrödinger equation was nevertheless correct. Thanks to this relation, the complex motion equation can be integrated in terms of a generalized Schrödinger equation that read

$$S_0 \left(\mathcal{D}_{jk} \partial_j \partial_k \psi + i \frac{\partial}{\partial t} \psi \right) - \phi \psi = 0. \tag{5.265}$$

Then by taking the imaginary and real parts of this equation after introducing $P = |\psi|^2$, one recovers respectively a continuity equation,

$$\frac{\partial P}{\partial t} + \mathrm{div}(P\Omega) = 0, \tag{5.266}$$

and an energy equation,

$$-\frac{\partial S}{\partial t} = \frac{1}{2} I_{jk} \Omega_j \Omega_k + \phi + Q, \tag{5.267}$$

where the first term is the standard kinetic energy, and where Q is a generalization of the quantum potential to the tensorial case,

$$Q = -S_0 \frac{\mathcal{D}_{jk} \partial_j \partial_k \sqrt{P}}{\sqrt{P}}. \tag{5.268}$$

This therefore ensures also in this case the Born interpretation of the wave function, according to which P gives the density of probability of the various values of the position angles.

We recover, when $S_0 = \hbar$, the standard quantum mechanical equation that can be constructed from the rotational motion Hamiltonian. This equation has been validated by experiment, since it has already been applied to the quantization of the motion of molecules taken in their whole as solid bodies. This result is an additional confirmation of the ability of the scale relativity method to recover standard quantum mechanical equations, and, as we shall see, to also suggest generalizations of these equations.

5.11.6. *Dissipative systems: first hints*

One can generalize the Euler–Lagrange equations to dissipative systems thanks to the introduction of a dissipation function f (see e.g. [265]).

The quantum-covariant form of these equations reads [362]

$$\frac{\hat{d}}{dt}\frac{\partial \pounds}{\partial \mathcal{V}_i} = \frac{\partial \pounds}{\partial x_i} - \frac{\partial f}{\partial \mathcal{V}_i}, \tag{5.269}$$

where f is linked to the energy dissipation by the equation $f = -d\mathcal{E}/2dt$. This becomes in the Newtonian case:

$$m\frac{\hat{d}}{dt}\mathcal{V} = -\nabla\phi - \frac{\partial f}{\partial \mathcal{V}} = -\nabla\phi - \sum_j k_{ij}\mathcal{V}_j. \tag{5.270}$$

We shall only consider here briefly the simplified isotropic case,

$$f = kv, \tag{5.271}$$

and its complex generalization,

$$\mathcal{F} = k\mathcal{V}. \tag{5.272}$$

We obtain a new generalized equation [362],

$$\mathcal{D}^2\Delta\psi + i\mathcal{D}\frac{\partial\psi}{\partial t} - \frac{\phi}{2m}\psi + i\frac{k}{m}\psi\ln\psi = 0, \tag{5.273}$$

which is a nonlinear Schrödinger equation. This equation remains scale-invariant under the transformation $\psi \to \rho\psi$, up to an arbitrary energy term. It corresponds to a perturbed Hamiltonian $H = H_0 + V$, with the operator V such that $V\psi = -i(k/m)\psi\ln\psi$. The standard methods of perturbation theory in quantum mechanics can then be used to look for the solutions of this equation. The same problem has also been considered in [6] with equivalent results and in [221] along with its soliton solutions.

5.11.7. *Field equations*

As it is well-known, the deep unity of physics manifests itself by the fact that field equations can also be given the form of Euler–Lagrange equations. The potentials play the role of the generalized coordinates, the fields play the role of the time derivatives of coordinates and the coordinates play the role of time, namely,

$$x \leftrightarrow \phi,$$

$$t \leftrightarrow x,$$

$$v = \frac{dx}{dt} \leftrightarrow F = \frac{d\phi}{dx}. \tag{5.274}$$

Once this substitution is made, field equations take the same form as the equations of motion. However, the vectorial nature of the position coordinates, compared with the scalar nature of the time variable, implies that it is far more difficult to obtain a scale-relativistic version of these equations. In what follows, we shall therefore work, as a first step, with only one space-time variable x and with a scalar potential. The many coordinate and vectorial or tensorial field cases remain an open problem.

Note also that this problem is not equivalent to a second quantization of the field. It corresponds to a situation in which the information on the field and on the potentials is lost in such a way that one can no longer know the potentials themselves, but only a density of probability of their values.

Classically, one defines a Lagrange function $L(\phi, F, x)$, then an action S from this Lagrange function. The action principle leads to field equations that take the form of Euler–Lagrange equations,

$$\frac{d}{dx}\frac{\partial L}{\partial F} = \frac{\partial L}{\partial \phi}. \tag{5.275}$$

For example, the Lagrange equation constructed from $L = \frac{1}{2}F^2 - k\rho\phi$ is the Poisson field equation, $d^2\phi/dx^2 = -k\rho$ (where ρ is the density of the sources of the field, $k = 4\pi G$ in the gravitational case and $k = 4\pi$ is the case of a constant electric field). This well-known structure of present physical theories allows us to apply the scale relativity method to fields themselves [362].

Consider indeed a field potential $\Phi(x)$ whose evolution with the coordinate x is a fractal function. We therefore describe its elementary variations as the sum of a classical (differentiable) part and of a fractal stochastic (nondifferentiable) fluctuation, namely,

$$d\Phi = d\phi + d\xi_\phi, \tag{5.276}$$

such that $\langle d\xi_\phi \rangle = 0$ and $\langle (d\xi_\phi)^2 \rangle = 2\mathcal{D}_\phi dx$.

In the same way as for the description of motion, the combined effect of the fractal fluctuations and of the passage to complex numbers due to the breaking of the $(dx \leftrightarrow -dx)$ reflection invariance leads to define a complex field $\mathcal{F}$, then a complex quantum-covariant derivative that reads

$$\frac{\hat{d}}{dx} = \frac{\partial}{\partial x} + \mathcal{F}\frac{\partial}{\partial \phi} - i\mathcal{D}_\phi\frac{\partial^2}{\partial \phi^2}. \tag{5.277}$$

The Lagrange function become itself a complex function, $\mathcal{L} = \mathcal{L}(\phi, \mathcal{F}, x)$, so that, using this covariant derivative, Eq. (5.275) becomes complex

Euler–Lagrange equations,

$$\frac{\hat{d}}{dx}\frac{\partial \pounds}{\partial \mathcal{F}} = \frac{\partial \pounds}{\partial \phi}. \tag{5.278}$$

From the action $\mathcal{S}$, itself become complex, one defines a wave function

$$\Psi = e^{i\mathcal{S}/2\mathcal{D}_\phi}, \tag{5.279}$$

which is related to the complex field as

$$\mathcal{F} = -2i\mathcal{D}_\phi \frac{\partial}{\partial \phi}\ln \Psi. \tag{5.280}$$

Let us now specialize again to the Poisson-like case, corresponding to a Lagrange function $\pounds = \frac{1}{2}\mathcal{F}^2 - \mathcal{U}(\phi)$ (this is the field analog of the Newtonian case for motion). In this case, the field equation can be integrated under the form of a generalized Schrödinger equation for the probability amplitude of the potential $\Psi(\phi)$, namely,

$$\mathcal{D}_\phi^2 \frac{\partial^2}{\partial \phi^2}\Psi + i\mathcal{D}_\phi \frac{\partial}{\partial x}\Psi = \frac{1}{2}\mathcal{U}\Psi. \tag{5.281}$$

In the case of the standard Poisson equation, $\mathcal{U}$ is given by $\mathcal{U} = -k\rho\phi$ and it can be replaced by $\phi\partial^2\phi/\partial x^2$.

Finally, under such a description, one gives up the possibility to strictly know the value of the potential ϕ at any point or instant and one instead introduces a probability amplitude $\Psi(\phi)$, which is such that the probability of a given value of ϕ is given by $P(\Phi) = |\Psi|^2(\Phi)$.

Problems and Exercises

Open Problem 6: Generalize the construction of a Schrödinger-type equation for fields to vectorial coordinates and to vectorial or tensorial fields. ∎

Open Problem 7: Apply the scale relativity approach to a "quantization" of thermodynamics.

Hints: One may use the symplectic structure of thermodynamics [443]. This formal structure (the Poisson bracket structure) is the same as that of classical mechanics. Indeed, the state equations are analogous to Hamilton–Jacobi equations, while the couples of thermodynamics variables, such as (S, V), where S is the entropy and V the volume, are analogous to coordinates and $(T, -P)$, where T is the temperature and P the pressure to their conjugate moments. A change of variable may change this identification, which is not absolute, but in any case we keep two couples of conjugate variables.

The first step in the application of the scale relativity methods to thermodynamics amounts to generalizing the form of these thermodynamical variables to explicitly scale dependent functions, $S(\delta x, \delta t)$, $V(\delta x, \delta t)$, $T(\delta x, \delta t)$, $P(\delta x, \delta t)$, the same being true of density, $\rho = \rho(\delta x, \delta t)$. Note that explicitly scale-dependent temperatures, pressures and densities may be a necessary ingredient of a more complete description of many multi-scale systems, in particular turbulent ones.

The second step amounts to introducing fractal fluctuations and local irreversibility, and therefore fractal fluid-like paths in infinite number for these thermodynamical variables. This leads to a probabilistic view in which they are no longer defined in a deterministic way, but only by a probability amplitude $\psi(S, V)$ and its Fourier transform $\varphi(T, -P)$, whose squared modulus give the probability density $P(S, V)$ and $P(T, P)$, and which is expected to be solution of a Schrödinger-type equation (new form of the state-Hamilton–Jacobi equation). ∎

5.12. Foundation of Quantum Mechanics on the Principle of Relativity

5.12.1. *Analysis of the problem*

We are now in a position to found the laws of quantum mechanics on the principle of relativity in the framework of a both motion and scale relativity theory [399]. To this end, we do not need to recover all the various properties, features and results of the quantum theory, which would be an impossible task owing to its wide development since now nearly hundred years. We may on the contrary benefit from its axiomatic present basis to shunt the question. Namely, it is sufficient to show that we are now able to derive all the postulates of quantum mechanics from the principle of relativity. This is the subject of the present section.

Bjorken and Drell [53] have collected the axioms of quantum mechanics in terms of six items (some of which actually containing several statements), to which one should add the seventh axiom of wave function collapse (von Neumann postulate). Let us briefly recall them hereafter (see also [267, 106, 60, 474, 319]):

(1) (i) Existence of a complex state function $\psi(q_i, t)$; (ii) probability interpretation (Born postulate): $P = |\psi|^2$.

(2) Correspondence principle, $p_i \to -i\hbar\partial/\partial q_i$, etc.
(3) Eigenstate postulate, $\Omega\psi_n = \omega_n\psi_n$.
(4) Expansion postulate, $\psi = \sum_n a_n\psi_n$.
(5) Projection postulate, $P(\omega_n) = |a_n|^2$.
(6) Schrödinger equation $i\hbar\partial\psi/\partial t = H\psi$.
(7) Von Neumann postulate.

However, if one considers these postulates in more detail, it appears that they do not really have all the same status. Actually, the set of statements we find in the literature as "postulates", "axioms" or "principles" of quantum mechanics can be split into three subsets: (i) main postulates which cannot be derived from more fundamental ones (in the framework of the standard quantum theory), (ii) secondary postulates that are often presented as basic ones, but which can actually be derived from the main ones and (iii) statements called "principles" (often for historical reasons), which are now generally recognized to be mere consequences of the postulates. In fact, the main postulates have already been proved in the previous sections, so that we shall only in what follows remind the main steps of their derivation, then we shall show how the other "postulates" derive from them.

In the present analysis, we use a formulation of the postulates within a coordinate realization of the state function, since it is in this representation that their scale relativistic derivation is the most straightforward. Their momentum realization can be obtained by the same Fourier transforms, which are used in standard quantum mechanics. The same is true of the Dirac representation, which is another mathematical formulation of the same theory. Indeed, it can follow from the definition of the wave functions as vectors of an Hilbert space upon which Hermitian operators act, these operators representing the observables that correspond to classical dynamical quantities.

5.12.2. *Main postulates*

Complex state function

Each physical system is described by a state function, which determines all that can be known about the system. The coordinate realization of this state function, the wave function $\psi(r, t)$, is an equivalence class of complex functions of all the classical degrees of freedom generically noted r, of the time t. More generally, it may also contain additional degrees of freedom, such as spin s which are considered as intrinsically quantum-mechanical.

The emergence of spin and of other quantum numbers, such as isospin from the nondifferentiable geometry will be considered in the two next chapters. Two wave functions represent the same state if they differ only by a phase factor. This part of the "postulate" can be derived from the Born postulate, since, in this interpretation, probabilities are defined by the squared norm of the complex wave function and therefore two wave functions differing only by a phase factor represent the same state. The wave function has to be finite and single valued throughout position space. It must also be a continuous and continuously differentiable function. This postulate has recently been relaxed by Berry's discovery [47] of the possibility of nondifferentiable wave functions in quantum mechanics. We have reached the same conclusion in the scale relativity framework [403] (see Sec. 5.8).

Scale relativistic derivation: In the scale relativity theory, the wave function is identified with the action, after a simple change of variable. The action becomes complex because of the two-valuedness of mean derivatives that comes from the symmetry breaking ($dt \leftrightarrow -dt$), itself a direct consequence of nondifferentiability. The use of complex numbers to represent this two-valuedness is a simplifying and covariant choice (see Sec. 5.4.1). Spin and the Pauli equation may also be derived in the scale relativity framework, as a consequence of a new two-valuedness coming from the breaking of the symmetry $dx \leftrightarrow -dx$. Even though it comes under the nonrelativistic theory, this case will be considered in Chapter 6 about relativistic quantum mechanics since, as in standard QM, one may show that its origin is fundamentally relativistic.

Correspondence principle

To every dynamical variable of classical mechanics, corresponds in quantum mechanics a linear, Hermitian operator, which when operating upon the wave function associated with a definite value of that observable (the eigenstate associated to a definite eigenvalue) yields this value times the wave function. The more common operators occuring in quantum mechanics for a single particle are listed below and are constructed using the position and momentum operators.

More generally, the operator associated with the observable A, which describes a classically defined physical variable is obtained by replacing in the "properly symmetrized" expression of this variable the above operators for r and p. This symmetrization rule is added to ensure that the

Position	$r(x, y, z)$	multiply by $r(x, y, z)$
Momentum	$p(p_x, p_y, p_z)$	$-i\hbar\nabla$
Kinetic energy	$T = p^2/2m$	$-(\hbar^2/2m)\nabla^2$
Potential energy	$V(r)$	multiply by $V(r)$
Total energy	$E = T + V$	$i\hbar\partial/\partial t = V(r) - (\hbar^2/2m)\Delta$
Angular momentum	(l_x, l_y, l_z)	$-i\hbar r \times \nabla$

operators be Hermitian and, therefore that the measurement results are real numbers.

However, the symmetrization (or Hermitization) recipe is not unique. As an example, the quantum-mechanical analog of the classical product $(px)^2$ can be either $(p^2x^2 + x^2p^2)/2$ or $\{(xp + px)/2\}^2$ [60, 319]. No correspondence rule with classical mechanics is able to resolve such ambiguities, since they come from the noncommutativity of operators. The different choices yield corrections of the order of some $\hbar$ power and, in the end, it is experiment that decides, which is the correct operator. This is clearly one of the main weaknesses of the axiomatic foundation of quantum mechanics [319], since the ambiguity begins with second orders and, therefore, concerns the construction of the Hamiltonian itself.

Scale relativistic derivation: as a consequence of the two-valuedness issued from nondifferentiability, the various physical quantities, in particular the conservative quantities, become complex. The fundamental classical relations that relate the action, considered as a function of the coordinates, to the conservative quantities have therefore complex counterparts. After performing the change of variable, which introduces the wave function from the action, these relations become equalities that link the physical quantities (whose real parts are, at the classical limit, the classical variables), to the action of operators on the wave function (see Sec. 5.6.3). We recover the quantum mechanical operators under their correct form, including in particular $p^2 \to -\hbar^2\Delta$ in the Hamiltonian (Sec. 5.6.7) and its tensorial generalization (Sec. 5.11.5), which yet comes under the ambiguous cases of standard quantum mechanics [319].

Schrödinger equation

The time evolution of the wave function of a non-relativistic physical system is given by the time-dependent Schrödinger equation

$$i\hbar\frac{\partial\psi}{\partial t} = \hat{H}\psi, \tag{5.282}$$

where the Hamiltonian $\hat{H}$ is a linear Hermitian operator, whose expression is constructed from the correspondence principle.

Scale relativistic derivation: the derivation of the Schrödinger equation has been given in Sec. 5.6. It is obtained by (i) constructing a covariant derivative that generalizes the total time derivative in a fractal and non-differentiable space, then by (ii) writing the motion equation as a geodesic equation; (iii) integrating it and writing it in terms of the wave function, which is another expression for the complex action. More generally, one may write, in terms of this "quantum-covariant" derivative, a covariant complex Euler–Lagrange equation derived from the stationary action principle, which is again integrated as the Schrödinger equation, including an exterior potential. The same result can be directly obtained fom the energy equation, which can itself be written in a covariant form (Sec. 5.6.7).

Von Neumann's postulate

If a measurement of the observable A yields some value a_i, the wave function of the system just after the measurement is the corresponding eigenstate ψ_i in the case that a_i is degenerate, the wave function is the projection of ψ onto the degenerate subspace.

Scale relativistic derivation: We have already developed this derivation in Sec. 5.7.2. Let us recall the argument and complete it. The von Neumann postulate is a direct consequence of our identification of a "particle" or of an ensemble of particles with families of geodesics of a fractal and nondifferentiable space-time. The various properties that characterize a quantum state are therefore considered as geometric properties (in position and scale space) of the geodesics, as described by their complex velocity field and therefore by the wave function that plays a role equivalent to a velocity potential. This includes the properties of quantization, which are at the origin of the quantum and particle view of quantum mechanics and which are a consequence of the properties of the equations of quantum mechanics.

Namely, a quantized (or not quantized) energy or momentum is considered as a global conservative geometric property of the geodesic fluid. Now, in a detection process the particle view seems to be supported by the localisation and unicity of the detection. But we consider that this may be a very consequence of the quantization (which prevents a splitting of the energy) and of the interaction process needed for the detection. For example, in a photon by photon experiment, the detection of a photon on a screen means that it has been absorbed by an electron of an atom of the screen, and that the geodesics are, therefore, concentrated in a zone of the order of the size of the atom.

Any measurement, interaction or simply knowledge about the system can be attributed to the geodesics themselves. In other words, the more general set of geodesics, which served to the description of the system before the measurement or knowledge acquisition (possibly without interacting with the system from the view point of the variables considered) is instantaneously reduced to the geodesic sub-set which corresponds to the new state (see also Sec. 5.7.2). For example, the various results of a two-slit experiment — one or two slits opened, detection behind a slit, detection by spin-flip that does not interact with the position and momentum which yields a pure "which-way information", quantum eraser, etc. [175, 484] — can be recovered in the geodesics interpretation (see [353], Sec. 5.5 and Fig. C8 and [360] Sec. 8.2).

Born's postulate: probabilistic interpretation of the wave function

The squared norm of the wave function $|\psi|^2$ is interpreted as the probability of the system of having values (r, s) at time t. This interpretation requires that the sum of the contributions $|\psi|^2$ for all values of (r, s) at time t be finite, i.e. the physically acceptable wave functions are square integrable. More specifically, if $\psi(r, s, t)$ is the wave function of a single particle, $\psi^*(r, s, t)\psi(r, s, t)dr$ is the probability that the particle lies in the volume element dr located at r at time t. Because of this interpretation and since the total probability of finding a single particle at any position is 1, the wave function of this particle must fulfill the normalization condition

$$\int_{-\infty}^{\infty} \psi^*(r, s, t)\psi(r, s, t)dr = 1. \tag{5.283}$$

Scale relativistic derivation: it has been given in Sec. 5.7.3. This derivation relies on the ability in the fluid representation of the geodesic motion, where the two-valuedness is expressed in terms of $P = |\psi|^2$ and of the real velocity field V, to write the equations of dynamics under the form of a Euler equation and of a continuity equation. As recalled in Sec. 5.7.3, the $\text{div}(PV)$ term in the continuity equation finds directly its origin from the Laplacian operator in the Hamiltonian, that itself comes from the free kinetic energy term. The quantum potential in the Euler equation has exactly the same origin.

Therefore, we have found here a potential difference between standard quantum mechanics and the quantum laws deduced from the scale relativity principles. Indeed, in the axiomatic foundation of quantum mechanics, $P = |\psi|^2$ is a universal postulate, which is set totally independently of the form of the Hamiltonian, while we find it, in the scale relativity theory, to be a direct consequence of the quadratic nature of its kinetic energy part. Although the form of the kinetic energy operator seems to be rather general and preserved in generalizations, such as in the rotational motion of solids (see Sec. 5.11.5 hereafter). One may nevertheless wonder whether (possibly effective) other forms could be found in some situations, which could allow to put both theories to the test and to distinguish between them.

5.12.3. *Secondary postulates*

One can find other statements in the literature, which are sometimes presented as "postulates". Let us examine some of them below and show how they are mere consequences of the above listed main postulates, so that we can consider them as demonstrated in the scale relativity framework [399].

Superposition principle

Quantum superposition is the application of the superposition principle to quantum mechanics. It states that a linear combination of state functions of a given physical system is a state function of this system. This principle follows from the linearity of the $\hat{H}$ operator in the Schrödinger equation, which is therefore a linear second order differential equation to which this principle applies.

Eigenvalues and eigenfunctions

Any measurement of an observable A will give as a result one of the eigenvalues a of the associated operator $\hat{A}$, which satisfy the equation

$$\hat{A}\psi = a\psi. \tag{5.284}$$

Proof: The correspondence principle allows one to associate to every observable an Hermitian operator acting on the wave function. Since these operators are Hermitian, their eigenvalues are real numbers, and such is the result of any measurement. This is a sufficient condition to state that any measurement of an observable A will give one of the eigenvalues a as a result of the associated operator $\hat{A}$. Now, we need to prove that it is also a necessary condition.

We consider the Hamiltonian operator first, assuming that the classical definition of its kinetic energy part involves only terms, which are quadratic in the velocity. The Schrödinger equation can be written

$$ i\hbar\frac{\partial\psi}{\partial t} = \hat{H}\psi. \tag{5.285} $$

First let us limit ourselves to the case where the potential ϕ is everywhere zero (free particle). The above Eq. (5.285) becomes

$$ i\hbar\frac{\partial}{\partial t}\psi(r,t) = -\frac{\hbar^2}{2m}\Delta\psi(r,t). \tag{5.286} $$

This differential equation has solutions of the kind

$$ \psi(r,t) = Ae^{i(k.r-\omega t)}, \tag{5.287} $$

where A is a constant and k and ω verify

$$ \omega = \frac{\hbar k^2}{2m}. \tag{5.288} $$

We now apply to the expression of ψ in Eq. (5.287) the operators $\hat{P} = -i\hbar\nabla$ and $\hat{E} = i\hbar\partial/\partial t$ and we obtain

$$ \hat{P}\psi(r,t) = -i\hbar\nabla\psi(r,t) = \hbar k\psi(r,t) \tag{5.289} $$

and

$$ \hat{E}\psi(r,t) = i\hbar\frac{\partial}{\partial t}\psi(r,t) = \hbar\omega\psi(r,t). \tag{5.290} $$

An inspection of Eqs. (5.288) to (5.290) shows that the eigenvalues $\hbar k$ of $\hat{P}$ and $\hbar\omega$ of $\hat{E}$ are related in the same way as the momentum p and the energy E in classical physics, i.e. $E = p^2/2m$. Owing to the correspondence principle, we can assimilate $\hbar k$ to a momentum p and $\hbar\omega$ to an energy E, thus recovering the de Broglie relations $p = \hbar k$ and the Einstein relation

$E = \hbar\omega$. Moreover, since Eqs. (5.289) and (5.290) can be rewritten

$$\hat{P}\psi(r,t) = p\psi(r,t) \tag{5.291}$$

and

$$\hat{E}\psi(r,t) = E\psi(r,t), \tag{5.292}$$

we have shown that, in the case where the classical definition of the (free) Hamiltonian writes $H = P^2/2m$, the measurement results of the P and E observables are eigenvalues of the corresponding Hermitian operators.

As regard the position operator, this property is straightforward since the application of the correspondence principle yields that to r corresponds the operator "multiply by r". We, therefore, readily obtain

$$\hat{R}\psi = r\psi, \tag{5.293}$$

which implies that r is actually the eigenvalue of $\hat{R}$ obtained when measuring the position.

Implementing the correspondence principle, these results can easily be generalized to all other observables, which are functions of r, p and E.

Expectation value

For a system described by a normalized wave function ψ, the expectation value of an observable A is given by

$$\langle A \rangle = \int_{-\infty}^{\infty} \psi^* \hat{A}\psi \, dr. \tag{5.294}$$

This statement follows from the probabilistic interpretation attached to ψ, i.e. from Born's postulate (for a demonstration see, e.g. [67]).

Expansion in eigenfunctions

The set of eigenfunctions of an operator $\hat{A}$ forms a complete set of linearly independent functions. An arbitrary state ψ can therefore be expanded in the complete set of eigenfunctions of $\hat{A}$ ($\hat{A}\psi_n = a_n\psi_n$), i.e. as

$$\psi = \sum_n c_n \psi_n, \tag{5.295}$$

where the sum may go to infinity. For the case where the eigenvalue spectrum is discrete and non-degenerate and where the system is in the normalized state ψ, the probability to obtain as a result of a measurement of A the eigenvalue a_n is $|c_n|^2$. This statement can be straightforwardly generalized to the degenerate and continuous spectrum cases.

Another more general expression may be given of this postulate: an arbitrary wave function can be expanded in a complete orthonormal set of eigenfunctions ψ_n of a set of commuting operators A_n. It reads

$$\psi = \sum_n c_n \psi_n, \tag{5.296}$$

while the statement of orthonormality reads

$$\sum_s \int \psi_n^*(r, s, t) \psi_m(r, s, t) dr = \delta_{nm}, \tag{5.297}$$

where δ_{nm} is the Kronecker symbol.

Proof: Hermitian operators are known to exhibit the two following properties: (i) two eigenvectors of an Hermitian operator corresponding to two different eigenvalues are orthogonal; (ii) in an Hilbert space with finite dimension N, an Hermitian operator always possesses N linearly independent eigenvectors. This implies that, in such a finite dimensional space, it is always possible to construct a base with the eigenvectors of an Hermitian operator and to expand any wave function in this base. However, when the Hilbert space is infinite this is no more mandatorily the case. This is the reason why one introduces the observable tool. An Hermitian operator is defined as an observable if its set of orthonormal eigenvectors is complete, i.e. if it determines a complete base for the Hilbert space (see e.g. [106]).

The probabilistic interpretation attached to the wave function (Born's postulate) implies that, for a system described by a normalized wave function ψ, the expectation value of an observable A is given by

$$\langle A \rangle = \int_{-\infty}^{\infty} \psi^* \hat{A} \psi \, dr = \langle \psi | \hat{A} | \psi \rangle. \tag{5.298}$$

Expanding ψ in a complete eigenfunction set of A or in a complete eigenfunction set of commuting operators, $\psi = \sum_n c_n \psi_n$, where the c_n's are complex numbers, yields

$$\langle A \rangle = \sum_m \sum_n c_m^* c_n \langle \psi_m | \hat{A} | \psi_n \rangle = \sum_m \sum_n c_m^* c_n a_n \langle \psi_m | \psi_n \rangle = \sum_n |c_n|^2 a_n, \tag{5.299}$$

since, from orthonormality, $\langle \psi_m | \psi_n \rangle = \delta_{mn}$.

Assuming that ψ is normalized, i.e. $\langle\psi|\psi\rangle = 1$, we can write

$$\sum_n |c_n|^2 = 1. \tag{5.300}$$

From the eigenvalue secondary postulate, the results of measurements of an observable A are the eigenvalues a_n of $\hat{A}$. Since the average value obtained from series of measurements of a large number of identically prepared systems i.e. all prepared in the same state ψ, is the expectation value $\langle A \rangle$, we are led, following the Born postulate, to identify the quantity

$$P_n = |c_n|^2 = |\langle\psi_n|\psi\rangle|^2 \tag{5.301}$$

with the probability that, in a given measurement of A, the value a_n would be obtained [67].

Note that, in deriving these results we have implicitly included the degeneracy index in the summations. A generalization to a degenerate set of eigenvalues and to a continous spectrum is straightforward [67].

It is worth stressing here again that this secondary postulate is readily derived from Born's postulate and the superposition principle, which is itself a mere consequence of the linearity of the $\hat{H}$ operator and of the Schrödinger equation. Therefore it is not an actual founding postulate even if it is often presented as such in the literature (see e.g. [106]).

Probability conservation

The probability conservation is a consequence of the Hermitian property of $\hat{H}$ [106]. It first implies that the norm of the state function is time independent and it also implies a local probability conservation, which can be written, e.g. for a single particle without spin and with normalized wave function ψ, as

$$\frac{\partial}{\partial t}\rho(r,t) + \mathrm{div}\,J(r,t) = 0, \tag{5.302}$$

where

$$J(r,t) = \frac{1}{m}\Re\left[\psi^*\left(\frac{\hbar}{i}\nabla\psi\right)\right]. \tag{5.303}$$

Reduction of the wave packet or projection hypothesis

This statement does not need to be postulated since it can be deduced from other postulates (see e.g. [35]). It is actually implicitly contained in von Neumann's postulate.

5.12.4. *Derived principles*

Heisenberg's uncertainty principle

If P and Q are two conjugate observables such that their commutator equals $i\hbar$, it is easy to show that their standard deviations ΔP and ΔQ satisfy the relation

$$\Delta P \Delta Q \geq \frac{\hbar}{2}, \qquad (5.304)$$

whatever the state function of the system [267, 319, 106, 67]. This applies to any couple of linear (but not mandatorily Hermitian) operators and, in particular, to the couples of conjugate variables: position and momentum, time and energy. Moreover, we have recalled at the beginning of this chapter (Sec. 5.2) how one may obtain generalized Heisenberg relations for any couple of variables [178].

The spin-statistic theorem

When a system is composed of many identical particles, its physical states can only be described by state functions, which are either completely antisymmetric (fermions) or completely symmetric (bosons) with respect to permutations of these particles, or, identically, by wave functions that change sign in a spatial reflection (fermions) or that remain unchanged in such a transformation (bosons). All half-spin particles are fermions and all integer-spin particles are bosons (see the next chapter for the construction of spin and spinors from the nondifferentiable geometry and for the derivation of the Dirac and Pauli equations in the scale relativity theory).

Demonstrations of this theorem have been proposed in the framework of field quantum theory as originating from very general assumptions. The usual proof can be summarized as follows: one first shows that if one quantizes fermionic fields (which are related to half-integer spin particles) with anticommutators one gets a consistent theory, while if one uses commutators, it is not the case; the exact opposite happens with bosonic fields, which correspond to integer spin particles, one has to quantize them with commutators instead of anticommutators, otherwise one gets an inconsistent theory. Then, one shows that the anti-commutators are related to the anti-symmetry of the wave functions in the exchange of two particles [267]. However, this proof has been claimed to be incomplete [106], but more complete ones have subsequently been proposed [177, 547].

The Pauli exclusion principle

Two identical fermions cannot be in the same quantum state. This is a mere consequence of the spin-statistic theorem.

This closes the list of quantum mechanical "postulates" and "principles". As we have seen, the main ones can be derived from the first principles of the scale relativity approach, which are themselves reformulations and extension of the standard relativity principles (up to now applied to position, orientation and motion, and now to scales). Then the secondary ones and the derived principles can be proved in the framework so constructed.

Now, despite this success, one cannot yet claim that we have founded the true quantum mechanics, which is really experimented in nature. This would be actually impossible (and this impossibility also applies to the present axiomatic theory of quantum mechanics), since all we can do is to attempt to falsify the theory by experiments. Moreover, the fact that we recover the postulates of standard quantum mechanics does not prove that some other consequences of the founding principles of the new theory could not contradict it. If inconsistencies were to appear, they could either be at variance with already known experimental results, in which case this would rule out the application in its present form of the theory to standard quantum mechanics, or, more interestingly, it could be used to put the scale relativity theory and also standard QM to the test. These questions will be considered in more detail in Chapter 10.

Problems and Exercises

Open Problem 8: Find the scale relativistic analog of a density matrix description of a quantum system.

Hint: The introduction of a density matrix in quantum mechanics corresponds to a situation where the knowledge of the state of a system is incomplete. One assumes that a system, described by coordinates $\{x\}$, is a part of a closed system, which is in a state given by the wave function $\psi(q, x)$. Since, in general, this wave function cannot be decomposed under a product of functions of x and q separately, the subsystem under consideration has no wave function. It may, however, be described in the most complete possible way by a density matrix that reads (see e.g. [267, 474])

$$\rho(x', x, t) = \int \psi^*(q, x', t)\psi(q, x, t)dq. \tag{5.305}$$

Its motion equation is, in the Schrödinger picture,

$$i\hbar\frac{d\rho}{dt} = [H, \rho].$$ (5.306)

The density matrix or "density operator" describes a statistical state instead of a pure state, and it is, therefore, the quantum analog of the classical density function. The classical equation that corresponds to Eq. (5.306) is known as Liouville's theorem [474]. It reads

$$\frac{\partial\rho}{\partial t} = \{H, \rho\}.$$ (5.307)

A possible way to tackle this problem in the scale relativity framework consists of deriving a generalized version of this equation in a nondifferentiable space-time, then of finding the form of the velocity fields and of their relation to the density matrix. ∎

Exercise 15 Develop the Heisenberg representation and the matrix formulation of quantum mechanics from its scale relativistic foundation. Show that one may recover the various commutation relations of standard quantum mechanics, and that the origin of the noncommutativity can be found in the second order term of the total covariant derivative and in the subsequent difference with the first order Leibniz rule.

Hint: Found the Schrödinger picture from the scale relativity principles, then use the equivalence between the Schrödinger and Heisenberg representations. Then translate the various matrix tools of quantum mechanics in terms of the geometric tools of scale relativity (see e.g. [447]). ∎

5.13. Quantum Entanglement in Scale Relativity

The term "entanglement" has been coined by Schrödinger to describe a quantum phenomenon he considered to be "the characteristic trait of quantum mechanics [...] that enforces its entire departure from classical lines of thought" [480]. Quantum entanglement, therefore, deserves a specific study, even though, having now derived the basic postulates of quantum mechanics from the principle of relativity including states of position, orientation, motion and scale of the reference system, it is naturally accounted for in the scale relativity approach since it relies on these postulates and is not separated from them. Our aim in this section is then to show how the scale relativity and nondifferentiable space-time description allows one to get an intuitive geometric view of entangled quantum states.

Let us briefly recall the nature of quantum entanglement. Schrödinger defines it in the following way [480].

"When two systems, of which we know the states by their respective representatives, enter into temporary physical interaction due to known forces between them, and when after a time of mutual influence the systems separate again, then they can no longer be described in the same way as before, viz. by endowing each of them with a representative of its own. I would not call that *one* but rather *the* characteristic trait of quantum mechanics, the one that enforces its entire departure from classical lines of thought. By the interaction the two representatives [the quantum states] have become entangled."

A typical example of an entangled system is given by a pair of spin one-half particles, which are created in an anticorrelated state such that $s_1 + s_2 = 0$, and which are emitted in opposite directions in space. If the spin of particle 1 is measured to be $s_1 = +1/2$ (respectively $-1/2$), the value of the spin of particle 2 becomes instantaneously $s_2 = -1/2$ (respectively $+1/2$), as can be proved by an effective measurement. This is achieved even when the particles are so distant and the two measurements so close in time that if there had been some transfer of information from one particle to the other, it would have to travel faster than light, which is excluded by special relativity. The puzzle comes from the fact that, before the measurement, neither s_1 nor s_2 do have determined values. Moreover, this undetermination of the individual spins is not a matter of incomplete knowledge about the system (as in 'hidden variables' theories), they actually do not exist as individual and intrinsic properties of each of the particles taken separately. This has been proved experimentally [32] by the explicit violation of the Bell inequalities [41], which would hold in the hidden parameter case.

The essence of entangled states is, therefore, the existence of global states for an ensemble of "particles" that would be classically considered to be separated systems. It therefore points out the fundamental non locality and inseparability properties of quantum systems.

Let us now show how quantum entanglement naturally comes under the scale relativity geometric description. In the relativistic viewpoint developed here, this behavior has nothing special nor incomprehensible. Indeed, in theories of relativity, the various physical quantities do not exist as intrinsic properties of objects, but only as relative properties of pairs, which cannot be attributed to any of the members of the pair. Consider for example two identical classical bodies with relative motion in empty space. The velocity is a property belonging to neither of the bodies taken individually, only the inter-velocity between the bodies does have a physical meaning. The same is therefore true of the momentum and of the kinetic energy. The individual energy is not an intrinsic property of a body, since it depends on the reference system. The total kinetic energy of the two bodies is clearly something that lies neither in the first body, nor in the second, nor in both of them, since it also depends on the reference system. It is not only a non local property, but even a non localized property. This is a general property of conservative quantities, whose status is that of global invariants of a system, not of localized characteristics intrinsic to an object.

In order to illustrate this point, let us consider a classical restricted three-body problem ($m_1 = m_2 \ll M$). For example, the relative motion of two Saturn satellites of similar masses on close orbits is known to be chaotic [444]. Beyond some rather short interval of time, it becomes therefore impredictible. As regards now the motion of their center of gravity, this is a two-body problem ($m \ll M$) which is not chaotic. From any initial conditions, the position of the center of gravity can then be predicted with a good precision. Assume now that the two satellites have been observed at some date, then have no longer been followed. There is no way to predict their position from theory after time intervals larger than the predictability horizon. But if one now observes one of the satellites, the position of the other can be immediately derived from the theoretical knowledge of the position of their center of gravity, and the second satellite will be immediately discovered at the expected position. There is no need for any information to go from one satellite to the other: the conservation law of the center of gravity is simply true at any time.

Even though this example from classical chaos shares some common points with an EPR experiment [160], there is an essential difference. Namely, this is a hidden parameter situation where, despite the absence of knowledge of the positions of the satellites, these positions do exist, as could have been checked by actual observations, which would not have perturbed the position state of the satellites. Therefore, in this case the probabilities remain classical and Bell's inequalities are satisfied.

On the contrary, there is no existence of the spins s_1 and s_2 separately before their measurement in the quantum entanglement case, while the conservation law $s_1 + s_2 = 0$ is true during the whole experiment, even though the determined physical quantity $s = s_1 + s_2$ cannot be attributed to any one of the particles individually and, more generally, cannot be considered as being localized.

All these features are easily understandable in terms of the geometric pictures of the scale relativity theory (see Sec. 5.10). Indeed, in this approach, we give up the concept of "particles" that would follow "trajectories". We identify the quantum mechanical "particles", described in standard quantum mechanics by a wave function, with the set of geodesics of the fractal space that corresponds to given conservative quantities. The complex action $\mathcal{S}$, and therefore the wave function $\psi = \exp(i\mathcal{S}/\mathcal{S}_0)$, which is a re-expression of this action, is unique for all the particles of the ensemble, i.e. for the whole set of geodesics. But when coming back to the velocity fields of (on) these geodesics, they may be partly separated in subsets characterized by different geometric properties, in particular, the mass, according to the relation $\mathcal{V}_k = -2i\mathcal{D}_k\nabla_k \ln \psi$, thus defining different "particles" from the same wavefunction. This is the basic understanding of quantum entanglement in the scale relativity framework: although different particles may be defined as such subsets of the whole geodesics ensemble, some of their properties may not be separated and may belong to the ensemble, but without existing at the level of the individual particles (geodesic subsets).

The conservative quantities, which define the particles are identified to purely geometric properties of the geodesics. For example, the mass is a manifestation of the fractal fluctuations, namely, $m = \hbar \, dt/\langle d\xi^2 \rangle$. Particles of different masses mean subsets of geodesics characterized by different mean fractal fluctuation amplitudes, $\langle d\xi_k^2 \rangle = (\hbar/m_k)dt$. Their velocity fields writes, in terms of these fluctuations, $\mathcal{V}_k = -i(\langle d\xi_k^2 \rangle/dt)\nabla_k \ln \psi$. The energy-momentum is given by the transition from scale dependence to effective scale independence in scale space, identified with the de Broglie scale, namely, $p^\mu = \hbar/\lambda^\mu$. The spin is an intrinsic angular momentum of the fractal geodesics (see Chapter 6). The charges are the conservative quantities that appear, according to Noether's theorem, from their internal scale symmetries (see Chapter 7). The existence of quantas, upon which the concept of "particle" is based, is just a consequence of the quantization of these geometric properties, which is itself derived from the properties of the geodesic equation.

Due to nondifferentiability, the set of geodesics contains always an infinity of fractal curves and fills space like a fluid (in analogy with Feynman's path integral approach [176]). There is no way to specify or to choose a given geodesic among their infinite number, since all of them minimize the proper time and are, therefore, physically equivalent and indistinguishable. The wave function is just another expression for the fluid velocity field, since it can actually be identified with a potential of velocity. A single trajectory has no meaning in the theory. Even in the case of the detection of a "particle" in a given region of space, which seems to support the "particle" interpretation of quanta, this region can never be reduced to a point, since any measurement is accompanied by its finite resolution.

This description is therefore fundamentally non local, and properties like entanglement and indistinguishability of identical particle (Sec. 5.10), etc. naturally follow.

5.14. Generalized Schrödinger Equations Founded on Extended Scale Laws

In the whole of the present chapter, we have considered only the simplest possible scale law underlying the construction of the motion equation. Namely, all the previous work relies on the description of the fractal fluctuation as

$$d\xi^2 = 2\mathcal{D}\zeta^2|dt|, \tag{5.308}$$

which expresses the fractal dimension $D_F = 2$ of the paths. This dimension corresponds to a full loss of information for the elementary displacements, which are neither correlated nor anticorrelated, in agreement with Feynman's discovery that the typical quantum mechanical paths (those which mainly contribute to the path integral) are of fractal dimension 2 [176]. It is therefore only under this condition that one may recover standard quantum mechanics.

Even in this case, we have shown that the theory so constructed was wider than standard QM, since it relies on a parameter $S_0 = 2m\mathcal{D}$, which only needs to be constant for the system under consideration to obtain a self-consistent theory. The case of standard quantum mechanics is strictly recovered only when S_0 takes the microscopic value $S_0 = \hbar$, so that this result already opens the possibility to have, for some particular systems and under some particular conditions of fractality and irreversibility, macroscopic quantum-like behaviors. This possibility has been briefly considered in Sec. 5.11 and it will be the subject of Chapter 10 and of many proposals of applications (Part III).

However, we have shown in Chapter 4 that a constant fractal dimension was a very special case, which we have called "scale-Galilean", of scale laws coming under the principle of scale relativity. In particular, we have suggested that the scale Galilean-like laws of constant fractal dimensions may be a large scale limit of more general special scale relativity laws taking a log-Lorentz form. In such a framework, one should make the difference between an invariant fractal dimension that keeps the value $D_F = 2$, thus still ensuring the independence between the successive elementary displacements and the apparent fractal dimension (that we have called "djinn"), which is a projected quantity and, therefore, becomes variable with scale (in analogy with the difference between the invariant proper time and the time coordinate in special motion relativity).

We shall now recall how one can relax some of the assumptions that lead to the standard quantum mechanics and obtain in this way generalized Schrödinger equations that go beyond this theory [358, 360]. This approach also allows one to understand why the fractal to nonfractal transition (identified here with the quantum to classical transition) is so fast, and why the fractal dimension jumps directly from $D_F = 1$ to $D_F = 2$. This results from the fact that the generalized Schrödinger equation one may construct for $1 < D_F < 2$ is degenerate and unphysical, in the case when one would want to apply it to an all-scale description.

Let us come back to the origin of our description. We still assume that space is continuous and we relax the usual hypothesis that it is differentiable. The geodesics of a nondifferentiable continuum are also expected to be continuous, nondifferentiable and in infinite number between any couple of points. This can be expressed by describing the position vector of a particle by a finite, continuous fractal $x(t, dt)$, explicitly dependent on the time differential element dt. Nondifferentiability also implies that the variation of the position vector between $t - dt$ and t and between t and $t + dt$ is described by two *a priori* different processes:

$$x(t + dt, dt) - x(t, dt) = v_+(x, t)dt + \zeta_+(t, dt)(dt/\tau_o)^{1/D_F}, \quad (5.309)$$

$$x(t, dt) - x(t - dt, dt) = v_-(x, t)dt + \zeta_-(t, dt)(dt/\tau_o)^{1/D_F}, \quad (5.310)$$

where D_F is the fractal dimension of the paths. Under this form, the variables $\zeta_\pm$ are finite velocity fluctuations described by stochastic variables. The last terms in these equations can also be written in terms of fluctuations $d\xi_{\pm i}$, which are of zero mean, $\langle d\xi_{\pm i} \rangle = 0$, and which satisfy

$$\langle d\xi_{\pm i} d\xi_{\pm j} \rangle = 2\mathcal{D}_{ij}(dt^2)^{1/D_F}. \quad (5.311)$$

Up to now, we have mainly considered the case $\mathcal{D}_{ij} = \mathcal{D}\delta_{ij}$, with $\mathcal{D}$ constant and related to the Compton length of the particle, i.e. it reads in the standard QM case,

$$\mathcal{D} = \frac{\hbar}{2m} = c\lambda_c/2. \tag{5.312}$$

The case of a tensorial $\mathcal{D}_{ij}$ has also been considered for the quantization of the rotational motion of solids (Sec. 5.11.5).

But it is clear from the above form of the fractal fluctuations that two generalizations of this expression may be considered: (i) fractal dimensions different from $D_F = 2$, either constant, or variable (see Chapter 4); (ii) fluctuation tensor $\mathcal{D}_{ij}$ no longer constant, but varying with position, time, and/or scale. In particular, such a case must be considered in applications of this approach to chaotic dynamics on time scale longer than the predictability horizon (see Part III).

The general problem of tackling these generalizations properly and in full detail and of studying their properties and consequences lies outside the scope of the present book. It is actually a huge domain of research, as can be seen from the various proposals of generalized scale laws made in Chapter 4. For each of these laws, considered as describing the underlying fractal structure of geodesics, one would need to reconsider and to generalize the whole construction of the present chapter.

Such a problem is left open for the moment. We shall instead consider a perturbative approach, in which one can still use the above construction as a basis and then describe the generalized behavior in terms of small differences.

We shall, therefore, only consider the case where the constant or variable fractal dimension D_F does not depart too much from $D_F = 2$. Indeed, in this case its deviation from 2 can be approximated in terms of an explicit scale dependence on the time resolution, as first noticed by Mandelbrot and Van Ness [304]. Namely we write the fractal fluctuation under the form

$$\langle d\xi_{\pm i}d\xi_{\pm j}\rangle = \pm 2\mathcal{D}_0(x,t)\delta_{ij}dt \times \left(\frac{|dt|}{\tau_o}\right)^{\frac{2}{D_F}-1}, \tag{5.313}$$

where τ_o is some characteristic time scale. Hence the effect of $D_F \neq 2$ can be dealt with in terms of a generalized, scale-dependent, parameter

$$\mathcal{D}(x,t,dt) = \mathcal{D}_0(x,t)\left(\frac{|dt|}{\tau_o}\right)^{\frac{2}{D_F}-1}. \tag{5.314}$$

The various steps of the construction of the covariant derivative are therefore left unchanged, provided the fractal dimension remains smaller than $D_F = 3$, which is implicit in the perturbative approach adopted here. Indeed, when it reaches 3, a new contribution from a third order derivative term is expected (see Sec. 5.4.3). The complex total derivative operator is found to be given by the same expression as in the case $D_F = 2$, namely

$$\frac{\hat{d}}{dt} = \frac{\partial}{\partial t} + \mathcal{V} \cdot \nabla - i\mathcal{D}(x, t, dt)\Delta, \qquad (5.315)$$

but with the parameter $\mathcal{D}$ being now a function of position, time and scale. This can be easily generalized to the complete covariant derivative by replacing the differentiable part $\mathcal{V}$ of the complex velocity field by the full velocity field $\tilde{\mathcal{V}}$ and to a tensorial parameter by replacing $\mathcal{D}\Delta$ by $\sum_{jk} \mathcal{D}_{jk}\partial_j\partial_k$. Let us now specialize our study.

Scale dependent fractal fluctuation parameter

We first consider a parameter $\mathcal{D}$, which is scale-dependent, but does not depend on space and time coordinates, namely, $\mathcal{D} = \mathcal{D}(dt) = \mathcal{D}_0(dt/\tau_o)^{(2/D_F)-1}$. This case includes the scale-dependent effect of a fractal dimension different from 2.

It also includes, more generally, variable fractal dimensions, such as in special scale relativity, where it takes a log-Lorentz form (see Sec. 4.4.5), namely,

$$D_F(dt) = 1 + \frac{1}{\sqrt{1 - \ln^2(\tau_o/dt)/\ln^2(\tau_o/\tau_{\mathbb{P}})}}. \qquad (5.316)$$

It jumps from $D_F = 1$ to $D_F = 2$ at the transition scale $\tau_o = \hbar/mc^2$, then it increases toward the small time scales, becoming divergent when the time interval dt tends to the Planck time scale $\tau_{\mathbb{P}}$. In this case the fluctuation parameter reads

$$\mathcal{D}(dt) = \mathcal{D}_0 \left(\frac{dt}{\tau_o}\right)^{\frac{2}{D_F(dt)}-1}, \qquad (5.317)$$

i.e.

$$\mathcal{D}(dt) = \mathcal{D}_0 \left(\frac{dt}{\tau_o}\right)^{\frac{-1+\sqrt{1-\mathbb{V}^2/\mathbb{C}^2}}{1+\sqrt{1-\mathbb{V}^2/\mathbb{C}^2}}}, \qquad (5.318)$$

where $\mathbb{V}(dt) = \ln(\tau_o/dt)$ and $\mathbb{C} = \ln(\tau_o/\tau_{\mathbb{P}})$. Expanding this relation to lowest order (when $\mathbb{V} \ll \mathbb{C}$, i.e. when the time scale remains far larger than the Planck time scale), one finds

$$\mathcal{D}(dt) = \mathcal{D}_0 \left(1 + \frac{1}{4} \frac{\mathbb{V}(dt)^3}{\mathbb{C}^2} + \cdots \right). \tag{5.319}$$

The scale dependence of the fluctuation parameter does not interfere with the various steps of the demonstration of the Schrödinger equation. The wave function may still be defined as

$$\psi(x, t, dt) = e^{\mathcal{S}/2m\mathcal{D}(dt)}. \tag{5.320}$$

It simply becomes an explicit function of dt, in continuity with the case when we used the full velocity field to define it, which already led to the construction of a fractal and nondifferentiable wave function. The next steps of the demonstration of the Schrödinger equation are preserved, so that one finally obtains

$$\mathcal{D}^2(dt)\Delta\psi + i\mathcal{D}(dt)\frac{\partial\psi}{\partial t} = \frac{\phi}{2m}\psi. \tag{5.321}$$

The form of the Schrödinger equation is preserved, but with a scale dependent $\hbar$-like parameter. This is possible provided the wave function ψ is itself a scale dependent function, possibly fractal, i.e. scale divergent, which is ensured if one considers that it describes the complete geodesic velocity field and not only its mean.

The behavior of this equation is in agreement with the underlying stochastic process being no longer Markovian for $D_F \neq 2$. The "ultraviolet" behavior ($dt \ll \tau_o$) and "infrared" behavior ($dt \gg \tau_o$) are reversed between the cases ($D_F < 2$) and ($D_F > 2$), but anyway they correspond to only two possible asymptotic behaviors, namely, $\mathcal{D} \to \infty$ and $\mathcal{D} \to 0$. Let us consider these asymptotic behaviors in more detail.

(i) $\mathcal{D} \to \infty$ [UV $D_F > 2$; IR $D_F < 2$]: in this case Eq. (5.321) is reduced to $\Delta\psi = 0$, i.e. to the equation of a sourceless, stationarity probability amplitude. Whatever the field described by the potential ϕ, it is no longer "felt" by the particle: this is nothing but the property of asymptotic freedom, which is already provided in quantum field theories by non-Abelian fields. This remark is particularly relevant for $D_F > 2$, since in this case this is the UV behavior, i.e. toward small length-scale and high energy, while we have been led, in the special scale relativity case, to introduce a scale-dependent generalized fractal dimension $D_F(dt) > 2$ (Eq. (5.319)) for virtual quantum trajectories considered at time scales smaller than the Einstein scale $\tau_o = \hbar/mc^2$.

(ii) $\mathcal{D} \to 0$ [UV $D_F < 2$; IR $D_F > 2$]: in this case Eq. (5.321) becomes completely degenerate ($\phi\psi = 0$). Physics seems to be impossible under such a regime. This result is also in agreement with what is known about the quantum-classical transition. Indeed one finds, when trying to translate the behavior of typical quantum paths in terms of fractal properties [353], that their fractal dimension quickly jumps from $D_F = 2$ (quantum) to $D_F = 1$ (classical) when the resolution scale becomes larger than the de Broglie scale. This fast transition actually prevents the domain $1 < D_F < 2$ to be achieved in nature (concerning the description of fractal space-time).

However the conclusion that we reached in [358] must be now somewhat moderated. It applies, strictly, to D_F very different from 2 and then indeed to the quantum/classical transition. But if one considers only a small perturbation to $D_F = 2$, then the asymptotic, degenerate behavior described above occurs only at very small scales, while one obtains at intermediate scales a new behavior of the generalized Schrödinger equation, namely, an explicit scale-dependence of this equation and of its solutions.

Finally, since the scale dependent parameter $\mathcal{D}$ is given by $\mathcal{D}(dt) = \mathcal{D}_0(dt/\tau_o)^{2/D_F-1}$, one sees that one recovers the standard Schrödinger equation for $D_F = 2$, as expected. This proves that the special value $D_F = 2$ of the fractal dimension of paths plays a critical role, since it suppresses an explicit scale dependence from the Schrödinger equation. The scale dependence therefore becomes hidden in the equation for this value and it reappears in a way, which is only indirect through, e.g. the Heisenberg relations.

Fractal fluctuation parameter function of coordinates

In the case where the parameter $\mathcal{D}$ is a function of space and time coordinates, one can no longer follow the various steps of the demonstration of the Schrödinger equation in the same manner as previously. Indeed, this parameter intervenes in the construction of the wave function, which was up to now defined as $\psi = e^{iS/2m\mathcal{D}}$. This is no longer possible. We shall, therefore, once again adopt a perturbative approach. Namely, we define an average value of the diffusion coefficient, $\langle \mathcal{D} \rangle$. We assume that this mean value is constant respectively to variables x and t, but that it may include an explicit scale-dependence in terms of δt and we write

$$\mathcal{D}(x,t) = \langle\mathcal{D}\rangle + \delta\mathcal{D}(x,t). \tag{5.322}$$

Then we introduce the complex function ψ from the relation

$$\psi = e^{i\mathcal{S}/2m\langle\mathcal{D}\rangle}. \tag{5.323}$$

Then ψ is linked to the complex velocity by the relation

$$\mathcal{V} = -2i\langle\mathcal{D}\rangle\nabla\ln\psi, \tag{5.324}$$

so that the equation of dynamics reads

$$2im\langle\mathcal{D}\rangle\frac{\hat{d}}{dt}(\nabla\ln\psi) = \nabla\phi. \tag{5.325}$$

After expansion, it can be written under the form [358, 360]

$$\nabla\left\{\frac{\phi}{2m\langle\mathcal{D}\rangle} - \frac{1}{\psi}\left(\mathcal{D}\Delta\psi + i\frac{\partial\psi}{\partial t}\right) + \delta\mathcal{D}(\nabla\ln\psi)^2\right\} = -\nabla(\delta\mathcal{D})\Delta\ln\psi. \tag{5.326}$$

The general study of this equation is left open to future studies. We shall only consider here the special cases when it can be integrated as a (possible nonlinear) Schrödinger equation, i.e. $\nabla(\delta\mathcal{D}) = 0$ or $\nabla(\delta\mathcal{D}) \ll 1$.

In this case, that corresponds either to a slowly varying diffusion coefficient in the domain considered, or, at the limit, to a parameter $\mathcal{D}$ depending on time but not on position, the right-hand side of Eq. (5.326) vanishes, so that it may be integrated, yielding

$$\mathcal{D}(x,t)\Delta\psi + i\frac{\partial\psi}{\partial t} = \left\{\frac{\phi}{2m\langle\mathcal{D}\rangle} + \delta\mathcal{D}(x,t)(\nabla\ln\psi)^2\right\}\psi, \tag{5.327}$$

up to a constant of integration that can be absorbed in a redefinition of the potential ϕ. This equation has kept the form of a nonlinear Schrödinger equation, but with a variable coefficient $\mathcal{D}$, which generalizes $\hbar/2m$, in factor of the Laplacian term implying that the Born relation $P = |\psi|^2$ may not be always fulfilled for such an equation.

Let us specialize again and assume that $\delta\mathcal{D}/\mathcal{D} \ll 1$. Then the effect of the term $\delta\mathcal{D}\psi(\nabla\ln\psi)^2$, which is in addition to the standard Schrödinger

Scale Relativity and Fractal Space-Time

equation, and the effect of $\mathcal{D}$ being a function of x and t can be treated perturbatively. One obtains the Schrödinger equation

$$\langle \mathcal{D} \rangle^2 \Delta \psi + i \langle \mathcal{D} \rangle \frac{\partial \psi}{\partial t} = \left\{ \frac{\phi(x,t)}{2m} - \langle \mathcal{D} \rangle [\Delta \ln \psi]_0 \delta \mathcal{D}(x,t) \right\} \psi, \qquad (5.328)$$

for which the Born postulate is now ensured. Hence the effect of a variable parameter $\mathcal{D}$ amounts to adding new terms to the potential in the standard Schrödinger equation. In particular, in the case where there is no applied external potential ϕ, one sees that the varying part of the fluctuation parameter, $\delta \mathcal{D}(x,t)$, plays the role of such a potential up to a multiplicative constant. This is therefore a first example of the appearance of an apparent scalar potential from the fractal geometry itself (in analogy with Newton's potential appearing in general relativity as a manifestation of the curved geometry of space-time). Such a behavior anticipates on the construction of gauge fields from scale relativity and fractal space-time, which will be the subject of Chapter 7.

Chapter 6

FROM FRACTAL SPACE-TIME TO RELATIVISTIC
QUANTUM MECHANICS

6.1. Introduction

In Chapter 5 we considered a mere fractal space, in which time was not yet affected by the fractal geometry. As we have seen in detail, the underlying nondifferentiability and fractality of space allows one to set the foundations of non-relativistic quantum mechanics. This means that, in that case, time plays the role of an invariant scalar, in agreement with Galilean motion relativity.

However, if one now considers increasing velocities and energies, one skips both to a special relativistic description of motion and to decreasing scales. The de Broglie quantum to classical transition, which was well approximated in the non-relativistic case by $\lambda_{\mathrm{dB}} = \hbar/mv$ should be replaced by its more correct expression, $\lambda_{\mathrm{dB}} = \hbar/p = \hbar\sqrt{1 - v^2/c^2}/mv$. When the resolution of space measurements reaches the Compton scale, $\delta x \sim \lambda_c = \hbar/mc$ and when that of time measurements reaches the Einstein scale, $\delta t \sim \tau_{\mathrm{E}} = \hbar/mc^2$, the energy fluctuations become of the order of particle masses and can therefore create virtual particle-antiparticle pairs.

This pair creation can be interpreted [420, 349, 353] as a transition to the fractality of time itself. Indeed, when $\delta t < \tau_{\mathrm{E}}$, the proper time becomes fractal and scale divergent. As a consequence, in the laboratory coordinate system, the particle is seen as running backward in time on time scales smaller than $\hbar/2mc^2$ (the factor 2 comes from the mass of the pair). The reason for the fundamentally different status of time with respect to space from the view point of scales is the existence of mass in the energy relation

$$E^2 = p^2c^2 + m^2c^4. \tag{6.1}$$

253

Indeed, the fractal to nonfractal transition, which corresponds to a quantum to classical transition, is fully given by the four-dimensional de Broglie scale $\lambda_\mu = \hbar/p^\mu$, and its time expression is $\tau_{dB} = \hbar/E = \hbar/\sqrt{p^2c^2 + m^2c^4}$, which is therefore always smaller than $\tau_E = \hbar/mc^2$ for a massive particle. This explains why, when going from the large scales to the small scales, one first encounters a fractal space without fractal time, then at smaller scales a full fractal space-time that gives rise to (motion) relativistic quantum mechanics.

We shall not come back anymore on this behavior here, which has been studied in detail in [353]. The aim of the present chapter is to generalize to relativistic quantum mechanics the foundation of quantum tools and quantum laws from the nondifferentiable geometry and the principle of relativity. We shall first derive the Klein–Gordon equation as the integral of a geodesic equation acting on a complex wave function. Then we shall show that a new two-valuedness of derivatives must be introduced, as a consequence of the breaking of the discrete symmetry under the transformation $dx^\mu \leftrightarrow -dx^\mu$ – we are now dealing with full four-vectors — (in addition to the symmetry breaking under the transformation $ds \leftrightarrow -ds$, which leads to a complex representation of physical quantities). This new two-valuedness is naturally described in terms of quaternions and biquaternions, which are equivalent to spinors and bispinors. Then we show that these bispinors are solutions of the Dirac equation. We finally consider Pauli spinors and the Pauli equation, although they correspond to a nonrelativistic situation. The reason for this choice is that, in the scale relativity approach as in standard quantum mechanics, one can prove that the spin is of fundamentally relativistic origin, and that the Pauli description can be recovered only as a nonrelativistic limit of the Dirac description, while a direct nonrelativistic construction, which would not provide us with the correct equation, seems to be forbidden.

6.2. Klein–Gordon Equation

6.2.1. *Geodesic equation*

Most elements of the nonrelativistic approach, as described in Chapter 5, remain correct, but now the time differential element dt is replaced by the proper time differential element ds. Not only space, but the full space-time continuum, is considered to be nondifferentiable and, therefore, fractal. We chose a metric signature $(+, -, -, -)$ and we consider only the critical case $D_F = 2$. The elementary displacement along geodesics now writes

$$dX_\pm^\mu = dx_\pm^\mu + d\xi_\pm^\mu. \tag{6.2}$$

Due to the breaking of the reflection symmetry $(ds \leftrightarrow -ds)$ issued from non-differentiability, we still define two "classical" derivatives, d_+/ds and d_-/ds, which, once applied to x^μ, yield two classical four-velocities,[a]

$$\frac{d_+}{ds}x^\mu(s) = v_+^\mu; \quad \frac{d_-}{ds}x^\mu(s) = v_-^\mu. \tag{6.3}$$

These two derivatives can be combined in terms of a complex derivative operator

$$\frac{\widehat{d}}{ds} = \frac{(d_+ + d_-) - i\,(d_+ - d_-)}{2\,ds}, \tag{6.4}$$

which, when applied to the position vector, yields a complex four-velocity

$$\mathcal{V}^\mu = \frac{\widehat{d}}{ds}x^\mu = V^\mu - i\,U^\mu = \frac{v_+^\mu + v_-^\mu}{2} - i\,\frac{v_+^\mu - v_-^\mu}{2}. \tag{6.5}$$

We are, once again, led to a stochastic description, due to the infinite number of geodesics of the fractal space-time. This forces us to consider the question of the definition of a Lorentz-covariant stochasticity in space-time. This problem has been addressed by several authors in the framework of a relativistic generalization of Nelson's stochastic quantum mechanics. Two mutually independent fluctuation fields, $d\xi_\pm^\mu(s)$, are defined, with zero expectation value ($\langle d\xi_\pm^\mu \rangle = 0$) and such that

$$\langle d\xi_\pm^\mu\, d\xi_\pm^\nu \rangle = \mp\lambda\,\eta^{\mu\nu} ds. \tag{6.6}$$

The constant λ is another writing — up to constants — for the coefficient $2\mathcal{D} = \lambda c$, with $\mathcal{D} = \hbar/2m$ in the standard quantum case, in which it is the Compton length of the particle. This process only makes sense in $\mathbb{R}^4$, i.e. the "metric" $\eta^{\mu\nu}$ should be positive definite. Indeed, the fractal fluctuations are of the same nature as uncertainties and "errors", so that the space and

---

[a]In this chapter, we take into account only the classical (differentiable) part of the velocity field. All the results obtained can be generalized to the full velocity field $\widetilde{\mathcal{V}}^\mu = \mathcal{V}^\mu + \mathcal{W}^\mu$, including the fractal (nondifferentiable) fluctuation of zero mean $\mathcal{W}^\mu$. As in the non-relativistic case, this generalization leads to the existence of fractal wavefunctions, which are solutions of the relativistic quantum equations (Klein–Gordon and Dirac equations).

the time fluctuations add quadratically. The sign corresponds to a choice of space-like fluctuations.

Dohrn and Guerra [143] introduce the above "Brownian metric" and a kinetic metric $g_{\mu\nu}$ and obtain a compatibility condition between them, which reads $g_{\mu\nu}\eta^{\mu\alpha}\eta^{\nu\beta} = g^{\alpha\beta}$. An equivalent method was developed by Zastawniak [550], who introduces, in addition to the covariant drifts v_+^μ and v_-^μ, new drifts b_+^μ and b_-^μ (note that our notations are different from his). Serva [489] gives up Markov processes and considers a covariant process, which belongs to a larger class, known as "Bernstein processes". In the end, all these proposals are equivalent and amount to transforming what would have been a Laplacian operator in an Euclidean four-dimensional space $\mathbb{R}^4$ into a Dalembertian.

In what follows we assume a Minkowskian metric for the underlying classical space-time and we adopt Einstein's convention about the sum over up and down indices. The two $(+)$ and $(-)$ derivatives of a function $f[x(s), s]$ can be written

$$d_\pm f/ds = \left(\frac{\partial}{\partial s} + v_\pm^\mu \, \partial_\mu \mp \frac{1}{2} \lambda \partial^\mu \partial_\mu \right) f. \qquad (6.7)$$

The situation is different from the non-relativistic case. Indeed, the various three-dimensional physical quantities are usually explicit functions of the time t, while in four dimensions they do not depend explicitly on the proper time s. In what follows, we shall therefore only consider s-stationary functions, so that the partial derivative $\partial/\partial s$, which would have been the equivalent of $\partial/\partial t$, disappears from the above expression. Finally, the covariant time derivative operator reduces to [356, 360]

$$\frac{\widehat{d}}{ds} = \left(\mathcal{V}^\mu + \frac{1}{2} i \lambda \partial^\mu \right) \partial_\mu. \qquad (6.8)$$

By introducing the velocity operator [387]

$$\widehat{\mathcal{V}^\mu} = \mathcal{V}^\mu + i \frac{\lambda}{2} \partial^\mu, \qquad (6.9)$$

it can be written under the first order form

$$\frac{\widehat{d}}{ds} = \widehat{\mathcal{V}^\mu} \, \partial_\mu. \qquad (6.10)$$

The next steps of the derivation of the motion equation remain similar to the fractal space case. We assume that the system under consideration can be characterized by an action $\mathcal{S}$, which is complex because the four-velocity is now complex. The same reasoning as in classical mechanics leads us to write

$$dS = -mc\, V_\mu\, dx^\mu \tag{6.11}$$

(see [446] for another equivalent choice). The least-action principle applied on this action yields the equations of motion of a free particle that takes the form of a geodesic equation

$$\frac{\widehat{d}}{ds} V_\alpha = 0. \tag{6.12}$$

Such a form is also directly obtained from the "strong covariance" principle and the generalized equivalence principle. We can also write the variation of the action as a functional of coordinates. We obtain the usual result (but here generalized to complex quantities):

$$\delta S = -mc\, V_\mu\, \delta x^\mu \Rightarrow \mathcal{P}_\mu = mc\, V_\mu = -\partial_\mu S, \tag{6.13}$$

where $\mathcal{P}_\mu$ is now a complex four-momentum. As in the nonrelativistic case, the wave function is introduced as being nothing but a re-expression of the action:

$$\psi = e^{iS/mc\lambda}, \tag{6.14}$$

so that the complex velocity field reads in terms of the wave function

$$V_\mu = i\lambda\, \partial_\mu (\ln\psi). \tag{6.15}$$

The equations of motion become

$$\frac{\widehat{d}}{ds} V_\alpha = \left(V^\mu + \frac{1}{2}i\lambda\, \partial^\mu\right)\partial_\mu V_\alpha = 0. \tag{6.16}$$

Introducing the wave function it reads

$$\left(\partial^\mu \ln\psi + \frac{1}{2}\partial^\mu\right)\partial_\mu\partial_\alpha \ln\psi = 0. \tag{6.17}$$

Now, since $\partial_\alpha(\partial_\mu \ln\psi\, \partial^\mu \ln\psi) = 2\,\partial^\mu \ln\psi\, \partial_\alpha\partial_\mu \ln\psi$, and since

$$\partial_\alpha(\partial_\mu\partial^\mu \ln\psi + \partial_\mu \ln\psi\, \partial^\mu \ln\psi) = \partial_\alpha\left(\frac{\partial_\mu\partial^\mu\psi}{\psi}\right), \tag{6.18}$$

we finally find that it takes the compact form

$$\partial_\alpha\left(\frac{\partial_\mu\partial^\mu\psi}{\psi}\right) = 0. \tag{6.19}$$

Therefore the equation of motion can finally be integrated in terms of the Klein–Gordon equation for a free particle,

$$\lambda^2 \, \partial^\mu \partial_\mu \psi + \psi = 0, \tag{6.20}$$

where $\lambda = \hbar/mc$ is its Compton length. The integration constant $-1/\lambda^2$ is chosen so as to ensure the identification of $\varrho = |\psi|^2$ with a probability density for the particle and to recover the non-relativistic limit. The proof is quite similar to that of the Schrödinger equation, with the three-dimensional Laplacian replaced by a four-dimensional Dalembertian, and the partial derivative $\partial \psi / \partial t$ having no equivalence, because the relativistic physical quantities are not explicit functions of the proper time s.

As shown by Zastawniak [550] and as can be easily recovered from the definition (6.5), the quadratic invariant of special motion-relativity, $v^\mu v_\mu = 1$, is naturally generalized as

$$\mathcal{V}^\mu \mathcal{V}_\mu^\dagger = 1, \tag{6.21}$$

where $\mathcal{V}_\mu^\dagger$ is the complex conjugate of $\mathcal{V}_\mu$. This ensures the covariance, i.e. the invariance of the form of equations, of the theory at this level.

6.2.2. *Quadratic invariant, Leibniz rule and complex velocity operator*

It has been recalled by Pissondes [446] that the square of the complex four-velocity is no longer equal to unity, since it is now complex. It can be derived directly from (6.18) after accounting for the Klein–Gordon equation. One obtains the generalized energy (or quadratic invariant) equation:

$$\mathcal{V}_\mu \mathcal{V}^\mu + i\lambda \partial_\mu \mathcal{V}^\mu = 1, \tag{6.22}$$

which is the relativistic analog of the non-relativistic energy equation $\mathcal{E} = \frac{1}{2} m \mathcal{V}^2 - im\mathcal{D}\nabla \mathcal{V}$ of Chapter 5 (the change of sign comes from the choice of the metric signature).

Now, taking the gradient of this equation, one obtains

$$\partial_\alpha (\mathcal{V}_\mu \mathcal{V}^\mu + i\lambda \partial_\mu \mathcal{V}^\mu) = 0 \Rightarrow \left(\mathcal{V}^\mu + \frac{1}{2} i\lambda \partial^\mu \right) \partial_\alpha \mathcal{V}_\mu = 0, \tag{6.23}$$

which is equivalent to Eq. (6.16) in the case of free motion, since, in the absence of external field, $\partial_\alpha \mathcal{V}_\mu = \partial_\mu \mathcal{V}_\alpha$.

Clearly, the new form of the quadratic invariant comes only under "weak covariance". Pissondes has therefore addressed the problem of implementing the strong covariance, i.e. of keeping the free, Galilean form of the equations of physics even in the new, more complicated situation, at all levels of the description. As already recalled in Chapter 5, the additional terms in the various equations find their origin in the very definition of the "quantum-covariant" total derivative operator. Indeed, it contains partial derivatives of first order (namely, $\mathcal{V}^\mu \partial_\mu$), but also second order ($\frac{1}{2} i \lambda \partial^\mu \partial_\mu$). Therefore, when one is led to compute quantities like $\widehat{d}(fg)/dt = 0$, the Leibniz rule to use becomes a linear combination of the first order and second order Leibniz rules. There is no problem, provided one always come back to the definition of the covariant total derivative. Some inconsistency would appear only if one, in contradiction with this definition, would want to use only the first order Leibniz rule $d(fg) = f\,dg + g\,df$. Indeed, one finds

$$\frac{\widehat{d}}{ds}(fg) = f\,\frac{\widehat{dg}}{ds} + g\,\frac{\widehat{df}}{ds} + i\lambda\,\partial^\mu f\,\partial_\mu g. \tag{6.24}$$

Pissondes attempted to find a formal tool in terms of which the form of the first order Leibniz rule would be preserved. He introduced the following "symmetric product"

$$f \circ \frac{\widehat{dg}}{ds} = f\,\frac{\widehat{dg}}{ds} + i\,\frac{\lambda}{2}\,\partial^\mu f\,\partial_\mu g, \tag{6.25}$$

and he showed that, using this product, the covariance can be fully implemented. In particular, one recovers the form of the derivative of a product, $\widehat{d}(fg) = f \circ \widehat{dg} + g \circ \widehat{df}$, and the standard decomposition in terms of partial derivatives, $\widehat{df} = \partial_\mu f \circ \widehat{dx^\mu}$.

However, one of the problem with this tool is that it depends on the two functions f and g. We shall therefore use another equivalent tool, which has the advantage to depend only on one function. We have already defined a complex velocity operator

$$\widehat{\mathcal{V}^\mu} = \mathcal{V}^\mu + i\,\frac{\lambda}{2}\,\partial^\mu, \tag{6.26}$$

so that the covariant derivative can be written in terms of an operator product that keeps the standard, first order form,

$$\frac{\widehat{d}}{ds} = \widehat{\mathcal{V}^\mu}\,\partial_\mu\,. \tag{6.27}$$

More generally, one defines the operator

$$\frac{\widehat{dg}}{ds} = \frac{\widehat{dg}}{ds} + i\,\frac{\lambda}{2}\,\partial^\mu g\,\partial_\mu, \tag{6.28}$$

which has the advantage to depend only on the function g. The covariant derivative of a product now writes

$$\frac{\widehat{d(fg)}}{ds} = \frac{\widehat{\widehat{df}}}{ds}\, g + \frac{\widehat{\widehat{dg}}}{ds}\, f, \tag{6.29}$$

i.e. one recovers the form of the first order Leibniz rule. Since $\widehat{f}g \neq \widehat{g}f$, one is led to define a symmetrized product, following Pissondes [447]. One defines $\dot{f} = \widehat{d}f/ds$, then one obtains

$$\dot{f} \otimes \dot{g} = \widehat{\dot{f}}g + \widehat{\dot{g}}f - \dot{f}\dot{g}\,. \tag{6.30}$$

This product is now commutative, $\dot{f} \otimes \dot{g} = \dot{g} \otimes \dot{f}$, and in its terms the standard expression for the square of the velocity is recovered, namely,

$$\mathcal{V}^\mu \otimes \mathcal{V}_\mu = 1. \tag{6.31}$$

The introduction of such a tool, that may appear formal in the case of free motion, becomes particularly useful in the presence of an electromagnetic field. This point will be further developed in Chapter 7. We shall show that the introduction of a new level of complexity in the description of a relativistic fractal space-time, namely, the account of resolutions that become functions of coordinates, leads to a new geometric theory of gauge fields, including the U(1) electromagnetic field. We find that the complex velocity is given in this case by

$$\mathcal{V}^\mu = i\lambda\, D^\mu \ln\psi = i\lambda\, \partial^\mu \ln\psi - \frac{e}{mc^2} A^\mu, \tag{6.32}$$

where A^μ is a field of dilations of internal resolutions that can be identified with an electromagnetic field.

Inserting this expression in Eq. (6.23) yields the standard Klein–Gordon equation with electromagnetic field [446]

$$\left(i\hbar\,\partial_\mu - \frac{e}{c}A_\mu\right)\left(i\hbar\,\partial^\mu - \frac{e}{c}A^\mu\right)\psi = m^2 c^2 \psi. \tag{6.33}$$

6.2.3. *Quantum potential*

Let us now establish the expression for the quantum potential in the relativistic case [392]. As we shall see, it is now obtained as a manifestation of the fractality of the full space-time, i.e. of the fractal dimension 2 of the four space-time coordinates.

The calculations are similar to the non-relativistic case. We decompose the complex velocity in terms of its real and imaginary parts, $\mathcal{V}_\alpha = V_\alpha - iU_\alpha$, so that the geodesic equation becomes

$$\left\{ V^\mu - i\left(U^\mu - \frac{\lambda}{2}\partial^\mu\right)\right\}\partial_\mu\,(V_\alpha - iU_\alpha) = 0, \tag{6.34}$$

i.e.

$$\left\{V^\mu\partial_\mu V_\alpha - \left(U^\mu - \frac{\lambda}{2}\partial^\mu\right)\partial_\mu U_\alpha\right\} - i\left\{\left(U^\mu - \frac{\lambda}{2}\partial^\mu\right)\partial_\mu V_\alpha + V^\mu\partial_\mu U_\alpha\right\} = 0. \tag{6.35}$$

The real part of this equation takes the form of a relativistic Euler–Newton equation of dynamics

$$\frac{dV_\alpha}{ds} = V^\mu\partial_\mu V_\alpha = \left(U^\mu - \frac{\lambda}{2}\partial^\mu\right)\partial_\mu U_\alpha. \tag{6.36}$$

Therefore, the relativistic case is similar to the non-relativistic one, since a generalized force also appears in the right-hand side of this equation. Let us now prove that it also derives from a potential. We set

$$\psi = \sqrt{P}\,e^{i\theta}. \tag{6.37}$$

Using the expression for U_α in terms of the modulus $\sqrt{P}$ of the wave function,

$$U_\alpha = -\lambda\,\partial_\alpha \ln\sqrt{P}, \tag{6.38}$$

we may write the force under the form

$$\frac{F_\alpha}{mc^2} = -\lambda\,\partial^\mu \ln\sqrt{P}\,\partial_\mu(-\lambda\,\partial_\alpha \ln\sqrt{P}) + \frac{\lambda^2}{2}\partial^\mu\partial_\mu\partial_\alpha \ln\sqrt{P}$$
$$= \lambda^2\left(\partial^\mu \ln\sqrt{P}\,\partial_\mu\partial_\alpha \ln\sqrt{P} + \frac{1}{2}\partial^\mu\partial_\mu\partial_\alpha \ln\sqrt{P}\right). \tag{6.39}$$

Since $\partial^\mu\partial_\mu\partial_\alpha = \partial_\alpha\partial^\mu\partial_\mu$ commutes and since $\partial_\alpha(\partial^\mu \ln f\,\partial_\mu \ln f) = 2\,\partial^\mu \ln f\,\partial_\alpha\partial_\mu \ln f$, we obtain

$$\frac{F_\alpha}{mc^2} = \frac{1}{2}\,\lambda^2\,\partial_\alpha\left(\partial^\mu \ln\sqrt{P}\,\partial_\mu \ln\sqrt{P} + \partial^\mu\partial_\mu \ln\sqrt{P}\right). \tag{6.40}$$

We can now make use of the remarkable identity (6.18) applied to $\sqrt{P}$,

$$\partial^\mu \ln\sqrt{P}\,\partial_\mu \ln\sqrt{P} + \partial^\mu\partial_\mu \ln\sqrt{P} = \frac{\partial^\mu\partial_\mu\sqrt{P}}{\sqrt{P}}, \tag{6.41}$$

and we finally obtain

$$\frac{dV_\alpha}{ds} = \frac{1}{2}\lambda^2\,\partial_\alpha\left(\frac{\partial^\mu\partial_\mu\sqrt{P}}{\sqrt{P}}\right). \tag{6.42}$$

Therefore, as in the non-relativistic case, the force derives from a potential energy

$$Q_R = \frac{1}{2}mc^2\,\lambda^2\,\frac{\partial^\mu\partial_\mu\sqrt{P}}{\sqrt{P}}, \tag{6.43}$$

that can also be expressed in terms of the velocity field U as

$$Q_R = \frac{1}{2}mc^2\left(U^\mu U_\mu - \lambda\,\partial^\mu U_\mu\right). \tag{6.44}$$

This is the relativistic expression for the quantum potential, in which the Laplacian of the non-relativistic case is replaced, once again, by a d'Alembertian. Its origin can therefore be directly traced back to the second order terms of the covariant derivative, which themselves come from the expression for the stochastic fractal fluctuation of fractal dimension 2. It is then established as a genuine manifestation of the fractal geometry of space-time.

At the non-relativistic limit $(c \to \infty)$, the Dalembertian operator $\partial^\mu\partial_\mu = (\partial^2/c^2\partial t^2 - \Delta)$ is reduced to $-\Delta$, and since $\lambda = 2\mathcal{D}/c$, we recover the nonrelativistic potential energy $Q = -2m\mathcal{D}^2\Delta\sqrt{P}/\sqrt{P}$. Note the correction to the potential introduced in [446], which is twice this potential and therefore cannot agree with the nonrelativistic limit.

6.2.4. *Invariants and energy balance*

We have seen that the four-dimensional energy equation $u^\mu u_\mu = 1$ is generalized in terms of the complex velocity under the form $\mathcal{V}^\mu\mathcal{V}_\mu + i\lambda\partial^\mu\mathcal{V}_\mu = 1$ [446]. Let us show that the additional term is itself a manifestation of the scalar field Q_R, which takes its origin in the fractal and nondifferentiable geometry. Start again with the geodesic equation

$$\frac{\widehat{d}\mathcal{V}_\alpha}{ds} = \left(\mathcal{V}^\mu + i\frac{\lambda}{2}\partial^\mu\right)\partial_\mu\mathcal{V}_\alpha = 0. \tag{6.45}$$

Introducing the wave function by the relation $\mathcal{V}_\alpha = i\lambda\,\partial_\alpha\ln\psi$, it becomes

$$\frac{\widehat{d}\mathcal{V}_\alpha}{ds} = -\frac{\lambda^2}{2}\,\partial_\alpha\left(\partial^\mu\ln\psi\,\partial_\mu\ln\psi + \partial^\mu\partial_\mu\ln\psi\right) = \frac{1}{2}\partial_\alpha\left(-\lambda^2\,\frac{\partial^\mu\partial_\mu\psi}{\psi}\right) = 0. \tag{6.46}$$

Under its right-hand form, this equation is integrated in terms of the Klein–Gordon equation

$$\lambda^2 \, \partial^\mu \partial_\mu \psi + \psi = 0. \tag{6.47}$$

Under its left hand form, the integral writes

$$-\lambda^2 (\partial^\mu \ln \psi \, \partial_\mu \ln \psi + \partial^\mu \partial_\mu \ln \psi) = 1. \tag{6.48}$$

Reintroducing in this equation the complex velocity field, it becomes

$$\mathcal{V}^\mu \mathcal{V}_\mu + i\lambda \partial^\mu \mathcal{V}_\mu = 1, \tag{6.49}$$

which is therefore another form taken by the Klein–Gordon equation, as expected from the fact that it is the quantum equivalent of the Hamilton–Jacobi or energy equation. Let us now separate the real and imaginary parts of this equation. We obtain:

$$V^\mu V_\mu - (U^\mu U_\mu - \lambda \, \partial^\mu U_\mu) = 1, \quad 2\, V^\mu U_\mu - \lambda \, \partial^\mu V_\mu = 0. \tag{6.50}$$

Then the energy balance writes, in terms of the additional potential energy Q_R

$$V^\mu V_\mu = 1 + 2\, \frac{Q_R}{mc^2}. \tag{6.51}$$

Let us show that we actually expect such a relation for the quadratic invariant in presence of an external potential ϕ. The energy relation writes in this case $(E - \phi)^2 = p^2 c^2 + m^2 c^4$, i.e. $E^2 - p^2 c^2 = m^2 c^4 + 2E\phi - \phi^2$. Introducing the rest frame energy by writing $E = mc^2 + E'$, we obtain

$$V^\mu V_\mu = \frac{E^2 - p^2 c^2}{m^2 c^4} = 1 + 2\, \frac{\phi}{mc^2} + \left(2\, \frac{E'}{mc^2}\, \frac{\phi}{mc^2} - \frac{\phi^2}{m^2 c^4} \right). \tag{6.52}$$

This justifies the relativistic factor 2 in equation (6.51) and supports the interpretation of Q_R in terms of a potential, at least at the level of the leading terms.

Now, concerning the additional terms, it should remain clear that this is only an approximate field theory description in terms of the manifestations of the fractal and nondifferentiable geometry of space-time. Therefore we expect the field theory description to be the first order approximation in the same manner as, in general relativity, the approximate description of the metric potentials in terms of a Newtonian potential.

In particular, in the non-relativistic limit $c \to \infty$ the last two terms of equation (6.52) vanish and we recover the non-relativistic energy equation (5.210) of Sec. 5, which is therefore exact in this case.

Problems and Exercises

Exercise 16 Generalize the theory to a zero mass particle.

Hint: Start from the complex invariant

$$\mathcal{V}^\mu \mathcal{V}_\mu + i\lambda \partial^\mu \mathcal{V}_\mu = 1, \tag{6.53}$$

with $\lambda = \hbar/mc$. It becomes, in terms of the complex four-momentum $\mathcal{P}^\mu = mc\mathcal{V}^\mu$,

$$\mathcal{P}^\mu \mathcal{P}_\mu + i\hbar \, \partial^\mu \mathcal{P}_\mu = m^2 c^2. \tag{6.54}$$

This expression can therefore be easily written in the zero mass case. It reads [447]

$$\mathcal{P}^\mu \mathcal{P}_\mu + i\hbar \, \partial^\mu \mathcal{P}_\mu = 0. \tag{6.55}$$

The complex momentum is related to the complex action by $\mathcal{P}^\mu = -\partial^\mu \mathcal{S}$, while the wave function is $\psi = \exp(i\mathcal{S}/\hbar)$. The above energy invariant, expressed in terms of the wave function, becomes the Klein–Gordon equation, $\hbar^2 \partial^\mu \partial_\mu \psi + m^2 c^2 \psi = 0$, which reads in the zero mass case

$$\partial^\mu \partial_\mu \psi = 0, \tag{6.56}$$

as expected for a wave propagating at the speed of light c. ∎

6.3. Dirac Spinors and the Dirac Equation

6.3.1. *Introduction*

One of the main results of the scale relativity theory is its ability to provide a physical origin for the complex nature of the wave function in quantum mechanics. Indeed, we have seen that it is a direct consequence of the non-differentiable geometry of space-time, which involves a symmetry breaking of the reflection invariance $ds \leftrightarrow -ds$, and therefore a two-valuedness of the velocity vector, which is accounted for in a covariant way by the use of complex numbers.

The next question that may be asked in the scale relativistic framework is that of the origin of spin. This is a crucial point for the theory, since the spin is considered, in the standard approach, to have no classical counterpart and to be of pure quantum origin. Now, the passage from a spin 0 wave function to a one-half spinor amounts to a new doubling of the variables. Namely, a Pauli spinor is described by a doublet (ψ^1, ψ^2) of complex wave functions, while the probability density is $P = |\psi^1|^2 + |\psi^2|^2$.

When going to the relativistic case, a new doubling is needed, which leads to Dirac bispinors.

This suggests that, once again, a discrete symmetry breaking should be at work, which could be of the same nature of the algebra doubling which gives rise to the complex wave function (see Sec. 5.4.1). The nondifferentiability of space-time indeed implies a new symmetry breaking, namely, that of the reflection of the space differential element, $dx^\mu \leftrightarrow -dx^\mu$. An additional two-valuedness of derivatives and geodesic velocity fields should therefore be introduced, which is naturally described in terms of quaternions (equivalent to spinors), then biquaternions (equivalent to bispinors) [92, 93].

This question has already been raised in other apparented frameworks. A step in this direction has been made by Gaveau *et al.* [197], who have generalized Nelson's stochastic mechanics [331] to the motion-relativistic case in (1+1) dimensions. However, an analytic continuation was needed to obtain the Dirac equation in such a framework, and, furthermore, stochastic mechanics is now known to be in contradiction with standard quantum mechanics as concerns multitime correlations in repeated measurements [526].

In the framework of a fractal space-time viewed as a geometric analog of quantum mechanics [420, 344], Ord has developed a generalized version of the Feynman chessboard model, which allows him to recover the Dirac equation in (3+1) dimensions, without analytical continuation [421, 316, 422, 423]. Our approach also involves a fractal space-time, but in a different way, since Ord's description remains of a fundamentally statistical nature. Here, the fractality of the space-time continuum is derived from its nondifferentiability, it is constrained by the principle of scale relativity and the Dirac equation is derived as an integral of the geodesic equation. This is therefore not a stochastic approach in its essence, even though stochastic variables must be introduced as a consequence of the new geometry, so it does not come under the contradictions encountered by stochastic mechanics.

6.3.2. *Reflection symmetry breaking of space differentials*

The discrete symmetry breaking $ds \leftrightarrow -ds$ is not the last word about the new structures implied by the nondifferentiable geometry. The total derivative of a physical quantity also involves partial derivatives with respect to the space variables, $\partial/\partial x^\mu$. Once again, from the very definition of derivatives, the discrete symmetry under the reflection $dx^\mu \leftrightarrow -dx^\mu$ should also be broken at a more profound level of description. Therefore, we expect the possible appearance of a new two-valuedness of the generalized velocity.

At this level one should also account for parity violation (as in the standard quantum theory, see e.g. [268]). Finally, the three discrete symmetry breakings

$$ds \leftrightarrow -ds \qquad dx^\mu \leftrightarrow -dx^\mu \qquad x^\mu \leftrightarrow -x^\mu$$

can be accounted for by the introduction of a biquaternionic velocity. It has been subsequently shown by Célérier [91–93] that one can derive in this way the Dirac equation, namely as an integral of a geodesic equation. In other words, this means that, as we shall now show, this new two-valuedness is at the origin of the bispinor nature of the electron wave function.

6.3.3. *Biquaternionic Klein–Gordon equation*

It has been known for long that the Dirac equation proceeds from the Klein–Gordon equation when it is written in a quaternionic form [264, 111]. However no physical reason was known for this important mathematical property. We have proposed [93] to introduce, as a consequence of the nondifferentiable geometry, a biquaternionic covariant derivative operator, leading to the definition of a biquaternionic velocity and of a biquaternionic wave-function, which we use to derive the Klein–Gordon equation in a biquaternionic form. We use the quaternionic formalism (see Sec. 6.3.4), as introduced by Hamilton [220], and further developed by Conway [111, 112] (see also Synge [504] and Scheffers [472]).

6.3.3.1. *Multi-valuedness of derivative*

Let us now reconsider the above construction of the scale relativity mathematical tool in the more general case involving the subsequent breaking of the symmetries[b]:

$$ds \leftrightarrow -ds,$$
$$dx^\mu \leftrightarrow -dx^\mu,$$
$$x^\mu \leftrightarrow -x^\mu.$$

---

[b]We consider here only the reversal of the vector dx^μ taken as a whole. However, additional two-valuednesses can be added in future works, by taking into account separately the breakings under the reflection symmetries on the four space-time coordinates, $dx \leftrightarrow -dx$, $dy \leftrightarrow -dy$, $dz \leftrightarrow -dz$, $dt \leftrightarrow -dt$.

We have, up to now, considered only the effect of nondifferentiability on the total derivative d/ds. Now the velocity fields of the geodesic bundles are functions of the coordinates, so we are led to analyse also the physical meaning of the partial derivatives $\partial/\partial x$ (we use only one coordinate variable in order to simplify the writing) in the decomposition $d/ds = \partial/\partial s + (dx/ds)\,\partial/\partial x$. Strictly speaking, $\partial f/\partial x$ does not exist in the nondifferentiable case. We are therefore once again led to introduce fractal functions $f(x, dx)$, explicitly dependent on the coordinate resolution, whose derivative is undefined only at the unobservable limit $dx \to 0$. As a consequence of the very construction of the derivative, there are two definitions of the partial derivative of a fractal function instead of one, namely,

$$\frac{\partial f}{\partial x_+}(x, dx) = \frac{f(x + dx, dx) - f(x, dx)}{dx}, \tag{6.57}$$

$$\frac{\partial f}{\partial x_-}(x, dx) = \frac{f(x, dx) - f(x - dx, dx)}{dx}. \tag{6.58}$$

They are transformed one into the other under the reflection $dx \leftrightarrow -dx$.

The additional $(x^\mu \leftrightarrow -x^\mu)$ symmetry corresponds to the parity P and time-reversal T symmetries, which breaking is already taken into account in the standard definition of Dirac spinors in terms of pairs of Pauli spinors [268].

Let us now consider the consequences of these symmetry breakings on the definition of the total derivative and more generally on the velocity fields. The symmetry breaking under the transformation $ds \to -ds$ leads one to define two total derivatives, d_+/ds and d_-/ds. The additional discrete symmetry breaking under the transformation $dx^\mu \to -dx^\mu$ leads one to a new two-valuedness of each one of these derivatives, and therefore to four total derivatives, d_{++}/ds, d_{+-}/ds, d_{-+}/ds and d_{--}/ds. Finally, under a parity transformation $X^\mu \to -X^\mu$, these four derivatives are not left unchanged and become respectively $\tilde{d}_{++}/ds$, $\tilde{d}_{+-}/ds$, $\tilde{d}_{-+}/ds$ and $\tilde{d}_{--}/ds$.

As concerns the velocity field, it is a function of the fractal coordinates, which are now considered to depend on the four resolutions dx^μ, i.e. $v = v[x^\mu(s, dx^\mu), s]$, with $\mu = 0$ to 3 (we keep here the s dependence for generality, but, as we shall see, the velocity field and therefore the wave function are not explicit functions of the invariant proper time s). This implies a new two-valuedness, since in general $v[x^\mu(s, dx^\mu), s] \neq v[x^\mu(s, -dx^\mu), s]$. Combined with the two-valuedness, which comes from the symmetry breaking under the transformation $ds \leftrightarrow -ds$ and with the parity transformation, this yields eight velocity fields $\{v^\mu_{++}, v^\mu_{+-}, v^\mu_{-+}, v^\mu_{--}\}$ and $\{\tilde{v}^\mu_{++}, \tilde{v}^\mu_{+-}, \tilde{v}^\mu_{-+}, \tilde{v}^\mu_{--}\}$ [93].

As in the nonrelativistic case, each of these components is itself a fractal function, which can be decomposed in terms of a differentiable ("classical") mean part and a nondifferentiable (fractal, explicitly scale-dependent) part of zero mean. Since there is no reason for the classical parts to be equal, they come under the three succesive doublings and constitute also a multiplet of eigth velocity fields.

Finally, for each of these cases, one finds a total derivative

$$\frac{d}{ds}{}_{\pm\pm} = \frac{\partial}{\partial s} + v^\mu_{\pm\pm}\partial_\mu \mp \frac{\lambda}{2}\partial^\mu\partial_\mu \qquad (6.59)$$

and its tilde counterpart, where, now and in what follows, only the classical part of the velocity field is taken into account.

6.3.3.2. *Origin of biquaternions and bispinors*

In analogy with the introduction of complex numbers in the non-relativistic case, we shall now combine the components of the velocity multiplet in terms of a unique biquaternionic number. As recalled in Sec. 5.4.1, the motivation for this choice comes from the fact that it gives the best solution to the problem of algebra doubling, which is posed from the mathematical point of view.

A last point to be justified is the use of complex quaternions (biquaternions) for describing the new algebra doublings that lead to bispinors and to the Dirac equation [93]. One could think that the argument given in Sec. 5.4.1 implies the use of Graves–Cayley octonions (and therefore the giving up of associativity) in the case of three successive doublings as now considered. However, these three doublings are not on the same footing from a physical point of view:

(i) The first two-valuedness comes from a discrete symmetry breaking at the level of the differential invariant, namely, dt in the case of a fractal space (yielding the Schrödinger equation) and ds in the case of a fractal space-time (yielding the Klein–Gordon equation). This means that it has an effect on the total derivatives d/dt and d/ds. This two-valuedness is achieved by the introduction of complex variables.

(ii) The second two-valuedness (differential parity and time reversal violation) comes from a new discrete symmetry breaking (expected as a manifestation of nondifferentiability) on the space-time differential element $dx^\mu \leftrightarrow -dx^\mu$. It is subsequent to the first two-valuedness, since it has an effect on the partial derivative $\partial_\mu = \partial/\partial x^\mu$ that intervenes in the complex covariant derivative operator

$$\frac{\widehat{d}}{ds} = \left(\mathcal{V}^\mu + i\,\frac{\lambda}{2}\,\partial^\mu \right) \partial_\mu \,. \tag{6.60}$$

(iii) The third two-valuedness is a standard effect of parity ("*P*") and time reversal ("*T*") in the motion-relativistic situation, which is not specific of the present approach and is already used in the standard construction of Dirac spinors [268]. It does not lead to a real information doubling, since Dirac spinors still have only three degrees of freedom as Pauli spinors do. Therefore, from the second and third doublings, quaternions can be introduced, which affects variables that are already complex due to the first, more fundamental, doubling. This finally leads to the complex quaternionic tool, which is used here.

It is worth noting that these symmetry breakings are effective only at the level of the underlying geometric description of the nondifferentiable fractal space-time. The effect of introducing a two-valuedness of variables in terms of double symmetrical processes amounts to recovering symmetry (in particular, time reversibility) in terms of the bi-process, and therefore in terms of the quantum tools, which are built from it. However, one may wonder whether this recovering could be partially incomplete, as indicated by the existence of P and CP (and therefore T) violations in elementary particle physics.

Finally, one may remark that such a process of complexification of the physical description tool by algebra doubling could be expected to be continued by including additional discrete quantum numbers, such as isospin, hypercharge, color, etc. Such a new and general extension of complex and quaternionic numbers is naturally accounted for by the mean of Clifford algebrae (see [24, 17, 25] and other colloquia of this series). This allows one to bring a justification from first principles of the fact, which was recognized long time ago [230], that Clifford algebra provides a very useful tool for description of geometry and physics, as more recently examplified by the development of the concept of Clifford space [89]. In the end, one may wonder whether associativity itself should not be given up in a more general framework, but this would obviously render the subsequent developments of mathematical physics very complicated: we shall not consider this possibility further in the present book.

6.3.3.3. *Biquaternionic covariant derivative*

The eight velocity fields have then been combined by Célérier [93] in terms of a complex quaternionic velocity field,

$$
\mathcal{V}^\mu = \frac{1}{2}(v^\mu_{++} + \tilde{v}^\mu_{--}) - \frac{i}{2}(v^\mu_{++} - \tilde{v}^\mu_{--})
$$
$$
+ \left[\frac{1}{2}(v^\mu_{+-} + v^\mu_{-+}) - \frac{i}{2}(v^\mu_{+-} - \tilde{v}^\mu_{++})\right] e_1
$$
$$
+ \left[\frac{1}{2}(v^\mu_{--} + \tilde{v}^\mu_{+-}) - \frac{i}{2}(v^\mu_{--} - \tilde{v}^\mu_{-+})\right] e_2
$$
$$
+ \left[\frac{1}{2}(v^\mu_{-+} + \tilde{v}^\mu_{++}) - \frac{i}{2}(\tilde{v}^\mu_{-+} + \tilde{v}^\mu_{+-})\right] e_3, \tag{6.61}
$$

where e_1, e_2 and e_3 are the quaternion units. The freedom in the choice of the actual expression for $\mathcal{V}^\mu$ is constrained by the requirements to recover the complex velocity in the zero spin limit and a real classical velocity at the classical limit. This leads to various possible choices, which actually are physically equivalent, as we shall see. In what follows, we shall indeed consider another choice when looking for Pauli spinors at the non-relativistic limit.

The biquaternionic velocity defined thus corresponds to a biquaternionic derivative operator $\widehat{d}/ds$, similarly defined, and yielding, when applied to the position vector X^μ, the corresponding velocity. For instance, the derivative operator attached to the velocity in Eq. (6.61) reads

$$
\frac{\widehat{d}}{ds} = \frac{1}{2}\left(\frac{d}{ds}{}_{++} + \frac{\tilde{d}}{ds}{}_{--}\right) - \frac{i}{2}\left(\frac{d}{ds}{}_{++} - \frac{\tilde{d}}{ds}{}_{--}\right)
$$
$$
+ \left[\frac{1}{2}\left(\frac{d}{ds}{}_{+-} + \frac{\tilde{d}}{ds}{}_{-+}\right) - \frac{i}{2}\left(\frac{d}{ds}{}_{+-} - \frac{\tilde{d}}{ds}{}_{++}\right)\right] e_1
$$
$$
+ \left[\frac{1}{2}\left(\frac{d}{ds}{}_{--} + \frac{\tilde{d}}{ds}{}_{+-}\right) - \frac{i}{2}\left(\frac{d}{ds}{}_{--} - \frac{\tilde{d}}{ds}{}_{-+}\right)\right] e_2
$$
$$
+ \left[\frac{1}{2}\left(\frac{d}{ds}{}_{-+} + \frac{\tilde{d}}{ds}{}_{++}\right) - \frac{i}{2}\left(\frac{\tilde{d}}{ds}{}_{-+} + \frac{\tilde{d}}{ds}{}_{+-}\right)\right] e_3. \tag{6.62}
$$

Note that we keep the same notation $\widehat{d}/ds$ for this operator since, as we shall see, it keeps the same form in terms of the biquaternionic velocity it already had in the complex case.

Substituting Eq. (6.59) and its tilde counterpart into Eq. (6.62), we obtain the expression for the biquaternionic proper-time derivative operator

$$\frac{\widehat{d}}{ds} = \{1 + e_1 + e_2 + (1 - i)e_3\}\frac{\partial}{\partial s} + \mathcal{V}^\mu \partial_\mu + i\frac{\lambda}{2}\partial^\mu \partial_\mu, \tag{6.63}$$

where the + sign in front of the Dalambertian still proceeds from the choice of the metric signature $(+, -, -, -)$.

Here, for generality, we have kept the $\partial/\partial s$ term. The constant in factor of this term depends on the choice made for recombining the velocity octuplet components. This actually poses no physical problem, since the various physical functions are not explicitly depending on s, so that this term vanishes in all cases. Therefore we reach the remarkable result that the form finally obtained for the biquaternionic derivative operator is unchanged with respect to the previous complex operator, namely

$$\frac{\widehat{d}}{ds} = \left(\mathcal{V}^\mu + i\frac{\lambda}{2}\partial^\mu\right)\partial_\mu, \tag{6.64}$$

where $\mathcal{V}^\mu$ is now a complex quaternionic velocity field. This same form is obtained whatever the definition chosen for constructing the biquaternionic velocity, so that the different choices allowed for this definition can be considered to merely correspond to different mathematical representations leading to the same physical result.

This is another manifestation of the covariance of the scale relativity description, i.e. of its ability to keep the simplest possible form of equations in more and more complicated situations. Finally, it is easy to check that this operator, applied to the position vector X^μ, gives back the biquaternionic velocity $\mathcal{V}^\mu$ of Eq. (6.61).

6.3.4. *Quaternionic calculus*

Before going on with the physics, let us give a brief reminder, following Célérier [91], about the elementary properties of quaternionic arithmetic and analysis, which will be of use in what follows [91].

6.3.4.1. *Definitions and algebraic properties*

A biquaternion $\phi = (\phi_0, \phi_1, \phi_2, \phi_3)$ is an ordered quadruplet of complex numbers. The ϕ_i's are the components of ϕ. The equality of two quaternions is equivalent to the equality of their corresponding components: $\phi = \psi$ if and only if $\phi_i = \psi_i$, with $i = 0, 1, 2, 3$.

The multiplication of a biquaternion by a complex number α and the addition and substraction of quaternions are defined by

$$
\begin{aligned}
\alpha\phi &= (\alpha\phi_0, \alpha\phi_1, \alpha\phi_2, \alpha\phi_3), \\
\phi + \psi &= (\phi_0 + \psi_0, \phi_1 + \psi_1, \phi_2 + \psi_2, \phi_3 + \psi_3), \\
\phi - \psi &= \phi + (-1)\psi.
\end{aligned}
\tag{6.65}
$$

Addition of quaternions is commutative and associative. The null quaternion $\phi = 0$, i.e. the neutral element for addition, writes $\phi = (0, 0, 0, 0)$. Multiplication of quaternions by complex numbers is commutative, associative and distributive. The quaternionic product $\phi\psi$ of two quaternions is itself a quaternion. The product is distributive and satisfies, for any complex number α,

$$
(\alpha\phi)\psi = \phi(\alpha\psi) = \alpha(\phi\psi).
\tag{6.66}
$$

Any quaternion can be decomposed as

$$
\phi = \phi_0 + e_1\phi_1 + e_2\phi_2 + e_3\phi_3,
\tag{6.67}
$$

where Hamilton's imaginary units e_i satisfy the associative but noncommutative algebra

$$
e_i e_j = -\delta_{ij} + \sum_{k=1}^{3} \epsilon_{ijk} e_k, \qquad i, j = 1, 2, 3
\tag{6.68}
$$

where ϵ_{ijk} is the usual completely antisymmetric three-index tensor with $\epsilon_{123} = 1$. From these rules, it is easy to establish that the product of two arbitrary quaternions is

$$
\begin{aligned}
\phi\psi = (&\phi_0\psi_0 - \phi_1\psi_1 - \phi_2\psi_2 - \phi_3\psi_3, \\
&\phi_0\psi_1 + \phi_1\psi_0 + \phi_2\psi_3 - \phi_3\psi_2, \\
&\phi_0\psi_2 + \phi_2\psi_0 + \phi_3\psi_1 - \phi_1\psi_3, \\
&\phi_0\psi_3 + \phi_3\psi_0 + \phi_1\psi_2 - \phi_2\psi_1).
\end{aligned}
\tag{6.69}
$$

This product is in general noncommutative, but is associative.

The adjoint of a quaternion, also named quaternion conjugate or Hamilton conjugate, is defined as

$$
\phi \to \bar{\phi} = (\phi_0, -\phi_1, -\phi_2, -\phi_3).
\tag{6.70}
$$

One also defines the scalar product of quaternions

$$\phi \cdot \psi = \frac{1}{2}(\phi\bar{\psi} + \psi\bar{\phi}) = \frac{1}{2}(\bar{\phi}\psi + \bar{\psi}\phi) = \phi_0\psi_0 + \phi_1\psi_1 + \phi_2\psi_2 + \phi_3\psi_3. \quad (6.71)$$

The norm of a quaternion is, therefore,

$$\phi \cdot \bar{\phi} = \phi\bar{\phi} = \phi_0^2 + \phi_1^2 + \phi_2^2 + \phi_3^2 = \bar{\phi}\phi = \phi \cdot \bar{\phi}. \quad (6.72)$$

The norm and scalar product are numbers, in general complex, the norm being of course the self-scalar product, but also the self-quaternionic product. The norm of a product is the product of the norms. When $\phi\bar{\phi} = 0$, ϕ is said to be null or singular. If $\phi\bar{\phi} = 1$, ϕ is unimodular.

The reciprocal or inverse of any non null quaternion ϕ is defined as

$$\phi^{-1} = \frac{\bar{\phi}}{\phi\bar{\phi}}. \quad (6.73)$$

It possesses the properties:

$$(\phi\psi)^{-1} = \psi^{-1}\phi^{-1},$$
$$\phi\phi^{-1} = \phi^{-1}\phi = 1. \quad (6.74)$$

The latter property is a key tool of the derivation of the biquaternionic Klein–Gordon equation that follows.

6.3.4.2. *Quaternions and Conway matrices*

A connection between quaternions and the Dirac–Eddington matrices has first been brought to light by Conway [111, 112]. It gives rise to the definition of a set of sixteen Conway matrices, a subsample of which is used in Sec. 6.3.7 to derive the Dirac equation from the Klein–Gordon equation. We give below the main steps of their construction.

Let us consider any two quaternions a and b, and the transformation

$$\phi \rightarrow \psi = a\phi b, \quad (6.75)$$

which is linear for the elements $(\phi_0, \phi_1, \phi_2, \phi_3)$ and can therefore be written in matrix form

$$\psi = M(a, b)\phi, \quad (6.76)$$

where M is a 4×4 matrix, while ϕ and ψ are 4×1 column matrices. The notation indicates that $M(a, b)$ is determined by a and b.

Consider now a couple of linear transformations and their resultant

$$\psi = a\phi b, \qquad \chi = c\psi d, \qquad \chi = ca\phi bd, \tag{6.77}$$

they can be written in matrix form

$$\psi = M(a,b)\phi, \qquad \chi = M(c,d)\psi, \qquad \chi = M(c,d)M(a,b)\phi, \tag{6.78}$$

therefore allowing us to write

$$M(c,d)M(a,b) = M(ca,bd). \tag{6.79}$$

In particular, for $a = c$ and $b = d$, Eq. (6.79) becomes

$$M(a,b)^2 = M(a^2,b^2). \tag{6.80}$$

If we substitute $a = b = -1$ in Eq. (6.75), we obtain the identical transformation and we are thus allowed to write $M(-1,-1) = I$, which is the unit matrix. Hence, by Eq. (6.80), we have the following theorem: "If a and b are any two quaternions satisfying

$$a^2 = b^2 = -1, \tag{6.81}$$

then, $M(a,b)^2 = I$. In other words, the matrix generated by a and b is a square root of unity."

Conway proposes the suggestive notation

$$M(a,b) = a(\ )b. \tag{6.82}$$

Now, any of the four quaternions e_1, e_2, e_3, i satisfies Eq. (6.81). We have therefore the following sixteen matrices, each a square root of unity:

$$
\begin{array}{cccc}
e_1(\ )e_1 & e_1(\ )e_2 & e_1(\ )e_3 & e_1(\ )i \\
e_2(\ )e_1 & e_2(\ )e_2 & e_2(\ )e_3 & e_2(\ )i \\
e_3(\ )e_1 & e_3(\ )e_2 & e_3(\ )e_3 & e_3(\ )i \\
i(\ )e_1 & i(\ )e_2 & i(\ )e_3 & i(\ )i.
\end{array}
$$

It is easy to calculate explicitly these matrices. The whole set of the sixteen Conway matrices can be found in, e.g. [504]. We only give below the four matrices used in Sec. 6.3.7

$$
e_3(\ )e_2 = \begin{pmatrix} 0 & 0 & 0 & -1 \\ 0 & 0 & -1 & 0 \\ 0 & -1 & 0 & 0 \\ -1 & 0 & 0 & 0 \end{pmatrix}, \qquad
e_1(\ )i = \begin{pmatrix} 0 & 0 & 0 & i \\ 0 & 0 & -i & 0 \\ 0 & i & 0 & 0 \\ -i & 0 & 0 & 0 \end{pmatrix},
$$

$$e_3(\quad)e_1 = \begin{pmatrix} 0 & 0 & -1 & 0 \\ 0 & 0 & 0 & 1 \\ -1 & 0 & 0 & 0 \\ 0 & 1 & 0 & 0 \end{pmatrix}, \quad e_3(\quad)e_3 = \begin{pmatrix} 1 & 0 & 0 & 0 \\ 0 & 1 & 0 & 0 \\ 0 & 0 & -1 & 0 \\ 0 & 0 & 0 & -1 \end{pmatrix}.$$

6.3.4.3. *Quaternionic derivation*

Owing to the noncommutativity of the quaternionic product, the functions of a quaternionic variable, which possess derivatives, i.e. for which the right and left derivatives are equal, are only the constant and linear functions. In fact, this is a general property shared by every number system with a noncommutative product [472].

However, a biquaternionic function ψ defined as a field in the four-dimensional space-time (namely, its quaternionic components $(\psi_0^\nu, \psi_1^\nu, \psi_2^\nu, \psi_3^\nu)$ of each space-time component ψ^ν are complex functions of the coordinates X^μ), can nevertheless be derived with respect to these coordinates. This is an important point since we shall be naturally led to introduce a biquaternionic wave function that will be subsequently shown to be equivalent to a Dirac spinor. For instance, $\partial_\mu \psi^\nu$ writes

$$\partial_\mu \psi^\nu = \frac{\partial \psi_0^\nu}{\partial X^\mu}(X^0, X^1, X^2, X^3) + e_1 \frac{\partial \psi_1^\nu}{\partial X^\mu}(X^0, X^1, X^2, X^3)$$
$$+ e_2 \frac{\partial \psi_2^\nu}{\partial X^\mu}(X^0, X^1, X^2, X^3) + e_3 \frac{\partial \psi_3^\nu}{\partial X^\mu}(X^0, X^1, X^2, X^3). \quad (6.83)$$

We are therefore allowed to apply all the ordinary complex derivation rules (with respect to the coordinates) to the components of this function.

On the contrary, any relation involving an application of ordinary derivation rules with respect to the quaternionic variable itself, to a quaternionic function other than the constant and linear functions, would be uncorrect. As we shall see, the following calculations leading to the quaternionic Klein–Gordon and Dirac equations are all along in perfect agreement with this constraint.

6.3.5. *Bispinors as biquaternionic wave functions*

Since $\mathcal{V}^\mu$ is now biquaternionic, the Lagrange function is also biquaternionic and, therefore, the same is true of the action $\mathcal{S}$. The stationary action principle applied on this action, or equivalently the generalized equivalence principle, as well as the strong covariance principle, all lead to write the equation of motion in the free case, i.e. in the absence of an exterior field, under the form of a differential geodesic equation

$$\frac{\hat{d}\mathcal{V}^\mu}{ds} = 0. \tag{6.84}$$

The elementary variation of the action, considered as a functional of the coordinates, keeps the usual form

$$\delta\mathcal{S} = -mc\,\mathcal{V}_\mu\,\delta x^\mu. \tag{6.85}$$

We thus obtain the biquaternionic four-momentum, as

$$\mathcal{P}_\mu = mc\mathcal{V}_\mu = -\partial_\mu\mathcal{S}. \tag{6.86}$$

The form of all these expressions is therefore conserved with respect to the complex case.

Concerning the introduction of the wave function, one should now be cautious, because in the complex case it involved a logarithm, which is not well defined in the quaternionic case. However, the relation between the gradient of the action and the complex wave function, $\partial_\mu\mathcal{S} = -iS_0\partial_\mu \ln\psi$ may also be written in an equivalent way as $\partial_\mu\mathcal{S} = -iS_0\psi^{-1}\partial_\mu\psi$. Under this form it becomes quite generalizable, thanks to the introduction of the inverse biquaternion ψ^{-1}. A biquaternionic wave function is therefore introduced as a re-expression of the biquaternionic action, namely

$$\psi^{-1}\partial_\mu\psi = \frac{i}{S_0}\,\partial_\mu\mathcal{S}, \tag{6.87}$$

knowing that one implicitly uses the quaternionic product between ψ^{-1} and $\partial_\mu\psi$. The biquaternionic four-velocity is derived from Eq. (6.86), as

$$\mathcal{V}_\mu = i\,\frac{S_0}{mc}\,\psi^{-1}\partial_\mu\psi. \tag{6.88}$$

This is the biquaternionic generalization of the definition used in the Schrödinger case: $\psi = e^{iS/S_0}$. Note that we could have chosen, for the definition of the wave function in Eq. (6.87), a commuted expression in the left-hand side, i.e. $(\partial_\mu\psi)\psi^{-1}$ instead of $\psi^{-1}\partial_\mu\psi$. But with this reversed choice, owing to the noncommutativity of the quaternionic product, we could not obtain the motion equation as a vanishing four-gradient, as in the Eq. (6.95), which follows in next section. Therefore, we retain the above simplest choice, as yielding an equation which can be integrated.

Note that this noncommutativity is of a more profound level than that already involved by the fractal and nondifferentiable description regarding the quantum operatorial tool. It is now at the level of the fractal space-time itself, which therefore fundamentally comes under Connes's noncommutative geometry [109, 276]. Moreover, this noncommutativity might be considered as a key for a future better understanding of the parity and CP violations, which will not be developed here.

Problems and Exercises

Open Problem 9: Could the noncommutativity involved in the non-equivalence between the choices $(\partial_\mu \psi)\psi^{-1}$ and $\psi^{-1}\partial_\mu \psi$ for the writing of the equation of motion and its integration help to understand the CP violation in elementary particle physics? ∎

6.3.6. *Free particle biquaternionic Klein–Gordon equation*

The equation of motion, Eq. (6.84), reads

$$\left(\mathcal{V}^\nu \partial_\nu + i \frac{\lambda}{2} \partial^\nu \partial_\nu \right) \mathcal{V}_\mu = 0. \tag{6.89}$$

After replacing $\mathcal{V}_\mu$ (respectively $\mathcal{V}^\nu$) by its expression given in Eq. (6.88), we obtain

$$i \frac{S_0}{mc} \left(i \frac{S_0}{mc} \psi^{-1} \partial^\nu \psi \, \partial_\nu + i \frac{\lambda}{2} \partial^\nu \partial_\nu \right) (\psi^{-1} \partial_\mu \psi) = 0. \tag{6.90}$$

By taking the complex limit (neglecting the additional biquaternionic components) and then the non relativistic limit, one can prove that the constant S_0 is given by

$$S_0 = mc\lambda, \tag{6.91}$$

i.e. S_0 identifies to $\hbar$ and λ to the Compton length in the standard quantum mechanics case. Then, the above equation can be simplified in such a way that it no longer involves any constant, namely,

$$\psi^{-1}\partial^\nu \psi \, \partial_\nu (\psi^{-1}\partial_\mu \psi) + \frac{1}{2} \partial^\nu \partial_\nu (\psi^{-1}\partial_\mu \psi) = 0. \tag{6.92}$$

Now, from the definition of the inverse of a quaternion,

$$\psi\psi^{-1} = \psi^{-1}\psi = 1, \tag{6.93}$$

we know that ψ and ψ^{-1} commute. But, as we have already seen, this is not necessarily the case for ψ and $\partial_\mu \psi^{-1}$ nor for ψ^{-1} and $\partial_\mu \psi$ and their contravariant counterparts. However, when we derive Eq. (6.93) with respect to the coordinates, we obtain

$$\psi \, \partial_\mu \psi^{-1} = -(\partial_\mu \psi)\psi^{-1},$$
$$\psi^{-1}\partial_\mu \psi = -(\partial_\mu \psi^{-1})\psi, \tag{6.94}$$

and identical formulae for the contravariant analogues.

Developing Eq. (6.92), using Eqs. (6.94) and the property $\partial^\nu \partial_\nu \partial_\mu = \partial_\mu \partial^\nu \partial_\nu$, we obtain, after some calculations,

$$\partial_\mu[(\partial^\nu \partial_\nu \psi)\psi^{-1}] = 0. \tag{6.95}$$

This four-gradient can be integrated as

$$(\partial^\nu \partial_\nu \psi)\psi^{-1} + C = 0, \tag{6.96}$$

of which we take the right product by ψ to finally obtain

$$\partial^\nu \partial_\nu \psi + C\psi = 0. \tag{6.97}$$

The integration constant can be identified by taking the complex and non-relativistic limit. It is given by $C = 1/\lambda^2$, i.e. in the case of standard quantum mechanics, $C = m^2 c^2/\hbar^2$. We therefore recognize the Klein–Gordon equation for a free particle with a mass m,

$$\hbar^2 \, \partial^\nu \partial_\nu \psi + m^2 c^2 \, \psi = 0. \tag{6.98}$$

But, in this equation, ψ is now a biquaternion, i.e. as we shall see, a Dirac bispinor.

6.3.7. *Derivation of the Dirac equation*

We now use a long-known property of the quaternionic formalism, which allows one to obtain the Dirac equation for a free particle as a mere square root of the Klein–Gordon operator [264, 111, 230].

We first develop the Klein–Gordon equation as

$$\frac{1}{c^2}\frac{\partial^2 \psi}{\partial t^2} = \frac{\partial^2 \psi}{\partial x^2} + \frac{\partial^2 \psi}{\partial y^2} + \frac{\partial^2 \psi}{\partial z^2} - \frac{m^2 c^2}{\hbar^2}\psi. \tag{6.99}$$

Thanks to the property of the quaternionic and complex imaginary units $e_1^2 = e_2^2 = e_3^2 = i^2 = -1$, Eq. (6.99) can be written under the form

$$\frac{1}{c^2}\frac{\partial^2 \psi}{\partial t^2} = e_3^2 \frac{\partial^2 \psi}{\partial x^2} e_2^2 + i e_1^2 \frac{\partial^2 \psi}{\partial y^2} i + e_3^2 \frac{\partial^2 \psi}{\partial z^2} e_1^2 + i^2 \frac{m^2 c^2}{\hbar^2} e_3^2 \psi e_3^2. \tag{6.100}$$

Célérier [91] has then taken advantage of the anticommutative property of the quaternionic units ($e_i e_j = -e_j e_i$ for $i \neq j$) to add to the right-hand side of Eq. (6.100) six vanishing couples of terms which are rearranged as

$$
\frac{1}{c}\frac{\partial}{\partial t}\left(\frac{1}{c}\frac{\partial\psi}{\partial t}\right) = e_3\frac{\partial}{\partial x}\left(e_3\frac{\partial\psi}{\partial x}e_2 + e_1\frac{\partial\psi}{\partial y}i + e_3\frac{\partial\psi}{\partial z}e_1 - i\frac{mc}{\hbar}e_3\psi e_3\right)e_2
$$

$$
+ e_1\frac{\partial}{\partial y}\left(e_3\frac{\partial\psi}{\partial x}e_2 + e_1\frac{\partial\psi}{\partial y}i + e_3\frac{\partial\psi}{\partial z}e_1 - i\frac{mc}{\hbar}e_3\psi e_3\right)i
$$

$$
+ e_3\frac{\partial}{\partial z}\left(e_3\frac{\partial\psi}{\partial x}e_2 + e_1\frac{\partial\psi}{\partial y}i + e_3\frac{\partial\psi}{\partial z}e_1 - i\frac{mc}{\hbar}e_3\psi e_3\right)e_1
$$

$$
- i\frac{mc}{\hbar}e_3\left(e_3\frac{\partial\psi}{\partial x}e_2 + e_1\frac{\partial\psi}{\partial y}i + e_3\frac{\partial\psi}{\partial z}e_1 - i\frac{mc}{\hbar}e_3\psi e_3\right)e_3.
$$

$$(6.101)$$

This form is remarkable, since it directly exhibits the factorization of the biquaternionic Klein–Gordon equation as a squared equation. (More precisely, it may be formally written as $(T^2 - S^2)\psi = 0$, i.e. $(T + S)(T - S)\psi = 0$). Indeed, it is obtained by applying twice to the biquaternionic wavefunction ψ the operator

$$
\frac{1}{c}\frac{\partial}{\partial t} = e_3\frac{\partial}{\partial x}e_2 + e_1\frac{\partial}{\partial y}i + e_3\frac{\partial}{\partial z}e_1 - i\frac{mc}{\hbar}e_3(\ )e_3. \qquad (6.102)
$$

The three first Conway matrices $e_3(\ )e_2$, $e_1(\ )i$ and $e_3(\ )e_1$ (see Sec. 6.3.4 and [504]), appearing in the right-hand side of Eq. (6.102), can be written in the compact form as $-\alpha^k$, with

$$
\alpha^k = \begin{pmatrix} 0 & \sigma_k \\ \sigma_k & 0 \end{pmatrix},
$$

where the σ_k's are the three Pauli matrices, while the fourth Conway matrix

$$
e_3(\ )e_3 = \begin{pmatrix} 1 & 0 & 0 & 0 \\ 0 & 1 & 0 & 0 \\ 0 & 0 & -1 & 0 \\ 0 & 0 & 0 & -1 \end{pmatrix}
$$

is recognized as the Dirac β matrix. We can therefore finally write Eq. (6.102) as the non-covariant Dirac equation for a free fermion

$$
\frac{1}{c}\frac{\partial\psi}{\partial t} = -\alpha^k\frac{\partial\psi}{\partial x^k} - i\frac{mc}{\hbar}\beta\psi. \qquad (6.103)
$$

The covariant form, in the Dirac representation,

$$
(i\hbar\gamma^\mu\partial_\mu - mc)\,\psi = 0, \qquad (6.104)
$$

can be recovered by applying $ie_3(\ )e_3$ to Eq. (6.103).

More generally, as concerns the Dirac equation in the presence of an electromagnetic field, we shall see in Chapter 7 that one can also construct such a U(1) field as a manifestation of the fractal geometry of space-time. This is done by including the effects of the geometry in the very definition of a new covariant derivative in a way similar to the general relativistic construction. This covariant derivative identifies with the standard covariant derivative of quantum electrodynamics, which was a mere rule in this framework but is now demonstrated in scale relativity, as we shall see, namely, $i\hbar D_\mu = i\hbar\partial_\mu - eA_\mu$. We can therefore write the Dirac equation for an electron of mass m and charge e, in an "external" electromagnetic field A_μ, as

$$[\gamma^\mu (i\hbar\partial_\mu - eA_\mu) - mc]\,\psi = 0\,, \tag{6.105}$$

where the wave function is still a complex quaternion.

The isomorphism, which can be established between the quaternionic and spinorial algebrae through the multiplication rules applying to the Pauli spin matrices, allows one to identify the wave function ψ to a Dirac spinor. Indeed, spinors and quaternions are both representations of the SL(2,C) group (see [462] for a detailed discussion of the spinorial properties of biquaternions).

In conclusion, we have shown that both Dirac spinors and the Dirac equation can be obtained as manifestations of the nondifferentiable and fractal geometry of space-time. Such a space-time is essentially non local, since the "particles" are identified with bundles of geodesics. Therefore we recover the non locality of the wave function of standard quantum mechanics. Moreover, having now derived the Dirac equation and the bispinor nature of the wave function exactly in its standard quantum mechanical form (there is no missing or additionnal variable in the nondifferentiable space-time representation, but only a change of variables with the same number of degrees of freedom), some profound aspects of quantum mechanics, such as the EPR paradox [160] and the breaking of Bell inequalities [41, 32], whose subtleties are best expressed when it is applied to spin, are also recovered in the new framework.

Problems and Exercises

Open Problem 10: We have accounted here only for the reversal of the vector dx^μ taken as a whole, $dx^\mu \leftrightarrow -dx^\mu$. Is it possible to obtain a more complete description by taking into account the three additional two-valuednesses, which come from considering independent breakings under the reflection symmetries on the four space-time coordinates separately, $dx \leftrightarrow -dx$, $dy \leftrightarrow -dy$, $dz \leftrightarrow -dz$, $dt \leftrightarrow -dt$?

Open Problem 11: In this chapter, we have taken into account only the classical differentiable part of the velocity field $\mathcal{V}^\mu$ and, therefore, introduced only a standard differentiable wave function ψ. Is it possible to generalize the description by working with the full velocity field $\mathcal{V}^\mu + \mathcal{W}^\mu$, including its nondifferentiable part of zero mean ($\langle \mathcal{W}^\mu \rangle = 0$) as made for the nonrelativistic case in Sec. 5.8? Do we recover in terms of quaternions and biquaternions the fundamental relation $\langle \mathcal{W}^2 \rangle = 0$ of Sec. 5.4.1? Is the Dirac equation still valid in this case, i.e. it would have fractal nondifferentiable solutions?

6.4. Pauli Spinors and Pauli Equation

It is well-known that the Pauli equation can be obtained as a non-relativistic approximation of the Dirac equation (see e.g. [268]). Two of the components of the Dirac bispinor become negligible when $v \ll c$, so that they become Pauli spinors (i.e. in the above representation the biquaternions are reduced to quaternions) and the Dirac equation is transformed in a Schrödinger equation for these spinors with a magnetic dipole additional term. Such an equation is but the Pauli equation. In this operation, the resulting magnetic moment of the electron is found to be twice the value that is classically expected.

6.4.1. *Attempt from reflection invariance breaking in fractal space*

According to the above results, one could expect that spinors and the Pauli equation could be easily understood in the scale-relativity framework as a manifestation of the fractality of space (but not time), while the symmetry breaking of space differential elements is nevertheless at work. If this were the case, it would however be possible to obtain them directly in the non-relativistic case as the manifestation of a nondifferentiable space.

Rather interestingly, such an attempt fails [96]. Indeed, it amounts to constructing four three-dimensional velocity fields $v_{\pm\pm}^k$ from the two symmetry breakings on the reflections $(dt \to -dt)$ and $(dx^k \to -dx^k)$. It has been proposed by Célérier [96] to combine them under a quaternionic velocity field in the following way:

$$\mathcal{V}^k = \frac{1}{2}(v^k_{++} + v^k_{--}) - \frac{i}{2}(v^k_{++} - v^k_{--}) + j\left[\frac{1}{2}(v^k_{+-} + v^k_{-+}) + \frac{i}{2}(v^k_{+-} - v^k_{-+})\right],$$
(6.106)

where i, j and $k = ij = -ji$ are the quaternion units, which we have written e_1, e_2 and e_3 in the complex quaternionic case, since it includes another unit i different from e_1.

The zero-spin case, which corresponds to no $(dx^k \rightarrow -dx^k)$ discrete symmetry breaking, can be easily recovered by making $j = 0$ in this expression. The quaternionic velocity so defined corresponds to a quaternionic derivative operator $\widehat{d}/dt$ similarly defined, which reads

$$\frac{\widehat{d}}{dt} = \frac{\partial}{\partial t} + \mathcal{V} \cdot \nabla - i\mathcal{D}\Delta,$$
(6.107)

as in the Schrödinger case, but with $\mathcal{V}$ now being a quaternion. Then a quaternionic wave function can be introduced through the relation

$$\mathcal{V} = i\frac{S_0}{m}\psi^{-1}\nabla\psi,$$
(6.108)

where ψ^{-1} is the inverse quaternion of ψ. Finally the quaternionic geodesic equation, $\widehat{d}\mathcal{V}/dt = 0$, expressed in terms of this wave function, becomes after some calculations

$$\frac{\partial}{\partial t}\left(\psi^{-1} \cdot \nabla\psi\right) - 2i\mathcal{D}\nabla(\Delta\psi \cdot \psi^{-1}) = 0.$$
(6.109)

The presence of the partial time derivative in this equation totally changes the situation with respect to the cases previously studied. In the complex Schrödinger case, it poses no problem since $\psi^{-1} \cdot \nabla\psi = \nabla \ln\psi$ is a gradient, while in the Klein–Gordon and Dirac cases its equivalent $\partial/\partial s$ is absent. Now, since ψ is a quaternion, $\psi^{-1} \cdot \nabla\psi$ is not a gradient in general, and the above equation cannot be integrated under the form of a Schrödinger equation.

It remains a third-order equation, the physical meaning of which is not obvious. Since in the scale relativity theory the meaning of the function ψ is not set as an axiom (contrary to standard quantum mechanics) but is instead deduced from the developments of the formalism (concerning, in particular, its status of wave function subjected to the superposition principle and the proof of Born's postulate, see Chapter 5), the impossibility to put this equation in the Schrödinger form also deprives such a function ψ of its physical meaning.

This apparent failure is actually a success of the scale relativity theory. Indeed, if it had been possible to integrate it, the resulting equation would have been physically wrong, since, in the presence of a magnetic field, one would not have obtained the correct magnetic moment of the electron.

One could require a genuine fundamental physical theory, not only to be able to derive from physical principles the correct equations of physics, but also to ensure that some physical mechanism would always prevent an equation to be written when it is unphysical. Such a requirement is clearly impossible when the foundation of a physical theory remains axiomatic. This is the case of standard quantum mechanics, where no physical principle prevents *a priori* from writing a Pauli equation with the wrong (non-relativistic) magnetic moment. Such an equation is rejected, not by principle, but by its disagreement with experiment and with the non-relativistic limit of the Dirac equation.

But the attempt of the scale relativity theory to found quantum mechanical laws on first principles allows one to come back to this question, since we have shown that it was indeed impossible to directly write a non-relativistic equation for spin-1/2 particles, and that it could, therefore, only be derived as a non-relativistic limit of the relativistic equation. Let us now discuss in detail how one obtains this limit in terms of the scale relativity description tools.

6.4.2. *Pauli quaternions as limits of Dirac biquaternions*

For the derivation of the Dirac equation in the framework of scale relativity that was developed in [93] and in the hereabove sections, we have been led to make a choice about the way to combine into a biquaternion $\mathcal{V}^\mu$ the eight non-degenerate components of the classical part of the velocity field, $v^\mu_{\pm\pm}$ and $\tilde{v}^\mu_{\pm\pm}$. The final physics does not depend on this choice, but it may be more or less easy to deal with in the calculations. Célérier has therefore been led to propose in [96] a more symmetrical expression,

$$
\begin{aligned}
\mathcal{V}^\mu = {}& \frac{1}{2}(v^\mu_{++} + v^\mu_{--}) - \frac{i}{2}(v^\mu_{++} - v^\mu_{--}) \\
& + \left[\frac{1}{2}(v^\mu_{+-} + v^\mu_{-+}) - \frac{i}{2}(v^\mu_{+-} - v^\mu_{-+}) \right] e_1 \\
& + \left[\frac{1}{2}(\tilde{v}^\mu_{++} + \tilde{v}^\mu_{--}) - \frac{i}{2}(\tilde{v}^\mu_{++} - \tilde{v}^\mu_{--}) \right] e_2 \\
& + \left[\frac{1}{2}(\tilde{v}^\mu_{+-} + \tilde{v}^\mu_{-+}) - \frac{i}{2}(\tilde{v}^\mu_{+-} - \tilde{v}^\mu_{-+}) \right] e_3.
\end{aligned}
\tag{6.110}
$$

Here i denotes the imaginary element used to write the complex components of the biquaternions as sums of a real and an imaginary part and the e_n, with $n = 1, 2, 3$, correspond to the i, j, k imaginary elements often used to write linear expressions of quaternions as a direct generalization of complex numbers. We can therefore write any biquaternionic wavefunction, solution of the Dirac equation, as

$$\psi = \psi_0 + \psi_1 e_1 + \psi_2 e_2 + \psi_3 e_3, \tag{6.111}$$

i.e. in function of its eight real components,

$$\psi = (\eta_0 + i\kappa_0) + (\eta_1 + i\kappa_1)e_1 + (\eta_2 + i\kappa_2)e_2 + (\eta_3 + i\kappa_3)e_3. \tag{6.112}$$

Since ψ is assimilated to a Dirac bi-spinor, it must have the same properties. In particular, it must be a unit quaternion, i.e. a quaternion with a unit norm, which is written as

$$\eta_0^2 + \kappa_0^2 + \eta_1^2 + \kappa_1^2 + \eta_2^2 + \kappa_2^2 + \eta_3^2 + \kappa_3^2 = 1. \tag{6.113}$$

The components of the biquaternionic four-velocity can be calculated from the relation (6.88)

$$\mathcal{V}_\mu = i \frac{S_0}{mc} \psi^{-1} \partial_\mu \psi. \tag{6.114}$$

Then by taking linear combinations, Célérier [96] has obtained the $v_{\pm\pm}^\mu$ velocity fields in terms of the wave function components

$$
\begin{aligned}
v_{++}^\mu = -\frac{S_0}{mc} \big(& \eta_0 \partial^\mu \kappa_0 + \kappa_0 \partial^\mu \eta_0 + \eta_1 \partial^\mu \kappa_1 + \kappa_1 \partial^\mu \eta_1 + \eta_2 \partial^\mu \kappa_2 + \kappa_2 \partial^\mu \eta_2 \\
& + \eta_3 \partial^\mu \kappa_3 + \kappa_3 \partial^\mu \eta_3 + \eta_0 \partial^\mu \eta_0 - \kappa_0 \partial^\mu \kappa_0 + \eta_1 \partial^\mu \eta_1 - \kappa_1 \partial^\mu \kappa_1 \\
& + \eta_2 \partial^\mu \eta_2 - \kappa_2 \partial^\mu \kappa_2 + \eta_3 \partial^\mu \eta_3 - \kappa_3 \partial^\mu \kappa_3 \big),
\end{aligned} \tag{6.115}
$$

$$
\begin{aligned}
v_{--}^\mu = -\frac{S_0}{mc} \big(& \eta_0 \partial^\mu \kappa_0 + \kappa_0 \partial^\mu \eta_0 + \eta_1 \partial^\mu \kappa_1 + \kappa_1 \partial^\mu \eta_1 + \eta_2 \partial^\mu \kappa_2 \\
& + \kappa_2 \partial^\mu \eta_2 + \eta_3 \partial^\mu \kappa_3 + \kappa_3 \partial^\mu \eta_3 - \eta_0 \partial^\mu \eta_0 + \kappa_0 \partial^\mu \kappa_0 - \eta_1 \partial^\mu \eta_1 \\
& + \kappa_1 \partial^\mu \kappa_1 - \eta_2 \partial^\mu \eta_2 + \kappa_2 \partial^\mu \kappa_2 - \eta_3 \partial^\mu \eta_3 + \kappa_3 \partial^\mu \kappa_3 \big),
\end{aligned} \tag{6.116}
$$

$$
\begin{aligned}
v_{+-}^\mu = -\frac{S_0}{mc} \big(& \eta_0 \partial^\mu \kappa_1 + \kappa_0 \partial^\mu \eta_1 - \eta_1 \partial^\mu \kappa_0 - \kappa_1 \partial^\mu \eta_0 - \eta_2 \partial^\mu \kappa_3 - \kappa_2 \partial^\mu \eta_3 \\
& + \eta_3 \partial^\mu \kappa_2 + \kappa_3 \partial^\mu \eta_2 + \eta_0 \partial^\mu \eta_1 - \kappa_0 \partial^\mu \kappa_1 - \eta_1 \partial^\mu \eta_0 \\
& + \kappa_1 \partial^\mu \kappa_0 - \eta_2 \partial^\mu \eta_3 + \kappa_2 \partial^\mu \kappa_3 + \eta_3 \partial^\mu \eta_2 - \kappa_3 \partial^\mu \kappa_2 \big),
\end{aligned} \tag{6.117}
$$

$$v^\mu_{-+} = -\frac{S_0}{mc}\big(\eta_0\partial^\mu\kappa_1 + \kappa_0\partial^\mu\eta_1 - \eta_1\partial^\mu\kappa_0 - \kappa_1\partial^\mu\eta_0 - \eta_2\partial^\mu\kappa_3 - \kappa_2\partial^\mu\eta_3$$
$$+ \eta_3\partial^\mu\kappa_2 + \kappa_3\partial^\mu\eta_2 - \eta_0\partial^\mu\eta_1 + \kappa_0\partial^\mu\kappa_1 + \eta_1\partial^\mu\eta_0 - \kappa_1\partial^\mu\kappa_0$$
$$+ \eta_2\partial^\mu\eta_3 - \kappa_2\partial^\mu\kappa_3 - \eta_3\partial^\mu\eta_2 + \kappa_3\partial^\mu\kappa_2\big), \tag{6.118}$$

$$\tilde{v}^\mu_{++} = -\frac{S_0}{mc}\big(\eta_0\partial^\mu\kappa_2 + \kappa_0\partial^\mu\eta_2 + \eta_1\partial^\mu\kappa_3 + \kappa_1\partial^\mu\eta_3 - \eta_2\partial^\mu\kappa_0 - \kappa_2\partial^\mu\eta_0$$
$$- \eta_3\partial^\mu\kappa_1 - \kappa_3\partial^\mu\eta_1 + \eta_0\partial^\mu\eta_2 - \kappa_0\partial^\mu\kappa_2 + \eta_1\partial^\mu\eta_3$$
$$- \kappa_1\partial^\mu\kappa_3 - \eta_2\partial^\mu\eta_0 + \kappa_2\partial^\mu\kappa_0 - \eta_3\partial^\mu\eta_1 + \kappa_3\partial^\mu\kappa_1\big), \tag{6.119}$$

$$\tilde{v}^\mu_{--} = -\frac{S_0}{mc}\big(\eta_0\partial^\mu\kappa_2 + \kappa_0\partial^\mu\eta_2 + \eta_1\partial^\mu\kappa_3 + \kappa_1\partial^\mu\eta_3 - \eta_2\partial^\mu\kappa_0$$
$$- \kappa_2\partial^\mu\eta_0 - \eta_3\partial^\mu\kappa_1 - \kappa_3\partial^\mu\eta_1 - \eta_0\partial^\mu\eta_2 + \kappa_0\partial^\mu\kappa_2 - \eta_1\partial^\mu\eta_3$$
$$+ \kappa_1\partial^\mu\kappa_3 + \eta_2\partial^\mu\eta_0 - \kappa_2\partial^\mu\kappa_0 + \eta_3\partial^\mu\eta_1 - \kappa_3\partial^\mu\kappa_1\big), \tag{6.120}$$

$$\tilde{v}^\mu_{+-} = -\frac{S_0}{mc}\big(\eta_0\partial^\mu\kappa_3 + \kappa_0\partial^\mu\eta_3 - \eta_1\partial^\mu\kappa_2 - \kappa_1\partial^\mu\eta_2 + \eta_2\partial^\mu\kappa_1$$
$$+ \kappa_2\partial^\mu\eta_1 - \eta_3\partial^\mu\kappa_0 - \kappa_3\partial^\mu\eta_0 + \eta_0\partial^\mu\eta_3 - \kappa_0\partial^\mu\kappa_3 - \eta_1\partial^\mu\eta_2$$
$$+ \kappa_1\partial^\mu\kappa_2 + \eta_2\partial^\mu\eta_1 - \kappa_2\partial^\mu\kappa_1 - \eta_3\partial^\mu\eta_0 + \kappa_3\partial^\mu\kappa_0\big) \tag{6.121}$$

and

$$\tilde{v}^\mu_{-+} = -\frac{S_0}{mc}\big(\eta_0\partial^\mu\kappa_3 + \kappa_0\partial^\mu\eta_3 - \eta_1\partial^\mu\kappa_2 - \kappa_1\partial^\mu\eta_2 + \eta_2\partial^\mu\kappa_1$$
$$+ \kappa_2\partial^\mu\eta_1 - \eta_3\partial^\mu\kappa_0 - \kappa_3\partial^\mu\eta_0 - \eta_0\partial^\mu\eta_3 + \kappa_0\partial^\mu\kappa_3 + \eta_1\partial^\mu\eta_2$$
$$- \kappa_1\partial^\mu\kappa_2 - \eta_2\partial^\mu\eta_1 + \kappa_2\partial^\mu\kappa_1 + \eta_3\partial^\mu\eta_0 - \kappa_3\partial^\mu\kappa_0\big). \tag{6.122}$$

(Note the change of notation with respect to [96], where the η_n's were denoted ϕ_n's and the κ_n's, χ_n: its motivation is not to create confusion with the small and large terms of a Dirac spinor, usually called ϕ and χ, that we now introduce.)

We are now in a position to take the non-relativistic limit of these velocity fields. Recall that a Dirac spinor $\psi = \psi_0 + \psi_1 e_1 + \psi_2 e_2 + \psi_3 e_3$ can be decomposed into two spinors ϕ and χ,

$$\psi = \begin{pmatrix} \phi \\ \chi \end{pmatrix}. \tag{6.123}$$

In the quaternionic representation, this amounts to shifting to a symplectic form of ψ, where

$$\phi = \psi_0 + e_1\psi_1, \tag{6.124}$$

and

$$\chi = \psi_2 - e_1 \, \psi_3, \tag{6.125}$$

so that the spinor reads $\psi = \phi + e_2 \, \chi$ since $e_3 = e_1 e_2$.

In the non-relativistic limit, the rest energy mc^2 becomes dominant and can be eliminated from the non-relativistic spinor by making the transformation $\psi \to \psi \, e^{imc^2 t/\hbar}$. Under this approximation, the components χ of the spinor become much smaller than ϕ. The substitution of this non-relativistic solution into the Dirac equation (6.105) yields, after some calculations (see e.g. [268]), a relation between the small and the large component that reads in the Dirac representation

$$\chi = \frac{\sigma \cdot \left(i\hbar \nabla - \frac{e}{c} \mathbf{A} \right)}{2mc} \, \phi, \tag{6.126}$$

i.e. $\chi \approx (v/c)\phi$, which justifies the approximation $\chi \ll \phi$. The equation for ϕ becomes

$$i\hbar \frac{\partial \phi}{\partial t} = \left[\frac{1}{2m} \left(i\hbar \nabla - \frac{e}{c} \mathbf{A} \right)^2 - \frac{e\hbar}{2mc} \sigma \cdot \mathbf{B} + eA_0 \right] \phi. \tag{6.127}$$

We recognize here the Pauli equation for the theory of spin in non-relativistic quantum mechanics with $i\hbar \nabla$ replacing the momentum operator $\hat{p}$ in accordance with the correspondence principle. As it is well-known, one of the main results of this equation (when it is derived from the Dirac equation as we did) is to yield the correct gyromagnetic factor $g = 2$ of the electron.

We can now go back to the scale relativity framework and translate the decomposition of a Dirac spinor in its small and large components into the corresponding property of the velocity fields. From Eqs. (6.124) and (6.125), one sees that the small components correspond to terms with an index 2 or 3. Since these components do not appear in the final Pauli equation, they vanish in the non-relativistic limit. By looking at equations (6.115)–(6.122), we then find, as a consequence, that all the tilde velocity fields vanish.

Now, these velocity fields had been introduced as a consequence of a doubling due to the breaking of parity, which is specific of the relativistic case. This is on accordance with the known fact that one passes from Pauli spinors to Dirac bispinors precisely because of a relativistic effect. Namely, some components of a Pauli spinor, after the action of a Lorentz boost, are transformed into the components of another "punctuated" spinor and reciprocally, so that one recovers invariance in terms of a couple of spinors, which is a Dirac bispinor. This means that there is actually no new information in a bispinor, this doubling being actually a kind of mirror effect.

Finally the only non-zero velocity fields are the four four-velocities v_{++}, v_{+-}, v_{-+} and v_{--}.

Now the Pauli spinor ϕ is obtained from the Dirac bi-spinor ψ by keeping only the large components of indices 0 and 1, namely (after the $e^{imc^2t/\hbar}$ transformation),

$$\phi = (\psi_0 + e_1\psi_1) = (\eta_0 + i\kappa_0) + (\eta_1 + i\kappa_1)e_1. \qquad (6.128)$$

This implies that we obtain the quaternionic velocity corresponding to the non-relativistic limit, first, by neglecting the small components in the expression of the biquaternionic four-velocity of equation (6.110), which gives

$$\begin{aligned}
\mathcal{V}^\mu = \frac{\mathcal{S}_0}{mc}\Big\{ &-(\eta_0\partial^\mu\kappa_0 + \kappa_0\partial^\mu\eta_0 + \eta_1\partial^\mu\kappa_1 + \kappa_1\partial^\mu\eta_1) \\
&+ (\eta_0\partial^\mu\eta_0 - \kappa_0\partial^\mu\kappa_0 + \eta_1\partial^\mu\eta_1 - \kappa_1\partial^\mu\kappa_1)i \\
&+ \big[-\eta_0\partial^\mu\kappa_1 - \kappa_0\partial^\mu\eta_1 + \eta_1\partial^\mu\kappa_0 + \kappa_1\partial^\mu\eta_0 \\
&+ (\eta_0\partial^\mu\eta_1 - \kappa_0\partial^\mu\kappa_1 - \eta_1\partial^\mu\eta_0 + \kappa_1\partial^\mu\kappa_0)i\,\big]e_1\Big\}, \qquad (6.129)
\end{aligned}$$

then, by distinguishing its component $\mathcal{V}^0$ from its three spatial components. As expected, one finds that it is equal to the velocity of light c [96]. One finally finds that the resulting three-dimensional velocity reads

$$\begin{aligned}
\mathcal{V}^k = \frac{1}{2}\left(v_{++}^k + v_{--}^k\right) - \frac{i}{2}\left(v_{++}^k - v_{--}^k\right) \\
+ \left[\frac{1}{2}\left(v_{+-}^k + v_{-+}^k\right) - \frac{i}{2}\left(v_{+-}^k - v_{-+}^k\right)\right]e_1. \qquad (6.130)
\end{aligned}$$

Since i and e_1 are both roots of -1, one recovers the standard quaternionic notation and the expression of Eq. (6.106) by setting $j = e_1$ and $k = i\,e_1$. We have therefore obtained the velocity corresponding to the non-relativistic case from the mere degeneracy of the biquaternionic relativistic one, in the symplectic form of a real quaternion. This degeneracy yields the non-relativistic velocity as a three-vector naturally derived from the four-velocity of the relativistic case.

In this way one obtains the correct Pauli equation for the spinor wave function constructed from this velocity field, thus proving the fundamentally relativistic and quantum nature of spin, which we attribute in scale relativity to the nondifferentiability of the quantum space-time geometry and not only of the quantum space.

6.4.3. *Spin as internal angular momentum of fractal geodesics*

6.4.3.1. *Geometric models of spin*

The geometric description of quantum physics brought by the scale-relativity and fractal space-time approach allows one to give a physical picture of what the spin actually is. Recall that the spin has been considered, since its discovery, as a physical quantity of pure quantum origin having no classical counterpart. Indeed, assuming an extension of the electron of the order of its classical radius $r_e = \alpha^{-1} \lambda_c$, where $\alpha = 1/137.036...$ is the fine structure constant and $\lambda_c = \hbar/mc$ is its Compton length, an angular momentum $\hbar/2$ would involve velocity of rotation of its surface of order $\alpha^{-1} c$, which is clearly excluded by special relativity.

The scale relativity theory allows one to suggest a new solution to this fundamental problem. This solution remains non-classical (owing to the fact that an everywhere nondifferentiable space-time is non-classical, as proved by the quantum-mechanical-type behavior of its geodesics), but it is however a geometric solution.

As recalled in this chapter, in the scale relativistic framework, the complex nature of the wave function and the existence of spin have both a common origin, namely, the fundamental two-valuedness of the derivative coming from nondifferentiability.

A model for the emergence of a spin-like internal angular momentum (that was, however, not yet quantized in units of $\hbar/2$) in fractal spiral curves of fractal dimension 2 had been proposed a long time ago [349, 353]. Note that this kind of fractal spiral curve has recently known a renewal of interest under the name "hyperhelices" [180, 517].

Let us briefly recall here the argument (see Fig. 6.1). One considers a fractal helice path of fractal dimension $D_F = 2$. The angular momentum $L_z = mr^2 \dot{\varphi}$ of a point mass following this path should classically vanish for $r \to 0$. But in the fractal spiral model, when a scale factor q^{-1} is applied to the radius r, the number of turns and therefore the rotation velocity is multiplied by a factor $p = q^{D_F}$, so that the angular momentum becomes multiplied by a factor $p \times q^{-2} = q^{D_F - 2}$. It therefore remains defined at the infinitely small limit $q^{-n} \to 0$ in the special case $D_F = 2$. In other words, the rotation velocity $\dot{\varphi} \to \infty$ when the radius $r \to 0$ in such a way that the

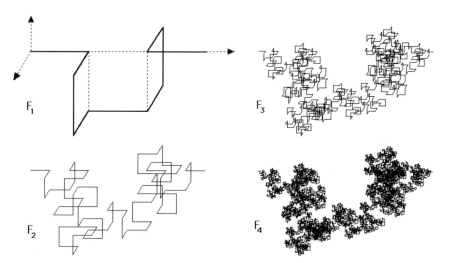

Fig. 6.1: **Model of fractal hyperhelice.** First four iterations on one period of an early model of infinitely spiral fractal curve ("hyperhelice"), from [349]. Its generator being made of nine segments of length $1/3$, its fractal dimension is $D_F = \log 9/\log 3 = 2$. The spin of such a curve, whose fundamental period is a de Broglie wavelength $\lambda_{dB} = 2\pi\hbar/mv$, is in the particular case shown on the figure $\sigma = 0.42\,\hbar$.

product $r^2\dot{\varphi}$ remains finite when $D_F = 2$, while it is vanishing for $D_F < 2$ and divergent for $D_F > 2$.

This model of spin therefore uses in an essential way the scale dependence of fractal geometry, which allows one to deal with vanishing and infinite quantities in a new manner. While in the standard differentiable approach, the encounter of a zero or infinite quantity usually leads one to stop a calculation, the explicitly scale-dependent tools of the scale relativity theory allows to go beyond such an obstacle and to prove the existence of finite and measurable quantities, which would remain undefined with standard methods.

Note that the second order terms in the quantum covariant total derivative, which are, in this framework, the basis of the Heisenberg relations, are exactly of the same nature: namely, $d\xi^2/dt$ is a differential element of first order in the differentiable theory, so that it should classically vanish. But in the fractal dimension 2 case, $d\xi^2/dt = (d\xi/dt)^2 \times dt = (2\mathcal{D}/dt) \times dt = 2\mathcal{D}$ is now finite since the fractal velocity $d\xi/dt$ is now formally infinite (at the limit $dt \to 0$).

This result solves the problem of the apparent impossibility to define a spin in a geometric way both for an extended object and for a point-like object and provides another proof of the critical character of the value $D_F = 2$ for the fractal dimension of quantum particle paths [176].

It is also remarkable that the existence of spiral structures at all scales is also one of the elements of description of spinors in the framework of Ord's reformulation of the Feynman relativistic chessboard model in terms of spiral paths [421].

6.4.3.2. *Numerical simulations of exact solutions*

We can now go beyond these fractal models of the spin and give a geometric physical picture of its nature [96] based on explicit solutions of the Pauli or Dirac equation, since the fractal velocity fields of the nondifferentiable space-time geodesics can be derived from these solutions. It is remarkable that this now exact geometric description, whose spin is quantized in units of $\hbar/2$, supports the main features of the previous rough fractal models.

In order to exhibit this picture, we shall perform numerical simulations of the stochastic differential equations that have been set at the origin of the description. Recall that we have decomposed the elementary displacements in a fractal space-time in terms of a classical (differentiable) part and of a fractal (nondifferentiable) part,

$$dX_{\pm\pm} = v_{\pm\pm}\, dt + d\xi_{\pm\pm}, \qquad (6.131)$$

where the geometric fractal fluctuation is replaced by a stochastic variable such that $\langle d\xi^2\rangle/c\,dt = \lambda_c$ and $\langle d\xi\rangle = 0$. Then the velocity fields $v_{\pm\pm}$, after they have been recombined as a unique biquaternionic velocity field, are solution of a bispinorial geodesic equation $\hat{d}\mathcal{V}/ds = 0$, which can be integrated in terms of the Dirac equation, whose non-relativistic limit is finally the Pauli equation. Therefore, solving the Pauli equation for a given physical problem yields a quaternionic wave function

$$\psi = \eta_0 + i\kappa_0 + (\eta_1 + i\kappa_1)e_1, \qquad (6.132)$$

from which the velocity fields can be derived, for example,

$$\begin{aligned}
v^\mu_{++} = -\frac{S_0}{mc}\big(&\eta_0\,\partial^\mu\kappa_0 + \kappa_0\,\partial^\mu\eta_0 + \eta_1\,\partial^\mu\kappa_1 + \kappa_1\,\partial^\mu\eta_1 \\
&+ \eta_0\,\partial^\mu\eta_0 - \kappa_0\,\partial^\mu\kappa_0 + \eta_1\,\partial^\mu\eta_1 - \kappa_1\,\partial^\mu\kappa_1\big).
\end{aligned} \qquad (6.133)$$

Then one can finally construct various realizations of the geodesics by performing numerical integrations of the stochastic differential equation

(see more details in Sec. 10.5)

$$dX_{++} = v_{++}\, dt + \zeta \sqrt{\lambda c}\, dt, \tag{6.134}$$

in which the explicit form of v_{++}, given by Eq. (6.133), is inserted, and where ζ is a normalized stochastic variable such that $\langle \zeta^2 \rangle = 1$ and $\langle \zeta \rangle = 0$.

A general form of a spinor wave function has been given by Cohen-Tannoudji *et al* [106], namely, for a spin $1/2$ particle,

$$|\psi\rangle = \cos(\theta/2)\, e^{-i\phi/2}\, |+\rangle + \sin(\theta/2)\, e^{i\phi/2}\, |-\rangle. \tag{6.135}$$

A simplified case has been studied by Dezael [137], who has considered the spinor

$$\psi = A_0\, e^{-\frac{i}{\hbar}(\overrightarrow{p_0}.\overrightarrow{r} - E_0 t + \sigma_0 \phi)} + A_1\, e^{-\frac{i}{\hbar}(\overrightarrow{p_1}.\overrightarrow{r} - E_1 t + \sigma_1 \phi)}. \tag{6.136}$$

From this expression, the biquaternionic velocity field, given by $m\overrightarrow{V} \simeq i\hbar \psi^{-1}\overrightarrow{\nabla}\psi$ can be derived. It must be such that $\overrightarrow{v}_{-+} = -\overrightarrow{v}_{++}$ in the nonrelativistic approximation considered here. This is possible only provided $\overrightarrow{p_0} = \overrightarrow{p_1}$, and $\sigma_0 = \sigma_1$ [137]. These conditions greatly simplify the expression of the biquaternionic velocity, which is reduced to

$$\overrightarrow{V} \simeq \overrightarrow{v}_{++} \simeq \frac{1}{m} \left(\overrightarrow{p_0} + \frac{\sigma_0}{r \sin\theta} \overrightarrow{u_\phi} \right), \tag{6.137}$$

where $\overrightarrow{u_\phi}$ is the unitary vector associated to the rotation by ϕ around $\overrightarrow{p_0}$. Dezael finally obtains the following differential equations, now written in Cartesian coordinates:

$$\begin{cases} \dot{x} = -\dfrac{\sigma_0}{m}\dfrac{y}{x^2 + y^2}, \\[2mm] \dot{y} = \dfrac{\sigma_0}{m}\dfrac{x}{x^2 + y^2}, \\[2mm] \dot{z} = \dfrac{p_0}{m}. \end{cases} \tag{6.138}$$

These equations clearly describe a class of spiral motions such that $mr^2\dot{\varphi} = \sigma_0$, where r can take any value, so that $r \to 0$ implies $\dot{\varphi} \to \infty$, in agreement with the properties of the early spiral models [349] briefly described hereabove.

Then one may finally carry over this expression for the classical velocity field into the stochastic differential equation (6.134) and integrate it numerically. A typical example of the spiral fractal paths obtained in this process is shown in Fig. 6.2.

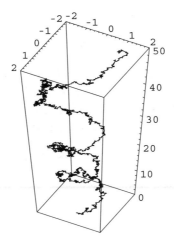

Fig. 6.2: Spinorial geodesic of a fractal space. Numerical simulation of a typical spinorial geodesic of a fractal space. This is one realization among an infinity of possible realizations of the solutions of the stochastic differential equation (6.134). We have chosen a large value for the spin, $\sigma = 5\,\hbar$, in order to render the spiral shape fully apparent. Note that a "same" given curve may be plotted at an infinity of possible resolution values dt, and that there is an infinity of such fractal geodesics (see e.g. numerical simulations in Sec. 10.5).

Problems and Exercises

Exercise 17 Generalize the above results to any value of the spin, in particular to spin one in the massive and non massive (electromagnetic) cases.

Hints: The wave function of a spin s particle is a symmetric spinor of rank $n = 2s$. One may then use the standard quantum mechanical result according to which a particle of spin s is equivalent to a system of $n = 2s$ particles of spin $1/2$, oriented in such a way that the total spin be s [267], to translate this description in terms of quaternions, then of the velocity fields of a fluid of geodesics.   ∎

Exercise 18 Develop the quantization of wave fields (*second quantization*) in the scale relativity framework.

Hints: Use the connection of this problem with the many-particle Schrödinger equation (see [474], p. 504), as developed in Sec. 5.10. Define the creation, destruction and number operators and give their geometric meaning.

In the spin one particle case, one may use the Proca equation (see e.g. [268]), which amounts to a Klein–Gordon equation for a spin 1 spinor, then the quantization is performed by the introduction of the creation-destruction operators. The electromagnetic case is then obtained by making $m = 0$.   ∎

Exercise 19 (proposed by M.N. Célérier). Derive Eq. (6.95) with the definition (6.87) of the biquaternionic wave function. Then show that the inverse choice $\partial_\mu \psi \, \psi^{-1} = (i/S_0)\partial_\mu S$ leads to an equation of motion, which cannot be put under a gradient form and, therefore, cannot be interpreted as a Klein–Gordon equation, since it remains a third order equation, the physical meaning of which is not obvious.

Hints: Use Eqs. (6.94) and the property $\partial^\nu \partial_\nu \partial_\mu = \partial_\mu \partial^\nu \partial_\nu$.   ∎

6.5. Generalization of the Klein–Gordon and Dirac Equations

The generalization of the Klein–Gordon and Dirac equations to scale laws more general than the constant fractal dimension $D_F = 2$ case considered up to now in this chapter is straightforward, following the same lines as in Sec. 5.14 for the generalization of the Schrödinger equation.

Indeed, all the relativistic construction that lead to the Klein–Gordon, Dirac, then Pauli equations, relies on the basic description of the fractal fluctuations generalized to four dimensions given in Eq. (6.6). This description is easily generalizable to a fractal dimension different from 2, and even possibly variable, as

$$\langle d\xi_\pm^\mu \, d\xi_\pm^\nu \rangle = \mp \lambda_D \, \eta^{\mu\nu} ds^{2/D_F}, \tag{6.139}$$

where λ_D identifies with the Compton scale λ when $D_F = 2$.

As in the nonrelativistic case, one may therefore incorporate the difference with the $D_F = 2$ situation by defining a scale dependent generalized Compton scale,

$$\widetilde{\lambda} = \lambda \left(\frac{|ds|}{\lambda} \right)^{(2/D_F)-1}, \tag{6.140}$$

in terms of which one recovers the form of the $D_F = 2$ case,

$$\langle d\xi_\pm^\mu \, d\xi_\pm^\nu \rangle = \mp \widetilde{\lambda} \, \eta^{\mu\nu} ds. \tag{6.141}$$

In other words, this is equivalent to the definition of a generalized Planck's "constant" that becomes explicitly scale dependent,

$$\widetilde{\hbar} = \hbar \left(\frac{|ds|}{\lambda} \right)^{(2/D_F)-1}. \tag{6.142}$$

The subsequent steps of the derivation of the Klein–Gordon equation are preserved, so that one obtains a scale-dependent generalized equation,

$$\widetilde{\lambda}^2 \, \partial^\mu \partial_\mu \psi + \psi = 0, \tag{6.143}$$

which can be solved in terms of an explicitly scale dependent (even possibly fractal and nondifferentiable) wave function.

Since the Dirac equation is naturally obtained from the factorization of the biquaternionic Klein–Gordon equation, it is also easily generalized as

$$\left(i \widetilde{\hbar} \gamma^\mu \partial_\mu - mc \right) \psi = 0. \tag{6.144}$$

This generalization becomes particularly relevant in elementary particle and high energy physics, at energies far larger than the mass energy of a particle, for which we have suggested [352, 353, 360] that the log-Lorentzian scale laws of the special scale relativity theory (see Sec. 4.4), that accounts for the genuine nature of the Planck length-scale as being invariant under dilations, should generalize the "Galilean" law of constant fractal dimension 2.

Such an explicit scale dependence of the generalized Dirac equation is actually not new. Indeed, the account of radiative corrections have also led to the introduction of electromagnetic form factors for the electron in quantum field theory (see, e.g. [268, 57]), which depend on the energy-momentum scale (being even logarithmically divergent) and enter in the Dirac equation. We shall consider these questions in more detail in Chapter 11, showing in particular that these two scale dependences (radiative corrections and scale-relativistic corrections) can be combined to yield a model for the emergence of the muon and the origin of its mass (Sec. 11.1.3.3).

Chapter 7

GEOMETRIC THEORY OF GAUGE FIELDS IN SCALE RELATIVITY

7.1. Introduction

Let us now review another important field of application of the fractal space-time/scale-relativity theory. In the previous chapters, we have (i) constructed from first physical principles the laws of scale transformation, which characterize the fractal structures of the geodesics of a nondifferentiable space-time; (ii) studied the consequences on motion of these fractal structures and found that they give rise to quantum mechanics both in the non-relativistic and relativistic cases.

In this description, the resolution variables $\mathbb{V} = \ln(\lambda/\varepsilon)$ can take all the values of the scale-space, but, as a first step, they do not themselves vary in function of other variables. Then we have considered the new situation of "scale dynamics", in which "scale-accelerations" are defined, so that the resolutions may vary with the "djinn" (variable scale-dimension).

We shall now consider the next step of the construction of the scale relativity theory. It consists of generalizing the scale variables ε to explicit functions of the space-time coordinates, $\varepsilon = \varepsilon(x, y, z, t)$. This means that the resolutions become themselves a field. Such a case can be described as a coupling between motion and scales, but it also comes under a "general scale-relativistic" description in which scale and motion are treated on the same footing. This approach provides us with a new interpretation of gauge transformations and therefore with a geometric interpretation of the nature of gauge fields, first in the Abelian case [356, 360, 384], then in the non-Abelian case [385, 394, 409].

In the present physical theory, one still does not really understand the nature of the electric charge and of the electromagnetic field. As recalled by Landau ([266], Chapter 16), in the classical theory the very existence of the charge e and of the electromagnetic four-potential A_μ are ultimately derived from experimental data. Moreover, the form of the action for a particle in an electromagnetic field cannot be chosen only from general considerations and it is, therefore, merely postulated. In other words, the Lorentz force must be added to Maxwell's equations in today's theory of electromagnetism.

This is to be compared to the status of gravitation in Einstein's theory [158]. The "charge" for gravitation is the energy-momentum itself, which is fully understood as the conservative quantities that take their origin in the fundamental symmetries of space-time (following Noether's theorem). Then the gravitational field is understood as the manifestation of a geometric property of space-time, namely, its curvature, which is itself self-imposed from the principle of general relativity. Finally, there is no need to add an independent equation of trajectories to the theory, since it is identified with the equation of geodesics, which is completely determined by the knowledge of the space-time geometry.

In comparison, the present state of the foundation of the classical electromagnetic theory looks far less satisfactory. This remains a standard field theory, not a geometric theory based on the principle of relativity. Up to now, the various attempts of foundation of electromagnetism on a space-time approach (Kaluza–Klein, see [291], Weyl [535], Dirac [140], etc.) have failed to yield new results that would have allowed to validate or refute them.

More recently, the quantum theory of electromagnetism and of the electron has added a new and essential stone in our understanding of the nature of charge. Indeed, in its framework, gauge invariance becomes deeply related to phase invariance of the wave function. The electric charge conservation is therefore directly related to the gauge symmetry. Such an understanding has led to important progress. In particular, the extension of the approach to non-Abelian gauge theories has allowed to incorporate the weak and strong field into the same scheme (see e.g. [7]).

However, despite this progress, the lack of a fundamental understanding of the nature of the gauge transformations and of the "arbitrary" gauge functions has up to now prevented from reaching the final goal of gauge theories: a genuine understanding of the nature of charge, of the origin of charge quantization and, as a consequence, the ability to theoretically predict its quantized value.

The theory of scale relativity, thanks to its generalization of the geometry of space-time to continuous but nondifferentiable geometries, allows one to reconsider this problem.

In this chapter, we first develop the classical theory of electrodynamics in the scale-relativity theory. The gauge fields can be constructed as manifestations of the dilations of the scale variables induced by space-time displacements. The theory also allows one to give a geometric meaning to the charges, which are defined as the conservative quantities that are built from the new scale symmetries. Then the equation of motion of a charged particle in an electromagnetic field is constructed in terms of a geodesic equation in a fractal space-time.

We then approach the quantum theory of electrodynamics by combining the various tools of the scale relativity approach, which have given rise to the quantum description on one hand and to classical electrodynamics on the other. The covariant derivative of quantum electrodynamics (QED) is constructed in a geometric way. Then we write a doubly covariant free-like geodesic equation, which can be finally integrated in terms of the Klein–Gordon and Dirac equations for a particle in an electromagnetic field.

As a consequence of this approach, since the Planck length-scale becomes invariant under dilations in the framework of the log-Lorentzian laws of special scale-relativity (Sec. 4.4), the quantization of charges is ensured because the possible scale ratios become limited. Then one theoretically predicts the existence of relations between coupling constants and mass scales, whose validity is supported by experimental data.

Then we move to non-Abelian gauge theories by considering more general transformations of the scale variables. Instead of working with only a global dilation, we take into account their general definition, which is actually tensorial, (see Sec. 3.6) and therefore their general transformations in the "scale space". This yields a geometric basis for non-Abelian gauge transformations. The gauge fields naturally appear as a new geometric contribution to the total variation of the action involving the scale variables. A generalized action is identified with the scale-relativistic invariant. The gauge charges emerge as the generators of the scale transformation group, i.e. they are identified with the conservative quantities, conjugates of the scale variables through the action, which find their origin in the now more complete symmetries of the "scale-space". In this way we found in a geometric way and recover the expression for the covariant derivative of gauge theories. Adding the requirement that under the scale transformations the fermion multiplets and the boson fields transform such that the derived Lagrangian remains invariant, we obtain Yang–Mills gauge theories as a consequence of scale symmetries issued from a geometric space-time description.

7.2. Statement of the Problem

Consider an electron in an electromagnetic potential A_μ. In the standard theory, it is well-known that this potential is invariant under a gauge tansformation $A_\mu \rightarrow A'_\mu = A_\mu + \partial_\mu \chi(x, y, z, t)$. Let us consider the wave function of an electron of well-defined energy, momentum, spin and charge. It may be written under the form

$$\psi = \psi_0 \exp\left\{ \frac{i}{\hbar}(px - Et + \sigma\varphi + e\chi) \right\}. \tag{7.1}$$

Its phase contains the usual products of fundamental quantities (space position, time, angle) and of their conjugate quantities (momentum, energy, angular momentum). They are related through Noether's theorem. Namely, the conjugate variables are the conservative quantities that originate from the space-time symmetries. This means that our knowledge of what are the energy, the momentum and the angular momentum and of their physical properties is founded on our knowledge of the nature of space, time and its transformations (translations and rotations).

This is true already in the classical theory, but there is something more in the quantum theory. There, the conservative quantities become quantized when the basic variables on which they are founded are limited. Concerning energy-momentum, this means that it is quantized only in some specific circumstances, e.g. bound states in atoms for which $r > 0$ in spherical coordinates. The case of the angular momentum is instructive. Indeed, its differences are quantized in an universal way in units of $\hbar$ because angles differences can not exceed 2π.

In comparison, the last term in the phase of Eq. (7.1) keeps a special status in today's standard theory. The gauge function $\chi(x, y, z, t)$ remains arbitrary, while it is clear from a comparison with the other terms that the meaning of charge e and the reason for its universal quantization can be obtained only by understanding the physical meaning of χ and why it is universally limited, since it is nothing but the quantity conjugate to the charge. As we shall see in what follows, the identification of χ with a resolution scale factor $\ln \varrho$ allows one to suggest solutions to these problems in the special scale-relativity framework [356].

7.3. Classical Electrodynamics in Scale Relativity

7.3.1. *Scale covariant derivative*

The theory of scale relativity allows one to get new insights about the nature of the electromagnetic field, of the electric charge, and about the physical meaning of gauge invariance [356, 360]. Consider indeed an electron or any other charged particle. In scale relativity, we identify the particle with a family of fractal paths, described as the geodesics of a nondifferentiable space-time. These paths are characterized by internal structures, which are fractal, i.e. explicitly dependent on one (or several) scale variable(s) that we have named "resolution" (by extension of the concept of resolution of a measurement apparatus).

Now, consider any one of these structures, lying at some (relative) resolution ε smaller that the Compton length of the particle (i.e. such that $\varepsilon < \lambda$) for a given relative position of the particle. In a displacement of the particle, the relativity of scales implies that the resolution at which this given structure appears in the new position will *a priori* be different from the initial one. Indeed, if the whole internal fractal structure of the electron was rigidly fixed, this would mean an absolute character of the scale space and a description of the fractal set of geodesics in terms of fractal rigid objects. Such a description would be clearly physically irrelevant and contradictory with the principle of relativity of scales.

This reasoning is quite analogous to Einstein's reasoning in general motion relativity, according to which, in a parallel transport a vector cannot remain identical to itself, since, in this case, space would be absolute, in contradiction with the principle of relativity.

Therefore we expect, in general, the occurrence of dilations of resolutions induced by translations (see Fig. 7.1). In other words, the scale variation includes a purely geometric increase, which reads

$$\delta\left(\ln\frac{\lambda}{\varepsilon}\right) = \frac{1}{q}A_\mu dx^\mu, \tag{7.2}$$

i.e.

$$q\frac{\delta\varepsilon}{\varepsilon} = -A_\mu dx^\mu. \tag{7.3}$$

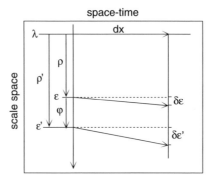

Fig. 7.1: **Scale dilations induced by space-time displacements.** Under an infinitesimal space-time displacement dx^μ of a fractal geodesic, a structure, which was lying at a relative scale $\varepsilon = \rho\lambda$ with respect to the Compton scale λ, has no reason to *a priori* remain at the same scale, because the scale space is not absolute. One therefore expects the appearance of a change of scale $\delta\varepsilon$ of such a structure under a displacement. If one starts form another structure at another scale $\varepsilon' = \rho'\lambda$, one expects another change of scale $\delta\varepsilon'$. The corresponding dilation field leads to introduce a vectorial potential having the properties of the electromagnetic potential, while the transformation from ε to ε' yields the basis for the gauge transformations and gauge invariance.

In this expression, the elementary dilation is written as $\delta\ln(\lambda/\varepsilon) = -\delta\varepsilon/\varepsilon$: this is justified by the Gell-Mann–Levy method, from which the dilation operator is found to take the form $\tilde{D} = \varepsilon\partial/\partial\varepsilon = \partial/\partial\ln\varepsilon$. Since the elementary displacement in space-time δx^μ is a contravariant four-vector and since $\delta\varepsilon/\varepsilon$ is a scalar, one must introduce a four-index quantity A_μ, which is itself a covariant four-vector, from the application of Einstein's rule about the summation over up and down indices. The constant q measures the amplitude of the scale-motion coupling. It can be subsequently identified with the active electric charge that intervenes in the potential. The choice of this form ensures that the dimensionality of A_μ be CL^{-1}, where C is the electric charge unit (e.g. $\varphi = q/r$ for a Coulomb potential).

In analogy with Einstein's construction of general relativity, we substract the new geometric effect from the total effect, in order to recover a purely inertial regime (see e.g. [266]). This can be expressed in terms of a scale covariant derivative. Namely, we set

$$\eta = q\ln(\lambda/\varepsilon), \tag{7.4}$$

i.e. $\eta = -q\mathbb{V}$ in terms of the "scale velocity" $\mathbb{V} = \ln(\lambda/\varepsilon)$ and we define

$$D_\mu\eta = \partial_\mu\eta - A_\mu. \tag{7.5}$$

7.3.2. *Gauge transformation*

Let us go on with the "field" A_μ of resolution dilations. If one wants such a field to be physical, it must be defined whatever the initial scale from which we started. Moreover, a scale being always defined relatively to another reference scale, the principle of scale relativity implies that a scale can change for two different but equivalent reasons: the scale can change while the scale of reference is kept fixed or because of a change of the reference scale itself.

We consider for the moment Galilean scale-relativity, in which the product of two dilations by factors ϱ and ϱ' is $\varrho'' = \varrho\varrho'$. Starting from another relative scale $\varepsilon' = \varrho\varepsilon$ (see Fig. 7.1), where the scale ratio ϱ may be any function of coordinates, i.e. $\varrho = \varrho(x, y, z, t)$, we are led to write

$$q\frac{\delta\varepsilon'}{\varepsilon'} = -A'_\mu dx^\mu. \tag{7.6}$$

Therefore we must introduce another four-vector A'_μ for the same displacement dx^μ of the same physical "object", and more generally an infinity of them depending on all the possible scales internal to it. This may seem to condemn such a physical quantity to have no possible physical meaning. But if one calculates A'_μ as a function of the previous field A_μ, one finds

$$A'_\mu = A_\mu + q\partial_\mu \ln \frac{\varepsilon'}{\varepsilon}, \tag{7.7}$$

i.e. by setting $\varepsilon'/\varepsilon = \varrho(x, y, z, t)$, one sees that the four-vector A_μ depends on the relative "state of scale" of the reference system, given by the relative "scale velocity" $\mathbb{V} = \ln \varrho(x, y, z, t)$. The transformation from A_μ to A'_μ finally reads

$$A'_\mu = A_\mu + q\partial_\mu \ln \varrho(x, y, z, t). \tag{7.8}$$

Therefore the four-vector A_μ is defined up to the gradient of a function of coordinates, which is exactly the well-known (but up to now misunderstood) gauge dependence property of the electromagnetic potential.

As a consequence we have suggested [356, 360, 394] to identify A_μ with an electromagnetic four-potential and Eq. (7.7) with its gauge transformation relation that writes in the standard way

$$A'_\mu = A_\mu + q\partial_\mu \chi(x, y, z, t). \tag{7.9}$$

It finally leads to gauge invariance because the potential is not directly observable, but only the field $F_{\mu\nu} = \partial_\mu A_\nu - \partial_\nu A_\mu$, which remains invariant under a gauge transformation.

The gauge function χ, in this relation, is usually considered as devoid of physical meaning. This is no longer the case here, since it is now identified with the logarithm of a scale ratio $\chi = \ln \varrho(x, y, z, t)$ between internal structures of the electron geodesics at scales smaller than its Compton length.

Our interpretation of the nature of the gauge function is compatible with its inobservability. Indeed, such a scale ratio is impossible to measure explicitly, since it would mean to make two measurements of two different relative scales smaller than the electron Compton length. But the very first measurement with resolution ε would change the state of the electron. Indeed, just after the measurement, its de Broglie length would become of order $\lambda_{\mathrm{dB}} \approx \varepsilon$ (see e.g. [353]), so that the second scale ε' would not be measured on the "same" electron, i.e. it would be in a different state. Therefore the ratio ϱ between the scales ε' and ε is expected to remain a virtual quantity.

However, even whether it cannot be directly measured, the gauge function has indirect consequences, so that the knowledge of its nature finally plays an important role. Indeed, it allows one to demonstrate the quantization of the electron charge and to relate its value to that of its mass [356, 360, 384, 400].

7.3.3. Definition of the electric charge

The fundamental new feature of scale relativity with respect to the standard view is the fractal nature of space-time, i.e. its explicit dependence on the relative resolution scale, which is characterized by $\ln \varrho = \ln(\lambda/\varepsilon)$ in the simplified case of a global dilation. In other words, the space of positions and instants must be completed by a space of scales.

Consider the action S for the electron. In the framework of a space-time theory based on a relativity principle, as it is the case here, it should be given directly by the length invariant s, i.e. $dS = -mc\,ds$. This relation ensures that the stationary action principle $\delta \int dS = 0$ becomes identical with a geodesics (Fermat) principle $\delta \int ds = 0$. Now the fractality of the geodesical curves to which the electron is identified means that, while S is an invariant with respect to space-time changes of the coordinate system, it is however a function of the scale variable, $S = S(\eta)$, at scales smaller than the Compton scale λ.

Therefore we expect the action to be an explicit function of $\eta = q \ln \varrho$, so that its differential reads

$$dS = \frac{\partial S}{\partial \eta} d\eta = \frac{\partial S}{\partial \eta}(D\eta + A_\mu dx^\mu), \qquad (7.10)$$

and we obtain

$$\partial_\mu S = D_\mu S + \frac{\partial S}{\partial \eta} A_\mu. \qquad (7.11)$$

What is the meaning of the derivative $-\partial S/\partial \eta$? Noether's theorem tells us that universal conservative quantities must emerge from the symmetries of the underlying space variables. Moreover, when considering the action as a function of coordinates at the upper limit of integration in the action integral, one finds that the conservative quantities are given by $p_k = -\partial_k S$. Now, space-time is completed in the scale-relativity framework by a scale space. Therefore, from the uniformity of the new scale variable $\ln \varrho$, a new conservative quantity can be constructed under the form of the derivative of the action with respect to scale transformations [356], namely,

$$\frac{e}{c} = -\frac{\partial S}{\partial \eta} = -\frac{\partial S}{q \partial \ln \varrho}. \qquad (7.12)$$

It can clearly be identified with the "passive" electric charge: in other words, the electric charge has now acquired a geometric meaning. It is now defined as the conservative quantity that arises from the uniformity of the scale space. The above choice is motivated by the expected symmetry of the "active" and "passive" charges in the final potential energy and by the fact that the action has the dimensionality of an angular momentum $[ML^2T^{-1}]$, while the squared charge dimensionality is $[ML^3T^{-2}]$.

One may also remark that in the case of an equality of the active and passive charge, $e = q$, and of their quantization, the above relation becomes a new definition for the fine structure constant $\alpha = e^2/\hbar c = -\partial(S/\hbar)/\partial \ln \varrho$.

Finally, the known form of the particle-field coupling term in the action (see next section),

$$S_{\text{pf}} = \int -\frac{e}{c} A_\mu dx^\mu \qquad (7.13)$$

appears as deriving from scale relativity principles, while it was merely postulated in the standard theory. This is an essential result since the Lorentz force can be deduced from the combination of this coupling term and of the free particle term, while the Maxwell equations derive from its combination with the field term [266]. The derivation of this action coupling term therefore allows one to reach a more profound foundation of electromagnetism, moreover of geometric nature.

7.3.4. Derivation of the Lorentz force from a geodesic equation

7.3.4.1. *Generalized invariant proper time*

Let us now write the action for a particle in an electromagnetic field under the form of the sum of its free term and of the particle-potential coupling term, namely,

$$S = S_{\mathrm{p}} + S_{\mathrm{pf}} = -\int mc\, ds - \int \frac{e}{c} A_\mu dx^\mu. \tag{7.14}$$

This action is not yet complete, since it should also contain the field term. This third term will be introduced after the construction of the electromagnetic field $F_{\mu\nu}$, which we now consider.

In the new framework, the two above terms in the action have both acquired a (geo)metric meaning. Indeed, as in any relativity theory, the action is now equivalent to the space-time invariant, $dS = -mc\, ds$, and the least action principle is therefore equivalent to a geodesic principle. The total elementary "length", i.e. its proper time in four dimensions, of a fractal curve reads

$$ds_{\mathrm{tot}} = (g_{\mu\nu} dx^\mu dx^\nu)^{1/2} + \frac{e}{mc^2} A_\mu dx^\mu. \tag{7.15}$$

The new meaning of this expression in the scale-relativity/fractal space-time description is as follows. The length of a fractal path can increase because its extremity has changed as a consequence of the motion of the particle. This is expressed by the first term, which is common to fractal and non-fractal paths, but also because of possible internal dilations (this is expressed by the second term). Finally the full length increase becomes distributed on the two terms: in the extreme cases, one can have a purely internal length increase that will have no counterpart in terms of space displacement and will therefore be equivalent to a potential energy or one can also consider an unfolding of the fractal path, in which this potential energy becomes manifested in terms of motion.

We shall therefore use the action-geodesic principle by postulating that the real paths are given by an optimization of the full invariant proper length, i.e. by the generalized fractal geodesic equation

$$\delta \int ds_{\text{tot}} = 0. \tag{7.16}$$

7.3.4.2. *Nature of the electromagnetic field*

Although we have recovered here the standard variational principle of classical electromagnetism, we shall nevertheless develop it again hereafter, since the new geometric interpretation will allow us to give a new meaning to the electromagnetic field. We shall see that it can be identified with a fractal space-time connection that defines the covariant derivative of a vector. This amounts to following Einstein's initial derivation of the motion equation in his relativistic theory of gravitation [158]. Indeed, in this derivation, the definition of the covariant derivative and of the Christoffel symbols directly proceeds from the geodesics-least action principle.

The variation of the invariant proper time reads

$$\delta s_{\text{tot}} = \int \left(\frac{dx_\mu d\delta x^\mu}{ds} + \frac{e}{mc^2} A_\mu d\delta x^\mu + \frac{e}{mc^2} \delta A_\nu dx^\nu \right) = 0. \tag{7.17}$$

After integration by parts and vanishing of the integrated terms (because the integral is varied for fixed values of the coordinates at its borns) it becomes

$$\int \left(du_\mu \delta x^\mu + \frac{e}{mc^2} dA_\mu \delta x^\mu - \frac{e}{mc^2} \delta A_\nu dx^\nu \right) = 0. \tag{7.18}$$

Finally, since $\delta A_\nu = (\partial A_\nu / \partial x^\mu) \delta x^\mu$ and $dA_\mu = (\partial A_\mu / \partial x^\nu) dx^\nu$, we obtain

$$\int \left\{ \frac{du_\mu}{ds} - \frac{e}{mc^2} \left(\frac{\partial A_\nu}{\partial x^\mu} - \frac{\partial A_\mu}{\partial x^\nu} \right) u^\nu \right\} \delta x^\mu ds = 0. \tag{7.19}$$

All the terms additional to the inertial ones now have a geometric origin. This leads us to define the covariant differential of the velocity as

$$Du_\alpha = du_\alpha - \frac{e}{mc^2} F_{\alpha\mu} dx^\mu, \tag{7.20}$$

in terms of a "connection"

$$F_{\alpha\mu} = \frac{\partial A_\mu}{\partial x^\alpha} - \frac{\partial A_\alpha}{\partial x^\mu}, \tag{7.21}$$

which can be identified with the electromagnetic field. It is easy to verify that this field is gauge invariant, i.e. in our framework, that it is now independent of the internal scale structure of the charged particle.

It may also be directly obtained from the commutator relation (in analogy with the definition of the Riemann tensor in Einstein's general relativity):

$$-q(\partial_\mu D_\nu - \partial_\nu D_\mu)\ln(\lambda/\varepsilon) = (\partial_\mu A_\nu - \partial_\nu A_\mu) = F_{\mu\nu}. \tag{7.22}$$

The first group of Maxwell equations directly derives from this expression, namely,

$$\partial_\mu F_{\nu\rho} + \partial_\nu F_{\rho\mu} + \partial_\rho F_{\mu\nu} = 0. \tag{7.23}$$

We are now able to define the covariant partial derivative of a resolution-dependent vector as

$$D_\mu B_\alpha = \partial_\mu B_\alpha + \frac{e}{mc^2}F_{\mu\alpha}. \tag{7.24}$$

In analogy with the construction of Einstein's covariant derivative in general relativity (see e.g. [266]), it amounts to subtracting from the total variation of the vector the new variation of geometric origin $(e/mc^2)F_{\alpha\mu}$, in order to let only its inertial part. While in motion general relativity the geometric variation of a vector is a consequence of curvature and manifests itself in terms of gravitation, in scale relativity it is a consequence of fractality and it manifests itself in terms of electromagnetism, i.e. of an Abelian gauge field (and more generally in terms of non-Abelian gauge fields, see [394] and Sec. 7.5).

Having defined and constructed the electromagnetic field, we are finally in a position to write the full action. Indeed, the field term of the action is naturally given by the square of the electromagnetic tensor, since it is the only scalar that satisfies its expected properties (see [266]). Therefore, the total action can be written as the sum of the free particle action S_p, field action S_f and particle-field coupling action S_pf,

$$S = S_\mathrm{p} + S_\mathrm{pf} + S_\mathrm{f} = -\int mc\,ds - \int \frac{e}{c}A_\mu dx^\mu$$
$$-\frac{1}{16\pi c}\int F_{\mu\nu}F^{\mu\nu}d\Omega. \tag{7.25}$$

The variational principle applied on the two last terms of the full action including the field action, after generalization to the current of several charges, yields Maxwell's equations [266],

$$\partial_\mu F^{\mu\nu} = -\frac{4\pi}{c}j^\nu. \tag{7.26}$$

7.3.4.3. *Geodesic equation from covariance principle*

By applying a generalized strong covariance principle that extends the covariance principle of general relativity, the motion equation of electrodynamics is established in terms of a geodesic equation that keeps, in terms of the covariant derivative, the form of the Galilean equation of free motion, namely

$$\frac{Du_\alpha}{ds} = u^\mu D_\mu u_\alpha = 0. \tag{7.27}$$

By expanding the expression of the covariant derivative, it reads

$$\frac{du_\alpha}{ds} - \frac{e}{mc^2} F_{\alpha\mu} u^\mu = 0, \tag{7.28}$$

in which we recognize the Lorentz equation of motion, that is now derived from first principles instead of being set as in the standard theory of classical electromagnetism or derived from the assumed particle-field action term, which is equivalent at the principle level.

Its meaning is that a charged particle is actually locally in free fall in a space-time, which is subjected to dilations and contractions at small scales (as a consequence of the presence of other charges). These local dilations and contractions of the internal structures of the geodesics that describe a "particle" manifest themselves at the macroscopic scales in terms of accelerations, which have been interpreted up to now as the effect of a force. In the new approach, this force or field can be understood as a mere manifestation of the local expansion and contraction properties of the fractal geometry.

It should be remarked that the electromagnetic field is now no longer separated from its source (the electron). The potential A^μ becomes a property of the "electron" geometry itself, since it is identified with the ability of its fractal structures to contract and dilate. Therefore the full four-momentum expression of the electron has now two terms,

$$P^\mu = mcu^\mu + \frac{e}{c} A^\mu, \tag{7.29}$$

which are, as we have previously seen, respective manifestations of its motion and scale properties. The new interpretation of the motion of an "electron" submitted to an "electromagnetic field" is that its state of motion changes because there is a transfer of energy-momentum toward (or from) its fractal structures.

One of the specificity of electromagnetism with respect to gravitation is easily accounted for by this view. Indeed, while gravitation is always attractive as regards the effect of active gravitational masses [483] (the repulsive effect of the cosmological constant is of a different nature, being equivalent to a negative pressure), electromagnetism can be attractive, repulsive or neutral when acting on a uncharged particle. This was one of the most difficult problems in the research of a geometric theory of electromagnetism. Indeed, general relativity suggested that, in a space-time theory, there should be an universality of the effects of the space-time geometry on the various bodies. Once space-time is curved, all particles, massive or not (i.e. gravitationally charged or not) feel this curvature, as demonstrated by the light deviation around the Sun.

On the contrary, the effect of the electromagnetic force depends on both the active and passive charges. How can this be reconciled with a geometric theory of space-time?

In the scale relativity interpretation this problem can be solved. Electromagnetism is not directly related to the fractal geometry itself, but instead to its scale variations, which can amount to local expansion, contraction or staticity of the fractal space-time that describes a charged or an uncharged particle. Another particle (i.e. another set of fractal geodesics) will also be characterized by the same properties. If it does not own fractal structures allowing dilations or contraction (i.e. if it is uncharged), such a particle will not "feel" the contraction or expansion of the other one. If both are able to expand or contract, energy-momentum can be transfered from one set of geodesics to the other, and one will see an attraction or a repulsion depending on the way the energy is transferred.

In a more exact description, one deals with a two-particle space-time, while the field is a wave of dilation that propagates at the velocity of light, in agreement with the solutions of the vacuum Maxwell equations, from one set of geodesics to the other. This is analogous to gravitational waves propagating between two gravitational bodies in general relativity. One therefore finally recovers in this way the attractive, repulsive and neutral character of the electromagnetic force.

7.3.4.4. *Geodesic equation from energy equation*

Before concluding this section, let us give another equivalent derivation of the Lorentz equation. Let us start from the energy equation

$$ds^2 = dx^\mu dx_\mu \Rightarrow u^\mu u_\mu = 1 \tag{7.30}$$

and take its partial derivative

$$\partial_\alpha(u^\mu u_\mu) = 0 \Rightarrow u^\mu \partial_\alpha u_\mu = 0. \tag{7.31}$$

As we shall see, this is nothing but the electrodynamics equation, under a form that is remarkable since there is no explicit presence of the field in it. In the free case it is indistinguishable from the free motion equation, since $\partial_\alpha u_\mu - \partial_\mu u_\alpha = 0$. In the presence of an electromagnetic field this commutation relation becomes wrong. Indeed, since $mcu^\mu + (e/c)A^\mu = -\partial_\mu S$ is a gradient, we have

$$\partial_\alpha u_\mu - \partial_\mu u_\alpha = \frac{e}{mc^2}(\partial_\mu A_\alpha - \partial_\alpha A_\mu), \tag{7.32}$$

so that

$$u^\mu \partial_\alpha u_\mu = u^\mu \left(\partial_\mu u_\alpha - \frac{e}{mc^2}F_{\alpha\mu}\right) = \frac{du_\alpha}{ds} - \frac{e}{mc^2}F_{\alpha\mu}u_\mu = 0. \tag{7.33}$$

Therefore the free-form equation $u^\mu \partial_\alpha u_\mu = 0$ is valid in both cases, without and with the presence of an electromagnetic field [446]. However, in the standard theory of electromagnetism in which the charges, the potential and their coupling are given from experiment without being derived from first principles, this is a mere result of the identification of the electromagnetic four-potential with an energy-momentum difference. In the scale-relativity framework, the nature of charges, of the field and the expression for their coupling are derived from a geometric description of space-time, so that this result acquires its full geometric and covariant meaning.

In conclusion of this section, let us highlight again that the progress made here respectively to the standard classical electromagnetic theory is that the Lorentz force and the Maxwell equations are derived in the scale relativity theory as being both manifestations of the fractal geometry of space-time, instead of being independently constructed. Moreover, a new physical meaning has been given to the electric charge and to gauge transformations. As a consequence of this interpretation, the charge of the electron is quantized in the special scale relativity framework and its quantized value can be related to the ratio of its mass over the Planck mass (see next section).

7.3.5. *Link with Weyl–Dirac theory*

Before generalizing the approach to quantum electrodynamics, let us notice that it shares some features with the Weyl–Dirac theory of electromagnetism [535, 140], but with new and essential differences. The Weyl theory considers scale transformations of the line element, $ds \to ds' = \rho\,ds$, but without specifying any fundamental cause for this dilation. The variation of ds should therefore exist at all scales, in contradiction with the observed invariance of the mass of the electron, and therefore of its Compton wavelength.

In the scale relativity proposal, the change of the line element comes from the fractal geometry of space-time, and it is therefore a consequence of the dilation of resolution. Moreover, the explicit effects of the dependence on resolutions is observable only below the fractal-nonfractal transition, which is identified in rest frame with the Einstein time-scale $\tau_E = \hbar/mc^2$ of the particle.

This solves the problem encountered in the Weyl theory and ensures the invariance of the observed electron mass, at least to lowest order. Indeed, in quantum electrodynamics accounting for radiative corrections, the self-energy of the electron is actually logarithmically dependent on the energy scale [530] (see [353] Chapter 6 for a detailed analysis of this behavior in the scale relativity framework).

7.4. Quantum Electrodynamics

7.4.1. *Introduction*

The problem posed by the foundation of quantum electrodynamics in the theory of scale relativity is far more difficult. Indeed, both the quantum properties and the electromagnetic properties are expected to be generated by the nondifferentiable and fractal geometry of space-time. One should therefore combine the quantum-covariant derivative that describes the induced effects of fractality and non-differentiability on motion and the scale-covariant derivative that describes the scale-motion coupling, i.e. the nonlinear effects of coordinate-dependent resolutions in terms of one unique covariant tool.

As in the classical case, our aim here is to succeed writing the Klein–Gordon and Dirac equations, including an electromagnetic field, in terms of a free geodesic equation that keeps the inertial Galilean form, $DV/ds = 0$, where now both the quantum behavior and the existence of the field are generated by the doubly covariant derivative.

7.4.2. *QED-covariant derivative*

Let us recall how one can recover the standard QED-covariant derivative in the scale-relativity approach. We consider again a generalized action that depends both on motion and scale variables. In the scale-relativistic quantum description, the four-velocity is now complex, so that the action writes $\mathcal{S} = \mathcal{S}(x^\mu, \mathcal{V}^\mu, \ln \varrho)$. This action gives the fundamental meaning of the wave function, defined as

$$\psi = e^{i\mathcal{S}/\hbar}. \tag{7.34}$$

Since the action is a complex number and becomes a complex quaternion in the generalized case that leads to the demonstration of the Dirac equation [93], this expression contains a phase and a modulus (the later becomes in the end a square-root of probability density).

The decomposition of the action performed in the framework of the classical theory still holds and now becomes (for an electron of charge e)

$$d\mathcal{S} = -i\hbar \, d\ln \psi = -mc\mathcal{V}_\mu \, dx^\mu - \frac{e}{c} A_\mu \, dx^\mu. \tag{7.35}$$

Equation (7.35) allows one to define a generalized complex energy-momentum,

$$\mathcal{P}^\mu = mc\mathcal{V}^\mu + \frac{e}{c} A^\mu. \tag{7.36}$$

This leads to a new expression for the relation between the complex velocity and the wave function,

$$\mathcal{V}_\mu = i\lambda \, \partial_\mu(\ln \psi) - \frac{e}{mc^2} A_\mu, \tag{7.37}$$

where $\lambda = \hbar/mc$ is the Compton length of the particle.

This relation is a covariant generalization of the free particle identity $\mathcal{V}_\mu = i\lambda \partial_\mu(\ln \psi)$. It leads to introduce a scale-covariant derivative as

$$\mathcal{V}_\mu = i\lambda \, D_\mu(\ln \psi). \tag{7.38}$$

Since we can write Eq. (7.37) under the form $mc\mathcal{V}_\mu \psi = [i\hbar\partial_\mu - (e/c)A_\mu]\psi$, we recognize the standard QED-covariant derivative operator when it is acting on the wave function ψ:

$$-i\hbar \, D_\mu = -i\hbar \, \partial_\mu + \frac{e}{c} A_\mu. \tag{7.39}$$

This provides one with an understanding of the nature and origin of the QED-covariant derivative from first principles, while it was merely set as a rule devoid of geometric meaning in standard quantum field theory.

This covariant derivative is directly related to the one introduced in the classical framework. Indeed, the classical covariant derivative was written $D_\mu = \partial_\mu + (1/q)A_\mu$ acting on ϱ, while $\psi = \psi_0 \exp[(i/\hbar)(eq/c) \ln \varrho]$. We therefore recover the expression (7.39) acting on ψ.

7.4.3. *Electromagnetic KG equation from a geodesic equation*

We are now able to combine the quantum-covariant and the scale-covariant derivatives in terms of a common tool. Recall that the quantum-covariance has been fully implemented by the use of a covariant complex velocity operator,

$$\widehat{\mathcal{V}^\mu} = \mathcal{V}^\mu + i(\lambda/2)\partial^\mu, \qquad (7.40)$$

in terms of which the decomposition of the quantum-covariant derivative keeps the first order Leibniz rule form, $\widehat{d}/ds = \widehat{\mathcal{V}^\mu}\partial_\mu$ (see also [446] on this point).

Then the classical electromagnetic field has been constructed thanks to a scale-covariant derivative D_μ that manifests the expansion-contraction properties of the fractal space-time and of its geodesics, such that the equation of motion reads $Du_\alpha/ds = u^\mu D_\mu u_\alpha = 0$.

One can therefore combine both tools and define a scale and quantum doubly-covariant derivative,

$$\frac{\widehat{D}}{ds} = \widehat{\mathcal{V}^\mu} D_\mu, \qquad (7.41)$$

in terms of which we can finally write an inertial-like, strongly covariant geodesic equation,

$$\frac{\widehat{D}}{ds}\mathcal{V}_\alpha = 0. \qquad (7.42)$$

We shall now prove that this extremely simple, free-form equation gives, after integration, the Klein–Gordon equation in the presence of an electromagnetic field. In other words, this means that it contains all the quantum terms and all the field terms, which are both generated through the double scale-covariance, which manifests the non-differentiability and the fractality of space-time.

As a first step, let us show that it allows one to obtain a quantum-covariant form of the Lorentz equation of dynamics. By successively developing the covariant derivatives, it becomes

$$\frac{\widehat{D}}{ds}\mathcal{V}_\alpha = \widehat{\mathcal{V}^\mu} D_\mu \mathcal{V}_\alpha = \widehat{\mathcal{V}^\mu}\left(\partial_\mu \mathcal{V}_\alpha + \frac{e}{mc^2}F_{\mu\alpha}\right) = 0, \qquad (7.43)$$

and since $\widehat{\mathcal{V}^\mu}\partial_\mu\mathcal{V}_\alpha = \widehat{d}\mathcal{V}_\alpha/ds$, we obtain

$$mc\frac{\widehat{d}}{ds}\mathcal{V}_\alpha = \frac{e}{c}\widehat{\mathcal{V}^\mu}F_{\alpha\mu}. \tag{7.44}$$

This equation has exactly the form of the classical Lorentz equation of dynamics, although this is a quantum equation whose integral is the Klein–Gordon equation with electromagnetic field. It is equivalent to that written by Pissondes in the scale-relativity framework by using a symmetric product [446], but in this previous work there was no justification of the existence of an electromagnetic field, which was included by using the standard QED-covariant derivative. The additional point here is that the theory generates both the field and the quantum behavior.

Let us go on from Eq. (7.43) and now develop the complex velocity operator. We obtain

$$\left(\mathcal{V}^\mu + i\frac{\lambda}{2}\partial^\mu\right)\left(\partial_\mu\mathcal{V}_\alpha + \frac{e}{mc^2}F_{\mu\alpha}\right) = 0. \tag{7.45}$$

While $\partial_\mu\mathcal{V}_\alpha - \partial_\alpha\mathcal{V}_\mu = 0$ in the free case, this is no longer true in the presence of an electromagnetic field. As in the classical situation it becomes

$$\partial_\mu\mathcal{V}_\alpha - \partial_\alpha\mathcal{V}_\mu = -\frac{e}{mc^2}F_{\mu\alpha}. \tag{7.46}$$

We therefore obtain a form of the equation in which the indices are exchanged with respect to the free case, namely,

$$\left(\mathcal{V}^\mu + i\frac{\lambda}{2}\partial^\mu\right)\partial_\alpha\mathcal{V}_\mu = \widehat{\mathcal{V}^\mu}\partial_\alpha\mathcal{V}_\mu = 0. \tag{7.47}$$

We now replace $\mathcal{V}^\mu$ by its covariant form and we obtain

$$\left(i\lambda\partial^\mu(\ln\psi) - \frac{e}{mc^2}A^\mu + i\frac{\lambda}{2}\partial^\mu\right)\partial_\alpha\left(i\lambda\partial_\mu(\ln\psi) - \frac{e}{mc^2}A_\mu\right) = 0. \tag{7.48}$$

After integration, one finds that this equation becomes the Klein–Gordon equation for a particle in an electromagnetic field,

$$\left(i\hbar\partial_\mu - \frac{e}{c}A_\mu\right)\left(i\hbar\partial^\mu - \frac{e}{c}A^\mu\right)\psi = m^2c^2\psi. \tag{7.49}$$

The expression (7.47) of the motion equation can also be obtained by differentiating the energy equation [446],

$$\partial_\alpha(\mathcal{V}^\mu \mathcal{V}_\mu + i\lambda\partial^\mu \mathcal{V}_\mu) = 0 \Rightarrow \widehat{\mathcal{V}^\mu}\partial_\alpha \mathcal{V}_\mu = 0. \tag{7.50}$$

This equation is valid without and with electromagnetic field, in the same way as its classical equivalent, $u^\mu \partial_\alpha u_\mu = 0$.

The Dirac equation in an electromagnetic field can subsequently be derived by the same method that has yielded the free Dirac equation from the Klein–Gordon one [98]. As we have seen in Chapter 6, the symmetry breaking under the reflection $dx^\mu \leftrightarrow -dx^\mu$, issued from nondifferentiability, leads to introduce an additional two-valuedness, which can be described by quaternionic and biquaternionic wave functions, to which the above derivation can be generalized (being anyway cautious with the noncommutativity of quaternions and its implications).

Problems and Exercises

Exercise 20 Generalize the above derivation to quaternionic (spinor) wavefunctions and biquaternionic (bispinor) wave functions.

Hints: One must be cautious in this derivation with the order under which the two (quantum and scale) covariant derivatives are applied. Indeed, the additional two-valuedness leads to the emergence of spin, which is no longer disconnected from the field. Indeed, through its link to the magnetic moment, it plays at some level the role of a charge. Therefore the quantum covariant derivative is no longer restricted to the emergence of the quantum behavior, it has also some effects on the field itself (see a solution of this problem in [98]).   ■

7.4.4. *Nature of the electric charge (quantum theory)*

In a gauge transformation $A'_\mu = A_\mu + e\partial_\mu\chi$, the wave function of an electron of charge e, becomes

$$\psi' = \psi\exp\left\{\frac{i}{\hbar} \times \frac{e}{c} \times e\chi\right\}. \tag{7.51}$$

We have reinterpreted in the previous sections the gauge transformation as a scale transformation of resolution, $\varepsilon \rightarrow \varepsilon'$, yielding an identification of the gauge function with a scale ratio, $\chi(x,y,z,t) = \ln\varrho = \ln(\varepsilon/\varepsilon')$, which is a function of space-time coordinates. In such an interpretation, the specific property that characterizes a charged particle is the explicit scale-dependence on resolution of its action, then of its wave function. The net result is that the electron wave function reads

$$\psi' = \psi \exp\left\{ i\frac{e^2}{\hbar c} \ln \varrho \right\}. \tag{7.52}$$

Since, by definition (in the system of units where the permittivity of vacuum is one),

$$e^2 = 4\pi \alpha\, \hbar c, \tag{7.53}$$

where α is the fine structure constant, Eq. (7.52) becomes [356]

$$\psi' = \psi \times e^{i4\pi\alpha \ln \varrho}. \tag{7.54}$$

This result supports the solution given in Sec. 7.3.3 to the problem of the nature of the electric charge in the classical theory. Indeed, considering now the wave function of the electron as an explicitly resolution-dependent function, we can write the scale differential equation of which it is solution as

$$-i\hbar \frac{\partial \psi}{\partial\left(\frac{e}{c}\ln \varrho\right)} = e\psi. \tag{7.55}$$

We recognize in $\tilde{D} = -i(\hbar c/e)\partial/\partial \ln \varrho$ a dilatation operator. Equation (7.55) can then be read as an eigenvalue equation issued from an extension of the correspondence principle (but here, demonstrated),

$$\tilde{D}\psi = e\psi. \tag{7.56}$$

This is the quantum expression of the above classical suggestion [356, 360], according to which the electric charge can be understood as the conservative quantity that comes from the new scale symmetry, namely, from the uniformity of the resolution variable $\ln(\varepsilon/\lambda)$.

7.4.5. *Charge quantization and mass-coupling relations*

While the results of the scale relativity theory described in the previous sections mainly deal with a new interpretation of the nature of the electromagnetic field, of the electric charge and of gauge invariance, we now arrive at one of the main consequences of this approach. As we shall see, it allows one to establish the universality of the quantization of charge and to theoretically predict the existence of fundamental relations between mass scales and coupling constants.

In the previous section we have recalled our suggestion [356, 360] of understanding the nature of the electric charge as being the eigenvalue of the dilation operator corresponding to resolution transformations (internal to the geodesics identified with the "particle"). We have written the wave function of a charged particle under the form of Eq. (7.55).

Let us now consider in more detail the nature of the scale factor ϱ in this expression. This factor describes the ratio of two relative resolution scales ε and ε' that correspond to structures of the fractal geodesical paths that we identify with the charged particle. However, this particle is not structured at all scales, but only at scales smaller than its Compton length, $ds < \lambda_C = \hbar/mc$. We can therefore take this upper limit as a reference scale and write

$$\psi' = \exp\left\{i4\pi\alpha \ln \frac{\lambda_C}{\varepsilon}\right\}\psi. \tag{7.57}$$

In the case of Galilean scale-relativity, such a relation leads to no new result, since ε can tend to zero, so that $\ln(\lambda_C/\varepsilon)$ is unlimited. But in the framework of special scale-relativity, scale laws take a log-Lorentzian form below a Galilean to Lorentzian transition scale, which we have suggested to identify with the Compton scale λ_C [352, 353]. The Planck length $l_{\mathbb{P}}$ becomes a minimal, unreachable scale, invariant under dilations, so that $\ln(\lambda_C/\varepsilon)$ becomes limited by $\mathbb{C} = \ln(\lambda_C/l_{\mathbb{P}})$, i.e.

$$\ln \frac{\lambda_C}{\varepsilon} < \mathbb{C} = \ln \frac{\lambda_C}{l_{\mathbb{P}}}. \tag{7.58}$$

This implies a quantization of the charge, which amounts to the relation $4\pi\alpha\mathbb{C} = 2k\pi$, i.e. [356]

$$\alpha\mathbb{C} = \frac{1}{2}k, \tag{7.59}$$

where k is integer. Since $\mathbb{C} = \ln(\lambda_C/l_{\mathbb{P}})$, it is equal to $\mathbb{C} = \ln(m_{\mathbb{P}}/m)$, where $m_{\mathbb{P}}$ is the Planck mass and m the particle mass, so that Eq. (7.59) finally amounts to a relation between a charged particle mass scale and the electromagnetic coupling constant, which is a dimensionless version of the square of the quantized charge. In this context, the existence of a well-defined and stable mass for a particle is a direct consequence of the quantization of its charge, which comes itself from the nature of the Planck length-scale as a limiting horizon.

The application of this approach to an attempt at understanding the experimental values of the mass and the charge of the real electron will be given in Chapter 11 (Sec. 11.1.3.2).

7.5. Non-Abelian Gauge Theories

7.5.1. *Scale-relativistic description*

7.5.1.1. *Introduction*

Let us now generalize the electromagnetic ($U(1)$, Abelian) description to a geometric foundation of non-Abelian gauge theories, based upon the scale relativity first principles [385, 394].

In the above Abelian description, we have considered only one resolution variable ε, or, in other words, only global dilations of the internal resolutions (as in a conformal group approach). However, we know that this is a very simplified description. Indeed, we have recalled in Chapter 3 that one may, more generally, characterize the various coordinates by different resolutions $(\varepsilon_x, \varepsilon_y, \varepsilon_z, \varepsilon_t)$ (in the same way as, in an actual position measurement, one may have different error bars on the various coordinates). Finally, the genuine nature of the resolutions is that of an "error" matrix or resolution matrix, i.e. of a tensor $\varepsilon_{\alpha\beta} = \rho_{\alpha\beta}\varepsilon_\alpha\varepsilon_\beta$, since it may also involve correlation coefficients (see Sec. 3.6).

We therefore consider that the fractal structures of the "particle" (i.e. of a family of geodesics of a non-differentiable space-time) are now described in terms of several scale variables that generalize the single resolution variable ε. For simplicity we omit the reference unit λ and we assume that the various indices can be gathered into one common index, i.e. $\eta_\alpha = \eta_\alpha(x, y, z, t)$, where $\alpha = 0$ to N.

In the simplest case (no correlation), $\eta_\alpha = \varepsilon_\alpha$, where ε_α correspond to the resolutions ε_α of the space-time coordinates $X_\alpha(\alpha = 0$ to 3, i.e. $N = 4$). When the full resolution matrix is taken into account in a n-dimensional space, one gets $N = n(n+1)/2$ since the resolution tensor is symmetric, i.e. $N = 6$ for space, $N = 10$ for space-time and $N = 15$ in special scale relativity when the resolutions are combined with a variable fractal dimension ("djinn") treated as the fifth dimension of a "space-time-djinn".

The scale variables may not be directly identified with the resolution matrix itself. Indeed, in analogy with general relativity (of motion), various transformations can be applied to these variables. They should now be considered as general "scale-space coordinates", in analogy with, e.g. time which becomes a "time coordinate" in Einstein's general relativity. For example, the transformation $\varepsilon_\alpha \to \ln \varepsilon_\alpha$ may be particularly relevant for them, as we have seen in the one-dimensional case.

In order to keep this generality, we shall therefore not be more specific about the choice of the scale variables in what follows. Our aim here is mainly to show that one may relate in a general way the scale relativistic tools to the standard description of current gauge theories. A full analysis of the possible final gauge groups is outside the scope of the present book. However, as we shall see, even at this preliminary stage of the analysis, one can show that in any case this full gauge group contains at least an SU(2) subset, and more generally an SU(5) subgroup. For example, there is a SU(2) subgroup of three-dimensional rotations in scale space, which is a good candidate for the isospin SU(2) transformation group. The theory therefore provides the elements to recover the standard U(1) × SU(2) electroweak theory (see Sec. 11.3 for new proposals about the nature of the Higgs field) and more generally the U(1) × SU(2) × SU(3) standard model and the SU(5) minimal grand unification theory. Recall that the various problems, which led to the rejection of SU(5) as the grand unification group are solved in the special scale relativistic framework, see [352, 353, 360] and Chapter 11.

7.5.1.2. *General scale transformations*

Let us consider infinitesimal transformations of the scale variables. The transformation law on the η_α can be written in a linear way as

$$\eta'_\alpha = \eta_\alpha + \delta\eta_\alpha = (\delta_{\alpha\beta} + \delta\theta_{\alpha\beta})\eta^\beta, \qquad (7.60)$$

where $\delta_{\alpha\beta}$ is the Kronecker symbol. This means that the fractality of space-time involves a geometric contribution

$$\delta\eta_\alpha = \delta\theta_{\alpha\beta}\eta^\beta. \qquad (7.61)$$

The $\delta\theta_{\alpha\beta}$ are the infinitesimal transformation tensor. It is noticeable that, in the general case when the scale variables are described by a two-index tensor, $\eta_{\alpha_1\alpha_2}$, it becomes a four-index tensor, $\delta\theta_{\alpha_1\alpha_2\beta_1\beta_2}$.

Let us now assume that the η_α's are functions of the standard space-time coordinates, $\eta_\alpha(x, y, z, t)$. This leads us to generalize the scale-covariant derivative previously defined in the electromagnetic case. The total variation of the resolution variables becomes the sum of the inertial variation, described by the covariant derivative and of the new geometric contribution, i.e.

$$d\eta_\alpha = D\eta_\alpha - \delta\eta_\alpha. \tag{7.62}$$

Note that we choose from now on to write the new geometric contribution with a minus sign, $-\delta\eta_\alpha = -\eta^\beta\delta\theta_{\alpha\beta}$, in order to recover the covariant derivative of gauge theories in its usual form at the end of the derivation. This is actually an inessential sign ambiguity, which is just a matter of definition.

The total variation therefore reads

$$d\eta_\alpha = D\eta_\alpha - \eta^\beta\delta\theta_{\alpha\beta} = D\eta_\alpha - \eta^\beta W^\mu_{\alpha\beta}dx_\mu. \tag{7.63}$$

This covariant derivative is now similar to that of GR, i.e. it amounts to subtracting the new geometric part in order to keep only the inertial part, for which the motion equation will therefore take a geodesical, free-like form. This is different from the case of the quantum-covariant derivative, which includes the effects of non-differentiability by adding new terms in the total derivative.

Recall that in the Abelian case, which corresponds to a unique global dilation, this expression can be simplified since $d\eta/\eta = d\ln\eta = d\chi$. This is no longer the case in the generalized non-Abelian case now considered, since one deals with different indices for $d\eta_\alpha$ in the left-hand side of this equation and η^β in the right-hand side. In this new situation we are led to introduce "gauge field potentials" $W^\mu_{\alpha\beta}$ that enter naturally in the geometrical frame of Eq. (7.63). These potentials are linked to the scale transformations as follows:

$$\delta\theta_{\alpha\beta} = W^\mu_{\alpha\beta}dx_\mu. \tag{7.64}$$

One should remain cautious about this expression and keep in mind that these potentials find their origin in a covariant derivative process and are therefore not gradients (this is expressed by the use of a difference sign $\delta\theta_{\alpha\beta}$ instead of $d\theta_{\alpha\beta}$). They formalize the coupling between displacements in space-time and transformations of the scale variables and play in Eq. (7.63) a role analogous to the one played in General Relativity by the Christoffel symbols. It is also important to notice that the $W^\mu_{\alpha\beta}$ introduced at this level of the analysis do not include charges. They are functions of the space and time coordinates only. They will therefore be habilitated to describe fields in a geometric way, in a manner that is partially independent of the sources of the field. This is a necessary choice because the present method generates, as we shall see, not only the fields but also the charges, from, respectively, the scale transformations and the scale symmetries of the dynamical fractal space-time.

7.5.1.3. *Multiplets*

After having written the transformation law of the basic variables (the η_α's), we are now led to describe how various physical quantities transform under these η_α transformations. These new transformation laws are expected to depend on the nature of the objects to transform, e.g. vectors, tensors, spinors, etc., which implies to jump to group representations.

In the case where the particle is a spin 1/2 fermion, it has been recalled in Chapter 6 that the relation between the velocity and the spinor fields reads

$$\mathcal{V}_\mu = i\lambda\psi^{-1}\partial_\mu\psi, \qquad (7.65)$$

where $\mathcal{V}_\mu$ and ψ are complex quaternions (equivalent to Dirac bispinors) and where the constant $\lambda = \hbar/mc$ is the Compton length of the particle.

However, bispinors are not a general enough description for fermions subjected to a general gauge field. Indeed, we consider here a generalized group of transformations, which therefore involves, anticipating on what follows, generalized charges. As a consequence of these new charges, whose existence will be fully justified below and their form specified, the very nature of the fermions is expected to become more complicated. Experiments do confirm this expectation. They have indeed shown that new degrees of freedom must be added in order to represent the weak isospin, the hypercharge and the color.

In order to account in a general way for this more complicated description, we shall simply replace the biquaternionic wave functions by multiplets ψ_k, where each component is a Dirac bispinor. This is a general method already employed in standard quantum mechanics, which is not different in its essence from the passage from real variables to complex, then to quaternionic variables. The specific nature of these multiplets becomes known *a posteriori* from the very way they interact with the other mathematical tools of the description.

As already remarked [360], when the scale variables become multiplets, the same is true of the charges. As we shall see in what follows, in the present approach it is at the level of the construction of the charges that the gauge set generators enter.[a]

In this case when the wave functions become biquaternionic multiplets, the multi-valued velocity becomes a biquaternionic matrix [385, 394],

$$\mathcal{V}_{jk}^{\mu} = i\lambda\psi_j^{-1}\partial^{\mu}\psi_k. \tag{7.66}$$

The biquaternionic (therefore noncommutative) nature of the wave function [93] (which is equivalent to Dirac bispinors) plays here an essential role, as previously announced. Indeed, it leads to write the velocity field as $\psi^{-1}\partial^{\mu}\psi$ instead of $\partial^{\mu}\ln\psi$ in the complex case, so that its generalization to multiplets involves two indices instead of one. The general structure of Yang–Mills theories and the correct construction of non-Abelian charges will be obtained thanks to this result, and could not be reached at all in its absence.

Therefore the action becomes also a tensorial two-index biquaternionic quantity, that is a function of the space-time coordinates, of the biquaternionic velocity field multiplets, and of the scale-space coordinates (generalized resolutions), namely,

$$dS_{jk} = dS_{jk}(x^{\mu}, \mathcal{V}_{jk}^{\mu}, \eta_{\alpha}). \tag{7.67}$$

In the absence of a field, it is linked to the generalized velocity (and therefore to the spinor multiplet) by the relation,

$$\partial^{\mu}S_{jk} = -mc\mathcal{V}_{jk}^{\mu} = -i\hbar\psi_j^{-1}\partial^{\mu}\psi_k. \tag{7.68}$$

7.5.1.4. *Non-Abelian gauge charges*

Now, in the presence of a field, i.e. when the second-order effects of the fractal geometry appearing in the right hand side of Eq. (7.63) are included — the field is no longer applied in an external way, but it now

---

[a]We speak here of "set" instead of "group" because there is no *a priori* physical argument from which the gauge set should be a group. Our general approach amounts to attempting to determine the nature of this set from physical principles rather than *a priori* posing it as a postulate. It could, for example, be a semi-group or a groupoid, i.e. some group axioms may lack, or it may on the contrary need another fundamental axiom. An example of such a situation is the special relativity set of transformation, which can finally be proved to be a (reflexive) group from only two axioms (internal law and reflection invariance), see [352, 353] and Chapter 4.

emerges in an internal way —, using the complete expression for $\partial^\mu \eta_\alpha$,

$$\partial^\mu \eta_\alpha = D^\mu \eta_\alpha - W^\mu_{\alpha\beta} \eta^\beta, \tag{7.69}$$

we are led to write a relation that generalizes Eq. (7.11) to the non-Abelian case,

$$\partial^\mu S_{jk} = \frac{\partial S_{jk}}{\partial \eta_\alpha} \partial^\mu \eta_\alpha = \frac{\partial S_{jk}}{\partial \eta_\alpha} \left(D^\mu \eta_\alpha - W^\mu_{\alpha\beta} \eta^\beta \right). \tag{7.70}$$

Thus we obtain

$$\partial^\mu S_{jk} = D^\mu S_{jk} - \eta^\beta \frac{\partial S_{jk}}{\partial \eta_\alpha} W^\mu_{\alpha\beta}. \tag{7.71}$$

This is a quite remarkable expression, since we note that, by the simple use of the partial differential calculus and of the index rules, new quantities appear in factor of the fields that have all the required properties to be identified with the gauge charges. We are therefore finally led to define a general set of scale transformations, which forms a group, and whose generators are [394]

$$T^{\alpha\beta} = \eta^\beta \partial^\alpha, \tag{7.72}$$

where we use the compact notation $\partial^\alpha = \partial/\partial\eta_\alpha$. Since each index of the scale variables represents two indices due to their tensorial nature, these generators read in a more complete way $T^{\alpha_1\alpha_2\beta_1\beta_2} = \eta^{\beta_1\beta_2} \partial^{\alpha_1\alpha_2}$.

This finally yields the generalized charges [385, 394],

$$\frac{\tilde{g}}{c} t^{\alpha\beta}_{jk} = \eta^\beta \frac{\partial S_{jk}}{\partial \eta_\alpha}, \tag{7.73}$$

which have the right particular structure of charges in non-Abelian gauge theories (see e.g. [7]). This group is submitted to a unitarity condition, since, when it is applied to the wave functions, $\psi\psi^\dagger$ must be conserved.

7.5.1.5. *Rotations in scale space*

In order to enlighten the meaning of the new definition we have obtained for the charges, we consider in the present section a subsample of the possible scale transformations on intrinsic scale variables: namely, those which are built from the antisymmetric part of the gauge set that can, therefore, be identified as "rotations" in the scale space. In this case the infinitesimal transformation is such that

$$\delta\theta_{\alpha\beta} = -\delta\theta_{\beta\alpha} \Rightarrow W^{\mu}_{\alpha\beta} = -W^{\mu}_{\beta\alpha}. \tag{7.74}$$

Therefore, reversing the indices in Eq. (7.71), we may write

$$\partial_{\mu}S_{jk} = D_{\mu}S_{jk} - \eta^{\alpha}\frac{\partial S_{jk}}{\partial\eta_{\beta}}W^{\mu}_{\beta\alpha}. \tag{7.75}$$

Taking the half-sum of Eqs. (7.71) and (7.75) we finally obtain

$$\partial_{\mu}S_{jk} = D_{\mu}S_{jk} - \frac{1}{2}\left(\eta^{\beta}\frac{\partial S_{jk}}{\partial\eta_{\alpha}} - \eta^{\alpha}\frac{\partial S_{jk}}{\partial\eta_{\beta}}\right)W^{\mu}_{\alpha\beta}. \tag{7.76}$$

This leads to define the new charges,

$$\frac{\tilde{g}}{c}t^{\alpha\beta}_{jk} = \frac{\partial S_{jk}}{\partial\theta_{\alpha\beta}} = \frac{1}{2}\left(\eta^{\beta}\frac{\partial S_{jk}}{\partial\eta_{\alpha}} - \eta^{\alpha}\frac{\partial S_{jk}}{\partial\eta_{\beta}}\right). \tag{7.77}$$

One recognizes here a definition similar to that of the angular momentum, i.e. of the conservative quantity that finds its origin in the isotropy of space; but the space under consideration is here the "scale space". Therefore the charges of the gauge fields are identified, in this interpretation, with "scale-angular momenta".

The subgroup of transformations corresponding to these generalized charges is, in three dimensions, a SO(3) group related to a SU(2) group by the homomorphism, which associates to two distinct 2×2 unitary matrices of opposite sign the same rotation. We are therefore naturally led to define a "scale-spin", which we propose to identify to the simplest non-Abelian charge in the current standard model: the weak isospin.

Coupling this SU(2) representation of the rotations in a three dimensional sub-"scale-space" to the U(1) representation of the global scale dilations (that describes the electromagnetism process) analyzed in Sec. 7.4, we are therefore able to give a physical geometric meaning to the transformation group corresponding to the U(1)×SU(2) representation of the standard electroweak theory [527, 471].

It is worth stressing here that the group of three-dimensional rotations in scale space is only a subgroup of an at least four-dimensional rotation group (one scale variable for each space-time coordinate). It is therefore at least SO(4), since the metric signature of the four dimensional scale space is $(+,+,+,+)$, and, more precisely, its universal covering group SU(2)×SU(2). However, the full group is expected to be far larger, since the tensorial nature of the resolution variables leads, in four dimensions, to $N = 10$ components, and therefore to at least an SO(10)-type gauge group. This is a promising result since the SO(10) group is presently one of the best candidates for a grand unification group beyond the standard model [435], namely, it includes as subgroup SU(5), and therefore U(1)×SU(2)×SU(3), and allows for extensions, such as the possibility of neutrino masses.

7.5.2. *Yang–Mills theory with the scale relativity tools*

7.5.2.1. *Simplified notation*

For the subsequent developments, we shall simplify again the notations and use only one index $a = (\alpha, \beta)$ for the scale transformations (in accordance with the standard notation in present gauge field theories). This index runs on the gauge group parameters, now written θ_a. For example in three dimensions, this means that we replace the three rotations $\theta_{23}, \theta_{31}, \theta_{12}$ respectively by $\theta_1, \theta_2, \theta_3$. But since the scale variables were themselves two-index tensors, this means that we have now compacted four indices into one.

We obtain the following more compact form for the complete action,

$$dS_{jk} = \left(D_\mu S_{jk} - \frac{\tilde{g}}{c} t_{jk}^a W_{a\mu} \right) dx^\mu, \tag{7.78}$$

and therefore

$$D^\mu S_{jk} = -i\hbar \, \psi_j^{-1} D^\mu \psi_k = -i\hbar \, \psi_j^{-1} \partial^\mu \psi_k + \frac{\tilde{g}}{c} t_{jk}^a W_a^\mu. \tag{7.79}$$

7.5.2.2. *Scale relativistic tools for Yang–Mills theory*

The previous equations have used new geometric concepts that are specific of the scale relativity approach, namely, (i) the scale variables η_α, (ii) the biquaternionic velocity matrix $\mathcal{V}_{jk}^\mu$ and (iii) its associated action S_{jk}. The standard algebraic concepts of quantum field theories, specifically, the fermionic field ψ, the bosonic field W_a^μ, the charges g, the gauge group generators t_{jk}^a and the gauge-covariant derivative D_μ are now all of them derived from these new geometric tools.

Let us show that we are thus able to recover the basic relations of standard non-Abelian gauge theories (see e.g. [7]). From Eq. (7.79), we first obtain the standard form for the covariant partial derivative, now acting on the wave function multiplets,

$$D^\mu \psi_k = \partial^\mu \psi_k + i\frac{\tilde{g}}{\hbar c} t_k^{ja} W_a^\mu \psi_j. \tag{7.80}$$

The ψ_k's do not commute one with each other since they are biquaternionic quantities, but this is the case neither of t_k^{ja} nor of W_a^μ, so that ψ_j can be put to the right as in the standard way of writing. From the multiplet point of view (index j), this means that we simply exchange the lines and columns.

Now introducing a dimensionless coupling constant α_g and a dimensionless charge g, such that

$$g^2 = 4\pi\alpha_g = \frac{\tilde{g}^2}{\hbar c}, \tag{7.81}$$

and redefining the dimensionality of the gauge field (namely, we replace $W_a^\mu/\sqrt{\hbar c}$ by W_a^μ), the covariant derivative may be more simply written under its standard form,

$$D^\mu \psi_k = \partial^\mu \psi_k + i g t_k^{ja} W_a^\mu \psi_j, \tag{7.82}$$

where all the three new contributions, g, t_k^{ja} and W_a^μ have been constructed from the origin with the tools of the scale relativity theory and have been given a geometric meaning.

In the simplified case of a fermion singlet, it reads

$$D^\mu = \partial^\mu + i g t^a W_a^\mu. \tag{7.83}$$

Let us now derive the laws of gauge transformation for the fermion field. Consider a transformation θ_a of the scale variables. As we shall now see, the θ_a's can be identified with the standard parameters of a non-Abelian gauge transformation. Indeed, using the above remark about the exchange of lines and columns, Eq. (7.68) becomes

$$-i\hbar\, \partial^\mu \psi_k = \partial^\mu S_k^j \psi_j. \tag{7.84}$$

This allows us to recover by a different way Eq. (7.82), from which we obtain the standard form of the transformed fermion multiplet in the case of an infinitesimal gauge transformation $\delta\theta_a$,

$$\psi_k' = \left(\delta_k^j - i g t_k^{ja} \delta\theta_a\right)\psi_j. \tag{7.85}$$

7.5.2.3. *Yang–Mills theories*

We now have at our disposal all the tools of quantum gauge theories. The subsequent developments are standard ones in terms of these tools. One introduces the commutator of the matrices t_a, which have *a priori* no reason to commute, under the form:

$$t_a t_b - t_b t_a = i f^c_{ab} t_c. \tag{7.86}$$

Therefore the t_a's are identified with the generators of the gauge group and the $i f^c_{ab}$'s with the structure constants of its associated Lie algebra. The non-commutativity of the generators and the requirement of the full Lagrangian invariance under the scale transformations finally imply the appearance of an additional term in the gauge transformation law of the boson fields. We obtain this additional term by the standard method recalled below.

We replace into the Lagrangian of the fermionic field the partial derivative ∂_μ by its covariant counterpart D_μ of Eq. (7.83). The development of the covariant derivative leads to the appearance of two terms, a free particle one and a fermion-boson coupling term,

$$\mathcal{L} = \bar{\psi}(i\gamma^\mu \partial_\mu - m)\psi - g\bar{\psi}\gamma^\mu t_a W^a_\mu \psi. \tag{7.87}$$

Let us now consider an infinitesimal scale transformation of the fermion field,

$$\psi \to \psi e^{-ig\delta\theta^b t_b}. \tag{7.88}$$

The requirement of the full Lagrangian invariance under this transformation involves also the coupling term. Let us consider the transformation of this term, except for the W_μ contribution,

$$\bar{\psi}\gamma^\mu t_a \psi \to \bar{\psi} e^{ig\delta\theta^b t_b} \gamma^\mu t_a \psi e^{-ig\delta\theta^b t_b}. \tag{7.89}$$

Accounting for the fact that this is an infinitesimal transformation, it becomes

$$\bar{\psi}(1 + ig\delta\theta^b t_b)\gamma^\mu t_a \psi(1 - ig\delta\theta^b t_b) = \bar{\psi}\gamma^\mu t_a \psi + ig\bar{\psi}\gamma^\mu \delta\theta^b (t_b t_a - t_a t_b)\psi. \tag{7.90}$$

We replace the commutator $t_b t_a - t_a t_b$ by its expression in Eq. (7.86), and we obtain

$$\bar{\psi}\gamma^\mu t_a \psi \to \bar{\psi}\gamma^\mu t_a \psi - g\bar{\psi}\gamma^\mu \delta\theta^b f^c_{ba} t_c \psi. \tag{7.91}$$

Then the requirement of invariance could be fullfilled only provided the transformation of the field W_μ^a itself involves a new term (in addition to the Abelian term $\partial_\mu \delta\theta^a$), i.e.,

$$W_\mu^a \to W_\mu^a + \delta W_\mu^a. \tag{7.92}$$

The transformation of the full coupling term now reads

$$\bar\psi \gamma^\mu t_a W_\mu^a \psi \to \left[(\bar\psi \gamma^\mu t_a \psi) - g f_{ba}^c \delta\theta^b (\bar\psi \gamma^\mu t_c \psi)\right][W_\mu^a + \delta W_\mu^a]. \tag{7.93}$$

Neglecting the second order term in the elementary variations and using the fact that we can interchange the running indices, we see that this expression is invariant provided

$$\bar\psi \gamma^\mu \left\{ t_a \left[\delta W_\mu^a - g f_{bc}^a \delta\theta^b W_\mu^c\right]\right\}\psi = 0. \tag{7.94}$$

A general solution, independent of the t_a's, to the requirement of the Lagrangian invariance in the non-Abelian case is therefore

$$\delta W_\mu^a = g f_{bc}^a \delta\theta^b W_\mu^c. \tag{7.95}$$

Finally, under an infinitesimal scale transformation $\delta\theta^b$, the non-Abelian gauge boson field W_μ^a transforms as

$$W_\mu^a \to W_\mu'^a = W_\mu^a + \partial_\mu \delta\theta^a + g f_{bc}^a \delta\theta^b W_\mu^c. \tag{7.96}$$

We recognize here once again a standard transformation of non-Abelian gauge theories, which is now derived from the basic transformations on the η_a's of Eq. (7.63).

We can finish as usual the development of standard Yang–Mills theories. The gauge field self-coupling term, $-\frac{1}{4}F_{\mu\nu}F^{\mu\nu}$, is retained as the simplest invariant scalar that can be added to the Lagrangian. It is defined as follows.

First, one defines the Yang–Mills field,

$$A_\mu \equiv t_a W_\mu^a, \tag{7.97}$$

which yields the covariant derivative of Eq. (7.83) under the standard form,

$$D_\mu = \partial_\mu + ig A_\mu. \tag{7.98}$$

Then, one establishes the analogue of the Faraday tensor of electromagnetism, by defining

$$F_{\mu\nu}^a \equiv \partial_\mu W_\nu^a - \partial_\nu W_\mu^a - g f_{bc}^a W_\mu^b W_\nu^c. \tag{7.99}$$

and

$$F_{\mu\nu} \equiv t_a F_{\mu\nu}^a, \tag{7.100}$$

which gives

$$F_{\mu\nu} = \partial_\mu A_\nu - \partial_\nu A_\mu + ig[A_\mu, A_\nu]. \tag{7.101}$$

One adds to the Lagrangian density $\mathcal{L}$ a kinetic term for the free Yang–Mills gauge field,

$$\mathcal{L}_A = -\frac{1}{4} F_{\mu\nu} F^{\mu\nu}. \tag{7.102}$$

This form is justified by the same reasons as in the standard theory (namely, it must be a scalar and constructed from the fields and not from the potentials, which are gauge dependent). The Euler–Lagrange equations therefore read

$$\partial_\mu F^{\mu\nu} + ig[A_\mu, F^{\mu\nu}] = 0. \tag{7.103}$$

Introducing the Yang–Mills derivative operator,

$$\nabla_\mu = \partial_\mu + ig[A_\mu, \ ], \tag{7.104}$$

one finally obtains the standard Yang–Mills equations, which generalize to the non-Abelian case the source-free Maxwell equation,

$$\nabla_\mu F^{\mu\nu} = 0. \tag{7.105}$$

It is therefore finally a complete and fully consistent gauge theory that may be obtained as a consequence of scale symmetries issued from a geometric fractal and nondifferentiable space-time description.

7.5.3. *Discussion*

We leave a full discussion of the nature of the gauge group (which emerges from the new geometric approach) and of its consequences for elementary particle physics open for future works.

We just want here to discuss the conclusion reached in Sec. 7.5.1.5, according to which this group is expected to be at least a SU(5) group, which was for long the best candidate for a grand unification theory (GUT) [198]. However, the group SU(5) was later dismissed as a possible unifying group because the GUT predictions for the weak mixing angle and the proton lifetime were found to contradict experimental results. We recall that here this problem is set in a completely different way and can be solved in the special scale relativity framework [352, 353], so that it is possible to reconsider working in such a simplest scheme (and more generally in the larger group SO(10), of which SU(5) is a subgroup).

The value of the weak mixing angle at unification scale under SU(5) (and under some other unification groups) is $\sin^2 \theta_w(m_{\text{GUT}}) = 3/8$. By running it down to the W/Z scale using the solutions to the renormalization group equations, one theoretically predicts $\sin^2 \theta_w(m_Z) = 0.210$ (see detail and references for this prediction in, e.g. [353]), while recent determinations yield (in the modified minimal subtraction scheme) $\hat{s}_Z^2 = 0.23113(15)$ [433] (where the number between parentheses is, as usual, the error on the last digits), which excludes the theoretical prediction in a statistically significant way.

Another drawback, which prevented SU(5) from being retained as a relevant gauge group for grand unified theory in the standard model, is its incompatibility with the bounds on the proton lifetime t_p as they are constrained by, e.g. the Super-Kamiokande data, which impose $t_p > 10^{33}$ yrs [433, 435]. These constraints are sufficient to rule out non supersymmetric (SUSY) or minimal SUSY SU(5) GUTs.

However, in the special scale relativity framework, these issues are set in a fundamentally different way. Indeed, the laws of dilation have a log-Lorentzian form as a direct manifestation of the principle of scale relativity (see Chapter 4), so that a generalized Compton relation between length-scales and mass scales is established [352, 353, 360] that reads, when taking as reference the Z boson scale,

$$\ln \frac{m}{m_Z} = \frac{\ln(\lambda_Z/\lambda)}{\sqrt{1 - \ln^2(\lambda_Z/\lambda)/\ln^2(\lambda_Z/l_{\mathbb{P}})}}, \tag{7.106}$$

where $l_{\mathbb{P}}$ is the Planck length-scale. A major consequence of this new structure of space-time is that the Planck length-scale becomes invariant under dilations and now plays the role devoted to the zero point. The Planck mass scale is no longer its inverse. From Eq. (7.106) one finds that the Planck mass-scale corresponds in the new framework to a length-scale λ_G given by $\ln(\lambda_Z/\lambda_G) = \ln(m_{\mathbb{P}}/m_Z)/\sqrt{2}$, which is nothing but the grand unification scale [352, 353], at which the three gauge couplings and also the gravitational coupling converge toward about the same value (see Fig. 11.4).

Since the effects of gravitation become dominant at that scale, a full unification of the gravitational field with the gauge fields is needed from this energy scale and beyond. In the scale and motion relativity framework, gravitation manifests the effects of the curvature of space-time, while gauge fields manifest the effects of its fractality. Now, when reaching the Planck energy, the curvature has increased so much that it becomes indistinguishable from the fractal fluctuations. The quantum, gravitational and gauge field properties become mixed in a unique extremely complicated geometric behavior that remains to be understood and to be described.

Such a Planck energy-scale theory cannot, in this purely geometric approach, be reduced to a quantum gravity theory, since, in the same way as the quantum behavior breaks the classical gravitational description, the gravitational behavior can also be shown to break the standard quantum description [349, 352]). It is therefore a completely new description, expected to ask for completely new concepts, which is needed at the Planck energy scale.

When going down to lower energies ($E < E_{\mathbb{P}}$), the unified field is spontaneously broken by the rapid decrease of curvature that separates the gravitational field and the gauge fields. Under such a scenario, the SU(5) group would be valid only on a small scale-range around the unification scale and would subsequently be broken in $U(1) \times SU(2) \times SU(3)$ by a cascade effect. One therefore recovers the quantum numbers of hypercharge, isospin and color and the Dirac spinor multiplets corresponding to each of the subgroups in the above direct product.

Due to the presence of gravitation at the unification scale, one expects the appearance of threshhold effects, so that there is no reason for the three U(1), SU(2) and SU(3) low energy running couplings to converge at exactly the same point. This is for example already the case for quantum electrodynamics at the electron scale, where it is known that the asymptotic value of the running electromagnetic coupling constant does not converge toward the Compton scale of the electron (i.e. toward the electron mass scale), but instead toward an energy about 4.1 times larger (see e.g., [268, 353]). It is quite possible for the SU(3) and gravitational field to separate first (see [360, 383] and Chapter 11 for a theoretical predictions of the strong coupling constant issued from such an analysis), then at slighly smaller energy the U(1) and SU(2) fields (see Fig. 11.4). This relaxes the constraint on the weak mixing angle and allows to render its experimental value consistent with the theoretical prediction [360].

As regards the proton lifetime, it is given in a SU(5) GUT by $t_p \propto m_G^4/(\alpha_G^2 m_p^5)$, where α_G is the GUT coupling constant, m_G the GUT mass scale and m_p the proton mass. The standard GUT predicts $m_G \sim 10^{15}$ Gev, which gives $t_p \sim 10^{31}$ yrs. Now, in special scale relativity, $m_G = m_{\mathbb{P}}$

[352, 353, 360]. This multiplies by a factor $(10^4)^4 = 10^{16}$ the predicted proton lifetime, which therefore becomes compatible with the experimental data. All these points and other possible consequences of the scale relativity theory for elementary particle physics will be treated in more detail in Chapter 11.

7.6. Conclusion and Future Prospect

In this chapter, we have attempted to reach an understanding from first principles, in terms of a geometric space-time description, of the nature of gauge transformations. Let us indeed recall the fundamental difference between the situation of transformations in the standard gauge theories and transformations whose geometric meaning is known, such as e.g. rotations in space or Lorentz transformations.

Consider indeed rotations in three-dimensional space. When various physical objects are subjected to various deformations (dilations and contractions) due to these rotations, we nevertheless know that the basic cause for these deformations is rotation. This knowledge could be considered as superfluous, for example, one can simulate on a computer screen any rotation by simply simulating all its effects on the projected two-dimensional variables. However, this does not correspond to a true understanding of their nature. The simulation is well done provided one sees the object "turning" on the screen, not only as if some of its sides were dilating and other contracting. As a consequence, this can be generalized to rotations in a four-dimensional space, which can be experienced by the same method.

We also know from the very beginning what Lorentz transformations are space-time rotations of the coordinates, i.e. in the case of an infinitesimal transformation, (i) $dx'^\alpha = (1 + \omega^\alpha_\beta)dx^\beta$. Then, once this basic definition is given, one can consider the effect of these transformations on various physical quantities ψ. This involves the consideration of representations of the Lorentz group adapted to the nature of the physical object under consideration, i.e. (ii) $\psi' = \left(1 + \frac{1}{2}\omega^{\alpha\beta}\sigma_{\alpha\beta}\right)\psi$ (see e.g. [528]).

If one compares this situation to that of the standard theory of gauge transformations, there was, up to now, no equivalent of the basic defining transformation (i), and the gauge group was directly defined through its action on the various physical objects (ii).

It is just an equivalent of the defining transformation (i) that we propose in the scale-relativity framework. In other words, we have given a geometric meaning to the gauge transformations. We now interpret them as scale transformations in the scale space. These transformations apply to the fractal structures, which characterize the geodesics of a fractal space-time (identified with a particle) at scales smaller than its Compton length. This last point allows one to solve the problem encountered by the Weyl theory.

Then the gauge fields are understood as the manifestation of a general-scale-relativistic effect, i.e. as the geometric effects of dilations and contractions of the internal fractal structures in scale space that are induced by the displacements in space-time. In other words, they correspond to coupling terms between motion and scale. Finally the charges are identified with the conservative quantities that find their origin in the symmetries of the new scale variables.

We are now provided with a theory where the gauge group is no more defined through its only action on the physical objects, as in the standard framework, but as the transformation group of the scale variables, and where the boson fields and the charges are given a physical meaning. We have established the following correspondences between the standard gauge theory items and the scale relativistic tools [394]:

- gauge transformations ↔ scale transformations in scale-space,
- internal gauge space ↔ local scale space,
- gauge fields ↔ manifestations of the fractal and scale-relativistic geometry of space-time (analogues of the Christoffel symbols issuing from the curvature of space-time in General Relativity),
- gauge charges ↔ conservative quantities, conjugate to the scale variables, originating from the symmetries of the scale space and generators of the scale transformation group.

On the basis of the first stones recalled here, a huge work of construction remains to be done, which may be traced (in a non exhaustive way) in the following open problems.

Problems and Exercises

Open Problem 12: Generalize the metric description of the scale space to fully tensorial scale variables, then develop the corresponding non-Abelian gauge theory while keeping all tensorial indices. ∎

Open Problem 13: Achieve the second quantization of gauge fields in the scale-relativity framework. ∎

Open Problem 14: Identify the full group of transformation in the non-Abelian case and decompose it in sub-groups. Identify the associated various charges with their geometric meaning. ∎

Open Problem 15: Generalize, with these definitions, the mass-coupling relations and their application to possible theoretical predictions of particle masses. ∎

Open Problem 16: Develop the new approach to the Higgs boson field according to which it would be a part of the full gauge group, separated at low energy from the field by a spontaneous (scale and motion) symmetry breaking (see Chapter 11 and [378]). Develop the consequences of this approach for the understanding of the symmetry breaking mechanism of the electroweak theory. ∎

Open Problem 17: Connect the scale relativistic description with the renormalization group approach, in particular with regard to the variation of the running couplings and of the particle masses in function of the scale. ∎

Chapter 8

QUANTUM-TYPE MECHANICS IN SCALE SPACE

8.1. Motivation

Let us now consider a new tentative development of the scale relativity theory. Recall that this theory is founded on the giving up of the hypothesis of differentiability of space-time coordinates. We reached the conclusion that the problem of dealing with nondifferentiable coordinates (under the meaning of class C0 coordinates, which are continuous and therefore can be differentiated, but which have no derivative in the standard way) could be circumvented by replacing them by fractal functions, i.e. explicit functions of the resolutions. These functions are defined in the space of resolutions, which we have called scale space. The advantage of this approach is that it sends the problem of nondifferentiability to infinity in the scale space, i.e. $\ln(\lambda/\varepsilon) \to \infty$ when $\varepsilon \to 0$.

In such a framework, standard physics should be completed by scale laws allowing to determine the physically relevant functions of resolution. We have suggested that these fundamental scale laws be written in terms of differential equations, which amounts to defining a differential fractal "generator" (Chapter 4).

Then the effects induced by these internal scale laws on the dynamics can be studied. We have found that the simplest possible scale laws that are consistent (i) with the principle of scale relativity and (ii) with the standard laws of motion and displacements, lead to a quantum-type mechanics as regards the laws of motion, i.e. the laws of displacement in space-time (Chapters 5 and 6).

Another generalization of the description has consisted of considering resolution variables that may themselves be functions of the space-time coordinates. This has led to a new geometric foundation of gauge field theories as being manifestations of the fractal and nondifferentiable geometry of space-time (Chapter 7).

However, the description of a nondifferentiable space-time, even at this level of generality, can be further generalized. Indeed, the choice to write the transformation laws in the scale space in terms of standard differential equations means that we have implicitly assumed the scale space to be differentiable and "classical". As we have seen in detail, the introduction of such a differentiable internal scale space of resolution variables allows nondifferentiability in standard space-time of positions and instants. This is once again a mere hypothesis that can now be relaxed at a more profound level of description of the theory.

We therefore have been led [388, 400] to tentatively consider a new extension of the theory of scale relativity in which the scale space is nondifferentiable, which should be followed by the construction of the motion laws based on such a scale space description. In order to solve the problem, we use the same method that has been built for dealing with nondifferentiability in space-time.

We have proved the basic theorem according to which continuity and nondifferentiability of space-time implies the existence of internal scale variables (Chapter 3). In the same way, continuity and nondifferentiability of the scale space leads one to explore a new level of still more inner structures that may be its manifestation. In analogy with the previous case, we expect the nondifferentiability of the scale space and its subsequent explicit dependence on new "superscale" variables to give rise to scale laws that take a quantum-like form instead of a classical one. The very difficult question of finding the "super-quantum" motion laws in position space that could be constructed on the basis of internal scale structures that are themselves described by quantum scale laws is left open to future studies. We have suggested [388, 400, 402] that such a future theory could be a kind of "third quantization".

8.2. Schrödinger Equation in Terms of Resolutions

Recall that for the construction of classical scale differential equations we have mainly considered two representations: (i) the logarithms of resolution are fundamental variables; (ii) the main new variable is the "scale time" or "djinn" and the resolutions are deduced as derivatives.

These two possibilities are also to be considered for the new present attempt to construct quantum scale-laws. The first one, which we shall only briefly study here, consists in the introduction of a "scale wave function", which is an explicit function of scale and space-time variables, $\psi[\ln \varepsilon(x,t), x, t]$. In the simplified case where it depends only on the time variable, one may write a Schrödinger equation acting in the space of resolution variables,

$$\mathcal{D}_\varepsilon^2 \frac{\partial^2 \psi}{(\partial \ln \varepsilon)^2} + i \mathcal{D}_\varepsilon \frac{\partial \psi}{\partial t} - \frac{1}{2} \phi_\varepsilon \, \psi = 0. \tag{8.1}$$

This is the quantum equivalent of the classical stationary wave equation giving rise to a log-periodic behavior (see Sec. 4.3.1). It is also related to the scale relativistic re-interpretation of gauge invariance (Chapter 7), in which the resolutions become "fields" depending on space and time variables, so that the wave function becomes a function of the variables $\ln \varepsilon$. However, in the case of gauge theories, only the phase was resolution-dependent, while now the modulus of the scale wave function depends on the resolution scale. This means that the solutions of such an equation give the probability of presence of a structure at some relative scale (in the scale space), and that time-dependent solutions describe the propagation of quantum waves in scale space.

This equation can be generalized to a more general description of the scale variables, which accounts for their tensorial nature, $\varepsilon_{\mu\nu} = \rho_{\mu\nu} \varepsilon_\mu \varepsilon_\nu$. One introduces a wavefunction $\psi(\varepsilon_{\mu\nu})$, and a Laplacian operator $\partial_{\mu\nu} \partial^{\mu\nu}$ is constructed on these variables. The Schrödinger equation in scale space therefore reads

$$\mathcal{D}_\varepsilon^2 \, \partial_{\mu\nu} \partial^{\mu\nu} \psi + i \mathcal{D}_\varepsilon \frac{\partial \psi}{\partial t} - \frac{1}{2} \phi_\varepsilon \, \psi = 0. \tag{8.2}$$

Indeed, since $\varepsilon_{\mu\nu} \varepsilon^{\mu\nu} = $ cst, the two operators $\partial^2 / \partial \varepsilon_{\mu\nu} \partial \varepsilon^{\mu\nu}$ and $\partial^2 / \partial \ln \varepsilon_{\mu\nu} \partial \ln \varepsilon^{\mu\nu}$ are equal up to a multiplicative numerical constant.

8.3. Schrödinger Equation in Terms of the "djinn"

Let us now consider the second representation in which the "djinn", which is a variable fractal dimension, has become the primary variable. Start with the general Euler–Lagrange form given to scale laws in Sec. 4.3.3 after introduction of the "djinn" τ,

$$\frac{d}{d\tau} \frac{\partial \tilde{L}}{\partial \mathbb{V}} = \frac{\partial \tilde{L}}{\partial \ln \mathcal{L}} \,, \tag{8.3}$$

where we recall that $\mathcal{L}$ is a fractal coordinate, τ is the "djinn" that generalizes to a variable fifth dimension the fractal dimension (minus the constant topological dimension), $\tilde{L}$ is the scale Lagrange function and $\mathbb{V} = \ln(\lambda/\varepsilon)$ is the "scale velocity".

In the Newtonian case, this equation becomes

$$\frac{d^2 \ln \mathcal{L}}{d\tau^2} = -\frac{\partial \Phi_S}{\partial \ln \mathcal{L}}. \tag{8.4}$$

Since the scale space is now generalized to a nondifferentiable and fractal geometry, the various elements of the new description can be used:

(i) Infinity of trajectories, leading to introduce a scale velocity field $\mathbb{V} = \mathbb{V}(\ln \mathcal{L}(\tau), \tau)$;

(ii) Decomposition of the derivative of the fractal coordinate in terms of a "classical part" and a "fractal part", described by a stochastic variable such that $\langle d\xi_s^2 \rangle = 2\mathcal{D}_s d\tau$;

(iii) Introduction of the two-valuedness of this derivative because of the symmetry breaking of the reflection invariance under the exchange $(d\tau \leftrightarrow -d\tau)$, leading to construct a complex scale velocity $\tilde{\mathbb{V}}$ based on this two-valuedness;

(iv) Construction of a new total covariant derivative with respect to the "djinn", which reads

$$\frac{\widehat{d}}{d\tau} = \frac{\partial}{\partial \tau} + \tilde{\mathbb{V}} \frac{\partial}{\partial \ln \mathcal{L}} - i \mathcal{D}_s \frac{\partial^2}{(\partial \ln \mathcal{L})^2}. \tag{8.5}$$

(v) Introduction of a wave function as a re-expression of the action, which is now complex, $\Psi_s(\ln \mathcal{L}) = \exp(i\mathcal{S}_s/2\mathcal{D}_s)$;

(vi) Transformation and integration of the above Newtonian scale-dynamics equation under the form of a Schrödinger equation now acting on scale variables:

$$\mathcal{D}_s^2 \frac{\partial^2 \Psi_s}{(\partial \ln \mathcal{L})^2} + i \mathcal{D}_s \frac{\partial \Psi_s}{\partial \tau} - \frac{1}{2} \Phi_s \Psi_s = 0. \tag{8.6}$$

8.4. Complexergy

In order to understand the meaning of this new Schrödinger equation in more detail, let us once again review the various levels of evolution of the concept of physical fractals adapted to a geometric description of a nondifferentiable space-time.

The first level in the definition of fractals is Mandelbrot's concept of "fractal objects" [305, 306].

The second step has consisted in a shift from the concept of fractal objects to scale relativistic fractals. Namely, the scales at which the fractal structures appear are no longer defined in an absolute way. Only scale ratios do have a physical meaning, not absolute scales.

In the third step, which is achieved in the new scale relativistic interpretation of gauge transformations of Chapter 7, we consider fractal structures (still defined in a relative way) that are no longer static. The scale ratios between structures become a field that may vary from place to place and with time.

The next level (in the present state of the theory) is given by the solutions of the above Schrödinger equation acting in scale space. The Fourier transform of these solutions will provide probability amplitudes for the possible values of the logarithms of scale ratios, $\Psi_s(\ln \varrho)$. Then $|\Psi_s|^2(\ln \varrho)$ gives the probability density of these values. Depending on the scale field and on the boundary conditions (in scale space), peaks of probability density will be obtained, this meaning that some specific scale ratios become more probable than others. Therefore, such solutions now describe quantum probabilistic fractal structures, defined in a relative way. The statement about these fractals is no longer that they own given structures at some (relative) scales, but that there is a given probability for two structures to be related by a given scale ratio.

Concerning the solutions of the scale Schrödinger equation themselves (before Fourier transform), they provide probability densities for the position on the fractal coordinate (or fractal length) $\ln \mathcal{L}$. This means that, instead of having a unique and determined $\mathcal{L}(\ln \varepsilon)$ dependence (as in the well-known example of the length of the Britain coast between two points of this coast), an infinite family of possible behaviors is defined, which self-organizes in such a way that some values of $\ln \mathcal{L}$ become more probable than others.

Let us now consider the stationary scale Schrödinger equation. It reads

$$2\mathcal{D}_s^2 \frac{\partial^2 \Psi_s}{(\partial \ln \mathcal{L})^2} + (\mathbb{E} - \Phi_s)\Psi_s = 0. \tag{8.7}$$

The "stationarity" of this equation means that it does no longer depend on the "djinn" (or scale time) τ. A new important quantity appears in this last representation. It is the conservative quantity, which, according to Noether's theorem, must emerge from the uniformity of the new "djinn" variable [353]. It is defined in analogy with the definition of energy from the time symmetry in classical mechanics, in terms of the scale-Lagrange function $\tilde{L}$ and of the resolution $\mathbb{V} = \ln(\lambda/\varepsilon)$, as

$$\mathbb{E} = \mathbb{V}\frac{\partial \tilde{L}}{\partial \mathbb{V}} - \tilde{L}. \tag{8.8}$$

This new fundamental prime integral is therefore a "scale energy", i.e. it is the equivalent for scale of what energy is for motion. It has first been introduced in [352, 353], and refered to as "complexergy" in [388].

A more complete understanding of the meaning of this new description can be reached by considering an explicit example, e.g. the case of a scale harmonic oscillator potential well. Recall that such a potential (in the repulsive case) has already been considered as an example of new scale-dynamical laws (see Sec. 4.3.6) and has yielded a confinement-like behavior. We shall now consider the quantum version of the effect of such a scale force (now in the attractive case). The stationary Schrödinger equation reads in this case

$$2\mathcal{D}_s^2 \frac{\partial^2 \Psi_s}{(\partial \ln \mathcal{L})^2} + \left\{ \mathbb{E} - \frac{1}{2}\omega^2(\ln \mathcal{L})^2 \right\} \Psi_s = 0. \tag{8.9}$$

The behavior of its solutions strongly depends on the complexergy $\mathbb{E}$, which can take only quantized values.

As we shall now see, the behavior of the above equation suggests an interpretation for this conservative quantity and allows one to link it to the complexity of the system under consideration. Indeed, let us consider the momentum solutions of the above scale Schrödinger equation, i.e. scale momentum wave functions $a[\ln(\lambda/\varepsilon)]$. Recall that the main variable is now $\ln \mathcal{L}$ and that the scale momentum is the resolution, $\ln \rho = \ln(\lambda/\varepsilon) = d\ln \mathcal{L}/d\tau$ (since we take here a scale mass $\mu = 1$). The squared modulus of the wave function yields the probability density of the possible values of resolution ratios.

The complexergy is quantized as a consequence of the presence of the harmonic oscillator field. The various solutions (plotted in Fig. 8.1 as concerns the three first levels) depend on the quantized values of the complexergy, that read, in terms of the quantum number n

$$\mathbb{E}_n = 2\mathcal{D}_s \omega \left(n + \frac{1}{2} \right), \tag{8.10}$$

where ω is the frequency. As can be seen in Fig. 8.1, the solution of minimal complexergy shows a unique peak in the probability distribution

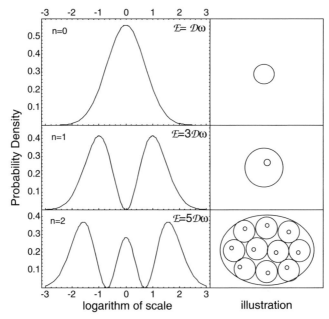

Fig. 8.1: **Solutions of a scale-Schrödinger equation (harmonic oscillator).** They provide a probability density for a structure to exist at a given (relative) scale. These solutions can be interpreted as describing systems characterized by an increasing number of hierarchical levels, as illustrated in the right-hand side of the figure. For example, living systems, such as prokaryotes, eukaryotes and simple multicellular organisms, have respectively one characteristic scale (cell size), two characteristic scales (nucleus and cell) and three characteristic scales (nucleus, cell and organism). One obtains a similar result, e.g. by solving a Schrödinger equation in resolution space under box limiting conditions (see Fig. 8.2).

of the $\ln(\lambda/\varepsilon)$ values. This can be interpreted as describing a system characterized by a single, more probable relative scale. Now, when the complexergy increases, the number of probability peaks $(n + 1)$ increases. Since these peaks are regularly distributed in terms of $\ln \varepsilon$, i.e. probabilistic log-periodicity, they are distributed in scale (ε) as powers of a given unitary ratio ρ^n, so that they can be interpreted as describing a system characterized by a hierarchy of imbricated levels of organization. A remarkable feature of this self-organization process is that the minimal complexergy, which yields the simplest structure with only one hierarchy level, is not vanishing. This "vacuum complexergy" is the scale analog of the vacuum energy of standard quantum mechanics.

Such a hierarchy of organization levels is one of the criteria that define complexity. Therefore, increasing complexergy corresponds to increasing complexity, which is one of the justifications for the chosen name for the new conservative quantity.

More generally, one can remark that the "djinn" is universally limited from below ($\tau > 0$). This implies that the complexergy is universally quantized, and that we expect the existence of discretized levels of hierarchy of organization in nature, as is actually observed, as well at microphysical scales (quarks, nucleons, nuclei, atoms, molecules) as at mesoscopic scales (chromosomes, proteins, nucleus, cell, tissue, organs, organisms then social organization in living systems) and astonomical scales (stars, galaxies, clusters of galaxies, superclusters) instead of a continuous hierarchy.

Let us consider another example that leads to essentially the same conclusion. In the above situation, the structuring comes from the field, namely, an attractive harmonic oscillator potential in scale space. But this kind of behavior is general in such a quantum-type theory.

Assume that $\ln \mathcal{L}$ is limited at lower and upper scales and that the system is free (no applied force). This is the scale equivalent of the well-known problem of a free quantum particle in a box. The solution of the Schrödinger equation is in this case ("scale box") a log-periodic law of probability that takes a sinusoidal form (see Fig. 8.2)

$$P = a \sin^2(b \ln \mathcal{L}). \tag{8.11}$$

A similar result can also be directly obtained from the Schrödinger equation (8.1) written in terms of the resolutions $\ln \varepsilon$ (considered in this case as main variables). While the corresponding classical equation would describe, e.g. a continuous evolution with time of the characteristic size of a system, the quantum approach yields a punctuated evolution.

Indeed, if the system is led to increase its energy with time, as is the case of many living systems, its evolution is predicted to proceed from one entangled hierarchy of organization to a more complicated one by punctuated jumps rather than by a continuous process, since only the quantized states are in equilibrium.

Some examples of application of this new approach to various sciences [388, 33, 402] will be suggested in Part III of this book. In particular it may yield an explanation for the discrete scale invariance describable by log-periodic laws observed in many biological systems (see Chapter 14).

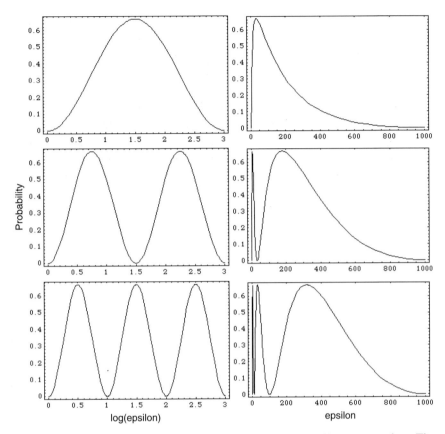

Fig. 8.2: **Log-periodic solutions of the free scale-Schrödinger equation.** The left figures gives, for $n = 0$ to 3, the probability density $P(\ln \varepsilon)$ in function of $\ln \varepsilon$, and the right figures the corresponding probabilities in function of ε, in order to show the scale hierarchy of the probability peaks.

Problems and Exercises

Open Problem 18: Construct the "superquantum" motion laws in position space based on a quantum underlying scale space.

Hints: We recall that standard quantum mechanics has been obtained as a manifestation, at the level of motion laws, of differentiable laws of scale transformation. In the simplest case, the scale dependance (and divergence) has been described by a fluctuation $d\xi^2 = 2\mathcal{D}\zeta^2 dt$, of fractal dimension $D_F = 2$ (Chapter 5). Generalizations of quantum mechanical equations based on more complicated scale laws, special scale-relativistic laws, scale dynamics laws, etc., have been given in Sec. 5.14. They lead to generalized Schrödinger equations, which become explicitly fractal and scale dependent. Their solutions are themselves generalizations of the nondifferentiable and fractal solutions, which have been found for the standard Schrödinger equation in Sec. 5.8.

But, here, a new step is taken forward with the suggestion that physical systems may exist in which scale laws become of a quantum nature. Instead of having a classical description of the fractal fluctuations $d\xi$, they become characterized by a probability amplitude $\psi(d\xi)$, from which a probability density $|\psi|^2$ of having a structure at a given (relative) value can be derived.

The problem posed is therefore to construct the new "superquantum" motion laws, which would be based on such quantum-type scale laws. ■

Part III: Applications

Chapter 9

INTRODUCTION

Part II of this book has been devoted to the foundation of the theory of scale relativity, under its three aspects of (i) scale laws finding their origin in the continuous and nondifferentiable geometry, (ii) motion laws of a quantum nature being based on these scale laws and (iii) gauge fields emerging as manifestations of the fractal geometry of space-time.

We now consider, in this Part III, the applications of the theory to various sciences. In Chapter 10 we describe some applications to physics, with the aim to prepare possible future laboratory experiments that could allow one to put some aspects of the theory to the test.

In Chapter 11 we mainly consider the possible consequences of the special scale relativity theory, in which the Planck length scale is considered to be a minimum scale invariant under dilations, for elementary particle and high energy physics. In particular, we describe the present status of the comparison between some theoretical predictions, which had been made in this framework, with the ever improving experimental values.

In Chapter 12, we study the cosmological consequences of the theory, mainly of its special scale relativity version, in which the cosmic length scale defined by the cosmological constant is considered to be a maximum scale invariant under dilations. In particular, the question of the meaning and value of the cosmological constant is analysed in detail, and its theoretical expectation [353] is compared with various determinations of its now known observational value.

Then, in Chapter 13, we apply the scale relativity approach to astrophysics, in particular to the question of the formation and evolution of gravitational structures in the Universe. The three conditions that lead to obtaining a generalized Schrödinger-like form for the equations of motion can be shown to apply, as a good approximation, to many astrophysical systems on many scales, including to the formation of planetary systems. We develop the kind of theoretical expectations one may derive in this framework, then we review the positive results obtained in the comparison of these predictions with the observational data, with a special emphasis on the universal structures unveiled in extrasolar planetary systems.

Finally the last Chapter 14 is devoted to a brief review of some applications of the theory to sciences of life and human sciences and geosciences, then by a prospect about its contribution to a possible future construction and development of a genuine theoretical systems biology founded on first principles.

Chapter 10

APPLICATIONS TO LABORATORY
AND EARTH SCALE PHYSICS

10.1. Introduction

This chapter is devoted to the study of various applications of the theory of scale relativity in physics, in particular in laboratory physics and more generally to physics at human scale.

We first study in more detail the quantum to classical (and classical to quantum) transitions in the scale relativity framework. By "transition", we mean here two different though related concepts:

(i) The conceptual transition from classical laws to quantum laws. This amounts to answering the question: what are the necessary and sufficient conditions that transform classical laws into quantum laws? The theory of scale relativity, by its ability to write a geodesic equation, which includes both classical and quantum laws, offers a unique occasion to analyse in detail how one passes from the quantum to the classical realm and reciprocally.

(ii) The effective transition of a physical system, which is quantum under some conditions and becomes classical under other conditions.

Then, we show that a more profound understanding of the nature of this transition allows us to propose new laboratory experiments, which may (i) test the predictions of the theory, and (ii) lead to new kinds of macroscopic quantum-type laboratory experiments and technology. Here and in the following chapters, the application of the same approach to natural systems will also be considered.

10.2. Quantum-Classical Transition

In Chapter 5, we have seen in detail that the nondifferentiable and fractal geometry can be translated into three properties (briefly recalled in the next section), which in their turn give rise to three terms in the covariant total derivative (in the spinless non-relativistic case). The full transition of a system from the classical to the quantum regime becomes effective only provided all of the three new properties have fulfilled.

This is a new and unexpected result, since one could think that the specific condition that transforms a classical system into a quantum one would be the discrete symmetry breaking of the reflection $dt \to -dt$, which is a completely new feature respectively to standard physics, a direct consequence of nondifferentiability and the genuine origin of the emergence of complex numbers in the quantum mechanical description [393]. However, when none, one or two of any of the three properties is fulfilled, one recovers various forms of classical systems (deterministic, fluid-like, fractal fluids, etc.). Quantum systems are obtained only when all the three conditions are present.

10.2.1. *Consequences of nondifferentiable geometry*

In order to analyse in greater detail the nature of the quantum to classical transition, recall that the Schrödinger equation is obtained, in the scale relativity approach, on the basis of three fundamental conditions:

(i) The trajectories are in infinite number. This condition leads one to use a statistical, fluid-like description, in which the velocity $v(t)$ is replaced by a velocity field $v[x(t), t]$. The fundamental cause for this undeterminism of trajectories is the nondifferentiability of space-time, which implies its fractality, and the subsequent identification of the trajectories with its geodesics.

(ii) The trajectories are fractal curves (of fractal dimension $D_F = 2$ in the critical case that leads to standard quantum mechanics). This comes directly from the fact that space-time itself is nondifferentiable and therefore fractal, which implies the fractality of its geodesics. This leads to generalize the concept of velocity to fractal velocity fields.

(iii) The invariance under the reflection transformation $(dt \leftrightarrow -dt)$ is broken. Therefore, while in its standard definition, the velocity does not exist any longer since the coordinate $x(t)$ is nondifferentiable, we replace $x(t)$ by a fractal coordinate $x(t, \delta t)$, and the standard velocity is replaced by

$$v_+[x(t, \delta t), t, \delta t] = \frac{x(t + \delta t, \delta t) - x(t, \delta t)}{\delta t}, \tag{10.1}$$

$$v_-[x(t, \delta t), t, \delta t] = \frac{x(t, \delta t) - x(t - \delta t, \delta t)}{\delta t}. \tag{10.2}$$

Such a twin process plays also a central role in Ord's statistical approach to the fractal space-time description [423]. In the scale relativity approach, it is a consequence of the nondifferentiable geometry itself, which supersedes the fractal geometry (since the fractality is a consequence of continuity and nondifferentiability).

A new generalization of the velocity is involved in this fundamental symmetry breaking. We now deal with two fractal velocity fields,

$$V_+ = v_+[x(t), t] + w_+[x(t, dt), t, dt], \tag{10.3}$$

$$V_- = v_-[x(t), t] + w_-[x(t, dt), t, dt], \tag{10.4}$$

each of them decomposed in terms of a "classical part" (v_+, v_-), which is differentiable and independent of resolution, and of a "fractal part" (w_+, w_-), explicitly dependent on the resolution interval dt and divergent at the limit $dt \to 0$.

10.2.2. *The triple transition from classical to quantum*

Before continuing, recall that these three conditions are only the minimal consequences of nondifferentiability and fractality, (which also include, in particular, the appearance of the spin and of gauge fields). We consider here only the simplest transition, i.e. the transition from a classical regime to a nonrelativistic quantum mechanical regime (or reversely). However, since even in this simplest case the nondifferentiability and the fractality of a space-time continuum manifest themselves under three consequences, we expect this transition to be already a complicated one.

Indeed, it is a combination of the passage from a deterministic velocity on a given trajectory to a velocity field (defined on the infinity of potential trajectories), then to a fractal velocity field, i.e. that is explicitly dependent on the resolution interval, and finally to a twin fractal velocity field, namely,

$$v(t) \to v(x(t), t) \to V[x(t, dt), t, dt]$$
$$\to \{V_+[x(t, dt), t, dt], V_-[x(t, dt), t, dt]\}. \tag{10.5}$$

Reversely, as we shall now see, the transition of a system from the quantum to the classical regime becomes effective provided one among the three new properties have disappeared (either in a change of scale or in a change of the transition scale, i.e. of velocity, temperature, etc.).

10.2.3. *Total derivative in nondifferentiable geometry*

The fundamental mathematical tool of scale relativity amounts to including the new effects into the writing of a more complete expression for the total time derivative [353]. The three conditions imply the appearance of three additional terms:

(i) The condition (1), which tells us that the various physical quantities are functions of $x(t)$ and t, implies to replace d/dt by the standard "Eulerian" total derivative

$$\frac{d}{dt} = \frac{\partial}{\partial t} + v \cdot \nabla. \tag{10.6}$$

(ii) The condition (2), which tells us that the fractal fluctuations $d\xi_k$ are now differential elements of order $1/2$, (which corresponds to fractal dimension 2 of the paths), leads one to introduce terms of second order in the total derivative. The Taylor expansion up to order two of the derivative of a physical quantity f reads

$$\frac{df}{dt} = \frac{\partial f}{\partial t} + \frac{\partial f}{\partial X_k}\frac{dX_k}{dt} + \frac{1}{2}\frac{\partial^2 f}{\partial X_j \partial X_k}\frac{dX_j \, dX_k}{dt}. \tag{10.7}$$

The term $dX_j \, dX_k/dt$ is usually infinitesimal, but when the fractal dimension is $D_F = 2$, its "classical part" reduces to $\langle d\xi_j \, d\xi_k \rangle/dt$ and it is therefore finite. The last term amounts to a Laplacian and one obtains

$$\left\langle \frac{df}{dt} \right\rangle = \left(\frac{\partial}{\partial t} + v \cdot \nabla + \frac{1}{2}\lambda\Delta \right) f. \tag{10.8}$$

In other words, the effect of the second condition is to add second order derivative terms in differential equations.

(iii) The condition (3), i.e. the two-valuedness of the velocity field that finds its origin in the nondifferentiability, leads to a two-valued velocity field

(v_+, v_-) represented by a unique complex velocity field

$$\mathcal{V} = V - iU = \frac{v_+ + v_-}{2} - i\frac{v_+ - v_-}{2}.$$ (10.9)

One introduces a twin classical derivative, $(d_+ f / dt, d_- f / dt)$, from which one defines a complex total derivative operator

$$\frac{\widehat{d}}{dt} = \frac{1}{2}\left(\frac{d_+}{dt} + \frac{d_-}{dt}\right) - \frac{i}{2}\left(\frac{d_+}{dt} - \frac{d_-}{dt}\right),$$ (10.10)

such as $\widehat{d}\,x / dt = \mathcal{V}$.

Finally, when combining the effect of condition (2) (that leads to introduce second order terms in differential equations) and of condition (3) (that leads to jump from a real to a complex description), the complex total derivative reads

$$\frac{\widehat{d}}{dt} = \frac{\partial}{\partial t} + \mathcal{V} \cdot \nabla - i\frac{\lambda}{2}\Delta.$$ (10.11)

Finally, since $\mathcal{V} = V - iU$, the three minimal consequences of the nondifferentiable and fractal geometry are expressed by the appearance in the total derivative of three additionnal terms, namely, $V \cdot \nabla$, $-iU \cdot \nabla$ and $-i(\lambda/2)\Delta$, so that it reads

$$\frac{\widehat{d}}{dt} = \frac{\partial}{\partial t} + V \cdot \nabla - iU\nabla - i\frac{\lambda}{2}\Delta,$$ (10.12)

where λ is the generalized Compton scale (assuming $c = 1$, the coefficient of Chapter 5, $\mathcal{D} = \lambda/2$, i.e. $\hbar/2m$ in standard quantum mechanics).

10.2.4. *Transition from Schrödinger to Euler–Newton equation*

As shown in many previous works [353, 360, 93] and in Chapter 5, if one now writes Newton's equation of dynamics which is nothing but Einstein's equation of geodesics in the Newtonian limit when the potential is a gravitational one, in terms of the above total derivative, namely,

$$m\frac{\widehat{d\mathcal{V}}}{dt} + \nabla\phi = 0, \tag{10.13}$$

one obtains after integration a Schrödinger equation

$$\frac{1}{2}\lambda^2\Delta\psi + i\lambda\frac{\partial}{\partial t}\psi - \frac{\phi}{m}\psi = 0, \tag{10.14}$$

where $\psi = \exp(iS/S_0)$ is a mere redefinition of the action S.

Recall also that a third form of these equations can be obtained by separating the real and imaginary parts of the Schrödinger equation and by taking as a new couple of variables the real part V of the complex velocity $\mathcal{V}$ and the squared modulus of the wave function $P = |\psi|^2$. We obtain a generalized Euler–Newton equation including a "quantum potential" and a continuity equation:

$$\left(\frac{\partial}{\partial t} + V \cdot \nabla\right)V = -\nabla\left(\frac{\phi}{m} - \frac{1}{2}\lambda^2\frac{\Delta\sqrt{P}}{\sqrt{P}}\right), \tag{10.15}$$

$$\frac{\partial P}{\partial t} + \operatorname{div}(PV) = 0. \tag{10.16}$$

The full transition from the quantum to the classical regime can now be made clear. Equation (10.13) is both classical and quantum. When all of the three terms are present, its integral is the Schrödinger equation. Therefore a full study of the classical to quantum transition in the scale relativity framework involves an analysis of the physical conditions under which these terms vanish. Such an analysis, concerning in particular its relation to decoherence and the role played by the de Broglie scale and by the thermal de Broglie scale, has been initiated in previous works ([353] Sec. 5.7, [355] Sec. 4.5).

Let us consider in more detail the various possible intermediate cases, when only some of the three additional terms are present, but not all of them.

(0) When the three additional terms vanish, one recovers the standard deterministic motion equation of classical mechanics, $m\, dv/dt = -\nabla\phi$. This result can be compared to the decoherence approach, which allows an understanding of the vanishing of the non-diagonal terms of a density matrix, but not a transition to a deterministic description, since the description tool remains probabilistic in its essence. In the scale relativity theory, the passage to a statistical and

probabilistic description is not set at a foundation level, but is only a very consequence of the nondifferentiability, which implies an infinite number of geodesics. In the differentiable and large scale limit of low resolution, one recovers an apparent classical deterministic trajectory.

(i) When only the first condition is fulfilled, there is an infinity of paths, but these paths are not fractal, which corresponds to the limit $\lambda \to 0$, and there is no two-valuedness of velocity, which corresponds to keeping a real velocity field. One easily verifies in the fluid representation of the equations (Eqs. (10.15), (10.16)) that in this case one recovers classical hydrodynamical-like equations.

(ii) When only the second condition (fractality of paths) is fulfilled, this corresponds to deterministic motion on a fractal curve. This classical problem of fractal motion is very interesting by itself and is at the heart of a large part of the multiple works involving fractals and fractal objects (see [305, 306], [416, 295] and many other series). The various differential methods outlined in Chapter 4 apply to this case.

(iii) The case when only the doubling of the velocity is present is special. Indeed, it would correspond to a situation where the velocity remains classical and finite (i.e. non scale dependent), but nevertheless two-valued, and therefore describable by complex numbers. The corresponding classical theory would be a complex general relativity.

Such a case could be interpreted as a more complete, partially virtual, description of reversible motion. For example, there is no way to characterize in an absolute way the direction of the motion of the Earth around the Sun. If we take v as initial condition, the opposite speed $-v$ is as well habilitated to be an initial condition for its description. A two-valued velocity could be a way to account for this elementary discrete undeterminism of classical mechanics.

Let us now consider the cases when two of the three conditions are fulfilled.

10.2.4.1. *Fractal velocity field: new form of diffusion equations*

Assume that only the two first conditions: (i) infinity of paths, and (ii) fractality of paths with fractal dimension $D_F = 2$, are fulfilled. This means that we are now describing a standard diffusion process of the Brownian motion type. It shows that one of the key conditions for obtaining a genuine quantum-type behavior is the differential irreversibility condition (3), i.e. the symmetry breaking of the reflection invariance under the transformation $(dt \leftrightarrow -dt)$. Let us demonstrate this important point by studying what happens when this condition is released [393].

The elementary displacements on each trajectory are decomposed as $dX = dx + d\xi$, where

$$dx = V(x(t), t) \, dt, \quad \langle d\xi^2 \rangle = \lambda \, dt, \tag{10.17}$$

i.e. $\lambda = 2\mathcal{D}$ is twice the diffusion coefficient in a diffusion interpretation of such a process. The velocity field $V(x(t), t)$ is now real. The effect of such a behavior on the dynamics can be described in terms of a covariant derivative that writes

$$\frac{D}{dt} = \frac{\partial}{\partial t} + V \cdot \nabla + \frac{1}{2}\lambda \Delta. \tag{10.18}$$

Contrary to what happens when the differential irreversibility condition is assumed, the second order contribution $(1/2)\lambda \Delta$ is now real instead of imaginary. We jump from the quantum case to this reduced situation by making the replacement $-i\lambda \to \lambda$ and, therefore, $\lambda^2 \to -\lambda^2$. In terms of this total derivative operator, Newton's fundamental equation of dynamics keeps its usual form,

$$m\frac{D}{dt}V = -\nabla\phi, \tag{10.19}$$

where ϕ is a potential energy. Since it preserves the form of equations, one can identify the derivative operator D/dt with a "covariant" derivative.

One can define a Lagrange function $L(x, V, t)$ and an action S such that $dS = L \, dt$, which are both real, since there is no longer any two-valuedness of the velocity vector. Let us now set

$$\varphi = e^{S/S_0}, \tag{10.20}$$

where S_0 must be introduced for dimensional reasons. The function φ is a real function, which plays a role similar to that played by the complex wave function ψ. When one considers the action as a function of coordinates, one obtains $v = \nabla S/m$, so one can replace V in Eq. (10.19) by

$$V = \frac{S_0}{m}\nabla \ln \varphi. \tag{10.21}$$

Therefore Eq. (10.19) becomes

$$-S_0 \left[\frac{\partial}{\partial t}(\nabla \ln \varphi) + \left(\frac{S_0}{m}(\nabla \ln \varphi \cdot \nabla)(\nabla \ln \varphi) + \frac{\lambda}{2}\Delta(\nabla \ln \varphi) \right) \right] = \nabla\phi. \tag{10.22}$$

Under the condition $S_0 = \lambda m$, this expression can be greatly simplified thanks to the identity [353]

$$2\nabla \ln \varphi \cdot \nabla(\nabla \ln \varphi) + \Delta(\nabla \ln \varphi) = \nabla \left(\frac{\Delta \varphi}{\varphi} \right). \tag{10.23}$$

We obtain:

$$\nabla \left(\lambda \frac{\partial \ln \varphi}{\partial t} + \frac{1}{2} \lambda^2 \frac{\Delta \varphi}{\varphi} \right) = -\frac{\nabla \phi}{m}. \tag{10.24}$$

Therefore we find a general prime integral of the motion equations, which reads

$$\frac{1}{2} \lambda^2 \Delta \varphi + \lambda \frac{\partial \varphi}{\partial t} = \left(\frac{A(t) - \phi}{m} \right) \varphi. \tag{10.25}$$

Though this equation may look like a Schrödinger equation, it has actually very different properties. The function φ is real, (while the wave function ψ was complex), the sign of the potential is reversed, and the integration function $A(t)$ does not vanish since φ has no phase. As a consequence, such an equation is not at all structuring, contrary to the standard Schrödinger equation. It remains a diffusion equation, without counterterms allowing stationary solutions.

This result gives the proof that a quantum-type behavior cannot be obtained from fractality alone. In order to obtain a genuine quantum behavior, one must definitely add, to the two first conditions — (1) fluid description of the velocity field of the path family and (2) fractality of the paths — the third condition (3) of two-valuedness of the velocity field issued from nondifferentiability.

Let us conclude this section with an application of this result to standard hydrodynamics. If one takes a negative value for the coefficient λ and sets $\nu = -\lambda/2$ and $m = 1$, Eq. (10.19) becomes the Navier–Stokes equation with a coefficient of viscosity ν. This form of the Navier–Stokes equation is obtained in every situations where $\nabla p/\rho$ is a gradient (achieved when there exists an univocal link between the pressure p and the density ρ, in particular in the isentropic case where $\nabla p/\rho = \nabla w$, where w is the enthalpy by unit of mass, see [269]). Therefore Eq. (10.25), which becomes

$$\nu^2 \Delta \varphi - \nu \frac{\partial \varphi}{\partial t} = \left(\frac{A(t) - \phi}{2} \right) \varphi, \tag{10.26}$$

is a prime integral of the Navier–Stokes equation in the case of a potential (irrotational) motion.

10.2.4.2. *Complex velocity field: new formulation of fluid mechanics*

Assume now that conditions (1: infinity of paths) and (3: differential time symmetry breaking) are fulfilled, while the fractal part of the velocity field is absent. In the previous section, we concluded that the two first prescriptions that describe a fluid of fractal geodesics were insufficient to obtain a quantum-type behavior. The third condition of discrete scale symmetry breaking seems to play an essential role, since it leads to a complex representation and therefore to the complex nature of the wave function. This remark led us to conclude [393] that it was the key condition for the transition to quantum laws.

This, however, is true only provided it is accompanied by the two other conditions. It is finally the full combination of the three conditions that leads to a quantum-type behavior.

Let us indeed consider a system, which can be described by

(i) a velocity field $v[x(t), t]$ instead of a deterministic velocity $v(t)$;
(ii) a two-valuedness of this velocity field (v_+, v_-), leading to introduce a complex velocity field $\mathcal{V} = (v_+ + v_-)/2 - i(v_+ - v_-)/2 = V - iU$.

The condition of fractality of the paths, which is described in terms of the introduction of a fluctuation $d\xi = \eta\sqrt{2\mathcal{D}\,dt}$ is no longer assumed here. Such a system is equivalent to the standard scale relativity system $dX_\pm = v_\pm\, dt + \eta_\pm\sqrt{2\mathcal{D}\,dt}$ in the limit $\mathcal{D} \to 0$, i.e. in the zero fractal fluctuation limit. The covariant derivative is then reduced to $\hat{d}/dt = \partial/\partial t + \mathcal{V} \cdot \nabla$ in this case.

Therefore, the various equivalent representations of the motion equations (geodesics, Schrödinger and fluid mechanics) remain valid, but they are now considered under the limit $\mathcal{D} \to 0$. The geodesic-dynamics equation,

$$\left(\frac{\partial}{\partial t} + \mathcal{V} \cdot \nabla\right)\mathcal{V} = -\nabla\phi, \tag{10.27}$$

can be integrated in terms of a Schrödinger-type equation

$$\mathcal{D}^2\Delta\psi + i\mathcal{D}\frac{\partial}{\partial t}\psi - \frac{\phi}{2}\psi = 0, \tag{10.28}$$

and then reformulated in terms of a Euler + continuity equation system,

$$\left(\frac{\partial}{\partial t} + V \cdot \nabla\right) V = -\nabla\phi + 2\mathcal{D}^2 \nabla \left(\frac{\Delta\sqrt{P}}{\sqrt{P}}\right), \tag{10.29}$$

$$\frac{\partial P}{\partial t} + \text{div}(PV) = 0. \tag{10.30}$$

In the limit $\mathcal{D} \to 0$, we see that the quantum potential vanishes and that this system becomes the standard system of equations of fluid mechanics (for zero pressure, or, more generally, with a pressure term inserted in the potential ϕ term, see Sec. 10.6):

$$\left(\frac{\partial}{\partial t} + V \cdot \nabla\right) V = -\nabla\phi, \tag{10.31}$$

$$\frac{\partial P}{\partial t} + \text{div}(PV) = 0. \tag{10.32}$$

As a consequence of this result, the doubling $v \to (v_+, v_-)$, which seemed to be a radical change in physics owing to the fact that it led to jump from real number to complex numbers in the description, takes a completely different meaning here. Indeed, complex numbers appear here as a simple representation of the transition to a statistical description in terms, not only of a velocity field V but also of a probability density P. In the absence of the fractality condition, they therefore describe a system that remains perfectly classical.

The interest of such a representation, even for a standard classical fluid description, is that it allows one to obtain from the same single geodesic-like (covariant) equation both the Euler dynamics equation and the continuity equation, which must be established in a separate way in the standard approach, which corresponds to setting only the prescription of an infinity of paths. Note also that, here, P is a probability density, and becomes equivalent to a matter density ρ in the limit of a continuous fluid, (for which all particles would be subjected to the same probability density P). It is therefore more general than standard fluid mechanics, since it also includes fluid approximations of the motion of an ensemble of individual particles, such as, for example, in the description of gravitational systems in astrophysics and cosmology, see Chapter 13.

We conclude from these detailed analyses of the various possible combinations (see Exercise 21 for the last one) of conditions — infinite number of paths, fractality of paths, two-valuedness of velocity field that the quantum behavior is obtained as a consequence of the three conditions together, while relaxing anyone of these conditions implies that the system under consideration remains classical.

Problems and Exercises

Exercise 21 Finally assume that conditions (2: fractality of path) and (3: two-valuedness of velocity) are fulfilled. Show that this case of complex velocity on a single deterministic trajectory, leads to no physically meaningful situation. ∎

10.3. Macroscopic Quantum-like Theories

10.3.1. *Analysis of the problem*

As we have seen in the previous sections, the quantum-type laws are obtained provided all three conditions issued from nondifferentiability (velocity field, fractality and two-valuedness) are simultaneously fulfilled. This is clearly the case of standard quantum mechanics in the microphysical domain, for which the description of elementary displacements as

$$dX_\pm = v_\pm\, dt + \eta_\pm \sqrt{2\mathcal{D}\, dt}, \quad \langle \eta_\pm \rangle = 0, \quad \langle \eta_\pm^2 \rangle = 1, \qquad (10.33)$$

can be considered as valid for any time interval $dt \to 0$, without any lower limit. This means that there is, in this case a complete and genuine nondifferentiability of space-time. This is a necessary condition for the violation of Bell's inequalities, namely, there should be no classical physics underlying quantum physics, i.e. no classical microscopic theory of quantum mechanics. This excludes many interpretations of quantum mechanics in terms of sub-quantum medium, underlying classical diffusion processes, etc., such as those postulated, e.g. by Bohm and Vigier [58, 59, 61] or Nelson's stochastic mechanics [331]. In the scale relativity description, there is no longer any separation between a "microscopic" description and an emergent "macroscopic" description (at the level of the wave function), since both are accounted for in the double scale space and position space representation.

Recall in this regard that the interpretation of the Planck length-scale and time-scale as a limiting lower scale in the special scale relativity framework does not change anything to this point. Indeed, it is understood as an horizon, not as a wall nor a cut-off. Only scale ratios have a physical meaning, not the individual scales themselves. An infinite number of successive dilations would still be needed to reach the Planck scale from any given finite scale in the new special scale relativistic framework. In other words, the Planck length-scale has, in this framework, all the properties formerly attributed to the zero point (recall that it no longer corresponds, in this case, to the Planck energy-scale).

An essential question can be asked concerning the applications of the scale relativity approach to realms other than the standard quantum theory: what are the conditions under which the description given by Eq. (10.33) can be considered as a valid approximation? This question and its answer concerns two types of applications, which we shall review in the following chapters: (i) the artificial reproduction of the conditions that lead to a quantum-type behavior in order to possibly obtain and develop new kinds of macroscopic quantum-type experiments and technology; (ii) the research and identification of natural systems, which have spontaneously developed such a behavior.

Strictly, Eq. (10.33) is valid for all values of dt only in the case of standard quantum mechanics and, therefore, only for $\mathcal{D} = \hbar/2m$. This means that for possible other quantum-type systems based on a different macroscopic constant $\mathcal{D}$, it should be valid only on a finite scale interval of the resolution variable dt or dx. The question that naturally arises and that we want to investigate in the present section is: what are the conditions on the width of this scaling interval under which the theory remains valid?

In the scale relativistic foundation of standard quantum mechanics, the relation $d\xi = \eta\sqrt{2\mathcal{D}|dt|}$ is considered to be valid at all scales. The de Broglie transition to the classical realm is but an effective transition, which comes from the domination of the classical term $dx = v\,dt$ over the fractal term $d\xi$ at scales $\delta x > \hbar/mv$. One does not consider that this fractal part has disappeared, even at macroscopic scales, but that it is only masked by the classical motion. This result agrees with the existence of macroscopic quantum effects, such as superconductivity, which manifest themselves precisely when the de Broglie or thermal de Broglie length $\hbar/m\langle v^2\rangle^{1/2}$ becomes macroscopic because the motion of the particles decreases. In other words, in the scale relativity framework, space-time is fractal (in terms of $S_0 = \hbar$) at all scales.

This conclusion allows us to make a new proposal to put the theory to the test. We suggest to make precision measurements at a relative scale up to now considered to be fully classical, in order to exhibit the underlying masked fractal/quantum behavior (see Fig. 10.1). It is known from Ehrenfest theorem that classical quantities remain at the heart of the quantum world as averages (as expected in the scale relativity approach from the fact that the fractal nondifferentiable part is of zero mean, so that

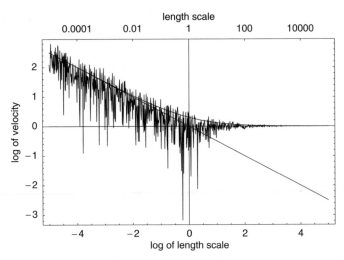

Fig. 10.1: **Differentiable and fractal parts of the velocity.** Representation of the total velocity $V = v + w$, including its classical-like, differentiable part v (horizontal line) and its fractal, stochastic and nondifferentiable part w as a function of the logarithm of the space resolution $\log \delta x$. The inclined line corresponds to $\langle w^2 \rangle^{1/2}$ and the curve to the sum $v + \langle w^2 \rangle^{1/2}$. The transition scale is the de Broglie scale. It is remarkable that fractal, i.e. quantum, fluctuations, though dominated by the classical part at scales larger than the de Broglie scale, i.e. in the classical regime, do not disappear at large scales but are only masked by the classical velocity. This suggests one should perform a test of the scale relativity theory by a measurement of expected very faint quantum effects at scales $\varepsilon \gg \lambda_{\mathrm{dB}}$, where the standard view of the quantum to classical transition (wave packet collapse) considers that the system has become totally classical.

$\langle dX \rangle = dx$). Reversely, the scale relativity theory suggests that quantum properties may also remain at the heart of the classical world.

We have insisted in Chapter 5 on the fact that the whole mathematical structure of the theory does not rely on the constant $\hbar$ and that it could be preserved for any value of this constant (and more generally of $\mathcal{D}$), even macroscopic. Moreover, we are now in a new situation regarding the nature of "quantum objects" and of "quantum properties". In the framework of standard quantum mechanics, where the foundation was axiomatic, the word "quantum" refered to the ensemble of the quantum properties, which could only be taken as a whole. One could not separate properties like a wave function solution of a Schrödinger or Dirac equation, the indiscernability of identical particles, the Pauli principle, the EPR paradox and the violation of Bell's inequalities, etc. The standard quantum theory is understood as including the whole ensemble of quantum properties, and other theories such as hidden parameter theories that contradict Bell's theorem are rejected as being unphysical.

We agree with this position as concerns the standard quantum theory based on $\hbar$, which we have identified with the manifestation of a truly nondifferentiable continuous geometry (i.e. without any lower limit to the fractal structures implied by such a geometry). The ability of the scale relativity theory to find a more profound origin to the various axioms of quantum mechanics also allows one to reconsider this conclusion for systems other than molecules, atoms, nuclei and elementary particles to which the full $\hbar$-based standard quantum mechanics is known to apply with great precision.

Indeed, in the scale-relativity framework, it is quite possible that only a sub-class of the conditions that yield the full quantum theory be achieved for some systems, which could therefore be described by a quantum-like tool. In particular, we have suggested that some chaotic systems could be described, beyond their horizon of predictability, by wave functions solutions of a Schrödinger equation based on a constant different from $\hbar$ (more generally, on a fractal parameter $\mathcal{D}$ different from $\hbar/2m$) and satisfying Born's postulate (see [353], Chapter 7.2, [357, 362] and Chapter 13). In such a case, other typical properties of genuine quantum systems, in particular those, which are linked to elementarity [531], such as undiscernability of identical particles or the EPR paradox, become irrelevant for such systems.

For such partly quantum macroscopic theories, the nondifferentiability is only approximate. We have indeed shown that true nondifferentiability and continuity imply scale divergence when $dx \to 0$ or $dt \to 0$, without any lower limit to this divergence, and we have used this scale dependence both as a signature and as a tool of description, in terms of the fractal and nondifferentiable part of differentials, $d\xi = \eta a |dt|^{1/D_F}$, where $\langle \eta \rangle = 0$, $\langle \eta^2 \rangle = 1$, while the dimensionality of the constant a is $[a] = [L\,T^{-1/D_F}]$. Reversely, if such a relation is true for a given system on a large range of time scales dt, which can be considered as microscopic respectively to its typical integrated time scale t, the scale relativity description leading to a Schrödinger equation may be a good approximation, even if one recovers a nonfractal behavior at still smaller time scales.

We shall therefore now attempt to characterize the conditions on the scale range under which a macroscopic Schrödinger equation can be derived as a valid approximation.

10.3.2. *Width of the scaling interval*

In the fully nondifferentiable case, we have seen that there is no transition, neither upper nor lower (i.e. toward the large and small scales), to the scale dependence described as $d\xi = \eta\sqrt{2\mathcal{D}|dt|}$ (when the fractal dimension takes the critical value $D_F = 2$).

The combination of this fractal term with the classical term $dx = v\,dt$ yields an effective transition at a generalized de Broglie scale λ_{dB}. Let us recall the argument. Since the classical space and time differentials are of the same order, the fractal fluctuation reads in terms of the space differential $d\xi = \eta\sqrt{\lambda_{\mathrm{dB}}|dx|}$, where λ_{dB} is defined from this relation. By comparing with the previous expression, one finds $\lambda_{\mathrm{dB}}\,dx = 2\mathcal{D}dt$, so that

$$\lambda_{\mathrm{dB}} = \frac{2\mathcal{D}}{dx/dt} = \frac{2\mathcal{D}}{v}. \tag{10.34}$$

In the standard quantum case, $\mathcal{D} = \hbar/2m$, and one recovers the usual de Broglie nonrelativistic length $\hbar/mv$. Therefore the full displacement $dX = dx + d\xi$ reads

$$dX = dx\left\{1 + \eta\left(\frac{\lambda_{\mathrm{dB}}}{dx}\right)^{1/2}\right\}, \tag{10.35}$$

in which one easily identifies λ_{dB} with the fractal-nonfractal transition.

This transition is a spontaneous transition, which corresponds to no real symmetry breaking, since the classical and the fractal terms are considered to exist at all scales. It is therefore relative to the conditions and even to the observer, depending on the mass, temperature (in the thermal de Broglie case) and relative velocity, and it is sent to infinity (i.e. it disappears) in the limit $v \to 0$.

In the macroscopic quantum-like theory, it should be combined with one or possibly two transitions, which are of a different nature. For example, in the application of the scale relativity approach to chaotic systems, e.g. in the case of planetesimal motions in a protoplanetary disk, the motion becomes Brownian-like (therefore of fractal dimension 2) beyond the mean-free path, i.e. beyond the predictability horizon $t_H \approx 20\,t_L$, where t_L is the inverse of the Lyapunov exponent ([353], Chapter 7.2), while it is classical and scale-independent at smaller time-scales where the deterministic theory is still effective to predict a precise trajectory. In the application to physics inside a fractal medium, there are two transition scales, a lower one given by the smallest structures of the medium that participates in the fractal geometry, and an upper one given by the largest structure, which may be the size of the fractal object in case it does not replicates itself under a periodic or pseudo-periodic pattern.

A description of such a double transition and its origin as a solution of a scale differential Eq. [363] has been given in Sec. 4.2.6. It is given by

$$\mathcal{L} = \mathcal{L}_0 \left(\frac{1 + (\lambda_0/\varepsilon)^{D_F - 1}}{1 + (\lambda_1/\varepsilon)^{D_F - 1}} \right), \tag{10.36}$$

where $\mathcal{L}$ is the length of a curve, which is scale dependent on the scale range $[\lambda_1, \lambda_0]$ and independent of the scale variable ε for $\varepsilon < \lambda_1$ and $\varepsilon > \lambda_0$, while D_F is an effective fractal dimension, which is constant and different from 1 only on the same scale range.

We can now apply such a solution to the description of a partly fractal velocity, which is an explicit function of the time differential dt, under the form

$$V = v \frac{1 + \eta \left(\tau_1/dt \right)^{1/2}}{1 + \eta \left(\tau_2/dt \right)^{1/2}}, \tag{10.37}$$

where we have chosen here $D_F = 2$. On the time scales range $\tau_2 < dt < \tau_1$, one recovers the standard scale relativistic description $V = v[1 + \eta(\tau_1/dt)^{1/2}]$, which underlies the foundation of quantum mechanics in our framework.

In case when the upper transition is the generalized de Broglie (relative) scale, one may identify τ_1 with $2\mathcal{D}$. The additional lower transition is equivalent to defining a generalized fractal fluctuation part such that

$$d\xi^2 = 2\eta^2 \times \widetilde{\mathcal{D}}(dt) \times |dt|, \tag{10.38}$$

where

$$\widetilde{\mathcal{D}}(dt) = \frac{\mathcal{D}}{1 + \eta^2(\tau_2/dt)}. \tag{10.39}$$

An example of such a scale dependent effective parameter $\mathcal{D}(dt)$ (equivalent to a variable Planck "constant" $\hbar$ since $\mathcal{D} = \hbar/2m$ in standard quantum mechanics) is given in Fig. 10.2.

This allows one to use the method of Sec. 5.14, in which the construction of a Schrödinger equation is generalized to a scale-dependent parameter $\mathcal{D}(dt)$. Provided the two other properties underlying its demonstration are preserved, at least as an approximation (infinity or large number of

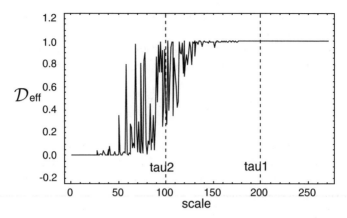

Fig. 10.2: **Effective scale-dependent parameter $\mathcal{D}$.** Example of scale-dependent effective fractal fluctuation parameter $\mathcal{D}$ (that generalizes $\hbar/2m$) in the case of the existence of a lower scale transition τ_2 (see text).

trajectories and two-valuedness of the velocity field), one obtains

$$\widetilde{\mathcal{D}}^2(dt)\Delta\psi + i\widetilde{\mathcal{D}}(dt)\frac{\partial\psi}{\partial t} = \frac{\phi}{2m}\psi. \qquad (10.40)$$

The parameter $\widetilde{\mathcal{D}}$ is constant at scales larger than τ_2. It strongly fluctuates around the time scale τ_2, then it vanishes for $dt \ll \tau_2$. This means that such a system is described by classical physics when $dt \ll \tau_2$, and by a partly quantum-like physics, i.e. a Schrödinger regime, when $dt > \tau_2$, up to the generalized de Broglie scale τ_1 beyond which one recovers again classical mechanics through a standard quantum to classical transition.

Such systems are therefore characterized by multiple scales of time and multiple scales of length, in a way that shares some common features with the renormalization group approach [537, 538, 539, 540], as emphasized in ([353] Chapter 6), but with also different properties. Indeed, the change of scale applies here on the very variable that plays the role of time and space differential elements, and, moreover, its consequences can be a change of physics as radical as going from a classical-like physics to a quantum-like physics and reversely.

The examination of Fig. 10.2 shows that the non constancy of $\mathcal{D}$ and its fractal fluctuations is already effective at scales far larger than τ_2. This shows, in agreement with the behavior of the effective fractal dimension shown in Fig. 10.3 of Sec. 4.2.6, that a necessary condition for the possibility of existence of a macroscopic Schrödinger regime for mesoscopic and macroscopic fractal systems is a large enough scale range

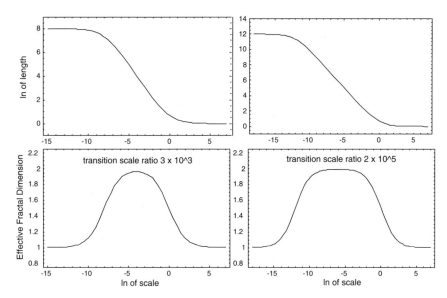

Fig. 10.3: **Effective fractal dimension between two transitions.** Effective fractal dimension between two transition scales: variation with scale of the length and effective fractal dimension of a curve, which is scale-dependent on a finite range of scales. In the left figures, the ratio between the two transition length-scales is 3×10^3. Although, according to the top figure, there seems to exist a scale range on which a constant fractal dimension is established, i.e. a constant slope in a log-log plot, the drawing of the effective fractal dimension in the down figure shows that it is never constant and barely reaches the underlying value $D_F = 2$. This is a consequence of the non null width (in scale space) of the transitions. In the left figures, the ratio between the two transition length-scales is enlarged to 2×10^5. In this case, the effective fractal dimension takes the expected constant value $D_F = 2$ on a scale range of about two decades.

over which the fractal behavior is ensured. Even a ratio such as $\tau_1/\tau_2 = 3 \times 10^3$ is insufficient (see Fig. 10.3), since one finds that the effective fractal dimension $D_F = 2$ is hardly established in such a case. A scale ratio of $\approx 10^5$ between the upper and lower transitions seems to be a minimum. This conclusion holds as well for time-scales as for length-scales, as can be shown from the two equivalent relations $d\xi^2 = 2\eta^2 \mathcal{D}\, dt$ and $d\xi^2 = \eta^2 \lambda_{\mathrm{dB}}\, dx$.

Such a scale ratio of $\approx 10^5$ is actually hard to achieve in laboratory experiments and concerns only large scale Earth physics, but it is on the contrary common at astronomical scales, where a scaling behavior over many scales is known to exist for several systems: protoplanetary disks, interstellar medium and distribution of galaxies. This explains why the first manifestations of such a macroscopic Schrödinger physics have been found in large number in the astronomical domain (see [123] and references therein and Chapter 13). In a laboratory experiment, for an upper scale of 10 cm, one would need to build a multiscale fractal structure down to the micrometer in order to reach such a ratio.

Even the Brownian motion of microscopic particles in a fluid, which is itself at the micron scale, does not reach a large enough scale range, since the molecular collisions are at the nanometer scales, so that the scale ratio is only some 10^3. Moreover, the dynamics is of the Langevin rather than Newtonian type, which prevents from obtaining a Schrödinger form for the motion equation. On the contrary, the astronomical Brownian motion of a planetesimal or of a proto-planet in a protoplanetary disk due to the gravitational encounters with the other bodies has all the expected characteristics at long time scales (very large number of potential trajectories, fractal dimension 2 and irreversibility) to be described by the new macroscopic quantum-like theory ([353] Chapter 7.2).

Another possible domain of application is fluid mechanics, meteorology and climate studies. But here too, only long time-scales and large length-scales are concerned. For example, if one estimates the lower scale at about 1 mm, this means that the Schrödinger-type approach could not be effective before scales of ≈ 100 m, which is beyond present laboratory experiments but could concern the atmosphere. The same is true for time scales: this approach holds only beyond the predictability horizon, which is about seven days for meteorological studies, which means that it would be more adapted to long-time scale studies of the climate. But it should also be clear that for such a difficult problem, which includes that of a theory of turbulence, the question should be set again from the very beginning by using the scale relativity methods, which we intend to do in the forthcoming years.

10.3.3. *Application to physics in fractal media*

An example of possible laboratory and technological application of such concepts could be the achievement of a medium, which would be fractal on several decades of scales ([353], Chapter 7.3), followed by the study of the propagation of waves and particles in such a medium and of its global properties. We have suggested that, provided the necessary and sufficient conditions analyzed above be achieved, the objects in such a medium could have macroscopic quantum-fluid-type properties, not based on the Planck constant $\hbar$, but on a new macroscopic constant specific to the system.

The idea underlying this proposal is that the relation of a medium ("container") to the objects it contains is similar to the relation of a space (or space-time) to its objects. The difference is in the universality of space-time geometry, while the geometry of a medium is a constraint on the motion of the objects it contains only in the limit where they remain inside it and linked to it. This kind of conceptual transport between fundamental physics and physics of materials has already been effective: a remarkable example is the analogy between superconductivity and the Higgs mechanism in the electroweak theory [527, 471]. Hence, in a superconducting material, the existence of the bosonic field created by the Cooper pairs change the properties of the free electrons, in particular their mass and charge. In the same way the existence of the Higgs field and of its $\lambda\phi^4$ potential allows the W and Z particles and the Higgs boson itself to acquire mass in the Weinberg–Salam–Glashow electroweak theory.

Here we suggest an analogy that goes in the reverse way and which applies to an even more profound level of theoretical physics. Recall indeed that field theories have evolved from the concept of force, to that of field and potential, then finally to the concept of space-time in Einstein's theory (and now, provided it is validated, in scale relativity theory). In the construction of the electroweak theory, properties of materials (superconductors) were used to infer analogous properties of gauge fields. Conversely, we have suggested to apply the theoretical developments of physics in a fractal space-time to new experiments and new technologies in fractal media [350, 353], including possible new macroscopic quantum-like properties and behaviors.

This proposal includes the possibility that such media partly subjected to a macroscopic Schrödinger regime and, therefore, to quantum-type potentials do already exist in nature. An anticipated case of this would be nothing else than living systems,(see [353, Chapter 7.3], [373, 377, 388, 33, 402] and Chapter 14 of this book), but the same approach could also be applied to the development of artificial "life", i.e. of inanimate systems characterized by properties usually attributed to living systems, such as growth, replication, etc.

Other kinds of applications include the study of the propagation of waves in geological media, in order to provide better solutions to the geophysical inversion problem [226] of various fractal systems in geography (tank basins [412], hydrological lattice, porous media such as karsts [311, 114]) multiphoton ionization by means of fractal geometry [426], etc.

10.4. Quantum Laws as Anti-diffusion Laws

10.4.1. *Introduction*

By writing the equations of motion under the form of a macroscopic Schrödinger equation, we have actually obtained a theory of self-organization.

Indeed, a Schrödinger-type equation is characterized by the existence of stationary solutions yielding well-defined peaks of probability linked to quantization laws, themselves a consequence of the boundary or environmental conditions, of the forces applied and of the symmetries of the system. It has therefore been suggested [362] to interpret these peaks of probability density as a tendency for the system to develop structures. In other words, the theory can be used to predict, not a fully deterministic organisation, but rather the most probable structures among the infinity of other close possibilities. Such a feature of this approach is compatible with the large variability that characterizes living systems (see Chapter 14). Moreover, it is a genuine theory of self-organization, since the structuring comes from the interior instead of being a consequence of an exterior force, but the particular organization obtained in a given situation is related to the environmental conditions. The natural tendency for the system to self-organize comes from the quantum-like nature of the laws, which, in the scale relativity framework, is itself a manifestation of relativity and covariance, i.e. of the non-absolute existence of the various properties defining a physical or biological system.

An important question concerning the scale relativity approach and the quantum formalism issued from it is its relation to diffusion processes. It could seem paradoxical that, starting from a description of infinitesimal displacements made in terms of stochastic processes, one obtains in the end a theory of self-organization.

In fact the principle of relativity (under its form of geodesic principle) can be seen as actually recombining the initial twin velocity process in terms of a genuine anti-diffusion. This means that the self-organization properties of the scale relativity theory lead to an entropy decrease, i.e. to a neguentropy, the source of which has been unsuccessfully searched for long for the foundation of life sciences.

The question of the relation between quantum laws and diffusion processes has been asked for long. The formal analogy that exists between the classical diffusion equation

$$D\Delta P - \frac{\partial P}{\partial t} = 0 \tag{10.41}$$

and the quantum, free Schrödinger-type equation,

$$\mathcal{D}\Delta\psi + i\frac{\partial\psi}{\partial t} = 0, \tag{10.42}$$

(where $\mathcal{D} = \hbar/2m$ in standard quantum mechanics), and also the parallelism between the undeterminisms of the quantum mechanical behavior and of stochastic processes, has led many authors to attempt understanding the quantum behavior in terms of real particle trajectories and diffusion process [132, 58, 59, 172, 532, 60, 331].

All these attempts claim to recover the quantum behavior in terms of a classical Markov–Wiener process describing a diffusion process in a classical space or space-time. The cause of this diffusion is therefore attributed to an external agent interacting with the particle, which is described depending on the authors as either hypothetical new particles [532], or a quantum mechanical potential [58, 59], or a subquantum medium [60], or a background field [331, 332]. Anyway, in every case, the assumption of the existence of a diffusion process leads to understanding the quantum mechanical system in terms of its interaction with something exterior to it.

Actually, it has been shown that these various attempts to recover the quantum behaviour in terms of classical diffusion processes have a fundamental flaw [205, 526, 362, 374]. Because of the complex-number nature of quantum mechanics, it is indeed necessary to introduce two (forward and backward) diffusion processes, and one of them can be shown to actually have no physical meaning (see what follows).[a]

Anyway, the differences between the above diffusion and quantum equations (real in one case, complex in the other, acting on probability P in one case and on a wave function $\psi = \sqrt{P} \times e^{i\theta}$ in the other) and between the behaviors of the processes that are described by their solutions are, by far, larger than their similarities: they can even be viewed as exactly opposite. Indeed, while diffusion processes are the archetype for dissipative, non stationary, non-isentropic systems characterized by desorganization (typically, the entropy increase is proportional to time t and the spreading is proportional to $\sqrt{t}$), the Schrödinger equation exhibits, for given boundary,

---

[a]Note that in the scale relativity approach the elementary description is made in terms of a twin stochastic process, but it is not associated with a classical diffusion interpretation, since it is understood as a direct manifestation of the non-differentiable geometry.

field and symmetry conditions, stationary and quantized solutions, which can be viewed as an archetype for self-organization, e.g. atoms.

A typical example of such opposite behaviors is the diffusion in a box, which leads to an unstructured final state of constant density, while the corresponding Schrödinger process yields well-defined structures given by peaks of probability density, in particular a central peak with vanishing density on the walls in the fundamental level.

We shall show in this section that one can support this conclusion by showing that the diffusion equation can be given the same form as the quantum equation, but in terms of a "diffusion potential", which is exactly the opposite of a quantum potential. Note that the relation between scale covariance, the Schrödinger equation and the hydrodynamical picture of diffusion-type processes has also been recently studied in [68, 192].

10.4.2. *Diffusion potential as anti-quantum potential*

10.4.2.1. *Equations for the velocity fields*

Let us first recall the equations that govern the various velocity fields in the scale relativity description, in particular the two velocity fields v_+ and v_-, which will play an important role in the comparison with classical diffusion processes.

As we have seen in Chapter 5, the imaginary part of the Schrödinger equation takes the form of an equation of continuity in terms of the velocity field V (defined as the real part of the complex velocity field $\mathcal{V}$) and of the square of the modulus of the wave function $P = |\psi|^2$,

$$\frac{\partial P}{\partial t} + \mathrm{div}(PV) = 0. \tag{10.43}$$

The imaginary part of the complex velocity field is given by [353]

$$U = \mathcal{D}\nabla \ln P = \frac{\mathcal{D}\nabla P}{P} \Rightarrow UP = \mathcal{D}\nabla P, \tag{10.44}$$

and therefore one obtains

$$\mathrm{div}(PU) = \mathcal{D}\Delta P. \tag{10.45}$$

We can now reintroduce the two velocity fields of the initial twin process, namely

$$v_+ = V + U = V + \mathcal{D}\nabla \ln P, \quad v_- = V - U = V - \mathcal{D}\nabla \ln P. \tag{10.46}$$

The sum of Eqs. (10.43) and (10.45) yields an equation for P and v_+ that is nothing but a Fokker–Planck equation,

$$\frac{\partial P}{\partial t} + \text{div}(Pv_+) = \mathcal{D}\Delta P, \qquad (10.47)$$

while their difference yields for v_-:

$$\frac{\partial P}{\partial t} + \text{div}(Pv_-) = -\mathcal{D}\Delta P. \qquad (10.48)$$

Therefore the v_+ process, which is described in an equivalent way by a stochastic differential equation $d_+ X = v_+ dt + \eta_+ \sqrt{2\mathcal{D}dt}$ and by a standard Fokker–Planck equation, may look, at this level of the description, as a standard classical diffusion process. But actually this is not the case, since the way to calculate the probability is fundamentally non classical.

Concerning the v_- process, Eq. (10.48), called a "backward Fokker–Planck" equation by Nelson [331], is set as a founding equation in stochastic mechanics, though it actually corresponds to no existing stochastic classical process, neither Markovian nor non-Markovian [205, 362], even at this simple level of description.

On the contrary, in the scale relativity framework none of these equations are set as founding stones, but they are derived from a geometric foundation, therefore they do not need to be interpreted *a posteriori*, and their similarity with the equations of classical stochastic processes is only superficial. It comes from the fundamental loss of information involved at the level of geodesics by a nondifferentiable and fractal geometry, given that the physics on a nondifferentiable manifold is in its essence non classical.

10.4.2.2. *Fluid representation of diffusion processes*

Let us now consider a classical diffusion process, such as e.g. a Brownian motion of small particles in a fluid. Such a process is described by a Fokker–Planck equation,

$$\frac{\partial P}{\partial t} + \text{div}(Pv) = D\Delta P, \qquad (10.49)$$

where D is the diffusion coefficient, P the probability density distribution of the particles and $v[x(t), t]$ is their mean velocity.

When there is no global motion of the diffusing fluid or particles ($v = 0$), the Fokker–Planck equation is reduced to the usual diffusion equation for the probability P,

$$\frac{\partial P}{\partial t} = D\Delta P. \tag{10.50}$$

This well-known equation holds in many situations, such as the Brownian motion of particles diffusing in a fluid, the diffusion of a fluid in another fluid (in this case P is replaced by the concentration of the diffusing fluid) and also the propagation of heat (in this case P is replaced by the temperature).

Conversely, when the diffusion coefficient vanishes, the Fokker–Planck equation is reduced to the continuity equation,

$$\frac{\partial P}{\partial t} + \text{div}(Pv) = 0. \tag{10.51}$$

10.4.2.3. *Continuity equation*

Let us now make, in the general case where v and D are *a priori* non vanishing, the change of variable

$$V = v - D\nabla \ln P, \tag{10.52}$$

which is inspired by the relation between V, v_+ and P in Eq. (10.46) (but now these variables describe quite classical processes).

We shall first prove that the new velocity field $V(x, y, z, t)$ is a solution of the standard continuity equation. Indeed, by using the Fokker–Planck equation and by replacing V by its above expression, we find

$$\frac{\partial P}{\partial t} + \text{div}(PV) = \{D\Delta P - \text{div}(Pv)\} + \text{div}(Pv) - D\,\text{div}(P\nabla \ln P). \tag{10.53}$$

Finally the various terms cancel each other and we obtain also the continuity equation for the velocity field V,

$$\frac{\partial P}{\partial t} + \text{div}(PV) = 0. \tag{10.54}$$

Therefore, the diffusion term has been absorbed in the re-definition of the velocity field.

10.4.2.4. *Euler equation and diffusion potential*

Let us now go one step further in such a fluid-like description of the diffusing motion. The question that we now want to address is: what is the form of the Euler equation for the velocity field V?

Case of vanishing mean velocity

Let us calculate the total time derivative of the velocity field V, first in the simplified case $v = 0$:

$$\frac{dV}{dt} = \left(\frac{\partial}{\partial t} + V \cdot \nabla\right) V = -D \frac{\partial}{\partial t} \nabla \ln P + D^2 (\nabla \ln P \cdot \nabla) \nabla \ln P.$$

$$(10.55)$$

Now, since $\partial \nabla \ln P / \partial t = \nabla \partial \ln P / \partial t = \nabla(P^{-1} \partial P / \partial t)$, we can make use of the diffusion equation so that we obtain

$$\left(\frac{\partial}{\partial t} + V \cdot \nabla\right) V = -D^2 \left\{ \nabla\left(\frac{\Delta P}{P}\right) - (\nabla \ln P \cdot \nabla) \nabla \ln P \right\}. \qquad (10.56)$$

In order to write this expression in a more compact form, we shall now use the fundamental remarkable identity, which has been proved in Sec. 5.6.4,

$$\frac{1}{\alpha} \nabla\left(\frac{\Delta R^\alpha}{R^\alpha}\right) = \Delta(\nabla \ln R) + 2\alpha(\nabla \ln R \cdot \nabla)(\nabla \ln R). \qquad (10.57)$$

By writing this remarkable identity for $R = P$ and $\alpha = 1$, we can replace $\nabla(\Delta P / P)$ by $\Delta(\nabla \ln P) + 2(\nabla \ln P \cdot \nabla) \nabla \ln P$, so that Eq. (10.56) becomes

$$\left(\frac{\partial}{\partial t} + V \cdot \nabla\right) V = -D^2 \{ \Delta(\nabla \ln P) + (\nabla \ln P \cdot \nabla) \nabla \ln P \}. \qquad (10.58)$$

The right-hand side of this equation comes again under the identity (10.57), but now for $\alpha = 1/2$. Therefore we finally obtain the following form for the Euler equation of the velocity field V:

$$\left(\frac{\partial}{\partial t} + V \cdot \nabla\right) V = -2D^2 \nabla\left(\frac{\Delta \sqrt{P}}{\sqrt{P}}\right). \qquad (10.59)$$

This is a new fundamental result [402]. Its comparison with the quantum result in the free case is striking:

$$\left(\frac{\partial}{\partial t} + V \cdot \nabla\right) V = +2\mathcal{D}^2 \nabla\left(\frac{\Delta \sqrt{P}}{\sqrt{P}}\right). \qquad (10.60)$$

This result is indeed remarkable for several reasons:

(i) It gives an equivalence between a standard fluid subjected to a force field and a diffusion process. However this force is very particular, since it is expressed in terms of the probability density at each point and instant.

(ii) The above "diffusion force" derives from a potential

$$\phi_{\mathrm{diff}} = 2D^2 \, \Delta\sqrt{P}/\sqrt{P}. \tag{10.61}$$

This expression introduces, in a striking way, a square root of probability in the description of what remains a totally classical diffusion process, while usually one encounters probabilities P in classical mechanics and probability amplitudes $\sqrt{P} \times e^{i\theta}$ in quantum mechanics.

(iii) But there is more: not only it is expressed in terms of the square root of probability, but this "diffusion potential" is exactly the opposite of the "quantum potential" $Q/m = -2\mathcal{D}^2\Delta\sqrt{P}/\sqrt{P}$.

The relation between quantum mechanics and diffusion processes is now enlighted in a new way, namely, they appear as exactly opposite. We conclude that it is, therefore, impossible to obtain quantum mechanics from a standard classical diffusion process, since it is now clear that quantum mechanics correspond to an "anti-diffusion" rather than to a diffusion. Reversely, one may suggest that confined quantum structures, such as those exemplified, e.g. by atoms in the $\mathcal{D} = \hbar/2m$ case, may be the best possibility of reversal of a diffusive process (see Sec. 10.4.3).

Case of non vanishing mean velocity

The Fokker–Planck Eq. (10.49) is common to the two cases, classical diffusion desorganization and quantum organization. How can this be reconciled with their antinomy?

To answer this question, let us now consider the general situation of a non vanishing mean velocity field v. In order to do this calculation we now introduce the indices in an explicit way. Equation (10.55) takes the form

$$\frac{\partial V^k}{\partial t} + V^j \partial_j V^k = \frac{\partial v^k}{\partial t} - D\,\partial^k \left(\frac{\partial P/\partial t}{P} \right)$$
$$+ (v^j - D\,\partial^j \ln P)\,\partial_j (v^k - D\,\partial^k \ln P). \tag{10.62}$$

Accounting for the Fokker–Planck equation it becomes

$$\frac{\partial V^k}{\partial t} + V^j \partial_j V^k = \left(\frac{\partial v^k}{\partial t} + v^j \partial_j v^k \right) - D\,\partial^k \left(\frac{D\Delta P - \partial_j P\,v^j - P\,\partial_j v^j}{P} \right)$$
$$- D\,v^j \partial_j \partial^k \ln P - D\,\partial^j \ln P\,\partial_j v^k + D^2\,\partial^j \ln P\,\partial_j \partial^k \ln P. \tag{10.63}$$

After some calculation we finally obtain

$$\frac{\partial V^k}{\partial t} + V^j \partial_j V^k = -2D^2 \partial^k \left(\frac{\partial_j \partial^j \sqrt{P}}{\sqrt{P}} \right) + \frac{dv^k}{dt}$$

$$+ D \{\partial^k \partial_j v^j + \partial_j \ln P (\partial^k v^j - \partial^j v^k)\}. \qquad (10.64)$$

In the case when v is potential, we set

$$v = \nabla \varphi, \qquad (10.65)$$

the last rotational term vanishes and the force in the right-hand side of this equation itself derives from a potential,

$$\Phi = 2D^2 \frac{\Delta \sqrt{P}}{\sqrt{P}} - D \Delta \varphi + \frac{\partial \varphi}{\partial t} + \frac{1}{2} (\nabla \varphi)^2. \qquad (10.66)$$

This is in particular the case of the scale relativistic description and of standard quantum mechanics, where $v = v_+$ is potential. The quantum potential plus possibly an external potential ϕ can therefore be obtained in this case provided

$$\frac{\partial \varphi}{\partial t} + \frac{1}{2} (\nabla \varphi)^2 - D \Delta \varphi = \phi - 4D^2 \frac{\Delta \sqrt{P}}{\sqrt{P}}. \qquad (10.67)$$

Under this condition, the Euler and continuity equations can be integrated under the form of a generalized Schrödinger equation,

$$\mathcal{D}^2 \Delta \psi + i \mathcal{D} \frac{\partial \psi}{\partial t} = \frac{1}{2} \phi \psi. \qquad (10.68)$$

Therefore the possible values of the velocity field $v = v_+$ differ fundamentally between the two situations (quantum *vs* diffusion). In particular $v_+ = U + V = 0$ is excluded in the quantum case, since it leads to the standard classical diffusion equation whose solutions spread as $\sqrt{t}$.

10.4.3. *Consequence for the diffusion inverse problem*

We hope that such a result may be useful in solving the diffusion inverse problem, for example in source reconstruction of atmospheric pollution [237] and possibly partial reversal of such a diffusion.

Assume indeed that we apply a force to particles diffusing in a fluid. Two extreme regimes can be considered. When the velocity is small enough, this force causes an acceleration. But for large enough velocities, the resistance force due to viscosity leads to reaching a constant velocity proportional to the force. Such a process is easily described in a simplified way in terms of the Stokes resistance force that reads for a sphere $F_R = u/b$ in terms of the mobility $b = 1/6\pi\eta R$, where R is the sphere radius, η the coefficient of viscosity and u the velocity (see e.g. [269] for a more general expression). The equation of motion in a constant force field F reads

$$m\frac{du}{dt} = F - \frac{1}{b}u, \tag{10.69}$$

and its solution writes

$$u = bF(1 - e^{-t/mb}). \tag{10.70}$$

We therefore recover the two extreme cases of Newtonian dynamics for small velocities ($u = Ft/m$) and dissipative motion for larger ones ($u = bF$), that we now consider hereafter.

10.4.3.1. *Newtonian dynamics*

Let us first consider this inverse problem in the case when the effect of a force is an acceleration.

(i) More specifically, assume that a diffusion process, e.g. a Brownian motion, described by the standard diffusion Eq. (10.50),

$$\frac{\partial P}{\partial t} = D\Delta P, \tag{10.71}$$

has led to some probability distribution $P(x, y, z, t_m)$ of the diffusing particles at time t_m. In an equivalent way, we have shown hereabove that the diffusion equation is equivalent to a Euler and continuity system in terms of a fluid velocity field V in the "diffusion potential" $\phi_{\text{diff}} = 2D^2\Delta\sqrt{P}/\sqrt{P}$, which is the opposite of a quantum potential,

$$\left(\frac{\partial}{\partial t} + V \cdot \nabla\right)V = -2D^2\nabla\left(\frac{\Delta\sqrt{P}}{\sqrt{P}}\right), \quad \frac{\partial P}{\partial t} + \text{div}(PV) = 0. \tag{10.72}$$

(ii) Assume now that we apply a force $F_1 = F_Q = 2D^2\nabla(\Delta\sqrt{P}/\sqrt{P})$ on the particles and that we are in the Newtonian dynamics regime. Assuming that the motion remains potential, this would result in a cancellation of the diffusion force and therefore in a purely fluid regime without any diffusion, described by the equations

$$\left(\frac{\partial}{\partial t} + V \cdot \nabla\right) V = 0, \quad \frac{\partial P}{\partial t} + \text{div}(PV) = 0. \tag{10.73}$$

It corresponds to $D = 0$, i.e. to $V = v$ and, therefore, to the reduction of the Fokker–Planck equation to the continuity equation.

(iii) In order to not only suppress, but even reverse this diffusion process, we now assume that we apply to the system a force $F_2 = 2F_Q = 4D^2 \nabla(\Delta\sqrt{P}/\sqrt{P})$. Under the Newtonian regime and potential motion assumed in this section, the effect of this force will be an acceleration dV/dt, so that the Euler and continuity equations now take the form (including an additional potential energy term ϕ for generality):

$$\left(\frac{\partial}{\partial t} + V \cdot \nabla\right) V = 2D^2 \nabla \left(\frac{\Delta\sqrt{P}}{\sqrt{P}} - \phi\right), \tag{10.74}$$

$$\frac{\partial P}{\partial t} + \text{div}(PV) = 0. \tag{10.75}$$

Let us show that, in the case when V describes a potential (irrotational) motion, this system of equation can be transformed under the form of a generalized Schrödinger equation, i.e. that the transformation between the Schrödinger and hydrodynamics form of the equations is reversible.

Let us first define a function $S(x, y, z, t)$ by the relation

$$V = \nabla S. \tag{10.76}$$

The Eq. (10.74) then takes the successive forms:

$$\frac{\partial}{\partial t}(\nabla S) + \frac{1}{2}\nabla(\nabla S)^2 + \nabla\left(\phi - 2D^2\frac{\Delta\sqrt{P}}{\sqrt{P}}\right) = 0, \tag{10.77}$$

$$\nabla\left(\frac{\partial S}{\partial t} + \frac{1}{2}(\nabla S)^2 + \phi - 2D^2\frac{\Delta\sqrt{P}}{\sqrt{P}}\right) = 0, \tag{10.78}$$

which can be integrated as

$$\frac{\partial S}{\partial t} + \frac{1}{2}(\nabla S)^2 + \phi + K - 2D^2\frac{\Delta\sqrt{P}}{\sqrt{P}} = 0, \tag{10.79}$$

where K is a constant that can be re-absorbed in a renormalization of the potential energy ϕ. One then combines this equation with the continuity equation to obtain

$$-\frac{\sqrt{P}}{2}\left(\frac{\partial S}{\partial t} + \frac{1}{2}(\nabla S)^2 + \Phi - 2D^2\frac{\Delta\sqrt{P}}{\sqrt{P}}\right) + i\frac{D}{2\sqrt{P}}\left(\frac{\partial P}{\partial t} + \mathrm{div}(P\nabla S)\right) = 0.$$

(10.80)

Let us set

$$\psi = \sqrt{P} \times e^{iS/2D}.$$

(10.81)

It corresponds to a relation between ψ, V and P that reads

$$-2iD\nabla\ln\psi = V - iD\nabla\ln P.$$

(10.82)

Then the Eq. (10.80), multiplied by $e^{iS/2D}$, becomes strictly identical to the following Schrödinger equation:

$$D^2\Delta\psi + iD\frac{\partial}{\partial t}\psi - \frac{\Phi}{2}\psi = 0.$$

(10.83)

The solutions $\psi = |\psi| \times \exp(i\theta)$ of this equation provide directly the velocity field and the probability density at each point and instant,

$$V = 2D\,\nabla\theta, \quad P = |\psi|^2.$$

(10.84)

In the above Schrödinger equation, it is the diffusion coefficient of the classical diffusion process that plays the role devoted to the ratio $\hbar/2m$ in standard quantum mechanics. The solutions of this equation are therefore expected to be subjected to a Heisenberg relation that reads

$$\Delta x \times \Delta v \geq D.$$

(10.85)

The final uncertainty in the reconstruction of the source is limited by the classical diffusion coefficient itself.

10.4.3.2. *Dissipative regime*

Let us now consider the inverse problem in the case when the effect of a force is a velocity instead of an acceleration.

(i) We start from a classical diffusive system described by the standard diffusion equation $\partial P/\partial t = D\Delta P$. In this case we have $v = 0$, so that

$$V = -D\nabla\ln P, \quad \left(\frac{\partial}{\partial t} + V \cdot \nabla\right)V = -2D^2\nabla(\Delta\sqrt{P}/\sqrt{P}).$$

(10.86)

(ii) Assume now that we apply a force

$$F_1 = \frac{D}{b} \nabla \ln P = T \nabla \ln P \tag{10.87}$$

on the diffusing particles that derives from the potential $\phi_1 = T \ln P$ (see also [548]), where the expression of this force in terms of the temperature T results from Einstein's formula for the diffusion coefficient, $D = bT$. Since we are in the regime where the velocity acquired by the particle is proportional to the force, it reads

$$v = bF_1 = D\nabla \ln P. \tag{10.88}$$

As a consequence the transformed velocity V vanishes in this case, and therefore also dV/dt, i.e.

$$V = v - D\nabla \ln P = 0, \quad \left(\frac{\partial}{\partial t} + V \cdot \nabla\right) V = 0. \tag{10.89}$$

Let us establish the form of the Fokker–Planck equation in this case. It reads

$$D\Delta P = \frac{\partial P}{\partial t} + \operatorname{div}(DP\nabla \ln P) = \frac{\partial P}{\partial t} + D\operatorname{div}(\nabla P), \tag{10.90}$$

and we finally obtain

$$\frac{\partial P}{\partial t} = 0. \tag{10.91}$$

This means that the system has become stationary and that the force F_1 has therefore stopped the diffusion.

(iii) Let us finally assume that we apply a doubled force

$$F_2 = 2\frac{D}{b} \nabla \ln P = 2T \nabla \ln P. \tag{10.92}$$

The particles acquire a velocity

$$v = bF_2 = 2D\nabla \ln P \tag{10.93}$$

and the Fokker–Planck equation becomes

$$D\Delta P = \frac{\partial P}{\partial t} + \operatorname{div}(2DP\nabla \ln P) = \frac{\partial P}{\partial t} + 2D\operatorname{div}(\nabla P), \tag{10.94}$$

so that it finally reads

$$-\frac{\partial P}{\partial t} = D\Delta P. \tag{10.95}$$

This is nothing but the classical diffusion equation for the reversed time $t \to -t$, whose absence of physical solution manifests the fundamental irreversibility of diffusion processes.

In this case the velocity field V writes

$$V = +D \nabla \ln P, \tag{10.96}$$

and its Euler equation becomes

$$\left(\frac{\partial}{\partial t} + V \cdot \nabla\right) V = \left(\frac{\partial}{\partial t} + D\nabla \ln P \cdot \nabla\right) D\nabla \ln P$$

$$= D\nabla \frac{\partial \ln P}{\partial t} + D^2 (\nabla \ln P \cdot \nabla)\nabla \ln P$$

$$= D^2 \left(-\nabla \frac{\Delta P}{P} + (\nabla \ln P \cdot \nabla)\nabla \ln P\right). \tag{10.97}$$

This is exactly Eq. (10.56), so that we finally obtain once again

$$\left(\frac{\partial}{\partial t} + V \cdot \nabla\right) V = -2D^2 \nabla \left(\frac{\Delta\sqrt{P}}{\sqrt{P}}\right), \tag{10.98}$$

which remains a diffusion equation. Therefore the doubling of the force in the dissipative Langevin regime has not reversed the diffusion, but through a double sign inversion, it has brought back the system to its original diffusive behavior.

10.4.4. *Consequence for negentropy*

The change of sign of the potential has therefore dramatic consequences (in the case of Newtonian dynamics), since in one case it yields a classical diffusion equation, which is known to lead to disorganization, irreversibility, entropy increase proportional to time t and spreading as $\sqrt{t}$, while in the other case it yields a Schrödinger equation that allows stationary solutions and therefore leads to structuring and self-organization (in dependence of the field, the limiting conditions and the symmetries).

This result leads one to a new view of the transition from the classical to the quantum organization. The Euler and continuity system of equations is characterized by three regimes instead of only two, in terms of the additional potential. Depending on the amplitude of this potential, given by $-\mathcal{D}^2$, 0 or $+\mathcal{D}^2$, one respectively obtains a quantum-type (Schrödinger) self-organized system, a classical hydrodynamic-like, isentropic system having weak capacity of organization or a diffusive system typical of entropy increase.

It is easy to verify that the Schrödingerian case is a negentropic system. Indeed, if one starts from a quantum-type organized structure described by a Schrödinger equation, or equivalently by a Euler and continuity system with quantum potential, and then abruptly cancels it, the structure is expected to dissipate itself and the entropy to increase. Reciprocally, the passage from the unstructured classical system to the organized quantum-like system therefore corresponds to a negative entropy or negentropy [481, 69], the value of which is easily computable by taking the opposite of the entropy increase from the structured to the unstructured system, $\sum_i P_i \ln P_i$.

10.5. Numerical Simulations of Fractal Geodesics

10.5.1. *Stochastic differential equations*

Let us consider in more detail the stochastic differential equations that have been constructed for describing the laws of motion in a nondifferentiable and fractal space. The two-valuedness of the velocity field can be expressed in terms of the doublet (v_+, v_-), or equivalently in terms of the real and imaginary parts (V, U) of the complex velocity $\mathcal{V} = V - iU$, which is related to the wave function $\psi = \sqrt{P}\, e^{iS/2m\mathcal{D}}$ as $\mathcal{V} = -2i\mathcal{D}\nabla \ln \psi$ [353].

Let us review the stochastic differential equations involving each of these velocity fields and the associated differential equations that link them to the probability density P (see Chapter 5):

$$dx = v_+ dt + d\xi_+, \quad \frac{\partial P}{\partial t} + \mathrm{div}(Pv_+) = \mathcal{D}\Delta P : \text{Fokker–Planck,}$$

(10.99)

$$dx = v_- dt + d\xi_-, \quad \frac{\partial P}{\partial t} + \mathrm{div}(Pv_-) = -\mathcal{D}\Delta P : \text{no classical process,}$$

(10.100)

$$dx = V dt + d\xi_V, \quad \frac{\partial P}{\partial t} + \mathrm{div}(PV) = 0 : \text{continuity equation,}$$

(10.101)

$$dx = U dt + d\xi_U, \quad \mathrm{div}(PU) = \mathcal{D}\Delta P : \text{no classical process,}$$

(10.102)

where $V = (v_+ + v_-)/2$, $U = (v_+ - v_-)/2$, $d\xi_V = (d\xi_+ + d\xi_-)/2$, $d\xi_U = (d\xi_+ - d\xi_-)/2$. We recall that the $d\xi_\pm = \zeta_\pm (2\mathcal{D}dt)^{1/2}$ are stochastic differential elements of fractional order (with $\langle \zeta_\pm \rangle = 0$ and $\langle (\zeta_\pm)^2 \rangle = 1$).

10.5.1.1. *Comments on stochastic quantum mechanics*

Recall that, in the scale relativity theory, these equations are derived from the fundamental equation of dynamics $m\,\widehat{d}\mathcal{V}/dt = -\nabla\phi$, so that their similarity with diffusion equations is only superficial. On the contrary, in stochastic approaches to quantum mechanics, they are usually set as founding equations and they are accompanied by their diffusion interpretation, which leads to contradictions with standard quantum mechanics [205, 526, 362].

The first of these equations for the v_+ process is indeed, at first sight, equivalent to a standard classical diffusion process, although, more profoundly, the computation of the v_+ velocity is not classical, involving the standard form of a stochastic differential equation and its corresponding Fokker–Planck, i.e. forward Kolmogorov equation [533, 254]. It has been called "forward Fokker–Planck" equation by Nelson in the framework of stochastic quantum mechanics [331, 332].

Now, concerning the v_- process, it has been introduced in stochastic quantum mechanics as a time reversed diffusion process, while its differential equation was called by Nelson "backward Fokker–Planck equation".

Recall however that the classical diffusion theory is founded on Markov processes, for which one obtains a Chapman–Kolmogorov equation of composition of probability densities f. It reads for continuous valued processes (for $s \le u \le t$) [533]

$$f(a, s; x, t) = \int_{-\infty}^{\infty} f(a, s; z, u) f(z, u; x, t) dz. \tag{10.103}$$

The impossibility to obtain a classical diffusion theory of quantum mechanics can be traced back to this equation, since its quantum equivalent is Feynman's equation for the composition of probability amplitudes ([176], p. 37) for events occuring in succession,

$$K(b, a) = \int_{x_c} K(b, c) K(c, a) dx_c \tag{10.104}$$

a rule summed up by Feynman as: *amplitudes for events occurring in succession in time multiply.* Such a multiplication of probability amplitudes, involving square roots of probability densities and, moreover, quantum phases, is therefore fundamentally in contradiction with the classical Chapman–Kolmogorov equation that involves products of probability densities. This underlies Wang–Liang's result according to which stochastic quantum mechanics contradicts standard quantum mechanics as regards multitime measurements [526].

Now, in classical diffusion theory, one derives from the Chapman–Kolmogorov equation a forward Kolmogorov equation (also called Fokker–Planck equation), and a backward Kolmogorov equation, which reduces to the same equation in the case of a diffusion process [533]. Therefore Nelson's so-called backward Fokker–Planck Eq. (10.100), which is set as a founding equation for stochastic quantum mechanics, does not correspond to any known classical process. The same is true of the last equations for the U velocity field.

In the scale relativity approach, on the contrary, the v_- velocity field is not defined as a backward (i.e. time-reversal) velocity, but comes from the discrete symmetry breaking in the transformation $dt \to -dt$ of the differential element dt considered and re-interpreted as an explicit scale variable. The Eq. (10.100) for v_- is derived as a fundamentally non-classical process, which manifests the non-differentiability of space and has therefore no classical equivalent. In the same way, the equations for the v_+ and V velocities, even if they look like classical diffusion or fluid equations, have not the same status as in classical physics. Indeed, they are not set as founding equations, but instead the stochasticity and these equations are derived as a very consequence of the non-differentiable and fractal geometry, whose effects cannot be reduced to them.

10.5.1.2. *Numerical simulations*

The equations for the velocity field V are particular among the others, since the differential equation involving P and V is not stochastic, but is instead the standard fluid-like continuity equation. Moreover, we know that it is completed by the Euler equation including a quantum potential.

However, this does not mean that we deal here with a classical fluid, since the way to compute the velocity field is fundamentally non-classical, involving the phase of a wave function, which is solution of a Schrödinger equation. Note also that, although the classical part of the velocity field V is not stochastic, one can more generally include the stochastic fractal term in a new definition of the velocity field $\widetilde{V} = V + d\xi_V/dt$ (see Sec. 5.8). Then one recovers a continuity equation, which derives from the imaginary part of the Schrödinger equation. The equations for $\widetilde{V}$ and P read in this case

$$dx = \widetilde{V}\,dt, \quad \frac{\partial P}{\partial t} + \mathrm{div}(P\widetilde{V}) = 0, \tag{10.105}$$

where $\widetilde{V}$ now includes divergent stochastic fluctuations.

We shall therefore consider two complementary ways to perform the numerical representation of the geodesics of a fractal space-time and to numerically derive the probablity density:

(i) Numerical integration of the differential stochastic equation

$$dx = Vdt + d\xi_V, \tag{10.106}$$

with and without the stochastic term.

(ii) Numerical integration of the differential stochastic equation

$$dx = v_+dt + \eta\,(2\mathcal{D}dt)^{1/2}, \tag{10.107}$$

ζ being a numerical stochastic variable, which is centered and normalized ($\langle\eta!\rangle = 0$ and $\langle(\eta)^2\rangle = 1$). For each system considered, the velocity field v_+ is computed from the relation

$$v_+ = U + V. \tag{10.108}$$

The velocity fields U and V are solutions of the quantum covariant equation of dynamics, $\widehat{d}(V - iU)/dt = -\nabla\phi/m$, which may be written under the form of hydrodynamical-like equations for two coupled fluids (Eqs. (10.112) and (10.113)). If possible, one may derive U and V directly by solving these equations without writing the Schrödinger equation, as done by Hermann [227]. One may also use the Schrödinger form of these equations to solve them in terms of an intermediary wave function $\psi = \sqrt{P}\,e^{i\theta}$ and then to derive the expression of the velocity fields as

$$U = \mathcal{D}\nabla\ln P, \quad V = 2\mathcal{D}\nabla\theta. \tag{10.109}$$

Under such an approach, v_+ and v_- are solutions of the differential equation

$$m\frac{\widehat{d}}{dt}\left(\frac{1-i}{2}v_+ + \frac{1+i}{2}v_-\right) = -\nabla\phi, \tag{10.110}$$

and the passage through the wave function description and the Schrödinger equation appears as a mere mathematical method of integration.

The stochastic variable η is represented by pseudo-random numbers such that $\langle\eta\rangle = 0$ and $\langle\eta^2\rangle = 1$ and satisfying to various probability distributions, for example to a Gaussian distribution. However, the whole formalism does not depend on the choice of this distribution. This is supported by the fact that the final probability density constructed from the geodesics does not depend on this choice.

Each particular example among the infinite number of possible geodesics is therefore represented in such a simulation as a random walk. This kind of simulation have already been performed in the framework of stochastic mechanics by McClendon and Rabitz [315], who have treated with positive results the Young two-slit problem and the tunneling problem. They have also found an absence of quantum chaos, in agreement with the scale relativistic view according to which the quantum properties already originate from an underlying infinite "chaos", which is understood as the effect of the nondifferentiable geometry [353, Chapter 5].

Since Eq. (10.107) is common to stochastic quantum mechanics and to the present scale relativity approach, this pioneering study and its results also apply to our case. However, the scale relativity view brings completely new features, which have been missed in it. Indeed, they start from the Langevin stochastic differential equation, re-written with our present notations as

$$dx(t) = v_+ dt + \sqrt{2\mathcal{D}}\, dw(t), \qquad (10.111)$$

where $\mathcal{D} = \hbar/2m$ and where $dw(t)$ is a stochastic variable. On the contrary, in the scale relativity approach:

(i) $\mathcal{D}$ is not reduced to its standard quantum expression and can take any macroscopic constant value specific of the system under consideration. For example, we shall see in Chapter 13 that when such an equation is applied to gravitational structuring at astronomical scales, one finds, in a Kepler potential around a mass M, that $\mathcal{D} = GM/2\alpha_g c$, where α_g is a "gravitational coupling constant" [3]. The paths thus obtained as geodesics are therefore not only typical of quantum mechanical "particles", but also of macroscopic trajectories of bodies under some conditions of fractality and irreversibility (e.g. the trajectory of a body in a protoplanetary disk).

(ii) The stochastic velocity fluctuation $W = \eta\,(2\mathcal{D}/dt)^{1/2}$ is an explicit function of the time interval dt, considered as a fundamental *variable*. This is an essential new point of our simulations compared to previous ones [315, 227], in which only one value of the time step size δt is taken (such that it is small with respect to the time scale of the system). In the scale relativity framework, we are led, on the contrary, to vary the size of the time step, at small time scales, in order to exhibit the fractal structure of the paths, but also up to long time scales of the order of the full integration time. This allows to put into evidence the de Broglie transition (see Fig. 10.6), which just comes from the different orders of the classical and fractal differential elements (respectively dt and $dt^{1/2}$).

It is clear that the potential number of examples of simulations is without any limit, considering the huge number of interesting quantum systems, which have been studied since the beginnings of quantum mechanics. We shall, therefore, give here very briefly only a few typical examples, leaving a more complete account open to future works.

10.5.2. *Examples of simulations*

10.5.2.1. *Quantum particle in a box*

A numerical simulation of a quantum particle in a box has been performed in the framework of the scale relativity approach in the pioneering work of Hermann [227]. He has shown, for this example, that the numerical prediction of quantum mechanical particle behavior could be obtained without using the Schrödinger equation. Moreover, since, in the scale relativity approach, the probability density is not defined at the beginning of the construction (contrary to stochastic mechanics), but is instead derived from the wave function, which is itself a manifestation of the geodesic velocity field, this result has provided a numerical support to the derivation of the Born postulate (see Sec. 5.7.3).

Hermann has started from the geodesic equation, $m\,\widehat{d}\mathcal{V}/dt = 0$, which becomes, in terms of the U and V velocity fields, equations of a coupled bi-fluid [353]:

$$\frac{\partial V}{\partial t} + (V \cdot \nabla)V = \mathcal{D}\Delta U + (U \cdot \nabla)U, \qquad (10.112)$$

$$\frac{\partial U}{\partial t} = -\{\mathcal{D}\Delta V + (V \cdot \nabla)U + (U \cdot \nabla)V\}. \qquad (10.113)$$

For the one-dimensional problem of a particle in a box, one has $V = 0$, so that the system of equations is reduced to

$$\mathcal{D}\Delta U + (U \cdot \nabla)U = 0, \quad \frac{\partial U}{\partial t} = 0. \qquad (10.114)$$

The solution of these equations is found by Hermann to be

$$U(x) = \frac{2\mathcal{D}n\pi}{a} \tan\left(-\frac{n\pi x}{a} + \frac{\pi}{2}\right), \qquad (10.115)$$

where a is the size of the box. Then, since $v_+ = U$ in this case, the numerical simulation proceeds by making several realizations of the stochastic process

$$dx(t) = \frac{2\mathcal{D}n\pi}{a} \tan\left(-\frac{n\pi x}{a} + \frac{\pi}{2}\right) dt + d\xi_+(t). \qquad (10.116)$$

Hermann has simplified the simulation by taking $2\mathcal{D}\, dt = 1$, which amounts, by counting time in terms of the number of time steps, to take $dt = \tau_E = 1$, where the Einstein time-scale $\tau_E = 2\mathcal{D}$ is equal to $\hbar/mc^2$ in the standard QM case (for $c = 1$). Such an elementary time scale, which is a characteristic of relativistic QM, is indeed far smaller than the typical time scales of the system, and it is therefore qualified to represent $dt \ll t$. However, by making this choice instead of varying the ratio dt/τ_E, the explicit scale dependence of the paths and the fractal-non fractal transition have not been investigated in this study (even though the fractality of the paths clearly appears in their drawing at such a resolution, see Fig. 10.7 of [227]).

Finally Hermann finds an excellent agreement between the QM theoretical prediction and the probability density obtained by plotting the histogram of the position distribution of the "particle" (for up to 5×10^8 steps). Moreover, the plot of typical paths thus obtained (Fig. 10.4) clearly show continuous "quantum jumps" of particles between different density peaks (in accordance with von Neumann's postulate and Feynman's path approach, that ensure the continuity of the path in space).

Let us now give some other examples of such numerical representations of the fractal space geodesics whose distribution leads to the quantum mechanical behavior. These simulations are similar to particle-by-particle

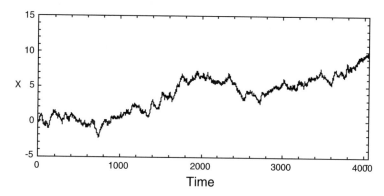

Fig. 10.4: **Typical path of a free particle.** Typical path of a free particle, obtained from a simulation of the stochastic differential Eq. (10.99) with a velocity field $v_+ = x/t$ and plotted for $2\mathcal{D} = 1$ and $dt = 0.01$.

experiments. This will be made apparent by taking only a small number of realisations N of the geodesics. As we shall see, even in this case the quantum-mechanical predictions are recovered in a fair way (within the expected $\sim \sqrt{N}$ fluctuations).

In these numerical simulations, we analytically establish the expressions for the velocity fields using Mathematica symbolic calculation abilities. We shall not give these expressions here, since they may be extremely complicated (they may take several pages in some cases: see some still rather simple examples in [315]).

10.5.2.2. *Young experiment: one-slit*

In a Young two-slit experiment, let us first consider the case when only one slit is open or when one knows by which slit the particle has passed (whatever be the way to obtain this knowledge). In this case one has to consider only the geodesics that come from this slit and go to the screen.

Examples of typical paths that connect the slit to the screen obtained in the simulation is given in Fig. 10.5. We do not give explicitly the expressions for the classical velocities, since, though they are analytic,

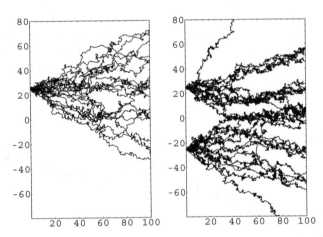

Fig. 10.5: **Fractal paths in the one- and two-slit cases.** Examples of individual fractal paths obtained in the numerical simulation of the one-slit (left) and two-slit (right) cases. The slit is at $(X = 0, Y = 25)$ in the one-slit case, and at $(X = 0, Y = \pm 25)$ in the two open slit case, X being the horizontal and Y the vertical coordinate. The screen lies at $X = 100$. The density distributions (Fig. 10.6 for the one-slit experiment, Fig. 10.8 in the two-slit experiment) is constructed from the distribution of the impact positions on the screen.

they are extremely long and complicated (see [315] for examples of such expressions).

The effect of varying the value of dt from small values to values of the order of or larger than the time-scale transition allows one to study the quantum to classical transition. As seen in Fig. 10.6, one indeed finds that the quantum probability peak becomes flattened out at long time scales in an increasing way when dt increases, i.e. the probability density tends to a constant value.

10.5.2.3. *Two-slit experiment*

The situation when the two slits are open while there is no way to know by which slit the "particle" has passed is described in the scale relativity framework by a fluid of geodesics that connect both of the slits to the screen. This is exemplified by the drawing in Fig. 10.7 of the square of the x component of the velocity field, which can be viewed as the equivalent of a fractal metric element (see Sec. 4.5). We have drawn $[(v_+)_x(x,y) + \eta\,(2\mathcal{D}/dt)^{1/2}]^2$ at three different time resolutions ($dt \to \infty$, $dt/2\mathcal{D} = 20$, $dt/2\mathcal{D} = 1/4$) in order to render manifest the de Broglie transition at the level of the fractal space. The metric oscillations, which manifest as the interference pattern on the screen can be clearly seen on the figure.

The numerical simulation of this situation by the stochastic differential Eq. (10.107) yields the expected interference pattern, even for a small number of geodesics incoming on the screen (see [315] for such simulations with a large number of trajectories).

But we may also consider the numerical integration of Eq. (10.106) that involves only the V velocity field. It can be made either with a fluctuation term $d\xi_V$ or without fluctuation (in this case the flow of geodesics is similar to a classical fluid), but with random initial conditions behind each hole. In each case we obtain the correct probability distribution (Fig. 10.8).

This is a remarkable result, compared with the simulation of the v_+ process. Indeed, the $v_+ = U + V$ process is linked to both the probability density by the relation $U = \mathcal{D}\nabla \ln P$ and to the phase of the wave function. On the contrary, the V velocity field depends only on the phase S, by the relation $V = 2\mathcal{D}\nabla S$ [353]. In other words, this means that by a numerical integration of the paths from the differential equation $dx = 2\mathcal{D}\nabla S\,dt$ and by constructing the histogram of their impact positions, one may recover the probability density, which is normally only given by the modulus of the wave function $P = |\psi|^2$.

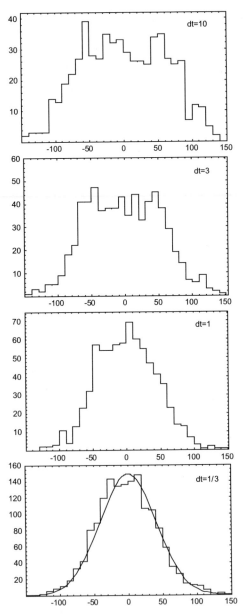

Fig. 10.6: **One-slit probability distribution at varying resolutions.** Numerical simulation of the distribution of impacts on a screen for a one-slit experiment (here the slit is at $X = 0$). The value of the variable dt is varied from $1/3$ to 10, while the transition time-scale is here at $\tau = 2$. One observes the transition from the expected quantum probability distribution (curve in the lower figure) to a flattened classical distribution of constant probability.

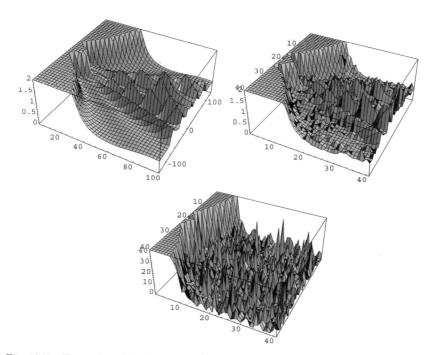

Fig. 10.7: **Fractal metric in a two-slit experiment.** In each figure, the two holes are to the left and the screen to the right. The top left figure shows the classical part of the effective metric element g_{xx}, given by the squared velocity field x component $(v_+)^2_x = (U_x + V_x)^2(x, y)$. The top right figure, corresponding to $(dt/2\mathcal{D}) = 20$, and the bottom figure, corresponding to $(dt/2\mathcal{D}) = 1/4$, include the fractal fluctuation at two different resolutions in order to illustrate the de Broglie transition.

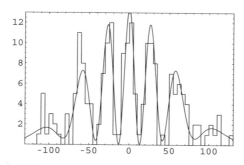

Fig. 10.8: **Numerical simulation of a two-slit experiment.** The figure gives the observed probability distribution of the impact positions of the simulated paths on the screen, described by the velocity field V alone (with random initial conditions behind each slit). It is compared with the interference pattern theoretically expected in standard quantum mechanics. This simulates a particle-by-particle Young experiment for a small number of particles. The difference between the theoretical prediction and the simulation agrees within the expected statistical fluctuations $\propto \sqrt{N}$.

This result seems to be contradictory with the current view according to which, in quantum mechanics, the phase carries a hidden information, which is additional to the observable one contained in the modulus. Indeed, here, we derive the observable quantity (P) from the phase alone.

In order to understand this result, one should however recall that the modulus and the phase of the wave function are calculated together, not as separated quantities, which would be solutions of different equations. This is also apparent in the equivalent Euler and continuity system with quantum potential, in which P and V appear in the two equations, which are therefore strongly coupled (see the next Section 10.6).

Moreover, recall that in the scale relativity approach, the doubling of the information in terms of the modulus $\sqrt{P}$ and the phase S, or equivalently in terms of the density P and of the potential velocity field V ultimately comes from a mirror effect at the infinitesimal level, namely, from the derivative two-valuedness issued from the reflection symmetry breaking under the discrete scale transformation $dt \rightarrow -dt$, which is itself a very consequence of nondifferentiability. This is reminiscent of the passage from Pauli spinors to Dirac bispinors described in scale relativity by a passage from quaternions to biquaternions, which also comes from the action of the discrete symmetry of parity in the (motion) relativistic case (see e.g. [268]). Hence, under mirror effects, it is clearly the same information that is seen in the "reality" and in the mirror, but, on the other hand, one must take into account the presence of a mirror in a room if one wants to obtain a correct and complete physical description.

We therefore suggest that the genuine nature of the quantum two-valuedness, as it is described by complex, then quaternionic numbers, is an elementary "mirror effect" which happens at all "points" of space-time (these "points" being, in addition, structured in scale).

10.5.2.4. *Hydrogen atom*

The same method can be applied to the simulation of electron geodesics in a hydrogen atom. We give in Fig. 10.9 an example of a geodesic for the state $n = 3$ and $l = n - 1 = 2$, and in Fig. 10.10 the histogram of the radial coordinate distribution of a unique geodesic during its time evolution. We clearly recover in this way the expected shape of the probability density given by Laguerre polynomials (see e.g. [267]).

10.5.2.5. *Harmonic oscillator*

Let us give the last example for a particle in a harmonic oscillator potential. In this case the expressions for the classical velocity fields remain simple:

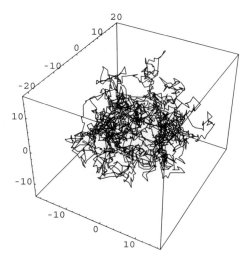

Fig. 10.9: **Numerical simulation of a geodesic in a hydrogen atom.** The fractal path is plotted here at the transition resolution ($dt = 1$ for $2\mathcal{D} = 1$ and $v = 1$).

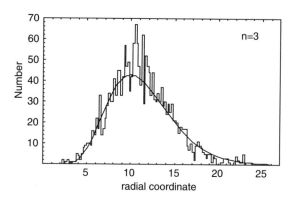

Fig. 10.10: **Numerical simulation of a hydrogen atom orbital.** A hydrogen atom orbital ($n = 3$) is simulated by the distribution of the radial coordinate on one of its geodesics (see Fig. 10.9), obtained by integration of the stochastic differential Eq. (10.107). The simulated probability distribution is compared with the theoretically expected Laguerre polynomial.

one finds $V = 0$, $U = -2\mathcal{D}(x, y, z)$ for the fundamental state $n = 0$ and $V = 0$, $U = 2\mathcal{D}(1/x - x, -y, -z)$ for the first excited state $n = 1$. Then we construct paths by a numerical integration of the stochastic differential equation $dx = v_+ dt + \zeta (2\mathcal{D}dt)^{1/2}$, where $v_+ = U + V$ and where ζ is simulated by a random function.

We also recover in this case the expected quantum probability distribution in a satisfactory way, as can be seen in Figs. 10.11 and 10.12. Similar

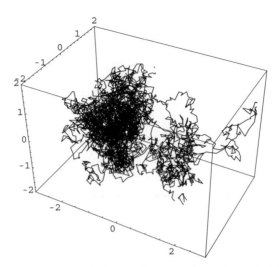

Fig. 10.11: **Quantum mechanical path (harmonic oscillator potential).** Numerical simulation of a quantum mechanical path in a three-dimensional harmonic oscillator potential ($dt = 1/100$, $2\mathcal{D} = 1$), for the first excited state $n = 1$. The dissymetry comes from the fact that there is only one path in this figure, with initial condition in the left lobe.

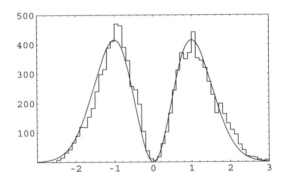

Fig. 10.12: **Probability distribution (harmonic oscillator potential).** Numerical simulation of the probability distribution in an harmonic oscillator potential ($dt = 1/100$, $2\mathcal{D} = 1$). The figure gives the histogram of the distribution of positions on the principal axis, for the time evolution of two paths following the stochastic differential Eq. (10.107), with initial conditions in each of the lobes. It is compared to the expected probability distribution in quantum mechanics (Hermite polynomial).

results have recently been obtained by other authors, who have developed, as in Hermann's work [227], numerical solutions of the scale relativity equations and obtained in this way the correct quantum probability densities without explicitly using the Schrödinger equation [11, 12].

10.6. Macroscopic Quantum Potentials

10.6.1. *Introduction*

We have shown in Chapter 5 that the scale relativity geodesic equation can either take a Schrödinger form or a fluid dynamics form with an added quantum potential, and that this remains true when the constant $\mathcal{D}$ is macroscopic instead of being reduced to its standard quantum mechanical expression $\hbar/2m$. This has opened the possibility of a new form of macroscopic quantum-type effects, different from the already known ones, such as superconductivity or Bose–Einstein condensates, which would no longer be constrained by the microscopic Planck constant $\hbar$. One is, therefore, led to consider the possible existence of classical systems, which would also be subjected to a quantum-like potential. Such systems would be expected to exhibit some quantum-like macroscopic properties specific to Schrödinger and nonlinear Schrödinger regimes (though certainly not every aspects of a genuine quantum system).

In the present section, we aim to study in more detail the properties of such macroscopic quantum potentials and of the systems in which they would intervene in order to prepare their possible use in future laboratory experiments and new technologies [403, 395, 402, 406]. We have shown (Sec. 5) that the system of equations (Euler equation and continuity equation) that describes a fluid in irrotational motion subjected to a generalized quantum potential, is equivalent to a generalized Schrödinger equation [353, 362, 374, 406]. Moreover, we show in this Chapter that, even in the presence of vorticity, it is also possible to obtain a Schrödinger-type equation, which includes the presence of a vectorial field, for a large class of systems.

Moreover, since the Euler and continuity system of equations can also be used as approximation in the description of many other types of physical, chemical, biological and other systems (chaotic mechanical and chemical systems, n particle dynamics, etc.), it is a full class of new experimental macroscopic quantum-like devices that can be considered.

10.6.2. *Schrödinger equation in fluid mechanics*

10.6.2.1. *From Schrödinger to Euler and continuity equations*

Let us recall once again a fundamental result that has been given in previous chapters, since it is at the heart of the new proposal presented now. By separating the real and imaginary parts of the generalized Schrödinger equation and by using a mixed representation of the motion equations in terms of (P, V), instead of (V, U) in the geodesic form and (P, θ) in the Schrödinger form, one obtains fluid dynamics-like equations, i.e. a Euler equation and a continuity equation (this is a generalization of the Madelung–Bohm transformation, but whose physical meaning is set from the very beginning instead of being *a posteriori* interpreted).

Let us recall this transformation in details, keeping all its steps, which may be useful in many situations (in particular when studying quantum potentials [406]). We first come back to the definition of the wave function by making explicit the probability and the phase,

$$\psi = \sqrt{P} \times e^{iS/S_0}. \tag{10.117}$$

The Schrödinger equation reads

$$S_0 \left(\mathcal{D} \Delta \psi + i \frac{\partial \psi}{\partial t} \right) = \phi \, \psi. \tag{10.118}$$

Under this form, in which the relation between S_0 and $\mathcal{D}$ is not yet explicited, i.e. $S_0 = 2m\mathcal{D}$ for each particle, it is easily generalizable to the many particle case, for which $\mathcal{D}\Delta$ is replaced by the sum on the particles $\sum_k \mathcal{D}_k \Delta_k$ (Sec. 5.10).

The time derivative of the wave function reads

$$\frac{\partial \psi}{\partial t} = \left(\frac{\partial \sqrt{P}}{\partial t} + \frac{i}{S_0} \sqrt{P} \frac{\partial S}{\partial t} \right) e^{iS/S_0}. \tag{10.119}$$

Its gradient reads

$$\nabla \psi = \left(\nabla \sqrt{P} + \frac{i}{S_0} \sqrt{P} \nabla S \right) e^{iS/S_0}. \tag{10.120}$$

We obtain for its second derivative

$$\Delta \psi = \left(\Delta \sqrt{P} - \frac{1}{S_0^2} \sqrt{P} \, (\nabla S)^2 + \frac{i}{S_0} \sqrt{P} \Delta S + \frac{2i}{S_0} \nabla \sqrt{P} \cdot \nabla S \right) e^{iS/S_0}. \tag{10.121}$$

Finally, by separating its real and imaginary parts, the Schrödinger equation takes the form (after simplification of the exponential term in factor)

$$\left(\mathcal{D} \Delta \sqrt{P} - \frac{\mathcal{D}}{S_0^2} \sqrt{P} \, (\nabla S)^2 - \frac{\phi}{S_0} \sqrt{P} - \frac{1}{S_0} \sqrt{P} \frac{\partial S}{\partial t} \right)$$

$$+ i \left(\frac{\mathcal{D}}{S_0} \sqrt{P} \Delta S + \frac{2\mathcal{D}}{S_0} \nabla \sqrt{P} \cdot \nabla S + \frac{\partial \sqrt{P}}{\partial t} \right) = 0. \tag{10.122}$$

Let us now introduce the generalized Compton relation $S_0 = 2m\mathcal{D}$ in this equation. We obtain:

$$-\frac{\sqrt{P}}{2m}\left(\frac{\partial S}{\partial t} + \frac{(\nabla S)^2}{2m} + \phi - 2m\mathcal{D}^2 \frac{\Delta\sqrt{P}}{\sqrt{P}}\right)$$

$$+ i\frac{\mathcal{D}}{2\sqrt{P}}\left\{\frac{\partial P}{\partial t} + \mathrm{div}\left(P\frac{\nabla S}{m}\right)\right\} = 0. \qquad (10.123)$$

The complex velocity $\mathcal{V} = V - iU$ being linked to the wave function by the relation $\mathcal{V} = -2i\mathcal{D}\nabla \ln \psi$, its real part is therefore given in terms of the phase by the relation [353]

$$V = \frac{\nabla S}{m}. \qquad (10.124)$$

Note that here this fundamental relation is derived (since V has been defined from the very beginning as the real part of the geodesic mean velocity field), while in the standard Madelung transformation V is defined from the above equation and it is therefore interpreted from it. In the scale relativity/nondifferentiable spacetime approach, the velocity field and the probability density characterize from the beginning of the description the bundle of potential geodesics.

By replacing in the above form of the Schrödinger equation $\nabla S/m$ by the real velocity field V, it reads

$$\left\{-\frac{\sqrt{P}}{2m}\left(\frac{\partial S}{\partial t} + \frac{1}{2}mV^2 + \phi + Q\right) + i\frac{\mathcal{D}}{2\sqrt{P}}\left(\frac{\partial P}{\partial t} + \mathrm{div}(PV)\right)\right\} e^{iS/2m\mathcal{D}} = 0,$$

$$(10.125)$$

so that we finally obtain a Euler-type equation and a continuity equation, namely,

$$\left(\frac{\partial}{\partial t} + V\cdot\nabla\right)V = -\nabla\left(\frac{\phi + Q}{m}\right), \qquad (10.126)$$

$$\frac{\partial P}{\partial t} + \mathrm{div}(PV) = 0, \qquad (10.127)$$

where a potential energy Q has emerged and which writes

$$Q = -2m\mathcal{D}^2 \frac{\Delta\sqrt{P}}{\sqrt{P}}. \qquad (10.128)$$

As already stressed, this scalar potential generalizes to a constant $\mathcal{D}$, which may be different from $\hbar/2m$ the quantum potential obtained in the

Madelung–Bohm transformation. The potential Q is now understood as a manifestation of the fractal geometry and the probability density is also interpreted as arising from the distribution of geodesics, so that the Born postulate is derived [93, 403], as can be verified by numerical simulations (see [227] and the previous Sec. 10.5).

More generally, in the many particle case (Sec. 5.10), the Schrödinger equation reads

$$S_0 \left(\sum_k \mathcal{D}_k \Delta_k \psi + i \frac{\partial \psi}{\partial t} \right) = \phi \psi, \tag{10.129}$$

where the index k runs on the particle rank and where the Laplacian operator Δ_k applies on the particle k coordinates. The wave function reads $\psi = \sqrt{P} e^{iS/S_0}$. In its term, the classical velocity field of each particle (k) geodesic is given by $V_k = \nabla_k S/m_k$. The energy, continuity and Euler equations take respectively the following forms

$$E = -\frac{\partial S}{\partial t} = \sum_k \frac{1}{2} m_k V_k^2 + \phi + Q, \tag{10.130}$$

$$\frac{\partial P}{\partial t} + \sum_k \nabla_k \cdot (P V_k) = 0, \tag{10.131}$$

$$m_k \left(\frac{\partial}{\partial t} + V_k \cdot \nabla_k \right) V_k = F_k, \tag{10.132}$$

with forces $F_k = -\nabla_k(\phi + Q)$, in terms of a quantum potential, which is generalized as

$$Q = -S_0 \frac{\sum_k \mathcal{D}_k \Delta_k \sqrt{P}}{\sqrt{P}}, \tag{10.133}$$

with $S_0 = 2 m_k \mathcal{D}_k$ for each particle of rank k (and $S_0 = \hbar$ in standard microscopic quantum mechanics).

10.6.2.2. *Inverse derivation: from Euler to Schrödinger equation (pressure-less potential motion)*

It is less well-known that the transformation from the generalized Schrödinger equation to the Euler and continuity equations with quantum potential is reversible (but see [506, 507]). Indeed, the Euler and continuity system reads in the pressure-less case

$$\left(\frac{\partial}{\partial t} + V \cdot \nabla\right) V = -\nabla \left(\phi - 2\mathcal{D}^2 \frac{\Delta\sqrt{\rho}}{\sqrt{\rho}}\right), \tag{10.134}$$

$$\frac{\partial \rho}{\partial t} + \operatorname{div}(\rho V) = 0. \tag{10.135}$$

Their form is similar to Eqs. (10.126) and (10.127), but with the probability density P replaced by the matter density ρ and with $m = 1$. Assume as the first step that the motion is irrotational (see the following section for the account of vorticity and pressure). Then we set

$$V = \nabla S. \tag{10.136}$$

Equation (10.134) takes the successive forms

$$\frac{\partial}{\partial t}(\nabla S) + \frac{1}{2}\nabla(\nabla S)^2 + \nabla\left(\phi - 2\mathcal{D}^2\frac{\Delta\sqrt{\rho}}{\sqrt{\rho}}\right) = 0, \tag{10.137}$$

$$\nabla\left(\frac{\partial S}{\partial t} + \frac{1}{2}(\nabla S)^2 + \phi - 2\mathcal{D}^2\frac{\Delta\sqrt{\rho}}{\sqrt{\rho}}\right) = 0, \tag{10.138}$$

which can be integrated as

$$\frac{\partial S}{\partial t} + \frac{1}{2}(\nabla S)^2 + \phi + K - 2\mathcal{D}^2\frac{\Delta\sqrt{\rho}}{\sqrt{\rho}} = 0, \tag{10.139}$$

where K is a constant that can be taken to be zero by a redefinition of the potential energy ϕ. Let us now combine this equation with the continuity equation as follows:

$$\left\{-\frac{1}{2}\sqrt{\rho}\left(\frac{\partial S}{\partial t} + \frac{1}{2}(\nabla S)^2 + \phi - 2\mathcal{D}^2\frac{\Delta\sqrt{\rho}}{\sqrt{\rho}}\right)\right.$$
$$\left. +i\frac{\mathcal{D}}{2\sqrt{\rho}}\left(\frac{\partial \rho}{\partial t} + \operatorname{div}(\rho\nabla S)\right)\right\} e^{iS/2\mathcal{D}} = 0. \tag{10.140}$$

We have therefore recovered the form (10.123) of the Schrödinger equation (with $m = 1$). Finally we set

$$\psi = \sqrt{\rho} \times e^{iS/2\mathcal{D}} \tag{10.141}$$

and the Eq. (10.140) is strictly identical to the following generalized Schrödinger equation:

$$\mathcal{D}^2\Delta\psi + i\mathcal{D}\frac{\partial}{\partial t}\psi - \frac{\phi}{2}\psi = 0. \tag{10.142}$$

Given the linearity of the equation obtained, one can normalize the modulus of ψ by replacing the matter density ρ by a probability density $P = \rho/M$, where M is the total mass of the fluid in the volume considered. These two representations are equivalent.

The imaginary part of this generalized Schrödinger equation amounts to the continuity equation and its real part to the energy equation that reads

$$E = -\frac{\partial S}{\partial t} = \frac{1}{2} m V^2 + \phi - 2\mathcal{D}^2 \frac{\Delta \sqrt{\rho}}{\sqrt{\rho}}. \qquad (10.143)$$

10.6.3. From Euler to Schrödinger: account of pressure

Consider now the Euler equations with a pressure term and a quantum potential term, namely,

$$\left(\frac{\partial}{\partial t} + V \cdot \nabla \right) V = -\nabla \phi - \frac{\nabla p}{\rho} + 2\mathcal{D}^2 \nabla \left(\frac{\Delta \sqrt{\rho}}{\sqrt{\rho}} \right). \qquad (10.144)$$

When $\nabla p / \rho = \nabla w$ is itself a gradient, which is the case of an isentropic fluid, and, more generally, of every case when there is an univocal link between pressure and density, e.g. a state equation [269], its combination with the continuity equation can be still integrated in terms of a Schrödinger-type equation [362],

$$\mathcal{D}^2 \Delta \psi + i \mathcal{D} \frac{\partial}{\partial t} \psi - \frac{\phi + w}{2} \psi = 0. \qquad (10.145)$$

This equation is the same as that obtained in Sec. 5.11.3, but the approach is somewhat different here. In Sec. 5.11.3 and in the other generalized Schrödinger-like equations considered in Chapter 5, the problem was to describe a fluid, which would be subjected to the fractal and small scale irreversibility described by the quantum covariant derivative, while now we consider a classical fluid, which would be subjected to a quantum-like potential.

In the next step the pressure term needs to be specified through a state equation, which can be chosen as taking the general form $p = k_p \rho^\gamma$.

In particular, in the acoustic approximation, the link between pressure and density writes $p - p_0 = c_s^2 (\rho - \rho_0)$, where c_s is the speed of sound in the fluid, so that $\nabla p / \rho = c_s^2 \nabla \ln \rho$. In this case, which corresponds to $\gamma = 1$, we obtain the nonlinear Schrödinger equation

$$\mathcal{D}^2 \Delta \psi + i \mathcal{D} \frac{\partial}{\partial t} \psi - k_p \psi \ln |\psi| = \frac{1}{2} \phi \psi, \qquad (10.146)$$

with $k_p = c_s^2$. When $\rho - \rho_0 \ll \rho_0$, one may use the additional approximation $c_s^2 \nabla \ln \rho \approx (c_s^2 / \rho_0) \nabla \rho$, and the equation obtained takes the form of the nonlinear Schrödinger equation which is encountered in the study of superfluids and of Bose–Einstein condensates (see e.g. [337, 173]) and which is also similar to the Ginzburg–Landau equation of superconductivity [270] (here in the absence of magnetic field),

$$\mathcal{D}^2 \Delta\psi + i\mathcal{D}\frac{\partial}{\partial t}\psi - \beta|\psi|^2\,\psi = \frac{1}{2}\,\phi\psi, \qquad (10.147)$$

with $\beta = c_s^2/2\rho_0$. In the highly compressible case the dominant pressure term is rather of the form $p \propto \rho^2$, so that $p/\rho \propto \rho = |\psi|^2$, and one still obtains a nonlinear Schrödinger equation of the same kind [337, 338].

10.6.4. *From Schrödinger equation in vectorial field to Euler and continuity equations*

Let us now consider a more general case. In the previous sections, only a scalar external field was taken into account. We shall now study the decomposition of the Schrödinger equation, which applies to a system subjected to a vectorial field (such as, e.g. a magnetic field). As we shall now show, it can also be generally decomposed in terms of a Euler-type equation and a continuity-type equation, with the external vectorial field playing a role similar to the rotational part of the velocity field. Thanks to this analogy, this decomposition applies actually to two different cases: (i) quantum fluids subjected to a magnetic field, such as in the Ginzburg–Landau equation of superconductivity; (ii) some fluids with a non-potential velocity field.

Start from the general form of the Schrödinger equation for a spinless particle subjected to a scalar field ϕ and to a vectorial field K_j (for example, an electromagnetic field):

$$\left\{\frac{1}{2}(-2i\mathcal{D}\nabla - K)^2 + \frac{\phi}{m}\right\}\psi = 2i\mathcal{D}\,\frac{\partial\psi}{\partial t}. \qquad (10.148)$$

In order to prepare the reverse derivation in which K actually represents the rotational part of the velocity field of the fluid under consideration, we have given here to the potential K a form in which it has the dimensionality of a velocity. In the case of an electromagnetic field, it is related to the vector potential A by the relation $K = (e/mc)A$. In the particular case when $\mathcal{D} = \hbar/2m$, one recovers the Schrödinger equation of standard quantum mechanics in the presence of a vectorial field,

$$\left\{\frac{1}{2m}(-i\hbar\nabla - mK)^2 + \phi\right\}\psi = i\hbar\frac{\partial\psi}{\partial t}. \qquad (10.149)$$

Note that this equation may be itself founded from the scale relativistic interpretation of gauge field theories according to which the field and the charges emerge as manifestations of the fractality of space-time [356, 394]. In this approach, the QED-covariant derivative $-i\hbar\tilde{\nabla} = -i\hbar\nabla - mK$ can be derived from geometric first principles, and therefore the electromagnetic Schrödinger equation can be established as the integral of a geodesic equation (see also [446]).

Let us expand the Hamiltonian. We obtain (reintroducing for clarity indices running from one to three)

$$-2\mathcal{D}^2\Delta\psi + 2i\mathcal{D}K_k\partial^k\psi + i\mathcal{D}(\partial_k K^k)\psi + \frac{1}{2}(K_k K^k)\psi + \frac{\phi}{m}\psi = 2i\mathcal{D}\frac{\partial\psi}{\partial t}.$$
(10.150)

We now express the wavefunction ψ in terms of its modulus and of its phase,

$$\psi = \sqrt{P} \times e^{i\theta}.$$
(10.151)

Therefore we have

$$\partial_k\psi = (\partial_k\sqrt{P} + i\sqrt{P}\,\partial_k\theta)e^{i\theta}, \quad \partial_t\psi = (\partial_t\sqrt{P} + i\sqrt{P}\,\partial_t\theta)e^{i\theta}, \quad (10.152)$$

$$\Delta\psi = \{(\partial_k\partial^k\sqrt{P} - \sqrt{P}\,\partial_k\theta\,\partial^k\theta) + i(2\,\partial_k\theta\,\partial^k\sqrt{P} + \sqrt{P}\,\partial_k\partial^k\theta)\}e^{i\theta}.$$
(10.153)

The Schrödinger equation becomes, after simplification of the $e^{i\theta}$ term in factor,

$$-2\mathcal{D}^2(\partial_k\partial^k\sqrt{P} - \sqrt{P}\,\partial_k\theta\,\partial^k\theta) - 2\mathcal{D}\sqrt{P}\,K_k\partial^k\theta + \left(\frac{1}{2}K_k K^k + \frac{\phi}{m}\right)\sqrt{P}$$

$$+ 2\mathcal{D}\sqrt{P}\,\partial_t\theta + i\{-2\mathcal{D}^2(\sqrt{P}\,\partial_k\partial^k\theta + 2\,\partial_k\theta\,\partial^k\sqrt{P}) + 2\mathcal{D}K_k\,\partial^k\sqrt{P}$$

$$+ \mathcal{D}(\partial_k K^k)\sqrt{P} - 2\mathcal{D}\,\partial_t\sqrt{P}\} = 0.$$
(10.154)

10.6.4.1. *Continuity equation*

Let us first consider the imaginary part of this equation. After multiplication by $2\sqrt{P}$ it becomes

$$-2\mathcal{D}\,\partial_t P - 2\mathcal{D}^2(2P\Delta\theta + 2\partial_k P\,\partial^k\theta) + 2\mathcal{D}(K_k\partial^k P + P\partial_k K^k) = 0.$$
$$(10.155)$$

Without the indices, it reads

$$\partial_t P + 2\mathcal{D}(P\Delta\theta + \nabla P \cdot \nabla\theta) - K \cdot \nabla P - P\nabla \cdot K = 0. \qquad (10.156)$$

Let us now introduce, as in the scalar field case, a potential motion velocity field

$$V = 2\mathcal{D}\nabla\theta. \qquad (10.157)$$

We obtain

$$\partial_t P + P\nabla \cdot V - P\nabla \cdot K + \nabla P \cdot V - \nabla P \cdot K = 0. \qquad (10.158)$$

This leads us to define a full "velocity field" as

$$v = V - K, \qquad (10.159)$$

in terms of which the above equation reads

$$\partial_t P + P\nabla \cdot v + \nabla P \cdot v = 0, \qquad (10.160)$$

and finally becomes the continuity equation

$$\frac{\partial P}{\partial t} + \mathrm{div}(P\,v) = 0, \qquad (10.161)$$

which is therefore generally valid, provided it is written in terms of the full velocity field $v = V - K$ instead of only the velocity field V, which is linked to the phase of the wave function.

10.6.4.2. *Energy equation*

Let us now consider the real part of Eq. (10.154). It reads

$$\sqrt{P}\left(2\mathcal{D}\,\partial_t\theta + 2\mathcal{D}^2\partial_k\theta\,\partial^k\theta - 2\mathcal{D}^2\frac{\Delta\sqrt{P}}{\sqrt{P}} + \frac{\phi}{m} - 2\mathcal{D}\,K_k\,\partial^k\theta + \frac{1}{2}K_k K^k\right) = 0.$$
$$(10.162)$$

We now use the equivalent notation $S = 2m\mathcal{D}\theta$ ($= \hbar\theta$ in the case of standard quantum mechanics), so that the wave function is now defined, like in previous sections, as

$$\psi = \sqrt{P} \times e^{iS/2m\mathcal{D}}. \tag{10.163}$$

We obtain

$$\sqrt{P}\left[\partial_t S + \frac{1}{2m}\partial_k S\,\partial^k S + \phi - 2m\mathcal{D}^2\frac{\Delta\sqrt{P}}{\sqrt{P}} - K_k\,\partial^k S + \frac{1}{2}mK_k K^k\right] = 0. \tag{10.164}$$

The potential part of the full velocity field now reads

$$V = \frac{\nabla S}{m}, \tag{10.165}$$

and we get the energy equation

$$\frac{\partial S}{\partial t} + \frac{1}{2}mV^2 + \frac{1}{2}mK^2 - mV\cdot K + \phi - 2m\mathcal{D}^2\frac{\Delta\sqrt{P}}{\sqrt{P}} = 0. \tag{10.166}$$

One recognizes, once again, the emergence of the full velocity field $v = V - K$ in this equation, where V is potential while K is rotational. In its terms the energy equation takes the same form as in the scalar field case, namely,

$$-\frac{\partial S}{\partial t} = \frac{1}{2}mv^2 + \phi - 2m\mathcal{D}^2\frac{\Delta\sqrt{P}}{\sqrt{P}}. \tag{10.167}$$

When the energy is conserved, $E = -\partial S/\partial t$. We therefore recover the same three contributions of kinetic energy $E_c = \frac{1}{2}mv^2$, exterior potential energy ϕ and quantum potential energy

$$Q = -2m\mathcal{D}^2\frac{\Delta\sqrt{P}}{\sqrt{P}}, \tag{10.168}$$

as in the previous case. The quantum potential also keeps its previous form exactly in this new (vectoriel field) situation.

Let us now take the gradient of the energy equation. One obtains

$$\frac{\partial V}{\partial t} + \frac{1}{2}\nabla(v^2) = -\nabla\left(\frac{\phi + Q}{m}\right). \tag{10.169}$$

In the potential case, $\frac{1}{2}\nabla(v^2) = (v\cdot\nabla)v$. But here, in the case of rotational motion, this relation leads to the introduction of a vorticity-like quantity, $\omega = \mathrm{curl}\,v$, i.e.

$$\omega_{\alpha k} = \partial_\alpha v_k - \partial_k v_\alpha = \partial_k K_\alpha - \partial_\alpha K_k. \tag{10.170}$$

Since K represents here a vector potential, $-\omega = \operatorname{curl} K$ therefore represents a magnetic-like field. In tensorial notation we have

$$\frac{1}{2}\partial_\alpha(v^k v_k) = v^k \partial_\alpha v_k = v^k \partial_k v_\alpha + v^k(\partial_\alpha v_k - \partial_k v_\alpha)$$
$$= v^k \partial_k v_\alpha + v^k \omega_{\alpha k}, \qquad (10.171)$$

i.e.

$$\frac{1}{2}\nabla(v^2) = (v \cdot \nabla)v + v \times \omega. \qquad (10.172)$$

Therefore, since $V = v + K$ and $\operatorname{curl} v = -\operatorname{curl} K$, one finally obtains the equation

$$\frac{\partial v}{\partial t} + (v \cdot \nabla)v = -\frac{\partial K}{\partial t} + v \times \operatorname{curl} K - \nabla\left(\frac{\phi + Q}{m}\right). \qquad (10.173)$$

One recognizes in the right-hand side of this equation the exact analog of a Lorentz force, to which is added the quantum force $-\nabla Q/m$. The term $-\partial K/\partial t$ is the analog of the magnetic contribution $-\partial A/c\,\partial t$ to the electric field $\mathcal{E} = -\partial A/c\,\partial t - \nabla\phi$, while $v \times \operatorname{curl} K$ is the analog of the magnetic force $(e/c)\,v \times \operatorname{curl} A$ (see e.g. [266]).

This equation has therefore exactly the form of the Euler equation that is expected for a fluid of velocity field v coupled to a scalar potential ϕ and to a vectorial potential K and subjected to an additional quantum potential Q. It agrees with the continuity equation, which is also written in terms of the full velocity field v.

10.6.4.3. *From Ginzburg–Landau equation to fluid equations with magnetic field and quantum potential*

Such an approach can be applied to the transformation of the Ginzburg–Landau equation of superconductivity into the classical equations for a fluid subjected to a magnetic field and to a quantum-like potential.

Let us start indeed from the Ginzburg–Landau equation of superconductivity [270] generalized to a coefficient $\mathcal{D}$, which may be different from $\hbar/2$,

$$\left(\mathcal{D}\nabla - i\frac{K}{2}\right)^2 \psi + \alpha\,\psi - \beta|\psi|^2\,\psi = 0, \qquad (10.174)$$

where $A = (mc/e)K$ is the magnetic vector potential.

From the previous decomposition, it is equivalent to the classical continuity and Euler equations of a fluid subjected both to a Lorentz force and to a quantum potential Q, namely (for $m = 1$)

$$\frac{\partial P}{\partial t} + \text{div}(P\,v) = 0, \tag{10.175}$$

$$\frac{\partial v}{\partial t} + v \cdot \nabla v = -\frac{\partial K}{\partial t} + v \times \text{curl}\,K - \nabla Q, \tag{10.176}$$

where $P = |\psi|^2$ and

$$Q = -2\mathcal{D}^2 \frac{\Delta\sqrt{P}}{\sqrt{P}}. \tag{10.177}$$

The reversibility of the transformation (see next Sec. 10.6.5) means that, if one applies to a classical charged fluid a classical force having exactly the form of the "quantum potential" Q (with a coefficient $\mathcal{D}$ no longer limited to the microscopic value $\hbar/2$), such a fluid would be described by the Ginzburg–Landau equation and it would, therefore, acquire some of the properties of a superconductor.

10.6.4.4. *Euler equation for special flows*

A more simple form of Euler equation may be recovered in rather general situations, as we shall now see.

When $v \times \text{curl}\,v = 0$, this means that v and $\text{curl}\,v$ are parallel, i.e. $\text{curl}\,v = \lambda v$ (Beltrami stream). In this case the Schrödinger in vectorial field equation takes the form of a standard Euler and continuity system of equations for a fluid subjected to a quantum-type potential $Q = -2m\mathcal{D}^2\,\Delta\sqrt{P}/\sqrt{P}$ and to a force $F_K = -\partial K/\partial t$, namely,

$$\frac{\partial v}{\partial t} + (v \cdot \nabla)v = F_K - \nabla\left(\frac{\phi + Q}{m}\right), \tag{10.178}$$

$$\frac{\partial P}{\partial t} + \text{div}(P\,v) = 0. \tag{10.179}$$

When $v \times \text{curl}\,v = \nabla\xi_f/m$, which corresponds to $\text{curl}(v \times \text{curl}\,v) = 0$, ξ_f plays the role of an additional scalar potential, and the Schrödinger equation in vectorial field may also be given the form of a standard Euler and continuity system of equations for a fluid subjected to a quantum-type force $F_Q = -\nabla Q$, with $Q = -2m\mathcal{D}^2\,\Delta\sqrt{P}/\sqrt{P}$, to a force $F_K = -\partial K/\partial t$

and to a total force $F = -\nabla(\xi_f + \phi)/m$, namely,

$$\frac{\partial v}{\partial t} + (v \cdot \nabla)v = F_K - \nabla\left(\frac{\phi + \xi_f + Q}{m}\right), \tag{10.180}$$

$$\frac{\partial P}{\partial t} + \text{div}(P\,v) = 0. \tag{10.181}$$

10.6.5. *Inverse problem: from Euler equation with vorticity to Schrödinger equation with vectorial field*

The previous calculations are reversible, and therefore they allow us to achieve a new result. The equations of the motion of a fluid including a rotational component subjected to a quantum-type potential can also be integrated in terms of a possibly nonlinear Schrödinger equation, the rotational part of the motion appearing in it under the same form as an external vectorial field.

Consider a classical non-viscous fluid subjected to a scalar potential ϕ and described by its velocity field $v(x, y, z, t)$ and its density $\varrho(x, y, z, t)$. These physical quantities are solutions of the Euler and continuity equations,

$$\left(\frac{\partial}{\partial t} + v \cdot \nabla\right)v = -\nabla\phi - \frac{\nabla p}{\varrho}. \tag{10.182}$$

$$\frac{\partial \varrho}{\partial t} + \text{div}(\varrho\,v) = 0. \tag{10.183}$$

In the case of an isoentropic fluid and more generally in all cases when there exists a univocal link between the pressure p and the density ϱ, $\nabla p/\varrho$ becomes a gradient [269], namely $\nabla p/\varrho = \nabla w$, where w is the enthalpy by mass unit in the isentropic case ($s = \text{cst}$). In this case we set

$$\nabla\phi + \frac{\nabla p}{\varrho} = \nabla(\phi + w) = \nabla\Phi, \tag{10.184}$$

and the Euler equation becomes

$$\left(\frac{\partial}{\partial t} + v \cdot \nabla\right)v = -\nabla\Phi. \tag{10.185}$$

Let us now assume that the classical fluid is subjected to an additional force

$$F_Q = -\nabla Q = 2\mathcal{D}^2\nabla\left(\frac{\Delta\sqrt{\varrho}}{\sqrt{\varrho}}\right), \tag{10.186}$$

so that the Euler and continuity equations read

$$\left(\frac{\partial}{\partial t} + v \cdot \nabla\right) v = -\nabla(w + \phi + Q), \tag{10.187}$$

$$\frac{\partial \varrho}{\partial t} + \mathrm{div}(\varrho\, v) = 0. \tag{10.188}$$

The Euler equation can be written under the form

$$\frac{\partial v}{\partial t} + \frac{1}{2}\nabla(v^2) - v \times \mathrm{curl}\, v = -\nabla(w + \phi + Q). \tag{10.189}$$

In this section, we specifically consider the case when the velocity field v is no longer potential. However, we can decompose it in terms of a potential (irrotational) contribution V and a rotational one K. Namely, we set

$$v = V - K, \quad V = \nabla S. \tag{10.190}$$

Then we build a "wavefunction" from the potential part only, by combining this function S and the density ϱ in terms of a complex function:

$$\psi = \sqrt{\varrho} \times e^{iS/2\mathcal{D}}. \tag{10.191}$$

Therefore $\partial v/\partial t = \partial \nabla S/\partial t - \partial K/\partial t = \nabla(\partial S/\partial t) - \partial K/\partial t$, so that the Euler equation now reads

$$\frac{\partial K}{\partial t} - v \times \mathrm{curl}\, K = \nabla\left(\frac{\partial S}{\partial t} + \frac{1}{2}v^2 + w + \phi + Q\right). \tag{10.192}$$

The scalar expression under the gradient is not vanishing in the general case. We call $-\chi(x, y, z, t)$ this function and we may therefore write a (formal) generalized energy equation,

$$-\frac{\partial S}{\partial t} = \frac{1}{2}v^2 + w + \phi + Q + \chi, \tag{10.193}$$

while Eq. (10.192) now writes

$$\frac{\partial K}{\partial t} - v \times \mathrm{curl}\, K = -\nabla\chi. \tag{10.194}$$

We have now recovered the conditions (energy equation and continuity equation), which lead to the construction of a nonlinear Schrödinger-type equation in terms of a complex linear combination of these two equations.

Therefore the whole calculation of the previous Sec. 10.6.4 can be reversed (with $m = 1$ and $P \propto \varrho$), so that we can integrate the Euler and continuity system in terms of a nonlinear Schrödinger-type equation including a vectorial field (analogous to the standard Schrödinger equation of a charged particle in a magnetic field),

$$\left(\mathcal{D}\nabla - i\frac{K}{2} \right)^2 \psi + i\mathcal{D}\frac{\partial \psi}{\partial t} = \left(\frac{w + \phi + \chi}{2} \right) \psi, \qquad (10.195)$$

where we recall that $\psi = \sqrt{\varrho} \times \exp(iS/2\mathcal{D})$. In general the pressure and, therefore, the enthalpy w is a function of the density $\rho = |\psi|^2$, which contributes to the nonlinearity of this equation. In the absence of vorticity, it is similar to the kind of nonlinear Schrödinger equation encountered in the study of superfluids and Bose–Einstein condensates (see e.g. [173, 337]).

Equation (10.195) can also be given an expanded form,

$$\mathcal{D}^2\Delta\psi + i\mathcal{D}\frac{\partial \psi}{\partial t} = \left\{ \frac{\phi + w + \chi}{2} + \frac{K^2}{4} + i\frac{\mathcal{D}}{2}\nabla \cdot K + i\mathcal{D}K \cdot \nabla \right\} \psi, \tag{10.196}$$

where the term between brackets in the right-hand side may be interpreted, when $K \cdot \nabla\psi$ is negligible, as a generalized potential energy.

In this Schrödinger equation, the rotational part K of the velocity field $v = V - K$ plays the role of an external vector potential, and therefore the vorticity $\omega = -\text{curl}\, K$ the role of the corresponding field. Its evolution equation is obtained by taking the curl of the Euler equation, namely,

$$\frac{\partial \omega}{\partial t} = \text{curl}(v \times \omega), \qquad (10.197)$$

which is equivalent to Eq. (10.194), but without the unknown function χ.

However, the situation here remains different and more complicated than the quantum mechanical Schrödinger equation in an external electromagnetic field of electric potential ϕ and vectorial potential K (up to constants), which is accompanied by the Maxwell equations for the external field. Here two terms are added, (i) the pressure p, expressed in terms of the enthalpy w, which may be known in function of the density ρ, i.e. $|\psi|^2$ through a state equation and lead to a nonlinear contribution; (ii) the unknown function $\chi(x, y, z, t)$. Therefore, except for the large class of flows for which $\chi = \text{cst}$, i.e.

$$\frac{\partial K}{\partial t} - v \times \text{curl}\, K = 0, \qquad (10.198)$$

in the general case this system of equations remains incomplete, since an equation for χ is lacking (see [285] for proposals concerning such a possible equation). Nevertheless, despite this situation, this result may

remain physically meaningful and useful, since the Schrödinger equation
and its solutions have general properties, which are valid whatever the
applied fields. It shows that the application of a quantum-like potential on
a fluid is sufficient to transform the energy and continuity equations into a
Schrödinger-like equation for a function ψ linked to the density by $\rho = |\psi|^2$,
even in the presence of vorticity.

Let us finally consider the stationary version of Eq. (10.195) in the
general case when the pressure terms reads $w = p/\rho \propto \rho$ (see Sec. 10.6.3).
We obtain

$$\left(\mathcal{D}\nabla - i\frac{K}{2}\right)^2 \psi + \alpha\,\psi - \beta|\psi|^2\,\psi = 0, \qquad (10.199)$$

with $\alpha = (E-\phi-\chi)/2$. This equation has exactly the form of the Ginzburg–
Landau equation of superconductivity [270], generalized to a coefficient $2\mathcal{D}$,
which may be different from $\hbar$. This result may be applied to the two cases
initially considered at the beginning of Sec. 10.6.4:

(i) The case where K represents the true vector potential of a magnetic
field. It may correspond to a classical charged fluid subjected to an
electromagnetic field and to a classical potential, which has been tuned
in order to give it the form of the quantum potential Q. As already
remarked in Sec. 10.6.4.3, the equations of motion of such a fluid
(continuity equation and Euler equation with a Lorentz force) may be
combined in terms of a single complex equation, which takes the form of
the Ginzburg–Landau equation of superconductivity. One may therefore hope such a fluid to acquire some of the properties of a quantum
fluid.

(ii) The case when K does not represent here an external magnetic field,
but a rotational part of the velocity field. This means that a nonlinear
Schrödinger form can also be given to the equation of motion of fluids
showing vorticity and subjected to an external potential Q having a
quantum-like form.

10.6.6. *From Navier–Stokes to nonlinear*
Schrödinger equation

Let us finally consider the general case of Navier–Stokes equations including
a viscosity term. The fluid mechanics equations including a quantum-type
potential read in this case

$$\left(\frac{\partial}{\partial t} + v \cdot \nabla\right) v = \nu \Delta v - \frac{\nabla p}{\varrho} - \nabla(\phi + Q), \qquad (10.200)$$

$$\frac{\partial \varrho}{\partial l} + \mathrm{div}(\varrho \, v) = 0, \qquad (10.201)$$

where the quantum-type potential energy is still given by

$$Q = -2\mathcal{D}^2 \frac{\Delta\sqrt{\varrho}}{\sqrt{\varrho}}. \qquad (10.202)$$

We set as in the previous section

$$v = V - K, \quad V = \nabla S, \quad \psi = \sqrt{\varrho} \times e^{iS/2\mathcal{D}}, \qquad (10.203)$$

i.e. V is the potential part of the full velocity field v. Therefore the viscosity term reads

$$\nu \Delta v = \nu \Delta(\nabla S - K) = \nu \nabla(\Delta S) - \nu \Delta K, \qquad (10.204)$$

and, assuming once again $\nabla p/\varrho = \nabla w$, the Navier–Stokes equation now takes the form

$$-\frac{\partial K}{\partial t} + \nu \Delta K - v \times \mathrm{curl}\, v = -\nabla\left(\frac{\partial S}{\partial t} - \nu \Delta S + \frac{1}{2}v^2 + w + \phi + Q\right). \qquad (10.205)$$

Combined with the continuity equation, it remains integrable in a Schrödinger equation, at least formally. Let us call χ the function between bracket on the right-hand side of this equation:

$$-\frac{\partial K}{\partial t} + \nu \Delta K - v \times \mathrm{curl}\, v = \nabla \chi. \qquad (10.206)$$

In this case one obtains an energy and a continuity equation that read

$$\frac{\partial S}{\partial t} - \nu \Delta S + \frac{1}{2}v^2 + w + \phi + \chi + Q = 0, \quad \frac{\partial \varrho}{\partial t} + \mathrm{div}(\varrho \, v) = 0, \qquad (10.207)$$

and which can be combined into the form of a nonlinear Schrödinger equation of the magnetic type,

$$\left(\mathcal{D}\nabla - i\frac{K}{2}\right)^2 \psi + i\mathcal{D}\frac{\partial \psi}{\partial t} = \frac{1}{2}(w + \phi + \chi - \nu \Delta S)\,\psi. \qquad (10.208)$$

The viscosity therefore leads to add a new nonlinear term in this nonlinear Schrödinger equation that depends on the phase $S/2\mathcal{D}$ of the wave function.

When the fluid motion is irrotational, the integration under the form of a nonlinear Schrödinger equation of the continuity and Navier–Stokes equations including a quantum potential is always possible.

10.6.7. *Schrödinger equation for the rotational motion of a solid*

10.6.7.1. *Introduction*

In the previous sections, a Schrödinger form has been obtained for the equations of motion and of continuity of a fluid subjected to a quantum potential. However, the method we used may be applied not only to a fluid but also to a mechanical system. Indeed, we have shown ([362], Sec. 5.11) that the scale relativity approach can be applied to the rotational motion of a solid, leading once again to a Schrödinger-type equation. Here, we give an improved demonstration of this Schrödinger equation [406], then, as in previous sections, we decompose it in terms of its real and imaginary parts and obtain a new generalized form of the quantum potential [406]. Reversely, the classical energy equation including this new quantum potential energy, combined with the continuity equation, yields a Schrödinger form for the equation of motion.

10.6.7.2. *Equation of rotational solid motion in scale relativity*

Let us briefly recall the results of [362, 365], in which a Schrödinger form was obtained for the equation of the rotational motion of a solid subjected to the three basic effects of a fractal and nondifferentiable space (namely, infinity of trajectories, fractal dimension 2 and reflection symmetry breaking of the time differential element).

The role of the variables (x, v, t) of translational motion is now played respectively by (φ, Ω, t), where φ stands for three rotational position angles (for example, Euler angles) and Ω for the angular velocity. We choose a contravariant notation φ^k for the angles, where the indices run from 1 to 3, and we adopt Einstein's convention for summation on upper and lower indices. The Euler–Lagrange equations for rotational motion classically read [265]

$$\frac{d}{dt} \frac{\partial L}{\partial \Omega^k} = \frac{\partial L}{\partial \varphi^k}, \tag{10.209}$$

in terms of a Lagrange function $L = (1/2)I_{ik}\Omega^i\Omega^k - \Phi$, where I_{ik} is the tensor of inertia of the solid body and Φ its potential energy in an exterior field. Therefore the angular momentum of the system is

$$M_i = \frac{\partial L}{\partial \Omega^i} = I_{ik}\Omega^k. \tag{10.210}$$

The torque is given by

$$K_i = \frac{\partial L}{\partial \varphi^i} = -\frac{\partial \Phi}{\partial \varphi^i}, \tag{10.211}$$

and the motion equations finally take the Newtonian form

$$\frac{dM_i}{dt} = K_i. \tag{10.212}$$

Let us now consider the generalized description of such a system in the scale relativity framework. Following the same road as for position coordinates, in the generalized situation when space-time is fractal, the angle differentials $d\varphi = dx_\varphi + d\xi_\varphi$ can be decomposed in terms of two contributions, a classical (differentiable) part dx_φ and a fractal fluctuation $d\xi_\varphi$, which is such that $\langle d\xi_\varphi \rangle = 0$ and

$$\langle d\xi_\varphi^j \, d\xi_\varphi^k \rangle = 2\mathcal{D}^{jk} \, dt, \tag{10.213}$$

where $\mathcal{D}^{jk}$ is now a tensor, which generalizes the scalar parameter $\mathcal{D}$ of the translational case. As we shall see in the following, this tensor is, up to a multiplicative constant, similar to a metric tensor.

In the same way as in the translational case, the breaking of reflection invariance $(dt \leftrightarrow -dt)$ on the time differential elements, which is a consequence of nondifferentiability, yields a two-valuedness of the angular velocity. This leads to introducing a complex angular velocity $\widetilde{\Omega}$, then a complex Lagrange function $\widetilde{L}(\varphi, \widetilde{\Omega}, t)$. The two effects of nondifferentiability and fractality of space can finally be combined in terms of a rotational quantum-covariant derivative [362, 365],

$$\frac{\widehat{d}}{dt} = \frac{\partial}{\partial t} + \widetilde{\Omega}^k \, \partial_k - i\,\mathcal{D}^{jk}\partial_j\partial_k, \tag{10.214}$$

where $\partial_k = \partial/\partial\varphi^k$. Using this quantum-covariant derivative, we may generalize to fractal motion the equation of rotational motion while keeping its classical form,

$$I_{jk} \frac{\widehat{d}\,\widetilde{\Omega}^k}{dt} = -\partial_j \Phi, \tag{10.215}$$

where I_{jk} is the tensor of inertia of the solid and Φ an externally added potential.

We then introduce a complex function, which will subsequently be identified with a wave function, as another expression for the complex action $\widetilde{S} = \int \widetilde{L}\, dt$,

$$\psi = e^{i\widetilde{S}/S_0}, \tag{10.216}$$

where S_0 is a real constant introduced for dimensional reasons. Now the complex angular momentum is, like in classical solid mechanics, linked to the complex action by the standard relation $\widetilde{M}_k = \partial \widetilde{S}/\partial \varphi^k$, so that one obtains

$$\widetilde{M}_\alpha = I_{\alpha k}\widetilde{\Omega}^k = -iS_0\, \partial_\alpha \ln \psi. \tag{10.217}$$

Let us therefore introduce the inverse of the tensor of inertia, $[I]^{-1} = I^{\alpha k}$, such that

$$I_{\alpha k} I^{k\beta} = \delta_\alpha^\beta. \tag{10.218}$$

This allows us to express the complex velocity field in terms of the wave function,

$$\widetilde{\Omega}^k = -iS_0\, I^{k\alpha}\, \partial_\alpha \ln \psi. \tag{10.219}$$

We can now replace the velocity field by this expression in the covariant derivative and in the rotational motion equations. We obtain

$$-iS_0 \left(\frac{\partial}{\partial t} - iS_0\, I^{k\beta}\partial_\beta \ln \psi\, \partial_k - i\, \mathcal{D}^{jk}\partial_j \partial_k \right) \partial_\alpha \ln \psi = K_\alpha, \tag{10.220}$$

which can be written as

$$-iS_0 \left(\partial_\alpha \frac{\partial}{\partial t} \ln \psi - i\{ S_0\, \partial_\beta \ln \psi\, I^{k\beta}\partial_k\, \partial_\alpha \ln \psi + \mathcal{D}^{jk}\partial_j \partial_k\, \partial_\alpha \ln \psi \} \right) = K_\alpha. \tag{10.221}$$

We have reversed in the second equation the places of $I^{k\beta}$ and ∂_β: this is possible since $I^{k\beta}$ is assumed to be constant. This reversal allows one to make appear the operator $I^{k\beta}\partial_k$. Provided the tensor of inertia plays the role of a metric tensor, we have $I^{k\beta}\partial_k = \partial^\beta$ and we recognize in the expression between brackets $\{\}$ a tensorial generalization of the expression, which was encountered in the translational motion case,

$$S_0 \, m^{-1} \, (\partial_\beta \ln \psi \, \partial^\beta) \partial_\alpha \ln \psi + \mathcal{D} \partial_k \partial^k \partial_\alpha \ln \psi. \tag{10.222}$$

Recall indeed that the relation

$$S_0 \, m^{-1} = 2\mathcal{D}, \tag{10.223}$$

which is nothing but a generalized Compton relation, allows one to transform this expression into a remarkable identity, which leads to the integration of the motion equation in terms of a Schrödinger equation.

Now we are able to generalize this result to the rotational motion case, despite the complication brought by the fact that the mass is replaced by the inertia tensor. Indeed, the inverse of the mass is replaced by the inverse tensor $[I]^{-1}$, and we can identify the fractal fluctuation tensor with this metric tensor up to a constant,

$$\mathcal{D}^{\alpha\beta} = \frac{S_0}{2} I^{\alpha\beta}, \tag{10.224}$$

i.e. in matrix form, $S_0[I]^{-1} = 2[\mathcal{D}]$. Note the correction to [362, 365] where an inverse relation between these quantities was erroneously given; the Schrödinger equation obtained in these papers nevertheless remains correct. This is a new tensorial generalization of the Compton relation. Moreover, this means that it is the inertia tensor itself, which is identified with a metric tensor and can be used to raise and lower the indices, e.g. $I^{jk}\partial_j\partial_k = \partial^k\partial_k$, while $\mathcal{D}^{jk}$ does the same but up to a constant, namely, $\mathcal{D}^{jk}\partial_j\partial_k = (S_0/2)\partial^k\partial_k$.

The existence of a similarity between the rotational diffusion term, $\widehat{M}_j D^{jk}\widehat{M}_k$, where $\widehat{M}$ denotes angular momentum operators and the corresponding quantum mechanical Hamiltonian $\widehat{M}_j I^{jk}\widehat{M}_k/2$ of a rigid body has already been remarked by Dale Favro [122] in his theory of rotational Brownian motion. Here we directly identify $\mathcal{D}^{jk}$ to $(S_0/2)I^{jk}$, but $\mathcal{D}^{jk}$, despite its stochastic definition, should not be confused with a standard diffusion coefficient.

The equation of motion now takes the form

$$-iS_0 \left(\partial_\alpha \frac{\partial}{\partial t} \ln \psi - i\frac{S_0}{2} \{ I^{k\beta}\partial_\beta \ln \psi \, \partial_k \partial_\alpha \ln \psi + I^{jk}\partial_j\partial_k\partial_\alpha \ln \psi \} \right) = K_\alpha,$$

$$\tag{10.225}$$

and, using the tensorial notation $I^{jk}\partial_j = \partial^k$, it can now be written as

$$-iS_0\left(\partial_\alpha\frac{\partial}{\partial t}\ln\psi - i\frac{S_0}{2}\{2\partial^k\ln\psi\,\partial_k\partial_\alpha\ln\psi + \partial^k\partial_k\partial_\alpha\ln\psi\}\right) = K_\alpha.$$

(10.226)

We recognize in this expression the remarkable identity of Sec. 5.6.4 so that we can simplify it under the form

$$-iS_0\left(\partial_\alpha\frac{\partial}{\partial t}\ln\psi - i\frac{S_0}{2}\partial_\alpha\frac{\partial_k\partial^k\psi}{\psi}\right) = -\partial_\alpha\Phi,$$

(10.227)

and finally write it globally as a gradient,

$$\partial_\alpha S_0\left\{\frac{(S_0/2)\partial_k\partial^k\psi + i\partial\psi/\partial t}{\psi}\right\} = \partial_\alpha\Phi.$$

(10.228)

This equation can therefore be integrated in the general case under the form of a new generalized Schrödinger equation that reads [362, 365, 406]

$$S_0\left(\mathcal{D}^{jk}\partial_j\partial_k\psi + i\frac{\partial}{\partial t}\psi\right) = \Phi\,\psi.$$

(10.229)

In terms of the inverse tensor of inertia, this rotational Schrödinger equation reads

$$\frac{1}{2}S_0^2\,I^{jk}\partial_j\partial_k\psi + i\,S_0\frac{\partial}{\partial t}\psi = \Phi\,\psi.$$

(10.230)

Since the tensor of inertia plays the role of a metric tensor, in particular for the lowering and raising of indices, it can finally be written as

$$\frac{1}{2}S_0^2\,\partial^k\partial_k\psi + i\,S_0\frac{\partial}{\partial t}\psi = \Phi\,\psi,$$

(10.231)

which keeps the form of the scalar case [353], while generalizing it.

The standard quantum case is recovered by identifying S_0 with $\hbar$, but, once again, all the mathematical structure of the equation and, therefore of its solutions, is preserved with a constant that can have any value, including a macroscopic one.

We may now conclude by returning to the fractal angular fluctuations that writes in terms of the inverse inertia tensor

$$\langle d\xi_\varphi^j\,d\xi_\varphi^k\rangle = S_0 I^{jk}\,dt.$$

(10.232)

We therefore gain a complete justification of the identification of the tensor of inertia with a metric tensor, since, owing to the fact that $I^{kj}I_{jk} = \delta_k^k = 3$, we obtain the invariant metric relation

$$dt = \frac{I_{jk}}{3S_0} \langle d\xi_\varphi^j \, d\xi_\varphi^k \rangle, \qquad (10.233)$$

where $S_0 = \hbar$ in the standard quantum case and where dt, which appears instead of its square dt^2 as a manifestation of the fractal dimension 2, is indeed the fundamental invariant since all this study is done in the framework of Galilean motion relativity.

10.6.7.3. *Fluid representation and newly generalized quantum potential*

Let us now give this Schrödinger equation a fluid mechanical form. This can be easily done by following the same steps as for the translational motion case, but now using tensorial derivative operators. We set

$$\psi = \sqrt{P} \times e^{iS/S_0} \qquad (10.234)$$

and we replace ψ by this expression in Eq. (10.231). The imaginary part of this equation reads

$$\frac{1}{2}\sqrt{P}\,\partial_k\partial^k S + \partial_k S\,\partial^k\sqrt{P} + \frac{\partial\sqrt{P}}{\partial t} = 0. \qquad (10.235)$$

i.e. after simplification,

$$\frac{\partial P}{\partial t} + \partial_k(P\partial^k S) = 0. \qquad (10.236)$$

Since S is the real part of the complex action, it is linked to the angular momentum M^k, which is itself the real part of the complex angular momentum and to the real angular velocity Ω^j by the relations

$$M_\alpha = I_{\alpha j}\Omega^j = \partial_\alpha S. \qquad (10.237)$$

Now, since $I^{k\alpha}I_{\alpha j} = \delta_j^k$, we find

$$\Omega^k = I^{k\alpha}I_{\alpha j}\Omega^j = I^{k\alpha}M_\alpha. \qquad (10.238)$$

Therefore,

$$\Omega^k = I^{k\alpha}\partial_\alpha S = \partial^k S. \qquad (10.239)$$

Finally, we find that the imaginary part of the rotational Schrödinger equation amounts, once again, to a continuity equation in the general case

$$\frac{\partial P}{\partial t} + \partial_k(P\Omega^k) = 0. \tag{10.240}$$

Note the correction to [362, 365], in which we concluded that a continuity equation could be obtained only in some particular reference systems: under the above tensorial form, it is correct in a generally covariant way. This means that the probability interpretation of $P = |\psi|^2$ is also generally ensured [93, 403].

The real part of Eq. (10.231) takes the form

$$\frac{\partial S}{\partial t} + \Phi - \frac{1}{2}S_0^2 \frac{\partial_k \partial^k \sqrt{P}}{\sqrt{P}} + \frac{1}{2}\partial_k S\, \partial^k S = 0. \tag{10.241}$$

Let us first consider the last term of this expression. It reads

$$\frac{1}{2}\partial_k S\, \partial^k S = \frac{1}{2}M_k\, \Omega^k = \frac{1}{2}I_{jk}\, \Omega^j\Omega^k = T_{\text{rot}}, \tag{10.242}$$

which is the classical expression of the rotational kinetic energy. In the conservative case, $E = -\partial S/\partial t$ is the total energy, so that we recover the standard energy equation,

$$E = \Phi + Q + T_{\text{rot}}, \tag{10.243}$$

but it now includes an additional potential energy that reads

$$Q = -S_0 \frac{\mathcal{D}^{jk}\partial_j\partial_k\sqrt{P}}{\sqrt{P}} = -\frac{1}{2}S_0^2 \frac{I^{jk}\partial_j\partial_k\sqrt{P}}{\sqrt{P}}. \tag{10.244}$$

This is a new generalization of the quantum potential in the rotational case.

A Euler-like equation including this quantum potential is simply obtained by taking the gradient of this equation,

$$\partial_\alpha\left(\frac{\partial S}{\partial t}\right) + \partial_\alpha\left(\frac{1}{2}\Omega^k M_k\right) = -\partial_\alpha(\Phi + Q). \tag{10.245}$$

We have $\partial_\alpha(\partial S/\partial t) = \partial(\partial_\alpha S)/\partial t = \partial M_\alpha/\partial t$, and since $M_\alpha = \partial_\alpha S$ is a gradient, $\partial_\alpha T_{\text{rot}} = \Omega^k \partial_\alpha M_k = \Omega^k \partial_k M_\alpha$, so that we finally obtain

$$\left(\frac{\partial}{\partial t} + \Omega^k \partial_k\right) M_\alpha = -\partial_\alpha(\Phi + Q), \tag{10.246}$$

which is indeed the expected generalization in terms of a Euler equation of the equation of dynamics $dM/dt = K$ when the rotational velocity becomes a velocity field $\Omega[\varphi(t), t]$.

10.6.7.4. *From Euler to Schrödinger equation*

Reversely, one may now consider a rotating body or an ensemble of rotating bodies, which are subjected to a fluctuating motion of rotation, such that the rotational velocity can be replaced, at least as an approximation, by a rotational velocity field $\Omega = \Omega[\varphi(t), t]$ [292].

Assume, moreover, that each body is subjected, in addition to the torque $-\partial_\alpha \Phi$ of a possible exterior field, to a quantum-like torque $K_Q = -\nabla_\varphi Q$, where the quantum potential Q is given by Eq. (10.244).

Such a system would be described by a Euler equation and a continuity equation that read

$$\left(\frac{\partial}{\partial t} + \Omega^k \partial_k \right) M_\alpha = -\partial_\alpha \left(\Phi - \frac{1}{2} S_0^2 \frac{I^{jk} \partial_j \partial_k \sqrt{P}}{\sqrt{P}} \right), \qquad (10.247)$$

$$\frac{\partial P}{\partial t} + \partial_k (P\Omega^k) = 0. \qquad (10.248)$$

After introducing the wave function $\psi = \sqrt{P} \times e^{iS/S_0}$, these equations can be recombined to yield a generalized Schrödinger equation,

$$\frac{1}{2} S_0^2 \, I^{jk} \partial_j \partial_k \psi + i \, S_0 \, \frac{\partial}{\partial t} \psi = \Phi \, \psi. \qquad (10.249)$$

Therefore this system would be expected to exhibit some quantum-type properties. Indeed, in the particular case $S_0 = \hbar$, we recover the standard Schrödinger equation of the quantum mechanical description of a rigid body which is used, e.g. for determining the rotational levels of molecules taken as a whole. In the macroscopic case, such an equation has been applied with encouraging results to the study of the probability distribution of the inclination and obliquity of chaotic astronomical bodies (see [365, 123] and Chapter 13).

10.6.8. *Application to laboratory experiments and new technologies*

These theoretical considerations can now be applied to the general conception of a new type of experiment that may, later, give rise to new "macroquantum" technologies. In these devices, a classical system describable by fluid mechanics-type equations would be led to exhibit a quantum-type behavior at macroscopic scales. Specifically, it would share the stabilizing properties of the solutions of the Schrödinger equation, or more generally of nonlinear Schrödinger equations [309, 283], in particular of the superfluid type [338, 407]. Such a system would benefit from the general existence of stationary solutions, which are obtained for quantized values of the conservative quantities, and depend on the limit conditions (instead of the initial conditions in the standard classical case). These solutions are generally characterized by the existence of probability peaks of the various variables that describe the system (positions, angles, . . .), i.e. by a tendency to structuring [362].

In these experiments, the constant would not be $\hbar$, which is still underlying known macroscopic quantum effects such as superfluidity, superconductivity, and Bose–Einstein condensates of cold atoms, but a macroscopic constant $S_0 = 2m\mathcal{D}$ which could be, in this framework, defined, changed and fine-tuned by the experimenter. One could even reverse the force, whose amplitude is proportional to S_0^2, and then observe the transformation of a self-organizing quantum-type process $(\mathcal{D}^2)$ into a weakly organized classical process $(\mathcal{D}^2 = 0)$, then into a desorganizing, entropy increasing, diffusion process $(-\mathcal{D}^2)$.

Note that such a proposal shares some similarities with that, made independently by Zak, for an expectation based intelligent control and reversible thermodynamics [548, 549]. Inspired just by the standard quantum potential itself, he has suggested to apply an "information potential" $-\alpha \ln \rho$ in order to stop and even reverse diffusion, but the reversal may reveal to be difficult and maybe impossible in the dissipative Langevin regime, see Sec. 10.4.3.2. Moreover, Zak suggested that more general potentials of that type could be at work in living systems, and he therefore tried to find them. This is consistent with our earlier proposal of application of the scale relativity approach to living systems [353, 373, 375, 377, 388], which naturally leads to a macroscopic quantum-type theory. But, in this context, the naturally controlling, regulating and self-organizing potential is known: it is but a macroscopic quantum-type potential.

The suggested experiment [403, 395] consists of applying to a classical system a potential that simulates a quantum potential, either directly (in the case when it is known theoretically or when a stationary regime has been reached), or by the mean of a three-step retroactive loop, involving a detector, a computer and an actuator:

(i) Detector: provide measurements with detectors of the quantity, which plays the role of the density in the continuity equation, which may be a matter density, a probability density, the height of the surface of a liquid, etc., $\rho(x, t) = \rho_{jn}$, at regular time interval $\{t_n\}$ on a grid at positions $\{x_j\}$.

(ii) Computer: compute the quantum potential $Q = -2\mathcal{D}^2\Delta\sqrt{\rho_n}/\sqrt{\rho_n}$ and the quantum force $(F_Q)_n = 2\mathcal{D}^2\nabla(\Delta\sqrt{\rho_n}/\sqrt{\rho_n})$ from these measurements at each time t_n.

(iii) Actuator: apply the new value of the potential or of the force to the system at each time t_n.

The continuous application of this loop would therefore simulate the presence of a quantum-type potential, so that the system would become describable by a Schrödinger equation.

We may also complete such an experiment by changing the sign of the amplitude $-\mathcal{D}^2$ of the quantum potential, which should lead to diffusion desorganization instead of quantum-type self-organization (see Sec. 10.4).

As we have already remarked, the advantage of such a proposal is that we are no longer constrained by the standard quantum value $\mathcal{D} = \hbar/2m$ that fixes the amplitude of the quantum force and that we can therefore give to it a macroscopic value, vary it, study its transition to zero (quantum to classical transition), etc.

Such a feedback loop involves: (1) a measurement device; (2) a computing device and (3) an actuating device. It is remarkable that this kind of feedback is typical of the fundamental functioning of living systems [388, 33, 402]. Indeed, even the simplest cell is able to direct itself in function of the gradient of some chemical quantity, which involves such a loop, even if it is at a still rough level.

Due to the large number of systems, which are describable by hydrodynamics equations, at least as an approximation, there is also a wide set of possible classical systems, which could be rendered (at least partially) quantum-like by this method. We stress once again the fact that this does not mean that they will share all the properties of genuine quantum systems of the microphysical scale. We mean here that they will be describable by a wave function ψ solution of a Schrödinger equation and such that $|\psi|^2$ yields a relevant measurable physical quantity.

Let us suggest a certainly non-exhaustive list.

(i) *Compressible fluids, gas or plasma.* In this case, the quantity to be measured and from which the quantum force is computed is the fluid matter density ρ, which is proportional to the probability density P. More generally, when the density is known to be related to another physical quantity, for example through a state equation like pressure or volume, but it may be temperature or velocity, it may be easier to measure this quantity and to derive the density to be inserted in the feedback loop.

A particularly interesting case could be the application to a plasma, for which the "detector" part (measurement of density) and "actuator" part (e.g. simulation of the quantum potential by using an electric field) of the retroactive loop would not be too hard to conceive with today's technology.

Numerical simulations of this kind of experiment applied to the example of an oscillatory nonspreading wave packet in an harmonic oscillator potental have given encouraging results, supporting the expected properties of self-organization, confinement and stabilization expected from the application of a quantum potential (see [395] and the following Sec. 10.6.9).

(ii) *Uncompressible fluids.* There are in this case several possibilities, among which:

(a) *Sound wave approximation.* In this case the quantum potential can be computed from the density variation $\rho' = \varrho - \varrho_0$ or from the pressure variation $p' = p - p_0 = \rho' c_s^2$, where c_s is the sound velocity in the fluid. Under this sound approximation, the quantum potential takes the simplified form

$$Q \approx -\frac{\mathcal{D}^2}{\varrho_0} \Delta \rho', \tag{10.250}$$

and the wave equation acquires a nonlinear term, i.e. it takes to lowest order a form already obtained in the study of superfluids [338],

$$\frac{\partial^2 \varrho'}{\partial t^2} - c_s^2 \Delta \varrho' = -\mathcal{D}^2 \Delta^2 \varrho'. \tag{10.251}$$

(b) *Gravity waves.* In the case of an uncompressible fluid in a basin, it is well-known that the variations of pressure involve variations of the height h of the fluid surface, which replaces the density in the continuity equation [269]. One may therefore conceive a two-dimensional experiment in which the quantum potential would be computed directly from the square root of this height, namely,

$$Q = -2\mathcal{D}^2 \frac{\partial^2 \sqrt{h}/\partial x^2 + \partial^2 \sqrt{h}/\partial y^2}{\sqrt{h}}. \tag{10.252}$$

A preliminary application to a simulation of quantum-like superfluid vortices has given positive results [406]. The application of the quantum force can be done by the action of a classical force at the surface (by wind, blower, an electromagnetic force for a charged liquid, varying pressure), by the action of current (the kinetic energy term $(1/2)v_z^2$ may play the role of a potential energy for the surface), or, more interestingly, one may simulate the quantum potential itself by a curved bottom (see [407] and the following Sec. 10.6.9).

(iii) *Mechanical systems.* In this case the quantum potential is computed from the measurement of a density of probability, while it is derived from the density of matter (or its equivalent) in fluid mechanics experiments.

   (a) *Translational motion.* The equations of an ideal fluid (Euler and continuity equation) are already currently used as an approximation of the behavior of a "gas" of particles, provided the particle mean free path is short and provided one can reduce the description of the interactions between the particles to a mean field approximation (plus possibly a pressure term). A typical example of this approach is the fluid description of the gas of galaxies in cosmology [438].

   In this case the suggested experiment consists of following the evolution of a large number of particles while subjecting each of them to a quantum-like potential computed in real time from the probability distribution of the whole ensemble.

   Another possibility, which would simulate the quantum mechanics of a single particle, consists in using a particle in chaotic motion and in computing the density of probability from many positions of the particle during a finite time interval.

   A third proposal, which would be another different simulation of the quantum mechanics of a single particle, consists in accompanying the single particle motion by a real time numerical simulation of a large number of virtual particles under the same conditions, then in computing the quantum potential from the density distribution of these virtual particles and finally in applying it to the real particle (then loop).

   (b) *Rotational motion.* As in the translational motion case, two main possible interpretations of the theoretical description tools and of their equations can be considered. These two interpretations lead, in their turn, to two different types of laboratory experiments and/or new technology with rotating bodies and, therefore, possibly, engines.

The first one consists of applying the description to an ensemble of bodies driven in a fluctuating rotational motion, for example in rotational Brownian motion [122], under equivalent conditions. In this case the density of probability, and therefore the quantum potential, can be determined at a given time from this ensemble.

The second possibility consists of applying the theory to a single body in fluctuating or chaotic rotational motion. In this case one may determine the density of probability and then deduce from it the quantum potential by measuring the values of the angles several times during a given time interval on which the body is expected to take all the possible values allowed by the probability density. In the case of a real laboratory experiment, this would involve three time scales, namely: (i) measurements of the angles on time intervals δt; (ii) determination of the probability density distribution from N_1 such measurements and calculation of the quantum potential on time intervals $\Delta t = N_1 \delta t$; (iii) repeated application of the quantum force on these time intervals during the total time of the experiment $t = N_2 \Delta t = N_1 N_2 \delta t$.

Each of these various cases, and yet other possible ones, will be considered in forthcoming specific works. We shall now describe in more detail two possible applications, one to the formation of a stable structure in a compressible fluid [395], and the other to the simulation of a superfluid-like quantized vortex in a finite height basin [407].

10.6.9. *Application to the oscillating wave packet*

10.6.9.1. *Introduction*

In the present section, we validate the concept of possible new macroscopic quantum-like experiment by numerical simulations of a fluid subjected to a generalized quantum force through a retroaction loop, as an anticipation of a future real laboratory experiment. The example chosen for this first attempt is the appearance of a non-spreading quantum-like oscillating wave packet in a compressible fluid (e.g. a plasma) subjected to an attractive harmonic oscillator potential. We find in these numerical simulations signatures of a quantum-like behavior, which are stable against various perturbations [395].

10.6.9.2. *The oscillating wave packet: theory*

As an example of application and as a preparation for a laboratory experiment, let us consider the simplified case of one-dimensional fluid motion in an external harmonic oscillator potential $\phi = (1/2)\omega^2 x^2$. This system is described by the two following equations:

$$\frac{\partial V}{\partial t} = -V\frac{\partial V}{\partial x} - \omega^2 x + 2D^2 \frac{\partial}{\partial x}\left(\frac{\partial^2 \sqrt{\rho}/\partial x^2}{\sqrt{\rho}}\right), \qquad (10.253)$$

$$\frac{\partial \ln \rho}{\partial t} = -\frac{\partial V}{\partial x} - V\frac{\partial \ln \rho}{\partial x}. \qquad (10.254)$$

Here we have written the continuity equation in terms of $\ln \rho$, which will be useful in the numerical simulations that follow. These two equations are equivalent to the one-dimensional generalized Schrödinger equation:

$$D^2 \frac{\partial^2 \psi}{\partial x^2} + iD\frac{\partial}{\partial t}\psi - \frac{1}{4}\omega^2 x^2 \psi = 0. \qquad (10.255)$$

It is well-known that it is possible to find a solution of this equation in the form of a wave packet whose center of gravity oscillates with the period of the classical motion and which shows no spreading with time [479, 474, 267]. Assuming that the maximal amount by which the center of gravity is displaced is a, the wave function reads in this case

$$\psi = \left(\frac{\omega}{2\pi D}\right)^{\frac{1}{4}} e^{-\frac{\omega}{4D}(x - a\cos\omega t)^2} \times e^{-i\left(\frac{1}{2}\omega t + \frac{\omega}{2D} ax\sin\omega t - \frac{\omega}{8D} a^2 \sin 2\omega t\right)}. \qquad (10.256)$$

Therefore the probability density reads

$$P = |\psi|^2 = \sqrt{\frac{\omega}{2\pi D}}\, e^{-\frac{\omega}{2D}(x - a\cos\omega t)^2}. \qquad (10.257)$$

This is an interesting case for a test of a genuine quantum behavior, since it involves a non vanishing phase in an essential way although this is a one-dimensional system. The velocity field is given by

$$V = -a\omega \sin\omega t, \qquad (10.258)$$

while the expression for the quantum potential is

$$Q(x, t) = D\omega - \frac{1}{2}\omega^2(x - a\cos\omega t)^2, \qquad (10.259)$$

so that the quantum force writes

$$F_Q = -\frac{\partial Q}{\partial x} = \omega^2(x - a\cos\omega t). \qquad (10.260)$$

Therefore the (varying) energy takes the form

$$E = \frac{1}{2}V^2 + \phi + Q = \mathcal{D}\omega + a\omega^2 x\cos\omega t - \frac{1}{2}a^2\omega^2\cos(2\omega t). \qquad (10.261)$$

When it is applied to the center of the packet $x = a\cos\omega t$, this expression becomes

$$E_c = \mathcal{D}\omega + \frac{1}{2}a^2\omega^2. \qquad (10.262)$$

We recognize in the second term, as expected, the energy of a classical pendulum. Concerning the first term, since standard quantum mechanics corresponds to the particular choice $\mathcal{D} = \hbar/2m$ (here with $m = 1$), the term $\mathcal{D}\omega$ is, very interestingly, nothing but the generalization of the vacuum energy for an harmonic oscillator, $\frac{1}{2}\hbar\omega$.

Therefore, we verify that the application of a quantum potential on the fluid has given to it some new properties of a quantum-like nature, such as a zero-point energy and the conservation of the shape of the wave packet.

10.6.9.3. *Toward a laboratory experiment*

In order to prepare a real laboratory experiment aiming at achieving such a new macroscopic quantum-like (super)fluid, we shall now present the result of numerical simulations of such an experiment. To this purpose these simulations are not based on the Schrödinger form of the equations, but instead on the classical Euler and continuity equations and on the application of a generalized quantum-like force.

As already remarked, the advantage of such a proposal is that one is no longer constrained by the standard quantum value $\mathcal{D} = \hbar/2m$ that fixes the amplitude of the quantum force, and that one can therefore give to it a macroscopic value, vary it, study its transition to zero (quantum to classical transition), etc.

10.6.9.4. *Iterative fitting simulation*

In this first simulation, we assume that the quantum force, which is a third derivative of the density, is not computed directly from the values of the density, but from a polynomial fit of the distribution of $\ln \rho$. In the special case considered here (the oscillating wave packet), we use a Gaussian fit of the density distribution (i.e. a second order polynomial fit to $\ln \rho$), so that we need to know only the mean and dispersion. More generally, one can decompose the distribution of $\ln \rho(x)$ into its successive moments. Therefore the density is written as

$$\rho_n(x) \propto \exp\left[-\frac{1}{2}\left(\frac{x - \bar{x}_n}{\sigma_n}\right)^2\right], \tag{10.263}$$

so that, once the mean and dispersion $\bar{x}_n$ and σ_n at time t_n are computed, the quantum force to be applied at each step (n) writes:

$$(F_Q)_n(x) = \frac{\mathcal{D}^2(x - \bar{x}_n)}{\sigma_n^4}. \tag{10.264}$$

Numerical simulation

The numerical simulation is performed by a simple Mathematica program, which reproduces the steps of the real experiment, specifically, at each time step t_n:

(i) It computes the mean and the dispersion of positions x according to the density distribution:

$$\bar{x} = \frac{\sum_j \rho(x_j)x_j}{\sum_j \rho(x_j)}, \tag{10.265}$$

$$\sigma^2 = \frac{\sum_j \rho(x_j)(x_j - \bar{x})^2}{\sum_j \rho(x_j)}. \tag{10.266}$$

(ii) It computes the force F_Q to be added in terms of these quantities as

$$(F_Q)_n(x) = \frac{\mathcal{D}^2(x - \bar{x}_n)}{\sigma_n^4 \, \delta x^3}, \tag{10.267}$$

where δx is the grid interval and intervenes here because we use finite differences.

(iii) It computes the logarithm of the density $\ln \rho$ and the velocity V at next time step t_{n+1} by transforming Eqs. (10.253, 10.254) into centered finite-difference equations (Forward Time Centered Space, FTCS scheme) using the Lax–Friedrichs method [454],

$$\ln \rho_j^{n+1} = \frac{\ln \rho_{j+1}^n + \ln \rho_{j-1}^n}{2} - \frac{\delta t}{2\,\delta x}\{(V_{j+1}^n - V_{j-1}^n)$$
$$+ V_j^n\,(\ln \rho_{j+1}^n - \ln \rho_{j-1}^n)\}, \tag{10.268}$$

$$V_j^{n+1} = \frac{V_{j+1}^n + V_{j-1}^n}{2} + \delta t\left(-V_j^n\,\frac{V_{j+1}^n - V_{j-1}^n}{2\,\delta x} + F_j^n + (F_Q)_j^n\right). \tag{10.269}$$

The lower index (j) is for space x and the upper one (n) is for time t; δt is the time step and $F(x) = -\omega^2 x$ is the external harmonic oscillator force. In the above Lax method, the terms $\ln \rho_j^n$ and V_j^n are replaced by their space average, which has the advantage to stabilize the FTCS scheme.

The initial conditions are given by the density distribution (Eq. 10.257) for $t = 0$.

Although this is a simple scheme in which we have not attempted at this stage to better control numerical error diffusion, it has given very encouraging results, since it has reproduced on several periods the expected motion of the quantum oscillating wave packet (see Fig. 10.13).

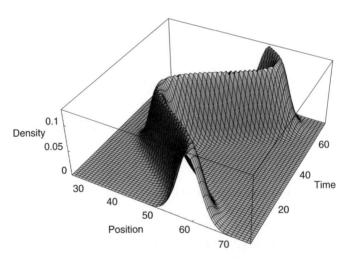

Fig. 10.13: **Fluid subjected to a quantum potential.** Result of the numerical integration of a Euler and continuity one-dimensional system with generalized quantum potential for the oscillating wave packet in an harmonic oscillator field. The quantum force applied on the fluid is calculated from a gaussian fit of the density distribution. The figure gives the density distribution obtained in function of position (space grid from 25 to 75) and time (time steps from 1 to 75, i.e. 1.2 period).

Perturbation of initial conditions

One of the possible shortcomings in the passage from the simulation to a real experiment may come from fluctuations in the initial conditions. Indeed, in the previous simulations, we have taken as initial density distribution that of the exact quantum wave packet. In order to be closer to a real experimental situation, we have therefore performed a new simulation similar to the previous one, but with an initial density distribution that is perturbed with respect to the Gaussian solution (Eq. 10.257): we have multiplied its values $\rho(x_j)$ at each point $\{x_j\}$ of the space grid by $\exp(\alpha)$, where α is random in the interval $[0, 1]$. A typical resulting initial density distribution is given in Fig. 10.14, followed by the distributions obtained on a full period (sub-figures 1 to 12) after application of the generalized quantum force.

Once again the result obtained is very enrouraging as concerns the possibility of performing a real laboratory experiment, since, despite the initial deformation, the wave packet remains stable during several periods. Moreover, not only the mean and dispersion of the evolving density distribution remain close to the ones expected for the quantum wave packet, but, as can be seen in Fig. 10.14, the initial perturbations have even been smoothed out.

General account of uncertainties

This encouraging result leads us to attempt a numerical simulation under far more difficult conditions: in order to simulate the various uncertainties and errors that may occur in a real experiment, in particular as concerns the density measurement, the application of the force, and physical effects not accounted in the simulation, such as pressure (see below), vorticity, etc., we have now added a fluctuation at each step of the retroactive loop. At each time step t_n, we have multiplied the density $\rho(x_j)$ at each point $\{x_j\}$ of the space grid by $\exp(\alpha)$, where α is random in the interval $[0, 1]$.

As can be seen in Fig. 10.15, despite the large errors added, the numerical simulation shows an oscillating wave packet, which, despite its large fluctuations, keeps its coherence. In particular, it keeps the values of the mean and dispersion (to about 5%) expected for the quantum solution on about $1/3$ of the period before the end of the simulation due to numerical errors.

Account of pressure

The account of pressure has been considered in Sec. 10.6.3. It leads to a nonlinear Schrödinger equation,

$$\mathcal{D}^2 \Delta\psi + i\mathcal{D}\frac{\partial}{\partial t}\psi - k_p \ln|\psi|\,\psi = \frac{1}{2}\phi\,\psi. \qquad (10.270)$$

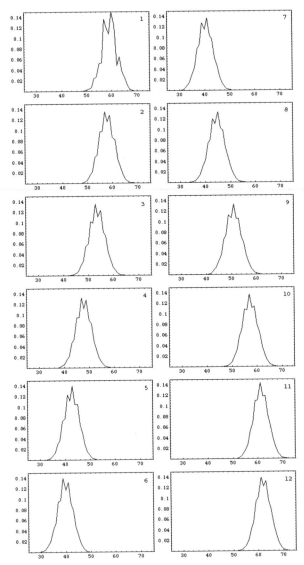

Fig. 10.14: **Fluid subjected to a quantum potential and to initial fluctuations.**
Result of the numerical integration of a Euler and continuity one-dimensional system with
added quantum potential for the oscillating wave packet. The conditions are the same as
in Fig. 10.13, except for the addition of a perturbation on the initial density distribution
(left top figure). The quantum force applied on the fluid is calculated from a Gaussian fit
of the density distribution. The successive figures give the density distribution obtained
in function of position (space grid from 25 to 75) and time (64 time steps corresponding
to one period, among which twelve of them, equally distributed, are shown).

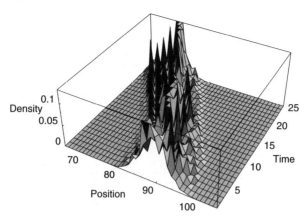

Fig. 10.15: **Fluid subjected to quantum potential and fluctuations.** The figure shows the result of the numerical integration of a Euler and continuity one-dimensional system with added generalized quantum potential for the oscillating wave packet. The conditions are the same as in Fig. 10.13, except for the addition of a perturbation on the density distribution at each time step of the simulation. The quantum force applied on the fluid is calculated from a Gaussian fit of the density distribution. The figure gives the density distribution obtained in function of position (space grid from 65 to 105) and time (time steps 1 to 25, which corresponds to almost half a period).

In the highly compressible case the dominant pressure term is rather $\propto \rho^2$, and the $\ln |\psi|$ term is replaced by $|\psi|^2$ in the nonlinear Schrödinger equation (see e.g. [338]).

The numerical integration is now performed by generalizing Eq. (10.269) as

$$
V_j^{n+1} = \frac{V_{j+1}^n + V_{j-1}^n}{2} + \delta t \left(-V_j^n \frac{V_{j+1}^n - V_{j-1}^n}{2\,\delta x} \right.
$$
$$
\left. + F_j^n + (F_Q)_j^n - k_p \frac{\ln \rho_{j+1}^n - \ln \rho_{j-1}^n}{2\delta x} \right). \qquad (10.271)
$$

The result is given in Figs. 10.16 and 10.17 for two different values of the pressure amplitude k_p. One finds that the addition of pressure leads to a small oscillatory spreading of the wave packet, but that its main superfluid-like features are preserved, since it nearly recovers its shape after half a period.

10.6.9.5. *Full finite differences simulation*

The success of this first simple simulation leads us to attempt a more direct feedback in which the quantum force is computed by finite differences from

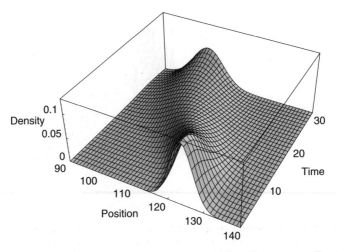

Fig. 10.16: **Fluid subjected to quantum potential and pressure.** Result of the numerical integration of a Euler and continuity one-dimensional system of equations with added generalized quantum potential and account of a pressure term, for the oscillating wave packet. The quantum force applied on the fluid is calculated from a Gaussian fit of the density distribution. The figure gives the probability density in function of position (space grid from 90 to 140) and time (time steps from 1 to 32). In this simulation (near half a period), the amplitude of the pressure term is $k_p = 5$.

the values of the density itself (while in the previous simulation we used an intermediate polynomial fit from which the force was analytically derived).

To this purpose we use a form of the generalized quantum potential and of the generalized quantum force according to which they can be expressed in terms of only $\nabla \ln P$ (or equivalently $\nabla \ln \rho$). Setting

$$H = \nabla \ln P, \tag{10.272}$$

we find:

$$Q = -\mathcal{D}^2 \left(\nabla \cdot H + \frac{1}{2} H^2 \right), \tag{10.273}$$

$$F_Q = -\nabla Q = \mathcal{D}^2 [\Delta H + (H \cdot \nabla) H]. \tag{10.274}$$

In one dimension, it reads

$$F_Q = \mathcal{D}^2 \left(\frac{\partial^3 \ln P}{\partial x^3} + \frac{\partial^2 \ln P}{\partial x^2} \frac{\partial \ln P}{\partial x} \right). \tag{10.275}$$

The numerical integration proceeds following the same lines as in the previous simulation, except for the first steps aiming at computing F_Q, which

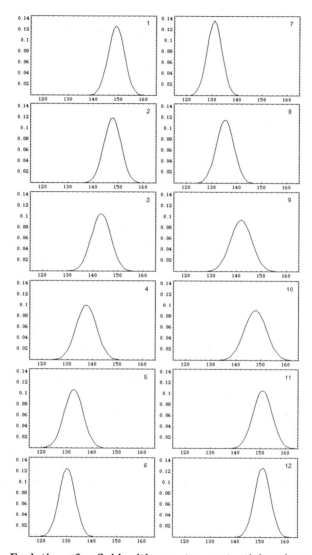

Fig. 10.17: **Evolution of a fluid with quantum potential and pressure.** The figure shows the result of the numerical integration of a Euler and continuity one-dimensional system of equations with added generalized quantum potential and account of a pressure term, for the oscillating wave packet. The quantum force applied on the fluid is calculated from a Gaussian fit of the density distribution. The figure gives the density distribution in function of the position (space grid from 120 to 160), for 12 equal time steps which cover a full period. In this simulation, the amplitude of the pressure term is $k_p = 1$. One sees that the effect of pressure amounts to an oscillating spreading of the wave packet, which nearly recovers its shape after half a period.

are replaced by a finite difference calculation according to Eq. (10.274). We calculate successively, for all values of the position index j,

$$H_j^n = \frac{\ln \rho_{j+1}^n - \ln \rho_{j-1}^n}{2 \, \delta x}, \qquad (10.276)$$

then similar relations for positions x_{j-1}, x_{j+1}, x_{j-2} and x_{j+2}, then

$$Q_{j-1}^n = -\mathcal{D}^2 \left\{ \frac{H_j^n - H_{j-2}^n}{2 \, \delta x} + \frac{1}{2}(H_{j-1}^n)^2 \right\}, \qquad (10.277)$$

then a similar relation for $Q(x_{j+1}, t_n)$, and finally

$$(F_Q)_j^n = \frac{Q_{j-1}^n - Q_{j+1}^n}{2 \, \delta x}. \qquad (10.278)$$

The calculation of $\ln \rho$ (from the continuity equation) and of V (from the Euler equation) are the same as previously. We have attempted to use other more precise formulas for the calculation of the second and third order derivatives in the expression of F_Q: this has led to essentially the same result.

Despite, once again, the roughness of the chosen integration method, the result obtained is satisfactory, since the motion of a quantum non-spreading oscillating wave packet has been reproduced on about $1/4$ of period before divergence due to the effect of computing errors (Fig. 10.18). This result has been obtained without using the Schrödinger equation, but instead an apparently "classical" hydrodynamic Euler/continuity system with an externally applied generalized quantum potential.

Adding a pressure term yields a similar result, i.e. reproduction of the motion of the wave packet on about $1/4$ of period before divergence due to the effect of computing errors, which confirms the result obtained with the Gauss fitting method, namely, a partial oscillatory spreading of the wave packet (Fig. 10.19).

10.6.9.6. *Discussion*

These preliminary simulations were intended to yield a first validation of the concept of a new kind of quantum-like macroscopic experiments based on the application to a classical system of a generalized quantum force [395, 406]. They have given positive results, since the expected quantum-type stable structure (here a non-spreading or weakly spreading oscillating wave packet) has been obtained during a reasonably long time of integration. These results, obtained by a rather simple integration method, are very

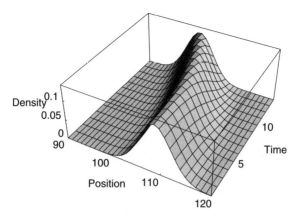

Fig. 10.18: **Fluid subjected to quantum potential (finite differences).** Result of the numerical integration of a Euler and continuity one-dimensional system with generalized quantum potential for the oscillating wave packet in an harmonic oscillator field. The quantum force applied on the fluid is directly calculated from the values of the density by finite differences. The density distribution obtained in function of position (space grid from 90 to 120) is shown on about 1/4 of period (14 time steps).

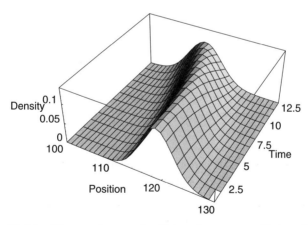

Fig. 10.19: **Fluid with quantum potential and pressure (finite differences).** Result of the numerical integration of a Euler and continuity one-dimensional system with generalized quantum potential for the oscillating wave packet in an harmonic oscillator field and account of pressure (1/4 of period before stop due to computing errors). The quantum force applied on the fluid is here directly calculated from the values of the density by finite differences. The figure gives the density in function of space position (grid from 100 to 130) and time (time steps from 1 to 13). A pressure term has been added ($k_p = 1$), whose effect is a small oscillatory spreading of the wave packet.

encouraging since they give the hope that a real laboratory experiment should be possible to achieve.

In the hydrodynamic case considered in this work, possible shortcomings are to be considered, such as the effects of finite compressibility, of vorticity, of viscosity at small scales, etc.

Here, we have attempted to obtain the first evaluation of these uncertainties by taking a pressure term into account, by adding large random fluctuations in the initial conditions, then by adding large fluctuations at each time steps of the simulation. The results obtained were again encouraging, since, despite the pressure term and the large fluctuations, the overall coherence of the wave packet and its period were preserved.

The present section was devoted to the description of a specific experiment restrained to a potential fluid. Concerning the question of vorticity, it is too wide to be discussed in detail here. Let us simply remark that adding a quantum potential to a fluid in potential motion is not expected to introduce vorticity. On the contrary, we hope that, in some situations, it will increase the irrotational part of a fluid. The example of the transition from a classical vortex to a quantum-like vortex is enlightening in this regard. Indeed, the classical vortex is potential in its outer region and rotational in the inner region which rotates like a solid body, while the quantum vortex (see e.g. [173]) is everywhere potential. We therefore expect that in the new experiment the application of a quantum-type potential will increase the size of the outer irrotational zone. A full treatment of the vorticity, by using a vectorial-field Schrödinger-like equation (Sec. 10.6.4) or more generally quaternionic wave functions (see [97] for their application to some particular cases of rotational flows) will be the subject of future works.

10.6.10. *Application to gravity waves and quantized vortex at the surface of a fluid*

Let us give another example of a possible macroscopic quantum-type experiment. We apply this concept to gravity waves at the surface of an incompressible fluid in a basin of finite height, with particular emphasis on the quantized votex [407]. We identify the height of the fluid in the basin with the square of the modulus of a complex wave function, which is solution of a nonlinear Schrödinger equation typical of superfluids. The quantum potential itself is therefore computed in terms of the fluid height.

We suggest three methods for applying this quantum potential to the fluid: (i) by the action of a force on the surface (pressure variations by wind or blower, electromagnetic force for a charged liquid); (ii) by a vertical current; (iii) by a curvature of the basin bottom. The effects of a curved bottom, of pressure or of a vertical current yield the quantum potential itself, while usually only the quantum force is accessible, so that such an experiment is expected to provide one with a macroscopic model of a quantum-type vacuum energy.

In addition, these results may provide explanations for the origin of some freak waves, which have already been satisfactorily described by nonlinear Schrödinger equations without clear understanding of the reason of this agreement [418]. Under this new interpretation, they would be the result of a local and rare generation of a quantum-like potential by either the wind force or vertical currents or a curved bottom or a combination of these effects [407].

10.6.10.1. *Gravity waves in a basin (flat bottom)*

Schrödinger-type equation

At first sight, gravity surface waves could be considered as unable to come under such a quantum potential description, since they correspond to the case of an incompressible fluid. The constancy of the density ρ prevents in this case to define a wave function from it (since ρ is expected to be the square of its modulus) and to define a quantum potential (since it is given by second derivatives of ρ).

However, we shall show that completely similar equations can be obtained for gravity waves by replacing the density by the height profile of the fluid surface. This allows us to suggest here a laboratory experiment in which a quantum force would be applied to the surface of the fluid and computed from the shape of this surface itself, thus forcing a superfluid-type behavior of the gravity waves. In some respect, e.g. the wave behavior, this case is actually very similar to the case of sound (that we shall consider in a separate work), except for its two-dimensional character. But it has also the advantage to allow the possibility of fluctuations of the surface which are not small with respect to the height of the basin, and which may even reach $h = 0$, therefore simulating with an uncompressible fluid the case of a perfectly compressible fluid.

We consider an incompressible fluid in potential (irrotational) motion, with free surface. A variation of internal pressure cannot yield a variation of density, so that it manifests itself in terms of a variation of its level in the basin. This implies a continuity equation in which the density is replaced by the fluid height h [269], namely,

$$\frac{\partial h}{\partial t} + \frac{\partial}{\partial x}(h\, v_x) + \frac{\partial}{\partial y}(h\, v_y) = 0. \tag{10.279}$$

We also assume that the velocities v_x and v_y do not depend on the z coordinate and that they derive from a potential φ:

$$v_x = v_x(x, y, t) = \frac{\partial \varphi}{\partial x}, \quad v_y = v_y(x, y, t) = \frac{\partial \varphi}{\partial y}. \tag{10.280}$$

Therefore the Euler equations for the two-dimensional velocity read

$$\frac{\partial v_x}{\partial t} + v_x \frac{\partial v_x}{\partial x} + v_y \frac{\partial v_x}{\partial y} = -\frac{1}{\rho}\frac{\partial p}{\partial x} - \frac{\partial Q}{\partial x}, \tag{10.281}$$

$$\frac{\partial v_y}{\partial t} + v_x \frac{\partial v_y}{\partial x} + v_y \frac{\partial v_y}{\partial y} = -\frac{1}{\rho}\frac{\partial p}{\partial y} - \frac{\partial Q}{\partial y}, \tag{10.282}$$

where Q is the quantum potential to be applied to the fluid. Its expression will be specified in the following. These equations can be completed by the energy equation (integral of the Euler equations):

$$\frac{\partial \varphi}{\partial t} + \frac{1}{2}(v_x^2 + v_y^2 + v_z^2) + Q + gz = -\frac{p}{\rho}, \tag{10.283}$$

where g is the gravity acceleration. At the surface of the fluid, the pressure is constant (given by the atmospheric pressure for an open basin), so that the right-hand side of this equation is a constant $-p_0/\rho$, which can be set to zero by a redefinition of the potential. Therefore, the energy equation reads on the surface $z = h(x, y, t)$

$$\frac{\partial \varphi}{\partial t} + \frac{1}{2}(v_x^2 + v_y^2 + v_z^2) + Q + gh = 0. \tag{10.284}$$

Assuming that $v_z = v_z(x, y, t)$, we can now separate the (x, y) behavior and the vertical behavior (z) by combining the gravity term and vertical velocity term under the form of a potential energy term

$$\phi(x, y, t) = gh + \frac{1}{2}v_z^2, \tag{10.285}$$

so that the two-dimensional continuity and energy equations now read

$$\frac{\partial h}{\partial t} + \frac{\partial}{\partial x}(h\, v_x) + \frac{\partial}{\partial y}(h\, v_y) = 0, \tag{10.286}$$

$$\frac{\partial \varphi}{\partial t} + \frac{1}{2}(v_x^2 + v_y^2) + Q + \phi = 0. \tag{10.287}$$

Therefore, by taking the derivative of the energy equation we find that the Euler equations for the two-dimensional velocity finally take the form

$$\frac{\partial v_x}{\partial t} + v_x \frac{\partial v_x}{\partial x} + v_y \frac{\partial v_x}{\partial y} = -g \frac{\partial \Phi}{\partial x} - \frac{\partial Q}{\partial x}, \tag{10.288}$$

$$\frac{\partial v_y}{\partial t} + v_x \frac{\partial v_y}{\partial x} + v_y \frac{\partial v_y}{\partial y} = -g \frac{\partial \Phi}{\partial y} - \frac{\partial Q}{\partial y}. \tag{10.289}$$

We recognize here the same form of the continuity and Euler equations that allowed us to combine them into a Schrödinger equation, but now with the density of matter ρ replaced by the basin height h.

This leads us to introducing a quantum-type potential of a new kind, *as a function of $\sqrt{h}$* instead of $\sqrt{\rho}$, given by

$$Q = -2\mathcal{D}^2 \frac{\Delta_2 \sqrt{h}}{\sqrt{h}}, \tag{10.290}$$

where $\Delta_2 = \partial^2/\partial x^2 + \partial^2/\partial y^2$. We are brought back to exactly the situation in which the real continuity equation and energy equation may be combined in terms of a unique complex Schrödinger-type equation. Therefore, we define a wave function $\psi(x, y, t)$ whose modulus is now the square root of the surface profile while its phase is related to the two-dimensional potential of velocity, as

$$\psi = \sqrt{h} \times e^{i\varphi/2\mathcal{D}}, \tag{10.291}$$

which is a solution of a generalized nonlinear Schrödinger equation,

$$\mathcal{D}^2 \Delta_2 \psi + i\mathcal{D} \frac{\partial \psi}{\partial t} = \frac{\phi}{2} \psi = \frac{1}{2} \left(g\,h + \frac{1}{2} v_z^2 \right) \psi. \tag{10.292}$$

The vertical velocity term is non negligible only when it is larger than a critical value given by

$$g\,h = \frac{1}{2} \left(\frac{dh}{dt} \right)^2, \tag{10.293}$$

which is but the free fall equation, of solution

$$h_{\text{crit}} = \frac{1}{2} g\,t^2, \quad (v_z)_{\text{crit}} = \pm g\,t. \tag{10.294}$$

It can therefore be neglected in many real experimental situations, so that we obtain the equation:

$$\mathcal{D}^2 \Delta_2 \psi + i\mathcal{D} \frac{\partial \psi}{\partial t} - \frac{1}{2} g|\psi|^2 \, \psi = 0. \tag{10.295}$$

This equation is a two-dimensional analog of the standard nonlinear Schrödinger equation of a superfluid (see e.g. [338, 173]). But this quantum fluid-type behavior can now be macroscopic from the very beginning, since the parameter $\mathcal{D}$ is no longer constrained to its standard quantum mechanical value $\mathcal{D} = \hbar/2$ as in the real superfluids known up to now.

This does not mean that all quantum properties of a genuine superfluid will manifest themselves in such an experiment. In particular, one does not expect to observe the property of superfluidity itself (i.e. vanishing viscosity), since the viscosity of the classical fluid is unaffected. But some quantum-like properties like quantization, increase of the zone of potential motion and possibly decrease of the effective viscosity, can nevertheless be expected.

Generation of a quantum potential

We have assumed that the fluid was subjected to a new form of quantum-like potential depending on the surface height, $Q = -2\mathcal{D}^2 \, \Delta_2 \sqrt{h}/\sqrt{h}$. How can we generate such a potential (or equivalently the corresponding force $F_Q = -\nabla Q$)? Let us come back to the two-dimensional energy equation, *a priori* without quantum potential. It can be decomposed in two parts:

$$\left(\frac{\partial \varphi}{\partial t} + \frac{1}{2}v_x^2 + \frac{1}{2}v_y^2 \right) + \left(\frac{1}{2}v_z^2 + gh + \frac{p}{\rho} \right) = 0. \tag{10.296}$$

The pressure term was previously assumed to be constant on the surface, but we may consider the more general situation of a variable pressure due to wind or a blower. The left part is the energy term, which participates in the construction of the two-dimensional Schrödinger equation, the right part plays the role of a potential energy. This expression therefore offers us three ways to generate a two-dimensional quantum potential: (1) by a vertical current, which would be such that $Q = v_z^2/2$; (2) by a variation of the fluid height created by a curved bottom, $Q = gh_b$ and (3) by a variation of pressure due to natural or artificial wind, $Q = p/\rho$.

When it is generated by the vertical kinetic energy term $Q = v_z^2/2 = -2\mathcal{D}^2\Delta_2\sqrt{h}/\sqrt{h}$ +cst and when the pressure is constant on the surface, we obtain Eq. (10.295) now in an exact way. We shall particularly develop here the curved bottom case, letting the other possibilities open to future studies.

Wave equation in the linear case

Let us briefly consider the linearized case when the amplitude of the wave remains small compared with the height of the basin. We assume that the bottom of the basin is flat, that the average height of the fluid is $h_0 = $ cst, and that the effective height at a point of coordinate (x, y) is

$$h(x, y, t) = h_0 + \zeta(x, y, t), \qquad (10.297)$$

with $\zeta \ll h_0$. In this case the quantum potential can be simplified to lowest order as

$$Q = -\frac{\mathcal{D}^2}{h_0} \Delta \zeta(x, y, t), \qquad (10.298)$$

which amounts to a two-dimensional Laplacian Δ_2, while the wave function becomes

$$\psi = \sqrt{h_0} \left(1 + \frac{1}{2} \frac{\zeta}{h_0} \right) \times e^{i\varphi/2\mathcal{D}}. \qquad (10.299)$$

In this case the equation of motion can be given the "classical" form of a wave equation, but including a source term coming from the quantum potential.

Before deriving this wave equation, let us first make a remark. Accounting for the fact that, at the surface of the fluid, we have

$$\frac{\partial h}{\partial t} = v_z = \frac{\partial \varphi}{\partial z}, \qquad (10.300)$$

and taking the derivative of the energy equation, we obtain (on the fluid surface):

$$\frac{\partial^2 \varphi}{\partial t^2} + g \frac{\partial \varphi}{\partial z} = -\frac{1}{2} \frac{\partial}{\partial t}(v^2) - \frac{\partial Q}{\partial t}. \qquad (10.301)$$

This shows that, contrary to the acoustic case approximation, there is no standard wave equation for the velocity potential.

However, a generalized wave equation can be constructed for the surface profile. Indeed, the continuity equation takes for $\zeta \ll h_0$ the simplified form

$$\frac{\partial \zeta}{\partial t} + h_0 \left(\frac{\partial v_x}{\partial x} + \frac{\partial v_y}{\partial y} \right) = 0. \qquad (10.302)$$

Taking its time derivative and introducing the velocity potential, we obtain

$$\frac{\partial^2 \zeta}{\partial t^2} + h_0 \Delta_2 \left(\frac{\partial \varphi}{\partial t} \right) = 0. \qquad (10.303)$$

Now the velocity potential is given by the energy equation

$$\frac{\partial \varphi}{\partial t} = -\left(\frac{1}{2} v^2 + Q + gh \right), \qquad (10.304)$$

so that we finally obtain the wave equation of a classical wave of propagation velocity $U = \sqrt{gh_0}$, but with an additional nonlinear term,

$$\frac{\partial^2 \zeta}{\partial t^2} - gh_0 \, \Delta\zeta = \Delta\left(\frac{v^2}{2} + Q\right). \tag{10.305}$$

By neglecting the v^2 term and by replacing the quantum potential by its linearized expression Eq. (10.298), it finally becomes

$$\frac{\partial^2 \zeta}{\partial t^2} - gh_0 \, \Delta\zeta = -\frac{\mathcal{D}^2}{h_0}\Delta^2\zeta. \tag{10.306}$$

This equation is structurally identical (after the reduction from three to two dimensions and the replacement of the density ρ by the surface profile ζ) to the wave equation directly obtained by linearization from the nonlinear Schrödinger equation of a superfluid [338].

10.6.10.2. *Application to a macroquantum vortex (flat bottom)*

A particularly relevant example of application of these ideas consists of applying them to the rotation of a fluid in a bucket. We shall now show that it is possible to transform a classical vortex into a "macroquantum" superfluid-like vortex by application of a quantum-like force, e.g. by the action of a wind at the surface.

Let us compare in detail the quantum-like stationary vortex with the classical vortex (see also [338]).

Classical vortex

Height-velocity relation

The Euler equations read in cylindric coordinates, in the case when the velocity field has no radial component, $v = v_\theta(r)$ and $\partial v_\theta/\partial t = 0$ (stationary flow),

$$\frac{v_\theta^2}{r} = \frac{1}{\rho}\frac{\partial p}{\partial r}, \tag{10.307}$$

$$0 = \frac{1}{\rho}\frac{\partial p}{\partial z} + g. \tag{10.308}$$

They are solved in terms of a pressure expression that reads

$$p = p_0 + \rho \int \frac{v_\theta^2}{r}\, dr - \rho g z. \tag{10.309}$$

Since the pressure is constant at the surface, one expects a surface profile given by

$$h = h_0 + \frac{1}{g} \int \frac{v_\theta^2}{r} \, dr. \tag{10.310}$$

Reversely, by derivating this expression with respect to r, we obtain a general relation for the velocity in terms of the profile,

$$v_\theta^2(r) = g \, r \, \frac{\partial h}{\partial r}. \tag{10.311}$$

Height and velocity field

The classical vortex is actually potential only in its outer regions. Let us call r_m the limit between the inner region where solid rotation $\propto r$ occurs and the outer region with rotation velocity $\propto 1/r$, and R the bucket radius. Let us also assume that the fluid rotates in a bucket, which is itself in rotation with a velocity $v_R = \Omega r_m^2 / R$. This choice has the advantage to preserve the exact $1/r$ rotation, as in an infinite radius bucket at rest. In the case when the rotation is such that the height is zero at the center, the velocity and the profile are in the inner region

$$v_\theta(\text{in}) = \Omega r, \quad h(\text{in}) = \frac{\Omega^2}{2g} r^2. \tag{10.312}$$

In the external region the motion is potential and such that

$$v_\theta(\text{ex}) = \frac{\Omega r_m^2}{r}, \quad h(\text{ex}) = h_0 \left\{ 1 - \left(\frac{r_0}{r} \right)^2 \right\}. \tag{10.313}$$

We verify that $v_\theta = \Omega r_m$ on the matching radius r_m from both the inner and outer relations. The radius r_0, which appears in the external profile corresponds to the extrapolation of this profile to zero heigth, $h = 0$. The matching condition yields

$$r_m^2 = 2 \, r_0^2. \tag{10.314}$$

The external profile gives the meaning of h_0. Namely, it is the fluid height extrapolated at infinite radius, so that it is given from the liquid height h_R at the bucket radius R by the relation

$$h_0 = \frac{h_R}{1 - (r_0/R)^2}. \tag{10.315}$$

The circulation $\Gamma = 2\pi\Omega r_m^2$ is given in this case ($h = 0$ at center) by

$$\Gamma = 4\pi\Omega r_0^2 = \frac{2\pi g h_0}{\Omega}, \tag{10.316}$$

so that one obtains the relation

$$r_0^2 = \frac{g h_0}{2\Omega^2}. \tag{10.317}$$

By matching the inner and outer profiles at radius r_m, one finds that $h_m = h_0/2 = \Omega^2 r_0^2/g$, which yields again the same relation. The corresponding profile is given in Fig. 10.20.

Quantum-like vortex

Generalized height-velocity relation

Let us now assume that we add a quantum potential Q, which is a function of r only, calculated from the surface profile and acting on the surface itself, e.g. by the action of the wind or of a blower or any other mean — electromagnetic field on a charged fluid, etc. Let us set

$$f = \sqrt{\frac{h}{h_0}}, \tag{10.318}$$

where h_0 is a reference height for the fluid in the basin (its precise definition will emerge in the following). The quantum potential reads

$$Q = -2\mathcal{D}^2 \frac{\Delta f}{f} = \frac{-2\mathcal{D}^2}{f}\left(\frac{\partial^2 f}{\partial r^2} + \frac{1}{r}\frac{\partial f}{\partial r}\right). \tag{10.319}$$

We still consider the purely circular velocity case. The Euler equations with quantum potential become

$$\frac{v_\theta^2}{r} = \frac{1}{\rho}\frac{\partial p}{\partial r} + \frac{\partial Q}{\partial r}, \tag{10.320}$$

$$\frac{1}{\rho}\frac{\partial p}{\partial z} + g = 0, \tag{10.321}$$

i.e.

$$\frac{v_\theta^2}{r} = \frac{1}{\rho}\frac{\partial p}{\partial r} - 2\mathcal{D}^2\frac{\partial}{\partial r}\left\{\frac{1}{f}\left(\frac{\partial^2 f}{\partial r^2} + \frac{1}{r}\frac{\partial f}{\partial r}\right)\right\}, \tag{10.322}$$

$$\frac{\partial p}{\partial z} = -g\rho. \tag{10.323}$$

This equation can be integrated as $p/\rho = p_0/\rho + g(h - z)$ since the pressure is constant on the free surface, so that one obtains, at the surface $z = h(r)$, the relation

$$\frac{1}{\rho}\frac{\partial p}{\partial r} = g\frac{\partial h}{\partial r}. \tag{10.324}$$

We can therefore easily generalize to the quantum potential case the previous classical height-velocity relation Eq. (10.311), namely,

$$v_\theta^2(r) = g\,r\frac{\partial}{\partial r}\left\{h - \frac{2\mathcal{D}^2}{g\sqrt{h}}\left(\frac{\partial^2\sqrt{h}}{\partial r^2} + \frac{1}{r}\frac{\partial\sqrt{h}}{\partial r}\right)\right\}. \tag{10.325}$$

Superfluid-like height and velocity field

Assume now that the fluid remains potential. In the case of a single vortex this implies that $v_\theta \propto 1/r$ (see e.g. [218]). Therefore the velocity reads

$$v_\theta = \frac{1}{r}\frac{\partial S}{\partial\theta} = \frac{2\mathcal{D}l}{r}, \tag{10.326}$$

with $S = 2\mathcal{D}\,l\,\theta$. Since $v_\theta^2/r = -\partial(2\mathcal{D}^2l^2/r^2)/\partial r$, one obtains, after integration, the equation

$$\frac{\partial^2 f}{\partial r^2} + \frac{1}{r}\frac{\partial f}{\partial r} + \left(\frac{1}{a^2} - \frac{l^2}{r^2}\right)f - \frac{gh_0}{2\mathcal{D}^2}f^3 = 0, \tag{10.327}$$

which has solutions only for integer values of l, i.e. for quantized values of the velocity. In this equation a is an integration constant whose value can be determined from the asymptotic infinite limit of the vortex. Indeed, when $r \to \infty$, we have by definition $h \to h_0$ so that $f \to 1$, while $\partial f/\partial r \to 0$ and $\partial^2 f/\partial r^2 \to 0$. Therefore we obtain

$$a^2 = \frac{2\mathcal{D}^2}{gh_0}. \tag{10.328}$$

Let us compare this equation with that of a vortex in a real quantum superfluid [419, 174, 211, 448]. A liquid Helium II superfluid can be described, under certain conditions, by a nonlinear Schrödinger equation

$$-(\hbar^2/2m)\nabla^2\psi + V_0|\psi|^2\psi = \mu\psi, \tag{10.329}$$

where V_0 characterizes a short range repulsive potential and μ is the chemical potential. The presence of a single quantized vortex in the fluid is characterized by a wave function of the type

$$\psi = f_l(r)\, e^{il\theta}, \tag{10.330}$$

where (r, θ) are the appropriate cylindrical coordinates and l an integer. The function $f_l(r)$ is then a solution of the equation [248]

$$\frac{\partial^2 f}{\partial r^2} + \frac{1}{r}\frac{\partial f}{\partial r} + \left(\frac{1}{a_s^2} - \frac{l^2}{r^2}\right) f - \frac{1}{a_s^2} f^3 = 0. \tag{10.331}$$

In this equation the fundamental distance scale, which governs the vortex scale is but the de Broglie length a_s, which is given by [248]

$$a_s^2 = \frac{\hbar^2}{2m\mu}. \tag{10.332}$$

Therefore the macroquantum vortex Eq. (10.327) has exactly the same form as this real quantum superfluid vortex equation. The distance scale a given by the relation $a^2 = 2\mathcal{D}^2/gh_0$, which characterizes the scale of the vortex, can be therefore identified, as expected, as a generalized de Broglie length. Indeed, we recover standard quantum mechanics in the special case $\mathcal{D} = \hbar/2m$, so that, for $m = 1$ which is relevant here, the gravity potential gh_0 in the expression of the de Broglie length

$$a = \mathcal{D}\sqrt{\frac{2}{gh_0}} \tag{10.333}$$

plays a role similar to that of the chemical potential μ in the real superfluid case.

The square of the parameter $\mathcal{D}$ characterizes the amplitude of the macroquantum potential Q. Reversely, once fixed the horizontal scale a of the experiment and its vertical scale h_0, the effective value of $\mathcal{D}$ is fixed to $\mathcal{D} = a\sqrt{gh_0/2}$. We shall therefore be able to simulate some of the properties of a genuine quantum vortex (usually constrained by the microscopic value of $\hbar$) in a now macroscopic experiment.

The quantum vortex profile is given by the solution of Eq. (10.327). In the central region, the solution is a Bessel function $f(r) \propto J_l(r/r_0)$ [248]. We give in Fig. 10.20 the result of a numerical integration of Eq. (10.327) in the case $l = 1$, which agrees with those of Kawatra [248] and Nore [338].

The comparison between the quantum-like and the classical vortex profiles (Fig. 10.20) shows significant differences. The quantum height profile is higher than the classical one in the central region and lower in the outer region, and they differ in both regions by a relative value that reaches $\approx 20\%$.

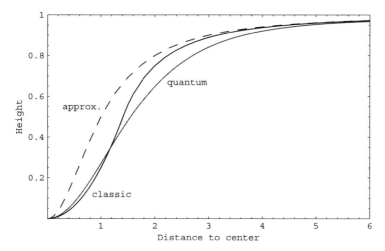

Fig. 10.20: **Classical and quantum vortex height profiles.** Comparison between the height profiles of a quantum superfluid vortex, its classical counterpart, and Fetter's approximation [173] to the quantum vortex (dashed line). The radial distance to the center r is in unit of $a = r_0$ (see text).

The quantum vortex velocity is given by $v_\theta(r) = 2\mathcal{D}l/r$ everywhere, and it is therefore quantized (see Fig. 10.21). The continuation of an $1/r$ velocity, i.e. of a potential motion in the inner region of the vortex and its quantization achieve two additional signatures of a quantum superfluid-like behavior, which can be put to the test in a real experiment, even though, due to viscosity, one does not expect it to be achieved for all values of the radius, contrary to the genuine quantum superfluid.

Quantum force

Let us finally give the expression of the quantum potential (see Fig. 10.22), and, therefore, that of the quantum force acting on the fluid surface. They can be directly established from Eq. (10.327) as a function of the surface profile. Since $2\mathcal{D}^2/a^2 = gh_0$, we obtain

$$Q = \frac{-2\mathcal{D}^2}{f}\left(\frac{\partial^2 f}{\partial r^2} + \frac{1}{r}\frac{\partial f}{\partial r}\right) = 2\mathcal{D}^2\left(\frac{1}{a^2} - \frac{l^2}{r^2}\right) - gh = gh_0\left(1 - l^2\frac{a^2}{r^2} - \frac{h}{h_0}\right),$$
(10.334)

and

$$F_Q = -\frac{\partial Q}{\partial r} = \frac{-4\mathcal{D}^2 l^2}{r^3} + g\frac{\partial h}{\partial r} = gh_0\left(-\frac{2a^2 l^2}{r^3} + \frac{1}{h_0}\frac{\partial h}{\partial r}\right).$$
(10.335)

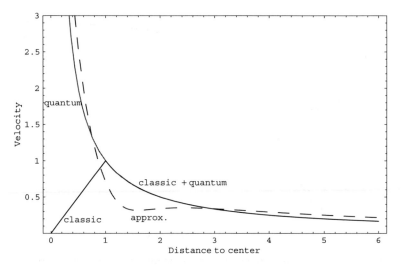

Fig. 10.21: **Classical and quantum vortex velocity fields.** Comparison between the velocities of a quantum superfluid vortex, its classical counterpart, and Fetter's approximation [173] to the quantum vortex (dashed line).

We note that the existence of the quantum potential cancels the gravity term gh in the total potential energy, which allows the velocity to remain everywhere proportional to $1/r$ in the quantum-type case.

10.6.10.3. *Generalization to a curved bottom*

Simulation of a quantum potential by a curved bottom

In the previous study, we have assumed that the bottom of the basin containing the liquid was flat. Let us now consider the more general case of a basin with a non-flat bottom.

Let $h_b = h_b(r, \theta, t)$ denote the height profile of the basin bottom, while $h_s(r, \theta, t)$ denotes the height of the fluid surface (measured with respect to a constant level). In this case the continuity equation no longer holds in terms of the visible surface height h_s, but instead in terms of the fluid height,

$$h_q = h_s - h_b. \tag{10.336}$$

It reads

$$\frac{\partial h_q}{\partial t} + \frac{\partial}{\partial x}(h_q \, v_x) + \frac{\partial}{\partial y}(h_q \, v_y) = 0. \tag{10.337}$$

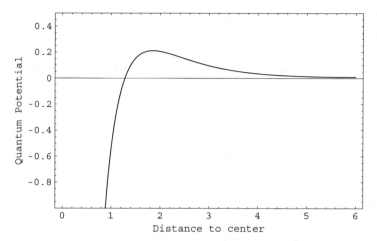

Fig. 10.22: **Form of the quantum potential for a quantized vortex.** Form of the quantum potential (for $l = 1$), which, added to the Euler equation, transforms a classical vortex into a superfluid quantum-like vortex. The distance to the center is in unit of the characteristic scale $a = \mathcal{D}\sqrt{2/gh_0}$.

This means that it is now $\sqrt{h_q}$ instead of $\sqrt{h}$ that can be called to play the role of a wave function modulus. When $v_z = 0$, the vertical Euler equation reads

$$\frac{1}{\rho}\frac{\partial p}{\partial z} + g = 0, \qquad (10.338)$$

so that we recover the same expression for the pressure as in the flat bottom case. Namely, let $p = p_0$ be the constant pressure at the free surface $z = h_s$, e.g. the atmospheric pressure, the pressure therefore reads [269]

$$\frac{p}{\rho} = \frac{p_0}{\rho} + g(h_s - z). \qquad (10.339)$$

The Euler equation therefore keeps the form of the flat bottom case. Namely, the pressure term in this equation remains $-\nabla p/\rho = -g\nabla h_s$, i.e. it still depends on the surface height h_s while the continuity equation now depends on the fluid height h_q.

But let us now decompose the surface height as $h_s = h_q + h_b$. The radial Euler equation (in cylindrical coordinates) becomes, in the absence of an additional external potential,

$$\frac{\partial v_r}{\partial t} = -v_r\frac{\partial v_r}{\partial r} + \frac{v_\theta^2}{r} - g\frac{\partial h_q}{\partial r} - g\frac{\partial h_b}{\partial r}. \qquad (10.340)$$

We have therefore recovered the previous form for the couple of Euler and continuity equations (now in cylindrical coordinates), but they are now written in terms of h_q, while the h_b term can be considered as an externally added potential. This result is remarkable since it means that the curved bottom profile $h_b(r, \theta, t)$ now plays the role of a potential (up to the gravity constant g),

$$Q = gh_b. \tag{10.341}$$

As we shall see in the following, this property can be used in at least two ways:

(i) In natural systems: search for natural sea bottom profiles, which could locally create a quantum-like wave.
(ii) Simulate the quantum potential by the profile of the basin bottom, according to the relation

$$h_b = -\frac{2\mathcal{D}^2}{g} \frac{\Delta_2 \sqrt{h_q}}{\sqrt{h_q}}, \tag{10.342}$$

in which the bottom height becomes a function of the fluid height $h_q = h - h_b$.

We can now, in the case of potential motion, build a wave function from the fluid height h_q and the velocity potential,

$$\psi = \sqrt{\frac{h_q}{h_0}} \times e^{i\varphi/2\mathcal{D}}, \tag{10.343}$$

so that the gravity term gh_q becomes a non linear term $g|\psi|^2$. In this case the Euler and continuity equations can be combined in terms of the nonlinear Schrödinger equation, which is typical of superfluids,

$$\mathcal{D}^2 \Delta_2 \psi + i\mathcal{D}\frac{\partial}{\partial t}\psi - \frac{1}{2}g|\psi|^2\psi = 0. \tag{10.344}$$

Application to a quantum-like vortex

Let us exemplify this result, as in the flat bottom case accompanied by a quantum-like force, by the study of a stationary vortex.

In such an experiment, the "classical" case now corresponds to a flat bottom $h_b = 0$, and therefore to a situation where the surface height is $h = h_q + h_b = h_q$. The "macroquantum" case involves a basin bottom

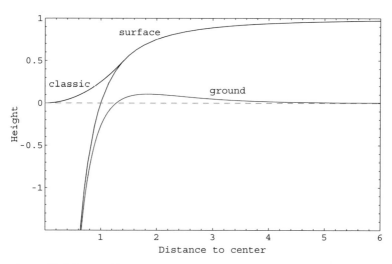

Fig. 10.23: **Height profiles in a macroquantized vortex experiment.** Profiles of the curved bottom h_b that simulates a quantum potential and of the corresponding surface h_s, compared with the "classical" profile (flat bottom). The quantum-like profile, given in Fig. 10.20 (for $l = 1$), which is a solution of a nonlinear Schrödinger equation of the superfluid kind, is the liquid height given by the difference $h_q = h_s - h_b$.

$h_b \neq 0$ that plays the role of a quantum potential, and therefore a surface height $h_s = h_q + h_b \neq h_q$ (see Fig. 10.23). One does not directly see $h_q = h_0 |\psi|^2 = h_s - h_b$, but it can be reconstructed from the measurement of h_s and h_b. Figure 10.20 therefore corresponds, in this case, to a comparison between the two shapes of $h_q = h_s - h_b$ in the "classical" ($h_b = 0$) and "quantum-like" cases ($h_b = Q/g$).

The various results of Sec. 10.6.10.2 still apply in this case, after having replaced h by h_q. In particular, the angular momentum remains quantized in terms of the quantum number $l = 0, 1, 2, \ldots$ and the velocity field is therefore also quantized as $v_\theta = 2\mathcal{D}l/r$.

Curved bottom height

Equation (10.327) now applies to $f_q = \sqrt{h_q/h_0}$ and reads

$$\frac{\partial^2 f_q}{\partial r^2} + \frac{1}{r} \frac{\partial f_q}{\partial r} + \left(\frac{1}{a^2} - \frac{l^2}{r^2} \right) f_q - \frac{1}{a^2} f_q^3 = 0, \tag{10.345}$$

where the characteristic scale a is still given by

$$a^2 = \frac{2\mathcal{D}^2}{gh_0}. \tag{10.346}$$

Let us now express the bottom height (see Figs. 10.22 and 10.23)

$$h_b = -\frac{2D^2}{g} \frac{\Delta_2 f_q}{f_q}, \tag{10.347}$$

which is the quantum potential up to the gravity constant g, in a dimensionless way. To this purpose, we replace the radial variable r by the dimensionless variable $x = r/a$, and we obtain, thanks to the above expression for a,

$$h_b = -h_0 \frac{1}{f_q} \left(\frac{\partial^2 f_q}{\partial x^2} + \frac{1}{x} \frac{\partial f_q}{\partial x} \right). \tag{10.348}$$

The expression in factor of h_0 is now dimensionless and depends only on the value of l.

Free surface height and liquid height

Let us now determine the free surface height profile in this "macroquantum" case. From the superfluid-like Eq. (10.345), one may write the bottom height as

$$h_b = -h_0 \frac{1}{f_q} \left(\frac{\partial^2 f_q}{\partial x^2} + \frac{1}{x} \frac{\partial f_q}{\partial x} \right) = -h_0 f_q^2 + h_0 \left(1 - l^2 \frac{a^2}{r^2} \right). \tag{10.349}$$

Finally, since $h_0 f_q^2 = h_q$ and $h_s = h_q + h_b$, one finds that the surface profile is given by

$$h_l(r) = h_0 \left(1 - l^2 \frac{a^2}{r^2} \right). \tag{10.350}$$

It is noticeable that one obtains the correct asymptotic height h_0 for $r \to \infty$, which means that the relation (Eq. 10.328) could have been established directly from this condition.

One therefore recovers the outer classical profile $h_s = h_0(1 - r_0^2/r^2)$, with two fundamental differences:

(i) The values of the characteristic scale r_0 are now quantized in agreement with the quantization of the differential rotational velocity $\Omega = 2Dl/r^2$, namely,

$$(r_0)_l = \sqrt{\frac{2}{gh_0}} \, Dl; \tag{10.351}$$

(ii) This profile is expected to continue in the inner region below the classical matching radius r_m. This agrees with the obtention of the same result for the velocity field (see Fig. 10.21), since this profile is typical of potential motion [218].

10.6.10.4. *Discussion and conclusion*

We have given the theoretical basis aiming at preparing a new kind of quantum-like laboratory experiment, in which the surface gravity waves of a classical liquid in a basin of finite height could be transformed to acquire some quantum-type properties typical of superfluids.

In the classical hydrodynamic case considered in this work, possible shortcomings are to be considered, such as the effects of vorticity, of viscosity at small scales, of uncertainties in the measurement of the height and in the application of the quantum-like force or in the deformation of the basin ground, etc.

In particular, while the quantum vortex (see e.g. [173]) is everywhere potential, the classical vortex is rotational in the inner region which rotates like a solid body. It is possibly potential only in its outer region (a zone that vanishes in the case of the rotating bucket). It is therefore clear that the superfluid-like properties cannot be implemented for all the region containing the fluid, but only in a sub-region. Indeed, the viscosity of the classical fluid implies the existence of a core radius inside, which solid rotation is expected and then vorticity. Therefore, although the fluid height is partly described by a superfluid equation, this does not mean that the specific property of superfluidity, namely, the vanishing of viscosity, is expected in such an experiment, since the microscopic classical nature of the fluid is not affected. However, the region of potential motion is expected to be increased toward the center in a larger domain than for the standard classical fluid, so that we could simulate in this way the decrease of an effective viscosity.

Future works will attempt to take into account these effects in more complete theoretical analyses and numerical simulations with improved integration schemes. Then real hydrodynamic laboratory experiments aiming at implementing the "macroquantum" behavior described here will be lead, in the hope to implement various quantum-type properties like quantization, non dissipation and self-organization.

In addition to applications to experimental devices, the concept of quantum-like potentials could also be applied to the understanding of "freak waves". They are extraordinary large water waves whose heights exceed by a factor larger than two the usual wave height, and which may have potentially devastating effects. It has been established that the nonlinear Schrödinger equation can describe many of the features of their dynamics, although it still remains unclear how they are generated in realistic oceanic conditions (see [418] and references therein). The results obtained here suggest new possibilities of understanding some of these freak waves, which may naturally arise for certain combinations of the surface height profile, currents and wind conditions. These combinations, rare but not impossible due to the large number of configurations on the whole surface of the Earth oceans, would simulate the appearance of a quantum-like potential and would therefore lead to quantum-like waves, whose height can be doubled respectively to classical waves because it is given by the square of the modulus of a wave function.

Problems and Exercises

Open Problem 19: In the case of the rotational motion of a solid body, write the equations of motion with quantum potential for a variable tensor of inertia $I_{ik} = I_{ik}(x, y, z, t)$. ∎

Open Problem 20: Develop experimental applications of the generalized fractal scale laws obtained in Chapter 4. ∎

Open Problem 21: Conceive experimental devices allowing one to implement, measure and apply the concept of "scale acceleration" defined in Chapter 4 (derivative of the scale velocity — or log of resolution — in terms of the "djinn"). Develop an experimental "scale dynamics" involving scale accelerations and scale forces. ∎

Open Problem 22: Develop experimental and technological applications of the geometric (Abelian and non-Abelian) gauge field theory described in Chapter 7. Find the generalized state of reference systems under which a gauge field can be suppressed or created. Achieve artificial charges and artificial fields. Use the properties of non-Abelian artificial fields (confinement, asymptotic freedom, etc.) for the realization of new technologies (batteries, energy storage). ∎

Chapter 11

APPLICATIONS TO ELEMENTARY PARTICLE
AND HIGH ENERGY PHYSICS

11.1. Implications of Fractal Space-Time
and Special Scale Relativity

Today's elementary particle and high energy physics relies on the implicit assumption of underlying "Galilean-like" scale laws. Namely, it is actually assumed without additional analysis that the concept of scale interval tending to zero does have a physical meaning and that the composition of two changes of scales by factors ϱ and ϱ' yields a total change of scale by a factor $\varrho'' = \varrho \times \varrho'$. Even though the Planck length-scale is generally considered to be a limit for known physics, in the standard model and in most of its extensions, the very existence of scales smaller than the Planck scale is nevertheless considered as physically meaningful.

The question of the nature of space-time geometry at Planck scale is nevertheless a subject of intense work (see e.g. [21, 274] and references therein). This is a central question for practically all theoretical attempts, including noncommutative geometry [109, 110], supersymmetry and superstrings theories [186, 208, 452], which themselves may include a noncommutative geometry [485] and for which the compactification scale is close to the Planck scale, and particularly for the theory of quantum gravity.

Indeed, after the progress achieved by Ashtekar [30, 31], the development of loop quantum gravity by Rovelli and Smolin [466, 468] led to the conclusion that the Planck scale could be a quantized minimal scale in Nature, involving also a quantization of surfaces and volumes [467].

This would mean that the Planck energy would also be a limit and/or an invariant energy scale [301], which contradicts the very existence of energies larger than the Planck energy (equivalent to a mesoscopic mass of 2×10^5 g).[a]

In this framework of attempts to construct a quantum gravity theory at the Planck scale, many works have pointed out the expected fractal structures and properties of the quantum space-time (see e.g. [18, 19, 46, 236], Kröger's review [258] and references therein). Although different from the scale relativity context, these works may be considered as complementary and the two approaches could well finally meet.

The main difference is that these quantum gravity studies assume the quantum laws to be set as fundamental laws. In such a framework, the fractal geometry of space-time at the Planck scale is a consequence of the quantum nature of physical laws, so that the fractality and the quantum nature co-exist as two different things.

In the scale relativity theory, there are not two things (in analogy with Einstein's general relativity theory in which gravitation is a manifestation of the curvature of space-time): the quantum laws are considered as manifestations of the fractality and nondifferentiability of space-time, so that they do not have to be added to the geometric description. This point is particularly relevant to address the Zeilinger and Svozil's attempt to measure the fractal dimension of space-time at quantum electrodynamics scales [551]. They concluded to its quasi non-fractality at these scales (they found a value $D = 4 - (5.3 \pm 2.5) \times 10^{-7}$ for the dimension of space-time), but, there too, they did not attribute the fractality to the quantum properties themselves.

Another difference between the two approaches is that, while the quantum gravity space-time is fractal only at the Planck scale, the scale relativity space-time is fractal at all scales, and this fractality becomes effective below the de Broglie scale (in the quantum domain). Anyway, the various approaches, including Connes' noncommutative geometry [109], meet on one fundamental point: the combination of geometry and quantum mechanics invariably leads to introduce fractal properties and nondifferentiability of space-time.

---

[a]This statement is only valid in the framework of standard dilation laws, which we have called "Galilean scale laws" by analogy, in which the product of two dilations by ρ and ρ' is $\rho'' = \rho \times \rho'$, and in which the Planck length-scale and time-scale are the same as the Planck energy-scale.

Over the last years, there has been significant research effort aimed at the development of a "Doubly-Special-Relativity" [20] (see a review in [21]), according to which the laws of physics involve a fundamental velocity scale c and a fundamental minimum length scale L_p, identified with the Planck length. These works have confirmed that "it is possible to formulate the Relativity postulates in a way that does not lead to inconsistencies in the case of space-times whose short-distance structure is governed by observer-independent scales of velocity and length, and that this new type of relativistic theories allows the introduction of a minimum length, without giving rise to a preferred class of inertial frames for the description of space-time structure".

The concept of a new relativity, in which the Planck length-scale would become a minimum invariant length, is exactly the founding idea of the special scale relativity theory [352], which has since been incorporated in other attempts of extended relativity theories [87, 88]. But, despite the similarity of aim and analysis, the main difference between the "Doubly-Special-Relativity" approach and the scale relativity one is that we have identified the question of defining an invariant length-scale as coming under a relativity of scales. Therefore the new group to be constructed is a multiplicative group that becomes additive only when working with the logarithms of scale ratios, which are definitely the physically relevant scale variables, as we have shown by applying the Gell-Mann–Levy method to the construction of the dilation operator (see Sec. 4.2.1).

Moreover, we have proved that the Lorentz transformation is the general solution to the special relativity problem of finding linear laws of coordinate transformation that come under the principle of relativity [352] (see Chapter 4). The existence of a minimal quantity invariant under the considered transformation (motion or scale) therefore does not have to be postulated, since it is a consequence of this proof. The theory of special scale relativity then suggests [352, 353, 355, 360] that the Planck length-scale be such an impassable and unreachable horizon scale, invariant under dilations. In this framework, the current elementary particle and high energy physics would be a (relatively) "large" scale approximation of a more profound theory, in which the laws of dilation take a log-Lorentzian structure. As a consequence, the Planck energy no longer corresponds to the Planck length and remains passable. This is similar to the velocity-momentum relation for which, in Galilean relativity laws, $p = mc$ incorrectly corresponds to $v = c$, while in the correct Einstein laws, $p = mc$ is obtained for $v = c/\sqrt{2}$.

In similarity with the theory of the relativity of motion, the passage from the Galilean to the Lorentzian-like theory may be expressed in terms of scale-relativistic "corrections". These corrections remain small at "large" scale (i.e. around the Compton scale of particles) and increase when going to smaller length scales (i.e. large energies) in the same way as motion-relativistic corrections increase when going to large speeds.

In this chapter, we shall not be exhaustive on all the proposals of applications of the scale relativity theory which have been suggested in elementary particles and high energy physics. We refer the reader to [353, 355, 360] for additional suggestions.[b]

Let us also quote another proposal, which could allow one to derive the mass and charge spectrum of fermions ([355] Sec. 6 and [360] Sec. 6.2). It was based on the observation that scale-relativistic corrections closely follow (up to a constant) the variation of the electric charge due to vacuum polarization by particle-antiparticle pairs of elementary fermions. It therefore allowed us to theoretically recover the observed R ratio of the e^+e^- annihilation cross sections into hadrons and muon-antimuon pairs, which amounts to lowest order to the sum of the square of elementary fermion charges.

We assume in this chapter that the reader is accustomed with some basic concepts, tools and methods of elementary particle physics as can be found in textbooks [219, 210, 7, 238, 502] in particular as regards the definition of coupling constants and their renormalization group equations and solutions. Indeed, the various tentative theoretical predictions recalled here and compared to experimental data are not, for most of them, predictions of the scale relativity theory alone, but mainly of the standard model itself and of its GUT extensions, in which one takes into account the new special scale relativistic relation between momentum-energy and space-time scales. A full theory in which generalized nonperturbative renormalization group equations would be derived as a more profound scale manifestation of space-time fractality and nondifferentiability remains to be constructed, so that the herebelow approach remains semi-phenomenological.

---

[b]One of these proposals ([349] and [353] Sec. 5.10) was a simple descriptive fractal model of the anomalous positron peaks, which were claimed at that time to be observed in heavy ion collisions at Darmstadt. However, these experimental results have not been confirmed in subsequent experiments [5], so that their observed fractality analysed in the model may have been a property of experimental errors instead of a genuine physical effect. Another possibility could be that they were a manifestation of the possible fractality and nondifferentiability of wavefunctions (Sec. 5.8).

11.1.1. *Special scale relativistic laws*

Let us briefly recall the results obtained in [352, 353] and given in Sec. 4.4, then apply them to the physical problem of high energy scale laws.

We define the resolution as $\varepsilon = \delta x = c\delta t$ and we set $\lambda_0 = c\tau_{dB} = \hbar c/E$. In rest frame, λ_0 is thus the Compton length of the system or particle under consideration. We have assumed [352] that the laws of dilation remain Galilean at scales larger than λ_0 and become "log-Lorentzian" at smaller scales. This assumption was based on the appearance of particle-antiparticle pairs at time-scales smaller than $\hbar/(2\,mc^2)$, which contribute to the self-energy of the particle through "radiative corrections". However, in the scale relativity view, we have identified these contributions to nothing else but the fractality of the space-time geodesics, which run backward in time below the Einstein-Compton scale, in accordance with the Wheeler–Feynman view of antiparticles as particles going backward in time (see detail in Chapter 6 of [353]). Now, as remarked by Landau [268], the very existence of these pairs suppresses the possibility of direct length or time interval measurement at scales smaller than the Compton length of the electron, and therefore affects also the physical meaning of what are space and time at very small scales.

11.1.1.1. *Law of composition of dilations*

Let us start from the transition scale λ_0, then let us make a scale transformation to a new scale ε. Then assume that we make a new scale transformation by a factor ϱ. At scales larger than λ_0, i.e. in the non-relativistic quantum-mechanical domain, one expects to obtain a new scale $\varrho\varepsilon$. But at scales smaller than λ_0, the "log-Lorentz" new law of composition of dilations reads, for $\varepsilon < \lambda_0$ and $\varepsilon' < \lambda_0$ (see Fig. 11.1),

$$\ln\frac{\varepsilon'}{\lambda_0} = \frac{\ln(\varepsilon/\lambda_0) + \ln\varrho}{1 + \ln\varrho\ln(\varepsilon/\lambda_0)/\ln^2(\lambda_0/l_{\mathbb{P}})}. \tag{11.1}$$

It must be understood to apply "inside" the particle, namely, in the scale relativity view, to geometric structures of the family of geodesics, at scale where not only space but also time is now fractal. Recall indeed that the Compton scale is just identified with the fundamental parameter λ_c that defines the fractal fluctuations, $d\xi^2 = \eta^2\lambda_c\,ds$, in the scale relativistic foundation of quantum mechanics, and that it plays also the role of a fundamental transition scale for the proper time s.

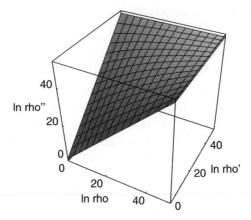

Fig. 11.1: **Log-Lorentzian composition law of dilations.** This figure illustrates the new law of composition of dilations in special scale relativity (Eq. 11.1). The x and y coordinates represent the logarithm of the scale factors ϱ and ϱ', and the z coordinate gives the logarithm of their product according to the log-Lorentzian composition law, $\ln \rho'' = (\ln \rho + \ln \rho')/(1 + \ln \rho \ln \rho'/\mathbb{C}^2)$. Here we have taken for the scale constant that plays for scale laws a role similar to that of c for motion laws the value $\mathbb{C} = \ln(\lambda_e/l_{\mathbb{P}}) \approx 51.528$. In this law, the product of two successive dilations is limited by the maximal value $e^{\mathbb{C}}$.

This relation introduces a fundamental length scale Λ (Sec. 4.4) that we have identified with the Planck length (currently $1.61605(10) \times 10^{-35}\,\mathrm{m}$),

$$\Lambda = l_{\mathbb{P}} = (\hbar G/c^3)^{1/2}. \tag{11.2}$$

As one can see from Eq. (11.1), if one starts from the scale $\varepsilon = l_{\mathbb{P}}$ and apply any dilatation or contraction ϱ, to this scale, one gets back the scale $\varepsilon' = l_{\mathbb{P}}$, whatever the initial value of λ_0, i.e. whatever the state of motion, since λ_0 is Lorentz-covariant under velocity transformations. In other words, the Planck length $l_{\mathbb{P}}$ and its time equivalent the Planck time $t_{\mathbb{P}} = l_{\mathbb{P}}/c$ are now interpreted as limiting lower length and time scales, impassable, unreachable and invariant under dilatations and contractions.

11.1.1.2. *Laws of scale transformation*

The length measured along a fractal coordinate, which was previously scale-dependent, for $\varepsilon < \lambda_0$, as

$$\ln \frac{\mathcal{L}}{\mathcal{L}_0} = \tau_0 \ln(\lambda_0/\varepsilon), \tag{11.3}$$

where the scale exponent τ_0 was constant, becomes in the new framework

$$\ln \frac{\mathcal{L}}{\mathcal{L}_0} = \frac{\tau_0 \ln(\lambda_0/\varepsilon)}{\sqrt{1 - \ln^2(\lambda_0/\varepsilon)/\ln^2(\lambda_0/l_{\mathbb{P}})}}. \qquad (11.4)$$

The main new feature of special scale relativity respectively to the previous fractal or scale-invariant approaches is that the scale exponent τ and therefore the fractal dimension $D_F = 1 + \tau$, which were previously constant $(D_F = 2, \tau_0 = 1)$, are now explicitly varying with scale, following the law

$$\tau(\varepsilon) = \frac{1}{\sqrt{1 - \ln^2(\lambda_0/\varepsilon)/\ln^2(\lambda_0/l_{\mathbb{P}})}}. \qquad (11.5)$$

The fractal dimension of geodesics has been shown in Secs. 5 and 6 to jump from $D_F = 1$ to $D_F = 2$ at the de Broglie scale $\lambda_{\mathrm{dB}} = \hbar/p$ of the particle.

Then we have suggested [352, 353] that the new scale of transition λ_0 from a scale-Galilean regime to a scale-Lorentzian new regime is the Compton scale of the particle, e.g. for the electron, $\lambda_0 = \lambda_e = \hbar/m_e c$. This suggestion was based on the remark that the standard concept of length and time radically changes beyond the Compton scale of the electron, because of the creation of electron-positron pairs by the measurement process itself at energies larger than $\hbar/mc^2$ [268].

In this framework, the fractal dimension is now varying beyond this scale, at first very slowly as

$$D_F(r) = 2\left(1 + \frac{1}{4}\frac{\mathbb{V}^2}{\mathbb{C}_0^2} + \cdots\right), \qquad (11.6)$$

where $\mathbb{V} = \ln(\lambda_0/\varepsilon)$ and $\mathbb{C}_0 = \ln(\lambda_0/l_{\mathbb{P}})$, then it tends toward infinity at very small scales when $\mathbb{V} \to \mathbb{C}_0$, i.e. $\varepsilon \to l_{\mathbb{P}}$.

We have called "djinn" the now variable scale exponent τ in the special scale relativistic framework, in which it plays the role of a fifth dimension (recall the change of notation with respect to previous works where it was often denoted δ). Indeed, the fact that the four fractal fluctuations $d\xi^\mu$ have positive definite metric allows to implement special scale relativity thanks to the introduction, at small scales $\varepsilon < \lambda_0$, of a five-dimensional Minkowskian "space-time-djinn" of signature $(+, -, -, -, -)$, whose fifth dimension is the variable scale exponent τ. The new fluctuation metric reads in this case

$$d\sigma^2 = \mathbb{C}_0^2 d\tau^2 - \Sigma_\mu \frac{(d\xi^\mu)^2}{\xi^2}. \tag{11.7}$$

This is to be compared to the achievement of Einstein-Poincaré special motion-relativity thanks to the introduction of a Minkowskian space-time [450, 321]. Hence motion in space can be described as rotation in space-time, and the full Lorentz invariance is finally implemented by jumping to a four-dimensional tensor description. Here the situation is similar, with the "scale time" τ playing for scale the same role as played by time for motion. There too, the scale transformations, i.e. the dilatations and contractions of space-time resolutions, are described as rotations in a five-dimensional Minkowskian fractal space-time.

From the experimental value of the electron mass and of the Planck mass [435], the constant $\mathbb{C}_0$ is found to take, for the electron, the value

$$\mathbb{C}_e = \ln \frac{m_P}{m_e} = 51.52797(7) \tag{11.8}$$

(the number into brackets is the uncertainty on the last digits).

11.1.1.3. *Scale relativistic mechanics*

A scale-relativistic mechanics has been developed (see [352], [353, Chapter 6] and Sec. 4.4.6) in analogy with motion-relativistic mechanics [266].

The special scale-relativistic invariant reads (for $\mathbb{C} = \ln(\lambda_0/l_\mathbb{P})$, λ_0 fixed and resolution $\varepsilon < \lambda_0$)

$$d\sigma^2 = \mathbb{C}^2 d\tau^2 - \frac{d\mathcal{L}^2}{\mathcal{L}^2}. \tag{11.9}$$

Under this form the new invariant depends on τ, which is not a directly measurable quantity. However it may also be expressed in terms of the measurement resolution ε, which is a measurable quantity,

$$d\sigma^2 = \left(\frac{\mathbb{C}^2}{\mathbb{V}^2} - 1\right) \frac{d\mathcal{L}^2}{\mathcal{L}^2}, \tag{11.10}$$

where $\mathbb{V} = \ln(\lambda_0/\varepsilon)$ and $\mathbb{C} = \ln(\lambda_0/l_\mathbb{P})$.

Assume now that physical scale laws emerge from a least action principle. We expect the action to be the integral over $d\tau$ of some Lagrange function $\widetilde{L} = \widetilde{L}(\ln \mathcal{L}, \mathbb{V}, \tau)$, in analogy with $L = L(x, v, t)$ in motion relativity. Moreover, we also expect its differential $d\widetilde{S} = \widetilde{L}\,d\tau$ to be given, up to some multiplicative constant, by the invariant $d\sigma$ [266].

If we denote as $\mathbb{V} = \ln(\lambda_0/\lambda)$ the relative scale state of the reference system, conservative quantities (i.e. prime integrals) $\mathcal{P} = \partial\widetilde{L}/\partial\mathbb{V}$ and $\mathcal{E} = \mathbb{V}\partial\widetilde{L}/\partial\mathbb{V} - \widetilde{L}$ are expected to emerge from the uniformity of $\ln\mathcal{L}$ and of τ respectively, according to Noether's theorem. Here "conservative" means that these quantities do not depend explicitly on the scale-time τ, which plays for scale the structural role played by time for motion.

Generalized Einstein–de Broglie relations

Let us consider first the uniformity of $\ln\mathcal{L}$. It implies the existence of a conservative quantity, a "scale momentum" $\mathcal{P}$, which is a function of the scale state $\ln(\lambda_0/\lambda)$,

$$\mathcal{P}(\lambda) = \mu\frac{\ln(\lambda_0/\lambda)}{\sqrt{1 - \frac{\ln^2(\lambda_0/\lambda)}{\ln^2(\lambda_0/l_{\mathbb{P}})}}}, \tag{11.11}$$

where μ is the constant, to be later determined, which comes from the fact that the action and the metrics invariant are equal only to some proportionality factor (this factor is equal to $-mc$ in motion relativity [266]). This relation is the scale-relativistic analog of the motion relativistic equation for momentum, $p = mv/\sqrt{1 - v^2/c^2}$, and it applies to space and time scale variables as well.

In order to understand the meaning of this result, we have first noted that physics must be invariant under the choice of the logarithm base. Then the form of Eq. (11.11) implies that $\mathcal{P}$ is itself a logarithm of some dimensionless quantity. Now Eq. (11.11) has been obtained from the uniformity of a space variable (in logarithm form since we are concerned with its scale dependence), from which the usual momentum also derives as a conservative quantity in classical mechanics. This has led us to write $\mathcal{P} = \ln(p/p_0)$.

The next step has consisted in finding the value of the constant μ by taking the limit $l_{\mathbb{P}} \to 0$, i.e. $\mathbb{C} \to \infty$ (this is the scale analog of taking the limit $c \to \infty$ to recover classical nonrelativistic mechanics from relativistic mechanics as regards motion laws). This limit should give back and, therefore, establish some fundamental equations of standard quantum mechanics. Indeed Eq. (11.11) becomes $p/p_0 = (\lambda_0/\lambda)^\mu$, and one recognizes in this equation the generalized Einstein–de Broglie relation for a fractal dimension $D_F \neq 2$.

As shown in Chapters 5 and 6, standard quantum mechanics is obtained only in the critical case $D_F = 2$, from which one therefore has $\mu = 1$. The relation becomes $p\lambda = p_0\lambda_0 = $ constant (for each of the four momentum components). This defines the Planck constant $\hbar$ and we obtain in the scale Galilean case the standard Einstein–de Broglie relation

$$p\lambda = p_0\lambda_0 = \hbar, \qquad (11.12)$$

which is generalized in the scale Lorentzian case as

$$\ln\frac{p}{p_0} = \frac{\ln(\lambda_0/\lambda)}{\sqrt{1 - \frac{\ln^2(\lambda_0/\lambda)}{\ln^2(\lambda_0/l_\mathbb{P})}}}. \qquad (11.13)$$

This is a fundamental result, since it provides us with an explanation of the Einstein–de Broglie relations, which were the first and most profound quantum mystery of the so-called wave-particle duality (why is the classical momentum also related to a wavelength?). Here the de Broglie relations are the scale relativistic analogs of Descartes's motion relation $p = mv$. This is understood in the scale and motion relativity framework by considering that a reference system is characterized both by a state of scale given by the relative quantity $\ln(\lambda_0/\lambda)$ and by a state of motion given by the relative velocity v. This explains why the same conservative quantity p can be related to both a wavelength and a velocity by similar relations, $\ln(p/p_0) = \mu\ln(\lambda_0/\lambda)$ and $p = mv$.

Schwarzschild and Compton relations

But there is something more here. Anticipating on the applications to cosmology in Chapter 12, one may remark that there are actually two ways to relate the scale conservative quantity $\mathcal{P}$ to a momentum ratio, and that we have considered hereabove only one of them, $\mathcal{P} = \ln(p_0/p)$. The other is the symmetric relation (in logarithm form) $\mathcal{P} = \ln(p/p_0)$, i.e. the inverse relation for momentum. Applied to the rest mass energy, the first relation becomes (in the standard Galilean $D_F = 2$ case) the Compton relation, which relates a wavelength $2\pi\lambda_c$ to an inertial mass,

$$\lambda_c = \frac{\hbar}{c}m^{-1}. \qquad (11.14)$$

The second relation is also remarkable, since it relates a mass to a length scale, now in a direct relation of proportionality. Such a relation, here established as a manifestation of the relativity of scales, already does exist from Einstein's general relativity of motion. It is the Schwarzschild relation that relates the Schwarzschild radius $r_s = 2\lambda_s$ to the active gravitational mass of a body,

$$\lambda_s = \frac{G}{c^2} m. \tag{11.15}$$

From the weak equivalence principle, the inertial and the passive gravitational masses are the same, and from the strong principle, both are the same as the active mass. The scale relativity approach therefore provides two relations for the same quantity, here the mass, that is linked to two inverse scales as two solutions of the same equation. The symmetry becomes even more clear if one expresses these two relations in terms of the Planck length and of the Planck mass [435]

$$m_{\mathbb{P}} = \left(\frac{\hbar c}{G} \right)^{1/2} = 2.17645(16) \times 10^{-8} \, \text{kg}. \tag{11.16}$$

One finds the symmetric "dual" relations [349, 351, 353]

$$\frac{\lambda_c}{l_{\mathbb{P}}} = \frac{m_{\mathbb{P}}}{m}, \quad \frac{\lambda_s}{l_{\mathbb{P}}} = \frac{m}{m_{\mathbb{P}}}, \tag{11.17}$$

which are, in the new framework, both expressions of the fundamental relation $\mathcal{P} = \mu \mathbb{V}$.

Consequence for the fundamental constants

It is well-known that one of the consequences of the special motion relativity theory has been to lead to a redefinition of the nature of the constants of physics. The very existence of space-time, of which space and time are but subspaces, leads to measure lengths and time with the same unit (as already done for the three dimensions of space), and therefore to define velocity as being dimensionless. In this context, the limit velocity c results from a simple projection effect from four dimensions to three and becomes the pure number $c = 1$, so that the question of its value loses any physical meaning. For this reason in today's state of physics where one keeps nevertheless meter and second as units, its value has been fixed to $c = 299792458 \, \text{m/s}$, and the meter is now derived from the second.

A similar consequence could come from the special scale relativity theory, if it were validated by experimental data. Indeed, in its framework the Planck time scale also becomes an invariant unit time, impassable and unreachable, resulting from a projection effect of five dimensions to four.

Moreover, the above mass-length relations validates the scale relativity of mass supported by Mach [297], according to which there is no absolute mass but only mass ratios. The inertial mass ratios are, according to Newton's initial definition, acceleration ratios. But one can now reach the more profound conclusion that mass ratios can be reduced to length ratios, and this for all forms of masses (this point, in connection with Mach's principle, will be developed in more detail in Sec. 12.4.4).

This means that one may definitely set the foundations of physics on the identification of the Planck length, time and mass with fundamental units $l_\mathbb{P} = 1, t_\mathbb{P} = 1$ and $m_\mathbb{P} = 1$, and that the question of their value (and therefore of their variation) loses any physical meaning. In such a framework the constants $\hbar$ and G should be deduced from these units as

$$\hbar = c l_\mathbb{P} m_\mathbb{P}, \quad G = c^2 l_\mathbb{P} m_\mathbb{P}^{-1}, \tag{11.18}$$

which are but the Compton and Schwarzschild relations for the Planck scales.

In this context the constants $\hbar$ and G have finally the same status: both of them are geometric constants that relate mass-energy-momentum to space-time. This doubling of constants simply comes from a fundamental symmetry of scale physics that leads to two equivalent but inverse mass-length relations. We therefore predict that any variation of the constants c, $\hbar$ and G is forbidden, since devoid of physical meaning.

Djinn uniformity and complexergy

Another prime integral, $\mathcal{E} = \mathbb{V} \partial L / \partial \mathbb{V} - L$, finds its origin in the uniformity of the "djinn" (scale time τ). This new quantity, which had no equivalent in standard physics, has been derived in [352, 353] and named "complexergy" in [388] (see Sec. 8.4). In the special scale relativistic case, this is the scale equivalent (from the viewpoint of the mathematical structure) of the relativistic expression for energy, $E = mc^2 / \sqrt{1 - v^2/c^2}$. It reads, for $\mu = 1$,

$$\mathcal{E} = \frac{\ln^2 \left(\frac{\lambda_0}{\Lambda} \right)}{\sqrt{1 - \frac{\ln^2 \left(\frac{\lambda_0}{\lambda} \right)}{\ln^2 \left(\frac{\lambda_0}{\Lambda} \right)}}}, \tag{11.19}$$

where $\Lambda = l_\mathbb{P}$ in the microphysical case.

Some applications of this concept to elementary particle physics are proposed in Sec. 11.2.

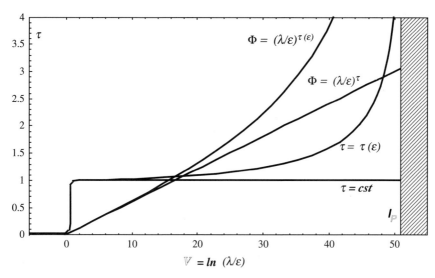

Fig. 11.2: **Transition from Galilean to Lorentzian scale relativity.** The fractal dimension $D_F = 1 + \tau$, that was constant at large length scale (i.e. low energy, here to the left of the figure) under Galilean scale laws, becomes scale dependent in the special scale relativity framework and now diverges toward small length-scales (i.e. high energy, here to the right of the figure) when resolution tends to the Planck scale, which now owns all the physical properties of the zero scale. A physical quantity, such as the length of a fractal curve, which was divergent as $(\lambda/\varepsilon)^\tau$ when $\varepsilon \to 0$, now diverges when ε approaches the Planck length scale.

11.1.1.4. *Special scale relativistic formulas*

Let us sum up various new formulas of special scale relativity (right-hand side formulas below) and compare them to the former standard laws (left-hand side formulas) that have been identified with the Galilean version of scale relativity (see [352, 353, 355, 360] and Fig. 11.2). In the former laws the various length and time scales encountered in various physical relations (de Broglie or Compton scale, space or time resolution in Heisenberg inequalities, length scale in the renormalization group equations, etc) were allowed to reach zero point. This is generally no longer allowed in the new framework, where the asymptotic zero point is replaced by the Planck space-time scale.

Compton relation:

$$\ln \frac{m}{m_0} = \ln \frac{\lambda_0}{\lambda}, \quad \text{with } \lambda_0 m_0 = \hbar/c; \quad \ln \frac{m}{m_0} = \frac{\ln(\lambda_0/\lambda)}{\sqrt{1 - \frac{\ln^2(\lambda_0/\lambda)}{\ln^2(\lambda_0/l_{\mathbb{P}})}}}. \tag{11.20}$$

Heisenberg's inequality:

$$\ln \frac{\delta p}{p_0} \geq \ln \frac{\lambda_0}{\delta x}, \quad \text{with } \lambda_0 p_0 = \hbar; \quad \ln \frac{\delta p}{p_0} \geq \frac{\ln(\lambda_0/\delta x)}{\sqrt{1 - \frac{\ln^2(\lambda_0/\delta x)}{\ln^2(\lambda_0/l_{\mathbb{P}})}}}. \tag{11.21}$$

Scale dependence of coupling constants:

$$\alpha^{-1}(r) = \alpha^{-1}(\lambda_0) + \beta_0 \ln \frac{\lambda_0}{r} + \cdots ; \quad \text{unchanged (in length scale)} \tag{11.22}$$

Scale invariant:

$$\tau = \tau_0 = \text{constant}; \quad d\sigma^2 = \mathbb{C}_0^2 d\tau^2 - \frac{d\mathcal{L}^2}{\mathcal{L}^2}, \quad \text{with } \mathbb{C}_0 = \ln \frac{\lambda_0}{l_{\mathbb{P}}}. \tag{11.23}$$

In these equations, valid at scales smaller than the Compton length of the particle λ_0 (or its Lorentz transform), the Planck length $l_{\mathbb{P}} = (\hbar G/c^3)^{1/2}$ is identified with a fundamental length-scale $\mathbf{\Lambda}$, that is invariant under dilations and contractions (in a way similar to the invariance of the velocity of light under motion transformations).

11.1.2. *Running coupling constants and fundamental mass scales*

11.1.2.1. *Theoretical prediction of mass scales*

As we shall now see, a theoretical prediction of mass-energy scales can be performed in the new framework (see Fig. 11.4) by various ways:

(i) The transitions from fractal (nondifferentiable) regime to classical (differentiable) regime can be identified with fundamental mass scales (in rest frame).

(ii) Inverse running couplings can be shown, theoretically and experimentally, to reach a critical value $4\pi^2$ at some specific scales (see next section): the electromagnetic coupling at infinite energy and the strong coupling at Planck energy satisfy the relation $\alpha^{-1}(E) = 4\pi^2$. The theoretical prediction of a new energy structure at $E = 3.27 \times 10^{20}$ eV can also be reached by solving this equation for the SU(2) weak coupling (see Sec. 11.4).

(iii) Some universal relations between masses and charges of elementary particles can be established. Such relations, which are demonstrated from the geometric interpretation of gauge transformations in the scale relativity framework (Chapter 7), apply to the mass and charge of the electron and also to the electroweak scale in combination with the critical coupling $1/4\pi^2$. They allow one to predict new energy scales when they are applied to the SU(3) and SU(2) couplings (respectively, $E = 4 \times 10^8$ GeV and $E = 76$ TeV, see Fig. 11.4).

(iv) A mechanism for generating the mass-charge spectrum of elementary fermions has been proposed [355, 360].

We shall now review in more detail these various relations by recalling how they are obtained, then we shall compare to the experimental data the quantitative predictions one can derive from them.

11.1.2.2. *Critical value $4\pi^2$ of inverse couplings*

Let us recall the simple argument [360] according to which the natural "bare" value of any coupling constant is given by the optimized value $\alpha_b = 1/4\pi^2$.

Assume that there exists a Coulomb-like force between two bodies. In the classical theory, this force is attributed to the fact that the bodies carry charges Q_1 and Q_2, and it writes

$$F = \frac{Q_1 Q_2}{4\pi r^2}. \tag{11.24}$$

The microscopic quantum theory, since the work of Einstein [156], interprets this process in terms of quanta. Namely, the two macroscopic bodies are themselves made of elementary substructures that carry quanta of charges e, so that $Q_1 = Z_1 e$ and $Q_2 = Z_2 e$. The force between two quanta of charge e now reads (considering here only its amplitude)

$$F = \frac{e^2}{4\pi r^2}. \tag{11.25}$$

It can be re-expressed in terms of a coupling constant α by introducing the Planck constant,

$$F = \frac{\alpha \hbar c}{r^2}, \tag{11.26}$$

so that one can make the identification $e^2 = 4\pi\alpha\hbar c$.

The second quantization theory states that the force itself is carried by the continous exchange of zero mass bosons between the two charges. Indeed the existence of a mass m for the intermediate exchanged particle would introduce a vanishing Yukawa term $\exp(-r/\lambda)$ in the expression of the force, where $\lambda = \hbar/mc$. The macroscopic, classical force therefore results from the average effect of the boson exchanges, i.e.

$$F = \left\langle \frac{\delta p}{\delta t} \right\rangle. \tag{11.27}$$

This classical force corresponds to the average variation of momentum over the time interval δt. Each charge quanta emits bosons isotropically, while only those, which are absorbed by the other quanta participate in the interaction. The three-dimensional nature of space therefore implies a geometric factor $1/4\pi$. We obtain

$$F = \frac{1}{4\pi} \frac{\delta p}{\delta t}. \tag{11.28}$$

Now the exchanged momentum comes under a Heisenberg inequality. It will therefore be optimized by the "saturating", minimal value of this inequality, which is of universal meaning. However, one should be cautious about the precise Heisenberg relation to be used here. We indeed do not look here for orders of magnitude, but for exact statements.

Fortunately, a general method making use of the concept of information entropy has been devised by Finkel [178] for constructing any exact Heisenberg relation between any couple of variables (see Sec. 5.2). It can be used to establish the Heisenberg relation, which is relevant here, relating the distance interval r and the momentum difference δp. While the usual Heisenberg relation for variances is saturated for the standard value $\hbar/2$, one finds in the case of intervals

$$\delta p \times r \geq \frac{\hbar}{\pi}. \tag{11.29}$$

The limiting value is precisely reached by the harmonic oscillator vacuum state. This is in accordance with the second quantization of the exchanged boson, which is just based (since Einstein's pioneering work in 1905) on a description in terms of elementary harmonic oscillators.

The last step of the demonstration makes use of the fact that the intermediate bosons are of null mass, so that they move at the velocity of light c, and therefore $\delta t = r/c$. We finally find for the optimized amplitude of the force

$$F = \frac{1}{4\pi^2} \frac{\hbar c}{r^2}. \tag{11.30}$$

The comparison with Eq. (11.26) yields [360]:

$$\alpha_b = \frac{1}{4\pi^2}. \tag{11.31}$$

This result may seem, at first sight, to be in contradiction with the experimental values of coupling constants. However, the reason why this value is not directly seen in the experimental measurement of coupling constants is two-fold:

(i) The couplings are actually explicitly scale-dependent (they are "running" with energy-scale), because of radiative corrections for the effects of fermion-antifermion pairs. Therefore the $4\pi^2$ value of couplings is expected to be reached only at some high energy particular and critical scales, to be determined.

(ii) In the standard model, the bare couplings (at infinite energy) remain either logarithmically divergent (QED) or vanishing (asymptotic freedom of non-Abelian fields). On the contrary, in the special scale-relativistic framework [352], the divergence problem of mass and charge left unsolved by the renormalization theory is solved, so that one becomes able to look at the bare values.

When accounting for this running character of couplings, the results obtained (see Fig. 11.4) support, as we shall see, the systematic occurrence of the critical value $4\pi^2$ for inverse couplings at several specific and fundamental scales.

11.1.2.3. *Mass-coupling relations*

One of the main results obtained in the new geometric theory of gauge fields (see Chapter 7) is the possible existence of relations between mass scales and couplings in the special scale relativistic case. Indeed, applying a gauge transformation to the electromagnetic field implies to change also the wave function of the electron, which becomes

$$\psi' = \psi e^{i4\pi\alpha \ln \rho} \tag{11.32}$$

where α is a coupling constant.

While, in Galilean scale relativity, the scale ratio ρ is unlimited, in the more general framework of special scale relativity it is limited by the ratio of the Planck scale over the Compton scale. This limitation implies the quantization of charge, following a general mass-charge relation (see Sec. 7.4.5 and [356]) that takes the form

$$\alpha \times \ln\left(\frac{m_{\mathbb{P}}}{m}\right) = \frac{k}{2},$$ (11.33)

where k is integer.

Accounting for the electroweak theory, according to which the electromagnetic coupling is only 3/8 of its high energy value (plus radiative corrections), one obtains

$$\frac{8}{3}\alpha_e \ln\left(\frac{m_{\mathbb{P}}}{m_e}\right) = 1,$$ (11.34)

where $\alpha_e = 1/137.036$ is the low energy fine structure constant and m_e is the electron mass. This relation is supported by the experimental data with a relative precision of 2×10^{-3} [360], becoming 10^{-4} when accounting for threshhold effects (see Sec. 11.1.3.2 for a more complete description). One may contemplate the possibility that the small remaining difference with the experimental value of the electron charge be of weak interaction origin, which could lead to a future expectation for the neutrino masses.

There actually remains an uncertainty, since the above relation corresponds to a particular choice of the value of the quantized number $k = 2$. It is interesting to consider the full spectrum, which defines only five characteristic scales. One obtains 8.5×10^{16} eV for $k = 1$, then very small energies corresponding to Compton lengths of 4.8 cm $(k = 3), 7 \times 10^6$ km $(k = 4$, typical solar system scales), 32.6 kpc $(k = 5$, typical galactic scales), while the following scales lie beyond the scale of the Universe.

11.1.2.4. *Generalized relation between mass scales and length scales*

It is clear that, in the special scale relativity framework, the new status of the Planck length-scale as a lowest unpassable scale must be universal. In particular, it applies also to the de Broglie and Compton relations themselves, while in their standard definition they may reach the zero length. We have given their generalized expressions in Sec. 11.1.1.4.

A fundamental consequence of these new relations for high energy physics is that the mass-energy scale and the length-time scale are no longer inverse as in standard quantum field theories, but they are now related by the special scale-relativistic generalized Compton formula that reads [352]

$$\ln \frac{m}{m_0} = \frac{\ln(\lambda_0/\lambda)}{\sqrt{1 - \ln^2(\lambda_0/\lambda)/\ln^2(\lambda_0/l_{\mathbb{P}})}}. \tag{11.35}$$

Therefore, the new mass-length relation can be given a covariant form, $m/m_0 = (\lambda_0/\lambda)^{\tau(\lambda)}$, which converges at large length-scale toward the standard relation since $\tau(\lambda_0) = 1$.

An important point to be emphasized is that this relation does not depend on the choice of the scale variables and of the fractal exponent that plays the role of the fifth dimension. We have indeed recalled in Chapter 4 that the variation with scale of a fractal coordinate $\mathcal{L}$ can be expressed either in terms of $\delta\mathcal{L}$ (this is a static view of the scale dependence), or in terms of a time parameter used to travel along the curve. The expression for the scale exponent in function of the fractal dimension differs between these two choices: the fractal length respectively reads $\mathcal{L} \propto \delta\mathcal{L}^{1-D_F}$ and $\mathcal{L} \propto \delta t^{(1/D_F)-1}$. As a consequence the generalization of scale laws to the log-Lorentz transformation laws of special scale relativity takes different forms for these two cases (see Sec. 4.4.5), in particular as regards the scale variation of the "djinn".

In this context, one should make the difference between underlying laws, which are not directly observable, and those, which have direct observational or experimental consequence. As already remarked, in the quantum relativistic domain a length interval loses its direct physical meaning [268], so that these coordinate-scale relations cannot be directly observed. The same is true of the law of dilation composition, which is only virtual in the quantum mechanical framework. Indeed, one may conceptualize the existence of a system having structures at many scales and then consider jumping from one scale to another, then from this second scale to the third. But in reality, such a three-scale structure and its description is only virtual, since it is unobservable for a given system, because any actual observation changes the system, so that the second change of scale cannot be made on the same initial system.

This virtuality does not deprive the coordinate-scale relations of physical meaning. For example it is shared by the phases of the wave function in quantum mechanics, which are not directly observable, but whose existence is necessary for a correct description. In the case of the special scale relativity framework, these virtual laws of scale transformation underlie the construction of conservative quantities, which leads to the above generalized Compton formula that relates mass-energy-scales to length-time scales, and which does not depend on the above choice. This relation does have direct consequences, since it changes the high energy structure of elementary particle physics, in a way that is both radical and observable at the scales of today's particle accelerators. In particular, one expects that future experiments to be achieved at the LHC at energies far larger than the W/Z energy of $\approx 100\,\mathrm{GeV}$ could manifest scale relativistic corrections, which are of order $\ln^3(\lambda_Z/r)$.

Concerning coupling constants, the fact that the lowest order terms of their Callan–Symanzik β-function are quadratic (i.e. their renormalization group equation reads $d\alpha/d\mathbb{V} = \beta_0 \alpha^2 + O(\alpha^3)$) implies that their variation with scale is unaffected at this order by scale-relativistic corrections (see more detail in [352, 353]), provided it is written in terms of length scale.

But this means that, when they are written in terms of momentum, the renormalization group equations and their solutions are now changed. The transformation between length and momentum is now performed by using Eq. (11.35).

The main consequence of this new high energy behavior is the solution to the divergence problem of electrodynamics and other gauge field theories [352] (see Fig. 11.3).

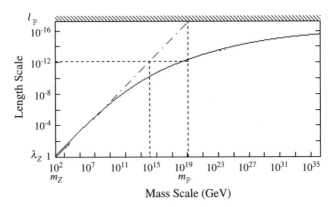

Fig. 11.3: **Mass-length relation in special scale relativity.** In the framework of log-Lorentzian scale transformations, the Planck length $l_\mathbb{P} = \sqrt{\hbar G/c^3}$ becomes a limit, which cannot be exceeded for all lengths in nature. The new relation is such that the Planck space-time scale corresponds to infinite energy-momentum. The Planck energy $m_\mathbb{P} \approx 1.22 \times 10^{19}$ GeV then allows one to define a new universal length scale (horizontal dashed line), which is found to be 10^{-12} times the Z boson scale. This length scale corresponds, in the previous standard theory (dot-dashed line), to an energy of 10^{14} GeV, i.e. just the value of the GUT scale deduced from the convergence of coupling constants.

11.1.3. *Theoretical predictions of standard model parameters*

In the new framework theoretical predictions of some of the free parameters of the standard model become possible. We have presented and checked such predictions in previous works [352, 353, 360, 383], by showing in particular how the agreement between the theoretical predictions and the experimental data improves with the precision improvement of measurements with time.

Indeed, in the recent years there has been an improvement of several experimental measurements [432, 433, 434, 435], so that it may now be interesting to check them again. The year 2006 values [435] for, respectively, top quark mass, Higgs boson mass, W and Z boson masses, strong coupling constant at Z scale, fine structure constant at Z scale, and $\sin^2 \theta$ of weak mixing angle at Z scale in the modified minimal substraction scheme (where it is defined through the SU(2) charge g and the U(1) charge g') are:

$$m_t = 174.2 \pm 3.3 \, \text{GeV}/c^2, \qquad m_H = 114 - 135 \, \text{GeV}/c^2,$$

$$m_W = 80.403 \pm 0.029 \, \text{GeV}/c^2, \quad m_Z = 91.1876 \pm 0.0021 \, \text{GeV}/c^2,$$

$$\alpha_3(m_Z) = 0.1176 \pm 0.0020, \qquad \alpha^{-1}(m_Z) = 128.91 \pm 0.02,$$

$$\hat{s}_Z^2 = g'^2/(g^2 + g'^2) = 0.23120 \pm 0.00015.$$

Scale of grand unification

Because of the new relation between length-scale and mass-scale, the theory yields a new fundamental scale, given by the length-scale corresponding to the Planck energy (see Fig. 11.3). This scale plays for scale laws the same role as played by the Compton scale for motion laws. Indeed, the Compton scale is $\hbar/mc$, while the velocity for which $p = mc$ is $c/\sqrt{2}$. This is a relative scale, which is defined relatively to a given Compton scale.

Let us take the Compton scale of the Z boson as a reference scale. In this case, the new scale is given, to lowest order, by the relation

$$\ln(\lambda_Z/\lambda_{\mathbb{P}}) = \mathbb{C}_Z/\sqrt{2}, \qquad (11.36)$$

where $\mathbb{C}_Z \approx \ln(m_{\mathbb{P}}/m_Z)$. Since $m_{\mathbb{P}}/m_Z \approx (10^{19} \, \text{GeV}/100 \, \text{GeV}) \approx 10^{17}$, the new length-scale linked to the Planck energy is therefore $\approx 10^{-17/\sqrt{2}} = 10^{-12}$ times smaller than the Z length-scale. In other words, this is but the GUT scale ($\approx 10^{14} \, \text{GeV}$ in the standard non scale relativistic theory) in the minimal standard model with "great desert hypothesis', i.e. under the assumption that there is no new elementary particle mass scale beyond the top quark.

This means that the GUT energy scale is now the Planck energy scale, so that the unification of the three gauge fields can no longer be separated from the question of unification with the gravitational field.

11.1.3.1. *Unification of gauge and gravitational fields*

As a consequence, the four fundamental couplings, U(1), SU(2), SU(3) and gravitational converge in the new framework toward about the same scale, which now corresponds to the Planck energy scale (Fig. 11.4). The GUT energy now being of the order of the Planck one ($\approx 10^{19}$ GeV), the predicted lifetime of the proton ($\propto m_{GUT}^4/m_p^5 \gg 10^{38}$ yrs) becomes compatible [352, 353, 360] with experimental results ($> 5.5 \times 10^{32}$ yrs) [435].

There is no strict convergence of the four couplings at Planck energy (see Fig. 11.4), but this may be understood by the threshold effects and the symmetry breaking mechanism. Indeed, at energies larger than the Planck energy, one expects the existence of a unique field, whose properties include, but also go beyond quantum, gauge field and gravitational field properties.

Recall indeed that in the same way as the quantum properties falsify the classical gravitational properties at the Planck energy scale the reverse is also true: namely, one can show that the quantum behavior is itself fundamentally affected by the general relativistic features [353], so that a quantum gravity theory could well be insufficient to really describe the new physics of the Planck scale. In the scale relativity framework, the problem amounts to describing a fractal and curved geometry where fractality and curvature become of the same order, so that the various manifestations of the geometry (gravitation, which manifests the curvature, quantum laws and gauge fields, which manifests the nondifferentiability and the fractality) are mixed and accounted for on the same footing. This reveals to be an extremely difficult task, since it does not mean to combine the motion and scale covariant derivatives after they have been separately constructed, but instead to construct a new covariance that accounts for all effects together. This problem is left open to future works.

Whatever the nature of this unique field, its inverse running coupling $\alpha_g^{-1} = (m_{\mathbb{P}}/m)^2 = \exp[2\ln(m_{\mathbb{P}}/m)]$ very rapidly increases when the energy becomes smaller than the Planck energy (see Figure 11.4). We have suggested [352, 353, 360] that when it reaches the critical value $4\pi^2$, the first symmetry breaking occurs, leading to the separation between the gravitational field (become negligible with respect to the other fields

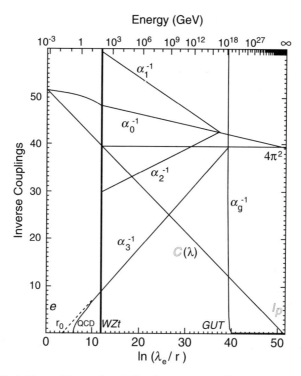

Fig. 11.4: **Variation with scale of the inverse couplings.** Variation with scale, in the special scale relativistic minimal standard model, of the inverse couplings of the fundamental interactions U(1), SU(2) and SU(3), of α_0^{-1}, which is 3/8 of the fine structure constant (electromagnetic coupling), and of the inverse gravitational coupling $\alpha_g^{-1} = (m_{\mathbb{P}}/m)^2$. The running couplings are no longer divergent or vanishing at infinite energy in the new framework. This scale relativistic diagram (mass-scale versus inverse coupling constants) shows well-defined structures and symmetries that are accounted for by mass/charge relations, and by the expected critical value $1/4\pi^2$ of coupling constants (see text). The unification scale, including gravitation, is now the Planck energy scale itself, which no longer corresponds to the Planck length scale (compare the upper and lower graduations).

at these scales), the SU(3) chromodynamics field and an electroweak field. Then, beyond a threshold (like, e.g. for the electromagnetic running coupling at the electron scale, see e.g. [268], [353] Sec. 6.2), the U(1) and SU(2) couplings would separate and run independently.

11.1.3.2. *Mass-charge relation for the electron*

We have shown in Sec. 7.4.5 and recalled in Sec. 11.1.2.3 that a general relation of the form $2\alpha \ln(m_{\mathbb{P}}/m) = k$ could be obtained between a particle mass m and the value of its coupling constant $\alpha(m)$ at corresponding energy.

In order to explicitly apply such a relation to the electron, we must account for the fact that we know from the electroweak theory that the electric charge is only a residual of a more general, high energy electroweak coupling. This coupling can be defined from the U(1) and SU(2) couplings as (see e.g. [353] and references therein)

$$\alpha_0^{-1} = \frac{3}{8}\alpha_2^{-1} + \frac{5}{8}\alpha_1^{-1}. \tag{11.37}$$

This coupling is such that $\alpha_0 = \alpha_1 = \alpha_2$ at the electroweak unification scale and it is related to the fine structure constant at the Z scale by the relation $\alpha = 3\alpha_0/8$. This means that, because the weak gauge bosons acquire mass through the Higgs mechanism, the interaction becomes transported at low energy only by the residual null mass photon. As a consequence the amplitude of the electromagnetic force abruptly falls by a factor $3/8$ at the Z scale. Therefore we have suggested that α_0 instead of α must be used in Eq. (7.59), $2\alpha\mathbb{C} = k$, for relating the electron mass to its charge.

We obtain, for $k = 2$ (see Sec. 11.1.2.3 for a discussion of other values of k), a mass-charge relation for the electron that reads [356, 360]

$$\ln \frac{m_\mathbb{P}}{m_e} = \frac{3}{8}\alpha^{-1}. \tag{11.38}$$

The electromagnetic coupling constant is known to become logarithmically dependent on the energy scales at energies larger than $m_e c^2$, i.e. at length-scales smaller than the Compton length of the electron, because of radiative correction, (equivalently, of electron-positron pair creation). We may therefore consider it as an explicit function of the running mass m, so that the above relation can be viewed as an equation for the unknown electron mass m,

$$\ln \frac{m_\mathbb{P}}{m} = \frac{3/8}{\alpha(m)}, \tag{11.39}$$

where $\alpha(m)$ is the running electromagnetic coupling (see Figs 11.5 and 11.6; see also Fig. 11.4 for a full view of the running coupling from Planck scale to electron scale). If this approach is correct, the solution of this equation should yield both $m = m_e$ and $\alpha_e = \alpha(m = m_e)$.

The existence of such a relation between the mass and the charge of the electron is supported by the experimental data. Indeed, using the known experimental values, the two members of this equation agree to within 0.2%.

One finds $\mathbb{C}_e = \ln(m_{\mathbb{P}}/m_e) = 51.528 \pm 0.001$, while $(3/8)\alpha_e^{-1} = 51.388$, where $\alpha_e = 1/137.036$ is the fine structure constant, i.e. the low energy electromagnetic coupling.

It also agrees within uncertainties with the theoretical prediction of the low energy fine structure constant, $\alpha^{-1} = 137.01 \pm 0.04$, which we have obtained in the special scale relativity framework from its expected bare value and the solution to its renormalization group equation [352, 353] (see Sec. 11.1.3.4 for a complete calculation, and Figs. 11.5 and 11.6).

The agreement is even stronger if one accounts for threshold effects at the electron scale, i.e. the measured fine structure constant and the running

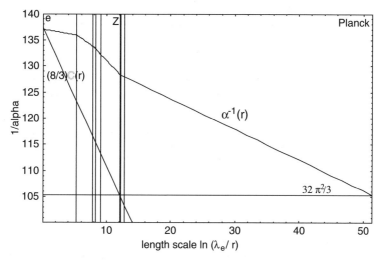

Fig. 11.5: **Running electromagnetic inverse coupling.** Variation in terms of the length-scale r of the running electromagnetic inverse coupling, between the Planck scale and the electron scale (see Fig. 11.4 for a larger view). Between the Planck and the Z boson scale, it is given by the effective running coupling $\alpha_2^{-1} + (5/3)\alpha_1^{-1}$, where α_1 in the U(1) coupling and α_2 the SU(2) coupling of the electroweak theory. In the special scale relativity framework, where the Planck length scale $l_{\mathbb{P}}$ is an invariant minimum scale in nature, it is given at $l_{\mathbb{P}}$ by the "bare" value $(8/3) \times 4\pi^2$ (see text). Between the Z scale and the electron scale (e), it is given by the running inverse fine structure constant $\alpha^{-1}(r)$, which is constant at scales larger than the electron Compton scale, but whose slope changes for each new contribution of particle-antiparticle pairs [thin vertical lines: respectively, from left to right, (e), $(\mu, u, d, s), (c), (\tau), (b), (t)$], and at the electroweak scale because of the mass acquired by the $W^{\pm}$ and Z bosons. The line $8\mathbb{C}(r)/3$, where $\mathbb{C}(r) = \ln(r/l_{\mathbb{P}})$, is a running special scale relativity "constant", in which the factor $8/3$ takes its origin in the electroweak symmetry breaking. The two curves cross precisely at the experimental value of the Compton scale of the electron (thus yielding its low energy mass and charge), as theoretically expected in the scale relativity theory (see a zoom on the crossing region in Fig. 11.6).

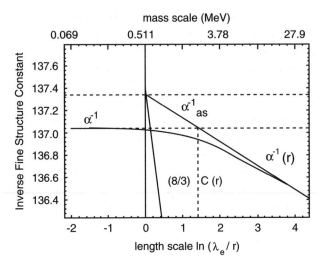

Fig. 11.6: **Mass-charge relation for the electron.** Observed convergence at the electron mass-scale (i.e. at the Compton length of the electron) of the asymptotic running electromagnetic inverse coupling $\alpha^{-1}(r)$ in terms of a running length-scale r and of the running scale-relativity constant $8\mathbb{C}(r)/3 = (8/3)\ln[m_{\mathbb{P}}/m(r)]$. This is an enlarged view by a factor ≈ 50 of Fig. 11.5. Such a convergence is theoretically expected in the framework of the scale-relativistic interpretation of gauge transformations that yields a mass-charge relation for the electron, which reads $(8/3)\alpha_e \ln(m_{\mathbb{P}}/m_e) = 1$. The final low energy fine structure constant differs from its asymptotic value by a small theshhold effect (see e.g. [268], [353] Sec. 6.2).

electron mass (at Bohr scale) differ from the limit of their asymptotic behavior, including radiative corrections (see Fig. 11.6).

Indeed, from the calculation of radiative corrections to the Coulomb law, one finds an asymptotic behavior valid for $r \ll \lambda_e$, where $\lambda_e = \hbar/m_e c$ is the Compton scale of the electron, [268, 353]

$$\alpha(r) = \alpha_e\left\{1 + \frac{2\alpha_e}{3\pi}\left[\ln\frac{\lambda_e}{r} - \left(\gamma + \frac{5}{6}\right)\right]\right\}, \qquad (11.40)$$

where α_e is the low energy fine structure constant and $\gamma = 0.577...$ is the Euler constant. The asymptotic running self-energy (mass) of the electron is given by [238, 353]:

$$m(r) = m_e\left\{1 + \frac{3\alpha_e}{2\pi}\left(\ln\frac{\lambda_e}{r} + \frac{1}{4}\right)\right\}. \qquad (11.41)$$

We now identify the scale relativity constant $\mathbb{C}_e = \ln(m_{\mathbb{P}}/m_e)$ with $(3/8)$ times the value of the asymptotic running fine structure constant at the scale where the asymptotic self-energy of the electron equals its mass. One obtains a new improved relation including threshold effects,

$$\ln \frac{m_{\mathbb{P}}}{m_e} = \frac{3}{8}\alpha_e^{-1} + \frac{1}{4\pi}\left(\gamma + \frac{13}{12}\right). \tag{11.42}$$

One finds $(3/8)\alpha^{-1}\{r(m = m_e)\} = 51.521(1)$ (see Fig. 11.6), which lies within 10^{-4} of the value of $\mathbb{C}_e = 51.528$.

Finally, by this process, the low energy values of the mass and of the charge of the electron have both been obtained from its postulated bare charge $1/2\pi$. This result can be interpreted as indicating that the mass of the electron is mainly of electromagnetic origin. It can therefore be used, reversely, to theoretically predict the electron mass from the experimental value of the fine structure constant. One finds $m_e(\text{th}) = 1.007m_e(\text{exp})$. Provided the approach is correct, this means either that the threshold corrections have not been fully accounted for or that there are other contributions to the electron self-energy, since it is experimentally known with an uncertainty of $0.3\,\text{ppm}$.

One may wonder whether the possible origin of the small remaining difference could be a weak interaction contribution. In this case, one may hope a future development of the same approach to yield a theoretical prediction of the electronic neutrino mass.

11.1.3.3. *Muon mass*

One of the main unsolved mysteries of elementary particle physics is the question of the origin and nature of the muon and tau particles, which exist in addition to the electron and more generally of the existence of three families of particles instead of one. Except for their larger mass, the muon and tau particles are quite similar to the electron (this is the $e - \mu - \tau$ universality). This is also true of the c and t quarks in addition to the u quark and of the s and b quarks, which are similar but more massive than the d quark.

In the scale relativity framework, this question can be addressed in terms of structures (in scale space) of the fractal geodesics to which particles are identified. The very first of these structures is the de Broglie scale, which is identified with the nonfractal to fractal transition in scale space. In particular, the Einstein–de Broglie time $\tau_E = \hbar/mc^2$ is understood as the nonfractal (at long time scales) to fractal (at small time scales) transition for the time coordinate. The Compton length $\lambda_C = c\tau_E = \hbar/mc$ and therefore the mass of the particle itself are understood as other expressions of this transition.

In an equivalent way, the mass is interpreted as yielding the amplitude of the fractal fluctuation on the geodesics,

$$d\xi^2 \sim \frac{\hbar}{m} ds, \tag{11.43}$$

here for the critical value $D_F = 2$ of the geodesics fractal dimension ("Galilean" scale relativity).

In special scale relativity, where the Galilean-like laws of dilation are generalized to log-Lorentzian laws, the fractal dimension becomes variable in function of length scale. The 'fractal fluctuations become (see Sec. 6.5)

$$d\xi^{D_F(d\xi)} \sim \frac{\hbar}{m} ds, \tag{11.44}$$

where the fractal dimension takes the value $D_F = 2$ at the Compton scale and then slowly increases toward the small scales. This new relation can then be put under the form of a "scale relativistic correction" to the previous "Galilean" relation, in terms of a scale varying effective mass.

This tentatively suggests a possible mechanism of generation of the mass and charge spectrum of elementary particles [355, 360]. It is based on the observation that the variation of the electron mass due to special scale-relativistic "corrections" is closely linked to its variation due to radiative corrections. Let us show that this mechanism may provide a hint on the origin of the muon and of its mass.

We take as scale variable $r = d\xi$ and we set

$$\mathbb{V} = \ln \frac{\lambda_e}{d\xi}, \tag{11.45}$$

where $\lambda_e = \hbar/m_e c$ is the Compton length of the electron and m_e its low energy mass.

When the fractal dimension is variable with scale, the mean fractal fluctuation can be written as

$$\frac{\langle d\xi^2 \rangle}{ds} = \frac{\hbar}{m_e} \left(\frac{ds}{\lambda_e} \right)^{\frac{2}{D_F(r)} - 1} \tag{11.46}$$

with $D_F(r) = 1 + \tau(r)$. In terms of the "djinn" τ it reads

$$\frac{\langle d\xi^2 \rangle}{ds} = \frac{\hbar}{m_e} \left(\frac{\lambda_e}{ds} \right)^{\frac{\tau-1}{\tau+1}}, \tag{11.47}$$

while the scales variables ds and $d\xi$ are linked by the relation

$$\ln\frac{\lambda_e}{ds} = (1+\tau)\ln\frac{\lambda_e}{d\xi}, \tag{11.48}$$

so that one obtains

$$\frac{\langle d\xi^2\rangle}{ds} = \frac{\hbar}{m_e}\left(\frac{\lambda_e}{d\xi}\right)^{\tau-1} = \frac{\hbar}{m_e}e^{\mathbb{V}(\tau-1)}. \tag{11.49}$$

In the special scale relativistic case, the "djinn" τ varies in terms of the scale variable as (see previous Sec. 11.1.1)

$$\tau(\mathbb{V}) = \frac{1}{\sqrt{1-\mathbb{V}^2/\mathbb{C}_e^2}}. \tag{11.50}$$

To lowest order it yields

$$\tau = 1 + \frac{1}{2}\frac{\mathbb{V}^2}{\mathbb{C}_e^2} \tag{11.51}$$

with $\mathbb{C}_e = \ln(\lambda_e/l_{\mathbb{P}})$. Then we finally obtain to lowest order in $\mathbb{V}$:

$$\frac{\langle d\xi^2\rangle}{ds} = \frac{\hbar}{m_e}\left(1+\frac{1}{2}\frac{\mathbb{V}^3}{\mathbb{C}_e^2}\right). \tag{11.52}$$

We can keep the same form for this expression as for the "Galilean" fractal fluctuation,

$$\frac{\langle d\xi^2\rangle}{ds} = \frac{\hbar}{m(\mathbb{V})}, \tag{11.53}$$

provided one introduces a scale dependent "running" effective mass $m(\mathbb{V})$ that reads to lowest order

$$\frac{m(\mathbb{V})}{m_e} = 1 - \frac{1}{2}\frac{\mathbb{V}^3}{\mathbb{C}_e^2}. \tag{11.54}$$

To this approximation, this relation achieves a kind of "scale relativistic correction" to the electron mass (see Fig. 11.7).

   More generally, if one assumes that the log-Lorentzian regime does not begin exactly at the Compton scale of the electron, one may introduce a zero point $\mathbb{V}_0$ in this relation:

$$\frac{m(\mathbb{V})}{m_e} = 1 - \frac{1}{2}\frac{(\mathbb{V}-\mathbb{V}_0)^3}{\mathbb{C}_e^2}. \tag{11.55}$$

One may indeed make the conjecture that the transition to the log-Lorentzian regime corresponds more precisely to electron-positron pair

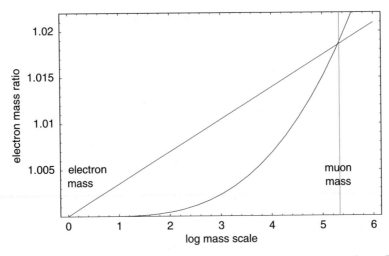

Fig. 11.7: **Electron mass: radiative versus scale relativistic corrections.** This figure shows the two contributions to the variation of the mass of the electron (over its low energy mass) in function of the energy scale (in logarithm). The straight line gives the radiative correction on the mass of the electron from the lowest order solution to its renormalization group equation. The curved line gives the special scale-relativistic correction to the electron mass (taking the mass scale $2m_e$ as zero point). The two corrections intersect at the muon mass and therefore cancel each other just for its value.

creations. It must be recalled that these pairs are re-interpreted, in the fractal geometric view, as parts of the geodesics that run backward in time and that they consequently ensure the transition from a fractal space to a fractal space-time (see Chapter 6). In this case, the transition would be for the mass scale $2m_e$, which corresponds to $\mathbb{V}_0 = \ln 2$.

In other respects, the electron mass is already known to vary with scale in quantum electrodynamics. To lowest order, the running self-mass of the electron, as given by radiative corrections [530] or equivalently by the solution to its renormalization group equation (see e.g. [238]) reads

$$\frac{m_{\mathrm{RC}}(\mathbb{V})}{m_e} = 1 + \frac{3\alpha_e}{2\pi}\mathbb{V}, \tag{11.56}$$

where α_e is the low energy fine structure constant. More profoundly, this scale dependence of the electron mass in quantum electrodynamics may itself be attributed to the fractality of space-time, i.e. to the structures of geodesics running backward in time for time scales smaller than $\hbar/m_e c^2$ [353, Chapter 6].

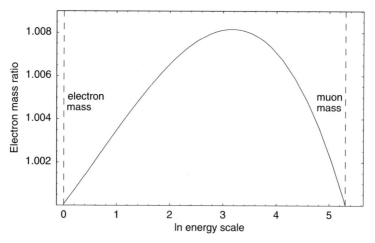

Fig. 11.8: **Generation of the muon mass.** This figure shows the full variation of the ratio $m(E)/m_e$, where $m(E)$ is the running mass of the electron and m_e its low energy mass, in function of the energy scale $\ln(E/m_e c^2)$ in the combined special scale relativity and quantum electrodynamics frameworks. The special scale relativistic correction (taking the electron-positron pair mass $2m_e$ as zero point), and the radiative correction to the electron mass cancel each other just for the value of the muon mass.

Therefore, the combination of quantum electrodynamics and of special scale relativity yields a full "correction" given by

$$\frac{m(\mathbb{V})}{m_e} = \frac{1 + (3\alpha_e/2\pi)\mathbb{V}}{1 + (\mathbb{V} - \mathbb{V}_0)^3/2\mathbb{C}_e^2}. \tag{11.57}$$

The intriguing result is that the two corrections cancel each other just for the mass scale of the muon (see Figs. 11.7 and 11.8). To the first approximation (for a zero point at the Compton scale of the electron, i.e. $\mathbb{V}_0 = 0$), the cancelation point is given by

$$\mathbb{V}_\mu = \sqrt{\frac{3\alpha_e}{\pi}}\,\mathbb{C}_e. \tag{11.58}$$

By identifying more precisely the transition to the log-Lorentz regime with the scale $2m_e$, we find that the mass scale where the two corrections cross is $\mathbb{V} = 5.306$, while the muon mass $m_\mu = 105.658369(9)\,\text{MeV}$ [435] corresponds to $\mathbb{V}_{\mu 0} = 5.332$. Moreover, this last value corresponds to the mass scale ratio of muon over electron mass, while $\mathbb{V}_\mu$ is defined from the ratios of their Compton lengths. In special scale relativity they are no longer the same, but linked by the relation

$$\mathbb{V}_\mu = \frac{\mathbb{V}_{\mu 0}}{\sqrt{1 + \mathbb{V}_{\mu 0}^2/\mathbb{C}_e^2}}, \qquad (11.59)$$

yielding $\mathbb{V}_\mu = 5.303$, whose relative difference with the cancellation point of the two corrections is only 6×10^{-4}.

More work is needed to decide if this result is meaningful and relevant for the understanding of the origin of the muon. In particular, the close agreement between the experimental mass and the obtained mass may be partly due to chance, since it depends on the choice of the zero points (numerical constants), which intervene in the various scale formula and variables (knowing that these numerical constants may also differ between the position and the momentum representations).

Problems and Exercises

Open Problem 23: Generalize the above mechanism of muon generation to a full generation mechanism of the mass and charge spectrum of elementary fermions (see [355, 360] for the tentative suggestion of such a mechanism).

Hint: This generation mechanism may be based on the full expression for the variation of the running electron coupling $\alpha(r)$ due to pairs of elementary fermions-antifermions,

$$\frac{\alpha_e}{\alpha(r)} = 1 - \frac{2\alpha_e}{3\pi} \left[\sum_{i=0}^{n} Q_i^2 \mathbb{V} - \sum_{i=0}^{n} (Q_i^2 \mathbb{V}_i) \right], \qquad (11.60)$$

where α_e ($\approx 1/137.036$) is the low energy fine structure constant, n is the number of elementary pairs of fermions of dimensionless charges $Q_i = q_i/e$ and of Compton lengths $\lambda_i = \hbar/m_i c > r$, and where we have set $\mathbb{V} = \ln(\lambda_e/r)$ and $\mathbb{V}_i = \ln(\lambda_e/\lambda_i)$. This formula is written to lowest order and neglects threshold effects. This relation means that the information about the masses of elementary charged fermions (contained in the Compton scales λ_i) is "coded" in the scale variation of the electric charge in terms of transitions scales where the slope abruptly changes, while the information about their charges is contained in the slopes themselves. Moreover, one can relate the running masses and the running charges through their renormalization group equations (see e.g. [353, Chapter 6]). For example, between the electron and muon scale one finds $m/m_e = (\alpha/\alpha_e)^{9/4}$.

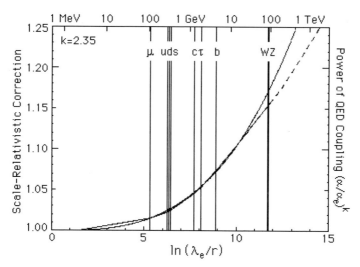

Fig. 11.9: **Special scale relativistic correction.** Comparison between the special scale-relativistic correction and the variation of the fine structure constant (to the power k) due to the pairs of elementary charged fermions.

The generation mechanism therefore consists of cancellating the radiative correction, which depends on the particle content, by the special scale relativity cubic V^3 correction (see Fig. 11.9). The points where the cancellations occur would yield the masses while the changes of slopes would yield the square of charges content of new particles at these mass scales.

∎

11.1.3.4. *Fine structure constant*

Coupling constants (charges) and self-energies (masses) have finite non-zero values at infinite energy in the new framework, while in the standard model they were either infinite (Abelian U(1) group) or null (asymptotic freedom of non-Abelian groups). Such a behavior of the standard theory prevented relating the "bare" (infinite energy) values of charges to their low energy values, while this is now possible in the scale relativistic standard model.

Let us denote by $\bar{\alpha} = \alpha^{-1}$ the inverse running couplings. We have defined [353] a formal QED inverse coupling,

$$\bar{\alpha}_0 = \frac{3}{8}\bar{\alpha}_2 + \frac{5}{8}\bar{\alpha}_1 = \frac{3}{8}\bar{\alpha}, \tag{11.61}$$

where $\bar{\alpha}_1$ and $\bar{\alpha}_2$ are respectively the U(1) and SU(2) inverse couplings. This definition relies on the fact that three among the four gauge bosons acquire mass through the Higgs mechanisms at low energy scales, so that the QED coupling jumps from α_0 to $\alpha = 3\alpha_0/8$ at the Z scale. This coupling is such that $\alpha_0 = \alpha_1 = \alpha_2$ at unification scale. It is therefore α_0 rather than α, which must be used in the mass-charge relation. One then obtains a relation for the electron (see Sec. 7.4.5) that reads, disregarding threshold corrections,

$$\ln \frac{m_{\mathbb{P}}}{m_e} = \frac{3}{8} \alpha^{-1}. \tag{11.62}$$

While, in the standard model, α_0 is logarithmically divergent, it becomes convergent in the special scale-relativistic framework. We have therefore suggested that the critical value $4\pi^2$ theoretically expected for bare couplings could apply to it, so that the expectation for the bare fine structure constant would be

$$\alpha_\infty^{-1} = \frac{32\pi^2}{3} = 105.276\ldots. \tag{11.63}$$

The difference between the infinite energy and the Z energy value of the fine structure constant can then be computed using the solution to the renormalization group equation for the running couplings that reads to lowest order (see [353, 360, 372], references therein and what follows for second order terms)

$$\alpha^{-1}(r) = \alpha^{-1}(\lambda_Z) - \left(\frac{5}{3\pi} + \frac{N_H}{6} \right) \ln \frac{\lambda_Z}{r} + \cdots, \tag{11.64}$$

where N_H is the number of Higgs doublets. By using the approximation $\mathbb{C}_Z \approx 4\pi^2$ in these equations, one obtains a simplified expression for its Z value for $N_H = 1$ Higgs doublet, [372, 383]

$$\alpha^{-1}(m_Z) = \frac{32\pi^2}{3} + \frac{22\pi}{3} + \frac{6}{\pi^2} + \cdots = 128.92, \tag{11.65}$$

which agrees within uncertainties with the recent determination $\alpha^{-1}(m_Z) = 128.91 \pm 0.02$ [435].

The difference between the Z value and the low energy value of α has been improved [432] to the value $\Delta\bar{\alpha}_{eZ} = 8.12 \pm 0.03$, and more recently $\Delta\bar{\alpha}_{eZ} = 8.126 \pm 0.020$ [435]. One obtains a prediction for the low energy fine structure constant from its scale-relativistic bare value $\alpha^{-1} = 137.04 \pm 0.045$ [383], now improved with the PDG2006 values to

$$\alpha^{-1} = 137.01 \pm 0.035, \tag{11.66}$$

which compares well with the experimental value $\alpha^{-1} = 137.036$ (to this order of precision). Note that the error comes here from the uncertainties in the coupling variation, not from the new theoretical expectation, which is exact.

Reversely, one may also run the electromagnetic coupling from the electron scale down to the Planck length-scale by using its renormalization group equation in order to put to the test the $4\pi^2$ conjecture for the inverse bare coupling. This procedure yields results that improve with time, thanks to the improvements of the experimental measurements [435] of the various parameters, which occur in this calculation (see what follows). The first attempt performed in 1991 had yielded [352] $\alpha_0^{-1}(\infty) = 40.0 \pm 0.5$, while with 2006 values one obtains $\alpha_0^{-1}(\infty) = 39.489 \pm 0.013$. The apparent convergence of these results toward $4\pi^2 = 39.478\ldots$ (see Fig. 11.10) supports the $4\pi^2$ conjecture about the bare inverse coupling.

A consequence of this result is that one theoretically predicts that the number of Higgs doublets, which contributes to $2.11N_H$ in the final value of $\bar{\alpha}$, must be $N_H = 1$.

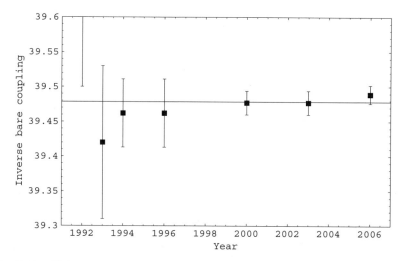

Fig. 11.10: **Values of bare electromagnetic coupling (at infinite energy).** Values of the inverse bare electromagnetic coupling $(3/8)\alpha^{-1}$ (at infinite energy), derived from the low energy fine structure constant and from the solution to the renormalization group equation for the running constant compared with the theoretical expectation for a bare inverse coupling, $\alpha_b^{-1} = 4\pi^2$ (see text and Fig. 11.4). The figure shows the results obtained in various publications since the 1991 theoretical prediction: [352] (40.0 ± 0.5 in 1992), [353] (1993), [355] (1994), [360] (1996), [372] (2000), [383] (2003) and the present one with 2006 values (unchanged in 2010).

Let us sum up the calculations that lead to these results. The running of the inverse fine structure constant from its infinite energy (i.e. Planck length-scale) value to its low energy (electron scale) value reads (see [353, 355] and references therein)

$$\bar{\alpha}(\lambda_e) = \bar{\alpha}(l_{\mathbb{P}}) + \Delta\bar{\alpha}^{(1)}_{\mathbb{P}Z} + \Delta\bar{\alpha}^{(2)}_{\mathbb{P}Z} + \Delta\bar{\alpha}^{L}_{Ze} + \Delta\bar{\alpha}^{h}_{Ze} + \Delta\bar{\alpha}^{\text{Sc-rel}} \quad (11.67)$$

where $\bar{\alpha}(l_{\mathbb{P}}) = \bar{\alpha}(E = \infty) = 32\pi^2/3$; $\Delta\bar{\alpha}^{(1)}_{\mathbb{P}Z}$ is the first order variation of the inverse coupling between the Planck length-scale and the Z boson length-scale, as given by the solution to its renormalization group equation,

$$\Delta\bar{\alpha}^{(1)}_{\mathbb{P}Z} = \frac{10 + N_H}{6\pi} \ln \frac{\lambda_Z}{l_{\mathbb{P}}} = \frac{10 + N_H}{6\pi} \mathbb{C}_Z = 23.01 + 2.11(N_H - 1).$$
$$(11.68)$$

For one Higgs doublet and taking $\mathbb{C}_Z \approx 4\pi^2$, it is approximated by $\Delta\bar{\alpha}^{(1)}_{\mathbb{P}Z} \approx 22\pi/3$ (see Eq. (11.65)).

$\Delta\bar{\alpha}^{(2)}_{\mathbb{P}Z}$ is its second order variation, which now depends on the three fundamental couplings α_1, α_2 and α_3, which may themselves be estimated thanks to their renormalization group equations (see Fig. 11.11)

$$\Delta\bar{\alpha}^{(2)}_{\mathbb{P}Z} = -\frac{104 + 9N_H}{6\pi(40 + N_H)} \ln\left\{1 - \frac{40 + N_H}{20\pi}\alpha_1(\lambda_Z)\ln\frac{\lambda_Z}{l_{\mathbb{P}}}\right\}$$
$$+ \frac{20 + 11N_H}{2\pi(20 - N_H)} \ln\left\{1 + \frac{20 - N_H}{12\pi}\alpha_2(\lambda_Z)\ln\frac{\lambda_Z}{l_{\mathbb{P}}}\right\}$$
$$+ \frac{20}{21\pi} \ln\left\{1 + \frac{7}{2\pi}\alpha_3(\lambda_Z)\ln\frac{\lambda_Z}{l_{\mathbb{P}}}\right\} = 0.767 \pm 0.030. \quad (11.69)$$

$\Delta\bar{\alpha}^{L}_{Ze}$ is the leptonic contribution to its variation between electron and Z scales,

$$\Delta\bar{\alpha}^{L}_{Ze} = \frac{2}{3\pi}\left\{\ln\left(\frac{m_Z}{m_e}\right) + \ln\left(\frac{m_Z}{m_\mu}\right) + \ln\left(\frac{m_Z}{m_\tau}\right) - \frac{5}{2}\right\} = 4.30578(5)$$
$$(11.70)$$

with the PDG 2006 values of the masses [435]. $\Delta\bar{\alpha}^{h}_{Ze}$ is the hadronic contribution to its variation between electron and Z scales, which can be precisely inferred from the experimental values of the ratio R of the cross sections $\sigma(e^+e^- \to \text{hadrons})/\sigma(e^+e^- \to \mu^+\mu^-)$ [78]. In [360] we quoted the

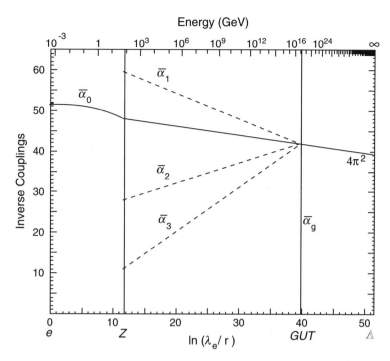

Fig. 11.11: **Variation with scale of the QED inverse coupling.** Variation of the QED inverse coupling between infinite energy scale (i.e. Planck length-scale in special scale-relativity, see up graduation), Z boson scale, and electron scale (see text). The running fine structure constant is given, at low energies, by $\alpha = 3\alpha_0/8$, because the weak bosons have acquired mass in the Higgs mechanism. To second order its variation between the Z and the Planck length scale depends on the other couplings (see text), but only an approximation of the $\alpha_1, \alpha_2,$ and α_3 couplings are needed in this calculation (dashed lines). (See Fig. 11.4 for a more precise description of their variation).

value $\Delta\bar{\alpha}_{Ze}^h = 3.94 \pm 0.12$, which is now greatly improved to [435]

$$\Delta\bar{\alpha}_{Ze}^h = 3.825 \pm 0.018. \tag{11.71}$$

Finally $\Delta\bar{\alpha}^{\text{Sc-rel}} = -0.18 \pm 0.01$ is the scale-relativistic correction, which comes from the fact that the length-scales and mass-scales of elementary particles are no longer directly inverse in the new framework [355]. Combining all these contributions, we have obtained $\bar{\alpha}(\lambda_e) = 137.08 + 2.11(N_H - 1) \pm 0.13$ [360], then more recently $137.04 + 2.11(N_H - 1) \pm 0.045$ [383] (the quoted uncertainty 0.03 was an underestimate). With the PDG2006 values, these estimates have been improved to

$$\bar{\alpha}(\lambda_e) = 137.006 + 2.11(N_H - 1) \pm 0.036, \tag{11.72}$$

which agrees within uncertainties with the experimental value 137.035 999 11(46) [435] for a number of Higgs doublets $N_H = 1$.

Note however that the quoted error may still be an underestimate. Indeed, (i) the precise choice of the reference scale in the determination of the constants $\mathbb{C}$ remains not fully clear and (ii) the QED coupling is runned up to the infinite energy (Planck length) scale, while we know that the physics is dramatically changed at the Planck energy scale, beyond which it is expected to be described by a "quantum gravity" theory which remains to be constructed. The running coupling between the Planck energy and the infinite energy scales should therefore not be considered as a real one, but as a theoretical purely mathematical extrapolation, which describes an ideal situation, the same being true of the ideal bare coupling $1/4\pi^2$.

Now, since the experimental value of the fine structure constant is far more precise than this partly theoretical expectation, it cannot be directly validated by its predictive power (e.g. about new unknown decimals, as in the strong coupling case studied in the next section). However, this result has an indirect consequence about the number of Higgs doublets, which is now strongly constrained in this framework. One indeed finds, by identifying the above result to the experimental inverse fine structure constant,

$$N_H = 1.018 \pm 0.020, \tag{11.73}$$

so one expects only one Higgs doublet, a prediction that the next generation of accelerators will certainly be able to validate or invalidate.

11.1.3.5. *QCD coupling*

We have made the conjecture [352, 353] that the SU(3) inverse coupling reaches the critical value $4\pi^2$ at unification scale, i.e. at an energy $m_{\mathbb{P}}c^2/2\pi$ in the special scale-relativistic modified standard model (see Fig. 11.4).

By the same method as for the fine structure constant, namely, by running the coupling from the Planck to the Z scale, this conjecture allows one to get a theoretical estimate for the value of the QCD coupling at Z scale. Indeed its renormalization group equation yields a variation of $\alpha_3 = \alpha_s$ with length scale given to first order, for a number n_f of quarks, by

$$\alpha_3^{-1}(r) = \alpha_3^{-1}(\lambda_Z) + \frac{33 - 2n_f}{6\pi} \ln \frac{\lambda_Z}{r}, \tag{11.74}$$

and to second order (for six quarks) by

$$\alpha_3^{-1}(r) = \alpha_3^{-1}(\lambda_Z) + \frac{7}{2\pi} \ln \frac{\lambda_Z}{r}$$

$$+ \frac{11}{4\pi(40 + N_H)} \ln \left\{ 1 - \frac{40 + N_H}{20\pi} \alpha_1(\lambda_Z) \ln \frac{\lambda_Z}{r} \right\}$$

$$- \frac{27}{4\pi(20 - N_H)} \ln \left\{ 1 + \frac{20 - N_H}{12\pi} \alpha_2(\lambda_Z) \ln \frac{\lambda_Z}{r} \right\}$$

$$+ \frac{13}{14\pi} \ln \left\{ 1 + \frac{7}{2\pi} \alpha_3(\lambda_Z) \ln \frac{\lambda_Z}{r} \right\}. \tag{11.75}$$

More complete solutions are now available thanks to the knowledge of the four-loop β function in QCD [464, 523], but the above solution is sufficient for our purpose owing to the other uncertainties in the calculation.

The variation with energy scale is obtained by making the transformation given by Eq. (11.35). This led to the expectation $\alpha_3(m_Z) = 0.1165 \pm 0.0005$ [352] (where the uncertainty actually was an underestimate) that compared well with the experimental value at that time, $\alpha_3(m_Z) = 0.112 \pm 0.010$.

This calculation has been more recently reconsidered [80], by using improved experimental values of the α_1 and α_2 couplings at Z scale, which intervene at second order, and by a better account of the top quark contribution. Indeed, its mass was unknown at the time of the first attempt in 1991, so that the running from Z scale to Planck scale was performed by assuming the contribution of six quarks on the whole scale range.

However, the now known mass of the top quark, $m_t = 174.2 \pm 3.3 \, \text{GeV}$ [435] is larger than the Z mass, so that only five quarks contribute to the running of the QCD coupling between Z scale and top quark scale, then six quarks between top and Planck scale. Moreover, the possibility of a threshold effect at top scale cannot be excluded.

One of the uncertainties comes from the determination of the constant $\mathbb{C}_Z = \ln(\lambda_Z / l_\mathbb{P})$. A first order estimate of this constant is given by $\ln(m_\mathbb{P} / m_Z) = 39.436$. However, in the special scale relativity framework, one should account for the fact that the mass scales and the length scales are no longer strictly inverse, and that only scale ratios do have a physical meaning. The Compton length of the Z boson λ_Z is therefore actually defined only with respect to a given reference scale. The electron Compton scale, which makes the transition between the low energy log-Galilean physics and the log-Lorentzian scale laws, should be taken as such a reference scale [352, 353].

Using the PDG2008 values [436]

$$m_e = 0.510998910(13)\,\text{MeV}/c^2, \quad m_Z = 91.1876(21)\,\text{GeV}/c^2, \quad (11.76)$$

and

$$m_{\mathbb{P}} = 1.22089(6) \times 10^{19}\,\text{GeV}/c^2, \quad (11.77)$$

one obtains

$$\mathbb{C}_e = \ln\frac{\lambda_e}{l_{\mathbb{P}}} = \ln\frac{m_{\mathbb{P}}}{m_e} = 51.52784(5) \quad (11.78)$$

and

$$\mathbb{C}_Z = \ln(\lambda_Z/l_{\mathbb{P}}) = \mathbb{C}_e - \ln\varrho_{Ze} = 39.7556(2), \quad (11.79)$$

where the scale ratio between the electron and Z boson lengths is now given by

$$\ln\varrho_{Ze} = \frac{\ln(m_Z/m_e)}{\sqrt{1 + \ln^2(m_Z/m_e)/\mathbb{C}_e^2}} = 11.7723(10). \quad (11.80)$$

With this value of $\mathbb{C}_Z$, one finds [80] the following results for the strong coupling value at Z scale, computed from the assumption that the running coupling crosses the gravitational coupling $\alpha_g = (m/m_{\mathbb{P}})^2$ at the critical value $1/4\pi^2$ and from the solution to its renormalization group equation between Planck and Z scale, assuming one Higgs doublet:

First order:	0.110557
With second order:	0.1156
Second order with account of top quark:	0.1166
With account of top Yukawa coupling:	0.1169.

The contribution of other Higgs doublets (of the order of -0.000006) and of the third and fourth order terms in the renormalization group equations ($+0.000005$) are negligible to this order of precision. The contribution of a possible top threshold effect (in position space) is estimated by Campagne [80] to be $+0.0008$. The final estimate can therefore be summarized as

$\alpha_s(m_Z) = 0.1169 - 0.1177$, i.e.

$$\alpha_s(m_Z)_{\text{pred}} = 0.1173 \pm 0.0004, \tag{11.81}$$

which agrees within uncertainties with our initial estimate 0.1165(5) [352]. This expectation is in good agreement with the recent experimental average [435]

$$\alpha_s(m_Z)_{\text{exp}} = 0.1176 \pm 0.0009, \tag{11.82}$$

where the quoted uncertainty is the error on the average (but note that the authors finally adopt, in their final result, a conservative uncertainty 0.002 to account for possible unknown uncertainties). In Fig. 11.12, we give the evolution of the measurement results of the strong coupling at Z scale, compared with this expectation.

Reversely, one may start from the Z experimental value of the strong coupling, $\alpha_s(m_Z) = 0.1176 \pm 0.0009$, then run it up to the Planck scale using the renormalization group equations and the scale relativistic corrections (see Fig. 11.4). One finds a difference 0.0916(8), and therefore $\alpha_s(\mathbb{P}) = 0.0260(12)$, which lies within less than one sigma of $1/4\pi^2 = 0.0253\ldots$, as expected.

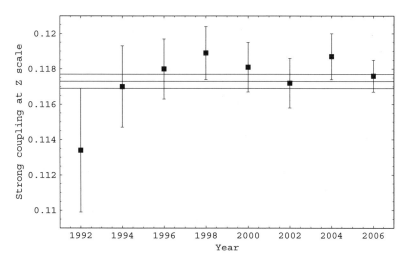

Fig. 11.12: **Values of $\alpha_s(M_Z)$ compared with theoretical expectation.** Measured values of $\alpha_s(M_Z)$ from 1992 to 2006 [428]-[435] compared with the expectation $\alpha_s(m_Z) = 0.1173 \pm 0.0004$ obtained by assuming that the inverse running coupling reaches the value $4\pi^2$ at Planck scale (see text and Fig. 11.4). The PDG2010 experimental value is yet improved to 0.1184 ± 0.0007. It lies within one standard deviation of the theoretical expectation with top threshold, 0.1177.

11.1.3.6. *Electroweak scale and solution to the hierarchy problem*

We have also proposed that the ratio of the WZ electroweak scale over the Planck scale is determined by the same number, i.e. by the bare inverse coupling $\bar{\alpha}_0(\infty) \approx 4\pi^2$ [353]. The large value of the ratio of GUT scale over electroweak unification scale is known as the hierarchy problem. This ratio is of the order of 10^{12} in the minimal grand unified theories where the unification energy is about 10^{14} GeV. It becomes 10^{14} in supersymmetric extensions where the unification scale becomes 10^{16} GeV. But in the special scale relativity framework, the GUT energy becomes the Planck energy, and therefore the hierarchy ratio becomes $m_{\mathbb{P}}/m_Z = 1.34 \times 10^{17}$.

A possible solution of this hierarchy problem could come from the existence of an expected mass-coupling relation

$$\alpha\mathbb{C} = 1, \tag{11.83}$$

applied to the critical bare charge at infinite energy $\alpha_b = 1/4\pi^2$. Since $\mathbb{C} = \ln(m_{\mathbb{P}}/m)$ to lowest order, such a relation yields a scale ratio

$$m_{\mathbb{P}}/m = e^{4\pi^2} = 1.397 \times 10^{17}, \tag{11.84}$$

which is remarkably close to the Planck over WZ scale ratio.

Indeed the relation $\ln(m_P/m) = \bar{\alpha}_b = 4\pi^2$ yields a mass scale $m_w = 87.393$ GeV, which lies between the W and Z boson masses (currently $m_Z = 91.1876(21)$ GeV/c^2, $m_W = 80.403(29)$ GeV/c^2, [435]). One may, more strictly, compute the new fundamental length scale λ_v given by

$$\ln \frac{\lambda_v}{l_{\mathbb{P}}} = 4\pi^2. \tag{11.85}$$

Since mass-scales and length-scales are no longer directly inverse in the scale-relativity framework, there is a "log-Lorentz" factor between them (when they are referred to low energy scales). Specifically, by taking the electron Compton scale as reference, the new mass-scale is given by

$$\ln \frac{m_v}{m_e} = \frac{\ln(\lambda_e/\lambda_v)}{\sqrt{1 - \ln^2(\lambda_e/\lambda_v)/\mathbb{C}_e^2}}. \tag{11.86}$$

With the present value of the gravitational constant $G = 6.6742(10) \times 10^{-11}$ MKSA [435], we obtain for the fundamental constant $\mathbb{C}_e = \ln(\lambda_e/l_{\mathbb{P}}) = 51.52785(9)$. Then the new theoretically predicted mass scale is

$$m_v = 123.210 \pm 0.021 \text{ GeV}/c^2, \tag{11.87}$$

which improves the previously obtained value $m_v = 123.23 \pm 0.09 \, \text{GeV}/c^2$ [383]. Such a mass scale may be linked with the vacuum expectation value v of the Higgs field, since one finds from the experimental data $v/\sqrt{2} = m_W/g = 123.11 \pm 0.03 \, \text{GeV}$ (where g is the SU(2) weak charge).

11.1.3.7. *Ratios of coupling constants*

In [360] we also attempted to apply the mass-charge relation to the SU(2) coupling α_2. To this purpose, we need to generalize the scale (i.e. gauge) transformations to no longer global dilations: one can instead consider different and independent dilations on the internal resolutions corresponding to the various coordinates. This generalization is just at the basis of the geometric construction of non-Abelian gauge fields. The group SU(2) corresponds to rotations in a three-dimensional scale-space. Therefore the phase term in a fermion field reads

$$\alpha_2 \left(\ln \frac{\varepsilon_x}{\lambda} + \ln \frac{\varepsilon_y}{\lambda} + \ln \frac{\varepsilon_z}{\lambda} \right) < 3\alpha_2 \ln \frac{l_{\mathbb{P}}}{\lambda}, \tag{11.88}$$

since the same coupling applies to the three variables and since all three resolutions are limited toward small scales by the Planck scale. We therefore obtain a generalized mass-coupling relation

$$3\alpha_{2Z} \mathbb{C}_Z = k, \tag{11.89}$$

with k integer. One finds that this relation is achieved at the Z scale for $k = 4$ [383], but, for the moment, we have no understanding of the origin of this precise value of k.

Reversely, one may therefore attempt to determine the SU(2) coupling at Z scale from the relation $\alpha_{2Z} \mathbb{C}_Z = 4/3$. One obtains $\alpha_{2Z}^{-1} = 29.8169 \pm 0.0002$. The experimental value is

$$\alpha_{2Z}^{-1} = \alpha_Z^{-1} \times \hat{s}_Z^2 = 29.802 \pm 0.027, \tag{11.90}$$

which lies within one standard deviation of the theoretical prediction.

More generally, the universality of mass-coupling relations for any gauge field implies that coupling ratios at Z scale should be integers. Current values ($\alpha_{1Z}^{-1} = 59.471 \pm 0.026; \alpha_{2Z}^{-1} = 29.802 \pm 0.027; \alpha_{3Z}^{-1} = 8.47 \pm 0.14$) [432] support this expectation within a 2σ confidence interval, since one obtains [383]

$$\left(\frac{\alpha_3}{\alpha_1} \right)_Z = 7.02 \pm 0.12; \quad \left(\frac{\alpha_2}{\alpha_1} \right)_Z = 1.996 \pm 0.002. \tag{11.91}$$

Now such a result is for the time being only incomplete and tentative, since there is no understanding of the integer values of these ratios, two for SU(2) and seven for SU(3).

Provided it is confirmed, the relation $\alpha_{2Z} = 2\alpha_{1Z}$ would fix the value of the weak angle at Z scale, or, in an equivalent way, the W/Z mass ratio. Namely, it yields

$$(\sin^2 \theta)_Z = \frac{3}{13}; \quad \frac{m_W}{m_Z} = \sqrt{\frac{10}{13}}. \tag{11.92}$$

This value, $3/13 = 0.230769\ldots$, is indeed close to the experimental values of $(\sin^2 \theta)_Z, \hat{s}_Z^2 = 0.23122(15)$ when it is defined from the charges and $s_{M_Z}^2 = 0.23108(5)$ when it is defined from M_Z by removing the top and Higgs dependence, which explains the smaller uncertainty, while the on-shell value is 0.2231 [435]. However, because of the experimental improvement, it now lies outside their confidence interval. This is not unexpected since, to this precision, the theoretical expectation also needs to be defined in a more precise way and would be subjected to small corrections depending on its definition.

In order to perform such a more precise prediction, one needs to justify the existence of a relation between the electroweak couplings at Z scale from a more complete theoretical approach. In today's unification theories, one expects new constraints only at large energies. But in the scale relativity framework, constraints may be set both at large and small scales [353]. We have suggested a more constrained electroweak theory (see [372, 376, 378] and Sec. 11.3) in which the Higgs field would actually be part of the total electroweak field, and which is expected to yield two additional constraints, the amplitude of the Higgs $\lambda\phi^4$ potential (and therefore the Higgs boson mass) and a relation between couplings. Additional work is needed in order to decide whether the relation obtained here semi-empirically would emerge in such a new framework.

11.2. Application of Quantum Mechanics in Scale Space

Let us now give first hints at attempts to apply the new quantum mechanics in scale space as outlined in Chapter 8, including the new concept of complexergy, to the physics of elementary particles.

This proposal led to the concept that fundamental scales in nature (more precisely, fundamental scale ratios) could be constrained by a Schrödinger-type equation acting in scale space. The solution of such an equation would therefore describe probability amplitudes for the scale at which a given structure may appear. In other words, fundamental scales could be defined by peaks of probabilities of their ratio with some reference scale. Such an approach naturally leads to the emergence of a hierarchy of scales, since the solutions give probability densities $P[\ln(\varepsilon/\lambda)]$.

The problem of the nature of the elementary particle masses seems to be a natural domain of application of such a tentative approach [388]. Indeed, there is an experimentally observed hierarchy of elementary particles, which are organized in terms of three known families, with mass increasing with the family quantum number. For example, a (e, μ, τ) universality is known among leptons: these three particles have exactly the same properties except for their mass and their family number. However, in the present standard model, there is no understanding of the nature of the families and no theoretical prediction of the values of the masses.

Hence, the experimental masses of charged leptons and of the "current quark masses" are [436]:

(i) $m_e = 0.510998910(13)\,\text{MeV}$; $m_\mu = 105.658367(4)\,\text{MeV}$;
  $m_\tau = 1776.84(17)\,\text{MeV}$;
(ii) $m_u = 1.5 - 3\,\text{MeV}$; $m_c = 1.25 \pm 0.09\,\text{GeV}$; $m_t = 174.2 \pm 3.3\,\text{GeV}$;
(iii) $m_d = 3 - 7\,\text{MeV}$; $m_s = 95 \pm 25\,\text{MeV}$; $m_b = 4.20 \pm 0.07\,\text{GeV}$.

Concerning neutrinos, there is now convincing evidence (through atmospheric, solar, reactor and accelerator data) that they can change from one flavor to another and that they are massive (see e.g. [435]). Square mass differences of $(m_{\nu_3}^2 - m_{\nu_2}^2)(\text{atm}) = (0.050 \pm 0.005)^2\,\text{eV}^2$ and $(m_{\nu_2}^2 - m_{\nu_1}^2)(\text{solar}) = \{(9.2 \pm 0.2) \times 10^{-3}\}^2\,\text{eV}^2$ have been found for neutrinos, which are approximate mixings $\nu_1 : (0, 1/2, 1/2)$, $\nu_2 : (1/3, 1/3, 1/3)$ and $\nu_3 : (2/3, 1/6, 1/6)$ of respectively $(\nu_e, \nu_\mu, \nu_\tau)$. This would lead to $m_{\nu_e} = 0.014\,\text{eV}$ and $m_{\nu_\mu} + m_{\nu_\tau} = 0.082\,\text{eV}$.

Basing oneself on the definition of complexergy as a new conservative quantity (which is to scale what energy is to motion) and on this mass hierarchy, one can suggest that the existence of particle families could be a manifestation of increasing complexergy, i.e. that the family quantum number would be nothing but a complexergy quantum number [388]. This would explain why the electron, muon and tau numbers are conserved in particle collisions, since such a fundamental conservative quantity (like energy) can be neither created nor destroyed. Under this view, the muon and tau particles would be excited states of the electron.

We have shown in Chapter 7 that the scale-relativistic re-interpretation of gauge transformations allowed one to suggest a relation between the mass of the electron and its electric charge (in terms of the fine structure constant). This result is compatible with the mass of the electron mainly being of electromagnetic origin. As already remarked, the Compton length of the electron and, therefore, its mass, i.e. of the less massive charged particle, naturally plays the role of a reference scale. More generally, the observed mass hierarchy between neutrinos, charged leptons and quarks also goes in this direction, suggesting that their masses are respectively of weak, electroweak and chromoelectroweak origin.

Although a full treatment of the problem must await a more advanced level of development of the theory, in which the "third quantization" description would be mixed with the gauge field one, some remarkable structures of the particle mass hierarchy already support such a view:

(i) The above values of quark and lepton masses are clearly organized in a hierarchical way. This suggests that their understanding is indeed to be searched in terms of structures of the scale-space, for example as manifestation of internal structures of iterated fractals [353].

(ii) With regards to the e, μ, τ leptons, we had already remarked in [353], in the framework of a fractal sef-similar model, that their mass ratios followed a power-law sequence, $m_\mu \approx 3 \times 4.1^3 \times m_e = 105.656\,\mathrm{MeV}$ and $m_\tau \approx 3 \times 4.1^5 \times m_e = 1776.1\,\mathrm{MeV}$. This empirical remark has already been predictive, since the mass derived for the τ from this empirical formula was in disagreement with its known experimental value at that time (1784 MeV), while more recent experimental determinations have given a mass of 1777 MeV [435], very close to the expectation. A possible way of research, not theoretically justified for the moment, is that such a law could correspond to the sequence $m_n/m_e = 3 \times 4.1^{1+2^n}$, in which case the next lepton would be heavy, with a mass $m_3 = 3 \times 4.1^9 m_e \approx 502\,\mathrm{GeV}$.

(iii) Concerning the quarks, the QCD gauge group is an SU(3) group. It is known that SU(3) is the dynamical symmetry group of a 3D harmonic oscillator field (see e.g. [474]). In the framework of a reinterpretation of gauge transformations as scale transformations on the internal scale variables, one is therefore led to link the SU(3) group to a scale harmonic oscillator field (see Chapter 4). In such a framework, the energy ratios are expected to be quantized as $\ln E \propto (3 + 2n)$. It may therefore be significant in this regard that the s/d mass ratio, which is far more precisely known than the individual masses since it can be directly determined from the pion and kaon masses, is found to be $m_s/m_d = 20.1$ [433], to be compared with $e^3 = 20.086$, which is the fundamental $(n = 0)$ level for scale ratios predicted by the above formula.

The suggestion of a link between a harmonic oscillator "scale field", which is a mere Newtonian-like way to describe a curved geometry in scale space, i.e. a geometry that deviates from the simple flatness of Galilean-type laws, is re-inforced by the following results and remarks:

(i) QCD is precisely characterized by the property of asymptotic freedom, which means that quarks become free at small scales, while the strong coupling constant increases toward large scales.

(ii) There is increasing evidence for an internal fractal structure of the proton, more generally of hadrons [280].

(iii) Free u and d quark masses, i.e. the masses they would have in the absence of confinement, are far smaller than their effective mass in the proton and neutron. This means that their Compton length $\lambda_c = \hbar/mc$ is larger than the confinement scale (of order 1 Fermi). This is exactly what is expected in the harmonic oscillator model of Sec. 4.3.6, in which the transition scale is pushed toward smaller length scales, i.e. larger energies (see Fig. 4.8).

(iv) Moreover, when viewed in terms of direct variables (X, Y, Z), a three-dimensional sphere in scale space $(\ln X, \ln Y, \ln Z)$ becomes a triad for large values of the variables (see Fig.11.13). Such a behavior may provide a very tentative model of color and quarks, in which the three quarks in hadrons would have no real separated existence, but could be identified with the three extremities of such a "three-tip string". Their appearance could therefore be the mere result of a change of reference system (i.e. of relativity), provided the genuine physical variables for the description of intra-hadron physics be the scale variables $(\ln X, \ln Y, \ln Z)$, in terms of which there is pure isotropy on the scale-sphere, while our measurement devices work in terms of the (X, Y, Z) variables.

11.3. Theoretical Expectation of Higgs Boson Mass

11.3.1. *Lowest order estimate*

Generalized scale relativity also allows us to suggest hints of a new version of the electroweak theory, which, as a consequence, would allow a theoretical

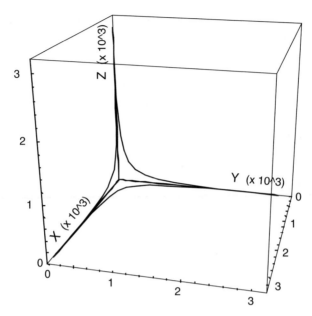

Fig. 11.13: **Three-tip string as sphere in scale space.** Three-dimensional sphere in scale space, $(\ln X)^2 + (\ln Y)^2 + (\ln Z)^2 = A^2$, plotted in terms of direct variables (X, Y, Z), for $A = 8$.

prediction of the value of the Higgs boson mass [372, 376, 378]. The summarized argument is as follows.

In today's electroweak scheme, the Higgs boson is considered to be separated from the electroweak field. Moreover, a more complete unification is mainly sought in terms of attempts of "grand" unifications with the strong field. However, one may wonder whether one could achieve a more tightly unified purely electroweak effective theory at intermediate energy. Recall indeed that in the present standard model, the weak and electromagnetic fields are mixed, but there remains four free parameters, which can, e.g. be taken to be the Higgs boson mass, the vacuum expectation value of the Higgs field and the Z and W masses. In the attempt sketched out hereafter, the Higgs field is assumed to be a part of the total field, so that only two free parameters would be left. As a consequence, the Higgs boson mass and the W/Z mass ratio could be derived in such a model.

Let us make a brief reminder about the structure of the present electroweak boson content. There is a SU(2) gauge field, involving three fields of null mass, i.e. $2 \times 3 = 6$ degrees of freedom (d.f.), a U(1) null mass field (2) and a Higgs boson complex doublet (4 d.f.), which makes a total of

12 degrees of freedom. Through the Glashow-Salam-Weinberg mechanism [201, 527, 471], three of the four components of the Higgs doublet become longitudinal components of the weak field, which therefore acquires mass ($3 \times 3 = 9$ d.f.), while the photon remains massless (2 d.f.), so that there remains a Higgs scalar, which is nowadays experimentally searched for (1 d.f.).

Let us now place ourselves in the framework of the scale relativistic geometric interpretation of non-Abelian gauge theories (Sec. 7.5). Instead of only one global dilation and scale variable as in the Abelian case, one considers four independant scale transformations on the four space-time resolutions, i.e. ($\varepsilon_x, \varepsilon_y, \varepsilon_z, \varepsilon_t$), which are intrinsic to the description and variable with space-time coordinates (recall that this remains a diagonal approximation of the very nature of resolutions that are tensorial).

This means that the scale space (i.e. here the gauge space) is at least four-dimensional, but note that this is not the final word on the subject, since this does not yet include the fifth "djinn" dimension τ that appears in special and general scale relativity. Moreover, the mixing relation between the B [U(1)] and W_3 [SU(2)] fields may also be interpreted as a spherical rotation in the full gauge (scale) space.

Therefore we expect the appearance of a six component antisymmetric tensor field (linked to the rotations in this space), corresponding in the simplest case to a SO(4) group (more generally, its universal covering group SU(2)×SU(2)). Such a zero mass field would yield 12 degrees of freedom by itself alone, so that it is able to include the 11 d.f. of the electromagnetic and weak fields, but also the degree of freedom of the residual Higgs field. But we are aware of the fact that additional degrees of freedom could be needed for such a construction to be self-consistent: we have seen in Sec. 7.5 that the full unifying gauge group (without gravitation) issued from the scale relativity framework is at least a SO(10) group. In future works we intend to reconsider the question in this extended framework.

We have tentatively explored the possibility that the Higgs boson appears as a separated scalar only as a low energy approximation, while in the new framework it would be one of the components of the unified field [372, 376, 378]. An analogy can be made with space-time physics, in which energy appears as scalar at low velocity, while it is ultimately a component of the energy-momentum four-vector. Here we consider a similar situation, but in the scale-space (i.e. in the gauge space), the large scale theory would be a Galilean approximation of special scale-relativity, involving a spontaneous symmetry breaking of the full gauge group as a mere result of the change of scale.

Such an attempt is supported by the form of the electroweak Lagrangian (we adopt here Aitchison's notations [7]). Its Higgs scalar boson part reads (in terms of the residual massive scalar σ):

$$L_H = \frac{1}{2}\partial_\mu\sigma\partial^\mu\sigma - \frac{1}{2}m_H^2\sigma^2 - \frac{1}{8}\lambda_a^2\sigma^4. \qquad (11.93)$$

We denote by λ_a the potential amplitude not to confuse with a Compton length. The vacuum expectation value $f = 246.22\,\text{GeV}$ [434] of the Higgs field is computed from the square and quartic terms, so that the Higgs boson mass is related to f and λ_a as

$$m_H = f\lambda_a. \qquad (11.94)$$

A prediction of the constant λ_a would therefore lead to a prediction of the Higgs mass. Now, a non-Abelian field reads in terms of its potential

$$F^{\alpha\mu\nu} = \partial^\mu W^{\alpha\nu} - \partial^\nu W^{\alpha\mu} - gc_{\beta\gamma}^\alpha W^{\beta\mu}W^{\gamma\nu}, \qquad (11.95)$$

where g is the (now unique) charge and $c_{\beta\gamma}^\alpha$ the structure coefficients of the Lie algebra associated to the gauge group. Its Lagrangian reads

$$L_W = -\frac{1}{4}F^{\mu\nu}F_{\mu\nu}. \qquad (11.96)$$

Therefore, it includes quartic W^4 terms coming from the W^2 terms in the field. Now our ansatz consists of identifying some of these W^4 terms, of coefficient $-\frac{1}{4}g^2(c_{\beta\gamma}^\alpha)^2$, with the Higgs boson ϕ^4 term of coefficient $-\frac{1}{2}\lambda_a^2$, where ϕ is the initial scalar doublet.

Let us first separate the six components of the total field in two sub-systems, $[W_1, W_2, W_3]$ and $[B_1, B_2, B_3]$, which is straighforward in a SU(2)×SU(2) group. The three W's can be identified with the standard SU(2) isospin field and, say, B_1 with the U(1)$_Y$ field, which is a subgroup of the second SU(2) group. Two vectorial fields remain, B_2^μ and B_3^μ. They will contribute in a non-vanishing way to the quartic term in the Lagrangian by their cross product. In the approximation (considered here) where their space components are negligible, we obtain

$$-B_{2\mu}B_{3\nu}B_3^\mu B_2^\nu = -\left[B_2^0 B_3^0\right]^2. \qquad (11.97)$$

Finally, we make the identification of these time components with the residual Higgs boson, $B_2^0 = B_3^0 = \sigma$. This allows a determination of the constant λ_a according to the relation

$$\lambda_a^2 = \frac{g^2 c^2}{2}, \tag{11.98}$$

where the squared Lie coefficient c^2 is equal to 1 in the case of an SU(2)×SU(2) group. Provided the global charge is identical to the SU(2) charge, and since the W mass is given by $m_W = gf/2$, one finally obtains a Higgs boson mass

$$m_H = \sqrt{2}\, m_W. \tag{11.99}$$

Using the recently precisely determined W boson mass, $m_W = 80.403 \pm 0.29\,\text{GeV}$ [436], we obtain a theoretical prediction [372, 376]:

$$m_H = \sqrt{2}\, m_W = 113.707 \pm 0.041 \text{ GeV}, \tag{11.100}$$

which is in agreement with current constraints (lower limit) and with a possible detection by the LEP at CERN with a mass of about $114\,\text{GeV}$ [433] (an analysis of 10 events by McNamara *et al.* [317] has yielded $m_H = 113.8 \pm 0.8$ GeV). More generally, one expects from Eq. (11.98) for $c^2 = k$ a relation $m_H = \sqrt{2k}\, m_W$, with k integer.

However, one should keep in mind that this result is obtained in the framework of an approximate theory. The SU(2)×SU(2) group is only a subgroup of the complete unified gauge group, from both the view points of the scale-relativity approach (other variables and other scale transformations should be taken into account) and of the standard quantum field theory approach, in which the unifying gauge group should also at least include as subgroup the SU(3) group of chromodynamics.

Although the self-consistency of this model then remains to be established, we hope that some of its ingredients could reveal to be useful in more complete attempts. In particular, the model is expected to provide a second constraint at low (Z) energy, which could be translated in terms of a relation between the U(1) and SU(2) couplings. As we have seen in the previous sections, the existence of such a relation, which reads $\alpha_2(M_Z) \approx 2\alpha_1(M_Z)$, is both supported by the experimental data and by the theoretical suggestion of mass-coupling relations issued from the new geometric construction of gauge fields in the scale relativity framework.

With regards to the above Higgs boson mass theoretical expectation, its value and the uncertainty about this value cannot be the last word, since one should also account for radiative corrections and for scale relativistic corrections.

11.3.2. *Radiative corrections*

Let us take a few steps towards a future and more precise estimate by calculating radiative corrections to the potential and, therefore, to the Higgs mass and to the coefficient of the ϕ^4 contribution. The full effective Higgs one-loop potential can be written as [490]

$$V = \frac{1}{2}\mu^2\phi^2 + \frac{1}{4}\lambda_s\phi^4 + B\phi^4 \ln\frac{\phi^2}{M^2} + V_s, \tag{11.101}$$

where

$$B = \frac{3}{64\pi^2}\left[\frac{1}{16}\left(3g^4 + 2g^2g'^2 + g'^4\right) - g_Y^4\right], \tag{11.102}$$

with $g_Y = \sqrt{2}m_t/f$. In these expressions, $f = 246.22\,\text{GeV}$ [435] is the mean value on the vacuum of the Higgs field (we keep Aitchison's notation [7] for this quantity), $m_t = 174.2 \pm 3.3\,\text{GeV}$ [435] is the top quark mass (the dominant contribution to this radiative correction). The SU(2) and U(1) charges are respectively $g = \sqrt{4\pi\alpha_{2Z}} = 0.649$ and $g' = \sqrt{(12/5)\pi\alpha_{1Z}} = 0.356$ at the Z boson scale. The scalar part of the effective potential reads

$$V_s = \frac{1}{64\pi^2}(\mu^2 + 3\lambda_s\phi^2)^2 \ln\frac{\mu^2 + 3\lambda_s\phi^2}{M^2} + \frac{3}{64\pi^2}(\mu^2 + \lambda_s\phi^2)^2 \ln\frac{\mu^2 + \lambda_s\phi^2}{M^2}. \tag{11.103}$$

Here we have adopted Sher's notation [490], which is related to the previous (Aitchison [7]) notation by

$$\lambda_s = \frac{1}{2}\lambda_a^2. \tag{11.104}$$

The one-loop corrected potential is compared to the zero-loop one in Fig. 11.14. The energy f is redefined from $\left[\frac{dV}{d\phi}\right]_{\phi=f} = 0$. Taking $M = f$, this yields $\mu^2 = -(\lambda_s + 2B)f^2$.

The mass of the Higgs boson is generally defined in terms of this more general form of the effective potential by

$$m_H^2 = \left[\frac{d^2V}{d\phi^2}\right]_{\phi=f} = \mu^2 + (3\lambda_s + 14B)f^2. \tag{11.105}$$

Neglecting the scalar potential, whose relative contribution amounts only to $\approx 5 \times 10^{-3}$, we finally obtain

$$m_H^2 = 2\lambda_s f^2\left(1 + 6\frac{B}{\lambda_s}\right), \tag{11.106}$$

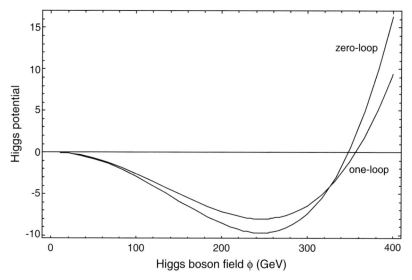

Fig. 11.14: **Higgs potential without and with radiative corrections.** Comparison between the standard zero-loop (uncorrected) $\lambda\phi^4$ Higgs potential and the potential corrected for radiative corrections to one loop.

which corresponds to a relative radiative correction of $\approx 3B/\lambda_s$ on m_H. With the above experimental values of the parameters, one finds $\sqrt{1 + 6B/\lambda_s} = 0.86$, yielding, for a zero-loop mass $(m_H)_0 = \sqrt{2\lambda_s} f = 113.73\,\mathrm{GeV}$, a Higgs mass corrected to one loop by an amount of 14%,

$$(m_H)_1 = 97.81 \text{ GeV.} \tag{11.107}$$

This value is compatible with the range deduced from precision electroweak measurements, $m_H = 95^{+45}_{-32}\,\mathrm{GeV}$, but it is excluded by the LEP200 experimental results, which have yielded a 1.7σ low significant detection at $114\,\mathrm{GeV}$ and a limit $(m_H)_{\mathrm{exp}} \geq 114\,\mathrm{GeV}$ to 95% confidence level [435]. From the second possible value in the sequence $\sqrt{2k}m_W, m_H = 2m_W = 160.84\,\mathrm{GeV}$, one would obtain $(m_H)_1 = 138.32\,\mathrm{GeV}$.

However, this is not the last word about this attempt of more precise prediction of the Higgs mass, since, to this order, the theoretical expectation should also be completed. Indeed, only the terms in the gauge field potential mimicking a $\lambda\phi^4$ term have been kept in our preliminary rough calculation, while other terms should also contribute to the effective potential. Moreover, a more complete account of radiative corrections can be obtained from the renormalization group [490], but the solution needs numerical integrations. These developments will be considered in future works.

Problems and Exercises

Open Problem 24: Establish the complete effective potential for an apparent Higgs boson under the above hypothesis that it is a low energy and large length scale manifestation of the unified gauge field itself, as a consequence of spontaneous scale symmetry breaking (under a SU(2)×SU(2) gauge group, or of a more general group). Deduce the predicted Higgs mass from this calculation.   ■

Open Problem 25: Find the radiative correction to the predicted Higgs boson mass from the full renormalization group approach [490].   ■

Open Problem 26: Find the additional relation that is expected in such a framework between the U(1) and SU(2) couplings.   ■

11.4. Ultra High Energy Cosmic Rays

11.4.1. *Energy showers at* 3×10^{20} *eV*

Ultra high energy cosmic rays observed beyond 10^{19} eV are still of unknown origin. The highest energy shower has been observed in 1991 by the Fly's Eye detector at $(3.2 \pm 0.9) \times 10^{20}$ eV [51, 52]. The second highest energy event is at 2.1×10^{20} eV. Such detections and more generally the highest energy spectrum of cosmic rays pose an important problem, because one expects a Greisen-Zatsepin-Kuzmin cutoff of the energy spectrum at $\approx 4 \times 10^{19}$ eV for travel distances larger than about 30 Mpc, due to pion photoproduction energy losses. Such a constraint strongly limits the possible astrophysical sources for such events, provided they are produced by accelerated known particles.

In order to circumvent this problem, it has been proposed that the highest energy cosmic rays originate from the decay of topological spacetime defects, such as cosmic strings or vortons [62]. Such theories would predict a continuing cosmic ray flux all the way up the grand unification mass scale (10^{23} eV in the minimal standard model, 10^{28} eV if the unification is at the Planck energy).

11.4.2. *Ultra-high energy structure in special scale-relativity*

A similar proposal can be made in the scale relativity framework [400, 390], but with another mass scale for the primary particle. The special scale relativity theory [352] introduces a generalized law of scale transformations of a log-Lorentz form. Moreover, as recalled at the beginning of this chapter, we expect the occurrence of new kinds of spacetime structures, linked in particular to mass-charge relation and to the critical value $4\pi^2$ of inverse couplings. One of these structures is given by the equation (see Fig. 11.4)

$$\alpha_2^{-1}(E) = 4\pi^2, \tag{11.108}$$

where α_2 is the SU(2) weak coupling and E is the energy-scale to be solved for. The scale dependence of the α_2 running coupling is given by the solution to its renormalization group equation, which, to first order, reads (see e.g. [353] Chapter 6.2 and references therein)

$$\alpha_2^{-1}(r) = \alpha_2^{-1}(\lambda_Z) + \left(\frac{5}{3\pi} - \frac{N_H}{12\pi}\right) \ln\left(\frac{\lambda_Z}{r}\right), \tag{11.109}$$

where λ_Z is the Compton length of the Z boson, N_H is the number of Higgs doublets and r a running length-scale. As recalled in the previous chapter, this solution remains correct in the special scale-relativity framework, provided it is written in terms of length-scale. Conversely, while one can replace $\ln(\lambda_Z/r)$ with $\ln(m/m_Z)$ in the standard model, this is no longer the case in special scale-relativity, since a log-Lorentz factor is now involved in the mass-scale to length-scale transformation,

$$\ln\left(\frac{\lambda_Z}{r}\right) = \frac{\ln(m/m_Z)}{\sqrt{1 + \ln^2(m/m_Z)/\ln^2(m_\mathbb{P}/m_Z)}}. \tag{11.110}$$

Solving Eq. (11.108) for $E = mc^2$ with the precise value of $\alpha_2^{-1}(m_Z) = 29.802 \pm 0.027$ [432] yields

$$E = (3.20 \pm 0.26) \times 10^{20}\,\text{eV}. \tag{11.111}$$

Including second order corrections in the renormalization group equation and accounting for the scale-relativistic correction on the fundamental constant $\mathbb{C}_Z = \ln(\lambda_Z/l_\mathbb{P}) = 39.756$, which differs by 1% from $\ln(m_\mathbb{P}/m_Z) = 39.436$, [353], one obtains an equivalent result,

$$E = (3.27 \pm 0.26) \times 10^{20}\,\text{eV}, \tag{11.112}$$

which agrees very closely with the maximal energy of cosmic rays observed at $(3.2 \pm 0.9) \times 10^{20}\,\text{eV}$ (see Fig. 11.15). Due to the large value of this energy, the agreement would remain remarkable even if the experimental error reveals to be underestimated.

Such a result, provided it is confirmed by observations, e.g. at the now available Pierre Auger observatory, allows one to put to the test the number

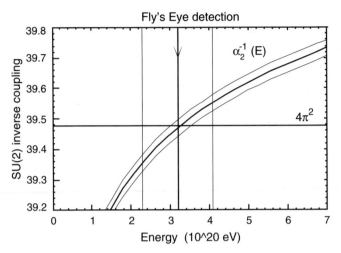

Fig. 11.15: **High energy particle expectation.** Theoretical prediction of a high energy "structure" (particle) from the weak interaction, which is solution of the equation $\alpha_2(E) = 1/4\pi^2$, where $\alpha_2(E)$ is the running SU(2) coupling and $1/4\pi^2$ is a critical value for couplings. From precise determinations of the coupling constant at Z scale [432], $\alpha_{2Z} = 29.802 \pm 0.027$ and using the solution to the renormalization group equation for the running coupling, one obtains $E = (3.27 \pm 0.26) \times 10^{20}$ eV, to be compared with the Fly's Eye detection at $(3.2 \pm 0.9) \times 10^{20}$ eV [52].

of Higgs doublets and the scale-relativistic log-Lorentz factor. Indeed, the predicted energy becomes 8.0×10^{19} for 0 Higgs doublet (excluded) and 1.7×10^{21} for 2 Higgs doublets, which will be excluded if no showers at energies larger than 3.2×10^{20} eV are observed. This would be a confirmation of our result concerning the running of the fine structure constant from its bare value to its low energy value, which depends on the number of Higgs doublets (see Sec. 11.1.3.4 and [360]):

$$\alpha^{-1} = 137.01 \pm 0.04 + 2.11(N_H - 1), \qquad (11.113)$$

which already strongly excludes $N_H \neq 1$.

This effect could also be used to discriminate between the Galilean scale relativistic (i.e. standard) and Lorentzian scale relativistic theory. Indeed, in the absence of a log-Lorentz factor, one obtains for one Higgs doublet $E = (2.0 \pm 0.12) \times 10^{19}$ eV, which is smaller than the Fly's Eye energy by a factor of 16 and lies below the GZK cutoff.

With additional Higgs doublets, one obtains $(1.9 \pm 0.1) \times 10^{20}$ eV for 3 doublets and 7.2×10^{20} eV for 4 doublets. The first value is already too low and the second will be excluded if the present limiting energy is not exceeded by future detections. This is therefore a new test of the log-Lorentz factor, that is added to other previous tests, such as the identification of the GUT scale with the Planck scale.

11.4.3. *Possible origin for the high energy cosmic rays*

Such a proposal, namely the decay of a new particle (the "Auger" particle) of mass $3.2 \times 10^{20}\,\text{eV}/\text{c}^2$, is falsifiable, since in this case one would not expect to find cosmic rays of energy far larger than the Fly's Eye value. Moreover, if it were confirmed, the distance limits set on the source of this event no longer apply. This reopens the possibility of a galactic source, as supported by the arrival direction that lies close to the galactic plane, at $b = 9.6$ deg, or of a very distant extragalactic source, e.g. 3C147, a QSO of redshift 0.545, which lies within the 1σ error box [164].

Chapter 12

APPLICATIONS TO COSMOLOGY

12.1. Introduction

The theory of scale relativity is founded on a generalization of the standard description of space-time and, as such, it can be expected to have cosmological consequences. The reader will find a good introduction to basic concepts in cosmology in standard references, such as [528, 438], to general relativity, on which today's cosmology is founded, in, e.g. [266, 528, 323] and also to the present challenges and essential issues of this science in [170].

As established in Chapter 4, the principle of relativity, when applied to scale transformations, leads to the suggestion of a scale-relativistic generalization of fundamental dilations laws [352].[a] These new special-scale-relativistic resolution transformations involve log-Lorentz factors and lead to the occurrence of a minimal, but also of a maximal length-scale in nature, which are both invariant under dilations. In the previous chapter we have studied the consequences for elementary particle physics of the suggested existence of a minimal length-scale, which replaces the zero from the viewpoint of its physical properties and which is identified with the Planck length $l_{\mathbb{P}}$. As regards the maximal scale, which replaces the infinite scale, it can be identified (as we shall see) with the cosmic scale $\mathbb{L} = \Lambda^{-1/2}$, where Λ is the cosmological constant.

The new interpretation of the Planck scale has several implications for the structure and history of the early Universe. In this chapter, we

---

[a]For example, the product of two scale ratios ρ and ρ' is generalized to $\rho'' = (\rho\,\rho')^{1/(1+\ln\rho\,\ln\rho'/\mathbb{C}^2)}$. One recovers the standard product $\rho'' = \rho\,\rho'$ in the limit $\mathbb{C} \to \infty$.

reconsider the questions of the origin of the universe, of the status of physical laws at very early times, of the horizon-causality problem and of fluctuations at recombination epoch.

The new interpretation of the cosmic scale has also consequences for our knowledge of the present universe, concerning in particular Mach's principle, the large number coincidence, the problem of the vacuum energy density and the nature and value of the cosmological constant (see Fig. 12.1). Scale-relativity also allows one to suggest a general solution to the missing mass (or so called "dark matter") problem and to expect the emergence of new fundamental energy scales [383].

12.2. Special Scale Relativity and Cosmology

It is well-known that the Galileo group is a degeneration of the more general Lorentz group of motion. As we have seen in Chapter 4, the same is true for scale laws, with an additional subtlety due to the effective scale symmetry breaking at the fractal-nonfractal transition.

The law of dilation composition in special scale relativity reads

$$\ln \varrho'' = \frac{\ln \varrho + \ln \varrho'}{1 + \ln \varrho \ln \varrho'/\mathbb{C}^2}. \tag{12.1}$$

Under this form the scale symmetry remains unbroken. Such a law corresponds, at the present epoch, only to the null mass limit. As we shall see hereafter, it is expected to apply in a universal way during the very first instants of the universe. This law assumes that at very high energy no static scale and no space or time unit can be defined, so that only pure contractions and dilations have physical meaning. In Eq. (12.1), there appears a universal purely numerical constant [353]

$$\mathbb{C} = \ln \mathbb{K}. \tag{12.2}$$

As we show in what follows, $\mathbb{K}$ can be identified with the ratio of the cosmic length $\mathbb{L} = \Lambda^{-1/2}$ (where Λ is the cosmological constant) over the Planck length $l_{\mathbb{P}}$, which leads to a numerical value $\mathbb{K} = 5.3 \times 10^{60}$ [353, 360, 383].

As we have seen in the previous chapters, the scale symmetry is broken in microphysics in an effective way by the mass of elementary particles, i.e. by the emergence of their Compton–de Broglie length,

$$\lambda_c = \frac{\hbar}{mc}, \quad \lambda_{dB} = \frac{\hbar}{mv} \tag{12.3}$$

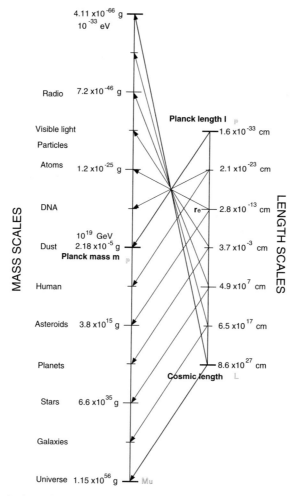

Fig. 12.1: **Relation of mass scales to length scales in Nature.** The range of mass scales in the Universe are linked to the range of length scales by two inverse relations. The first is a classical relation of proportionality, which can be established in a geometric way from Einstein's general relativity. The "geometry = matter" fundamental meaning of Einstein's field equations (see e.g. [323]) is translated into the Schwarzschild length, $r_s = Gm/c^2$ (the Schwarzschild radius being twice this expression). This relation can be written as $r_s/l_{\mathbb{P}} = m/m_{\mathbb{P}}$, where $l_{\mathbb{P}} = \sqrt{\hbar G/c^3}$ is the Planck length and $m_{\mathbb{P}} = \sqrt{\hbar c/G}$ is the Planck mass. The second relation is the Compton relation, originating from quantum theory, which also finds a geometric explanation in the present framework of the scale relativity theory. It reads $r_c = \hbar/mc$ and it can, therefore, be written as $r_c/l_{\mathbb{P}} = m_{\mathbb{P}}/m$. Due to the existence of these two inverse relations, while the full range of length scales and of time scales is 5.3×10^{60} (see [353] Sec. 7.1, and the present chapter), the full range of possible mass scales is doubled and, therefore, reaches 2.8×10^{121}.

and for extended objects by the thermal de Broglie length

$$\lambda_{th} = \frac{\hbar}{m\langle v^2\rangle^{1/2}} = \frac{\hbar}{\sqrt{2mkT}}. \tag{12.4}$$

The scale symmetry is also broken, in the cosmological realm, toward the relatively small scales by the emergence of static structures (galaxies, groups, cluster cores) of a typical size

$$\lambda_g \approx \frac{1}{3}\frac{GM}{\langle v^2\rangle}. \tag{12.5}$$

The effect of these two symmetry breakings is to separate the scale space into at least three domains, a quantum scale-dependent domain at microscopic scales, a classical scale-independent domain at intermediate scales and a cosmological scale-dependent domain toward the large scales [353], in which explicit scale-dependence and quantum-like laws take place again (see next chapter on gravitational structuring). The transitions between these domains is evidently relative to mass, velocity, energy, etc. and their typical scales, therefore, strongly depend on the type of cosmological constituent considered.

The consequence is that, in the two scale-dependent domains, these transition scales λ can be taken as references for scale ratios in the log-Lorentz transformation. As recalled in Sec. 4.4.5, one therefore does not deal any longer with dimensionless scale ratios, but with a new law involving dimensioned space and time intervals [352]. The law of composition of dilations now takes the form (instead of the "Galilean" form $\varepsilon' = \rho \times \varepsilon$):

$$\ln\frac{\varepsilon'}{\lambda} = \frac{\ln\rho + \ln(\varepsilon/\lambda)}{1 + \ln\rho\ln(\varepsilon/\lambda)/\ln^2(L/\lambda)}. \tag{12.6}$$

This means that we combine Galilean scale-relativity with Lorentzian scale-relativity on two different ranges of scales separated by the transition scale λ. While the dilation ratio from λ to ε is given by the Galilean ratio λ/ε (the same being true for λ over ε'), this is no longer the case of the dilation ratio ρ from ε to ε': namely, $\varepsilon'/\varepsilon \neq \rho$ in the above formula. In this log-Lorentzian law of composition of dilations, a dimensioned scale L appears, such that $\mathbb{C} = \ln(L/\lambda)$, which, as can be directly checked in the above expression, is invariant under dilations [352, 353]. It can be identified with:

(i) the Planck length-scale $l_\mathbb{P} = (\hbar G/c^3)^{1/2}$ toward the small scales,
(ii) the cosmic scale $\mathbb{L} = \Lambda^{-1/2}$ (where Λ is the cosmological constant) toward the large scales.

We also recall that the law of transformation of a fractal coordinate $\mathcal{L}$ in a scale change of the reference system by a dilation ρ is now given, in the asymptotic scale-dependent domain by

$$\ln(\mathcal{L}'/\mathcal{L}_0) = \frac{\ln(\mathcal{L}/\mathcal{L}_0) + \tau \ln \varrho}{\sqrt{1 - \ln^2 \varrho/\mathbb{C}^2}}, \tag{12.7}$$

while the scale exponent $\tau = D_F - D_T$, which plays the role of a fifth dimension (called "djinn"), transforms, in a scale transformation by a factor ρ, as

$$\tau(\varepsilon') = \frac{\tau(\varepsilon) + \ln \rho \ln(\mathcal{L}/\mathcal{L}_0)/\mathbb{C}^2}{\sqrt{1 - \ln^2 \rho/\mathbb{C}^2}}. \tag{12.8}$$

In such a framework, when crossing the transition scale in the scale space, one jumps from a regime dominated by the four classical variables dx^μ to a new regime dominated by the five fractal variables $d\xi^\alpha$, made of the four fractal fluctuations and of the "djinn" (the variable fractal dimension).

Note that the idea of a non-compactified fifth dimension has also been suggested in a different context in the last years [459] and its cosmological implications are currently actively studied. However this concept remains different from the scale relativity one, since it deals with the fourth space-like dimension. Moreover this extra dimension is postulated arbitrarily and therefore it still needs a direct detection or an indirect one through its possible effects, while the introduction of the "djinn" relies on the already established existence of the four resolutions for the four space-time coordinates, and on the description of their dilations and contractions as rotations in the five-dimensional "space-time-djinn".

12.3. Theoretical Prediction of the Space Dimensionality

Let us first consider a general consequence for the number of dimensions of space-time (and therefore for cosmology) of one of the results from the previous chapter. We have recovered in Sec. 11.1.2.2 the expression for the Coulomb force without referring to the dimensionality of space, but only to the exchange of zero mass bosons.

This provides us with a possible demonstration that the topological dimension of space is expected to be $D_T^S = 3$ at large scales. Indeed, the classical distance dependence of an infinite range force is $r^{1-D_T^S}$, while we have obtained here a r^{-2} dependence only from the Heisenberg relation.

More generally, a generalized Heisenberg relation can be demonstrated for any fractal dimension D_F of the geodesics of the fractal space-time [353]. It reads

$$\delta p \times \delta r^{D_F - 1} \geq (\hbar/2) \lambda_c^{D_F - 2}, \qquad (12.9)$$

where λ_c is the Compton length. We recover the standard Heisenberg relation for the crivital value $D_F = 2$ (see e.g. [360]), which is the universal Feynman value of the fractal dimension, which corresponds to standard quantum mechanics.

We therefore obtain a general relation between the topological dimension of space and the fractal dimension of geodesics,

$$D_T^S = D_F + 1, \qquad (12.10)$$

which yields $D_T^S = 3$ from $D_F = 2$. The special scale-relativistic transformation in which the effective value of D_F becomes a variable does not contradict this result, since the relevant fractal dimension in this case is the proper (or scale-covariant) dimension, which is nothing but the five-dimensional invariant itself, which keeps the value $\sigma = 2$.

Moreover, while from the fractal dimension $D_F = 2$ of quantum paths at low energy we recover a dimensionality of space $D_T^S = 3$, the expected increase of the fractal dimension at high energies in the special scale relativity framework leads to the possibility of extra topological dimensions, which are also a necessary ingredient of the superstrings approach [26, 208, 452].

12.4. Cosmological Implications of Scale Relativity

12.4.1. *New possible solution to the "dark matter" problem*

Before applying the scale relativity principle to the study of the global geometry of the Universe, let us briefly address one of the most important problems of today's cosmology, that of the so-called "missing mass" or "dark matter". This question will also be considered in the next chapter about gravitational structuring, in particular as concerns the flat rotation curves of spiral galaxies, but we shall give here a more general argument. We suggest that the various dynamical and gravitational lensing effects that are tentatively interpreted in the standard approach as necessitating the existence of large amounts of unseen matter could be a consequence of the fractality of space [123, 376, 383].

Let us start again from the three most basic new properties of a non-differentiable manifold (as compared with a Riemannian manifold) namely: (i) infinite number of geodesics; (ii) decomposition of each elementary displacement in terms of the sum of a differentiable part and of a nondifferentiable part of fractal dimension 2; (iii) two-valuedness of the velocity vector due to irreversibility in the reflection $dt \leftrightarrow -dt$. Accounting for these properties, the geodesic equation in a curved and fractal space can be integrated under the form of a Schrödinger equation [353, 362] that writes at the Newtonian limit

$$\mathcal{D}^2 \Delta\psi + i\mathcal{D}\frac{\partial}{\partial t}\psi = \frac{1}{2}\varphi\psi, \tag{12.11}$$

where φ is the Newtonian potential, which is a solution of the Poisson equation

$$\Delta\varphi = 4\pi G\rho. \tag{12.12}$$

From a description of motion in terms of an infinite family of geodesics, the meaning of $P = |\psi|^2$ imposes itself as giving the probability density of the particle positions, in agreement with Born's postulate (see Sec. 5.7). Now, by separating the real and imaginary parts of the Schrödinger equation and writing it in terms of P and the classical velocity V, one obtains respectively a generalized Euler–Newton equation and a continuity equation [374]:

$$\left(\frac{\partial}{\partial t} + V \cdot \nabla\right)V = -\nabla(\varphi + q), \tag{12.13}$$

$$\frac{\partial P}{\partial t} + \text{div}(PV) = 0. \tag{12.14}$$

This system of equations is equivalent to the classical one used in the standard approach of gravitational structure formation (see e.g. [438]), except for the appearance of an extra potential energy $Q = mq$ that writes (see Sec. 5.7.1 and Chapter 10)

$$Q = -2m\mathcal{D}^2\frac{\Delta\sqrt{P}}{\sqrt{P}}. \tag{12.15}$$

This potential is a manifestation of the fractality of space, in the same way as the Newtonian potential is a manifestation of space-time curvature.

Since the effects that have led to the proposal of missing mass (or of modified Newtonian gravity) are systematically the existence of a missing energy in the energy balance of a gravitational system (through dynamics, lensing or global cosmological energy balance), and since the scale relativity approach just provides one with an additional potential energy of fractal geometric origin (without perturbing the Newtonian dynamics that finds its origin from curvature), it seems very worth reinvestigating the dark matter question in the scale relativity context [376].

Indeed, let us come back to the Schrödinger form of these equations. Two extreme situations (and any intermediate between them) can be considered:

(i) The particles fill the probability density distribution, so that $\rho \propto P$. In this case the system of equations is a coupled Schrödinger–Poisson (Hartree) system [283], of the kind used to describe superconductivity. This case corresponds to a self-gravitating body, such as a cluster of galaxies. Markowich *et al.* [309] have recently demonstrated the general existence and nonlinear stability of steady states of the Schrödinger–Poisson system, with conserved total energy.

(ii) There are only very few test-particles, so that from the view-point of matter density, we deal with the vacuum. This case may correspond to the outer regions of spiral galaxies, (but it is more probable that they contain large quantities of baryonic dark matter). Therefore φ is a solution of $\Delta\varphi = 0$, i.e. $\varphi = -G\,\Sigma_i(M_i/r_i)$. The Schrödinger equation with such a potential has also general stationary solutions.

Therefore, in both cases, we can write a time-independent Schrödinger equation that takes the simplified form

$$2\mathcal{D}^2\Delta\psi + (\mathcal{E} - \varphi)\psi = 0, \qquad (12.16)$$

where $\mathcal{E} = E/m$ and φ is the steady-state solution for the potential. In the gravitational macroscopic case considered here, this equation is subjected to the principle of equivalence (contrary to the standard microscopic quantum mechanics case, where $\mathcal{D} = \hbar/2m$), and therefore it does not depend on the inertial mass m of the bodies. For this steady-state solution, one finds (in one dimension) the relation

$$\mathcal{E} = \varphi + q. \qquad (12.17)$$

Since $\mathcal{E} = E/m \propto v^2$, this relation expresses one of the main results of the scale-relativity-Schrödinger approach to gravitational structuration: (i) the expected (and now observationally supported [123]) universal quantization of velocities, i.e. the theoretical prediction that, in various gravitational systems, the probability distribution of the velocities will, in general, not be flat, but instead will have a tendency to show structures at values independent of the particular system [353] Chapter 7.2, [228, 357, 361, 364, 369, 374, 379]; (ii) the observed values of the velocity (e.g. rotation curves of galaxies, velocity dispersion in clusters of galaxies, etc.) is determined, not only by Newton's potential, but also by the new potential energy Q.

Our proposal, therefore, is that this potential, which is related to the baryonic matter density as Newton's potential, but not through a Poisson equation (Eq. 12.15), is at the origin of the various dynamical and lensing effects usually attributed to unseen additional mass. For example, in the Kepler problem, which appears in the outskirts of spiral galaxies, the additional potential writes $q = Q/m = \mathcal{E} + GM/r = -(GM/r_0)(1 - r_0/r)$, and we directly recover the result obtained in [123] for the fundamental energy level, but now whatever the state.

12.4.2. *Primeval density fluctuations*

Another fundamental consequence of the quantum gravitational theory concerns the question of density fluctuations at the recombination epoch. In the standard approach to the problem of structure formation, no structure can be formed from a strictly constant matter density. The standard model of formation assumes therefore that the present structures originate from the gravitational growth of early fluctuations that existed at the recombination epoch. These structures have actually been observed through their temperature signature in the Cosmic Microwave Background Radiation field, though at a low level of $\delta T/T \approx 2 \times 10^{-5}$.

However, in the standard approach these fluctuations should themselves be understood in terms of still earlier fluctuations. But the strong isotropization that prevails during the prerecombination phase poses a difficult problem. The standard solution to this problem consists of making the postulate that the Universe would have known an inflation phase, i.e. exponential expansion, which would have multiplied quantum fluctuations by an enormous factor.

We can suggest another solution to this problem in the scale-relativity framework, without any need for an inflationary phase. Indeed, as described in more detail in Chapter 13, the fractality of space-time at large scales involves a transformation of the motion equations into a macroscopic Schrödinger-type equation. As a direct consequence, there is a tendency to form structures at any epoch: these structures are described by probability density distributions given by the square of the modulus of the probability amplitudes, which are solutions of this gravitational Schrödinger equation [353, 362].

The classical approach, because of its deterministic and differentiable nature, predicts structures at a given epoch, which are the results of an evolution from previously existing structures, taken as initial conditions (in position and velocity). The new quantum-like approach is organized in a fundamentally different way. The loss of determinism of individual trajectories is compensated by a determinism of structures. At each epoch, stationary or steady-state solutions can be found in correspondence with the shape of the potential and the limiting and matching conditions. These structures do also evolve (as given by the time-dependent Poisson-Schrödinger system), in correspondence with the evolution of the environment.

Therefore, one expects the occurence of quantum-like fluctuations (structures) at the decoupling epoch. These structures would be, here, manifestations of the macroscopic quantum theory based on $\mathcal{D} = GM/2\alpha_g c$, which applies at this epoch, instead of the microscopic quantum theory for which $\mathcal{D} = \hbar/2m$ (and which applied at an earlier epoch). No inflation is needed in the new framework to obtain a scale invariant, quantum-like, spectrum at $z = 1000$ (and see Sec. 12.5.3 for an explanation, without inflation, of why causally disconnected regions in the Cosmic Microwave Background are at the same temperature).

12.4.3. Dimensionless physics

Let us briefly discuss some implications of special scale relativity with regards to the status of units [383] (see also Sec. 11.1.1.3) that may have important consequences for the very early primordial universe, as discussed in Sec. 12.5.2. We already know that special motion-relativity has changed the status of space and time units. Indeed, the very existence of space-time implies to use the same units for length and time intervals. This is achieved since 1985, when it was decided to have the unit of length derived from the unit of time and c fixed. Therefore, the genuine nature of velocities consists of dimensionless pure numbers always smaller than one.

A new step can be made in this direction using special scale relativity. Indeed, in its framework, every length (or time) interval is written in terms of its ratio over the Planck length (time)-scale. The Planck length and time scales thus appear as natural units of length and time intervals, whose genuine nature is found to be pure, dimensionless numbers larger than one.

In the end, this implies that space and time units do not really exist, since, in the same way as the limitation of three-velocity is a pure effect of projection from four-space-time to three-space, the Planck limit is the simple result of projection from five-dimensional to four-dimensional space. More generally, if one replaces the three fundamental constants G, $\hbar$ and c by their expressions in terms of the Planck-time, the Planck-length and the Planck-mass in any equation of physics, all quantities appear in these equations in terms of their ratio over the corresponding Planck units, which, ultimately, vanish from physics. Let us give a simple example.

The gravitational force is written since Newton as

$$F = \frac{-GMm}{r^2}. \tag{12.18}$$

The first step towards dimensionless physics consists of writing it in terms of quantum units, i.e. $\hbar c$, as currently done for the electric force. One obtains

$$F = -\hbar c \frac{(M/m_{\mathbb{P}})(m/m_{\mathbb{P}})}{r^2}. \tag{12.19}$$

This expression of the force, in which $M/m_{\mathbb{P}}$ and $m/m_{\mathbb{P}}$ appear analog to charges, unifies the microscopic and macroscopic forms. But by introducing a Planck unit of force according to its dimensionality, i.e. $f_{\mathbb{P}} = m_{\mathbb{P}} \, l_{\mathbb{P}} \, t_{\mathbb{P}}^{-2} = c^4 G^{-1}$, one can write Newton's force under the form

$$\frac{F}{f_{\mathbb{P}}} = -\frac{(M/m_{\mathbb{P}})(m/m_{\mathbb{P}})}{(r/l_{\mathbb{P}})^2}. \tag{12.20}$$

The (Galilean) scale relativity of mass and length becomes fully apparent in this expression. Each of the ratios is now a pure number ($f = F/f_{\mathbb{P}}$, $\mu = M/m_{\mathbb{P}}$, $\mu' = m/m_{\mathbb{P}}$, $\rho = r/l_{\mathbb{P}}$) and the Newtonian force has taken a dimensionless, constantless form,

$$f = -\frac{\mu \, \mu'}{\rho^2}. \tag{12.21}$$

Such an operation is possible on any physical expression and it takes sense through the interpretation of the Planck-scale as a scale invariant under dilations, making of the Planck-length and Planck-time the natural units of length and of time.

12.4.4. *Mach principle and the scale relativity of mass*

Concerning mass, let us be more specific about the status of the Planck mass. We have seen in the above expression that it appears as a natural mass unit simply by writing the Newton force in quantum units. But a more complete answer may be reached by an analysis of another important problem of physics, which is what Einstein called the "Mach principle" [158, 159].

As reminded in [353, 370, 400], there are actually three levels of "Mach principle".

(i) The first level amounts to determining how inertial systems are defined. Einstein's general relativity solves the problem thanks to the equivalence principle. Inertial systems are locally defined as systems, which are in free fall in the gravitational field, i.e. that follow geodesic motion in the curved space-time.

(ii) The second level expresses the hope that the amplitude of "inertial forces", i.e. of forces that appear precisely when motion is no longer inertial, could be determined from the gravitational field.

(iii) The third level of "Mach principle" considers that there are fundamental relations between elementary particle scales and cosmic scales, and it is therefore related to the Eddington–Dirac large number hypothesis (see [353] Chapter 7.1). We shall be more specific about it in the forthcoming Sec. 12.6 about the cosmological constant and the vacuum energy density.

We now concentrate on (ii). As analysed by Sciama [483], the implementation of this "Mach principle" makes sense only by considering inertial forces as a gravitational force of induction, i.e. as a very manifestation of the general relativity of motion (in analogy with the understanding by Einstein and Poincaré of the Faraday electromagnetic induction as a special relativity effect of change of the reference system). Sciama has shown that a simple mean to express this "Mach II" principle is to write that, if this principle is true, anybody should be in free motion in its own proper reference frame. Therefore the total energy of a body, which is the sum of its own energy and of its energy of interaction with all the other bodies in the Universe, should be zero in its rest frame, namely,

$$mc^2 + \sum_i \frac{-GmM_i}{r_i} = 0. \tag{12.22}$$

The sum of the (M_i/r_i)'s is convergent in finite Universes, such as spherical models, and in Universes having a horizon. In particular, as demonstrated in [353] Chapter 7.1 and recalled below, the spatially flat models, which are now favored by cosmological tests, are Machian.

In the convergent cases, this sum then defines a mass scale M_U and a length-scale R_U for the Universe through the relation

$$\sum_i \frac{M_i}{r_i} = 2\frac{M_U}{R_U},$$

(12.23)

where the factor 2 is conventional. Under this definition, the Mach II principle is finally expressed in terms of a Schwarzschild black hole-like condition for the whole Universe [159, 370, 483],

$$\frac{2G}{c^2}\frac{M_U}{R_U} = 1.$$

(12.24)

Note that the inertial mass of the body has disappeared from this equation. This means that this view of Mach's principle agrees with the weak principle of equivalence and predicts no anisotropy of inertia. This is different from Weinberg's conception [528], who considers that the absence of observed anisotropy of inertia (now measured to better than $\Delta m/m < 10^{-24}$) favors the equivalence principle against Mach's principle.

In the standard framework, there is a direct and fundamental opposition between Mach's principle and the strong equivalence principle. One indeed expects an anisotropy of active gravitational mass, even though this effect remains too small to be presently observable [483]. This problem is solved in the special scale relativity framework where the length $\mathbb{L}$ given by the cosmological constant becomes an horizon invariant under dilations that replaces the infinite length. Indeed, in this case any observer lies at the center of an apparent sphere of radius $\mathbb{L}$. Provided the Universe is infinite from the view point of its standard topology, this isotropic sphere would appear to be of infinite matter density and would therefore dominate any other local mass, no matter how large it may be.

Remember also that it is precisely the discovery by Einstein that the implementation of the Mach II principle involved a relation $M/R =$ constant for the whole Universe that lead him to introduce the cosmological constant in his field equations [159]. We shall come back to this point in the forthcoming section about the cosmological constant.

Einstein finally concluded that the Mach principle was implemented by the constraint that the Universe be described by a finite spherical model. However, infinite models can also be Machian. In particular, this is the case of the Einstein–de Sitter model [353]. Indeed, the integral over M/r should be made up to the horizon $R_U = c/H$, and the mass inside this horizon is

$M_U = \frac{4}{3}\pi\rho(c/H)^3$, so that the Schwarzschild relation becomes

$$\frac{2G}{c^2}\frac{M_U}{R_U} = \frac{8\pi G\rho}{3H^2} = 1. \tag{12.25}$$

which is exactly the space flatness condition $\Omega_M = 1$ of this model. This remains true with a cosmological constant, since the Schwarzschild relation reads in this case $2GM_U/c^2 R_U + \Lambda R_U^2/3 = 1$, which is exactly the flatness condition $\Omega_{\text{tot}} = \Omega_M + \Omega_\Lambda = (8\pi G\rho + \Lambda c^2)/3H^2 = 1$.

Let us now go on with the question of the nature of masses. The general relation Eq. (12.24) between lengths and masses has given the hope [37] that, ultimately, the concept of mass disappears from physics and be replaced by a relative combination of the interdistances between all bodies in the Universe. This hope was apparently deceived.

However, as argued in Sec. 11.1.1.3, this disappearance of mass (as a primary concept — it remains a derived concept of practical use) may actually be already the case, even though it has not yet been fully realized. Indeed, in Einstein's general relativity the gravitational mass is fully equivalent to the Schwarzschild radius. Let us define a Schwarzschild length $l_S = GM/c^2$ half the Schwarzschild radius. In the spirit previously described, introducing the Planck scales, the relation between the active mass and this length reads

$$\frac{M}{m_{\mathbb{P}}} = \frac{l_S}{l_{\mathbb{P}}}. \tag{12.26}$$

If we now look more closely at the various formulas where the active gravitational mass appears, we discover that it can be replaced everywhere by the Schwarzschild radius without any loss of meaning nor of generality. Actually, we discover an even more profound fact: the length carries the fundamental meaning and the mass is only a derived concept. For example, the g_{00} term of Schwarzschild metrics is naturally integrated as $g_{00} = 1 - 2l_S/r$, and l_S is only subsequently interpreted as giving the active gravitational mass. Even in Newton's theory, the very nature of the central mass in the Kepler problem is that it determines the natural length unit of the system. Hence, Kepler's third law between the semi-major axes a and the period T of planetary orbits reads

$$\left(\frac{a}{l_S}\right)^3 = \left(\frac{cT}{2\pi l_S}\right)^2. \tag{12.27}$$

This is also apparent from the fact that the product GM for bodies in the solar system (and therefore their Schwarzschild length) is known with a much higher precision than G and M separately.

Finally, such an analysis leads to writing Newton's force (Eq. 12.18) under another equivalent and highly significant dimensionless form,

$$\frac{F}{f_{\mathbb{P}}} = -\frac{r_1}{r} \times \frac{r_2}{r}, \qquad (12.28)$$

where $r_1 = GM/c^2$ and $r_2 = Gm/c^2$. Such a writing is fundamentally scale relativistic, since lengths appear here only through ratios.

Let us now consider the inertial mass. Since the introduction of matter waves by de Broglie, the inertial mass m of a particle is also fully equivalent to a length: it is associated to its Compton length λ_c,

$$\lambda_c = \frac{\hbar}{mc}. \qquad (12.29)$$

The situation is exactly similar to the gravitational, macroscopic case. Indeed, it is easy to verify that in any formula in which the inertial mass of a body appears, it can be replaced by its Compton length. For example, the energy of the fundamental level of the hydrogen atom writes, in dimensionless terms,

$$\frac{E}{E_{\mathbb{P}}} = -8\pi^2\alpha^2\frac{m_e}{m_{\mathbb{P}}} = -8\pi^2\alpha^2\frac{l_{\mathbb{P}}}{\lambda_e}, \qquad (12.30)$$

where $E_{\mathbb{P}} = m_{\mathbb{P}}\,c^2$ is the Planck energy. Therefore, we suggest that, like the gravitational mass, the inertial mass is but an intermediate concept, which is ultimately fated to vanish from physics.

Such a suggestion is supported by the scale relativistic explanation of the nature of the Compton length of a particle. Recall indeed that we identify "particles" with the geodesics family of a nondifferentiable fractal space-time, and that the Compton length is a geometric characterization of these geodesics (see Chapter 5). Mass ratios can be replaced everywhere by length ratios, so that the Planck mass itself might disappear from physics as a fundamental concept. The Planck length and time scales themselves appear as horizons resulting from a projection effect, representing a fundamental "1" playing for scales the same part as played by the velocity c for motion.

Finally, the equality of the inertial and active gravitational mass for a given body (in accordance with the strong equivalence principle) is translated in terms of an inverse relation between the Compton length and the Schwarzschild length,

$$\left(\frac{l_\mathbb{P}}{\lambda_c} = \frac{M_I}{m_\mathbb{P}} \right) = \left(\frac{M_A}{m_\mathbb{P}} = \frac{l_S}{l_\mathbb{P}} \right). \tag{12.31}$$

These scales are inverse one of the other, i.e. they are but symmetric elements in the scale space, namely, $\ln(l_S/l_\mathbb{P}) = -\ln(\lambda_c/l_\mathbb{P})$. Moreover, both of them are solutions of the search for conservative scale quantities, obtained from applying Noether's theorem in the framework of scale relativistic mechanics (see Sec. 11.1.1.3).

12.5. Consequences for the Early Universe

The history of the early Universe is nowadays described in correspondence with our knowledge of high energy particle physics. This is due to the fact that, when going toward the past, the primeval Universe encounters conditions of ever increasing density, pressure and temperature, so that it is described in terms of a gas of elementary particles of increasing energy. This means that the progress that is made in our understanding of elementary particles (transition scales and values of the coupling constants) has immediatly its counterpart in our view of the primeval universe. In what follows, we shall not attempt to give a complete new description of the scale relativistic history of the early Universe, but only point out some features by which it differs from the standard view.

12.5.1. *New formulation of the question of the origin*

The new meaning attributed to the Planck-length and the Planck-time drastically changes the standard view about the primeval singularity. The first new physical implication of cosmological importance is the non-existence of the zero instant as a meaningful physical concepts. The evolution of the Universe does not begin any more at the instant $t = 0$ (i.e. $\log(t/t_0) = -\infty$ for any $t_0 > 0$), but instead from the Planck time $t_\mathbb{P} = l_\mathbb{P}/c$ (see [353] Chapter 7.1). However this new structure should not be misinterpreted, since in the new theory, the Planck scale owns all the properties of the previous zero instant. This means that temperature, redshift, energy, density and all the quantities A, which were previously diverging as t^{-k}, are now diverging toward the past when t tends to $t_\mathbb{P}$ as

$$\log \frac{A}{A_0} = \frac{k \log(t_0/t)}{\sqrt{1 - \log^2(t_0/t)/\log^2(t_0/t_\mathbb{P})}}. \tag{12.32}$$

The scale factor of expansion of the Universe can itself no longer become smaller than the Planck length $l_\mathbb{P}$: the Universe starts asymptotically from the Planck scale. This solves the problem of the singularity, which actually is not part of the model in the standard general relativistic approach. For example, in the hyperbolic Robertson–Walker solution with trivial topology, the Universe is infinite at any instant $t \neq 0$, as small as it could be, while the spatial part of the metric vanishes at $t = 0$ since the scale factor is such that $a(0) = 0$. This kind of paradox no longer appears in the scale relativistic framework.

12.5.2. *Very early dimensionless phase of the Universe*

Actually, even such a description in terms of a limiting Planck scale remains misleading. Indeed, we have insisted from the very beginning on the fact that only scale ratios do have a physical meaning, so that any description in terms of a dimensioned scale assumes a static, invariant scale can be used as reference unit. For example, the expansion of the Universe is meaningful at the present epoch only because there does exist small scales length-units that are not carried away by the expansion, in particular the Bohr radius, which is linked to the line energies, which serve to measure the redshift.

At the present epoch, the static to nonstatic transition scale is given by the typical radius of self-gravitating extragalactic objects, such as galaxies and the centers of clusters of galaxies, i.e. $\lambda_g \approx GM/3\langle v^2 \rangle$. In the early Universe, static scales are given by scales, such as the QCD confinement scale, the Compton scale of the electron, then the Bohr scale, but they are effective only at epoch later than the phase transition to which they correspond, i.e. quark-hadron transition, e^+-e^- pair annihilation, recombination.

If one goes back still further in the past, one reaches the Planck energy scale. In the special scale-relativity framework, it is no longer equivalent to the Planck length-time scale that now corresponds to infinite energy. Moreover, we have found [352] that the Planck mass scale becomes naturally the scale of full unification between gravitation and the U(1), SU(2) and SU(3) fields (see Fig. 11.4). The X bosons that carry the unified field, therefore, have a Planck mass as their mass.

The GUT length-scale identifies with the Compton length of the X particles and is predicted to be $e^{4\pi^2/\sqrt{2}} = 10^{12}$ times smaller than the Z scale (see Chapter 11). It would be the minimal possible static scale in nature. Before the GUT transition, physics can therefore be only pure dimensionless physics. The Planck-length and the cosmic length lose their meaning (more precisely, in terms of evolution toward the future, they have no meaning yet), and only their ratio $\mathbb{K} = 5.3 \times 10^{60}$ is meaningful. During this very early era, unbroken scale relativity holds, and all the physical laws should be expressed in terms of pure, dimensionless numbers. The dimensioned physics as we know it now becomes possible only thanks to (and after) the Planck energy symmetry breaking.

12.5.3. *Solution to the horizon/causality problem*

One of the main results of scale relativity concerning the primeval Universe is its ability to solve the causality/horizon problem (see [353] and references therein). Let us recall the nature of this problem. When looking at two directions separated by a large angle, e.g. two opposite directions, we observe regions of the Universe, which, for a large enough redshift, may have never been connected in the past. The problem is particularly strong concerning the microwave background radiation, due to its high isotropy ($\delta T/T \approx 2 \times 10^{-5}$) and its early origin ($z \approx 1000$): at least twenty such independent regions would be observed in the framework of standard cosmology.

Such causally disconnected regions should behave as completely independent universes and it becomes very strange that no large fluctuation of the microwave background temperature is observed. The solution to this problem is usually searched for in the framework of inflationary cosmology. However one may remark that inflation is to some extent an ad hoc solution, in particular as concerns its cause (now unobservable scalar field, primordial black holes), that must be postulated additionally to the presently known content of the Universe. Moreover it does not solve the problem in principle: in its framework the presently observed regions of the universe would have been causally connected in the past, but this does not remain true in the distant future.

The horizon/causality problem is simply solved in the special scale-relativity framework without the need for an inflation phase, thanks to the new role played by the Planck lenth-time scale. It is identified with a limiting scale, invariant under dilations. This implies a causal connection of all points of the universe at the Planck epoch. This is due to the fact that, if one accounts for the scale-Lorentz factor, one finds that the light cones that rest on two arbitrary distant points flare when $t \to t_\mathbb{P}$ and finally always cross themselves in their past (see Fig. 12.2).

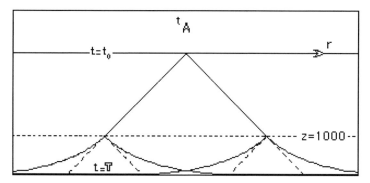

Fig. 12.2: **Flare of light cones in special scale relativity.** Illustration of the flare of light cones in special scale relativity, allowing causal connection of any couple of points in the universe.

12.6. The Nature of the Cosmological Constant and of "Dark Energy"

12.6.1. *The cosmological constant as an invariant cosmic scale*

In the same way as the minimal scale-relativistic invariant scale is naturally identified with the Planck length-scale [352], the maximum invariant scale can be identified with the scale of the cosmological constant, $\mathbb{L} = \Lambda^{-1/2}$ [353, 360, 370, 383].

In order to support such a proposal, let us reconsider Hawking's quantum cosmology calculation [224] in the special scale-relativity framework. In [529], Weinberg considered five different approaches to the solution of the cosmological constant problem and concluded that Hawking's result, according to which the cosmological constant should be zero, solved the problem.

Hawking's calculation [224] amounts to estimating the wave function of the Universe as a path integral, which is solution of the Wheeler–DeWitt equation [136, 536]. He finds that the expected distribution of values for the cosmological constant should show an enormous peak at $\Lambda_{\text{eff}} = 0$. Indeed, this quantum cosmology approach yields a probability density for the various values of the cosmological constant proportional to

$$P = e^{3\pi/G\Lambda}, \tag{12.33}$$

(in units where $\hbar = c = 1$), where Λ is the true effective cosmological constant that is measured in gravitational phenomena at long ranges. When a cosmological constant is included in the Schwarzschild metric around an active gravitational mass M, the time metric potential reads

$$g_{00} = 1 - \frac{2GM}{c^2 r} - \frac{\Lambda}{3} r^2. \tag{12.34}$$

Therefore, when $r \gg 2GM/c^2$, which is the case at cosmological scales, the length $r_H = \sqrt{3/\Lambda}$ appears as an horizon for the Schwarzschild variable r, which is defined as the distance to the central mass for which the circumference of a circle is $C = 2\pi r$. The same is true in the de Sitter metric, which is a good approximation to the present state of the universe, whose energy balance is now known to be dominated by the cosmological constant term (or "dark energy"). In terms of this horizon, the Hawking probability density reads

$$P = \exp\left(\frac{\pi r_H^2}{G}\right). \tag{12.35}$$

It is noticeable that πr_H^2 could be identified to a cross section. This probability density has an infinite peak for $\Lambda \to 0+$, i.e. for an horizon r_H pushed to infinity, so that this calculation predicts a zero effective cosmological constant in the standard framework [224, 529]. The subsequent observational evidence for a significantly nonzero cosmological constant (see what follows) seems to have invalidated this approach. Let us however reconsider this conclusion in the new scale relativistic framework.

The cosmological constant is fundamentally a curvature term, as can be seen in Einstein's field equations,

$$R_{\mu\nu} + \left(\Lambda - \frac{1}{2}R\right) g_{\mu\nu} = -\frac{8\pi G}{c^4} T_{\mu\nu}. \tag{12.36}$$

Therefore, in a framework where it becomes a variable, such a running cosmological constant is the square of the inverse of a running fundamental length L,

$$\Lambda = \frac{1}{L^2}. \tag{12.37}$$

The above Schwarzschild horizon is given by $r_H = \sqrt{3}\, L$. The infinite peak of Hawking's probability density for $\Lambda = 0$ therefore means an infinite peak for $L \to \infty$.

All this calculation implicitly assumes underlying Galilean dilation laws, in which an infinite length scale is assumed to have physical meaning. On the contrary, in scale relativity, the infinite length-scale is replaced by a maximal scale $\mathbb{L}$, which is unreachable and invariant under dilations.

The operation by which one considers an ever increasing running L is precisely such a dilation under which $\mathbb{L}$ is invariant. Consequently, the infinite scale in Hawking's calculation is replaced by the length-scale $\mathbb{L}$.

More specifically, in the scale-Galilean case, Hawking's probability density may be written under the form

$$P = \exp\left\{ \exp\left(2 \ln \frac{L}{\lambda} \right) \right\}, \tag{12.38}$$

where λ is a static reference length-scale ($\lambda \ll \mathbb{L}$). It becomes in the scale log-Lorentz case

$$P = \exp\left\{ \exp\left(\frac{2 \ln(L/\lambda)}{\sqrt{1 - \ln^2(L/\lambda)/\ln^2(\mathbb{L}/\lambda)}} \right) \right\}. \tag{12.39}$$

As can be easily checked in this expression, the probability density has an infinite peak for $L \to \mathbb{L}$, i.e. for a cosmological constant $\Lambda = 1/\mathbb{L}^2$.

Let us conclude this section by a relevant remark about units. As in the case of the Planck length scale, there actually remains an uncertainty about a possible multiplying numerical factor. In the case of the Planck length, it could be defined in terms of the Planck constant h or of the reduced Planck constant $\hbar$. The difference between these constants is that h defines a full wavelength, such as in the definition of the de Broglie wavelength, h/mv, i.e. it corresponds to a variation 2π of the phase, while $\hbar$ corresponds to a variation of the phase by 1 radian. This is the reason why we have called "de Broglie length" the scale $\lambda_{\mathrm{dB}} = \hbar/mv$ (while h/mv is the de Broglie *wave*length) and "Compton length" $\lambda = \hbar/mc$. A similar question occurs with the intervening of G, which could also be replaced by $8\pi G$ as in Einstein's equations. Whatever the choice made for the reference Planck scale, the same choice should be made for the scale to which it is referred, e.g. $\hbar$ in the Compton relation and in the Planck scale, and for the length, time and mass Planck scale, so that the numerical constants are expected to finally vanish (see Sec. 12.4.4). Ultimately, the final identification of the precise definition of the invariant scale will therefore come from a comparison of the theoretical predictions with the observational and experimental data, which, up to now, seems to confirm the choice of $\hbar$ and G. For this, see what follows.

As regards the scale of the cosmological constant, one could equivalently make the identification of the invariant maximal scale with the Schwarzschild–de Sitter horizon, $\mathbb{L} = \sqrt{3/\Lambda}$. There too, this question can be decided by the comparison with the observational data of the theoretical predictions in which the invariant scale $\mathbb{L}$ appears.

12.6.2. *Cosmological constant and slope of the correlation function*

It has been observed for long that the correlation functions of various classes of extragalactic objects (from galaxies to superclusters) were characterized by a power law variation in function of scale, with an exponent $\gamma = 1.8$ at scales ≈ 1–$10\,\mathrm{Mpc}$ [128], smaller than the value $\gamma = 2$ expected from the simplest models of hierarchical formation. At larger scales the question of the transition to uniformity is still discussed, some authors claiming that there is no transition and that the slope is $\gamma = 1$ everywhere (see [108] and references therein), which corresponds to a fractal dimension 2 of the matter distribution.

The scale relativity approach suggests a complementary view about this problem, and may reconcile the apparent contradictions between the observationally determined values of γ [353, 359, 360]. The value $\gamma = 2$ is nothing but what is expected in the framework of Galilean scale laws. It is the manifestation of a fractal dimension $\delta = 3 - \gamma = 1$. In scale relativity, one must adopt Lorentzian laws at large scales in order to ensure scale-covariance. The fractal dimension of the galaxy distribution, which is of zero topological dimension, being a dust of "particles", now becomes itself scale-varying and depends on the cosmological constant $\Lambda = 1/\mathbb{L}^2$ as

$$D_F(r) = \frac{D_0}{\sqrt{1 - \ln^2(r/\lambda_g)/\ln^2(\mathbb{L}/\lambda_g)}}, \qquad (12.40)$$

where λ_g is the typical static radius of the objects considered ($10\,\mathrm{kpc}$ for giant galaxies, 100 to $300\,\mathrm{kpc}$ for clusters...), and where D_0 is the asymptotic fractal dimension at scale λ_g. Several consequences and new predictions arise from this formula.

First, we expect $D_0 = 1$, then $\gamma = 3 - D_0 = 2$ at small scales. Several observations support this expectation. The flat rotation curves of galaxies imply halos in which mass varies as $M(r) = r^D$, with $D = 1$. Dwarf galaxies have been found to be autocorrelated at small scales (10–$200\,\mathrm{kpc}$) and characterized by a power $\gamma \approx 2.2$ [518]. The analysis of the CfA survey by Davis and Peebles [128] shows that, apart from fluctuations coming from deconvolution, the average γ is 2 between 10 and $300\,\mathrm{kpc}$, while it reaches its value 1.8 only between 1 and $10\,\mathrm{Mpc}$ (see their Fig. 3).

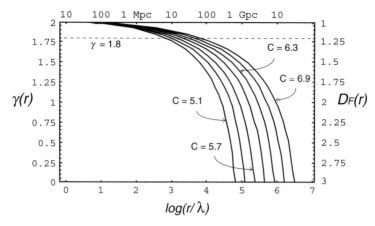

Fig. 12.3: **Cosmological variation of fractal dimension with scale.** Underlying variation with scale of the power $\gamma = 3 - D_F$ (where D_F is the fractal dimension) of the galaxy-galaxy autocorrelation function in special scale relativity, for various values of the constant $\mathbb{C} = log(\mathbb{L}/\lambda_g)$, where $\mathbb{L} = \Lambda^{-1/2}$, for $\lambda_g = 10$ kpc. This variation is a property of space, which should be completed by a knowledge of the organization of matter in such a fractal space.

Starting from the small scale value $\gamma = 2$ (i.e. $D_F = 1$) and using the above formula, one correctly expects a value of $\gamma = 1.8$, (i.e. $D_F = 1.2$) at a scale of $\approx 10\,\text{Mpc}$. One then predicts a fast variation of γ toward larger scales. Some early results [82] seemed to support such an expectation. More recent analyses of the SDSS survey have confirmed that the effective observed fractal dimension seems to be variable and to increase with increasing scale [95]. More detailed studies are needed to decide whether the increase is continuous or by stages, including the value $D_F = 2$, reached at scales of $\approx 20-40\,\text{Mpc}$. Such stages are quite possible in the hereabove model, since Eq. (12.40) refers to a variation with scale of the space fractal dimension, while the observed fractal dimension is that of objects in this space. The comparison with observation should therefore include a model of structure formation in such a fractal space whose fractal dimension increases with scale. In the next years, the analysis of large surveys like the Sloan Digital Sky Survey (see [2, 508] and references therein), which contains about 700 000 redshifts of galaxies could allow one to put the variable fractal dimension proposal to the test.

Conversely, provided the variation with scale of the fractal dimension be confirmed, one could use this variation as a direct measurement of the cosmological constant, since it depends on it in a fundamental way. Indeed, the scale $\mathbb{L}$ of the cosmological constant appears in this variation as an asymptotic horizon where the fractal dimension formally diverges. Although one clearly cannot scan such large scales (since it is the analog of the infinite scale in the standard theory), one may at least follow the scale variation up to the length scale where the fractal dimension reaches $D_F = 3$, which would appear as a transition to uniformity.

In Fig. 12.3, we have represented the function $\gamma(r) = 3 - D_F(r)$ for various values of $\mathbb{C} = \log(\mathbb{L}/\lambda_g)$, in the case of giant galaxies with $\lambda_g = 10\,\mathrm{kpc}$. The observed values of D_F (1.2 at 5–10 Mpc, 2 at 20–40 Mpc) are consistent with the estimate of the cosmic scale $\mathbb{L} = 2.8\,\mathrm{Gpc}$ from the vacuum energy density (see [353] and what follows). But a genuine estimate should account for the radii of galaxies in the considered sample, for the initial value of D_0 at small scales in these galaxies and for the effects of structure formation at each considered scale (in a hierarchical structure formation framework, as now privileged). Such a determination is expected to become possible in the coming years.

The transition to uniformity ($\gamma = 0$, $D_F = 3$) is reached only at very large scales (≈ 0.5 to 1 Gpc, depending on the galaxy population considered). This seemed to be supported by the suggestion that the COBE map is characterized by a low fractal dimension $D_F = 1.43 \pm 0.07$ [133]. Early results of analysis of the SDSS find a smaller transition scale at about $70\,h^{-1}\,\mathrm{Mpc}$ [235], i.e. about 100 Mpc, but this still may be an effect of the size limit of the catalog. These questions are certainly worth to be analysed further using the more recent WMAP, Planck and SDSS data.

12.6.3. *Cosmological constant and cosmic background radiation*

One of the recent unexplained fundamental observational effect that could support the new interpretation of the cosmological constant as defining a very large scale horizon invariant under dilations is the observed limit for angular correlation in the 3K cosmic background radiation (CBR) fluctuation unveiled by the COBE and WMAP data.

Indeed, the angular correlation function shows a definite cut-off at large angular scales (see e.g. Fig. 16 of [500]), which has been confirmed by the WMAP three year data [501]. Beyond an angle of about 60 degrees, the WMAP data no longer shows any correlation while the ΛCDM and other models predict correlations or anticorrelations at all angular scales.

This observational result is easily understood in the special scale relativity framework. Indeed, the cosmic scale $\mathbb{L}$ appears as an horizon from all points of the Universe and at any epoch. Therefore fluctuations could be correlated at the decoupling epoch $z = 1100$ only between points separated by a maximal distance $\mathbb{L}$. The sphere $z = 1100$ is itself now seen as very close to the horizon $\mathbb{L}$, so that the present observer and the two lastly correlated points nearly form an equilateral triangle. The predicted limit for angular correlations is therefore 60 degrees, as observed.

Note that a such a limit is expected for any finite cosmological model, either spherical, or subjected to topological conditions of closure [296], but at angular scales which can be different from 60 degrees.

Problems and Exercises

Open Problem 27: When the power spectrum of galaxies is compared with that of the 3K CBR, a strong difference of amplitude is found when trying to match them. However, they are observed at very different scales, though the gap between them begins to be filled both by the decrease of the minimal angular size of CBR missions (WMAP, Planck) and by the increase of size of galaxy surveys (SDSS, etc.). Study the possibility that the difference comes not from amplitude, but from the calculation of scales from angular observations. Indeed, in the special scale relativity framework, the infinite scale is replaced by the cosmic scale $\mathbb{L}$ and all scales becomes smaller than this cosmic scale according to the log-Lorentz relation. ■

Exercise 22 More generally, the transformation from momentum variables to position variables is performed by a Fourier transform. The same is true when transforming a correlation function into a power spectrum [438]. Now the Fourier transform implicitly assumes an underlying 'Galilean' structure of scale laws. Establish the form of a generalized Fourier transformation in special scale relativity.

Hint: In one dimension the standard Fourier transform reads

$$\phi(p) = \frac{1}{\sqrt{2\pi}} \int_{-\infty}^{+\infty} \psi(x)e^{ipx}\,dx. \tag{12.41}$$

Now the relation $px = p_0 x_0$ becomes in special scale relativity (for $p > 0$)

$$\ln\frac{x}{x_0} = \frac{\ln(p_0/p)}{\sqrt{1 + \ln^2(p_0/p)/\mathbb{C}^2}} = \ln\left(\frac{p_0}{p}\right)^{1/\sqrt{1+\ln^2(p_0/p)/\mathbb{C}^2}}. \tag{12.42}$$

so that one simply obtains a new transform

$$\Phi(p) = \phi\left[\left(\frac{p}{p_0}\right)^{1/\sqrt{1+\ln^2(p/p_0)/\mathbb{C}^2}}\right].$$
(12.43)

■

12.6.4. *Cosmological constant and Mach principle*

Let us show that the very existence of the cosmological constant actually solves in a general way the problem of whether Einstein's theory of general relativity is Machian or not.

We have recalled in Sec. 12.4.4 that the "Mach II" principle should be implemented in cosmology provided there exists a fundamental Schwarzschild-like relation $GM_U/c^2R_U = 1$ for the Universe as a whole. Einstein [159] has introduced the cosmological constant in the gravitational field equation precisely in order to make the final theory agree with the first principles from which it was constructed, including Mach's principle. Later, in 1924, it has been demonstrated by Cartan [84] that the Einstein tensor with a cosmological constant is the most general one that satisfy Einstein's criteria in his construction of the general relativity equations (in particular the null divergence condition).

Moreover, all cosmological solutions of Einstein's equations based on the cosmological principle of homogeneity and isotropy at large scale are characterized by a prime integral that reads $\rho R^3 = $ constant in dust models, which are valid for the present epoch. This means that all models are characterized by an invariant energy M_Uc^2, which can be neither created nor destroyed. Conversely, all cosmological models, which are solutions of the 1915 version of Einstein equations (without cosmological constant) are non-static, i.e. are characterized by a variable scale factor $R(t)$, so that M_U/R_U is variable. This was the problem Einstein wanted to solve in his 1917 paper by introducing a cosmological constant, and Eddington and Dirac by compensating the variation of R_U by a variation of G.

However, the solution proposed by Einstein was to construct a particular cosmological model for which $R_U = $ constant. Such a model is indeed allowed when a cosmological constant is present. However, it is unique, metastable and we know since 1929 that it contradicts the observation of the expansion of the Universe. Remember that when Einstein made this proposal in 1917, the Universe was still identified with the "gas" of stars, which is globally static and which is now recognized to be only our Milky

Way galaxy. Later, Einstein suggested that the solution was given by closed models, which are characterized by a maximal value of the scale factor. Anyway, none of these solutions to the Mach problem are satisfactory, since the initial question was a matter of first principle concerning the whole theory.

It is actually easy to show that general relativity in its whole is Machian, in agreement with Barbour and Bertotti's view [38, 39]. Indeed, as already recalled, the cosmological constant is a curvature, i.e. the inverse of the square of a cosmic length. Therefore, the introduction of a fundamentally invariant cosmological constant amounts to the introduction of an invariant length $\mathbb{L}$ [353]. This length is present in Einstein's equation whatever the model, even beyond the validity of the cosmological principle. As a consequence, any cosmological model being characterized by an invariant mass M_U, a universal Machian relation can be written whatever the model:

$$\frac{2G}{c^2}\frac{M_U}{\mathbb{L}} = \text{constant.} \tag{12.44}$$

Moreover, in the special scale relativity framework, in which the length-scale $\mathbb{L}$ is identified with the maximal possible resolution scale, and replaces the infinite scale, $\mathbb{L}$ is a Universal horizon, which exists independently from any model.

12.6.5. *Cosmological constant and vacuum energy density*

One of the most difficult open questions in present standard cosmology is the problem of the vacuum energy density and of its manifestation as an effective cosmological constant [83, 529]. This question dates back to Lemaitre's understanding that the cosmological constant term in Einstein's field equations is equivalent to the effects of a Lorentz-invariant vacuum [286]. Indeed, Einstein's equations read

$$R_{\mu\nu} - \frac{1}{2}Rg_{\mu\nu} = -\frac{8\pi G}{c^2}T_{\mu\nu} - \Lambda g_{\mu\nu}, \tag{12.45}$$

while the energy-momentum tensor reads for a perfect fluid

$$T_{\mu\nu} = \left(\rho + \frac{p}{c^2}\right)u_\mu u_\nu - \frac{p}{c^2}g_{\mu\nu}. \tag{12.46}$$

Lemaitre then remarked that a Lorentz-invariant vacuum is characterized by a negative pressure, which is exactly the opposite of its energy density, $p_v = -\rho_v c^2$. When they are inserted in the energy-momentum tensor, this pressure and this density cancel each other in the $u_\mu u_\nu$ term, while the negative pressure remains in the $g_{\mu\nu}$, thus playing the role of an effective positive cosmological constant Λ_{eff}, such that

$$\rho_v = -\frac{p_v}{c^2} = \Lambda_{\text{eff}} \frac{c^2}{8\pi G}. \tag{12.47}$$

Its effects are equivalent to those of a repulsive force. The Einstein equations therefore read

$$R_{\mu\nu} - \frac{1}{2}Rg_{\mu\nu} + \left(\Lambda + \frac{8\pi G}{c^2}\rho_v\right)g_{\mu\nu} = -\frac{8\pi G}{c^2}T_{\mu\nu}, \tag{12.48}$$

where Λ is the geometric cosmological constant and $(8\pi G/c^2)\rho_v$ is the effective cosmological constant coming from the vacuum energy density.

This remark has lead Lemaitre [286], then Zeldovich [552] to suggest that one could calculate the value of the cosmological constant from the quantum vacuum energy density. Actually, this amounts to assuming that the geometric contribution is vanishing and that the full value of the cosmological constant comes from vacuum energy. This hope was deceived, since the vacuum energy in standard quantum field theory is the sum of the zero point energies of all normal modes of the various fields and it is therefore infinite. If one imposes an arbitrary cut-off ν_{max} to the possible wave numbers, one finds the vacuum energy to vary as ν_{max}^4 [529]. By choosing, once again in an arbitrary way, $h\nu_{\text{max}} = E_{\mathbb{P}}$, the Planck energy, one finds a calculated vacuum energy density $\approx 10^{120}$ times larger than the observed cosmological constant.

Since 1998, definite observational proofs of the existence of such a cosmological contribution have been obtained [440, 463, 194, 129]. It even reveals to be the dominant one in the energy balance of the universe. But instead of considering this result as a measurement of the cosmological constant, this contribution has been called "dark energy" and considered to be of unknown origin. This expression is somewhat misleading, since the repulsive character of its effect is to be attributed to a negative pressure, while the energy term in the stress-energy tensor has disappeared for a state equation $p/\rho c^2 = w = -1$. Even if w was different from this cosmological constant-like value, the remaining energy contribution would be attractive. However, an advantage of this approach to the problem has been that the coefficient w of the state equation has entered into the set of measured quantities. The 2006 WMAP three year results have yielded $w = -0.97 \pm 0.08$ [501] and the SDSS data $w = -0.94 \pm 0.09$ [508]. This result has been confirmed and reinforced by the WMAP 5 years data, yielding $w = -0.97 \pm 0.06$ [233], which is a strong indication that the so-called "dark energy" is a cosmologial constant.

The scale relativity approach actually generalizes Zeldovich's [552] one, and allows to suggest a solution to the cosmological constant problem [353, 359, 360]. Moreover, this solution gives physical meaning in an exact way to the Eddington–Dirac large number coincidence (see also Sidharth [491] about this point), i.e, to the third form of Mach principle considered in Sec. 12.4.4.

The first step towards this solution consists of taking into account the well-established fact that the quantum vacuum is explicitly scale-dependent. As a consequence, the Planck value of the vacuum energy density, that gives rise to the 10^{120} discrepancy with observational limits, is only relevant at the Planck scale and becomes irrelevant at the cosmological scale.

We expect such a naturally scale dependent vacuum energy density ρ to be solution of a scale differential equation that reads

$$\frac{d\rho}{d\ln r} = \Gamma(\rho) = a + b\rho + O(\rho^2), \tag{12.49}$$

where ρ has been normalized to its Planck value, so that it is always <1, allowing us to perform a Taylor expansion of $\Gamma(\rho)$. This equation is solved as

$$\rho = \rho_c \left\{ 1 + \left(\frac{r_0}{r} \right)^{-b} \right\}, \tag{12.50}$$

where $\rho_c = -a/b$ can be identified with the cosmological energy density. We recover a combination of a power law behavior at small scales and of scale-independence at large scale (see Chapter 4), with a fractal to nonfractal transition about some scale r_0 that comes out as an integration constant.

Such a result can also be directly obtained from general relativity and quantum mechanics by writing that the final effective cosmological constant is the sum of a geometric, invariant term from the left-hand side of Einstein equations and of the quantum vacuum energy density contribution in the energy-momentum tensor, which is explicitly scale dependent,

$$\Lambda_{\text{tot}} = \Lambda_{\text{geom}} + \Lambda_{\text{quant}}(r). \tag{12.51}$$

Various contributions to the vacuum energy density can be considered [552]. The largest comes from the vacuum energy itself, which depends on scale as r^{-4} (as in the Casimir effect). However, when considering the various field contributions to the vacuum density, we may always choose $\langle E \rangle = 0$, i.e. the energy density of the vacuum can always be renormalized.

But consider now the gravitational self-energy of vacuum fluctuations. It writes:

$$E_g = \frac{G}{c^4} \frac{\langle E^2 \rangle}{r}. \tag{12.52}$$

The Heisenberg relation $\langle E^2 \rangle^{1/2} = \hbar c / r$ prevents from making $\langle E^2 \rangle = 0$, so that the gravitational self-energy cannot vanish. We find that $\langle E^2 \rangle \propto r^{-2}$, then $E_g \propto r^{-3}$, then finally a gravitational self-energy density $\rho_g \propto E_g / r^3 \propto r^{-6}$. Knowing that it is $\rho_{\mathbb{P}}$ at Planck scale $l_{\mathbb{P}}$, we therefore obtain the asymptotic high energy behavior:

$$\rho_g = \rho_{\mathbb{P}} \left(\frac{l_{\mathbb{P}}}{r} \right)^6, \tag{12.53}$$

From this relation we can make the identification $b = -6$ in Eq. (12.50). Therefore we obtain the complete behavior of the vacuum energy density,

$$\rho = \rho_c \left\{ 1 + \left(\frac{r_0}{r} \right)^6 \right\}, \tag{12.54}$$

where $\rho_c = \Lambda c^2 / 8\pi G$. We therefore obtain for the vacuum energy density corresponding to the geometric cosmological constant:

$$\rho_c = \rho_{\mathbb{P}} \left(\frac{l_{\mathbb{P}}}{r_0} \right)^6. \tag{12.55}$$

For $r \ll r_0$ the total energy density (12.54) is dominated by the scale-dependent term that decreases very rapidly with increasing scale, then becomes negligible beyond r_0. For $r \gg r_0$, it becomes dominated by the invariant term ρ_c, which is identified with the geometric cosmological constant term, i.e. $\rho_c = \Lambda c^2 / 8\pi G$. But now, the value of the geometric cosmological constant is given by the value reached by the quantum energy density at the transition scale r_0,

$$\Lambda = \frac{8\pi G}{c^2} \rho_{\mathbb{P}} \left(\frac{l_{\mathbb{P}}}{r_0} \right)^6, \tag{12.56}$$

so that the problem of determining the cosmological constant now amounts to determining this transition length-scale.

Note that all the above calculation has been performed in the standard quantum mechanical framework, i.e. from the scale relativity viewpoint, in

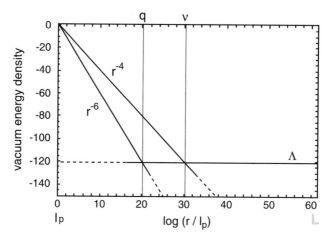

Fig. 12.4: **Scale variation of vacuum energy density.** Variation of the vacuum energy density [in $(l_\mathbb{P}/r)^4$] and of the gravitational self-energy density of quantum vacuum fluctuations [in $(l_\mathbb{P}/r)^6$] in the framework of Galilean scale relativity. We have suggested [353] that the r^{-6} gravitational self-energy contribution crosses the geometric cosmological constant contribution Λ at a scale of 70 MeV (electron "classical" radius and quark confinement transition denoted by q in the figure). This scale lies at the third of the full interval going from the Planck scale $l_\mathbb{P}$ to the cosmic scale $\mathbb{L}$ (in scale space described by logarithms), which validates the Eddington–Dirac large number relation. One derives a cosmological constant $\Lambda = 1/\mathbb{L}^2$, where $\mathbb{L} = 5.3 \times 10^{60} l_\mathbb{P}$. The vacuum energy contribution crosses Λ at the middle of the scale space (in log). From the previous determination of Λ, one finds that this is a mesoscopic scale ($\approx 40\,\mu$m), corresponding to an energy 5.3×10^{-3} eV, which is the typical neutrino mass needed to explain solar neutrino oscillations.

the framework of Galilean scale relativity. It is easy to generalize Eq. (12.54) in the case of log-Lorentzian dilation laws. One obtains

$$\rho = \rho_c \left\{ 1 + \left(\frac{r_0}{r} \right)^{6/\sqrt{1-\mathbb{V}^2/\mathbb{C}^2}} \right\}, \tag{12.57}$$

where $\mathbb{V} = \ln(r_0/r)$ and $\mathbb{C} = \ln(r_0/l_\mathbb{P})$. This generalization does not change the final result, since both formulae become equivalent at the scale r_0, which is the relevant scale with regards to the determination of the cosmological constant.

At this stage, we are already able to demonstrate the second Eddington–Dirac's large number relation [138, 139, 152, 528], according to which cosmological scales are given by the cube of typical elementary particle scales in Planck scale unit, and to write it in an exact way and in terms of invariant quantities, i.e. we do not need varying constants to implement it in this form.

Indeed, introducing the cosmic length-scale $\mathbb{L} = \Lambda^{-1/2}$, we get the relation

$$\mathbb{K} = \frac{\mathbb{L}}{l_{\mathbb{P}}} = \left(\frac{r_0}{l_{\mathbb{P}}}\right)^3 \equiv \left(\frac{m_{\mathbb{P}}}{m_0}\right)^3, \tag{12.58}$$

where $r_0 = \hbar/m_0 c$ is the Compton length associated with the mass scale m_0. Then, in this framework, the power 3 in the Eddington–Dirac relation is readily explained and understood as coming from the power 6 of the gravitational self-energy of vacuum fluctuations over the power 2 that connects the invariant impassable scale $\mathbb{L}$ to the cosmological constant (Fig. 12.4).

12.6.6. *Theoretical estimate of the cosmological constant*

The value of the cosmological constant can be estimated through this approach only provided the transition scale r_0 be identified. Our first suggestion ([353] Chapter 7.1) has been that this scale is given by the classical radius of the electron. We assumed that the mass of the electron is of pure electromagnetic origin and that the gravitational self-energy of the electron at scale r_0 is the smallest possible energy in the special scale relativity framework, namely, $\hbar c/\mathbb{L}$. This led to the relation

$$\alpha \frac{m_{\mathbb{P}}}{m_{\mathrm{e}}} = \mathbb{K}^{1/3}, \tag{12.59}$$

i.e. to a transition scale

$$r_0 = r_{\mathrm{e}} = \alpha \frac{\hbar}{mc}. \tag{12.60}$$

In this case the predicted cosmic length therefore reads [353]

$$\mathbb{L} = \left(\alpha \frac{m_{\mathbb{P}}}{m_{\mathrm{e}}}\right)^3 l_{\mathbb{P}}, \tag{12.61}$$

and the cosmological constant

$$\Lambda = \left(\frac{m_{\mathrm{e}}}{\alpha m_{\mathbb{P}}}\right)^6 \frac{1}{l_{\mathbb{P}}^2} = \frac{l_{\mathbb{P}}^4}{r_e^6}. \tag{12.62}$$

Let us now give another set of circumstancial evidences motivating the choice of $70\,\mathrm{MeV}$ for the transition scale r_0. The classical radius of the electron r_e actually defines the e^+-e^- annihilation cross section and the e^--e^- cross section $\sigma = \pi r_e^2$ at energy $m_e c^2$ [219]. This length corresponds to an energy $E_e = \hbar c / r_e = 70.02\,\text{MeV}$. This means that it yields the "size" of an electron viewed by another electron. Therefore, when two electrons are separated by a distance smaller than r_e, they can no longer be considered as different, independent objects.

The consequence of this property for the primeval Universe is that r_e should be a fundamental transition scale. When the Universe scale factor was so small that the interdistance between any couple of electrons was smaller than r_e, there was no existing genuine separated electron. Then, when the cooling and expansion of the Universe separated the electron by distances larger than r_e, the electrons that will later combine with the protons and form atoms appear for the first time as individual entities.

In subsequent works [360], we have remarked that this length scale also coincides very closely with fundamental scales, such as half the pion mass and the QCD scale, leading to the suggestion that the final value of the cosmological constant would be fixed by this fundamental transition during the primeval Universe era. Indeed, before the epoch of this quark-hadron transition, the expansion of the Universe applies to free quarks interdistances. When the distance between quarks reaches the size of hadrons, the quarks become confined, the nucleons are formed, and quarks are no longer subjected to the expansion. The radius of hadrons defines a static scale while the expansion of the Universe continues, but now between hadrons. Therefore the two constituents of atoms, nucleons and electrons finally appear at about the same epoch.

The end of this confinement process occurs when the lightest hadrons (the pions) are formed, since it corresponds to the largest possible interquark distance. The effective mass of quarks in the charged pion is $m_{\pi^\pm}/2 = 69.7851(2)\,\text{MeV}$ and in the neutral pion, $m_{\pi^0}/2 = 67.4883(3)\,\text{MeV}$. Their associated Compton lengths define therefore a transition for expansion to staticity in scale space.

Moreover, a calculation of the confinement QCD scale for six quark flavors also yields a value close to this energy. We have found $\lambda_{\text{QCD}} = 71 \pm 2\,\text{MeV}$ from the extrapolation of the running α_3 behavior beyond the Z scale (see Fig. 11.4). The diameter of nucleons corresponds to an energy $2 \times 64\,\text{MeV}$. One may wonder whether this coincidence of the classical electron radius and of several QCD scales could be a low energy manifestation of a strong-electroweak unification theory. Note in this context that there is another fundamental justification for relating the quark confinement scale and the cosmic scale $\mathbb{L}$ by a Dirac–Eddington large-number relation: it is the fact that, in the special scale-relativity framework, $\mathbb{L}$ is a confinement scale for the whole universe, while r_0 is a confinement scale for the subset of the Universe, which is contained inside hadrons.

Therefore we have suggested to identify the transition scale r_0 with the 70 MeV scale (see Fig. 12.5). By taking $r_0 = 70 \pm 2.5$ MeV, an interval, which incudes the various scales considered hereabove, one obtained from Eq. (12.58) [353, 360] $\mathbb{K} = (5.35492 \pm 0.57) \times 10^{60}$, allowing us to estimate a value of the cosmological constant of $\Lambda = (1.36 \pm 0.29) \times 10^{-56}\,\mathrm{cm}^{-2}$, i.e. a scaled cosmological constant $\Omega_\Lambda = (0.388 \pm 0.083)h^{-2}$, where $h = H_0/100\,\mathrm{km\,s}^{-1}\,\mathrm{Mpc}^{-1}$.

If one now assumes that the transition scale is precisely the classical radius of the electron, one obtains with the CODATA 2002 values of the fundamental constants a far more precise estimation,

$$\mathbb{K} = (5.3000 \pm 0.0012) \times 10^{60}, \tag{12.63}$$

i.e. $\mathbb{C}_U = \ln\mathbb{K} = 139.82281(22)$.

Using more recent values of the universal constants [436], we obtain a still increased precision:

$$\mathbb{K}_{\mathrm{pred}} = (5.2999 \pm 0.0008) \times 10^{60}, \tag{12.64}$$

i.e. $\mathbb{C}_{U\mathrm{pred}} = \ln\mathbb{K}_{\mathrm{pred}} = 139.82279(15)$, corresponding to

$$\Lambda_{\mathrm{pred}} = (1.36284 \pm 0.00027) \times 10^{-56}\,\mathrm{cm}^{-2}, \tag{12.65}$$

i.e. a scaled cosmological constant

$$\Omega_{\Lambda\mathrm{pred}} = (0.38875 \pm 0.00008)h^{-2}. \tag{12.66}$$

The invariant cosmic length scale is now

$$\mathbb{L}_{\mathrm{pred}} = (2.77604 \pm 0.00028)\,\mathrm{Gpc}, \tag{12.67}$$

i.e. $\mathbb{L}_{\mathrm{pred}} = (8.56598 \pm 0.00086) \times 10^{25}\,\mathrm{m}$. In time units, it corresponds to a fundamental invariant time scale:

$$\mathbb{T}_{\mathrm{pred}} = (9.05444 \pm 0.00091)\,\mathrm{Gyr}. \tag{12.68}$$

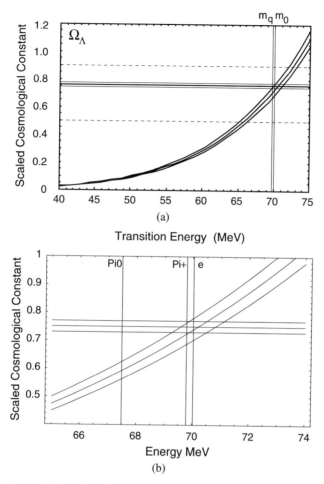

Fig. 12.5: **Expectation for the value of the cosmological constant.** These figures show the variation of the predicted value of the scaled cosmological constant $\Omega_\Lambda = \Lambda c^2/3H_0^2$ in function of the transition energy scale m_0. It is expected to depend on this transition scale and on H_0 as $\Omega_\Lambda \propto m_0^6 H_0^{-2}$ (see text). Top figure: the increasing curve corresponds to this power 6 dependence, calculated using the recently measured value of the scaled Hubble constant, $h = 0.73 \pm 0.019$. The two curves which flank it correspond to 1σ uncertainty on H_0, i.e. $h = 0.71$ and $h = 0.75$. The crossing of this expectation with the observational values of Ω_Λ (horizontal lines) yields the possible range of values for the transition scale. The dashed horizontal lines give the year 2000 observational range, $\Omega_\Lambda = 0.7 \pm 0.2$. The full horizontal line gives the 2006 value improved by a factor larger than 10, $\Omega_\Lambda = 0.761 \pm 0.017$ [508]. The values of the effective mass of quarks in charged pions (m_q) and of the mass-energy $m_0 = \alpha^{-1} m_e$ corresponding to the classical radius of the electron are shown as vertical lines. Bottom figure: zoom on the top figure using 2009 measured values from WMAP 5 years ($h = 0.725 \pm 0.019$, $\Omega_\Lambda = 0.750 \pm 0.020$ [328]). The π^+ quark scale and electron classical radius scale still lie in the error bars, while the π^0 quark scale is clearly excluded with the more recent data.

Most of the error in that case comes from the still poorly determined value of the gravitational constant G.[b]

Note on Newton's constant of gravitation

The value of the gravitational constant G has recently known a confused history. Indeed, various measurements during the years 1990's revealed to have much larger error bars than the 1986 accepted value, some of them being even incompatible with it. Recognizing this situation, the recommended value has been for some years $G = (6.673 \pm 0.010) \times 10^{-11} \, \mathrm{m^3 \, kg^{-1} \, s^{-2}}$, 12 times less precise than the 1986 value. Since 1997, several new measurements have been performed which, despite some remaining inconsistencies between some of them (in particular between the two last, most precise measurements, see Fig. 12.6), has nevertheless allowed to reduce again the error bar. The recommended CODATA 2002 value is $G = (6.6742 \pm 0.0010) \times 10^{-11} \, \mathrm{m^3 \, kg^{-1} \, s^{-2}}$ [326]. In 2008 ([436], the value has been improved to $G = (6.67428 \pm 0.00067) \times 10^{-11} \, \mathrm{m^3 \, kg^{-1} \, s^{-2}}$.

For most physical uses the imprecision of the value of G is of no consequence. In particular, in astronomy it is the product GM, which is far better known of the gravitational constant and of a mass M that occurs in most equations instead of G or M separately (for example, in the solar system it is given by $GM_\odot$, where $M_\odot$ is the mass of the Sun). However, in the theory of scale relativity the situation is different. As we have seen in the various applications of the theory to high energy and elementary particle physics in the previous chapter, and here to cosmology, the Planck scale plays a very fundamental role in the theoretical predictions. In its expression, which depends on c, $\hbar$ and G, $c = 299\,792\,458 \, \mathrm{m/s}$ is exact (as a consequence of Poincaré–Einstein motion-special relativity), $\hbar = 1.054\,571\,68(18) \times 10^{-34} \, \mathrm{J\,s}$ is known with a relative uncertainty of 1.7×10^{-7}, while $G = 6.6743(7) \times 10^{-11} \, \mathrm{m^3 \, kg^{-1} \, s^{-2}}$ is known only up to 1.0×10^{-5}. A precision measurement of G therefore becomes a high stake objective for the scale relativity framework.

12.6.7. *Validation by comparison with observational data*

Let us now compare our theoretical expectation with the measured values of the cosmological constant. This comparison, if it is made in terms

---

[b]Note that in [353] we gave a rough value (from the same theoretical formula), $\Omega_\Lambda = 0.36 \, h^{-2}$, since, at that time, we did not attempt to compute it with high precision. But the given source value of Λ is the same as given here and was already more precise (the imprecision lied only in a rough choice of the constants intervening in unit changes).

of Ω_Λ, also depends on the uncertainties on the Hubble constant. One may therefore compare the theoretical and observational values of the scaled, dimensionless cosmological constant, which corresponds to the reference value for the Hubble constant, $H_{100} = 100\,\mathrm{km/s.Mpc}$,

$$\Omega_\Lambda h^2 = \frac{\Lambda c^2}{3H_{100}^2}. \tag{12.69}$$

Early measurements

The history of the attempts to measure the cosmological constant is long (see Fig. 12.7).

Already in 1975 Gunn and Tinsley, in a paper entitled *An accelerating Universe* [215], concluded from a combination of various cosmological tests that "the most plausible cosmological models have a positive cosmological constant" and "will expand for ever", while "the Hubble diagram alone may dictate that Λ be non-zero". From various constraints, they found that $-1.5 < q_0 < -0.2$ and $0.02 < \Omega_M < 0.30$, which implied, from the

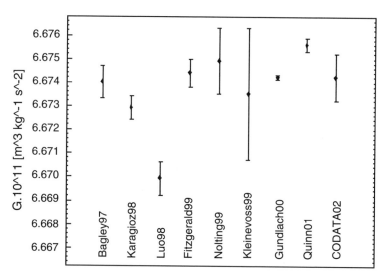

Fig. 12.6: **Measurements of the gravitational constant.** Values of the gravitational constant G measured since 1997 (see [214, 458] and references therein), followed by the CODATA 2002 recommended value, $G = (6.6742 \pm 0.0010) \times 10^{-11}\,\mathrm{m^3\,kg^{-1}\,s^{-2}}$ [326]. The more recent CODATA 2006 value is slighly improved at $G = (6.6743 \pm 0.0007) \times 10^{-11}\,\mathrm{m^3\,kg^{-1}\,s^{-2}}$ [436]. Systematic errors are unfortunately still present in these measurements, as shown by the 4σ discrepancy of the last two and most precise measurements. Provided these errors are understood, a factor 10 in the uncertainty is expected to be gained in the near future.

relation $\Omega_\Lambda = \Omega_M/2 - q_0$, an estimate $0.2 < \Omega_\Lambda < 1.3$, and therefore, with the range of Hubble constant values measured at this epoch, $H_0 = 50 - 80$, a reduced cosmological constant $0.08 < \Omega_\Lambda h^2 < 0.83$.

In 1982, an analysis of the infrared Hubble diagram of elliptical galaxies up to redshift $z \approx 1$ also gave an indication of a positive and large cosmological constant, $\Omega_\Lambda(\text{obs}) = 0.9 \pm 0.6$ [342]. With $H_0 = 50 - 80$, this yielded $0.08 < \Omega_\Lambda h^2 < 0.96$.

By the end of the eighties, the problem of the age of the Universe, i.e. of the discrepancy between the age directly measured (from globular clusters, age of the Galaxy, etc) and the expectation from cosmological models without cosmological constant had become dramatic. At that time, there were already suggestions that there was actually no age problem, and that this discrepancy was simply a proof of the existence of a positive cosmological constant and may lead to a measure of its value [347, 353]. Indeed, in a flat model ($k = 0$) with positive cosmological constant $0 < \Omega_\Lambda < 1$, the age of the Universe is such that

$$H_0\, t_0 = \frac{1}{3\sqrt{\Omega_\Lambda}} \ln \frac{1 + \sqrt{\Omega_\Lambda}}{1 - \sqrt{\Omega_\Lambda}}. \tag{12.70}$$

The age of the Universe was estimated at that time to be $t_0 = (16 \pm 3)$ Gyrs and the scaled Hubble constant $h = 0.75 \pm 0.10$. With these values, one obtains [347] $\Omega_\Lambda h^2 = 0.49 \pm 0.18$. In Fig. 12.7, we adopt a conservative error bar of ± 0.25 to account for the uncertainty on Ω_k at that time.

In the year 2000, the Hubble constant was determined to be $h = 0.71 \pm 0.07$. With this value, the theoretical prediction yielded a scaled cosmological constant $\Omega_\Lambda(\text{pred}) = 0.77 \pm 0.15$ [372, 383], which was already in good agreement with its measured value at that time from the Hubble diagram of type I supernovae [440, 463, 194], the angular power spectrum of the cosmic microwave radiation [129] and gravitational lensing [520], $\Omega_\Lambda(\text{obs}) = 0.7 \pm 0.2$, yielding $\Omega_\Lambda h^2 = 0.35 \pm 0.17$.

In 2003, the WMAP one year analysis had given $h = 0.71 \pm 0.03$ and $\Omega_\Lambda(\text{obs}) = 0.73 \pm 0.04$, yielding a highly improved value $\Omega_\Lambda h^2 = 0.368 \pm 0.051$.

Recent measurements

More recently, the WMAP three year analysis of 2006 [501] gave $h = 0.73 \pm 0.03$ and $\Omega_\Lambda(\text{obs}) = 0.72 \pm 0.03$. These results, combined with the recent Sloan (SDSS) data [508], yield, assuming $\Omega_{\text{tot}} = 1$ (as supported by its

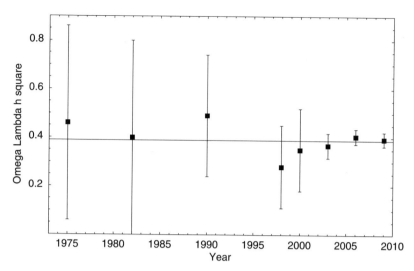

Fig. 12.7: **Evolution of the cosmological constant measurements.** Evolution of the measured values of the dimensionless cosmological constant $\Omega_\Lambda h^2 = \Lambda c^2/3H_{100}^2$, from 1975 to 2010 (see text), compared with the theoretical expectation (horizontal line), $\Lambda = l_\mathbb{P}^4/r_e^6$ [353], where $l_\mathbb{P}$ is the Planck length and r_e the "classical radius" of the electron. This expectation gives numerically $\Omega_\Lambda h^2(\text{pred}) = 0.38875 \pm 0.00008$.

WMAP determination, $\Omega_{\text{tot}} = 1.003 \pm 0.010$)

$$\Omega_\Lambda(\text{obs}) = \frac{\Lambda c^2}{3H_0^2} = 0.761 \pm 0.017, \quad h = 0.730 \pm 0.019. \quad (12.71)$$

Moreover, these recent results have also reinforced the cosmological constant interpretation of the "dark energy" with a measurement of the coefficient of the equation of state $w = -0.941 \pm 0.094$ [508], which encloses the value $w = -1$ expected for a cosmological constant.

With these values, one finds a still improved cosmological constant

$$\Omega_\Lambda h^2(\text{obs}) = 0.406 \pm 0.030, \quad (12.72)$$

which corresponds to a cosmic scale

$$\mathbb{L}(\text{obs}) = (2.72 \pm 0.10) \text{ Gpc}, \quad \text{i.e. } \mathbb{K}(\text{obs}) = (5.19 \pm 0.19) \times 10^{60}, \quad (12.73)$$

in good agreement with the theoretically estimated values, $\mathbb{L}(\text{pred}) = 2.7760(3)$ Gpc, and $\mathbb{K} = 5.3000(8) \times 10^{60}$.

The five-year WMAP data [233] have essentially confirmed these results. The WMAP data alone give $\Omega_\Lambda(\text{obs}) = 0.742 \pm 0.030$ and

$h = 0.719 \pm 0.026$ [148], yielding

$$\Omega_\Lambda h^2(\text{obs}) = 0.384 \pm 0.043, \tag{12.74}$$

very close to the expected value 0.389. Combined with other surveys, one obtains $\Omega_\Lambda h^2(\text{obs}) = 0.361 \pm 0.021$ [252] and $\Omega_\Lambda h^2(\text{obs}) = 0.394 \pm 0.029$ (wmap5 + 2df, [328]), among equivalent other determinations depending on the combined surveys and on the choice of the cosmological model.

The evolution of these experimental determinations is shown in Fig. 12.7 where they are compared with the theoretical expectation

$$\Omega_\Lambda h^2(\text{pred}) = 0.38875 \pm 0.00008. \tag{12.75}$$

The convergence of the observational values towards the theoretical estimate, despite an improvement of the precision by a factor of more than 20, is striking.

The difference between the theoretical and observed value is non significant. Moreover, the precision on the theoretical prediction remains about 400 times better than the observational error, so that the expected improvements of the cosmological parameters will allow to put again this prediction to the test in the near future. The theoretical value may also be improved thanks to improvements of the measurement of the constant of gravitation.

Another way to check the agreement between the theoretical expectation and the observational data consists of comparing the expected value of Ω_Λ deduced from both theory $(\Omega_\Lambda h^2)$ and observation (of h) to its observational value. One finds

$$\Omega_\Lambda(\text{pred}) = \Omega_\Lambda h^2(\text{pred}) \times h^{-2}(\text{obs}) = 0.729 \pm 0.020, \tag{12.76}$$

which is fully consistent with $\Omega_\Lambda(\text{obs}) = 0.726 \pm 0.015$ from the five-year WMAP data [252]. The error bars are now of the same order.

These values are now so precise that some of the conjectured transition scales, despite their proximity to the electron classical radius, can be excluded. The scale of transition predicted from the observed value of Ω_Λ and h is indeed

$$E_{\text{trans}} = \mathbb{K}^{1/3} E_\mathbb{P} = (70.53 \pm 0.89)\,\text{MeV}. \tag{12.77}$$

The π^0 scale at 67.49 Mev is now excluded at the 3.4 σ level and the proton scale at 64 MeV at the 7 σ level. The electron scale at 70.02 MeV and the $\pi^\pm$ scale both remain in the error bar. This energy is also still consistent with the QCD energy scale (extrapolated to 6 quarks) at 71 ± 3 MeV (see Fig. 12.5).

12.6.8. *New mass scales*

One of the new ways offered by the scale relativity approach for identifying fundamental mass scales is the interpretation of transitions between a fractal, nondifferentiable and scale-dependent regime and a classical, scale-independent and differentiable regime as being (in rest frame) Compton lengths $\hbar/mc$.

Hence the transition between the $1/r^6$ gravitational self-energy density of quantum fluctuations and the constant energy density linked to the geometric cosmological constant could be tentatively identified in the previous section with the effective quark mass in charged pions (i.e. the minimal possible effective quark mass) ≈ 70 MeV$/c^2$.

In this context, let us consider again the energy density contribution, that varies as $1/r^4$ (see Fig. 12.4). It crosses the cosmological constant contribution at a scale that is the "middle" of the Universe in scale space, given by

$$\frac{\lambda_s}{l_{\mathbb{P}}} = \left(\frac{\mathbb{L}}{l_{\mathbb{P}}}\right)^{1/2}. \tag{12.78}$$

This length scale takes the value $\lambda_s = 37.209(7) \, \mu$m from the hereabove determination of the cosmic scale $\mathbb{L} = 5.300(1) \times 10^{60} \, l_{\mathbb{P}}$. Interpreted as a Compton length, it corresponds to a mass [383]

$$m_s = \frac{m_{\mathbb{P}}}{\sqrt{\mathbb{L}/l_{\mathbb{P}}}} = (5.3032 \pm 0.0010) \times 10^{-3} \, \text{eV}. \tag{12.79}$$

Another argument pointing to the possible existence of such a fundamental mass scale in nature is given by the new quantum mechanics in scale space (Chapter 8). Indeed, the free Schrödinger equation applied to the whole scale space, which is limited by the Planck and cosmic scales yields solutions similar to that of a quantum particle in a box. The fundamental level solution is a sinus function that peaks at the middle of the scale range. Therefore, one expects a peak of probability for an energy structure linked to the whole Universe at all scales just at the above "middle" energy. It could be identified with the smallest particle mass in nature.

Is such a small mass scale relevant to some physical or astrophysical process? There is now convincing evidence that neutrinos have nonzero masses (see Sec. 11.2). Indeed, there are three reported indications that neutrinos oscillate and therefore have mass (atmospheric neutrinos, solar neutrinos and neutrinos from stopped π^+'s in the LSND experiment). The results of the three experiments can also be accomodated in a four neutrino scheme including a sterile neutrino, i.e. that does not participate in the normal weak interaction [432].

In particular, the solar neutrino observations are understood in terms of a squared mass difference $m_{\nu_2}^2 - m_{\nu_1}^2 \approx 10^{-5}\,\mathrm{eV}^2$. Provided the ν_1 be mainly a ν_e and $m_{\nu_e} \approx 0$ [432], this yields a mass $m_\nu \approx 3 \times 10^{-3}\,\mathrm{eV}$ for the other neutrino state. The above mass scale at 5.3×10^{-3} is therefore compatible with such a neutrino mass. It also agrees with the estimate $m_\nu \approx 10^{-8} m_e$ derived by Sidharth from another argument [492].

Chapter 13

APPLICATIONS TO GRAVITATIONAL STRUCTURING IN ASTROPHYSICS

13.1. Introduction

One of the main still open problems of today's cosmology is that of the formation and evolution of structures from the action of gravitation at the astronomical scales. Some years ago, J. Silk wrote [493]: "Galaxy formation theory is not in a very satisfactory state. This stems ultimately from our lack of any fundamental understanding of star formation. There is no robust theory for the detailed properties of galaxies." The same can be said of the formation of planetary systems, as it is now demonstrated by the discovery of extrasolar planetary systems with properties that were unexpected from the previously accepted models of formation (such as Jupiter-size planets very close to the central star and highly elliptic orbits) (see e.g. [169] and references therein). Moreover, at the scale of galaxies and at extragalactic scales, this question is strongly interconnected with that of "dark matter". The existence of large quantities of unseen matter is a necessary ingredient in the standard approach, since, in its absence, the self-gravitational attraction of visible matter is insufficient for galaxies to form within the known age of the universe. However, while the anomalous dynamical effects (flat rotation curves of spiral galaxies, velocity dispersion of clusters of galaxies, etc.) and gravitational lensing effects that the dark matter hypothesis attempts to explain are firmly established, dark matter itself escapes any detection.

In this chapter, we investigate how the scale relativity and fractal space-time approach can contribute to these open questions [353, 360, 123, 455]. Fractals are now widely applied in astrophysics as models for astronomical objects, population of objects, media and processes (see [170, 225] and references therein), but we are here concerned with a different question. Namely, we are to investigate the consequences of the fractal structures of space on astronomical objects.

After having briefly recalled the foundations of the theory, we apply it more specifically to a generalized theory of gravitation. In this new approach, space becomes not only curved but also fractal beyond some characteristic scale relative to the system under consideration. The induced effects on motion (in standard space) of the internal fractal structures of geodesics (in scale space), are to transform classical mechanics into a quantum-like mechanics, i.e. Newton's fundamental equation of dynamics into a Schrödinger-like equation [362]. Then we give the fundamental solutions of this macroscopic quantum-type equations, which are adapted to a large class of astrophysical situations (central potential, constant density, halos, ejection processes). Note that the use of a Schrödinger representation in astrophysics and cosmology has also been proposed in [505], at least as a method of resolution of hydrodynamics equations (by making $\hbar \to 0$) and in [107], including the account of a quantum potential (also previously introduced in [374]).

Finally the main body of the chapter aims at giving a large panorama of the various predicted effects and of their quantitative and statistically significant verifications in astrophysical data [123]. We describe structures, self-organized in terms of the same gravitational coupling constant, ranging from the scale of our Earth, the Solar System and extra-solar planetary systems, star-forming regions, planetary nebulae, galaxies, our Local Group of galaxies and clusters of galaxies, to large scale structures of the Universe.

13.2. Theory

13.2.1. *Origin of macroscopic Schrödinger mechanics at astronomical scales*

One of the properties of gravitation recognized for long is its scale invariance (see e.g. [437, 65, 85, 169]. It has been noticed by Laplace that "one of the remarkable properties of [Newtonian attraction] is that, if the dimensions of all the bodies in the universe, their mutual distances and their velocities were to increase or diminish proportionately, they would describe curves entirely similar to those which they at present describe [...]. The laws of nature therefore only permit us to observe relative dimensions" (quoted by Mandelbrot, [306, p. 419]). The last statement of this quotation clearly anticipates on the principle of scale relativity. The self-similarity of gravitational systems already plays an important role in their understanding and in their modelization [147]. It allows one to understand simply the existence of approximate scaling laws, such as the Titius-Bode empirical law of planetary distances in the solar system, which are just a manifestation of the scale invariance of the initial protoplanetary disk [206, 279].[a]

Here, we deal with a deeper level of description, in which one considers not only the global scale invariance of the standard coordinates, but also an explicit scale dependence of these coordinates on resolution variables, which comes under the scale relativity tools.

We have recalled in Chapter 5 that a nondifferentiable continuous manifold is fractal, so that its geodesics are characterized by at least three conditions: (i) infinite number, (ii) fractality and (iii) infinitesimal irreversibility. The effects of these conditions can then be combined in terms of a quantum-covariant derivative. Then the equation of geodesics, written in terms of this covariant derivative, can be integrated under the form of quantum mechanical equations.

We have also recalled in Chapter 10 that there may exist a large class of macroscopic systems, in particular those subjected to ergodic or almost ergodic chaos, which may come under the same description in an approximate or effective way (cf. [353] Chapter 7.2) when they are described at time scales larger than their predictability horizon. The conditions for such an approximation to hold have also been studied in Chapter 10: in particular, we have found that the scale range on which the fractality condition is valid must be at least of the order of 10^4–10^5. This constraint, in addition to the chaoticity, suggests that the astronomical realm may come under the scale relativity description, at least concerning some of its systems.

This does not mean that such systems would be described by the full quantum theory, which is clearly specific of the microphysical scales. But

---

[a]The Titius-Bode "law" of planetary distance is of the form $a + b \times c^n$, with $a = 0.4$ AU, $b = 0.3$ AU and $c = 2$ in its original version. It is partly inconsistent — Mercury corresponds to $n = -\infty$, Venus to $n = 0$, the Earth to $n = 1$, etc. It therefore "predicts" an infinity of orbits between Mercury and Venus and fails for the main asteroid belt and beyond Saturn. It has been shown by Hermann [228] that its agreement with the observed distances is not statistically significant. As we shall see in Sec. 13.3, in the scale relativity framework, the predicted law of distance is not a Titius-Bode-like power law but a more constrained and statistically significant quadratic law of the form $a_n = a_0 n^2$.

this means, more simply, that their probability density distribution can be, under some special conditions, calculated as the square of the modulus of a "wave function", which is solution of an equation of the Schrödinger form. In this equation the Planck constant is replaced by a macroscopic constant, which is specific to each considered system.

Let us briefly recall how the fundamental equation of dynamics can be integrated under the form of a Schrödinger equation. The three effects of nondifferentiability and fractality lead to construct a complex time-derivative operator that reads

$$\frac{\hat{d}}{dt} = \frac{\partial}{\partial t} + \mathcal{V} \cdot \nabla - i\mathcal{D}\Delta. \tag{13.1}$$

A complex probability amplitude can be introduced as a reformulation of the generalized complex action $\mathcal{S}$,

$$\psi = e^{i\mathcal{S}/2m\mathcal{D}}. \tag{13.2}$$

Finally the equivalence and strong covariance principles lead to write a geodesic equation in fractal space under the form of the equation of free Galilean motion,

$$\frac{\hat{d}}{dt}\mathcal{V} = 0. \tag{13.3}$$

After re-expression in terms of ψ and integration, it becomes the free Schrödinger equation

$$\mathcal{D}^2\Delta\psi + i\mathcal{D}\frac{\partial}{\partial t}\psi = 0. \tag{13.4}$$

13.2.2. *Generalized theory of gravitation*

Let us now consider the motion of a free particle in a curved space-time whose spatial part is also fractal beyond some temporal transition (e.g. the predictability horizon in case of strong chaos) and/or space transition (e.g. galaxy sizes). The equation of motion can be written, in the first order approximation, as a free-like geodesic equation that combines the general relativistic covariant derivative, which describes the effects of curvature and the scale-relativistic covariant derivative (Eq. 13.1), which describes the effects of fractality. In the Newtonian limit it reads

$$\frac{\widehat{D}}{dt}\mathcal{V} = \frac{\hat{d}}{dt}\mathcal{V} + \nabla\left(\frac{\phi}{m}\right) = 0, \tag{13.5}$$

where ϕ is Newton's potential energy. Once written in terms of the wave function ψ, this equation can be integrated under the form of a gravitational Schrödinger equation with a potential term

$$\mathcal{D}^2 \Delta \psi + i \mathcal{D} \frac{\partial}{\partial t} \psi = \frac{\phi}{2m} \psi. \tag{13.6}$$

Since the imaginary part of this equation is the equation of continuity, and the real part the Euler equation (with a quantum potential), $P = |\psi|^2$ can be interpreted as giving the probability density of the particle positions. For a Kepler potential and in the time-independent (stationary) case, the equation becomes [353, 354, 357]

$$2\mathcal{D}^2 \Delta \psi + \left(\frac{E}{m} + \frac{GM}{r} \right) \psi = 0. \tag{13.7}$$

Even though it takes this Schrödinger-like form, this equation is still in essence an equation of gravitation. It must therefore keep the fundamental properties it owns in Newton's and Einstein's theories, i.e. it must agree with the equivalence principle [361, 209, 3]. This implies that it must be independent of the mass of the test-particle. Moreover, GM provides the natural length and time unit of the system under consideration. This is a general property of the Newton-Kepler problem, as can be seen in Kepler's third law $(a/GM)^3 = (P/2\pi GM)^2$. As a consequence, the parameter $\mathcal{D}$ takes the form:

$$\mathcal{D} = \frac{GM}{2w}, \tag{13.8}$$

where w is a fundamental constant that has the dimension of a velocity [361]. Once divided by the velocity of light, it plays the role of a macroscopic gravitational coupling constants $\alpha_g = w/c$, as remarked by Agnese and Festa [3]. One of the main results of the application of the theory to astrophysical data is that the solutions of this gravitational Schrödinger equation are indeed characterized by an universal quantization of velocities in terms of a constant w_0 and its multiples and sub-multiples [361, 3].[b]

This particular form of the fractal fluctuation parameter $\mathcal{D}$ has consequences on the applicability of the scale relativity approach to the macroscopic gravitational situations considered here. Indeed, the wave function $\psi = \sqrt{P} e^{iS/S_0}$ is solution of a Schrödinger equation provided

---

[b]Its value is found to be $w_0 = 144.7 \pm 0.5$ km/s for the inner solar system and extrasolar planetary systems.

the generalized Compton relation $S_0 = 2m\mathcal{D}$ holds (see Chapter 5). Since $\mathcal{D} = GM/2w$ in the Kepler problem, this means that the parameter $S_0 = GMm/w$ now depends on the inertial mass m. The macroscopic theory can therefore be applied only to the one test-particle case, or to the many particle case provided they all have the same mass. This is not a problem in most applications considered in this chapter, since the idea is to look for the probability density of one given particle (for example, a planetesimal in a protoplanetary nebula) due to the action of the other bodies, which intervene in the exterior potential, and then to apply the obtained probability density to all the particles.

13.3. Theoretical Predictions

The generalized gravitational Schrödinger equation obtained above can now be used as the motion equation in the many systems, which come under the three conditions that underlie its demonstration: large number of possible trajectories, fractal dimension 2 of trajectories, and local irreversibility. It should be noted that these conditions amount to a loss of information about angles, position and time.

In general, the equations of evolution are the Schrödinger–Newton equation and the classical Poisson equation,

$$\mathcal{D}^2 \Delta\psi + i\mathcal{D}\frac{\partial\psi}{\partial t} - \frac{\varphi}{2}\psi = 0, \tag{13.9}$$

$$\Delta\varphi = 4\pi G\rho, \tag{13.10}$$

where φ is the potential and $\phi = m\varphi$ the potential energy.

By separating the real and imaginary parts of the Schrödinger equation we get respectively a generalized Euler–Newton equation (written here in terms of the Newtonian potential energy ϕ) and a continuity equation,

$$m\left(\frac{\partial}{\partial t} + V \cdot \nabla\right)V = -\nabla(\phi + Q), \tag{13.11}$$

$$\frac{\partial P}{\partial t} + \text{div}(PV) = 0. \tag{13.12}$$

This system of equations is equivalent to the classical one used in the standard approach to gravitational structure formation, except for the emergence of an extra potential energy term Q that writes

$$Q = -2m\mathcal{D}^2\frac{\Delta\sqrt{P}}{\sqrt{P}}. \tag{13.13}$$

In the case where the particles are assumed to fill the "orbitals" (for example, the planetesimals in a protoplanetary nebula), the density of matter becomes proportional to the density of probability, $\rho \propto P = |\psi|^2$,

$$\rho = m_0 P, \tag{13.14}$$

where m_0 is the total mass of the system. Then the Schrödinger and Poisson equations can be combined in terms of a single Hartree equation for matter alone [362],

$$\Delta \left(\mathcal{D}^2 \frac{\Delta \psi}{\psi} + i \mathcal{D} \frac{\partial}{\partial t} \ln \psi \right) - 2\pi G m_0 |\psi|^2 = 0. \tag{13.15}$$

Another situation occurs when the number of bodies is small. They follow at random one among the possible trajectories, so that $P = |\psi|^2$ is in this case a mere probability density instead of a matter density, while space remains essentially empty.

In what follows, we shall mainly study the simplified situation of a potential which is assumed to be globally unaffected by the structures that it contributes to form. Typical examples are the two-body problem (planetary systems in the Kepler potential of the star, binary systems in terms of reduced mass), the formation cosmological problem (particles embedded into a uniform density background), ejection processes (planetary nebulae, etc.).

13.3.1. *Kepler potential*

Let us first study the general Kepler problem. The classical potential $\varphi = -GM/r$ can be inserted in the Schrödinger equation. One obtains [353]

$$\mathcal{D}^2 \Delta \psi + i \mathcal{D} \frac{\partial \psi}{\partial t} + \frac{GM}{2r} \psi = 0. \tag{13.16}$$

We look for wave functions of the form $\psi = \psi(\mathbf{r}) \times \exp(iEt/2m\mathcal{D})$. This equation is similar to the quantum mechanical hydrogen atom equation after making the two substitutions $\hbar/2m \to \mathcal{D}$ and $e^2 \to GMm$, where m is the inertial mass of the test particle.

We have shown in Sec. 13.2.2 that the parameter $\mathcal{D}$ should take the form $GM/2w$, where w is a velocity. Let us give a complementary argument. As already remarked, contrary to the hydrogen atom case, the motion equation must come here under the principle of equivalence since it is a geodesic equation and therefore it must be independent of the inertial mass of the "particle". This implies that $\mathcal{D}$ does not depend on the inertial mass (while in standard QM it reduced to the inertial mass since in this case one had $\mathcal{D} = \hbar/2m$).

However, contrary to what happens in the classical theory, the Schrödinger-like equation of motion can be shown to be gauge invariant. If the potential $\phi = m\varphi$ is replaced by $\phi + GMm\partial\chi(t)/c\partial t$, where the factor GMm ensures a correct dimensionality, then Eq. (13.16) remains invariant provided ψ is replaced by $\psi e^{-i\alpha_g \chi}$, where the "Agnese constant" α_g [3] is a dimensionless coupling constant related to $\mathcal{D}$ by

$$\alpha_g \times 2\mathcal{D} = \frac{GM}{c}. \tag{13.17}$$

Therefore, by introducing a constant

$$w_0 = \alpha_g c, \tag{13.18}$$

one finds that, in the new macroscopic quantum-like theory, the equivalent of the Compton length $\mathcal{D}$ takes the form

$$\mathcal{D} = \frac{GM}{2\alpha_g c} = \frac{GM}{2w_0}. \tag{13.19}$$

The relation (13.18) is quite comparable with a similar relation in the hydrogen atom, where the average orbital velocity of the electron in the fundamental level is given by $v_e = \alpha c$, where α is the fine structure constant.

It is interesting, in this respect, to compare the hierarchical atomic and subatomic structure induced by electromagnetism and constructed from powers of the fine structure constant to the hierachical structure of planetary systems, which finds its origin in the new gravitational coupling (see Fig. 13.1).

Thus, we can use the standard solutions, which are expressed in terms of Laguerre polynomials (see e.g. [267, 474]). In spherical coordinates, the radial part and the angular part are separable,

$$\psi(\mathbf{r}) = \psi_{nl\bar{m}}(r, \theta, \phi) = R_{nl}(r) Y_l^{\bar{m}}(\theta, \phi). \tag{13.20}$$

We denote the "magnetic" quantum number by $\bar{m}$ in order to prevent a possible confusion with the inertial mass m.

The energy over mass ratio is quantized as

$$\frac{E_n}{m} = -\frac{G^2 M^2}{8\mathcal{D}^2 n^2} = -\frac{1}{2}\frac{w_0^2}{n^2}, \quad \text{for } n \in \mathbb{N}^*, \tag{13.21}$$

while the natural length unit is the Bohr radius $a_0 = 4\mathcal{D}^2/GM = GM/w_0^2$. This means that, instead of the energy, it is the velocity, which is

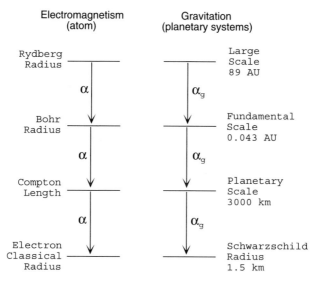

Fig. 13.1: **Comparison between QED and gravitational hierarchy.** Left: a hierarchy of scales involving three powers of the fine structure constant is induced by QED in atomic and subatomic physics. From the Compton length $\lambda_c = \hbar/mc$ and the inverse fine structure constant $\alpha^{-1} = 137.036$, one defines the Bohr radius $r_B = \alpha^{-1}\lambda_c$, which characterizes the minimal size of atoms, the Rydberg radius $r_R = \alpha^{-2}\lambda_c$, which is typical of their largest sizes, and the classical radius of the electron $r_e = \alpha\lambda_c$, which is typical of the electron size since it defines the electron-electron cross section $\sigma = \pi r_e^2$ at rest energy. Right: in the case of planetary systems, a similar hierarchy is found in terms of a gravitational coupling α_g, whose value in the inner solar system is such that $\alpha_g^{-1} = c/w_0 = 2070$ for $w_0 = 144.8\,\text{km/s}$. Starting from the Schwarzschild length $l_s = GM/c^2$ (half the Schwarzschild radius), one obtains a scale $r_p = \alpha_g^{-1}l_s \approx 3000\,\text{km}$, which is typical of planetary sizes, then the gravitational Bohr radius $a_0 = GM/w_0^2 = \alpha_g^{-2}l_s \approx 0.043\,\text{AU}$, which defines the fundamental level for inner-solar-system-like planetary systems, as confirmed by exoplanets (see Sec. 13.5), then a scale $\alpha_g^{-2}l_s \approx 89\,\text{AU}$, which is typical of the size of the outer solar system (Kuiper belt). By taking a three time smaller constant $\alpha_g^{-1} = 690$, the gravitational Bohr radius becomes the Sun radius $R_\odot = 0.00465\,\text{AU}$ and the outer scale becomes $3.3\,\text{AU}$, which is typical of the size of the inner solar system (asteroid belt between Mars and Jupiter).

fundamentally quantized in such a macroscopic quantum-type theory coming under the equivalence principle. In the Kepler problem, the typical velocity $v = (GM/a)^{1/2} = (2\pi GM/P)^{1/3}$ (where a is the semi-major axis and P the period) of orbiting objects is therefore given by the simple quantization law

$$v_n = \frac{w_0}{n}. \tag{13.22}$$

This essential result [353, 354, 357, 362] agrees with the fact that the fundamental constant has no longer the dimension of an angular momentum, as the Planck action constant of standard quantum mechanics, but of a velocity.

13.3.1.1. *Consequences for semi-major axes and eccentricity*

Let us consider particles (e.g. of gas, dust, planetesimals in a protoplanetary disk, etc.) involved in highly chaotic and irreversible motion in a central Kepler potential. At time-scales longer than the predictability horizon, the classical orbital elements, such as semi-major axes, eccentricities, inclinations, obliquities, etc. are no longer defined. However, the stationary Schrödinger equation (13.7) that describes their motion in terms of a probability amplitude ψ does have solutions, which are characterized by well-defined and quantized values of conservative quantities (prime integrals), such as energy E, angular momentum L, etc. Therefore we expect the particles to self-organize in the "orbitals" defined by these solutions, which are characterized by zones of higher probability density, then to form objects (e.g. planets) by accretion.

Owing to the conservation theorems, after the end of the formation process, the motion of the objects which have been formed remains given by the same values of the prime integrals, but it is either no longer chaotic or it is characterized by a far larger chaos time (inverse of the Lyapunov exponent). Then one recovers classical orbital elements (semi-major axis a, eccentricity e, etc.) linked to the conservative quantities by the classical relations, e.g. $E/m = -GM/2a$, $(L/m)^2 = GMa(1 - e^2)$, etc.

The theoretical prediction of the probability distribution of a given orbital element can therefore be obtained by looking for the quantum states of a conserved quantity, which is a direct indicator of the available observable. This is achieved by choosing the symmetry of the reference system in a way which is adapted to the observable. For example, the spherical symmetry solutions describe states of fixed E, L^2 and L_z. From these solutions we can recover the semi-major axis expectation, but not the eccentricity, since the definition of L^2 combines a and e (this comes in correction to the preliminary attempts of [353, 364] to derive the expected probability density for the eccentricity). The parabolic coordinate solutions describe states of fixed E, L_z and A_z, where A is the Runge–Lenz vector, which is a conservative quantity that expresses a dynamical symmetry specific of the Kepler problem (see e.g. [267]). This vector identifies with the major axis and is directed toward the perihelion while its modulus is the eccentricity itself.

The radial part of Kepler orbits in a spherical coordinate system are expressed in terms of Laguerre polynomials. They depend on two integer quantum numbers $n > 0$ and $l = 0$ to $n - 1$. Their average distance is given by (see e.g. [267]):

$$\langle r \rangle_{nl} = a_0 \left(\frac{3}{2} n^2 - \frac{1}{2} l(l+1) \right). \tag{13.23}$$

Semi-major axes

For the maximal value of the angular momentum ($l = n - 1$), the mean distance of the test particle becomes $\langle r \rangle_n = a_0(n^2 + n/2)$, while the probability peak lies at r_n (peak) $= a_0 n^2$. The energy being quantized as given in Eq. (13.21), the probability peaks for the semi-major axes lie at values

$$a_n = \frac{GM}{w_0^2} n^2 = a_0 n^2. \tag{13.24}$$

Eccentricities

The theoretical expectation for the eccentricity distribution is obtained in parabolic coordinates by taking as z axis the major axis of the orbit. One obtains for the projection of the Runge–Lenz vector on this axis [379]

$$A_z = e = \frac{k}{n}, \tag{13.25}$$

where the number k is an integer and varies from 0 to $n - 1$.

Let us demonstrate this result. In previous works [353, 364], we have considered the solutions of the gravitational Schrödinger equation in a spherical coordinate system. However, these solutions, which depend on the three quantum numbers n, l and $\bar{m}$, are not adapted to the problem of the eccentricity distribution. Indeed, the $1 - e^2$ dependence of L^2 has for consequence that large variations of e yield small variations of L, so that the angular momentum is actually a bad indicator of orbit circularity.

Moreover, one should account for the angle-angular momentum Heisenberg inequalities. In the astrophysical situation considered here, the planetesimals are described beyond their mean free path, i.e. $\Delta\theta \approx 1$ rd, and the Heisenberg relation writes $\Delta\theta \times \Delta(L/m) \approx \mathcal{D}$ (where $\mathcal{D}$ is the fractal fluctuation parameter, given by $\mathcal{D} = \hbar/2m$ in standard microscopic quantum mechanics). Therefore we obtain $\Delta l \approx 1/2$. Such an uncertainty confirms the above analysis, since it shows that the value of l is a bad indicator of eccentricity for small quantum numbers. In particular, it prevents from having strictly zero angular momentum in the fundamental level.

We have suggested [379, 391] a solution to this problem. A specific theoretical prediction of the eccentricity distribution can actually be performed by working in parabolic coordinates. Indeed, the wave functions in parabolic coordinates describe stationary states in which E, L_z and A_z do have determined values [267].

In the classical description (valid here at small time-scales for the planetesimals and at large time-scales, after the accretion process ended, for the planet), $\mathbf{A}$ is the Runge–Lenz vector, which is a constant of the motion in the Kepler problem (see e.g. [474, 267]):

$$\mathbf{A} = \mathbf{p} \times \mathbf{L} - \frac{\mathbf{r}}{r}. \tag{13.26}$$

The Lenz vector is oriented on the orbital major axis toward perihelion, while its modulus is nothing but the eccentricity itself, i.e. $A = e$.

Therefore, by taking the z axis along the major axis of the orbit, the permitted values of A_z in the quantum regime (planetesimal era) give the solution to this problem. This choice of the z axis is a key-point for getting this solution. Indeed, in such a quantum-like description of planetesimal orbits, one should be aware that the angular momentum vector and Runge–Lenz vector are orthogonal to each other. More generally, the choice of the coordinate system and of its symmetries must be done in accordance with the observable quantity that is studied.

The stationary states in parabolic coordinates are determined by three quantum numbers, n_1, n_2 and $\bar{m}$ (the L_z quantum number), which are combined together to give the main quantum number n as

$$n = n_1 + n_2 + |\bar{m}| + 1. \tag{13.27}$$

For $n_1 \neq n_2$, the orbitals are disymmetrical with respect to the (x, y) plane, but, since the z axis is along the major axis, this is simply a manifestation of the fact that the focus of the corresponding classical orbit is excentred (the case $n_1 = n_2$ corresponds to circular orbits). The energy levels, i.e. the possible values of the semi-major axis a (or equivalently of the period P) are unchanged with respect to the spherical coordinate solutions.

The number $|\bar{m}|$ can take values from 0 to $n - 1$, then the number n_1 can vary from 0 to $n - |\bar{m}| - 1$. The allowed values of A_z are therefore determined from the numbers n_1, n_2 and n by the formula

$$A_z = \frac{n_2 - n_1}{n}. \tag{13.28}$$

At the end of the accretion process of planetesimals (more generally, planet embryos or protoplanets) moving in stationary states characterized by these values of A_z, planets will be formed in orbits having eccentricities $e = A_z$ because of conservation laws. Therefore, one theoretically predicts the exoplanet eccentricity distribution should show peaks for quantized values given by the simple formula quoted hereabove, $e = k/n$, where the number k is integer and varies from 0 to $n-1$.

13.3.1.2. *Consequences for the angular distribution*

The angular part of the wave function is also quantized [360]. In spherical coordinates, the angular momenta are quantized as $(L/m)^2 = 4\mathcal{D}l(l+1)$ and its projected component as $L_z/m = 2\mathcal{D}\bar{m}$, where m is the mass and $\bar{m}$ is the third quantum number, which varies from $-l$ to l. The angular solutions are expressed in terms of the spherical harmonics $Y_l^{\bar{m}}(\theta, \phi)$, whose importance for morphogenesis will be pointed out in a forthcoming section (in particular for shapes formed through ejection processes).

13.3.1.3. *Consequences for dynamics*

The momentum solutions are separable in spherical coordinates and expressed by the function: $\Psi_{n,l,\bar{m}}(\mathbf{p}) = F_{n,l}(p) \times Y_l^{\bar{m}}(\vartheta, \phi)$. The functions $Y_l^{\bar{m}}(\vartheta, \phi)$ are standard spherical harmonics, the $F_{n,l}$ functions are proportional to the Gegenbauer functions $C_N^\nu(\chi)$. The momentum distribution is given by $|p \cdot F_{n,l}(p)|^2$. The mean square value of the observable $\mathbf{p}$ is given by [48]

$$\langle p^2 \rangle = \int_0^\infty p^2 |F_{n,l}|^2 p^2 \, dp = \left(\frac{p_0}{n}\right)^2, \tag{13.29}$$

where $p_0 = 2m\mathcal{D}/a_0$ is the Bohr momentum. We obtain

$$\langle p^2 \rangle = \left(\frac{GMm}{2n\mathcal{D}}\right)^2 = \left(\frac{mw_0}{n}\right)^2 \quad \text{for } v \ll c, \quad \langle v^2 \rangle = \left(\frac{w_0}{n}\right)^2, \tag{13.30}$$

since $\mathcal{D} = GM/2w_0$.

13.3.2. *Constant density: harmonic oscillator potential*

Let us now consider the gravitational potential given by a uniform mass density ρ [360]. The domain of application of this case is, evidently, cosmology in the first place (where ρ depends on time), but this can also apply as an approximation to the interior of extended bodies (stars in galaxies, galaxies in superclusters, etc.). Solving for the Poisson equation yields a harmonic oscillator gravitational potential $\varphi(r) = 2\pi G\rho r^2/3$, and the motion equation becomes

$$\mathcal{D}^2\Delta\psi + i\mathcal{D}\frac{\partial\psi}{\partial t} - \frac{\pi}{3}G\rho r^2\psi = 0. \tag{13.31}$$

This is the Schrödinger equation for a particle in a three-dimensional isotropic harmonic oscillator potential with frequency

$$\omega = 2\sqrt{\frac{\pi G\rho}{3}}. \tag{13.32}$$

The stationary solutions [360, 267] are expressed in terms of the Hermite polynomials $\mathcal{H}_n$,

$$R(x,y,z) \propto \exp\left(-\frac{1}{2}\frac{r^2}{r_0^2}\right)\mathcal{H}_{n_x}\left(\frac{x}{r_0}\right)\mathcal{H}_{n_y}\left(\frac{y}{r_0}\right)\mathcal{H}_{n_z}\left(\frac{z}{r_0}\right), \tag{13.33}$$

which depend on the characteristic scale

$$r_0 = \sqrt{\frac{2\mathcal{D}}{\omega}} = \mathcal{D}^{1/2}(\pi G\rho/3)^{-1/4}. \tag{13.34}$$

The energy over mass ratio is also quantized as

$$\frac{E_n}{m} = 4\mathcal{D}\sqrt{\frac{\pi G\rho}{3}}\left(n + \frac{3}{2}\right). \tag{13.35}$$

The main quantum number n is an addition of the three independent axial quantum numbers $n = n_x + n_y + n_z$.

Spatial consequences

The first Hermite polynomials are:

$$H_0 = 1; \quad H_1 = 2x; \quad H_2 = 4x^2 - 2; \quad H_3 = 8x^3 - 12x; \ldots. \tag{13.36}$$

The mode zero is a Gaussian of dispersion $\sigma_0 = r_0/\sqrt{2}$ (see Fig. 13.26). The positions of the peaks of mode n are given by the solutions to the equation

$$H_{n+1} - xH_n = 0. \tag{13.37}$$

The mode $n = 1$ is a binary structure (Fig. 13.26), whose peaks are situated at $x_{peak} = \pm r_0$.

The mode $n = 2$ (Fig. 13.26) has three peaks at $0, \pm\sqrt{5/2}r_0 \approx \pm1.58\,r_0$. For $n = 3$, one finds $x_{peak} = \pm0.602\,r_0$ and $\pm2.034\,r_0$. More generally, one can find an approximation for the position of the most extreme peak, by keeping only the two leading terms. One finds for small values of n ($n \leq 3$):

$$x_{max} = \sqrt{n(n+3)}\frac{r_0}{2} \approx \left(n + \frac{3}{2}\right)\frac{r_0}{2}. \tag{13.38}$$

One may predict, from these equations and their solutions, that matter will have a tendency to form structures [362] according to the various modes of the quantized three-dimensional isotropic harmonic oscillator [360, 90] whose dynamical symmetry group is SU(3). Depending on the conservative quantities and their associated quantum numbers (n_x, n_y, n_z), a simple or multiple (double, chain, trapeze) structure is obtained (see Fig. 13.26). This prediction of the theory can be checked in astrophysical data, since such morphologies are indeed found in the Universe on many scales (star formation zones, compact groups of galaxies, multiple clusters of galaxies). Moreover, quantitative predictions can be made, as for example the distance separation [360] of the extreme density peaks, which is given by the approximation $\Delta r_{max} = (n^2 + 3n)^{1/2}r_0$, etc.

Dynamical consequences

If one now considers the momentum representation rather than the position one, one predicts a distribution of velocities that is given by the same functions. Hence, the fundamental level is given by a Maxwellian distribution such that

$$P \propto e^{-(v/v_0)^2} \tag{13.39}$$

in terms of a characteristic velocity that reads

$$v_0 = \sqrt{2\mathcal{D}\omega} = 2\mathcal{D}^{1/2}(\pi G\rho/3)^{1/4}. \tag{13.40}$$

This result agrees with the classical description that leads to a Maxwellian distribution of velocities in self-gravitating systems, such as clusters of galaxies. But it also yields a new result, according to which the velocity dispersion depends on the underlying fractal fluctuation parameter $\mathcal{D}$.

More generally, one finds excited states with several probability peaks, whose most distant values are, for one-dimensional-type solutions $(n, 0, 0)$,

$$v_{max} = \sqrt{n(n+3)}\frac{v_0}{2}. \tag{13.41}$$

From this study, we note that the difference between the extreme velocity peaks $\Delta v_{max} = 2v_{max}$ is quantized in a quasi-linear way for small quantum numbers, since it is of the order of $\approx 2v_0$, $\approx 3v_0$, and $\approx 4v_0$ for the modes $n = 1, 2, 3$ respectively.

The main conclusion is the prediction that the various cosmological constituents of the Universe are expected to lie at preferential relative positions and move with preferential relative velocities, as described by the various structures implied by the quantization of the isotropic three-dimensional harmonic oscillator [360, 90].

13.3.3. Matching of potentials

13.3.3.1. Kepler and harmonic potentials

The above solutions only take into account the background potential, while a complete solution should also account for self-gravity of the forming structure. In the scale relativity framework applied to the formation of structure in (and from) a cosmological background density in expansion, this means to solve a nonlinear time-dependent Schrödinger–Poisson system, which is a difficult problem coming under numerical integration [283, 284].

However, some information about these more complete solutions can be obtained by remarking that, just outside the structure, which has been formed from the cosmological background of density ρ_b, the potential can be approximated by a Kepler potential, while it remains a harmonic oscillator potential at farther distance.

This leads one to consider the combination of a central body (Kepler potential) and of a constant density (harmonic oscillator potential). We know from standard quantum mechanics that the energy spectra of the Kepler (local) and background density (global) cases must be matched together [513]. The open question that will now be investigated is how and at which scale [90].

We want to find the most probable trajectories of a test particle in the gravitational field of a massive body (or equivalently the motion of a binary system in reduced coordinates), that is also subjected to the effect of the cosmological fluid and of an underlying fundamental chaos. The corresponding potential is equivalent to the sum of a Coulomb potential

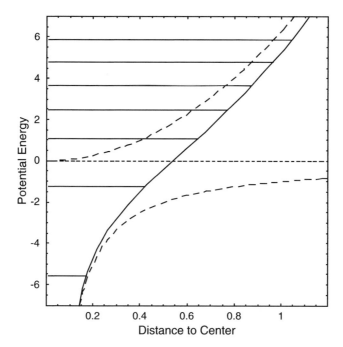

Fig. 13.2: **Matching of Kepler and harmonic oscillator potentials.** Sum of a Kepler potential (lower dashed curve) and of an harmonic oscillator potential (upper dashed curve) in terms of distance. A schematic representation of the quantized energy levels is given as horizontal lines.

and that of an harmonic oscillator. It is shown in Fig. 13.2. Energy is always quantized in such a potential well. In the stationary case, the equation for this problem writes

$$2\mathcal{D}^2 \Delta\psi + \left(\frac{E}{m} + \frac{GM}{r} - \frac{2\pi}{3} G\rho_b r^2 \right) \psi = 0. \qquad (13.42)$$

Since we still deal here with a symmetric central field, ψ can be decomposed as $\psi = R(r) Y_{lm}(\theta, \varphi)$, so that this problem can be reduced to the search of the radial function $\chi(r) = r R(r)$, which satisfies the equation

$$\frac{d^2\chi}{dr^2} + \left[\frac{1}{2\mathcal{D}^2} \left(\frac{E}{m} + \frac{GM}{r} - \frac{2\pi}{3} G\rho_b r^2 \right) - \frac{l(l+1)}{r^2} \right] \chi = 0. \qquad (13.43)$$

We shall not attempt here to find the exact solution to this problem. In the limits $r \to 0$ and $r \to \infty$, we recover respectively the Kepler and harmonic oscillator cases studied hereabove. We shall then only study how they match at intermediate distances.

Let us write the potential in the form $\phi = \frac{1}{2}m\omega^2 r^2 - \alpha/r$, where $\omega = \sqrt{4\pi G\rho_b/3}$ and $\alpha = GmM$ in the gravitational case considered here. It vanishes for a distance:

$$r_* = \left(\frac{2\alpha}{m\omega^2}\right)^{1/3} = \left(\frac{3M}{2\pi\rho_b}\right)^{1/3}. \tag{13.44}$$

One therefore obtains the relation

$$M = \frac{2\pi}{3}\rho_b r_*^3, \tag{13.45}$$

which means that $r_*/(2^{1/3})$ is the radius of the sphere of density ρ_b having the mass M (i.e. of the background zone from which the central mass has been formed, see Fig. 13.3).

For $r \gg r_*$, the energy is quantized as $E_{n_2}/m = \mathcal{D}\omega(2\,n_2 + 3)$, i.e. it tends toward an equidistant spacing for high values of the energy.

For $r \ll r_*$, the energy is quantized as $E_{n_1}/m = -(G^2 M^2/8\mathcal{D}^2)n_1^{-2}$.

Then the quantum number n_1 keeps its meaning only as long as the inequality $4\mathcal{D}^2 n_1{}^2/GM < r_*$ holds, i.e.

$$1 \leq n_1 < n_* = \frac{\sqrt{GMr_*}}{2\mathcal{D}}. \tag{13.46}$$

Therefore a Keplerian spectrum is ensured only provided $2\mathcal{D} \leq \sqrt{GMr_*}$. When $\mathcal{D} = GM/2w_0$ this condition becomes $r_* \geq a_0 = GM/w_0^2$, as expected since a_0 is the macroscopic gravitational analog of the Bohr radius.

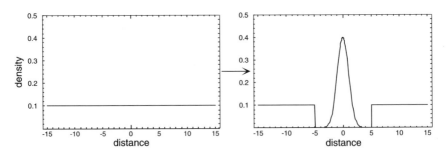

Fig. 13.3: **Formation of a structure from a background medium in a vacuole-type model.** The global harmonic oscillator potential defined by the background induces the formation of a local structure, in such a way that the average density remains constant, i.e. the matter of the structure is taken from the background.

Let us finally compute the difference between energy levels for this value of n_1. In standard quantum units, it reads

$$\Delta E_1 = \frac{m\alpha^2}{2\hbar^2}\left[\frac{1}{n_*{}^2} - \frac{1}{(n_* + 1)^2}\right] \approx \frac{m\alpha^2}{\hbar^2 n_*{}^3} = \hbar\omega = \Delta E_2. \qquad (13.47)$$

(One recovers the gravitational case in these expressions by making the substitution $\hbar \to 2m\mathcal{D}$). As expected, the inner and outer energy spectra are matched when the Kepler energy difference becomes of the same magnitude as the harmonic oscillator energy difference between levels, $\Delta E/m = (\hbar/m)\omega = 2\mathcal{D}\omega$. This matching occurs around the length-scale r_*, that characterizes the transition between the two regimes (see Fig 13.2).

13.3.3.2. *Vacuole ("Swiss-Cheese") model*

The above results may be easily generalized to a model with two levels of uniform density [90]. We consider a spherical, internal object of density ρ_i and radius r_i surrounded by a vacuum of radius r_e, digged out in a background of density ρ_e. The interest of such a two-level vacuole model is that one can solve exactly Einstein's equations for such a structure [153]: one describes the inner region by either an interior Schwarzschild metric or by a Friedmann one, the intermediate vacuum by a Schwarzschild metric, then the background by a standard Friedmann–Robertson–Walker model. The matching conditions on the hypersurfaces that separate the three domains express the continuity of the potentials and of their derivatives.

As remarked in Chapter 12, such models have already been extensively used to study gravitational lensing effects in an exact way [340, 341, 343, 346]. Here we shall consider only their Newtonian counterpart, for which the same matching conditions hold. Note also that they achieve a good approximation of the position distribution found for the zero mode harmonic oscillator, $\rho = \rho_m e^{-(r/r_0)^2}$, with $r_i \approx r_0$ and $\rho_i \approx \rho_m$.

The potential writes, respectively, in the three domains,

$$\phi_i = -E_i + \frac{1}{2}\omega_i{}^2 r^2; \quad \phi_s = -\frac{\alpha}{r}; \quad \phi_e = -E_e + \frac{1}{2}\omega_e{}^2 r^2. \qquad (13.48)$$

The matching conditions write

$$E_i = \frac{1}{2}\omega_i{}^2 r_i{}^2 + \frac{\alpha}{r_i}; \quad E_e = \frac{1}{2}\omega_e{}^2 r_e{}^2 + \frac{\alpha}{r_e}, \qquad (13.49)$$

$$\omega_i{}^2 r_i = \frac{\alpha}{r_i{}^2}; \quad \omega_e{}^2 r_e = \frac{\alpha}{r_e{}^2}. \qquad (13.50)$$

Equation (13.50) is nothing but the "null apparent mass condition" [153] that states that the matching is possible only provided the central mass is exactly the mass which has been digged out from the background. Indeed it writes, after coming back to the "gravitational variables",

$$M = \frac{4}{3}\pi\rho_i r_i{}^3 = \frac{4}{3}\pi\rho_e r_e{}^3. \tag{13.51}$$

Once the energy is quantized, we recover relations similar to those of the previously studied case, with slightly different numerical constants (remembering that in the macroscopic case, one replaces $\hbar$ by $2m\mathcal{D}$),

$$r_e = \left(\frac{\alpha}{m\omega_e{}^2}\right)^{1/3}; \quad n_{se} \approx \left(\frac{m\alpha r_e}{\hbar^2}\right)^{1/2}; \quad \Delta E_{se} \approx \frac{m\alpha^2}{\hbar^2 n_{se}{}^3} = \hbar\omega_e,$$

$$\tag{13.52}$$

and, provided $r_i \gtrsim \hbar^2/m\alpha$,

$$r_i = \left(\frac{\alpha}{m\omega_i{}^2}\right)^{1/3}; \quad n_{si} \approx \left(\frac{m\alpha r_i}{\hbar^2}\right)^{1/2}; \quad \Delta E_{si} \approx \frac{m\alpha^2}{\hbar^2 n_{si}{}^3} = \hbar\omega_i, \tag{13.53}$$

where n_{se} and n_{si} are respectively the largest and smallest quantum numbers in the intermediate vacuum region. When the above condition on r_i is not fulfilled, this means that there is no energy level in the inner region and that the last energy level is then the fundamental level of the vacuum region, $E = m\alpha^2/2\hbar^2 = -G^2M^2m/8\mathcal{D}^2$.

The general conclusion of this study is that, for hierarchized systems, we expect the inner and outer energy levels to be commensurable. The inner and outer "velocity quanta" are related by a relation of the kind

$$\frac{v_i}{v_e} \approx \frac{n_{si}}{n_{se}}. \tag{13.54}$$

This result is more general than the particular model considered here. Indeed, contrary to the atomic physics situation, a gravitational system is not described by a unique Schrödinger equation with unique constants, but more generally by a set of Schrödinger equations with self-organization constants that depend on the various subsystems at different scales. But the solutions of these equations must then be matched at the interface between the various regions, which involves relations between the constants. This is typically what is observed in hierarchical systems, such as our own solar system [361, 364], which can be shown to contain at least five related levels of hierarchy.

13.3.4. *Application to gravitational structuring in cosmology*

The application of the above approach to cosmology is now straighforward [360, 90]. We consider, as the first step, the simplified case when the structure formation is achieved on a small enough time scale so that the time dependence of the density could be neglected.

Let us analyse the conditions under which we can neglect the time-dependence of the density as a first approximation. The density of the universe varies in terms of redshift as $\rho = \rho_0(1 + z)^3$; denoting by τ the lookback time from the present epoch, we can write $z \approx H_0 r/c \approx H_0 \tau \approx K\tau/T_0$, where T_0 is the age of the Universe, and where $K = 1$ at the limit of the empty model and $K = 2/3$ for the Einstein–de Sitter model. Then $\rho \approx \rho_0(1 + 3K\tau/T_0)$, and the time dependence of the density can be neglected for time intervals small with respect to the full age of the Universe.

When considering very different epochs, different values of ρ must be taken, so that it implies that different kinds of structures are expected during the cosmological evolution, but also at the same epoch at various places where the density differs.

The various steps of the derivation of Sec. 13.3.2 are recovered and can now be expressed in terms of cosmological parameters. The standard fundamental equation of dynamics for the formation and evolution of structures reads [438]

$$\frac{d^2 r}{dt^2} = -\nabla\left(\frac{2\pi}{3}G\rho r^2\right), \tag{13.55}$$

where $\rho = \rho_b(1 + \delta(x, t))$, δ being the dimensionless density contrast. This equation is valid both in Newtonian cosmology and in Friedmann models in terms of the proper distance. Let us assume that $\delta \ll 1$, which is valid at the beginning of the formation of the structure, so that the potential remains dominated by the background density. Under the new approach, this equation can be transformed into a macroscopic Schrödinger equation that reads in the stationary case

$$2\mathcal{D}^2\Delta\psi + \left(\frac{E}{m} - \frac{2\pi}{3}G\rho_b r^2\right)\psi = 0. \tag{13.56}$$

The frequency of the harmonic oscillator is

$$\omega = \sqrt{\frac{4\pi G\rho_b}{3}} = \frac{1}{\sqrt{2}}H\Omega_M^{1/2}, \tag{13.57}$$

where Ω_M is the usual cosmological matter density parameter,

$$\Omega_M = \frac{8\pi G \rho_b}{3H^2}.$$ (13.58)

The solutions, which describe the structures at the moment of their formation, are those described in Sec. 13.3.2 (see Fig. 13.26). This kind of structures is indeed typical of those observed in formation regions at various scales (stars, clusters of stars, galaxies, clusters of galaxies). Their energy is quantized as

$$\frac{E_n}{m} = \mathcal{D}H\sqrt{2\Omega_M}\left(n + \frac{3}{2}\right).$$ (13.59)

The characteristic length is

$$r_0 = \sqrt{2\mathcal{D}/\omega} = 2^{3/4}\mathcal{D}^{1/2}H^{-1/2}\Omega_M^{-1/4},$$ (13.60)

and the characteristic velocity is

$$v_0 = 2^{1/4}\mathcal{D}^{1/2}H^{1/2}\Omega_M^{1/4}.$$ (13.61)

The next steps consist of taking into account the time variation of the background density, and the self-gravitating potential of the density contrast. This is a full subject of research in itself, which has been studied in more details by Ceccolini [90] (see the following exercices) and which we shall no longer consider here.

Problems and Exercises

Exercise 23 Write the nonlinear Schrödinger equation that describes the formation and evolution of structures in a fractal space, including the effects (i) of the expansion of the background Universe, (ii) of the cosmological constant and (iii) of the self-gravitation of the density contrast. ∎

Exercise 24 Show that the hereabove Schrödinger-type equation for cosmological structure formation can be decomposed in terms of exactly the same Euler and continuity fluid-like system of equations of the standard theory of formation [438] including an additional potential (usually attributed to "dark matter") [374]. ∎

Exercise 25 Show that, contrary to the standard approach, (i) this "dark potential" does not have to be arbitrarily added, but emerges naturally as a manifestation of the fractal geometry, in a way similar to the Newton potential manifesting the curved geometry [376, 90]; (ii) is not linked to the density of matter through the Poisson equation, but through a quantum-type potential equation. ∎

Open Problem 28: Solve the above equations for structure formation and evolution and show that this dark potential is a good candidate to explain the apparent missing nonbaryonic matter $\Omega_{nb} = 0.23$ at the cosmological level, while the visible matter yields $\Omega_v \approx 0.005$ and the baryonic dark matter $\Omega_b \approx 0.045$. ∎

13.3.5. *Ejection process*

In many ejection or growth processes (planetary nebulae, supernovae, star formation, ejection of matter from the Sun, as due e.g. to the infall of sungrazer comets, etc.), the observed ejection velocity seems to be constant in the first approximation (as a result of the cancellation between various dynamical effects). This particular behavior is consistent with a constant or null effective potential, i.e. it corresponds to the free motion case.

This means that this problem, when it is formulated in the scale relativity framework, becomes formally equivalent to a scattering process during elastic collisions [123, 124, 125, 126]. Indeed, recall that the collision of particles is described in quantum mechanics in terms of an incoming free particle plane wave and of outcoming free plane and spherical waves. The Schrödinger equation of a free particle reads

$$2m\mathcal{D}^2 \Delta\psi(\mathbf{r}) + E\psi(\mathbf{r}) = 0, \tag{13.62}$$

where $E = p^2/2m = 2m\mathcal{D}^2 k^2$ is the energy of this free particle. The stationary solutions of this equation in a spherical coordinate system are $\psi(\mathbf{r}) = R(r)Y_{l\bar{m}}(\theta, \phi)$. The equation of radial motion becomes [267]

$$R''(r) + \frac{2}{r}R'(r) + \left[k^2 - \frac{l(l+1)}{r^2}\right]R(r) = 0. \tag{13.63}$$

We keep the solution corresponding to a flow of central particles (ejection-scattering process), namely, the divergent spherical waves. The general solution is expressed in terms of the first order Hankel functions,

$$R(r) = +iA\sqrt{\frac{k\pi}{2r}}H^{(1)}_{l+\frac{1}{2}}(kr). \tag{13.64}$$

Radial consequences

$[R(r)]^2$ represents the spatial probability of presence for a particle ejected in a unit of time. But our aim is to know the evolution of the probability function for distances and times higher than the ejection area. The spatial probability density for a particle emitted at time (t_e) in an elementary spherical volume reads

$$dP(r, \theta, \phi, t, t_e) = [R(r - V_0(t - t_e))]^2 \frac{1}{r^2} r^2 dr [Y_{l\bar{m}}(\theta, \phi)]^2 \sin\theta \sin\phi \, d\theta \, d\phi.$$

$$(13.65)$$

Angular consequences

In a way similar to the interpretation of the spatial solutions, the angular solutions can also be interpreted in terms of a self-organized morphogenesis. Indeed, the matter is expected to be ejected with a higher probability along the angle values given by the peaks of probability density distribution. One may therefore associate quantized shapes, which can be spherical, plane, bipolar, etc. to each couple of quantum numbers $(l, \bar{m})$, i.e. to the quantized values of the angular momentum and of its projection on a given axis. These various possibilities will be considered in more detail in Sec. 13.7.

13.3.6. *Theoretical expectation of the gravitational coupling*

Let us conclude this theoretical part by an attempt to find a theoretical expectation of the value of the new fundamental gravitational coupling, which we suggest to name the "Agnese constant" α_g [3, 361].

We have recalled hereabove that, if the potential $\phi = m\varphi$ is replaced by $\phi + GMm\partial\chi(t)/c\partial t$, then Eq. (13.16) remains invariant provided ψ is replaced by $\psi e^{-i\alpha_g \chi}$. Now, in similarity with the electromagnetic case (see Chapter 7), we can interpret the arbitrary gauge function χ, up to some numerical constant, as the logarithm of a scale factor ρ in resolution space, i.e. $\chi = \ln \rho$. In the special scale relativity framework, such a scale factor is limited by the ratio of the maximal cosmic scale $\mathbb{L} = \Lambda^{-1/2}$ over the Planck scale $l_\mathbb{P}$ (see Chapter 12):

$$\ln \rho < \mathbb{C}_U = \ln(\mathbb{L}/l_\mathbb{P}). \qquad (13.66)$$

This limitation of χ in the phase of the wave function ψ implies a quantization of its conjugate quantity α_g, following the relation, for three independent scale transformations on the three space resolutions

(see Secs. 7.4.5 and 11.1.3.2)

$$\mu \times 3\alpha_g \mathbb{C}_U = \frac{k}{2},$$ (13.67)

where k is integer and where the numerical constant μ must be introduced to account for the overall geometry of the Universe and is therefore expected to involve powers of π factors, e.g. the volume of a spherical Universe is $V = 2\pi^2 R^3$. It is remarkable, in this respect, that the relation

$$\frac{3}{2}\pi^2 \alpha_g \mathbb{C}_U = 1$$ (13.68)

yields a value $\alpha_g^{-1} = 2070.10 \pm 0.15$, i.e. $w_0 = 144.82 \pm 0.01 \, \text{km/s}$. This theoretical expectation will be compared to the observational data in the next sections. It is in good agreement with the present observational determinations $w_0 = 144.7 \pm 0.5 \, \text{km/s}$ [361] of the inverse coupling at inner solar system scales. Namely, at these scales, the Agnese coupling is ≈ 15.1 times smaller than the QED coupling (i.e. the fine structure constant $1/137.036$). Reversely, from such a relation, if it were confirmed, a precise measurement of w_0 would provide one with a new way of determining the cosmological constant, since it enters in the definition of $\mathbb{C}_U$ [374].

The Agnese coupling constant $\alpha_g = 1/2070$ is only one particular constant, relevant to the inner solar system, among a sequence of constants with integer ratios. Another (not incompatible) way to understand these values and the reason for their multiplicity is based on the fundamental scale invariance of gravitation.

Contrarily to atomic physics, where the Planck constant $\hbar$, the mass m and the quantized charge e of electrons have constant values, which definitively fix the size of atoms in a unique way (at least under standard vacuum energy conditions, see [384]), the active gravitational masses in the astronomical realm are not quantized and can take values differing by factors up to 10^{40} and more (see Fig. 12.1). The same is true of length scales, which differ by factors reaching up to 10^{20}.

The gravitational potential $GM/r \sim E/m \sim v^2$, which is the newly quantized quantity in the macroscopic quantum-like theory, is observationally found to cover a far smaller range. For example, the Keplerian velocity at the Sun radius, $437.1 \approx 3 \, w_0 \, \text{km/s}$ is of the same order as the rotation velocity of our Galaxy at the Sun distance ($250 \, \text{km/s}$), the velocity of galaxies in groups and clusters or even our velocity with respect to the cosmic background radiation ($369 \pm 2 \, \text{km/s}$ [435]). This allows the various values of the constants w to remain in the same range. As we shall see, the value $w_0 = 144.7\,\text{km/s}$, which characterizes our own inner solar system and most of the exoplanetary systems discovered up to now, seems to be the most often observed, simply because values of the potential $GM/r \approx w_0^2$ are the most frequent in current astrophysical systems. The other observed values are multiple and submultiples, e.g. $3w_0$ and $9w_0$ in the intramercurial solar system, $w_0/5$ in the outer solar system, $w_0/35$ in the distant scattered Kuiper belt, $3w_0$ for pulsar planets, $2w_0$ for eclipsing double stars, etc.

We have seen that such a hierarchy can be explained and described by writing various Schrödinger equations with different constants

$$\mathcal{D}_{jk} = \frac{GM_j}{2w_k}, \tag{13.69}$$

for the various inner Keplerian cores of structures (when they exist), then, by matching the solutions between subsystems (described by different w_k for the same central mass M_j, then between different systems (see an explicit example with da Rocha's detailed study of our Local Group of galaxies [126]). But the theory of scale relativity also provides one with a theoretical way to tackle more directly this question. It amounts to using quantum mechanical-like laws in the scale space, and to write a Schrödinger equation for the scales of structures. The solution, as explained in Chapter 8, gives peaks of probability for such scales and therefore naturally describes a hierarchy of embedded structures.

These leads us to suggest the possibility that, ultimately, the set of values of constants w be matched to the velocity of light, i.e. that they read $w = c/\prod_i p_i^{k_i}$, where the p_i are prime numbers. In this regard, one may remark that the value of the inverse coupling obtained above can be decomposed as $2070 = 2 \times 3^2 \times 5 \times 23$, which includes the other observed values $2w_0$, $3w_0$, $5w_0$ and $9w_0$. A possible test of this suggestion would consist of searching for very high w values, such as $c/2$, $c/3$, etc. around compact objects (pulsars, galactic nuclei, etc.).

Let us now apply this theoretical approach to the description and understanding of structures at astronomical scales.

13.4. Solar System

13.4.1. *Planetary system formation*

Let us first apply the above theory to the formation of planetary systems. The geodesic equation from which we start can be given the form (after integration) of a Schrödinger equation, but also of fluid mechanics equations, which are those used in the standard approach to structure formation [438]. Therefore, the application of the fractal space approach to the formation of planetary systems is a mere extension of the standard models of formation (see, e.g. [65, 169] and references therein).

It may be remarked that this proposal [353, 354, 357], made before the discovery of exoplanets close to their stars, which have imposed migration theories, is nothing but a peculiar form of such migration theories. Indeed, these theories amount to taking into account the coupling of protoplanets with the disk, e.g. through a mean field approach, which allows the energy and, therefore, the semi-major axis of the planetary orbit not to remain constant. In the scale relativity generalization of standard formation scenarios, we simply consider that the main coupling of a protoplanet with the disk is due to the scattering effect of the closest encounters, which we describe by a Brownian motion. This agrees with recent results of numerical simulations according to which the migration is not monotonous but Brownian-like.

Disk formation

Consider a protoplanetary disk of bodies (planetesimals, planet embryos, protoplanets) during the formation of a planetary system. At the lowest order approximation, one may neglect the self-gravitation of the disk. The motion of each planetesimal in the central Kepler potential of its star comes under the conditions under which a gravitational Schrödinger equation can be written. The various bodies are subjected to numerous gravitational encounters that scatter them in such a way their motion becomes Brownian and highly chaotic. On time scales large with respect to their mean free path one completely loses the position, angle and time information about the individual trajectories. Beyond the horizon of predictability, the possible trajectories become in infinite number, fractal of dimension 2 and irreversible, and these three conditions allow one to give to the motion equation a Schrödinger form.

We therefore expect each body to be subjected to a probability density distribution, which is a solution of this Schrödinger equation. During the first phases of the formation process, when the disk is still made of dust (and later of planetesimals) and gas, the number of bodies to which this probability density applies is very large. Then we expect the set of bodies to fill the "orbitals" (see Fig. 13.4). At this stage, the probability density distribution $|\psi|^2$ derived from the geodesics-Schrödinger

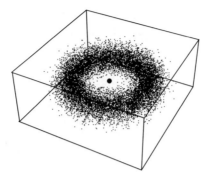

Fig. 13.4: **Solution of gravitational Schrödinger equation.** Example of solution of the gravitational Schrödinger equation for "particles" in a Kepler potential ($l = n - 1$). It is represented by a number density of points proportional to the probability density $P = |\psi|^2$.

equation therefore describes the matter density distribution of the initial disk or sub-disk. Examples of the general expected shapes for these density distributions are given in Fig. 13.5. In this figure, we have shown the proper solutions corresponding to given values of E, L^2 and L_z. We see that, in position space, the sub-disks corresponding to the various values of the (quantized) energy overlap, but they are actually well separated in phase space since the width of the energy distribution is very small for these solutions. Moreover, other solutions (less spread in position and more spread in energy-momentum) can be constructed as wave packets, satisfying a macroscopic Heisenberg relation $\Delta x \times \Delta p \sim 2m\mathcal{D}$. Whatever the width of these solutions, they remain characterized by peaks of probability ranging according to a n^2 distance law.

Wave packets

A subsequent step of the evolution of the system is the appearance of localized wavepackets (squeezed states) which move along the classical orbit. Such quasiclassical wavepackets, even on elliptic orbits, have been theoretically described in the standard quantum framework of atomic physics [76, 329, 188] and experimentally observed [546, 300]. In the macroscopic gravitational framework, they may describe the passage from a full disk to a concentrated planet embryo or protoplanet.

Moreover, these solutions are not the last word about the planetary formation in the scale relativity framework. Indeed, they correspond only to a first approximation in which only the dominant potential of the central star is taken into account. A more complete description should also account

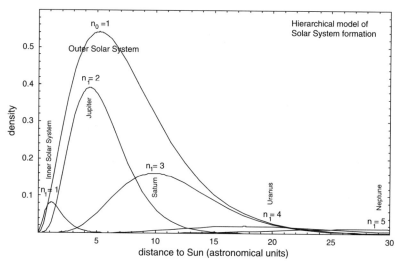

Fig. 13.5: **Hierarchical model of formation of the Solar System.** The density distribution of gas, dust and then planetesimals is given by solutions of a fractal geodesics equation (in the Kepler potential of the Sun), which can be transformed into a macroscopic Schrödinger equation. These solutions (generalized Laguerre polynomials) have the general form shown in the figure, involving a density peak depending on the energy of the system and vanishing density at small and large distances. This shape agrees with the mass distribution of planets in the inner and outer solar systems (see Fig. 13.7). The whole solar system is described by a fundamental level solution ($n_0 = 1$, peaking at Jupiter distance). It is then fragmented into the outer solar system, in which the inner solar system as a whole ranks $n_1 = 1$, Jupiter $n_1 = 2$, etc. Then again the inner solar system (peaked at Earth distance) fragments itself in terms of Mercury ($n_2 = 3$), Venus ($n_2 = 4$), Earth ($n_2 = 5$), Mars ($n_2 = 6$), etc.

for self-gravitation of the disk. In particular, in the wave packet phase, one expects the effects of self-gravitation to prevent spreading along the orbit and to finally lead to a concentrated object (protoplanet), which would finally accrete. In such a scenario the accretion stage would occur in a rather late phase, without any (or few) need for a sweeping process.

The final orbital elements of the planets formed in this way are expected to follow the laws that have been described in the previous section, i.e. most probable values of semi-major axis $\propto n^2$, of eccentricity $\propto k/n$, etc. In particular, in the absence of external perturbation, the law of conservation of energy leads to $a_n \propto n^2$, while the law of conservation of the center of gravity leads to a similar distance law given by the orbital mean distance, $r_n \propto n^2 + n/2$ (see in Sec. 13.5.3 an application of this case to the planetary system around the pulsar PSR 1257+12).

Dispersion of planet distances

The question of the expected dispersion of the distribution of planets around this n^2 law is a complex one. The answer to this question depends on the kind of system considered. It is, in general, not given by the width of the proper solutions shown in Fig. 13.5 (which scales as $n^{3/2}$ for circular orbits): indeed, this width is that of the initial distribution of planetesimals, while the planet position results from concentration and accretion of this distribution. This width therefore remains relevant only in the cases when the formation process has not been fully effective, for example when the disk of planetesimals has given rise to an asteroid belt. This is the case for the belt between Mars and Jupiter, which indeed covers the full $n = 8$ and $n = 9$ regions, whose probability peaks closely correspond to the most massive small planets of the belt, Ceres and Hygeia, or the Kuiper belt. When large planets are formed from the disk, the dispersion around their expected $\sim n^2$ distance may be the result of:

(i) an intrinsic macroquantum dispersion given by a generalized Heisenberg relation $\Delta E \Delta t \sim 2m\mathcal{D}$, while $E \propto 1/a$, where Δt is the typical time of fluctuation of the wavepackets. This case yields a small dispersion and holds only in the absence of additional perturbations: it is relevant, e.g. for the system of three planets around the pulsar PSR 1257+12, whose masses are of the order of the Earth mass (Sec. 13.5.3).

(ii) perturbations during the formation process and, after formation, during the evolution of the formed planetary system, due to interactions between the protoplanets and planets, coupling between the protoplanet and the disk (in addition to that accounted for in the Brownian motion accounted for in the scale relativity approach), and other possible physical effects (star wind, radiation pressure, magnetic fields, etc.). These interactions include both classical effects, such as resonances, which also intervene in a important way in the observed distribution of asteroids in the main belt and in the Kuiper belt, and macroquantum effects, which could be studied by including them in the macroscopic Schrödinger equation.

Hierarchic model

An important feature of the scale relativity approach is that it naturally leads to a hierarchy of structures [353, 361]. We have given hereabove general arguments for this result. Let us now singularize it to the case of a planetary system.

Consider a system of test-particles (e.g. planetesimals) in the dominant potential of the Sun. Their evolution on large time-scales is governed by Eq. (13.7), in terms of a constant $\mathcal{D}_j = GM/2w_j$. The particles then form a disk whose density distribution is given by the solution of the Schrödinger equation based on this constant. This distribution can then be fragmented in sub-structures still satisfying Eq. (13.7) (since the central potential remains dominant), but with a different constant w_{j+1}. We can iterate the reasoning on several hierarchy levels. The matching condition between the orbitals implies $w_{j+1} = k_j w_j$, with k_j integer.

Our own Solar System is indeed organized following such a hierarchy on at least five levels, from the Sun's radius to the Kuiper belt (see Figs. 13.5 and 13.6, and the following sections). In particular, the inner solar system as a whole can be identified with the fundamental level ($n_e = 1$) of the outer solar system, in which Jupiter is in $n_e = 2$, Saturn in $n_e = 3$, etc. Concerning the inner solar system, its peak of density is given by its most massive planet, the Earth, which is ranked $n_i = 5$ (Mercury ranks $n_i = 3$, Venus $n_i = 4$ and Mars $n_i = 6$). By matching the density peaks of these two

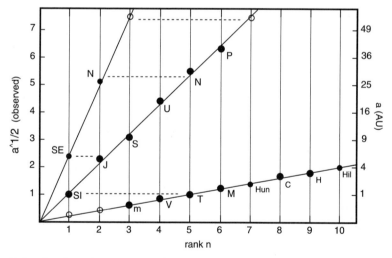

Fig. 13.6: **Semi-major axes of planets in the inner and outer solar system.** Comparison of the observed semi-major axes of planets in the inner and outer solar system with the predicted values, $a_n/GM = (n/w)^2$, with $w_{in} = 144\,\text{km/s}$ in the inner solar system and $w_{ex} = (144/5)\,\text{km/s}$ in the outer system. "SI" stands for the peak of the inner solar system, which corresponds to $n = 1$ in the outer system, and "SE" for the peak of the outer system (Jupiter at $n = 2$). Additional peaks are predicted beyond Pluto (see the Kuiper belt section hereafter) and before Mercury (see the section "Intramercurial structures").

embedded levels, we therefore expect the ratio of their coupling constants to be $w_i/w_e = 5$ (and therefore the ratio of their length scales to be ≈ 25, as observed).

Problems and Exercises

Open Problem 29: Develop in more details the theory of formation of planetary systems outlined in this section. In particular, solve the gravitational Schrödinger equation including self-gravitation of the disk for wave packet solutions [76, 329, 188] describing the concentration of planetesimals into proto planets.   ∎

13.4.2. *Planetary distances*

As expected in the hierarchical model of formation, the inner and outer solar systems are both organized in a similar way, in terms of semi-major axes distributed according to a n^2 law (see Fig. 13.6). For the inner system, the gravitational coupling constant is found to be $w_0 \sim 144\,\mathrm{km/s}$ [353, 357, 364]. This is easy to see from the orbital velocities of the planets: we expect to have $v_n = w_0/n$, so that an estimate $w_0 = nv_n$ can be obtained from each planet. From the known values of the mean orbital velocities: Mercury ($47.89 = 143.7/3\,\mathrm{km/s}$), Venus ($35.03 = 140.1/4\,\mathrm{km/s}$, the Earth ($29.79 = 148.9/5\,\mathrm{km/s}$) and Mars ($24.13 = 144.8/6\,\mathrm{km/s}$), one obtains a mean value of the coupling constant $w_0 = 144.4 \pm 1.8\,\mathrm{km/s}$. Accounting also for the mass peaks of the main asteroid belt (see Sec. 13.4.4), which lie around Ceres ($17.6 = 140.8/8\,\mathrm{km/s}$) and Hygiea ($16.5 = 148.5/9\,\mathrm{km/s}$), one obtains $w_0 = 144.5 \pm 1.5\,\mathrm{km/s}$.

Moreover, such a n^2 law is not adjustable (contrary to a scaling law of the Titius-Bode type), so that the ranks of the planets are fixed in a constrained way. This is easy to see from the square roots of the semi-major axes of planetary orbits, which are expected to be given by $\sqrt{a} = \sqrt{a_1}\,n$. One finds respectively for Mercury, Venus, the Earth and Mars $\sqrt{a} \approx 0.6$, 0.8, 1.0 and 1.2 (a in Astronomical Units). The difference is constant and equal to 0.2. Therefore Mercury, with a value of 0.6, is constrained to rank $n = 3$, Venus $n = 4$, the Earth $n = 5$ and Mars $n = 6$ (note that the region corresponding to the value $n = 7$ in the asteroid belt is nearly empty because of a 2:1 resonance with Jupiter).

As an additional consequence, the very first result of the theory has been to predict the existence of two additional probable zones for planetary orbits [353, 354, 357], at 0.043 AU ($n = 1$) and 0.17 AU ($n = 2$), for the solar system but also for extrasolar planetary systems. At the time of the prediction, no exoplanet was yet known. As we shall see, this prediction has now received strong support from the discovery of exoplanets and of structures in the very inner solar system. The same is true of the outer solar system beyond Pluto, which can now be checked using the Kuiper belt objects.

Let us briefly discuss other related approaches to the question of planetary distances (see more details in [357, 364, 374]). It has been remarked for long (see [437, p. 726]) that n^2 distance laws give a far better fit to the planetary distribution in the inner and outer solar systems than other proposed empirical laws, in particular the Titius-Bode law.

The suggestion to use the formalism of quantum mechanics for the treatment of macroscopic problems, in particular for understanding structures in the solar system, dates back to the beginnings of the quantum theory (see e.g. [240, 293] and [456] for additional references). However, these early attempts suffered from the lack of a convincing justification of the use of a quantum-like formalism and were hardly generalizable.

Later on, there were some attempts at developing a macroscopic quantum theory under the motivation of describing various effects of redshift quantization [209, 105]. The problem with these attempts is that they did not allow an understanding of the observed quantization and of its meaning, which was wrongly interpreted as of non-Doppler origin, while in the present approach we interpret them in terms of standard probability densities of normal velocities. The result according to which the solar system planets and the extra-solar planets (for which we are certain that we deal with true velocities) show the same structures supports our interpretation.

A third approach, better motivated theoretically, has been the suggestion to use Nelson's stochastic mechanics [331] as a description of the diffusion process in the protosolar nebula [8, 54]. The problem with such a suggestion is that Nelson's twin diffusion process is actually not a standard classical diffusion, it actually corresponds to no existing Markovian or non-Markovian process [205], so that a justification is lacking for its application to macroscopic systems.

More recently, it has been proposed by Reinisch to use a nonlinear Schrödinger equation [456], which yields a distance law different from n^2. He predicts probability peaks at 0.14, 0.38, 0.71, 1.14 and 1.67 AU for the inner solar system and similar extrasolar systems. This proposal yields a bad fit for the Earth (1 UA) and Mars (1.52 UA), but it is definitely excluded by the existence of a main peak of exoplanets at the $n = 1$ value predicted by the n^2 law, namely, $0.043 \, \mathrm{AU/M_{\odot}}$ (see Sec. 13.5).

We conclude this discussion by noting that Laskar [279] developed a simplified classical model of planetary accretion that also yields, among many other laws (most of them of a scaling form), a n^2-like distance law (of the form $\sqrt{a} = \mathrm{b}n + \mathrm{c}$) for a particular choice of the initial mass distribution. This is an interesting convergence of the standard approach with the generalized one considered here, although the final mass distribution corresponding to this law is at variance with that observed in the solar system. Moreover, one does not expect in this case a universal law, since it depends on very particular initial conditions in the protoplanetary disk. Therefore this classical approach is confronted to the difficult problem of explaining why, as well in the Solar system as in extrasolar systems, the constant c is zero, the ratio $\mathrm{b}/\sqrt{M}$ is a universal constant, and why this n^2 law holds even in the case of one single planet (see Sec. 13.5).

13.4.3. *Satellites and rings*

It has been shown by Hermann *et al.* [228] that the various systems of rings and satellites around the outer giant planets also come under the same n^2 law in a statistically significant way. Moreover the planet radii themselves belong to the sequence. Such a result is indeed expected, since a generalized Schrödinger equation can also be applied to the problem of the formation of the central bodies, and that matching conditions on the probability amplitude should be written between the interior and exterior solutions (similar to the Cauchy conditions in general relativity).

In a subsequent study, Antoine [27] has solved the Schrödinger equation in the case of cylindrical two-dimensional symmetry, and he has compared the solutions to the observed main density peaks of Saturn rings. In this case the energy becomes quantized as $E_n = E_0/(n+1/2)^2$. A similar result has been obtained by de Oliveira Neto *et al.* [134]. The remarkable fact discovered by Antoine is that the ring median distances follow this law with the same value $w_0 = 144 \, \mathrm{km/s}$ as the inner solar system. One indeed finds values of $n = 6.45, 7.57, 8.51, 9.46$ and 13.43, all of them close to $n + 1/2$ as expected. The probability to obtain such a result by chance (computed by folding the distribution on the expected period and by comparing to a heads or tails probability expectation) is smaller than 10^{-3}.

Problems and Exercises

Exercise 26 Solve the two-dimensional Schrödinger equation in a Kepler potential and show that in this case the quantization law is in $(n + 1/2)^2$ instead of n^2 in the three-dimensional case. Compare this theoretical expectation with the density distributions of the various flat rings in the solar system. ∎

13.4.4. *Planetary mass distribution according to distance*

The above theoretical approach provides a model for the mass distribution of matter according to radial distance in the solar planetary system [357, 364]. Indeed, as well the observed distribution of the whole system as that of the inner system, which stands globally as the first orbital of the outer one, agree with the predicted law of probability density. We have therefore used the hierarchical model described above (see Fig. 13.5) to predict the masses of the planets in units of Jupiter mass, in agreement with runaway models, in which the formation process is strongly dependent on the most massive planet. We have considered the whole initial disk as given by a fundamental level probability distribution, $P_1(r) \propto r^2 e^{-2r/a_J}$, peaking at Jupiter distance. Then we have considered that it is fragmented into sub-orbitals according to the expected planetary distances $r_n = r_1 n^2$ and to the probability density values $P_1(r_n)$ at these distances, integrated over an interval $\delta r = (2n+1)r_1$. The process has then be iterated for the inner solar system, which is in its whole the orbital $n = 1$ of the outer solar system. The result [364] is given in Fig. 13.7. The distribution according to distance of the masses of planets in the inner and outer solar system are in good agreement with such a model.

Only the mass of Neptune is much higher than expected. But even this discrepancy is easily explained by the existence of a larger system in which the mass peak of the whole planetary system (i.e. Jupiter) ranks $n = 1$, and in which Neptune ranks $n = 2$ (dashed line in Fig. 13.7; see also Fig. 13.6). One may also remark that under the local approximation of constant density, a first order account of self-gravitation can be done in the framework of a harmonic oscillator model. As we have seen in Sec. 13.3.2, the fundamental level solution of the harmonic oscillator Schrödinger equation yields one object, but the first excited level yields a pair of objects (see Fig. 13.26).

This provides a solution to the ancient unsolved problem of the high occurence of binary objects in astronomy at several scales (stars, clusters of

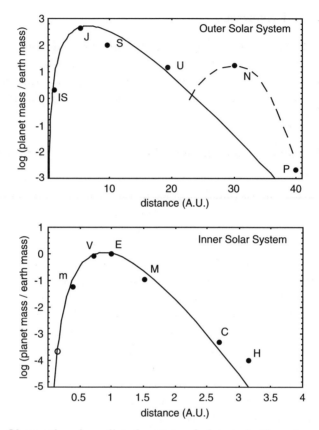

Fig. 13.7: **Observed and predicted masses of planets in the solar system.** IS stands for the inner solar system as a whole (two Earth masses), C and H for Ceres and Hygeia, which are the two main mass peaks in the asteroid belt between Mars and Jupiter. The $n = 2$ predicted mass at 0.17 AU in the inner solar system is only 10^{-4} of that of the Earth, which is probably too small for a planet to have formed. One therefore expects the existence of a remaining intramercurial asteroid ring orbiting the Sun around this distance (see Sec. 13.4.5.3).

stars, galaxies, clusters of galaxies). Applied to planetary system formation, this leads to the suggestion that planet pairs may be common during the formation era, which are most of the time unstable and therefore transient. The similarity of the masses of Uranus and Neptune could come under such an explanation. It is also noticeable in this context that the sum of the masses of the whole inner solar system is very precisely (to 1%) twice the mass of the Earth. The inner solar system, which is in its whole the fundamental level of the outer system, could then have known such an unstable binary phase, within which the final formation of the Earth-Moon system fits well.

Note finally that this model will certainly help solving another problem encountered by standard models of planetary formation. The accretion time of planetesimals, though acceptable for earth-like planets, becomes too large for giant planets. In the above framework, the initial distribution of planetesimals is no longer flat, but already peaked at about the final value of the planet positions, which should shorten the accretion process.

13.4.5. *Intramercurial structures*

13.4.5.1. *The sun*

Sun radius

Through Kepler's third law, the velocity $w = 3 \times 144.7 = 434.1\,\mathrm{km/s}$ is very closely the Keplerian velocity at the Sun radius, which is given by

$$w_\odot = \sqrt{\frac{GM_\odot}{R_\odot}} = 437.1\,\mathrm{km/s}, \qquad (13.70)$$

for $R_\odot = 0.00465$ AU. This result opens the possibility that the whole structure of the solar system be ultimately brought back to the Sun radius itself. Such a result is not unexpected in the scale-relativity approach. Indeed, the fundamental equation of stellar structure is the Euler equation, which can also be transformed in a Schrödinger equation, yielding a probability density distribution for star radii. The peak(s) of such a probability distribution would correspond to preferential values of the star radii. Matching conditions between the probability amplitude that describes the interior matter distribution (the Sun) and the exterior solution (the Solar System) during the formation epoch imply a matching of the positions of the probability peaks.

Solar cycle period

One can also apply the scale relativity approach to the organization of the solar plasma itself. One expects the distribution of the various relevant physical quantities that characterize the solar activity at the Sun surface (sun spot number, magnetic field, etc.) to be described by a wave function whose stationary solutions read

$$\psi = \psi_0 e^{iEt/2m\mathcal{D}}. \tag{13.71}$$

The energy E results from the rotational velocity and, to be complete, should also include the turbulent velocity, so that $E = m(v_{rot}^2 + v_{turb}^2)/2$. This means that we expect the solar surface activity to be subjected to a fundamental period (which is nothing but the macroscopic equivalent of a de Broglie period for the Sun) given by

$$\tau = \frac{2\pi m\mathcal{D}}{E} = \frac{4\pi\mathcal{D}}{v_{rot}^2 + v_{turb}^2}. \tag{13.72}$$

Now we can use our knowledge of the parameter $\mathcal{D}$ at the Sun radius, $\mathcal{D} = GM_\odot/2w_\odot$ to finally obtain [388, 396]

$$\tau = \frac{2\pi GM_\odot}{w_\odot(v_{rot}^2 + v_{turb}^2)}. \tag{13.73}$$

If the matter of the Sun was still rotating with its Keplerian velocity, one would have $v_{rot} = w_\odot$, so that one would recover (neglecting the turbulent velocity) the Keplerian period at the distance of the Sun radius, $P = 2\pi GM_\odot/w_\odot^3$. However, as many other Sun-like stars, the Sun has been subjected to an important angular momentum loss since its formation (see e.g. [473]). As a consequence, its rotational velocity is about 200 times smaller. The average sideral rotation period is 25.38 days, yielding a velocity of 2.01 km/s at equator [437]. The turbulent velocity has been found to be $v_{turb} = 1.4 \pm 0.2$ km/s [272]. Therefore we find numerically [388]

$$\tau = (10.2 \pm 1.0)\,\text{yrs}. \tag{13.74}$$

The observed value of the period of the Solar activity cycle, $\tau_{obs} = 11.0$ yrs, falls nicely close to this time scale.

This is an interesting result, owing to the fact that there is, up to now, no existing theoretical prediction of the value of the solar cycle period, except in terms of very rough order of magnitude [553]. It is generally considered that such a prediction should emerge from a model of the solar dynamo, which is not at all contradictory with the above scale relativity approach. Indeed, we contemplate the possibility of a future application of this approach to the description of stellar interiors (see also Sec. 13.8.6). Since stars are presently described in terms of hydrodynamics equations, these equations can be easily generalized to include the fractal fluctuation and irreversibility terms that lead to a Schrödinger form for the equation of dynamics (see Chapters 5 and 10). Knowing that the interior wave function and the surface wave function should ultimately be matched, as already remarked, one expect the interior and surface period to be related.

Moreover, since we now have at our disposal a simple and precise formula for a stellar cycle, which accounts for the solar period, the advantage is that it can be tested with other stars. The observation of the magnetic activity cycle of distant solar-like stars remains a difficult task, but it has now been performed on several stars [470]. A first attempt of testing formula (13.73) on these stars gives very encouraging results, as can be seen in Fig. 13.8. We have indeed obtained a satisfactory agreement between the observed and predicted periods, in a statistically significant way ($P \approx 10^{-4}$), despite the small number of objects.

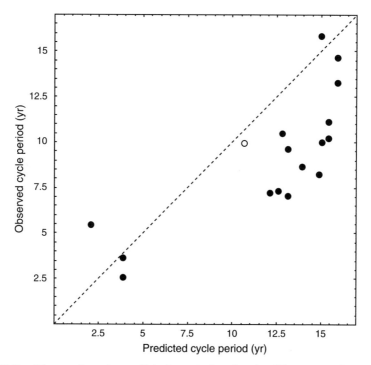

Fig. 13.8: **Observed and predicted periods of solar-like stars.** Comparison between the observed values of the period of solar-like star cycles (inactive stars with better determined behavior in Table 1 of [470]) and the predicted periods (see text). The open point is for the Sun. The correlation is significant at a probability level $P \approx 10^{-4}$ (Student variable $t \approx 5$).

Problems and Exercises

Open Problem 30: Apply the scale relativity approach (leading to transform the fundamental equation of dynamics into a macroscopic Schrödinger equation) to the problem of the internal structure of stars (including the Sun). ∎

Open Problem 31: Apply the scale relativity approach, which is equivalent to a macroscopic quantum-like hydrodynamics description including a macroscopic quantum potential, to the problem of the Sun magnetohydrodynamics (which underlies the question of Sun spots).

Hints: the numerous contributions intervening in the Sun magnetohydrodynamics equations may combine themselves to lead to chaotic and impredictible dynamics, then to the emergence of the fractal and irreversibility terms, which transform the classical-like equation into a macroscopic quantum-type equation. ∎

13.4.5.2. *Circumsolar dust*

On the basis of an intramercurial constant $w_{im} \approx 3 \times 144 = 432\,\mathrm{km/s}$, one expects probability density peaks lying at $4.09\,R_\odot$, $9.2\,R_\odot$, $16.32\,R_\odot$, etc. This can be checked by studying the density distribution of interplanetary dust [123].

The possible existence of intramercurial bodies is limited by dynamical constraints, such as the presence of Mercury, and thermodynamical constraints, such as sublimation of small bodies [364]. As a consequence asteroids can be found only in the zone 0.1–0.25 AU, but the inner zone 0.005–0.1 AU can yet be checked using the distribution of interplanetary dust particles in the ecliptic plane (originating from comets and asteroids) that produce the F-corona.

Since 1966, there have been several claims of detection during solar eclipses of IR thermal emission peaks from possible circumsolar dust rings (Peterson [442], MacQueen [298], Koutchmy [255], Mizutani *et al.* [324], Lena *et al.* [287]). These structures have been systematically observed at the same distance from the Sun during five eclipses between 1966 and 1983. MacQueen finds two radiance peaks at $4.1\,R_\odot$, which is equivalent to a Kepler velocity $v = 216\,\mathrm{km/s}$ and at $9.2\,R_\odot = 0.043\,\mathrm{AU}$, which corresponds very precisely to a Kepler velocity $v = 144\,\mathrm{km/s}$ (see Fig. 13.9), i.e. to the expected background level of the inner Solar System and of extrasolar planetary systems at the same scale (see Sec. 13.5 herebelow).

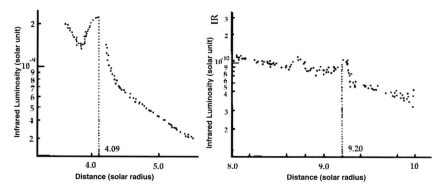

Fig. 13.9: **Dust density peaks near the Sun (adapted from [298]).** Two density peaks were observed at exactly the predicted distances 4.09 $R_\odot$, and 9.20 $R_\odot$ = 0.043 AU, which correspond respectively to Keplerian velocities of $432/2 = 216$ km/s and $432/3 = 144$ km/s. This last value is that of the main peak of exoplanets (see Sec. 13.5).

However, more sensitive observations during the 1991 solar eclipse [263, 234] did not confirm these detections. While Lamy *et al.* [263] conclude that the previous detections were in error, Hodapp *et al.* [234] argue that the earliest observations were credible and therefore that the observed structures were transient features, perhaps due to the injection of dust into near-solar space by a Sun-grazing comet. This last interpretation is quite in agreement with the scale-relativity approach. Indeed, the dynamics of the dust particles that are at the origin of the solar F-corona is determined by the Sun gravity, the Poynting–Robertson and corpuscular pressure drag, the radiation pressure force and the effect of sublimation [262]. The combination of these effects leads to chaotic motion with a small predictability time horizon, and, would, therefore come under the gravitational Schrödinger equation. The dust injected in the circumsolar space would thus accumulate during a finite time into the predicted high probability zones and finally spiral toward the Sun due to the Poynting–Robertson drag.

Therefore a possible independent test of the theory could consist of new real time IR observations during a forthcoming eclipse taking place just after the passage of large size sungrazer(s).

13.4.5.3. *Vulcanoid belt*

An attempt to test the prediction of the theory according to which one or several objects (most probably an asteroid belt) could exist around 0.17 AU

($n = 2$ for the inner solar system) from the Sun has been performed by Schumacher and Gay [477].

They have tried to detect vulcanoids by analysing the SOHO/LASCO images using automatic detection. Their conclusion [477] is that it is very difficult to detect such objects even if they do exist, because of the high level of noise coming from the solar corona, which remains after cleaning the images. Their result is that there is no object of diameter larger than 60 km around ≈ 0.17 AU. This lefts open plenty of possibilities, in particular the most probable one, i.e. an asteroid belt made of objects of size smaller than 10 km. Indeed, the prediction of the scale relativistic planetary model formation (see above) is that a total mass of 10^{-4} times the Earth mass is expected in this zone. It is probable that such a small total mass has not been able to accrete so that the matter remains discretized in terms of planetesimals. Moreover, recent numerical simulations [171] have shown that there should exist a stable zone between 0.1 and 0.2 AU, and, moreover, that some of the known Earth-crossing asteroids could well come from this zone.

In situ detection could be possible in the future using, e.g. the Solar Orbiter spacecraft, which will reach a distance to the Sun of 0.21 AU. Another suggestion [123] consists of searching for the possible perturbations that such an asteroid belt would induce on the motion of the Aten and Apollo Earth-crossing asteroids, in particular those which enter the very inner solar system. Six such objects are presently known having perihelion distances smaller than 0.17 AU (see http://cfa-www.harvard.edu/cfa/ps/mpc.html), among which 1995 CR, of perihelion $q = 0.119$ AU and inclination $i = 4.0°$, and 2000 BD19 of perihelion $q = 0.092$ AU, inclination $i = 25.7°$, which crosses the ecliptic at the expected belt distance of 0.174 AU.

Problems and Exercises

Open Problem 32: Describe the main asteroid belt between Mars and Jupiter by combining the macroscopic Schrödinger approach with the classical celestial mechanics approach. Apply this description to an expectation of the structure of the postulated vulcanoid belt around 0.17 AU.

Hints: the macroscopic Schrödinger equation approach yields the general shape of the density distribution for each quantized energy (i.e. semi-major axis) regions. For example, in the main belt it corresponds to $n = 7$ to 10.

But the classical celestial mechanics effects, acting on smaller time scales, perturb this macroquantum-type distribution. In particular, the $n = 7$ and $n = 10$ orbitals correspond to destructive resonances with Jupiter and are therefore nearly empty. More generally, the various resonances dig gaps in the asteroid distribution, such as the Kirkwood gaps, and must be accounted for in addition to the minima between the orbitals to obtain a full description of the belt.   ∎

13.4.5.4. *Sungrazer comets*

Parabolic orbit comets can be used to check deeper intramercurial structures [478, 123]. The sungrazers, in particular those recently observed by the SOHO satellite, enter in the very inner solar system. The eccentricity of these objects is very close to $e = 1$ (their orbits are quasi-parabolic). Therefore their perihelion distances become a direct indicator of angular momentum.

Angular momentum quantization

The square of the angular momentum is classically given, in term of semi-major axis a and eccentricity e, by the relation

$$(L/m)^2 = GMa(1 - e^2). \tag{13.75}$$

Now expressed in terms of perihelion distance $q = a(1 - e)$, this expression becomes

$$(L/m)^2 = GMq(1 + e). \tag{13.76}$$

In the case of quasi-parabolic objects as those considered here ($e \approx 1$), we finally obtain

$$(L/m)^2 = 2GMq. \tag{13.77}$$

Therefore, up to a quantity, which is a constant for a given system (Sun of mass $M_\odot$ and sungrazer comet of mass m), the perihelion distance identifies with the squared angular momentum. It is thus particularly easy to apply the Schrödinger-like approach to the problem of the perihelion distribution of sungrazer comets.

In particular, this is the case for the long period comets in the Solar System. In this case, the solutions correspond to states of fixed values of the squared modulus of the angular momentum, L^2 and of its z component L_z, which are quantized as

$$(L/m)^2 = \left(\frac{GM}{w}\right)^2 \ell(\ell + 1), \tag{13.78}$$

$$(L_z/m)^2 = \left(\frac{GM}{w}\right) m_{\mathrm{mag}}, \tag{13.79}$$

where ℓ runs from 0 to $n - 1$ and m_{mag} from $-\ell$ to $+\ell$.

Data

The sungrazing comets are a particularly interesting sample to test for the very inner Solar System, since their perihelion may be smaller than the Sun radius, and their velocity at perihelion may reach $600\,\mathrm{km/s}$. We therefore expect their probability distribution to be structured in terms of a new multiple of the constant $w_0 = 144\,\mathrm{km/s}$.

The analysis has been performed [478] on the sample of 333 sungrazers ($q < \approx 0.015\,\mathrm{AU}$) from the list updated by Biesecker *et al.* (2001 data) on the NASA web site [50].

Theoretical expectation for sungrazer perihelia

Let us now apply the theory to these particular objects. They have quasi-parabolic orbits, i.e. their eccentricity is $e = 1$. During their journey through the Solar System, their trajectories, when they are analysed at long time-scale resolution, have no determined perihelion distance, due to the multiple diffusions they suffer from the planets and from other bodies. The long time-scale description can therefore use the scale-relativity tool that transforms the geodesics equation into a Schrödinger equation. Then, from its solutions, we can make predictions about the expected probability distribution of angular momentum (Eq. 13.78). Now, at the small time-scales during which the sun-grazers are observed (namely, when they come close to the Sun), one recovers deterministic trajectories and determined values of the perihelion distance q. These values are linked to squared angular momentum through Eq. (13.77).

Therefore we finally expect the probability density distribution of observed perihelion distances to show peaks for the quantized values

$$q_\ell = \frac{\ell(\ell + 1)}{2}\frac{GM}{w^2}, \tag{13.80}$$

where ℓ takes only integer values.

Solving this equation for ℓ allows one to make correspond to each value of q an effective value $\hat{\ell}$ given by

$$\hat{\ell} = \frac{\sqrt{1 + 8q/a_1} - 1}{2},$$ (13.81)

where $a_1 = GM/w^2$ is the semi-major axis of the fundamental level for the hierarchic sub-structure determined by w.

The test we shall perform finally consists of checking that the distribution of the observables $\hat{\ell}$ (Eq. 13.81) show peaks for integer values.

Concerning the value of w (and of the corresponding fundamental radius a_1), we already know that the inner planetary system, which ranges from 0.4 AU (Mercury) to about 3 AU (asteroid belts) is characterized by the value $w_0 = 144$ km/s (i.e. $a = 0.043$ AU). However, the perihelia of the comets under consideration here lie at scales smaller than the inner Solar planetary system (0.0015 AU to 0.15 AU). As previously remarked, this leads us to make the conjecture that these comets characterize new substructures of the very inner Solar System, defined by integer multiples of $w_0 = 144$ km/s.

The sun radius already corresponds to such a substructure, with a Keplerian velocity of $w_\odot = 432$ km/s $= 3 \times 144$ km/s. Concerning the sungrazing comets, since they may have perihelia smaller than the Sun radius and may reach velocities of 600 km/s or more, one is led to consider a still more profound quantization level with a new integer factor (because of matching conditions on wave functions).

Results and statistical analysis

The histogram of the distribution of the perihelia of the 333 sungrazers is given in Fig. 13.10 and is compared with the probability peaks predicted from Eq. (13.81) for $w_1 = 3 \times 432 = 1296$ km/s (this value corresponds to 1/9 of the Sun radius). As can be seen in the figure, the density curve corresponds to a decreasing law with well defined peaks at the integer values of $\hat{\ell}$ ($\hat{\ell} = 4$, 5 and 6). There is also one sungrazer found at a $\hat{\ell}$ value close to 7. The radius of the sun corresponds to the value 3.72 of this variable, which means that the peak $\hat{\ell} = 4$ is not a cut-off peak, but has a real physical significance.

We have performed a statistical analysis to check the validity of this result. We separated the interval of the q values in two equal subsets corresponding respectively to the maxima and minima of probability density. Namely, we computed for each comet the effective value $\hat{\ell}$ of

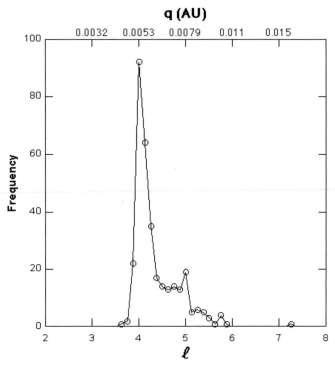

Fig. 13.10:   **Density peaks of sungrazer comets.** Distribution of the perihelions of sungrazer comets and comparison to the scale relativity expectation (integer values of l, see text), according to [478, 123].

ℓ, which is the solution of Eq. (13.80), as given in Eq. (13.81). The maxima corresponds to $k - 0.25 < \hat{\ell} < k + 0.25$ and the minima to $k + 0.25 < \hat{\ell} < k + 0.75$, with k integer. We find that 236 out of the 333 comets lie in the expected peaks of probability density ($+$) and only 94 in the minima ($-$). The probability to obtain such a result by chance is $P \approx \sum_{i=1}^{94} C_{333}^i 2^{-333} \approx 5.10^{-16}$, where C_j^i denotes the combination of j objects taken i by i (8 σ level of significance).

Conversely, one can use this sample for a new independent determination of w_0, by scanning the values of a_1 in Eq. (13.81), and looking for a maximum of the number of comets falling in the maxima as defined above. We find that the comet distribution is stable for w values in the interval $w_1 = 1300.5 - 1319.5 \,\mathrm{km/s}$, i.e., $w_1 = 9 \times (145.5 \pm 1.1) \,\mathrm{km/s}$. This new determination of the fundamental structure constant w_0 agrees within error bars with the average value previously obtained from several effects at different scales, $w_0 = 144.7 \pm 0.6 \,\mathrm{km/s}$ ([361]).

Discussion

The shape of the sungrazer perihelion distance distribution can be interpreted as follows. In the standard view (that we keep here), the Kreutz sungrazers come presumably from a comet that was broken on a previous passage, and the fragments are now hitting or coming very close to the Sun. Due to diffusion by various perturbation effects, the perihelion distances spread out (in this case, between ≈ 0.003 AU and ≈ 0.15 AU). Due to the information loss involved by such a chaotic dynamics, the classical approach is in general unable to be more specific.

The additional ingredient brought by the scale relativity approach is simply that the density function is expected to be non uniform and to exhibit peaks at some precise values. This does not mean that we can predict in a more precise way the perihelion of any given individual comet, which remains between 0.003 and 0.15 AU, but simply that some values are more probable than others. Moreover, these values depend only on the Sun mass and on the structure constant w_1, but they are independent of the comets themselves, which serve here only as test particles manifesting universal structures.

This result may play an important role in the interpretation of the theory and in determining its domain of application. Indeed, concerning the planetary distances, the scale relativity approach is considered to be applicable only during the formation era. Here, on the contrary, its applicability for the Kreutz sungrazer comets, which are of recent origin, lead to conclude that it can also be applied to the present motion of some solar system bodies. Actually, only very small perturbations are needed during their travel through the solar system to account for a small displacement of their perihelion distances from a hollow to a peak of the density distribution.

13.4.6. *Trans-Neptunian structures: Kuiper belt*

Many small and medium size bodies have now been discovered in the region around and beyond 40 AU, known as the Kuiper belt. Some of them being even larger than Pluto, this has now led to a change of nomenclature, in which Pluto and similar objects are no longer considered as main planets, but as "dwarf planets".

For the outer solar system, the w constant is close to 29 km/s, i.e. 144/5 km/s, where $n = 5$ is the rank of the mass peak in the inner solar system, given by the Earth position. This implies a quantization of the semi-major axis probability peaks according to the law $a_n = 1.1n^2$ AU. This law has been used to predict the distribution of Kuiper belt components [353, 364] before their discovery (see Fig. 13.6, adapted from [364]). Most Kuiper Belt Objects (KBOs) are found around 40 AU and they therefore agree with the $n = 6$ predicted peak whose Pluto is the dominant dwarf planet. At larger distance, another population of objects known as the scattered Kuiper belt objects (SKBOs) can be used to put the n^2 law to the test for $n > 6$. Their distribution is given in Fig. 13.11, and compared to these predictions [123].

A satisfactory agreement is found for these SKBOs, in particular concerning the main observed number density peak at about 57 AU, which agrees closely with the first expected trans-Plutonian peak ranking $n = 7$ [357]. Note that, as for the main asteroid belt between Mars and Jupiter, one fully recovers the observed number density distribution when also accounting for the resonances with the large planets, as in the standard

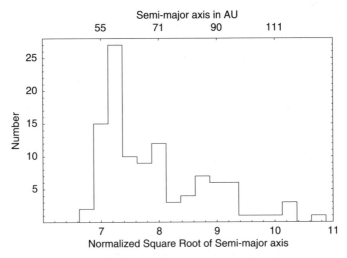

Fig. 13.11: **Scattered Kuiper Belt Objects.** This figure gives the distribution of the semi-major axis of Scattered Kuiper Belt Objects, compared with the theoretical expectation of probability density peaks for the outer solar system. One theoretically expects probability peaks for integer values of the variable $\sqrt{a/1.115}$, a being the semi-major axis of the orbit in AU. The data (Oct. 2009 list of SKBOs from the IAU Minor Planet Center, http://www.cfa.harvard.edu/iau/lists/ Centaurs.html) supports this expectation in a statistically significant way. To estimate the significance of this result, we have folded the distribution on the predicted period and compared the number of objects in the expected peak and hollow regions. In the absence of effect there should be a 50–50% probability of falling in the two regions (like in a heads or tails situation). The observed data yield a total of 76 SKBOs in the peak region against 38 in the hollow region. The probability to obtain such a difference by chance is only 2×10^{-4}.

celestial mechanics approach. The newly discovered dwarf planet Eris (2003 UB313), which was revealed to be larger than Pluto, has been found at a semi-major axis of 68 UA, which is close to the next predicted peak $n = 8$ at ≈ 70 UA.

13.4.7. *Sedna and the very distant Kuiper belt*

A new hierarchical level of the outer solar system can then be expected for the SKBO population, by taking their main peak at 57 AU as fundamental level. One therefore expects probability peaks at $a_n = 57n^2$ AU, i.e. 228 AU ($n = 2$), 513 AU ($n = 3$), 912 AU ($n = 4$), 1425 AU ($n = 5$), etc. [191]. It is therefore remarkable that the very distant dwarf planet Sedna (2003 VB12) has been discovered just with a semi-major axis of 509 AU, and that several additional small planets are now known to orbit according to the $n = 2$ and $n = 3$ predicted peaks, in particular 2002 GB32 $a = 217$ AU, 2000 CR105 $a = 220$ AU, 2001 FP185 $a = 224$ AU, and 2000 OO67 $a = 552$ AU (see Fig. 13.12). Two new small planets have been discovered even more recently, one close to the $n = 2$ peak, 2006 UL321 with $a = 261$ AU, and a new one just in the $n = 4$ peak, 2006 SQ372 with $a = 915$ AU [322], which matches fairly well the structures expected from the scale relativity approach.

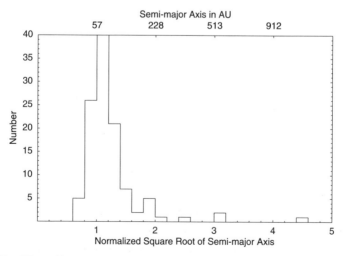

Fig. 13.12:   **Very distant Scattered Kuiper Belt Objects.** Histogram (2009 data) of the distribution of $n = (a/a_0)^{1/2}$, where a is the semi-major axis in AU, for the very outer solar system (Scattered Kuiper Belt Objects including the dwarf planet Sedna, which falls in the $n = 3$ peak). The fundamental level $a_0 = 57$ AU is the main SKBO peak and corresponds to a constant $w_{\rm SKBO} = 144/(5 \times 7) = 4.11$ km/s.

Problems and Exercises

Open Problem 33: Describe in more details the Kuiper belt density distribution by accounting for the resonances with the large planets, in particular with Neptune. ■

13.4.8. *Obliquities and inclinations of planets and satellites*

13.4.8.1. *Theoretical reminder*

The method of transformation of classical dynamics equations into Schrödinger equations under the information-loss conditions has also been applied to the equations of the rotational motion of solids (see [365, 362, 406] and Secs. 5.11.5 and 10.6.7.2). The planetary and satellite obliquities in the solar system are known to be most of the time chaotic and they therefore come under the Schrödinger description on time scales long with respect to the chaos time. In the case of a free rotational motion, we have obtained a generalized form of Schrödinger equation,

$$S_0 \left(\mathcal{D}_{jk} \partial_j \partial_k \psi + i \frac{\partial}{\partial t} \psi \right) = U \psi, \tag{13.82}$$

where U is a rotational potential energy and where the tensor $\mathcal{D}_{jk}$ is linked to the tensor of inertia I_{jk} of the body by a relation, which generalizes the Compton-like relation $S_0 = 2m\mathcal{D}$ [406] (with $S_0 = \hbar$ for standard microphysical quantum mechanics):

$$\mathcal{D}_{jk} = \frac{1}{2} S_0 I_{jk}. \tag{13.83}$$

We shall now apply this equation to our own Solar System and show that its solutions allow one to account for some observed characteristics of the distribution of planet obliquities and of the inclination of their orbits [365].

13.4.8.2. *Application to planet obliquities*

Let us look for the solutions of this generalized Schrödinger equation in the simplest case. Namely, being interested here in obliquities and inclinations, we treat either the planet or satellite or the couple planet/satellite-orbit as a spinning top. We consider in what follows only the problem of free rotational motion, focusing on the first Euler angle θ. It is only in that particular and simplified case that obliquities and inclinations can be treated together as solutions of the same equation. The scale-relativistic equation of evolution of the angle θ writes:

$$\frac{\tilde{d}^2\theta}{dt^2} = 0. \tag{13.84}$$

Looking for stationary solutions, Eq. 13.84 takes after integration the form of a one-dimensional time-independent Schrödinger equation,

$$\frac{d^2\psi}{d\theta^2} + A^2\psi = 0, \tag{13.85}$$

where $A = E/2I = \Omega/2\mathcal{D}$ since the rotational energy is $E = \frac{1}{2}I\Omega^2$.

The solutions, which have a peak at $\theta = 0$ (as observed and as predicted by the classical theory) and which satisfy the periodic condition $P(\theta+\pi) = P(\theta)$, write

$$P(\theta) = a\cos^2(n\theta), \tag{13.86}$$

with n integer.

13.4.8.3. *Comparison with observational data*

The peaks in the observed distribution of both obliquities and inclinations of planets and satellites in the Solar System agree remarkably well with the predicted periodicity of Eq. (13.86), in terms of a unique value $n = 7$ for the whole system. This can be seen in Fig. 13.13. The data is taken from Beatty and Chaikin [40], Encrenaz *et al.* [169] and Lang [273]. We mix values of the inclination relative to the planet equator with those relative to the orbital plane, since a common quantization law is predicted for all these values. Note that, at the still rough level of our analysis (free rotational motion description) our prediction concerns only the periodicity of the most probable values of the angles, but not the amplitude of the peaks. This result allows one to account for some very striking features shared by the observed inclinations and obliquities in our Solar System, in particular:

(i) The value 23.45° of the obliquity of the Earth is considered in the classical description to be the chance result of the last collision at the end of the formation epoch. This collision has probably given rise to the Earth-Moon couple, then the Earth obliquity has been stabilized to its value by the Moon [278]. Moreover, all of the terrestrial planets have probably experienced large, chaotic variations of their obliquities at some time in the past. In particular the obliquity of Mars is still in a large chaotic region [277].

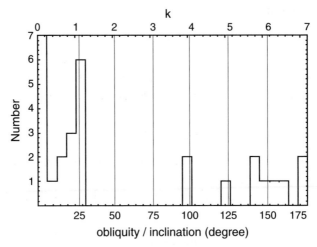

Fig. 13.13: Obliquity and inclination distribution of planets and satellites. Distribution of the obliquities and inclinations of planets and satellites in the Solar System, compared with the scale-relativity expectation (integer values of k).

This classical description is probably globally correct, but cannot explain why a large number of bodies in the Solar System have values of their obliquities and inclinations very close to the Earth value (see Fig. 13.13). Indeed the obliquities of Mars, Saturn and Neptune are respectively $23.98°$, $26.73°$, and $29.6°$, (mean $25.9 \pm 1.4°$ for 4 points), and the inclinations of Jupiter's satellites Leda, Himalia, Lysithea and Elara, and of Neptune's satellite Nereid are respectively $26.1°$, $27.6°$, $28.8°$, $24.8°$ and $27.5°$ (full mean $26.5 \pm 0.7°$ for 9 points). Both average values lie within one sigma of the first non-zero predicted probability peak at $180°/7 = 25.7°$ (see Fig. 13.13).

(ii) The obliquity of Uranus, which is heeled over its orbital plane, is actually $98°$, within $5°$ of the predicted peak at $\theta = 102.8°$ ($k = 4$). The inclination $\sim 100°$ of Charon's orbit around Pluto also agrees with this peak.

(iii) The retrograde rotations of Venus on itself ($\theta = 178°$), which is one of the puzzling feature of the Solar System [277] and of Saturn's satellite Phoebe on its orbit ($\theta = 177°$) also come under the same description, since they agree with the probability peak at $180°$. The same is true of the orbital inclination of four of the Jupiter satellites and of Neptune's satellite Triton, which yield a mean value $153.6° \pm 3.6°$, to be compared with the predicted peak $k = 6$ at $\theta = 154.3°$.

Finally, we can fold the whole observed distribution on the predicted period $\pi/7$ and then perform a χ^2 test: the probability of obtaining by chance the observed difference with a uniform distribution is $P \ll 10^{-4}$.

13.4.8.1. *Discussion*

These results do not contradict those obtained from standard celestial mechanics, which are valid below the predictivity horizon, while the scale relativity approach applies beyond the horizon. Both approaches are therefore complementary. Peaks of probability such as those expected here can be obtained from chaotic classical motion (case of Mars) at small time scales if, e.g. the time elapsed in the peaks is far larger than the time needed to jump from one peak to another. In the case of the Earth obliquity, the argument of its stabilization by the Moon [277] remains valid, but our theory adds a prediction of the value at which it can be stabilized (namely, around 25°).

Recall also that these results are obtained in the framework of a very simplified description (free rotational motion). It has however already predictive power. Indeed, two probability peaks at 54.4° and 77.1° were still empty at the time of this study [365] but they have been since filled by new observations. Hence the peak at 128.6° was empty until the measurement of Pluton's obliquity at 122.5°. In works to come, we shall try to improve the model and to answer questions that remain open, concerning in particular the origin of the empirical value $n = 7$.

Let us conclude by remarking that this proposal reinforces greatly the probability that a large number of Earth-like planets exist in the universe, stable enough for life to develop and survive. Indeed, it is expected that around solar-type stars, the orbit $n = 5$ will have a semi-major axis of $\sim 1\,\mathrm{AU}$, and that its obliquity can be $\sim 25°$ with high probability.

Problems and Exercises

Exercise 27 Show that some of the empty peaks of Fig. 13.13 are now filled by newly discovered satellites of the large planets. ∎

13.4.9. *Space debris around the Earth*

The implications of the scale relativity theory seem to be consistent with the observation of many quantized gravitational systems. Now it could be interesting to have not only observational, but also experimental validations of the scale relativity proposal.

One could suggest a gravitational experiment consisting of sending test particles in chaotic motion in a Keplerian gravitational potential. Actually such an experiment has already been done, even though it was not in purpose. Indeed, diffusing space debris in orbit around the Earth provides us with exactly this kind of experiment. A large class of these space debris has diffused through collision process so that the information about their initial orbits is lost. Thus, we can apply to these objects the Keplerian solutions allowed by the theory.

Starting from the best fit value of the fundamental gravitational velocity constant for inner solar system planets and exoplanets, $w_0 = 144.3$ km/s, and with the Earth mass of 5.977×10^{24} kg, we expect average orbiting distance given by: $\langle r \rangle_n = (GM/w_0^2)(n^2 + n/2) = 19.15 \times (n^2 + n/2)$ km. For $n = 18$, we find the Earth radius with a remarkable precision: $GM(n^2 + n/2)/w_0^2 = 6375$ km, while the equatorial radius of the Earth is 6378.160 km (while the next predicted level is 700 km higher). This result, though it needs further theoretical analysis, is theoretically founded for the same reasons as the connection of the Sun radius with the Solar System structures (see above section). Thus the mean distance of the space debris are predicted to be given by the next probability peaks at 718 km ($n = 19$), 1475 km ($n = 20$), 2269 km ($n = 21$), etc.

The available data (< 2000 km) [28] clearly show two density peaks at 850 km and 1475 km (see Fig. 13.14). The second peak is in total accordance with the prediction (however a more complete analysis is still needed to verify that it does not correspond to predetermined orbits). Concerning the first peak, it is necessary to take into account the dynamical braking of the earth atmosphere. This braking deviates particles toward the Earth in a region up to about 700 km. Moreover, the observed distribution should also be corrected for spikes that correspond to identified debris still orbiting about their original orbit. These corrections should be performed before reaching a firm conclusion about this test of the theory. More recent data, when compared to the earlier one, seem to confirm the diffusive character of the space debris motion, which therefore adds weight to the existence of peaks, in particular the 1475 km one, which should have vanished in the absence of a self-organizing underlying process.

Problems and Exercises

Exercise 28 Correct the observed distribution of space debris around the Earth for their interaction with the atmosphere. Compare the first peak of the corrected distribution with the expected peaks at altitude 718 km. ∎

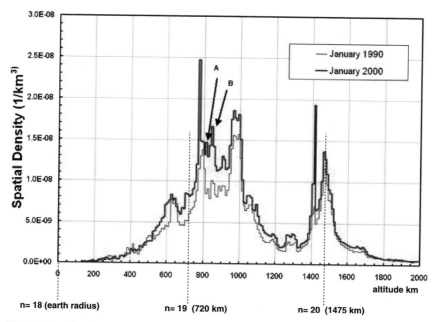

Fig. 13.14: **Density distribution of space debris (adapted from [28]).** Theoretically expected peaks lie at altitudes of 720 km and 1475 km.

Exercise 29 Compare the distribution of space debris around the Earth more distant than 2000 km with the expected distant peaks at 2269 km ($n = 21$), 3101 km ($n = 22$), etc. ∎

13.5. Extrasolar Planetary Systems

13.5.1. *Semi-major axis*

Until recently, one could claim (in 1994, [357]) that

> "one of the difficulty of theories of the Solar System formation and structures is, up to now, its uniqueness: we do not know whether an observed "law" is a peculiar configuration of our own system, or whether it is shared by all planetary systems in the universe. But we can expect such other systems to be discovered in the forthcoming years, and new informations to be obtained about the very distant solar system (Kuiper's belt, Oort cometary cloud...)."

This is now all changed since the discovery by Mayor and Queloz in 1995 of the first exoplanet around a solar-type star [314]. The number of observed extrasolar planetary systems has since regularly increased (more than 400 exoplanets are known at the beginning of 2010). These exoplanets now allow one to put to the test the universality of structures expected form the scale relativity approach to planetary formation [353, 357].

Concerning semi-major axes, the presently observed exoplanets fall in the distance range of the inner and intramercurial solar system (see Fig. 13.15). The n^2 law of the inner solar system can be tested by considering for each exoplanet the variable $4.83\sqrt{a/M}$, where a is the planet semi-major axis (in AU), M the star mass (in solar mass unit, $M_\odot$) and $w_0 = 144/v_{\text{Earth}} = 4.83$, which is the gravitational coupling $c\alpha_g = 144\,\text{km/s}$ in Earth velocity unit. The theoretical expectation is therefore that this variable should cluster around integer values (without any free parameter and performing no fit of the data, since we take the inner solar system constant).

The observed distribution of exoplanets is given in Fig. 13.15 and nicely supports this expectation. Moreover, the statistically significant agreement between the observational data and the inner solar system n^2 law (see Fig. 13.17) has been obtained since the very first observations of exoplanets [361, 3], and then has remained stable while the number and the diversity of exoplanets increased [361, 369, 374, 123, 191, 396, 411]. The probability to obtain such an agreement between the data and the theoretical prediction has also regularly increased and reaches 3×10^{-7} for the last fully studied sample [411] (see Figs. 13.16 and 13.17 for a description of two methods of statistical analysis, which yield consistent results).

A particularly intriguing result concerns the $n = 1$ and $n = 2$ orbitals at 144 and 72 km/s. No large planet is present in our own solar system in these density peaks, as expected from its mass distribution, which is peaked at Jupiter distance to the Sun, but we have given arguments according to which they may reveal themselves by transient dust and a possible asteroid ring, which remains to be discovered directly. As regards extrasolar systems, a large number of exoplanets have been found in these peaks, in particular in the $n = 1$ peak, which is the dominant one. The proximity of these exoplanets to their star (the so-called 51-Peg-like exoplanets) is a puzzle for standard theories of formation, while, in the scale relativity framework, these planets orbit preferentially around $a/M = 0.043$ or $0.17\,\text{AU/M}_\odot$.

Statistical significance of the periodicity

Let us give additional details about the determination of the statistical significance of the above results. We look for clustering around integer values of the ratio $144/v$, i.e. equivalently (from Kepler's third law), of $\tilde{n} = 4.83(a/M)^{1/2} = 4.83(P/M)^{1/3}$. In other words, we expect a periodicity

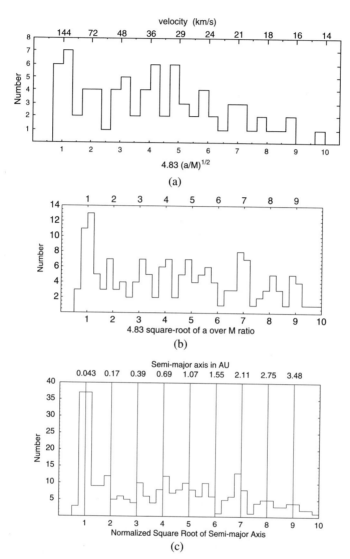

Fig. 13.15: **Distribution of exoplanet semi-major axes.** The figures compare this distribution with the predicted peaks (integer values of the variable $\tilde{n} = 4.83(a/M)^{1/2} = 4.83(P/M)^{1/3}$, where a is the semi-major axis in AU, P the period in year and M the mass of the central star in $M_\odot$). Mercury ($v = 47.9\,\mathrm{km/s}$), Venus ($v = 35.0\,\mathrm{km/s}$), the Earth ($v = 29.8\,\mathrm{km/s}$) and Mars ($v = 24.1\,\mathrm{km/s}$) lie respectively in the peaks $n = 3, 4, 5$ and 6. (a): year 2003 data (probability 10^{-4} to obtain by chance such a distribution). (b): year 2005 data (probability 4×10^{-5}). (c): year 2008 data (probability 4×10^{-7}, see Figs. 13.16 and 13.17). Data taken from [476].

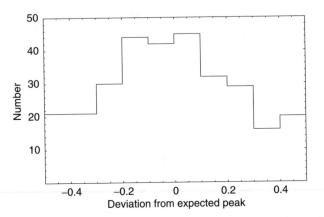

Fig. 13.16: **Folding of exoplanet semi-major axis distribution (2008 data).** The figure shows the histogram of the difference betwen $\tilde{n} = 4.83(a/M)^{1/2}$ (where a is the semi-major axis in AU and M the mass of the central star in $M_{\odot}$) and the nearest integer (predicted value of the probability peaks). In the absence of effect this distribution would be nearly flat. On the contrary, 193 planets out of 301 fall in the region $[-0.25, +0.25]$ and only 108 outside. The probability to obtain by chance such a deviation with the 50–50% distribution expected in the random case is found to be 4×10^{-7}.

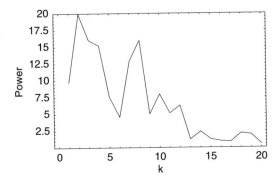

Fig. 13.17: **Power spectrum of exoplanet semi-major axis distribution.** We have taken all values of the variable $\tilde{n} = 4.83(a/M)^{1/2}$ (where a is the semi-major axis of the planet orbit in AU and M the mass of the central star in $M_{\odot}$), ranging between 0.7 and 8.7 (276 planets, 2008 data). Their power spectrum shows a highly significant peak at $k = 8$ with a power $p = 16.0$, therefore corresponding to the expected periodicity $\Delta\tilde{n} = 1$. The probability to obtain such a peak of power $p_0 = 16$ for a previously expected value of the period (that previously observed in the inner solar system) is $e^{-p_0} = 3 \times 10^{-7}$, corresponding to a 5σ statistical significance. This estimate supports the probability obtained with the folding method (Fig. 13.16).

with period $\Delta \tilde{n} = 1$ of this variable. The value $w_0 = 144\,\text{km/s}$ is taken from independent results (e.g. extragalactic data on binary galaxies, see Sec. 13.8.4). The period P of the planet orbit is the primarily measured quantity (most of the time from the induced motion of the star, through its radial velocity, or directly through observation of transit in front of the star). The quantity M is the mass of the star, which is now known with an uncertainty of about 5% for most of the stars [476] and is generally less than 10%.

A first method of statistical analysis of the periodicity consists in folding the distribution of $\tilde{n}$ on the expected period. This amounts to constructing the histogram of the difference $\delta n = \tilde{n} - n$ with respect to the closest integer [361, 374]. This results in Fig. 13.16 (for the 2008 data). Then we separate the δn domain into two equal intervals respectively of high expected probability, $[-0.25, +0.25]$ and low expected probability, $[0.25, 0.75]$. Among 301 points, 193 fall in the interval $[-0.25, +0.25]$ and 108 in $[0.25, 0.75]$, while an equal number of points would be expected in the absence of periodicity. The advantage of this method (based on the smallness of the period we are looking for) is that the existing biases in the exoplanet distribution (and also their real large scale distribution) affects very few the result and its significance. Indeed, the distribution obtained in Fig. 13.15 (bottom) contains many exoplanets recently discovered by the transit method [476], which are therefore preferentially close to their star (for the transit to be observable). This increases the relative number of planets with small distances to their star (51-Peg-like objects) respectively to the top and middle Fig. 13.15 in which the detection method was nearly exclusively radial velocity. This bias affects the overall distribution, in particular concerning the ratio of inner solar system-like planets (corresponding to $n = 3$ to 9) over intra-mercurial-like exoplanets ($n = 1$ and 2). But it does not explain why the exoplanets close to their star cluster in a very narrow way around a semi-major axis of $0.043\,\text{AU/M}_\odot$ ($n = 1$). Since we look for very small scale fluctuations in $\sqrt{a/M}$, the larger scale fluctuations due to biases and to the real distribution (seemingly showing an intramercurial peak and an inner solar system-like peak, as also supported by the significant power obtained for $k = 2$ in the power spectrum) cancel out each other in the folded distribution.

The probability that, for 301 trials, 108 events (or less) fall in a $1/2$ large interval and 193 in the complementary one is $p = \sum_{i=0}^{i=108} (301, i) \times 2^{-301} \approx 4 \times 10^{-7}$, where $(301, i)$ denotes a binomial coefficient. Therefore we can exclude at better than the 5σ level of statistical significance that such a result be obtained by chance. The power spectrum performed in Fig. 13.17 confirms this estimate.

Error analysis

One of the still open questions is that of the natural dispersion of exoplanets around the quantized values of $\sqrt{a/M} = (P/M)^{1/3}$, or, in other words, of the natural width of the probability peaks. This width is mixed with the various measurement uncertainties and systematic errors, and it can therefore be recovered from an error analysis.

The main source of error is about the parent star mass. The relative error is now less than $\approx \pm 10\%$ (reaching 5% for most stars), so that the uncertainty on $\tilde{n}$, which is proportional to $M^{-1/3}$, is less than $\approx 0.033n$. It is only $0.017n$ for most stars, and since n remains smaller than 10 (see Fig. 13.15), the measurement error on $\tilde{n}$ remains smaller than ≈ 0.17 (while looking for peaks separated by $\Delta\tilde{n} = 1$). This results therefore supports the observability of the peaks, even for the most distant planets to their star.

The inner Solar System, for which the planet positions are known with precision, and the exoplanet $n = 1$ orbital, for which the contribution of the error on the star mass remains small, allow us to obtain an observational estimate of the intrinsic dispersion on $\tilde{n}$. We get for the inner Solar System $\sigma = 0.121$ (7 objects), a value, which is confirmed by the $n = 1$ exoplanets, for which $\sigma = 0.124$ [374]. The correction of the other peaks for the measurement error yield the same result.

This value of the peak width of about 0.125 could be compared in the future with possible theoretical estimates, which remains to be done. The width should depend on the various conditions and perturbations that intervene during the accretion process and later, during the system evolution. In the case of the system of three small mass planets around the pulsar PSR B1257+12 [361], which can be shown to have remained stable since its formation (see Sec. 13.5.3), the width can be estimated from the quantum-type approach and conservation relations and takes a very small value.

Substructures at intramercurial distances

In the solar system, we have seen that there are hints of substructures at intramercurial distances $a < 0.4$ AU (see Sec. 13.4.5). These structures are organized according to a constant $w_\odot \sim 432 \, \text{km/s} = 3 \times 144 \, \text{km/s}$,

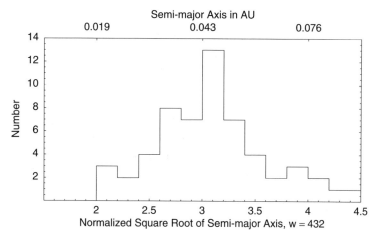

Fig. 13.18: **Substructures in exoplanet semi-major axis distribution.** The figure shows the histogram of the variable $\tilde{n} = 14.5 \times (a/M)^{1/2}$, which corresponds to a gravitational coupling constant $w_{\odot} = 432$ km/s (2008 data, error on $\tilde{n}$ smaller than 0.25). The main peak at 0.043 AU/M$_{\odot}$ lies now at $n = 3$ in this hierarchical sub-level. Expected new peaks in $n = 2$ and 4 [374] may begin to emerge.

which is nothing but the Keplerian velocity at the distance of the Sun radius (and more generally, the mean Keplerian velocity around solar-type stars [290]). The extrasolar planetary distribution can be also looked for such substructures, in particular for the existence of planets still closer to their star than the 51-Peg planets. In terms of the constant $w_{\odot}$, the 51-Peg planets at 0.043 AU/M$_{\odot}$ (corresponding to a Keplerian velocity of 144 km/s) now rank $n_{432} = 3$, so that one may look for probability peaks at $n_{432} = 2$, corresponding to a Keplerian velocity of 216 km/s [374]. The Fig. 13.18 shows a zoom on the short distance region of Fig. 13.15, which indicates the possible emergence in the data of such substructures.

Problems and Exercises

Exercise 30 Update the analysis of the semi-major axis distribution of exoplanets using recent data. Analyse the various biases in the extrasolar planet samples and show that the expected probability peaks in $\sqrt{a/M}$ are stable and statistically significant whatever the subsamples and methods of detection (mainly, radial velocity and transit). ∎

13.5.2. *Eccentricity*

The eccentricity distribution of the exoplanets can be studied with respect to the general Keplerian eccentricity solution, $e = k/n$ (see Sec. 13.3.1.1). The eccentricity distribution of the global sample (combining all the exoplanets and the inner solar system bodies) agrees with the predicted quantized distribution around integers $k = n \times e$ (see Fig. 13.19). The associated probability level (calculated by the folding method around the expected period) is 10^{-4} for the 2005 data [379, 191, 396]. When combining the eccentricity and semi-major axis distribution, the probability to find such a distribution by chance becomes smaller than 10^{-7} (see Fig. 13.20).

The discovery of the large eccentricities of many exoplanets was the second puzzle posed to standard models of formation. Indeed, in these models the accretion proceeds through sweeping, and the resulting orbits are quasi circular. On the contrary, in the scale relativity framework leading to a Schrödinger-like form of the energy equation, elliptical wave packet solutions of this equation can be obtained [329, 55]. They describe an elliptical disk of planetesimals, which localize and concentrate (as a quantum process helped and accelerated by self-gravitation) to finally form a protoplanet on a classical orbit having the eccentricity of the initial disk (given in the quantum-like regime by the Runge–Lenz invariant vector).

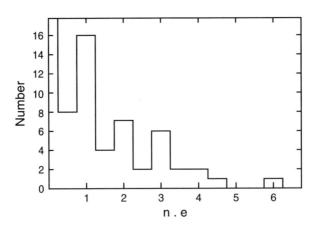

Fig. 13.19:   **Distribution of exoplanet eccentricities.** The figure shows the histogram of the product $n \times e$, i.e. of $4.83 \times e \times (a/M)^{1/2}$, where e is the planet orbit eccentricity, a its semi-major axis and M the star mass. We expect this variable to cluster around integers. This is supported by the observational data in a statistically highly significant way (2005 data).

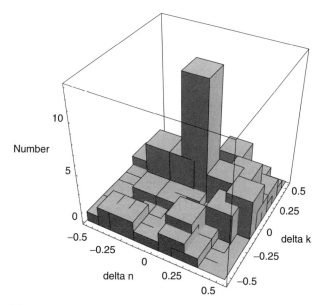

Fig. 13.20: **Exoplanet semi-major axes and eccentricities.** The figure shows the two-dimensional histogram of the two differences ($\delta n = \tilde{n} - n, \delta k = \tilde{k} - k$), where $\tilde{n} = 4.83(a/M)^{1/2}$, $\tilde{k} = e \times \tilde{n}$ and n and k are the nearest integers. In these expressions, a is the semi-major axis of the planet orbit, e its eccentricity, and M the star mass. In the scale relativity model of planetary formation, we expect these differences to cluster around the origin. This is supported by the observational data in a statistically highly significant way ($P \approx 10^{-7}$, 2005 data, subsample with errors on $\tilde{n}$ and $\tilde{k}$ smaller than 0.25).

The eccentricity may finally range from $e = 0$ to $e = 1 - 1/n$. Therefore only circular orbits are permitted on the fundamental level $n = 1$, but increasingly eccentric orbits are possible when n increases.

Problems and Exercises

Exercise 31 Update the analysis of the eccentricity distribution of exoplanets with recent data. Show that the eccentricity maximal value is $e_n = 1 - 1/n$ for a given value of the semi-major axis rank n.

Hint: We expect the eccentricity to show peaks of probability for values $e = k/n$, where k (which takes values from 0 to $n - 1$) and n, the rank of the semi-major axis distribution (peaking for $4.83(P/M)^{1/3}$), are both integers. Therefore, putting to the test the quantization of the probability peaks of the variable $\tilde{k} = 4.83 \times e \times (P/M)^{1/3}$ involves three measured quantities, the planet orbit eccentricity e, its period P and the star mass M.

While the relative error on the period is usually small and that on the mass is about 5 to 10%, the relative observational error on eccentricities may be larger [476]. The expected structure can then be studied only by selecting a subsample of exoplanets for which the error on $\tilde{k}$ is small enough (we have taken a limit of 0.25 in Fig. 13.20). ∎

Exercise 32 Develop in more details the model of planetary formation in terms of gravitational wave packets on possible elliptical orbits, which are solutions of the macroscopic Schrödinger equation in a Kepler potential (including self-gravitation of the disk). ∎

13.5.3. *Planetary system around the pulsar PSR 1257+12*

13.5.3.1. *Precision structuring of PSR 1257+12 planetary system*

The first extrasolar planetary system ever discovered has been the system of three planets found by Wolszczan around the pulsar PSR B1257+12 [543, 544]. Even though this star is not solar-like, this system deserves a special study [361, 366, 397], because (i) the pulse timing measurements allow a very precise determination of the orbital elements of the planets and (ii) their small masses (four Earth mass for two of them and the Moon mass for the third one) allow both the celestial mechanics models of evolution and the scale relativity model of formation to become also very precise. This system therefore stands out as a kind of ideal gravitational laboratory for studying the formation and evolution of planetary systems and putting models to the test.

The planets probably result from an accretion disk formed around the very compact star after the supernova explosion. Even though this is a secondary process, there is general agreement that the formation process should be similar to the standard picture [121].

We can therefore expect the purely gravitational formation process, described in the scale relativity approach by a macroscopic gravitational Schrödinger equation, to be still valid in this case. Moreover, the smallness of the planet masses implies very few perturbations between the subdisks from which the three planets have been formed and a negligible effect of self-gravitation, so that the theoretical predictions (based on conservation laws, in particular of the center of gravity of each subdisk) are expected to become very precise. The compacity of the star also suggests that planetary orbits be self-organized in terms of a smaller scale than the inner solar system (i.e. of a multiple of $w_0 = 144\,\mathrm{km/s}$).

On the one hand, the precision of pulse timing measurements has allowed to use the PSR B1257+12 system as a highly accurate probe of planetary dynamics. Indeed, it has been pointed out that the near $3:2$ resonance between the orbits of the two main planets should lead to precisely predictable and observable mutual gravitational perturbations [460, 302, 461, 303]. These non-Keplerian gravitational effects have been soon detected [544], and they have now been observed with high precision, yielding an irrefutable confirmation of the existence of planets around the pulsar and allowing a determination of the true masses and of the orbital inclinations of the planets [253].

On the other hand, the observed distances of the planets to the pulsar can be shown [361, 366] to be in very good agreement with the scale relativity theoretical expectations. To the first approximation, we expect the semi-major axes to show peaks of probability scaling as n^2, where n is integer. But, when perturbations to the simple one-object Kepler potential model are small and the trajectories are nearly circular (which is the case here, the present excentricities of the three planets being $e_A = 0$, $e_B = 0.0186 \pm 0.0002$ and $e_C = 0.0252 \pm 0.0002$ [253]), one may use the center of gravity conservation law to refine the theoretical expectation and state that, after the formation process the planets should lie at a distance given by the center of mass of the planetesimal wave packet, which scales as $n^2 + n/2$, i.e. $n^2(1 + 2/n)$ [188, 56]. This means a "correction" of 6 to 10% with respect to the simple n^2 law for the pulsar planets.

Already with the 1994 values of the orbital elements obtained by Wolszczan [544] from four years of observation, the agreement was excellent [361, 366]. The three observed planets A, B and C are indeed found to correspond to ranks $n = 5$, 7 and 8 with a high relative precision. With the n^2 law the agreement is already at the level of 1% and 0.3%. One finds:

$$(P_B/P_C)^{1/3} = 0.878 \text{ to be compared with } 7/8 = 0.875,$$

$$(P_A/P_C)^{1/3} = 0.636 \text{ to be compared with } 5/8 = 0.625.$$

With the $n^2 + n/2$ law, it becomes (see Fig. 13.21):

$$(P_B/P_C)^{1/3} = 0.8783 \text{ to be compared with } (52.5/68)^{1/2} = 0.8787,$$

$$(P_A/P_C)^{1/3} = 0.6366 \text{ to be compared with } (27.5/68)^{1/2} = 0.6359.$$

A factor of about 10 has been gained with this conservative law. The relative agreement between observation and theoretical expectation has become respectively 4×10^{-4} and 1.1×10^{-3} for the B/C and A/C ratios.

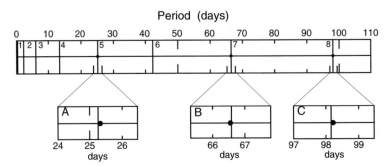

Fig. 13.21: **Periods of pulsar planets (1994 data).** The agreement between the observed periods and the predicted ones is so precise that we have made three zooms by a factor of 10 in order to show the differences (less than three hours for periods of several months).

Today the comparison between the data and the theoretical predictions can be considered again, since the determination of the orbital elements of the three planets has been greatly improved thanks to the increase to 13 years of the observing time and to the account of gravitational mutual effects [253]. This improvement of the data and of their fitting model has led to a change by 5σ of some of the derived orbital elements. This could have degraded the agreement with the scale relativity model. On the contrary, it has yielded a new improvement by a factor of about 20 on the A/C ratio. One now finds, for $P_A = (25.262 \pm 0.003)$ days, $P_B = (66.5419 \pm 0.0001)$ days and $P_C = (98.2114 \pm 0.0002)$ days [253]:

$$(P_B/P_C)^{1/3} = 0.8783 \text{ to be compared with } (52.5/68)^{1/2} = 0.8787,$$

$$(P_A/P_C)^{1/3} = 0.63597 \text{ to be compared with } (27.5/68)^{1/2} = 0.63593.$$

The relative agreement is now at the level of respectively 4×10^{-4} and 6×10^{-5} for the B/C and A/C ratios. The probability to find such a close agreement, accounting for all other possible fractional ratios with n up to 10, can be estimated to be less than 10^{-4} [366]. Moreover, using the standard pulsar mass $M = (1.4 \pm 0.1)M_\odot$ in the relation

$$P_n = \frac{2\pi GM}{w^3}(n^2 + n/2)^{3/2}, \tag{13.87}$$

one obtains for the coupling constant of this system $w = (2.96 \pm 0.07) \times 144\,\text{km/s}$, which is consistent with an expected integer multiple of the inner solar system constant w_0 by a factor of 3 and with the value of the Keplerian velocity $w_\odot = 435\,\text{km/s}$ at the Sun radius.

The precision of these results suggests to attempt improving the model, both concerning the formation as the evolution of the system. Indeed, the presently observed orbital elements result from the formation era followed by about one billion years of evolution, while the theoretical expectations correspond to the end of the formation. The comparison between the observational and theoretical orbital elements should therefore take the system evolution into account.

13.5.3.2. *Heliumoid model of the PSR1257+12 planetary system formation*

Let us address the problem of the formation of two planets from two subdisks in the scale relativity approach. The question to be solved is whether the self-gravitation and the mutual interaction of the subdisks leads to non negligible corrections to the previous one-body model.

Consider a star surrounded by a disk of planetesimals, which can itself be separated in two parts, (B) and (C). The central star is of mass M and the individual planetesimals are assumed to have equal mass μ. The total Hamiltonian reads

$$\hat{H} = \sum_{B,C} \left(-\frac{\tilde{\hbar}^2}{2\mu}\Delta - \frac{M\mu}{r} - \sum_B \frac{\mu^2}{r_{12}} - \sum_C \frac{\mu^2}{r_{12}} - \sum_{BC} \frac{\mu^2}{r_{12}} \right), \qquad (13.88)$$

where $\tilde{\hbar} = 2\mu\mathcal{D}$, with $\mathcal{D} = GM/2w_0$.

The three last terms are respectively the self potentials of the B and C rings and the potential of interaction between B and C, r_{12} being the interdistance between two planetesimals. Assume that the total number of planetesimals in B and C is respectively N_B and N_C. The masses of the B and C rings, and therefore of the planets B and C which will finally be formed from these rings, are

$$m_B = N_B\mu, \quad m_C = N_C\mu. \qquad (13.89)$$

The first two terms of the Hamiltonian then become

$$\sum_{B,C} \left(-\frac{\tilde{\hbar}^2}{2\mu}\Delta - \frac{M\mu}{r} \right) = -\frac{\tilde{\hbar}^2}{2\mu^2} (m_B\Delta_B + m_C\Delta_C) - \frac{Mm_B}{r_B} - \frac{Mm_C}{r_C}. \qquad (13.90)$$

The distribution of planetesimals in the rings B and C are given by solutions of the gravitational Schrödinger equation

$$\hat{H}\psi = E\psi, \qquad (13.91)$$

where $\hat{H}$ is given by Eq. (13.88). In the first approximation, since $m_B/M \ll 1$ and $m_C/M \ll 1$ (the ratios are $\approx 10^{-5}$ for planets B and C of the PSR1257+12 system and $\approx 10^{-7}$ for planet A), the potential is strongly dominated by the central body, and ψ_B and ψ_C are solutions of the equation

$$2\mathcal{D}^2 \Delta\psi + \left(\frac{E}{\mu} + \frac{M}{r} \right) \psi = 0. \tag{13.92}$$

In the second approximation, we can now use these solutions to derive the probability number density of planetesimals in the disks, $\rho_B = |\psi_B|^2$ and $\rho_C = |\psi_C|^2$. The self potential of ring B can now be written as a sum of the planetesimal interaction gravitational energy over all couples of planetesimals in this ring, namely,

$$-\sum_{i,j} \frac{\mu^2}{r_{12}} = -\sum_i N_B \mu^2 \int_B \frac{|\psi_B|^2}{r_{12}} dV_B = -m_B^2 I_{BB}. \tag{13.93}$$

This sum has therefore been reduced to the integral

$$I_{BB} = \int_B \frac{|\psi_{B1}|^2 |\psi_{B2}|^2}{r_{12}} dV_1 dV_2. \tag{13.94}$$

Similar calculations can be made for the self-potential of C and for the BC interaction term. Namely, we define the integrals

$$I_{CC} = \int_C \frac{|\psi_{C1}|^2 |\psi_{C2}|^2}{r_{12}} dV_1 dV_2, \tag{13.95}$$

$$I_{BC} = \int_{BC} \frac{|\psi_{B1}|^2 |\psi_{C2}|^2}{r_{12}} dV_1 dV_2. \tag{13.96}$$

Then the Hamiltonian now takes the form

$$\hat{H} = -\frac{\tilde{\hbar}^2}{2\mu^2}(m_B \Delta_B + m_C \Delta_C) - \frac{M m_B}{r_B}$$

$$- \frac{M m_C}{r_C} - m_B^2 I_{BB} - m_C^2 I_{CC} - m_B m_C I_{BC}. \tag{13.97}$$

The three last terms therefore also give the correction to the total energy of the system due to self-gravitation and to interactions of the two rings, which reads

$$E_{\text{tot}} = -\frac{1}{2}w_0^2\left(\frac{m_B}{n_B^2} + \frac{m_C}{n_C^2} + \frac{2m_B^2}{M}I_{BB} + \frac{2m_C^2}{M}I_{CC} + \frac{2m_B m_C}{M}I_{BC}\right),$$
(13.98)

where n_B and n_C are the main quantum numbers of the B and C wave functions (namely, $n_B = 7$ and $n_C = 8$ for the PSR 1257+12 system), and where the I integrals are now dimensionless (namely, the lengths are expressed in gravitational "Bohr" units, $r_{\text{Bohr}} = GM/w_0^2$).

To be complete, one adds the third planet A to the description, whose mass is yet ≈ 200 times smaller than that of planets B and C, so that the self-potential correction to its energy is negligible, while the interactions given by the integrals I_{AB} and I_{AC} may be relevant.

Finally the relative correction on the energy of the three planets is found to be

$$\frac{\Delta E_A}{E_A} = 2n_A^2\left(\frac{n_B^2}{n_A^2 + n_B^2}\frac{m_B}{M}I_{AB} + \frac{n_C^2}{n_A^2 + n_C^2}\frac{m_C}{M}I_{AC}\right), \quad (13.99)$$

$$\frac{\Delta E_B}{E_B} = 2n_B^2\left(\frac{m_B}{M}I_{BB} + \frac{n_C^2}{n_B^2 + n_C^2}\frac{m_C}{M}I_{BC}\right), \quad (13.100)$$

$$\frac{\Delta E_C}{E_C} = 2n_C^2\left(\frac{m_C}{M}I_{CC} + \frac{n_B^2}{n_B^2 + n_C^2}\frac{m_B}{M}I_{BC}\right), \quad (13.101)$$

where we have attributed the interaction contribution to each planet according to its energy, and where the self-gravity of planet A and its effects on planets B and C have been neglected.

The integrals I_{BB}, I_{CC} and I_{BC} are exactly those which are encountered in the helium and heliumoide quantum problem (see, e.g. [267]). Their numerical calculation for $n_A = 5$, $n_B = 7$ and $n_C = 8$, which are the values of the principal "graviquantum" number of the three planets of the PSR1257+12 system, and $l = n - 1$ (quasi circularity), yields:

$$I_{77} = 0.0173582, \quad I_{88} = 0.0134383, \quad I_{78} = 0.0221588,$$

$$I_{57} = 0.0377702, \quad I_{58} = 0.0307991.$$

Since the dimensionless energies are $E_n = -1/2n^2$, the perturbation can be expressed in terms of a correction on the expected main quantum numbers, $\delta n/n = -(1/2)\Delta E/E$. Using the measured values of the planet masses, $m_A = 0.020 \pm 0.002$, $m_B = 4.3 \pm 0.2$ and $m_C = 3.9 \pm 0.2$ Earth mass [253] and the standard pulsar mass of 1.4 $M_\odot$, one finds very small corrections,

$$\delta n_A = -0.000052, \quad \delta n_B = -0.000090, \quad \delta n_C = -0.000102.$$

However, in the absence of a precise knowledge of the mass of the pulsar, only ratios of quantum numbers (or equivalently, of periods, semi-major axes or energy) can be compared. The above corrections can then be expressed in another way. We fix as reference $n_C = 8$, then the expected values of the two other planets become: $(n_A)_{\mathrm{pred}} = 5.000012$, while the observed value is $(n_A)_{\mathrm{obs}} = 5.00029 \pm 0.00020$, and $(n_B)_{\mathrm{pred}} = 7.000000$, while the observed value is $(n_A)_{\mathrm{obs}} = 6.996987 \pm 0.000008$.

As expected at the beginning of this study from the smallness of the planet masses, the corrections are negligible for this system, since they remain smaller than the observational errors despite the precision of the data. However, a new possible improvement of the determination of the orbital elements in the future may render them meaningful. Moreover, all this calculation is also relevant for multiple systems with larger planetary masses, in particular Jupiter-like planets, in which case the corrections are no longer negligible.

Problems and Exercises

Exercise 33 Apply the above heliumoid model of planetary system formation to the known multiple planetary systems with large planetary masses (several Jupiter masses), in order to estimate the self-gravity and interaction corrections to be applied to the values of the expected probability peaks.   ∎

13.5.3.3. *Evolution of a 3:2 near resonant two-planet system*

Three-body perturbation theory

The general three-body perturbation equations (see Brouwer & Clemence [75]) in the Lagrange–Laplace theory have been applied by Malhotra [303] to the PSR 1257+12 planetary system. The gravitational interactions of low mass planets, such as the PSR 1257+12 planets, which are $\approx 10^5$ times smaller than the pulsar mass, can be analysed in terms of osculating ellipses, i.e. of orbits that are instantaneously elliptical, but whose orbital parameters are time dependent. For a 3 : 2 near commensurability between two planets (1) and (2), identified respectively with planet B and C of the

PSR 1257+12 planetary system, and assuming coplanar orbits, Malhotra found that the principal perturbation components in the interaction Hamiltonian are given by

$$\mathcal{H}' = -\frac{Gm_1 m_2}{a_2} \left\{ [\mathcal{P}(\psi, \alpha) - \alpha \cos \psi] + \frac{1}{2} A_1(\alpha)(e_1^2 + e_2^2) \right.$$
$$- A_2(\alpha) e_1 e_2 \cos(\omega_1 - \omega_2) + C_1(\alpha) e_1 \cos(\phi - \omega_1)$$
$$\left. + C_2(\alpha) e_2 \cos(\phi - \omega_2) \right\}, \tag{13.102}$$

where the planet masses are $m_1 \ll M_*$ and $m_2 \ll M_*$ (M_* being the central star mass), a_1 and a_2 are the osculating orbital semi-major axes, e_1 and e_2 the osculating eccentricities, $n_1 = 2\pi/P_1$ and $n_2 = 2\pi/P_2$ the mean orbital frequencies, and ω_1 and ω_2 the longitudes of periastron, and where (keeping Malhotra's notations)

$$\alpha = a_1/a_2,$$
$$\psi = \lambda_1 - \lambda_2,$$
$$\phi = 3\lambda_2 - 2\lambda_1,$$
$$\mathcal{P}(\psi, \alpha) = (1 - 2\alpha \cos \psi + \alpha^2)^{-1/2},$$
$$A_1(\alpha) = +\frac{1}{4} \alpha b_{3/2}^{(1)}(\alpha) = +2.50,$$
$$A_2(\alpha) = -\frac{1}{4} \alpha b_{3/2}^{(2)}(\alpha) = -2.19,$$
$$C_1(\alpha) = -\frac{1}{2} \left(6 + \alpha \frac{d}{d\alpha} \right) b_{1/2}^{(3)}(\alpha) = -2.13,$$
$$C_2(\alpha) = +\frac{1}{2} \left(5 + \alpha \frac{d}{d\alpha} \right) b_{1/2}^{(2)}(\alpha) = +2.59. \tag{13.103}$$

The λ_j are the mean longitudes. The $b(\alpha)$ are Laplace coefficients. The numerical values of the A_j and C_j coefficients are those, which correspond to the value $\alpha = a_{01}/a_{02} = 0.771416(2)$ observed for the B and C planet ratio in the PSR 1257+12 system [253].

Three types of significant perturbations are apparent in this Hamiltonian:

(i) those related to conjonctions of the planets, described by the term $[\mathcal{P}(\psi, \alpha) - \alpha \cos \psi]$,

(ii) secular effects responsible for the slow precession of the absides, described by the terms with coefficients A_1 and A_2, and

(iii) the effects of the $3:2$ near-commensurability of the mean motions, described by the remaining terms.

Malhotra wrote the first time derivative of the orbital elements as:

$$\frac{\dot{a}_1}{a_1} = +2\frac{m_2}{M_*}n_1\alpha\left[\frac{\partial}{\partial\psi}\mathcal{P}(\psi,\alpha) + \alpha\sin\psi + 2C_1(\alpha)e_1\sin(\phi - w_1)\right.$$

$$\left. + 2C_2(\alpha)e_2\sin(\phi - w_2)\right] \tag{13.104}$$

$$\frac{\dot{a}_2}{a_2} = -2\frac{m_1}{M_*}n_2\alpha\left[\frac{\partial}{\partial\psi}\mathcal{P}(\psi,\alpha) + \alpha\sin\psi + 3C_1(\alpha)e_1\sin(\phi - w_1)\right.$$

$$\left. + 3C_2(\alpha)e_2\sin(\phi - w_2)\right]. \tag{13.105}$$

Then, using the variables

$$h_j = e_j\sin w_j, \quad k_j = e_j\cos w_j, \tag{13.106}$$

for the two planets $j = 1, 2 = B, C$, the variations of the eccentricities and pericenters read

$$\dot{h}_1 = A_{11}k_1 + A_{12}k_2 + B_1\cos\phi,$$

$$\dot{h}_2 = A_{21}k_1 + A_{22}k_2 + B_2\cos\phi,$$

$$\dot{k}_1 = -A_{11}h_1 - A_{12}h_2 - B_1\sin\phi,$$

$$\dot{k}_2 = -A_{21}h_1 - A_{22}h_2 - B_2\sin\phi. \tag{13.107}$$

Setting

$$\mu_1 = \frac{m_1}{M_*}, \quad \mu_2 = \frac{m_2}{M_*}, \tag{13.108}$$

the coefficients of the matrix A_{ij} and the vector coefficients B_j read [303]

$$A_{11} = \mu_2 n_1\alpha A_1(\alpha), \quad A_{12} = \mu_2 n_1\alpha A_2(\alpha),$$

$$A_{21} = \mu_1 n_2 A_2(\alpha), \quad A_{22} = \mu_1 n_2 A_1(\alpha),$$

$$B_1 = \mu_2 n_1\alpha C_1(\alpha), \quad B_2 = \mu_1 n_2 C_2(\alpha). \tag{13.109}$$

Analytic solution

A partial analytical solution for this system of equations had been found by Rasio *et al.* [461] in a simplified case. They have treated the inner planet as a massless "test particle" perturbed by the outer planet, which was assumed to move on an unperturbed Keplerian orbit. To this approximation and neglecting the conjunctions (close encounters), they found that the system can be integrated in terms of a period-eccentricity relation that reads, for small eccentricities,

$$n_1 = n_*(1 + 3e_1^2), \tag{13.110}$$

where n_* is the value of the frequency for which $e_1 = 0$. However, this result holds only when $m_2 \gg m_1$, a condition which does not apply to the PSR 1257+12 system for which $m_1 \approx m_2$. In order to find a general analytical solution for the Malhotra system of equations, let us define a function that characterizes the conjunction effects on the semi-major axes variation,

$$\Gamma_c = \frac{\partial}{\partial \psi} \mathcal{P}(\psi, \alpha) + \alpha \sin \psi, \tag{13.111}$$

and a function that depends on eccentricities, namely,

$$\Gamma = C_1(\alpha)e_1 \sin(\phi - \omega_1) + C_2(\alpha)e_2 \sin(\phi - \omega_2). \tag{13.112}$$

In terms of these quantities, the semi-major axes equations now read

$$\dot{a}_1 = +2\mu_2 n_1 a_1 \alpha (\Gamma_c - 2\Gamma),$$
$$\dot{a}_2 = -2\mu_1 n_2 a_2 (\Gamma_c - 3\Gamma). \tag{13.113}$$

Therefore the two equations can be combined into a relation, which no longer depends on the eccentricities,

$$\frac{\dot{a}_1}{4\mu_2 n_1 a_1 \alpha} + \frac{\dot{a}_2}{6\mu_1 n_2 a_2} = \frac{1}{6}\Gamma_c. \tag{13.114}$$

We can now use Kepler's third law for the osculating orbits to write

$$n_1 a_1 \alpha = (GM_*)^{1/2} a_1^{1/2} a_2^{-1}, \quad n_2 a_2 = (GM_*)^{1/2} a_2^{-1/2}, \tag{13.115}$$

so that the above relation becomes

$$M_*^{1/2}\left(\frac{1}{2m_2}a_1^{-1/2}\dot{a}_1 + \frac{1}{3m_1}a_2^{-1/2}\dot{a}_2\right) = \frac{\Gamma_c}{3a_2}. \tag{13.116}$$

Prime integrals of the equations of motion

Under the approximation where the conjunction term is neglected, it leads
to a prime integral of the motion,

$$3m_1 a_1^{1/2} + 2m_2 a_2^{1/2} = Q_a, \tag{13.117}$$

where $Q_a = 3m_1 a_{01}^{1/2} + 2m_2 a_{02}^{1/2} = $ cst. This analytical result shows that
the main oscillations of the two planet orbits are strongly coupled and
are always in opposition. This effect agrees with the numerical integration
of Wolszczan [544, Fig. 2]. If we now take the conjunction term (label c)
into account, we can neglect the variation of a_2 in the right-hand side of
Eq. (13.116), and we obtain the integral

$$3\mu_1 a_1^{1/2} + 2\mu_2 a_2^{1/2} \left(1 - \frac{1}{2}\mu_1 \frac{2\pi}{P_2} \int \Gamma_c(t) dt\right) = Q_a. \tag{13.118}$$

For the system PSR1257+12, we have $\mu_1 \approx 10^{-5}$ in the correction term,
$2\pi/P_2 = 23.37 \, \mathrm{yr}^{-1}$, while the integral of Γ_c shows two components of
amplitudes 0.05 and 0.25 yrs, so that the correction to the fluctuations of
semi-major axes (which are themselves very small, of the order of 3×10^{-4})
remains smaller than 4×10^{-5} in proportion. Therefore the prime integral
of the motion given by Eq. (13.117) is valid to a very good approximation.

Let us now consider the eccentricity equations. From the relations $e_1^2 = h_1^2 + k_1^2$ and $e_2^2 = h_2^2 + k_2^2$, we derive

$$e_1 \dot{e}_1 = h_1 \dot{h}_1 + k_1 \dot{k}_1, \quad e_2 \dot{e}_2 = h_2 \dot{h}_2 + k_2 \dot{k}_2. \tag{13.119}$$

From Eqs. (13.107) we obtain the expressions

$$e_1 \dot{e}_1 = A_{12}(h_1 k_2 - k_1 h_2) + B_1(h_1 \cos\phi - k_1 \sin\phi),$$

$$e_2 \dot{e}_2 = A_{21}(h_2 k_1 - k_2 h_1) + B_2(h_2 \cos\phi - k_2 \sin\phi). \tag{13.120}$$

From the very definition of h_j and k_j, we find

$$h_1 k_2 - h_2 k_1 = e_1 e_2 \sin(\omega_1 - \omega_2), \tag{13.121}$$

so that the derivative of the eccentricities finally read

$$\dot{e}_1 = A_{12} e_2 \sin(\omega_1 - \omega_2) + B_1 \sin(\omega_1 - \phi),$$

$$\dot{e}_2 = -A_{21} e_1 \sin(\omega_1 - \omega_2) + B_2 \sin(\omega_2 - \phi). \tag{13.122}$$

We recognize in these expressions two contributions of respectively free
and forced oscillations (see [303, 202]). For the PSR 1257+12 system, the

free oscillations correspond to long-term motion (periods ~ 6200 yrs and ~ 92000 yrs [303, 202]), while the forced oscillation are on a shorter time scale (period $2\pi/(3n_2 - 2n_1) = 5.586$ yrs) and of smaller amplitude.

Let us consider only the long term motion characterized by the free oscillations. For them, the derivatives of the eccentricities read

$$\dot{e}_1 = A_{12}e_2 \sin(\omega_1 - \omega_2), \quad \dot{e}_2 = -A_{21}e_1 \sin(\omega_1 - \omega_2), \quad (13.123)$$

and we therefore find the relation

$$A_{21}e_1\dot{e}_1 + A_{12}e_2\dot{e}_2 = 0. \quad (13.124)$$

It can be integrated in terms of a new conservative quantity (valid for long-term motion),

$$A_{21}e_1^2 + A_{12}e_2^2 = \text{cst}. \quad (13.125)$$

With $A_{12} = \mu_2 n_1 \alpha A_2$, $A_{21} = \mu_1 n_2 A_2$, it yields a prime integral of long-term motion

$$\mu_1 n_2 e_1^2 + \mu_2 n_1 \alpha e_2^2 = \text{cst}. \quad (13.126)$$

For the PSR 1257+12 planetary system, the relative variation of the semi-major axes is $\approx 2 \times 10^{-4}$, while the eccentricities vary by $\approx \pm 50\%$ on the long term. With $\alpha = a_1/a_2$ and $(n_2/n_1)^2 = (a_1/a_2)^3$ from Kepler's third law, the semi-major axes can therefore be considered as constant to the first approximation, as also shown by numerical integration [202], so that this conservative quantity can be written $m_1 a_1^{1/2} e_1^2 + m_2 a_2^{1/2} e_2^2 = \text{cst}$, or equivalently

$$Q_e = e_1^2 + \frac{m_2}{m_1}\alpha^{-1/2}e_2^2 = \text{cst}. \quad (13.127)$$

This result once again shows that the motions of planets B and C are tightly coupled and that their eccentricities oscillate on the long term in exact opposition (see Fig. 1c of [202]).

Expression for the eccentricity function Γ

Let us come back to the full motion, including the free and forced osciilations of the eccentricities. From equations (13.122), we derive expressions for $\sin(\omega_1 - \phi)$ and $\sin(\omega_2 - \phi)$, which we may now insert in the expression for the quantity Γ (Eq. 13.112). This quantity describes the eccentricity-dependent contribution to the variation of the semi-major axes. We find

$$\Gamma = \frac{C_1}{B_1} e_1 \dot{e}_1 + \frac{C_2}{B_2} e_2 \dot{e}_2 + \left(\frac{C_2 A_{21}}{B_2} - \frac{C_1 A_{12}}{B_1} \right) e_1 e_2 \sin(\omega_1 - \omega_2). \quad (13.128)$$

When replacing the various coefficients by their expressions, one finds that $C_2 A_{21}/B_2 = C_1 A_{12}/B_1 = A_2$, so that the last term vanishes and Γ is finally given by

$$\Gamma = \frac{e_1 \dot{e}_1}{\mu_2 n_1 \alpha} + \frac{e_2 \dot{e}_2}{\mu_1 n_2}. \quad (13.129)$$

To the approximation (very good in this context) where n_1, n_2 and α are constant, this function can now be easily integrated as

$$S\Gamma = \int \Gamma(t) dt = \frac{1}{2\mu_2 n_1 \alpha} (e_1^2 + K e_2^2), \quad (13.130)$$

where we have set

$$K = \frac{m_2}{m_1} \alpha^{-1/2} = \frac{\mu_2}{\mu_1} \alpha^{-1/2}, \quad (13.131)$$

i.e. $K = (m_2/m_1)\sqrt{a_2/a_1}$. With the presently measured values of the masses of planets B and C, $m_B = (4.3 \pm 0.2) M_\oplus$ and $m_C = (3.9 \pm 0.2) M_\oplus$ [253], this parameter takes, for the PSR 1257+12 system, a value close to 1, namely,

$$K_{\text{PSR}} = 1.03 \pm 0.10. \quad (13.132)$$

We recognize in the term $e_1^2 + K e_2^2$ the hereabove long-term motion prime integral (Eq. 13.127).

Relation between semi-major axes and eccentricities

These results allow us to finally integrate analytically the semi-major axes equations in function of the eccentricities and to provide an exact expression for the incomplete Rasio *et al.* [461] formula. Let us first neglect the conjunction contribution. The equation for the semi-major axis of planet

(1) writes

$$\frac{\dot{a}_1}{a_1} = -4\mu_2 n_1 \alpha \Gamma(t). \tag{13.133}$$

Using the expression obtained for Γ, it writes

$$\frac{\dot{a}_1}{a_1} = -4(e_1 \dot{e}_1 + K e_2 \dot{e}_2). \tag{13.134}$$

This equation is easily integrated as

$$\ln \frac{a_1}{a_{1*}} = -2(e_1^2 + K e_2^2), \tag{13.135}$$

which may be approximated for small values of the eccentricities by

$$\frac{a_1}{a_{1*}} = 1 - 2(e_1^2 + K e_2^2), \tag{13.136}$$

where a_{1*} is the (virtual) value of the semi-major axis for which $e_1 = e_2 = 0$. This relation applies to the PSR1257+12 system, in which the eccentricities of planets B and C run between about 0.01 and 0.03. We may also express this result in terms of variations of the semi-major axis and eccentricities with respect to some initial conditions, a_{10}, e_{10} and e_{20},

$$a_1 = a_{10}\{1 - 2[(e_1^2 - e_{10}^2) + K(e_2^2 - e_{20}^2)]\}. \tag{13.137}$$

This result allows to recover and to generalize the Rasio *et al.* [461] formula, $a_1 = a_{1*}[1 - 2e_1^2]$, in which e_2 was assumed to be constant (the term $K(e_2^2 - e_{20}^2)$ vanishes in this case). However, it also shows that the correct result cannot be obtained as a mere "test particle" limit $m_1/m_2 \to 0$, since this increases the factor K instead of decreasing it.

Let us finally derive the expression for the variation of the semi-major axis of the outer planet. Its equation reads

$$\frac{\dot{a}_2}{a_2} = 6\mu_1 n_2 \Gamma(t) = \frac{6}{K}(e_1 \dot{e}_1 + K e_2 \dot{e}_2). \tag{13.138}$$

It is integrated, for small eccentricities, as

$$\frac{a_2}{a_{2*}} = 1 + \frac{3}{K}(e_1^2 + K e_2^2), \tag{13.139}$$

i.e. respectively to initial conditions,

$$a_2 = a_{20}\left\{1 + \frac{3}{K}[(e_1^2 - e_{10}^2) + K(e_2^2 - e_{20}^2)]\right\}. \tag{13.140}$$

We therefore find that the variation of the semi-major axes of the two planets (except for the small amplitude conjunction effects) depend only on the expression $e_1^2 + K e_2^2$, which we have found to be a prime integral of the long-term motion. This allows us to conclude (within the approximations used in these analytical calculations) that there is no long-term variation of the semi-major axes of planets B and C. This result fully agrees with the Gozdziewski *et al.* [202] numerical integrations of their long-term motion, which also derived their stability.

13.5.3.4. *Application to the PSR 1257+12 system evolution*

Semi-major axes, analytic solutions

The evolution of the semi-major axis of the orbit of planet A of the PSR 1251+12 system is mainly determined by the conjunction effects by planets B and C. An analytic expression can be obtained for this evolution [413]: the corresponding period evolution, which is given from Kepler's third law by $a_A^{3/2}$, is plotted in Fig. 13.22.

The evolution of the semi-major axis of planet B (i.e. planet 1 in the previous two-planet analysis) can be analytically integrated under the form:

$$a_B(t) = a_{B0} \left\{ 1 - 4\mu_C \frac{2\pi}{P_{C0}} \alpha \left(S\Gamma(t) - \frac{1}{2} S\Gamma_c(t) \right) \right\}, \qquad (13.141)$$

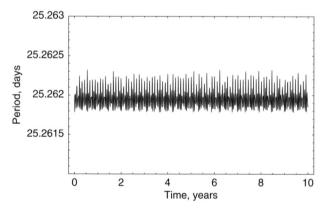

Fig. 13.22: **Evolution of planet A period.** We show in this figure the evolution over ten years of the period P_A of planet A in the PSR 1257+12 planetary system, plotted from its analytic expression. The fluctuations are mainly due to the conjunction effects by planets B and C. Except for this small oscillation, the period is stable on the long term (the motion of planet A is decoupled from the dynamics of planets B and C in the long-term scale [202]).

where

$$ST_c(t) = \int \Gamma_c(t)dt = \frac{\mathcal{P}(\psi(t)) - \mathcal{P}(\psi_0) - \alpha(\cos\psi(t) - \cos\psi_0)}{n_{B0} - n_{C0}}, \quad (13.142)$$

$$\mathcal{P}(\psi) = \left(1 + \alpha^2 - 2\alpha\cos\psi\right)^{-1/2}, \quad (13.143)$$

$$\psi(t) = (n_{B0} - n_{C0})t + \psi_0, \quad (13.144)$$

and

$$ST(t) = \int \Gamma(t)dt = \frac{1}{2\mu_C n_{B0}\alpha}[(e_B(t)^2 - e_{B0}^2) + K(e_C(t)^2 - e_{C0}^2)]. \quad (13.145)$$

The corresponding period evolution, which is given from Kepler's third law by $a_B^{3/2}$, is plotted in Fig. 13.23. It compares satisfactorily with the perturbation of orbital period derived from numerical integration of equations of motion by Wolszczan [544, Fig. 2].

Finally, the evolution of the semi-major axis of planet C (i.e. planet 2 in the previous two-planet analysis) can be analytically integrated under the form:

$$a_C(t) = a_{C0}\left\{1 + 6\mu_B\frac{2\pi}{P_{C0}}\left(ST(t) - \frac{1}{3}ST_c(t)\right)\right\}. \quad (13.146)$$

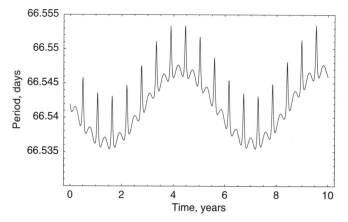

Fig. 13.23: **Evolution of PSR 1257+12 planet B period (see text).** The fluctuations are due to the mutual effects between planets B and C. Except for this small oscillation, the period is stable on the long term.

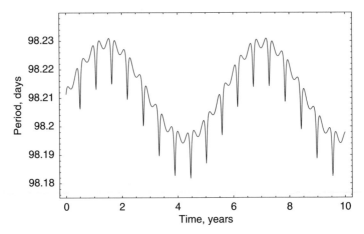

Fig. 13.24: **Evolution of PSR 1257+12 planet C period (see text).** The fluctuations are due to the mutual effects between planets B and C. Except for this small oscillation, the period is stable on the long term.

The corresponding period evolution, which is given from Kepler's third law by $a_C^{3/2}$, is plotted in Fig. 13.24. It also compares satisfactorily with the perturbation of orbital period derived from numerical integration of equations of motion by Wolszczan [544, Fig. 2]. The coupling of the two B and C orbits, which leads to motion in opposition of the two planets, is clear on Figs. 13.23 and 13.24.

13.5.3.5. *Consequences for the formation model*

Periods

We can now compare to the expectation of the wave packet formation model $(n^2 + n/2)$, not only the mean semi-major axis ratio of planets B and C as previously done, but the full variation with time of this ratio. It can be expressed in terms of the ratio of effective quantum numbers n_B and n_C, which is expected to be equal to $7/8$. The time evolution of the ratio $(n_B/n_C)/(7/8)$ is shown in Fig. 13.25.

An important point to notice is that this ratio, apart from very small fluctuations of about 3×10^{-4}, is stable on long time scales. This definitively justifies the validity of the comparison of today's periods with those expected from the formation model, despite the evolution of the system over billion years. Moreover, we can see in this figure that, when taking into account the mutual effects between planets B and C, the minimal difference

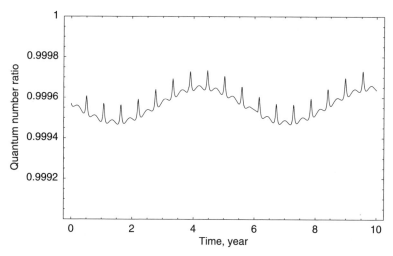

Fig. 13.25: **Evolution of planet B and C quantum number ratio.** Evolution over ten years of the ratio $[n_B(t)/n_C(t)]/[7/8]$ for the two planets B and C in the PSR 1257+12 planetary system. The minimal deviation from the expected value 1 reaches 2.5×10^{-4}.

between the value expected at the end of the formation era and the observed value has been decreased again by a factor of ~ 2, from 4.3×10^{-4} to about 2.5×10^{-4} (while it is 0.6×10^{-4} for the A/C ratio).

This high precision suggests to push further the model and to attempt to obtain a theoretical estimate of the width of the probability peaks. In the framework of the macroscopic quantum-type scale relativity approach, this width is given by a generalized Heisenberg relation $\Delta E \Delta t \approx 2m\mathcal{D}$, with $\mathcal{D} = GM/2w$. In this relation, the time fluctuation Δt may be estimated from the characteristic revival time of localized wave packets. These wave packets are characterized by a revival time $t_{\text{rev}} = 2nP/3$ where $P \propto n^3$ is the classical period and a superrevival time $t_{\text{sr}} = 3nt_{\text{rev}}/4$ [56]. Inserting this last value in the macroscopic Heisenberg relation, one obtains $\Delta E/E = 1/(\pi n^3)$, i.e. $\Delta n/n = 1/(2\pi n^3)$. For $n = 8$ this relation gives $\Delta n/n \approx 3 \times 10^{-4}$. Though this is a very preliminary estimate (one should take the self-gravitation of the wave packet into account), the fact that it is of the same order of size as the observed difference seems very encouraging.

Eccentricities

Concerning eccentricities, the precession and near-resonance effects between planets B and C imply an important relative variation of nearly a factor 3.

One finds, both in the numerical integration [202] and in the analytic solution [413], that the two planet eccentricities vary in opposition in the range 0.0125 to 0.0285 on a period of about 6200 years (plus a smaller component of period about 92000 years). It would therefore have no meaning to attempt obtaining a theoretical expectation of the individual eccentricities from the formation model. However, we have seen that the two eccentricities combine in terms of a conservative quantity (Eq. 13.127), $Q_e = e_B^2 + K e_C^2$. With the observational values $e_B = 0.0186 \pm 0.0002$, $e_C = 0.0252 \pm 0.0002$, $m_B = (4.3 \pm 0.2) M_\oplus$ and $m_C = (3.9 \pm 0.2) M_\oplus$ [253] leading to $K = 1.03 \pm 0.10$, one finds $Q_e(\text{obs}) = 0.00100 \pm 0.00002(e) \pm 0.00004(K)$.

Since this quantity is an invariant prime integral of the long term evolution of the system, it must have been fixed at the end of the formation era. One may therefore consider the possibility of deriving it as one among the possible quantized values in the macroscopic quantum-type model. In this aim, let us carry the model farther.

The planetesimal wave packets, being not only quantum-like wave packets but also gravitational structures, are expected to concentrate to form protoplanets, then the planets by final accretion. But during this concentration phase, the conditions under which the geodesic equation (i.e. the fundamental equation of dynamics) may be integrated under the form of a Schrödinger equation still apply. This results in the appearance of a scale factor f on the gravitational coupling constant, and therefore also on the main quantum numbers. This is another manifestation of the combination of the scale invariance of gravitation with a macroquantum Schrödinger description, which differs profoundly from the behavior of the microscopic atomic quantum regime whose scales are fixed by the constancy of the Planck constant $\hbar$. But the quantization of the solutions at each stage and the conservation of energy finally implies that the scale factor be integer (this process is quite similar to that yielding hierarchically embedded levels of organization in planetary systems, as verified in our Solar System, see [361] and Secs. 13.3.6 and 13.4).

Applied to the PSR 1257+12 planet formation, the initial ranks $n_B = 7$ and $n_C = 8$ are successively transformed into $n_B = 7f$ and $n_C = 8f$ with f increasing during the wave packet concentration. The coupling constant $w_\odot$ becomes correspondingly $f w_\odot$, then allowing the planet distances GMn^2/w^2 to remain the same. The width of the orbitals, on the contrary, being given by $\sigma_n \sim n^{3/2}$ [267], relatively decreases as $\sigma/a \sim f^{-1/2}$, as expected for such a concentration process.

Let us apply this process to the eccentricity quantization. We have seen that it is given by the amplitude of the Runge–Lenz vector [267], and is therefore expected to be quantized as $e_{kn} = k/n$, with $k = 0$ to $n - 1$. The first quantized value is $e = 0$. It yields a satisfying first approximation for the B and C planet eccentricities, $\langle e \rangle \approx 0.02$. To a better level of precision, the first excited value is obtained for $k = 1$, i.e. $e_B = 1/7f$ and $e_C = 1/8f$, with f integer. For $f = 6$, which corresponds to the beginning of the spatial separation of the orbitals, one obtains $Q_e = (1/42)^2 + K(1/48)^2 = 0.001014$ ($K = 1.03$ fixed), in fair agreement with the observed value 0.001000 ± 0.000018 (the agreement being preserved for all possible values of $K = 0.93$ to 1.13). Reversely, this result can be put to the test in the future since, if correct, it provides a value of K:

$$K = -\frac{e_B^2 - (1/42)^2}{e_C^2 - (1/48)^2} = 1.099 \pm 0.065, \tag{13.147}$$

from which an estimate of the mass ratio can be obtained,

$$\frac{m_C}{m_B} = 0.965 \pm 0.058, \tag{13.148}$$

more precise than the presently known value $m_C/m_B = 0.91 \pm 0.10$ [253]. A possible future improvement of the observational values of the eccentricities would still improve this estimate.

Problems and Exercises

Open Problem 34: Use the above model of formation and evolution to predict the distances and periods, in terms of probability density peaks, of other possible planets in the PSR1257+12 planetary system.

Hints: The $n^2 + n/2$ law yields in particular more probable short periods at 0.322 days ($n = 1$), 1.958 days ($n = 2$) and 5.96 days ($n = 3$). Wolszczan *et al.* [545] have obtained timing data for about 30 successive days. An analysis of the residuals of these data after substraction of the effects of the three known planets (the dispersion of these residual being still larger than the error bars) has yielded a marginal detection (at a significance level of 2.7 σ) of a planet with a period $P = 2.2$ days and a mass 0.035 $M_\oplus$, which is compatible with the $n = 2$ expectation [397]. The absence of planet for $n = 6$ may be well understood from the numerical integrations and simulations of Gozdziewski *et al.* [202], which show this particular zone to be unstable (although there is an extended stable zone between planets A and B). ∎

13.6. Galactic Structures

13.6.1. *Star formation*

Let us now consider the application of the new approach to some still unsolved fundamental problems in the standard theory of star formation. This concerns in particular the morphogenesis of the star forming regions. Indeed, as a first approximation, one can describe the interstellar medium from which a star forms in terms of an average constant density, i.e. of a three-dimensional isotropic harmonic oscillator potential. The solutions of the corresponding Schrödinger equation are given in Sec. 13.3.2 and illustrated in Fig. 13.26. For an increasing energy, these solutions describe single objects ($n = 0$), then binary structures ($n = 1$), then 3-chains and trapeze-like structures ($n = 2$), 4-objects chains ($n = 3$), etc. Such morphologies are naturally unstable and rapidly evolve just after their formation, since the potential is locally changed by the structuring itself.

It is remarkable that zones of star formations, such as O and B associations, are for long known to show in their central regions, in a systematic way, double structures (ex. the "butterfly" in N159 of the Large Magellanic Cloud), chain-like morphologies (ex. NGC 7510), and trapeze-like morphologies (ex. the Orion trapeze), etc. (see Fig. 13.27), as expected in the scale relativity approach [123].

Another morphological specificity of star formation at a smaller scale is the presence of disks associated with polar jets. As we have seen in Sec. 13.3.5 and as will be illustrated hereafter in Sec. 13.7 (see the cases ($l = 2, m = 0$) and ($l = 4, \bar{m} = 0$), this is precisely the result obtained when looking for the angular dependence of the solutions of the Schrödinger equation, when assuming that the matter and the gas has been preferentially ejected at angles given by the peaks of angular probability density. This vast subject can only be touched upon here and clearly merits to be developed elsewhere in more detail.

13.6.2. *Binary stars*

A crucial test of the theory consists of verifying that it applies to pure two-body systems. The formation of binary stars remains a puzzle, while more than 60% of the stars of our Galaxy are double. Conversely, in the new approach the formation of a double system is obtained very easily, since it corresponds to the solution $n = 1$ of the gravitational Schrödinger equation in an harmonic oscillator potential, i.e. a uniform density background (while

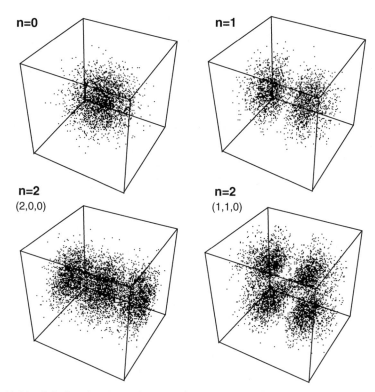

Fig. 13.26: **Modes in three-dimensional harmonic oscillator potential.** The three first modes of the solution of a Schrödinger equation for a particle in a three-dimensional harmonic oscillator potential, which correspond to the gravitational potential of a background of constant density (the mode $n = 2$ decays into two submodes). In the scale relativity approach, the geodesic equation can be integrated in terms of a Schrödinger equation, so that structures are formed even in a medium of strictly constant density. Depending on the value of the energy, discretized stationary solutions are found, that describe the formation of one object ($n = 0$), two objects ($n = 1$), etc. We have simulated these solutions by distributing points according to the probability density. The mode $n = 1$ corresponds to the formation of binary objects (binary stars, double galaxies, binary clusters of galaxies).

the fundamental solution $n = 0$ represents a single spherically symmetric system): see Figs. 13.26 and 13.27.

After its formation, the binary system will evolve according to its local Kepler potential. The binary Keplerian problem is solved, in terms of reduced coordinates, by the same equations as single objects in a central potential. This solution brings informations about the inter-velocities and

Fig. 13.27: **Morphologies of star clusters.** Top left: the globular cluster M13; top right: the "butterfly" nebula in M59 [229]; bottom left: the chain cluster NGC 7510; bottom right: the trapeze in Orion nebula.

the inter-distances between stars. The observed velocity is expected to be quantized as $v_n = w/n$ with w equal to 144 km/s or a multiple or submultiple (depending on the scale of the binary star).

As an example of application, eclipsing binaries are an interesting case of closeby systems, for which we therefore expect the gravitational constant to be a multiple of 144 km/s. It has indeed been found [369] that the average velocity of the 1048 eclipsing binaries in the Brancewicz and Dvorak catalog of eclipsing binaries [66] is $w = 289.4 \pm 3.0 = 2 \times (144.7 \pm 1.5)$ km/s (see Fig. 13.28). Moreover, a good fit of their interdistance distribution is given by the probability distribution of the fundamental Kepler orbital.

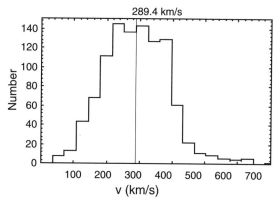

Fig. 13.28: **Velocity difference of eclipsing binary stars.** Distribution of the intervelocity between binary stars in the Brancewicz catalog of eclipsing binaries. The average velocity is $2 \times (144.7 \pm 1.5)$ km/s.

13.6.3. *High-velocity clouds*

High-velocity clouds (HVCs) are neutral gas clouds with anomalous observed velocities. They are grouped in large structures named complexes. Precise values for the distances of HVCs in complex A (chain A) have confirmed their positions in the halo [542]. A particular structure of such typical clouds has been observed by Pietz *et al.* [445] in the complex C. Velocity bridges have been identified, that seem to be in accordance with the expected Keplerian velocity distribution $(144/n \, \text{km} \cdot \text{s}^{-1})$ [123]. We give two examples of this effect in Fig. 13.29, which shows the mass density towards the line of sight in function of the radial velocity (the zero velocity is associated to the HI gas in the Galactic disk). Large bridges at $144 \, \text{km} \cdot \text{s}^{-1}$ with sub-structures near submultiples, such as $\pm 21 \, \text{km} \cdot \text{s}^{-1}$ $(1/7)$, $\pm 24 \, \text{km} \cdot \text{s}^{-1}$ $(1/6)$, $\pm 36 \, \text{km} \cdot \text{s}^{-1}$ $(1/4)$ and $-48 \, \text{km} \cdot \text{s}^{-1}$ $(1/3)$, are clearly apparent in these diagrams.

This distribution can be considered as a signature of a Keplerian interaction between the molecular clouds in the Galactic halo and the Galactic disc.

13.6.4. *Proper motion near the Galactic center*

The study of the velocity distribution of stars near galaxy centers could reveal to be particularly interesting in the context of testing the theory. Indeed there is increasing evidence that galaxies like ours host in their center compact masses (likely black holes) of several $10^6 \, M_\odot$, based on

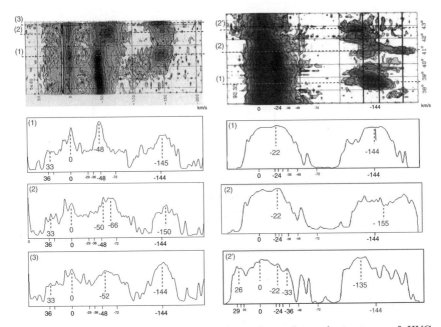

Fig. 13.29: **High-velocity clouds in the galaxy.** Internal structures of HVCs in complex C (centered on $b = 54° \cdot 6$ and $l = 92° \cdot 4$). Details reveal a Keplerian distribution in $144/n$ km $\cdot$ s^{-1}. (Adapted from [445]).

observation of Keplerian velocity-distance relations. Therefore, this could allow one to put the Keplerian quantization law to the test for large values of the velocities, then for large values of the coupling constant w, which should ultimately reach c (i.e. $\alpha_g = 1$).

We give here an example of such a work using observations by Eckart and Genzel [151] of the proper motions of 39 stars located between 0.04 and 0.4 pc from the Galactic center. They have found that these observations support the presence of a central mass of $(2.5 \pm 0.4) \times 10^6\ M_\odot$, which has been since confirmed by improved data. Though the velocities do not yet reach high values and the error bars are large (≈ 60 km/s), the observed distribution of velocity components, given in Fig. 13.30, is compatible with a w/n quantization matching with the 144 km/s sequence.

13.7. Planetary Nebulae

Planetary nebulae, despite their misleading names, are a general stage of evolution of low mass stars. They result from the ionization by the radiation of the central star of previously ejected outer envelopes.

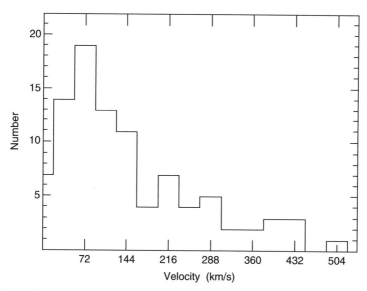

Fig. 13.30: **Stellar velocity near the galactic center (from the data of [151]).**
The diagram mixes the Right Ascension and Declination components of velocity derived
from proper motion, with radial components derived from spectral shift (for a fraction
of the sample). The main probability peak lies at $72 = 144/2$ km/s.

The standard theory to explain the formation of planetary nebulae
(PNe) is the interacting stellar wind (ISW) model. The simple cases of
spherical and elliptical PNe are easily reproduced by this model, but many
far more complicated morphologies have now been discovered, in particular
thanks to Space Telescope high resolution imaging. Despite many attempts
using numerical simulations made with a large number of initial conditions
and accounting for several physical effects (magnetic fields, companion star
perturbation, collimated outflows), no coherent and simple understanding
of these shapes has yet emerged.

Our approach to this problem is a scale-relativistic generalization of
the ISW model, which has been developed in collaboration with da Rocha
[123, 124, 125]. We account for the chaotic motion of the ejected material
and we simply replace the standard equation of dynamics used in the
model by the generalized one (written in terms of the covariant derivative).
It becomes a Schrödinger equation having well-defined angular solutions
$\psi(\theta, \phi)$. Their squared modulus $P = |\psi^2|$ is identified with a probability
distribution of angles that is characterized by the existence of maxima
and minima (see Figs. 13.31–13.33). This means that we automatically

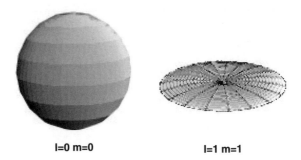

l=0 m=0   **l=1 m=1**

Fig. 13.31: **Predicted shapes of planetary nebulae.** Examples of expected quantized morphologies for ejection processes and planetary nebulae in function of the quantum numbers ($l = 0$ to $l = 1$).

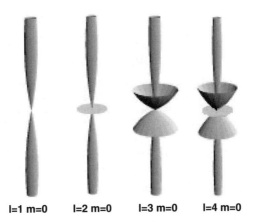

l=1 m=0   **l=2 m=0**   **l=3 m=0**   **l=4 m=0**

Fig. 13.32: **Predicted shapes of planetary nebulae.** Examples of expected quantized morphologies for ejection processes and planetary nebulae in function of the quantum numbers ($l = 1$ to $l = 4$). The shapes shown correspond to the most probable shapes, i.e. when matter is ejected along the probability peaks of angle values.

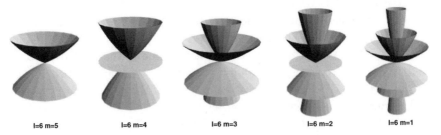

l=6 m=5   **l=6 m=4**   **l=6 m=3**   **l=6 m=2**   **l=6 m=1**

Fig. 13.33: **Predicted shapes of planetary nebulae.** Examples of expected quantized morphologies for ejection processes and planetary nebulae in function of the quantum numbers ($l = 6$). The shapes shown correspond to the most probable shapes, i.e. when matter is ejected along the probability peaks of angle values.

find, by this method, that the star has a tendency to eject matter along certain angle values that are far more probable than others. Therefore we are able to predict the existence of discretized possible morphologies, in correspondence with quantized values of the square angular momentum L^2 and of its projection L_z.

This approach is not contradictory with the standard ones. This is already clear from the fact that we use the same hydrodynamical equations as in the standard theory, except for the presence of an additional quantum-like potential. Moreover, the new result amounts to the finding that the various morphologies are quantized and related to quantized values of prime integrals. Now, for each individual PN with a given morphology, it remains needed to understand why specific values of L^2 and L_z are achieved, which may involve the existence of a companion star, etc. as in the standard approach. But the new point here is that the variability of shapes and their non-spherical symmetry (while the field is spherically symmetric) may now be simply understood in terms of states of a fundamental conservative quantity, the angular momentum.

Observational results indicate that the propagation velocity is nearly constant (this fact is already used in numerous simulations [113, 149]). This means that the PNe shells have an effective free Galilean motion. Therefore the problem of their structuration can be dealt with in terms of a free Schrödinger equation.

13.7.1. *Elementary morphologies*

In the context of scale relativity, the global formation of shapes is understood as a consequence of the geometry of geodesics, whose distribution is described by the generalized Schrödinger equation (13.62). Its solutions, $\psi_{nl\bar{m}}(r, \theta, \phi) = R_{nl}(r) \cdot Y_l^{\bar{m}}(\theta, \phi)$ have two separable parts, the radial part (13.64), which gives us information about matter density along the structure and the angular part, which imposes global shape specificity.

The model can also take into account perturbative terms, such as second order terms in the velocity power series expansion, external influences, the magnetic contribution in the ejection process, etc. Many hydrodynamic simulations neglect the magnetic force [149, 318], though PNe are expected to have strong magnetic fields, like the red giant stars and the white dwarfs [193]. The simplest description consists of introducing a poloidal magnetic field. With this particular geometry, the flow of ionized particles (because of the UV star radiation) should be deviated in the same direction, i.e. towards the axis of symmetry. Moreover, for particular $(l, \bar{m})$ values, bipolar structures naturally emerge. In these singular objects, a self-gravitating force appears. This force acts like the magnetic field and induces a constriction along the axis of symmetry of the PNe, which can be accounted for in a more complete description.

13.7.2. Quantized shapes

Let us give a synthesis of the different shapes allowed by this model [123, 124]. Note that an analogy of planetary nebulae shapes with spherical harmonics has also been noticed by Oldershaw, basing himself on discrete self-similarity arguments [417]. However, this analogy concerned the probability distribution representation in space, which has actually no direct morphological meaning. Here we use these probabilities to construct the most probable shapes by considering an ejection of matter along the peaks of probability for angles.

Three categories summarize all the possibilities:

(i) Spherical and elliptic: The basic spherical shape is obtained for the specific value of $(l = \bar{m} = 0)$ while for $([\bar{m} = \pm l]\forall l \geq 1)$, the PNe will evolve to an equatorial disc (Fig. 13.31). The tilting on the line of sight of this disc will induce a spherical or an elliptical shape for the PN.

(ii) Bipolar ejection: The general process upon which the whole description is based is an ejection process. Therefore, it is not surprising to find solutions describing bipolar jets (Fig. 13.32). For $([\bar{m} = 0]\forall l \geq 1)$, the density distribution is concentrated on the axis of the objects.

(iii) Bipolar shell: All the other solutions give bipolar shell structures (Fig. 13.33). The empirical relation $(l - \bar{m}) + 1$ constrains the number of internal structures. For example, $(l = 6, \bar{m} = 2)$ gives five structures (one disc and four shells). This brief presentation allows one to classify all the elementary shapes and gives a method for constraining the structure with the couple $(l, \bar{m})$.

13.7.3. Comparison with observations

The following four examples (Fig. 13.34) show the direct comparison between observed structures and predicted quantized shapes built with the Schrödinger model. We can see that many exotic shapes are naturally described in this framework.

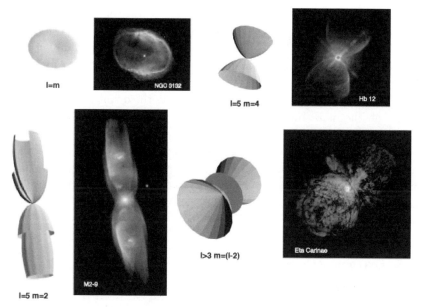

Fig. 13.34: **Predicted and observed planetary nebula shapes.** Comparison between predicted quantized shapes and typical observed planetary nebulae, plus the η Carinae nebula (STScI images, adapted from http://ad.usno.navy.mil/pne/caption. html).

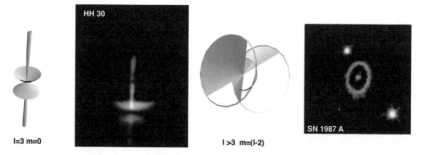

Fig. 13.35: **Comparison between predicted shapes and ejection processes.** Generalization of the theoretical description of ejection process. The left hand side represents a young star ejection state, and the right hand side, a supernova explosion (adapted from STScI images).

Moreover, the general solutions could be used in many ejection/ scattering cases other than PNe. In Fig. 13.35, the shapes of young star ejection states or supernovae explosions are positively compared with quantized solutions (also valid for inward motion). For example, in the case of the Supernova 1987A, whose remnant is still very young, one may deproject the image and measure the ejection angle θ at the origin. We have found $\theta = 41.2 \pm 1.0$ deg., which supports an identification with the case $l = 4$, $m = 2$ corresponding to the observed morphology, for which the predicted angle is 40.86 deg [124].

Let us conclude this section by emphasizing the quantitative predictive power of the new model, exemplified by the SN1987A case. Future extremely large telescope will allow detailed measurements of the velocity fields of planetary nebulae and similar objects, from which the values of the angular momenta could be deduced. These values are predicted to be quantized, in correspondence with quantized shapes and with quantized values of the angles (see tables in [124]), which could be put to the test in the coming years.

13.8. Extragalactic Structures

13.8.1. *"Dark matter" problem*

In order to explain the numerous effects, which are unexplainable by the sole action of the gravitational force of visible and baryonic matter (flat rotation curves of spiral galaxies, large velocity dispersion of clusters of galaxies, cosmological energy balance, gravitational lensing, formation and evolution of structures, etc.), one usually makes the hypothesis of the existence of huge quantities of invisible "dark" matter that would be the dominant mass density constituent of the Universe. However, despite decades of very active research, this missing matter continues to escape detection. Another suggestion was to modify Newton's gravity (the MOND hypothesis [320]), but such an ad-hoc hypothesis is hardly consistent with general relativity and with different observations at different scales.

Moreover, the dark matter problem is deeply connected with the problem of formation of galaxies and of large scale structures in the Universe. Indeed, in the standard approach to this problem, the growth of the very small $z = 1000$ fluctuations up to today's ($z = 0$) structures is impossible in the absence of a large amount of dark matter. However, as recalled in Silk's quotation at the beginning of this chapter, even with dark matter the theory of gravitational growth remains unsatisfactory.

The scale relativity approach may provide an original solution to both problems. Indeed, the fractal geometry of a nondifferentiable space-time solves the problem of formation at all scales (this is the subject of the whole chapter) and it also implies the appearance of a new scalar potential (Eq. 13.13), which manifests the fractality of space in the same way as Newton's potential manifests its curvature. We have suggested [376, 383, 389, 396] that this new potential energy may explain the anomalous dynamical effects, without needing any missing mass.

Let us sum up again how this new energy emerges from the fractal geometry. As we have shown in Chapter 5, the fundamental equation of dynamics in a fractal space, characterized by a fractal fluctuation such that $\langle d\xi^2 \rangle = 2\mathcal{D}\, dt$, reads $m\hat{d}\mathcal{V}/dt = -\nabla\phi$, with $\hat{d}/dt = \partial/\partial t + \mathcal{V} \cdot \nabla - i\mathcal{D}\Delta$ and $\mathcal{V} = V - iU$, and it can be integrated under the form a Schrödinger equation that reads in the time-independent case

$$2m\mathcal{D}^2\Delta\psi + (E - \phi)\psi = 0, \qquad (13.149)$$

where ϕ is the steady-state solution for the potential, given by $\Delta\phi = 4\pi G\rho$ in terms of the visible and baryonic dark matter density ρ. The real part of this equation is the energy balance equation, which reads

$$E = \phi + Q + \frac{1}{2}mV^2(x, y, z), \qquad (13.150)$$

where Q is an additional potential energy that has therefore naturally emerged from the fractal geometry. When there is a large quantity of matter, each particle is subjected to the probability density $P = |\psi|^2$, so that the density of matter becomes proportional to it, $\rho = m_0 P$, and the new potential energy is given by

$$Q = -2m\mathcal{D}^2\frac{\Delta\sqrt{\rho}}{\sqrt{\rho}}. \qquad (13.151)$$

We emphasize once again that the system of equations finally obtained is exactly the same as the standard system of structure formation [438]. It is a Euler and continuity equation system,

$$\left(\frac{\partial}{\partial t} + V \cdot \nabla\right)V = -\nabla\left(\frac{\phi + \phi_{\text{add}}}{m}\right), \quad \frac{\partial\rho}{\partial t} + \text{div}(\rho V) = 0, \qquad (13.152)$$

(plus possible additional terms of pressure, cosmological constant, other fields, etc.). But the essential difference is that the additional potential is currently assumed to be a dark matter potential $\phi_{\text{add}} = \phi_{\text{DM}}$, given by a Poisson equation $\Delta\phi_{\text{DM}} = 4\pi G\rho_{\text{DM}}$, in terms of a dark matter density ρ_{DM}, which is itself arbitrarily postulated and which has never been observed, while in the scale relativity proposal $\phi_{\text{add}} = Q$ is a non-Poissonian potential which is given by Eq. (13.151) in terms of the baryonic matter density.

13.8.2. *Flat rotation curves of spiral galaxies*

Let us exemplify this suggestion in the case of the flat rotation curves of spiral galaxies. The flat rotation curves observed in the outer regions of spiral galaxies is one of the main dynamical effects, which is at the root of the "missing mass" problem. Indeed, farther than the visible radius of galaxies, a very small quantity of matter is detected by all the possible methods used (images at all wavelengths, gravitational lensing, 21 cm radio observations, etc.). Therefore one expects the potential to be a Kepler potential $\phi \propto 1/r$ beyond this radius, and as a consequence one expects the velocity to decrease as $v \propto r^{-1/2}$, while all available observations show that the velocity remains nearly constant up to large distances for all spiral galaxies.

The formation of an isolated galaxy from a cosmological background of uniform density is obtained, in its first steps, as the fundamental level solution $n = 0$ of the Schrödinger equation with an harmonic oscillator potential (Eq. 13.31). Its subsequent evolution is expected to be a solution of a Hartree equation (13.15) [283, 284].

Once the galaxy is formed, let r_0 be its outer radius, beyond which the amount of visible matter becomes small. The potential energy at this point is given, in terms of the visible mass M of the galaxy, by

$$\phi_0 = -\frac{GMm}{r_0} = -mv_0^2. \tag{13.153}$$

The observational data tells us that the velocity in the exterior regions of the galaxy keeps the constant value v_0. From the virial theorem, we also know that the potential energy is proportional to the kinetic energy, so that it also keeps a constant value given by $\phi_0 = -GMm/r_0$. Therefore r_0 is the distance at which the rotation curve begins to be flat and v_0 is the corresponding constant velocity. In the standard approach, this flat rotation curve is in contradiction with the visible matter alone, from which one would expect to observe a variable Keplerian potential energy $\phi = -GMm/r$. This means that one observes an additional potential energy given by

$$Q_{\text{obs}} = -\frac{GMm}{r_0}\left(1 - \frac{r_0}{r}\right). \tag{13.154}$$

The regions exterior to the galaxy are described, in the scale relativity approach, by a Schrödinger equation with a Kepler potential energy $\phi = -GMm/r$, where M is still the sole visible mass, since we assume here no dark matter. The radial solution for the fundamental level is given by

$$\sqrt{P} = 2e^{-r/r_B}, \tag{13.155}$$

where $r_B = GM/w_0^2$ is the macroscopic Bohr radius of the galaxy.

It is now easy to compute the theoretically predicted form of the new potential (Eq. 13.13), knowing that $\mathcal{D} = GM/2w_0$,

$$Q_{\text{pred}} = -2m\mathcal{D}^2 \frac{\Delta\sqrt{P}}{\sqrt{P}} = -\frac{GMm}{2r_B}\left(1 - \frac{2r_B}{r}\right) = -\frac{1}{2}w_0^2\left(1 - \frac{2r_B}{r}\right). \tag{13.156}$$

We therefore obtain, without any added hypothesis, the observed form (Eq. 13.154) of the new potential term. Moreover the visible radius and the Bohr radius are now related, since the identification of the observed and predicted expressions yield $r_0 = 2r_B$. The constant velocity v_0 of the flat rotation curve is also linked to the fundamental gravitational coupling constant w_0 by the relation $w_0 = \sqrt{2}v_0$. This prediction is consistent with an analysis of the observed velocity distribution of spiral galaxies from the Persic–Salucci catalog [441], as shown in Fig. 13.36. The peak velocity is $142 \pm 2\,\text{km/s}$, while the average velocity is found to be

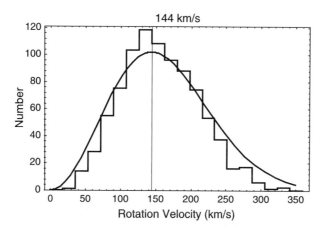

Fig. 13.36: **Distribution of rotation velocities of spiral galaxies.** Distribution of the outermost observed velocities in spiral galaxies (flat rotation curves) from the catalog of rotation curves for 967 spiral galaxies by Persic and Salucci [441]. The fitted continuous curve is proportional to $v^2 \exp(-(v/144)^2)$.

156 ± 2 km/s, so that $\sqrt{2}\langle v \rangle = 220 \pm 3$ km/s. This value, which we also obtain in the study of clusters of galaxies (see herebelow) is within 1σ of $3/2 \times 144.7$ km/s.

13.8.3. *Compact groups*

As remarked by Hickson [231], compact groups of galaxies provide the best environment for galaxy interactions to occur, so it is natural to study these systems in order to better understand the interaction process and its effects. But the existence of such systems is a puzzle, as they are unstable to gravitational interactions and mergers. A way out of this impasse is to assume that new compact groups are constantly forming. Finally Hickson reaches the conclusion, with now many other specialists, that "in compact groups such as those illustrated [in the Atlas of Compact Groups of Galaxies], we may be actually observing the process of galaxy formation".

It is therefore remarkable in this context that the kind of morphologies displayed in a systematic way by these compact groups is just chain and quadrilateral structures quite similar to those encountered in star forming regions (see Fig. 13.37). The basic solutions obtained from the gravitational Schrödinger equation for formation from a background of constant matter density (i.e. harmonic oscillator potential) once again explain these morphologies in a very simple way.

13.8.4. *Galaxy pairs*

In the same way as binary stars in our Galaxy, binary galaxies are very common structures in the Universe. For example our own Local Group of galaxies, which deserves a special study at the end of this chapter, is organized around a pair of giant spirals (Milky Way Galaxy and Andromeda nebula). This is easily accounted for in our framework, since double structures are the lowest energy solution (beyond the fundamental level that represents an isolated object) of the gravitational Schrödinger equation with a harmonic oscillator potential.

Moreover, binary galaxies are one of the first extragalactic systems for which a redshift quantization effect in terms of 144 km/s and its submultiples has been discovered by Tifft [510]. However, this effect has been interpreted by Tifft and other authors as an "anomalous redshift" of non-Doppler origin, whose existence would therefore question the whole foundation of cosmology.

Fig. 13.37: **Examples of galaxy grouping.** Examples of galaxy grouping: (top left) the giant elliptical galaxy M87 in the center of the Virgo cluster (photo AAT, D. Malin); (top right) the center of the Coma cluster; (bottom) two examples of typical compact groups of galaxies: the triplet HCG 14 [232] and a trapeze structure in the quintet HCG 40 (http://www.naoj.org/Science/press_release/9901/HCG40.html).

Today, one can disprove this anomalous redshift interpretation. Indeed, under such an interpretation, there should be a fundamental difference of behavior between motion deduced from extragalactic redshifts and the motion of planets. The discovery [353, 361, 360, 3], motivated by the scale relativity predictions, that the planets of our own Solar System and of extrasolar planetary systems do have velocities involved in the same sequence $v_n = (144/n)\,$km/s (namely, the velocities of Mercury, Venus, the Earth and Mars are respectively ≈ 48, 36, 29 and 24 km/s, which correspond to $n = 3$ to 6) has definitively excluded the anomalous redshift interpretation.

In the scale relativity approach, the observed peaks are simply peaks of probability density for standard velocities and Doppler redshifts, with a given width, quite similar to those predicted in a classical hydrodynamic approach. In this interpretation, there is no longer any contradiction with the foundation of cosmology, since, on the contrary, it confirms the extragalactic redshifts as being of Doppler and cosmological expansion origin.

Moreover, the probability peaks for the velocity values can be interpreted as a tendency for the system to structure. In the same way as there are well-established structures in the position space (stars, clusters of stars, galaxies, groups of galaxies, clusters of galaxies, large scale structures), the velocity probability peaks are simply the manifestation of structuration in the velocity space. In other words, as it is already well-known in classical mechanics, a full view of the structuring can be obtained in phase space.[c]

The application of the scale relativity approach to galaxy pairs comes, as for binary stars, under the Kepler case in reduced coordinates [360]. One therefore predicts probability peaks of the deprojected velocity difference between the members of the pair given by $v_n = w_0/n$. This prediction is in opposition with Tifft's empirical (and not theoretically justified) model according to which the quantization should be in terms of nv_0, with $v_0 = 12\,\text{km/s}$. A detailed study of this effect has been performed by da Rocha on several catalogs of galaxy pairs, reaching a total of almost 2000 pairs, which has yielded statistically significant results [126]. The comparison with the observational data favors the $1/n$ law, since the peaks expected from a linear law at, e.g. 60, 84, 96 km/s, etc. do not appear in these data.

Several methods of deprojection of the velocity difference (only the radial component is observed) and of the interdistance of binary galaxies (only the two transverse components are observed) have been developed [515, 516]. Indeed, if the velocity V has the distribution $P(V)$, v_r will have the distribution

$$p(v_r) = \int_{v_r}^{\infty} \frac{P(v)}{v}\,dv. \qquad (13.157)$$

For example, a uniform distribution of velocities will be projected as an affine decreasing distribution and a Dirac comb as a stair. Assuming isotropy, one may theoretically inverse this relation, and therefore recover

---

[c]This phase space is, in the macroscopic gravitational application of scale relativity, a (position, velocity) space instead of a (position, momentum) one, because of the combination of the Schrödinger form of the dynamics equation with the equivalence principle. It implies quantization relations for E/m, p/m, etc. instead of E or p in standard quantum mechanics, and therefore for the velocity v, whatever may be the spatial scale.

the original distribution from the derivative of the projected distribution (see Fig. 13.39),

$$P(v) = -v \left[\frac{dp(v_r)}{dv_r} \right]_{v_r}. \qquad (13.158)$$

One therefore expects, if there are velocity probability peaks as predicted, the projected velocity difference to show plateaux up to the corresponding velocity of the peak. The observed distribution obtained with large catalogs [126] confirms this expectation (see Fig. 13.39 a). The presence of such plateaux instead of redshift peaks invalidates the anomalous non-velocity redshift interpretation of these effects. After deprojection, these plateaux yield clear velocity peaks, in particular a significant one at a value compatible with $\approx 144\,\mathrm{km/s}$ (Fig. 13.39 b).

The results obtained confirm the existence of probability peaks both in the velocity space (see Figs. 13.38 and 13.39) and in the position space [515, 516] (more precisely, in the interdistance to mass ratio distribution), in agreement with the scale relativity prediction for the Kepler potential of the pairs in reduced coordinates. These peaks are correlated through Kepler's third law, which is a final demonstration of the Doppler and motion origin of the redshift differences in galaxy pairs.

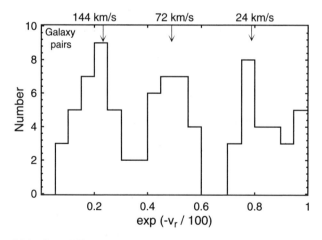

Fig. 13.38: Velocity difference between galaxies in pairs. Deprojection of the intervelocity distribution of galaxy pairs [516] from the Schneider–Salpeter catalog with precision redshifts [475]. The two main probability peaks are found to lie at 144 and 72 km/s.

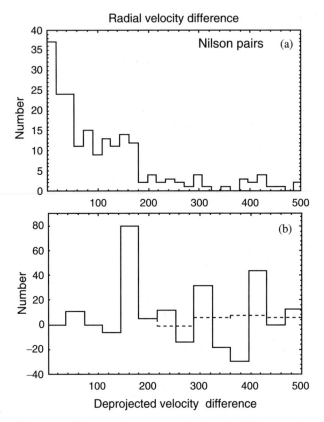

Fig. 13.39: **Velocity difference between galaxies in Nilson pairs.** We give in Fig. (a) the distribution of observed (projected) velocity differences between galaxies in a catalog of 596 pairs [534, 126] derived from the Nilson catalog [333] (here, the 256 pairs which are such that the error on the velocity difference is smaller than 36 km/s), and in Fig. (b) a deprojection of this distribution by taking its derivative (see text). The negative values at high velocity, which come from the fluctuations on small numbers, disappear with a larger bin (dashed line). The main probability peak is statistically significant and compatible with a velocity of 144 km/s.

13.8.5. *Local Group of galaxies*

A detailed analysis of our own Local Group of galaxies has been performed by da Rocha [123, 126], with remarkable results that we only summarize here. It is a particularly interesting system as concerns the application of the gravitational Schrödinger equation, since it is essentially made of two giant spiral galaxies, namely our own Milky Way galaxy (MW) and M31, surrounded by their dwarf companions. The gravitational potential is therefore dominated by a Kepler central potential around each giant galaxy, with the dwarf galaxies playing the role of test particles, while galaxies that lie at intermediate distance are subjected to the potential of the pair, which is similar, from the view point of the gravitational Schrödinger approach, to a diatomic molecular problem.

Moreover one disposes of both precision distances and velocity for the galaxies of our Local Group, so that the expected quantization law can be tested in position and velocity space. The relation between the two structurings involves Kepler's third law for the dwarf galaxies, which are close to either our Galaxy or M31, which may therefore also be validated, thus providing a new argument in favor of the Doppler nature of the redshifts.

The analogy of its structure (shared with several other loose groups of spirals) with old expanding stellar associations has been pointed out long ago by de Vaucouleurs [521]. As we have seen in previous sections, the theory of scale relativity allows one to understand the gravitational formations of such binary systems in terms of the first excited state ($n = 1$), which is solution of the gravitational Schrödinger equation for a constant background density (harmonic oscillator potential).

The use of the scale relativity approach in this case is also supported by the investigation of the motion of these galaxies in numerical simulations (see e.g. [519]) that has demonstrated the chaotic and violent past and future history of the Local Group. Moreover the loose character of this group implies a velocity field, which is locally dominated by the gravitation of the two giant spirals, but which is expected to join the Hubble expansion field in its outer regions. The expected quantization law is therefore rather complicated in this case, since it should correspond to a Kepler potential near M31 and MW, then to a two-body potential in an intermediate region and finally to a harmonic oscillator potential at the scale of the local supercluster. A possible redshift quantization in the Local Group in units of $72 \, \text{km/s}$ has already been suggested [29] (though, as already remarked, with a completely different interpretation from ours: we stress once again that the scale relativity view is a totally standard interpretation in terms of probability densities of velocities).

Observations show that there is a net age difference between both dominant galaxies and the rest of the dwarf galaxies in our Local Group [313]. One can also assert that the gas is isotropic in each subgroup and is subjected to the simple Keplerian potential of the dominant galaxies.

Thus, one can use the Keplerian solutions developed in Sec. 13.3.1. All the solutions should be constrained in order to agree with the initial system (spherical symmetry and isotropic subgroups). Spherical harmonics, $Y_l^{\bar{m}}(\theta, \phi)$, reveal an isotropic arrangement only for $l = 0$ and $\bar{m} = 0$.

Then the mean distance to the gravitational center is given by the formula $\langle r \rangle_n = (3GM/2w_0^2)n^2$. This equation assumes a quantization law in n^2 for the galactic distances with regard to the center of the dominant galaxies. The two dominant galaxies will therefore infer two different laws in two different domains (in this first Keplerian stage), with possibly two different constants w_0 that must be matched. A first step then consists of considering the main constant at $144\,\mathrm{km} \cdot \mathrm{s}^{-1}$ and the closest values ($288\ \mathrm{km} \cdot \mathrm{s}^{-1}$, $72\,\mathrm{km} \cdot \mathrm{s}^{-1}$). The mass M is the visible mass of the Milky Way $M = 7.2 \times 10^{10} M_\odot$ [126]. For M31, the observational data yields $M = 13.2 \times 10^{10} M_\odot$.

Furthermore, it is necessary to treat the case of the remote galaxies and of the NGC 3109 galaxy subgroup. From the theoretical point of view, this is an interesting problem since, assuming a global coherence of the double system in its outer regions described in terms of a global $\mathcal{D}$ value, it shares some common features with the Schrödinger equation written for molecules like H_2^+ (namely, a test particle subjected to two attractive centers). One finds that the wave function, solution of such a problem, is $\psi = a_1\psi_1(r_1) + a_2\psi_2(r_2)$, where ψ_1 and ψ_2 are solutions of the Schrödinger equations written using this global $\mathcal{D}$ for each individual Kepler potential. The global solution can be subsequently matched with the local solutions. Unfortunately, the case of the Local Group (asymmetrical gravitational double system) is more complicated because $\mathcal{D}_{MW} \neq \mathcal{D}_{M31}$ and $\psi_{MW} \neq \psi_{M31}$. Nevertheless, the molecular solution has the advantage to supply a simple presence probability, revealing the interference of the individual solutions: $\mathcal{P} = a_1^2\mathcal{P}_1^2 + a_2^2\mathcal{P}_2^2 + 2a_1a_2\sqrt{\mathcal{P}_1\mathcal{P}_2}\cos(\Delta\theta)$. Even if this solution can not be used as such in the more complex macroscopic gravitational case, it will be interesting to look at the configuration of the remote galaxies with regard to the laws that apply around M31 and the Milky Way.

13.8.5.1. *Structures in position space*

The precision on the distances and the radial velocities in Mateo's synthesis work [313] is sufficient for using them directly in this study. By the knowledge of the Sun vector, one calculates the galactocentric distances d. To verify the existence of a law of the form $d = d_0 \times n^2$, one analyzes the distribution of the observable $\tilde{n} = \sqrt{d/d_0}$. One expects this distribution to exhibit probability peaks around integers values (n).

Milky Way subgroup

The study of the data about the distances of the close-by galaxies reveals a minimum for a characteristic distance $d_0 = 5.50\,\mathrm{kpc}$. This result is in very good agreement with the theoretical prediction of $5.57\,\mathrm{kpc}$ for $w_0 = 288\,\mathrm{km \cdot s^{-1}}$. One therefore computes the values of $\tilde{n} = \sqrt{d/5.57}$, then the differences between them and the nearest integer $(\delta n = \tilde{n} - n)$. In the standard framework one expects these differences to be uniformly distributed between -0.5 and $+0.5$, while, in the present approach, one expects them to peak around zero.

The result, shown in the histograms below (Fig. 13.40), clearly favors the gravitational Schrödinger approach, despite the small number of objects.

Andromeda subgroup

The distance distribution of M31 companion galaxies is also in good agreement with a quantized n^2 distribution, for a value of $d_0 = 10.72\,\mathrm{kpc}$ (Fig. 13.41). Once again, the Keplerian model developed for $w_0 = 288\,\mathrm{km/s}$ (that gives an expected value $d_0 = 10.25\,\mathrm{kpc}$) is close to the observed law in $r = 10.72\,n^2\,\mathrm{kpc}$.

Remote galaxies subgroup

The configuration of the remote galaxies can also be compared with the M31 quantization law and with the MW quantization law (see the histograms (Fig. 13.42)).

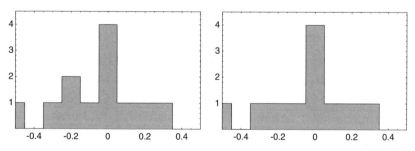

Fig. 13.40: **Histograms of the difference between the variable $\tilde{n} = \sqrt{d/5.57\,\mathrm{kpc}}$ and its nearest integer, for the Milky Way companion galaxies.** The right-hand side of the figure represents the MW subgroup, in which the galaxy Leo I has been excluded due to the large uncertainty on its distance.

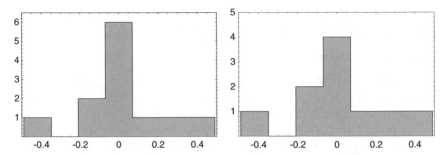

Fig. 13.41: **Histograms of the difference between the variable** $\tilde{n}$ = $\sqrt{d/10.72\,\mathrm{kpc}}$ **and its nearest integer, for M31 companion galaxies.** The right-hand side represents the M31 subgroup without And II and EGB0427 + 63, which are subjected to large uncertainties).

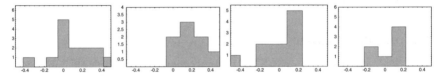

Fig. 13.42: **Histograms of the difference between the variable** $\tilde{n} = \sqrt{d/d_0}$ **and its nearest integer** n, **for the subgroup of remote galaxies.**

The histograms show maxima of the histograms close to zero, which means a higher probability for integer values of the variables, as expected. The galaxy distribution continues to agree with the quantization laws despite the larger distance values and the intervention of both potentials. This opens the possibility that the global solution be a linear combination of the solutions found independently for each subgroup (such as in a molecular-like case).

Final result for positions

Combining all the gravitational sub-systems of the Local Group, we can now draw a global histogram of the relative differences $\delta n = \tilde{n} - n$. The histograms obtained for these two samples differ significantly from a uniform distribution and peak at zero as expected (Fig. 13.43).

This behavior corresponds to a distribution of $\sqrt{d/d_0}$ that shows probability peaks for integer values. The probability to obtain such a distribution by chance is estimated to be $P = 2 \times 10^{-5}$. This value corresponds to a statistical meaning better than the 4σ level. This is supported by the limited sample, for which one obtains a similar result, $P = 5 \times 10^{-5}$.

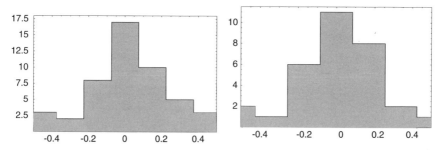

Fig. 13.43: **Distribution of distances of galaxies in our Local Group.** The figures give histograms of the difference between the variable $\tilde{n} = \sqrt{d/d_0}$ and its nearest integer n for all galaxies in the Local Group. The right-hand side histogram is for the limited sample.

13.8.5.2. *Structures in velocity space*

The velocity field of the Local Group provides an interesting test of the theory. One expects the velocity to show peaks of probability for $\langle v^2 \rangle = (w_0/n)^2$. The study of the spatial structures leads us to identify the constant $w_0 = 288$ km/s for the Local Group. A correction by the solar vector is performed to obtain galactocentric radial velocities.

As a first step one considers the subset surrounding the Milky Way. In this case the spatial distribution has been found to be given by the law $r = 5.50\,n^2$ kpc. The corresponding radial velocity law is given by $\langle v^2 \rangle = (w_0/n)^2$. The solutions in the position space and velocity space are equivalent with regard to the Schrödinger equation. Therefore, one associates to a given satellite energy a position and a velocity state given by the same main quantum number n.

The distance-velocity relation around the MW galaxy reads, in terms of the fundamental constant 288 km/s,

$$\frac{v}{288\,\text{km/s}} = \pm\sqrt{\frac{5.50\,\text{kpc}}{r}}. \tag{13.159}$$

As one can see in Fig. 13.44, the data are compatible with the model based on $w = 2 \times 144$ km/s, except for the deviation of some individual galaxies such as Leo I. The right hand side of the figure draws the same distance-velocity diagram completed with the remote galaxies (the Andromeda subgroup is not represented), and matched to the Hubble velocity field at large distances. However, due to the uncertainties on the velocities, which are strongly dependent on the choice of the solar vector, and the small number of objects, the present data in the Local Group does

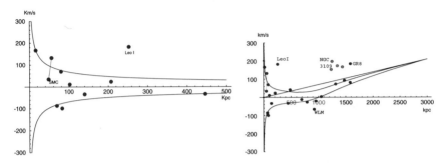

Fig. 13.44: **Distance velocity relation in the Local Group.** Left hand side: distribution of the galaxy velocities in the MW subgroup in function of their distances. The two curves represent the theoretical prediction from Kepler's third law and a velocity constant of 288 km/s. Right hand side: larger scale representation of the Local Group showing the reconnection of the local Keplerian velocity field to the outer Hubble field.

not allow to put to the test the expected velocity quantization, which must therefore await future more precise data.

13.8.6. *Clusters of galaxies*

Clusters of galaxies are the first gravitational systems in which the dark matter problem has been discovered by Zwicky [554, 555]. The Coma cluster of galaxies, which is the best studied rich cluster thanks to its relative proximity (100 Mpc) was also the first system in which probability peaks in the redshift distribution have been claimed [509], in units of ≈ 216 km/s (but with a wrong interpretation in terms of non-velocity redshifts). However, a new analysis using more recent and precise data fail to reproduce the effect (see Fig. 13.45). However, an analysis of the internal velocity distribution of several clusters reveals in a systematic way the existence of internal pseudo-periodic structures which are beyond statistical fluctuations. Two examples are given in Figs. 13.45 and 13.46.

The velocity dispersions σ of rich clusters of galaxies are of the order of 1000 km/s, which is about ten times larger than the values σ_0 expected from an application of the virial theorem to the visible mass of galaxies. Interpreted in terms of the existence of a large amount of dark matter, the ratio of dark over visible mass is

$$\frac{M}{M_0} = \left(\frac{\sigma}{\sigma_0}\right)^2 . \tag{13.160}$$

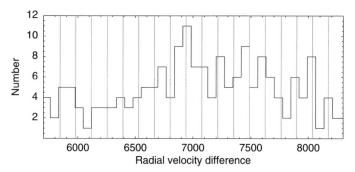

Fig. 13.45: **Radial velocity distribution in the Coma cluster.** Histogram of the radial velocity distribution (in km/s) of galaxies in the Coma cluster (214 galaxies with precision redshifts in the RC3 catalog within 6 degrees of cluster center). The distribution shows peaks with a period of 137 km/s (vertical lines). By folding the distribution on the period, one finds 136 galaxies of 214 in the peaks and 78 outside. The probability to obtain such a difference by chance is $P < 10^{-4}$. The power spectrum is consistent with this result: it yields a marginally significant peak of power $p = 7$ for a period ≈ 140 km/s.

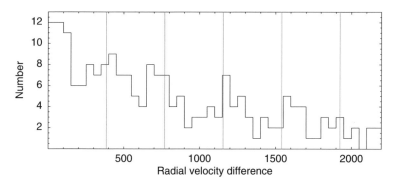

Fig. 13.46: **Radial velocity distribution in the cluster A 576.** Observed distribution of differences with the mean δv of radial velocities (in km/s) of galaxies in the A 576 cluster (206 galaxies with precision redshifts [325]). A periodicity of $w = 385$ km/s is found in this distribution (vertical lines). By folding it on the period, one finds that, 127 galaxies out of 206 lie in the peak and 79 outside. The probability for such a difference to happen by chance is $P = 4 \times 10^{-4}$. The power spectrum is consistent with this analysis: it yields a marginally significant peak of power $p = 8$ for a period ≈ 366 km/s.

In the scale-relativistic framework, the description of clusters of galaxies comes under the Hartree (Schrödinger–Poisson) equation applied to self-gravitating system. This problem is quite similar to that of the internal structure of stars (except for the additional nuclear energy that intervene in the equilibrium of stars). The question of the gravitational stability

or instability of the finite isothermal sphere has been asked for long (see
e.g. [103] and references therein). Even a classical treatment may lead to
a geometric hierarchy of scale, suggesting a fractal-like behavior of the
solutions [486, 487, 103]. One may easily construct a scale relativistic
generalization of the Emden/polytrope sphere, in terms of a Schrödinger
equation where the potential reads

$$\phi = k\rho^{\gamma}, \tag{13.161}$$

where the polytrope index is $n = 1/\gamma$, and where one recovers the Boltzman
case $\phi = k \ln(\rho_0/\rho)$ for $\gamma = 0$. One obtains a nonlinear Schrödinger equation

$$\mathcal{D}^2 \Delta \psi + i\mathcal{D}\frac{\partial \psi}{\partial t} - \frac{k}{2m}|\psi|^{2\gamma}\psi = 0. \tag{13.162}$$

It takes the form of the Ginzburg–Landau equation of superconductivity
in the case $\gamma = 1$ (see Chapter 10). A program of numerical resolution
of this class of equations for self-gravitating systems is under development
[283, 284].

Meanwhile, one can use the fact that the solution will be subjected
to a generalized Heisenberg relation that reads $\sigma_x \times \sigma_v \approx 2\mathcal{D} = GM/2w_0$,
where w_0 is found to be equal to $144\,\text{km/s}$ in several systems and at several
various scales. In this case, a typical velocity dispersion for large clusters of
$\approx 800\,\text{km/s}$ can be obtained for a small cluster mass $M \approx 5 \times 10^{13} M_\odot$ that
does not include excedentary non-baryonic dark matter, since we get the
numerical relation: $(1\,\text{Mpc}) \times (800\,\text{km/s}) = G \times (5 \times 10^{13} M_\odot)/2 \times (144\,\text{km/s})$
[389].

With regards to the possible internal structures in velocity space, they
should be given by $P_v = |a(v)|^2$, where $a(v)$ is the wave function in
velocity space. This wave function is given by the Fourier transform of the
position wave function $\psi(x)$, which is a solution of the Hartree equation.
Consider a Maxwellian velocity distribution, which is a solution of the
formation equation in a medium of average constant density, i.e. here, the
cosmological background, see Sec. 13.3.2, perturbed by periodic fluctuations
(Fig. 13.47a),

$$P_v = a(v)^2 = e^{-(v/v_0)^2}\left(1 + a\cos\frac{v}{v_1}\right)^2. \tag{13.163}$$

Such a velocity distribution is obtained by Fourier transform from a matter
distribution in space (bottom Fig. 13.47a) that includes the cluster (central
peak) but also a distant shell of matter around it. This is an interesting

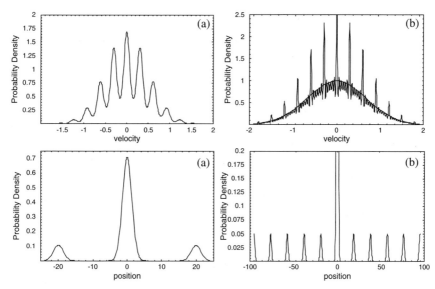

Fig. 13.47:   Idealized models of periodic velocity fluctuations inside clusters of galaxies. (a) A shell of matter around the cluster yields periodic fluctuations of velocity inside the cluster (the ratio between the density in the shell and in the central cluster has been increased for clarity). (b) A periodic repartition of matter outside the cluster reinforces the periodic velocity peaks.

result owing to the fact that the actual large scale distribution of galaxies shows such bubble-like structures and large voids (see [170] and references therein). Actually, the property of the Fourier transform means that the large scale space structures are in correspondence with small scale velocity structures (as summarized in the Heisenberg relation). This is a fascinating possibility, since if true, it would mean that the study of the small scale velocity structure of astrophysical objects would yield informations about the large scale universe. In particular, some observational evidence has been found according to which the large scale distribution of galaxies could be pseudo-periodic (see Sec. 13.8.8). Such a periodicity would re-inforce the possible peaks in the velocity distribution of clusters of galaxies, as shown in Fig. 13.47b.

A detailed application of the scale-relativity theory to these structures is still to be developed, since a global description in terms of a quantum-like statistical physics is also needed in this case. This will be the subject of forthcoming works.

13.8.7. *Local Supercluster*

The existence of a preferential value of $\approx 36 = 144/4 \, \text{km/s}$ for the inter-velocity of galaxies at the scale of the local supercluster was first suggested by Tifft and Cocke [511]. Croasdale [120], then Guthrie and Napier [216] found some support for this claim using spirals with accurately measured redshifts up to ≈ 1000 km/s. In a more recent work [217], they have confirmed the effect with galaxies reaching $\approx 2600 \, \text{km/s}$. A power spectrum analysis on their database has been performed [123] and has yielded a peak of power $p = 17$ corresponding to a preferential intervelocity of $37.5 \, \text{km/s}$, close to the expected value $144/4 = 36 \, \text{km/s}$. The probability to obtain such a peak by chance in the power spectrum for a prealably expected value (among 100 trials) is $P = 4 \times 10^{-6}$. A marginally significant peak (of probability 1%) is also found at $432 = 3 \times 144 \, \text{km/s}$, which is the solar and PSR 1257+12 value (see Fig 13.48).

Moreover, Guthrie and Napier remark that the phenomenon appears strongest for the galaxies linked by group membership. We confirm this result by a specific analysis of the group galaxies (Fig. 13.49), which show peaks of their intervelocity distribution at $\approx 144, 72, 36 \, \text{km/s}$ and possibly sub-levels (while $108 = 3 \times 36 \, \text{km/s}$ is absent), as expected for a local Kepler potential for which $v_n = w_0/n$. It is also supported by the study by Jacob

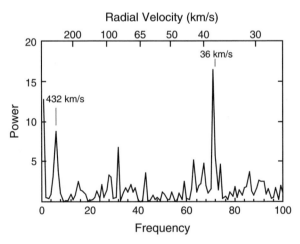

Fig. 13.48:   **Power spectrum of radial velocities in the local supercluster.** Power spectrum of the radial velocities of the 97 galaxies in the Guthrie–Napier sample with accurately determined redshifts, corrected for the optimum solar vector (219 km/s, 96 degree, −11 degree). The total range of velocities is 2665 km/s. Significant peaks are obtained for $v \approx 36 \, \text{km/s}$ and $v \approx 432 \, \text{km/s}$.

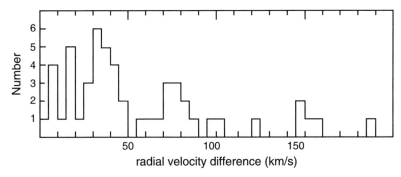

Fig. 13.49: **Distribution of velocity differences in groups of galaxies.** Distribution of closest intervelocities of galaxies, which are members of groups in the Guthrie–Napier accurate redshift sample, showing probability peaks around 144, 72 and 36 km/s.

[239] and Lefranc [282] of other independent samples of galaxies at the scale of the local supercluster, which have provided significant peaks at 48 and 36 km/s.

All these properties can be readily explained in the scale relativity framework. Indeed, the potential to be inserted in the gravitational Schrödinger equation for describing the structuration of these objects is, to the first order approximation, a combination of a three-dimensional harmonic oscillator potential describing the density background at the scale of the local supercluster and of local Kepler potentials in the grouping zones. The solution $n = 1$ of the harmonic oscillator leads one precisely to expect the existence of a preferential intervelocity depending on the density and on the gravitational coupling constant. Moreover, since these structures are understood as the equivalent in velocity space of the well-known structures in position space (galaxies, groups, clusters, etc.), we also expect their strengthening in groups and the local replacement of a linear quantization law by an inverse law.

13.8.8. *Very large scale structures*

The morphology of superclusters of galaxies is, like stellar associations, ensembles of star clusters and groups of galaxies, characterized by the existence of pairs, chains, etc. as expected from the solution of the Schrödinger equation in the cosmological potential (i.e. harmonic oscillator in de Sitter coordinates) produced by a uniform background matter density.

Redshift structures of the kind theoretically predicted by the scale relativity/macroscopic Schrödinger approach, i.e. probability peaks for some specific values of redshift differences, have also been detected on very large scales. Broadhurst *et al.* [74] have detected a pseudoperiodicity in units of $\approx 12800\,\mathrm{km/s}$ ($128\,h^{-1}\,\mathrm{Mpc}$) in the distribution of galaxies in two opposite cones directed toward the North and South Galactic poles, up to distances going beyond one $h^{-1}\,\mathrm{Gpc}$. This effect has since been confirmed in several redshift samples of galaxy and clusters of galaxies [271, 154, 469].

At even larger scales, the existence of peaks in the probability density of quasar redshifts has been pointed out by Burbidge and Burbidge [77] and subsequently confirmed by many authors. One finds peaks for redshift differences such that $\Delta \ln(1 + z) = 0.206$ [246].

The detailed understanding of such effects, since they involve cosmological scales reaching gigaparsecs and more, would need the development of a new level of the theory, namely, a scale relativistic generalization of the construction of global cosmological models, which are currently done in the framework of Einstein's general relativity. Such a wide enterprise goes beyond the frame of the present study, and is left open to future works.

Let us only briefly note that we expect the gravitational coupling constants $\alpha_g = w/c$ at different scales to be interrelated, and, ultimately, we expect the w's to be related to the maximal velocity c (i.e. $\alpha_g = 1$), while large scale structures in global cosmology are one of the astrophysical domains (with compact objects) where the velocities become close to the light velocity, allowing one to test this conjecture.

In this respect, the observed values of the large scale and small scale constants are consistent with such a matching. Indeed, the inverse coupling constant corresponding to the tentative theoretical value $w_0 = 144.82 \pm 0.01\,\mathrm{km/s}$ (Sec. 13.3.6), which agrees with the observed one 144.7 ± 0.5 [374], is $\alpha_g^{-1} = 2070.10 \pm 0.15$. We expect some other coupling constants applying to various scales to be multiples or submultiples of this value, and in the same time submultiples of c (but we cannot exclude other integer ratios), while the sequence of values derived from the decomposition in prime factors $2070 = 2 \times 3^2 \times 5 \times 23$ yields:

(i) $c/23 = 13034\,\mathrm{km/s}$: this value is in very close agreement with the Broadhurst *et al.* periodicity at $128\,h^{-1}$ Mpc, which is a very precise estimate since the power spectrum analysis has given a highly significant sharp peak $P(n) = 80$ while the comoving separation of pairs reaches $1.4\,h^{-1}$ Gpc [74]. The comoving distance between density peaks has been calculated by Broadhurst *et al.* in a flat model without cosmological constant, i.e. it is given in terms of redshift z by

$H_0 r/c = 2(1 - 1/\sqrt{1+z})$. Assuming a centered observer, one finds that the radial velocity difference between the two first peaks is therefore $\delta(cz) = 13076\,\mathrm{km/s}$, very close to $c/23$.

(ii) $c/5$, which corresponds to $\delta z = 0.2$, the possible quasar periodicity.

(iii) $2w_0 = c/(3^2 \times 5 \times 23) \approx 288\,\mathrm{km/s}$, $3w_0 = c/(2 \times 3 \times 5 \times 23) \approx 432\,\mathrm{km/s}$, $9w_0 = c/(2 \times 5 \times 23) \approx 1296\,\mathrm{km/s}$, values which have been identified (in addition to the main value $w_0 = 144\,\mathrm{km/s}$) in gravitational structures at all scales (sun, solar system, extrasolar planetary systems, binary stars, galaxy pairs, etc.).

(iv) $w_0/5$, which is the coupling constant of the outer solar system, $w_0/(5 \times 7)$, that of the very distant SKBOs, etc.

Problems and Exercises

Exercise 34 Apply the scale relativity/macroscopic Schrödinger theory to other examples of gravitational structuring, and compare the predicted structures to the observational data.

Hints: The conditions which underlie the fractal approach may apply to a very large ensemble of astrophysical systems over many scales, a small fraction of which has been considered here, most of the time in an incomplete way, so that much work remains to be done. The method consists in writing the linear or nonlinear Schrödinger equation for the gravitational potential (including when necessary the self-gravitation) of the system under consideration (interstellar medium, internal dynamics of the Galaxy, star clusters, galaxy groups, etc.) then in solving this equation (if necessary, numerically) for the various quantized quantities (in terms of symmetries and limiting conditions). ∎

Chapter 14

APPLICATIONS TO LIFE SCIENCES
AND OTHER SCIENCES

We shall finally conclude this book with a brief overview of possible applications of the scale relativity approach to sciences other than physical sciences, including life sciences, sciences of societies, historical [181] and geographical sciences [310, 182, 183, 412] and human sciences [512, 382, 398, 404, 525]. It will unfortunately be impossible to be exhaustive about both the already-obtained results and the various current developments, since this is a whole domain of research by itself, which is now wide open to investigation, and to which entire books [373, 408, 525], parts of review papers [377, 388, 400] and full review papers [33, 402] have been devoted.

14.1. Fractals in Biology

Self-similar fractal laws have already been used as models for the description of a large number of biological systems (lungs, blood network, brain, cells, vegetals, etc., see e.g. [416, 295], previous volumes, and references therein), as well as modelization tools that often manifest a fractal behavior, such as cellular automata (see [439] and references therein), arithmetic relators [330], etc.

The scale relativistic tools, which generalize scale invariance to scale covariance, may also be relevant for a description of behaviors and properties, which are typical of living systems (see Fig. 14.1). Some examples have been given in [375, 102, 373, 380, 86, 377] with regards to halieutics, log-periodic branching laws applied to species evolution, society evolution, embryogenesis and cell wall models.

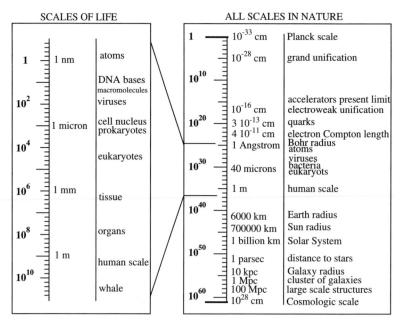

Fig. 14.1: **Physical and biological scales.** The range of biological scales (left), embedded in the range of physical scales (right).

Scale relativity, through its development of a macroscopic Schrödinger-type theory, may also provide a physical and geometric framework for the description of additional properties, such as formation, duplication, morphogenesis and imbrication of hierarchical levels of organization. This approach does not mean to dismiss the importance of chemical and biological laws in the determination of living systems, but on the contrary to attempt to establish a geometric foundation that could underlie them. It is hoped that the scale relativity tools and methods could participate in the future foundation of a genuine mathematical biology, which would have a predictive power analogous to that of mathematical physics since its foundation four centuries ago.

14.2. Applications of Log-Periodic Laws

14.2.1. *Species evolution*

14.2.1.1. *Introduction*

Let us first consider the application of log-periodic laws to the description of critical time evolution. Recall that a log-periodic generalization to scale invariance has been obtained in Sec. 4.3.1 as a solution to wave-like differential scale equations. Interpreted as a distribution of probability, such solutions therefore lead to a geometric law of progression of probability peaks for the occurence of events.

Several studies, in particular by D. Sornette *et al.*, have shown that many biological, natural, sociological and economic phenomena obey a log-periodic law of time evolution, such as can be found in some critical phenomena: earthquakes [495], stock market crashes [496], evolutionary leaps [102, 373, 380], long time scale evolution of western and other civilizations [373, 380, 213], world economy indices dynamics [241], embryogenesis [86], etc. Thus emerges the idea that this behavior, typical of temporal crisis, could be extremely widespread, as much in the organic world as in the inorganic one [497].

In the case of species evolution, one observes the occurrence of major evolutionary leaps leading to bifurcations among species, which proves the existence of punctuated evolution [203] in addition to the gradual one. The global pattern is assimilated to a "tree of life", whose bifurcations are identified to evolutionary leaps, and branch lengths to the time intervals between these major events [102]. As early recognized by Leonardo da Vinci, the branching of vegetal trees and rivers may be described at a first self-similar approximation by simply writing that the ratio of the lengths of two adjacent levels is constant in the mean. We have made a similar hypothesis for the time intervals between evolutionary leaps, namely, $(T_n - T_{n-1})/(T_{n+1} - T_n) = g$. Such a geometric progression yields a log-periodic acceleration for $g > 1$, a deceleration for $g < 1$ and a periodicity for $g = 1$. The events converge towards a critical time T_c, which can then be taken as reference (except when $g = 1$, which corresponds to an infinite critical time), yielding the following law for the event T_n in terms of the rank n:

$$T_n = T_c + (T_0 - T_c)g^{-n}, \tag{14.1}$$

where T_0 is any event in the lineage, n the rank of occurrence of a given event and g is the scale ratio between successive time intervals. Such a chronology is periodic in terms of logarithmic variables, i.e. $\log|T_n - T_c| = \log|T_0 - T_c| - n \log g$.

A statistically significant log-periodic acceleration has been found at various scales for global life evolution, for primates, for sauropod and theropod dinosaurs, for rodents and North American equids. A deceleration law was conversely found in a statistically significant way for echinoderms and for the first steps of rodents evolution (see Figs. 14.2 and 14.3 and more detail in [102, 373, 380]). One finds either an acceleration toward a critical date T_c or a deceleration from a critical date, depending on the considered lineage.

14.2.1.2. A fractal tree model

Constructing the evolutionary law

Let us consider a node in a tree where a branch divides into k sub-branches. Let us assume that the total cross-section before (level n) and after (level $n+1$) the node is preserved. If this section is bidimensional (as, for example, with conservation of sap flow), this is reflected in the relationship between radii: $k\, r_{n+1}^2 = r_n^2$. But a more general relationship can be considered by introducing a fractal dimension D: $k\, r_{n+1}^D = r_n^D$. If we now accept that the tree is fully self-similar (as a minimal simplifying assumption), the ratio of branch lengths will then be equal to the ratio of their radii, giving $g = k^{1/D}$. Since $g > 1$, the total length measured along a given lineage is therefore finite, since it is given by the converging infinite sum: $L_c = L_0(1 + g^{-1} + g^{-2} + \cdots) = gL_0/(g - 1)$. For a temporal tree, these "lengths" are given by the time interval between two evolutionary events: $L_n = T_{n+1} - T_n$. Convergence of the above series therefore means there is a critical time, T_c, marking the end of the evolutionary process for a given lineage (or its beginning in case of deceleration).

If we now take the final critical time T_c as the time origin, self-similarity is preserved, because the time interval ratios relative to this origin are still given by g^n. Finally, we recover the log-periodic law of Eq. (14.1), $T_n = T_c + (T_0 - T_c)g^{-n}$.

This law is dependent on two parameters only, g and T_c, which, of course, have no reason a priori to be constant for the entire tree of life. Note that g is not expected to be an absolute parameter, since it depends on the density of events chosen, i.e. on the adopted threshold in the choice of their importance (namely, if the number of events is doubled, g is replaced by $\sqrt{g}$). Only a maximal value of g, corresponding to the very major events, could possibly have a meaning. On the contrary, the value of T_c is expected to be a characteristic of a given lineage, and therefore not to depend (within error bars) on such a choice [380].

Methodology

Our method of statistical analysis of the fit between the data and this law consists of using Student's t variable associated with the correlation

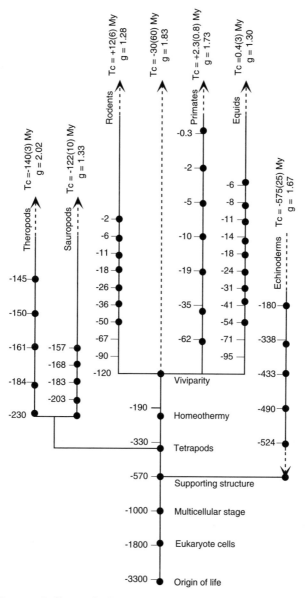

Fig. 14.2: **Log-periodic evolution on the "tree of life".** The dates of major evolutionary events of seven lineages (common evolution from life origin to viviparity, Theropod and Sauropod dinosaurs, Rodents, Equidae, Primates including Hominidae, and Echinoderms) are plotted as black points in terms of $\log(T_c - T)$, and compared with the numerical values from their corresponding log-periodic models (computed with their best-fit parameters). The adjusted critical times T_c and scale ratios g are indicated for each lineage [102, 373, 380]. The number between brackets is the uncertainty on T_c in million years.

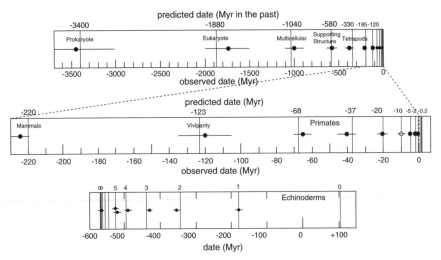

Fig. 14.3: Log-periodic acceleration and deceleration of three lineages (from [102, 373, 380]). The dates are here plotted in real time to show clearly the acceleration or deceleration. Top: main trunk of the 'tree of life', from the apparition of life to homeothermy and viviparity (last two large plotted points). The dates are best-fitted by an accelerating log-periodic law $T_n = T_c + (T_0 - T_c) \times g^{-n}$, with $T_c = -30 \pm 60$ Myr and $g = 1.83$ (for a total time interval of 3.5 Gyr). Middle: (zoom on the last 200 Myr of the top figure): comparison of the dates of the major leaps in the evolution of primates including the hominids plus the two preceding main events: apparition of mammals and viviparity, with an accelerating log-periodic law. The best fit gives $T_c = 2.0 \pm 0.4$ Myr and $g = 1.78 \pm 0.01$. For the 14 dates from the origin of life to the appearance of the Homo sapiens bauplan, one finds $T_c = 2.1 \pm 1.0$ Myr and $g = 1.76 \pm 0.01$. The probability is less than 10^{-4} to obtain such a fit by chance. Bottom: comparison with a decelerating log-periodic law of the dates of the major leaps in the evolution of echinoderms. The best fit yields $T_c = -575 \pm 25$ Myr and $g = 1.67 \pm 0.02$ (origin excluded). This means that the critical time from which the deceleration starts is their date of apparition (within uncertainties). All results are statistically highly significant (see the quoted references for details about the data and their analysis).

coefficient in the graph [event rank n, $\log(T_c - T_n)$] as the statistical estimator. When T_c is given, the law of Eq. (14.1) becomes linear when it is expressed in logarithm form. Therefore we vary continuously the values of T_c, and for each of these values we compute the values of g by a least-square fit, then we determine the associated t_{st} (Student). Then we construct the curve $t(T_c)$ (see examples in Figs. 14.10 and 14.12 below). The optimized value of T_c is given by the peak of this curve, and the error bar on T_c is estimated from it half-width at half-maximum. Finally, Monte-Carlo simulations have been made to calibrate this estimator and define the

Table 14.1: Values of the peak of the Student's t variable that corresponds to probability thresholds 1/100 (2.3 sigma), 1/1000 (3 sigma) and 1/10000, according to the number of dates in the sample.

n	4	5	6	7	8	9	10	12	14	16
$t(0.01\%)$	6860	293	82	93	66	69	70	53	62	72
$t(0.1\%)$	828	100	50	46	42	37	38	44	50	52
$t(1\%)$	85	34	28	27	27	26	30	32	37	40

associated probability, by applying the same analysis to dates chosen at random and arranged in chronological order (Table 14.1).

14.2.1.3. *Application to the evolution of species*

This log-periodic evolutionary law has been put to the test at various levels of analysis [102, 373, 380, 408]: (i) the tree of life at a global level, from the first appearance of life to viviparity [135]; (ii) sauropod and theropod dinosaurs postural structures [261, 488]; (iii) rodents families [99, 204]; (iv) the North American equid genera [204]; (v) primates bauplans, including the hominids [101]; (vi) Echinoderms groups [223].

In each case we find that a log-periodic law provides a satisfactory fit for the distribution of dates, with different values of the critical date T_c and of the scale ratio g for different lineages. The obtained behavior may be an acceleration or a deceleration depending on lineage and time scale. The results are statistically significant.

We give in what follows (see also Fig. 14.2) the adopted dates (in Myrs before present) for the major jumps of the studied lineages. The error bars are typically $\delta T/T \approx 10\%$ or less, i.e. $\delta \log(T_c - T) \approx 0.04$. Since we are interested here in pure chronology, if several events occur at the same date (within uncertainties), they are counted as one. Then we give the result of the least-square fit of the log-periodic model and the associated Student variable with its corresponding probability to be obtained by chance. For each lineage we include in the analysis the common ancestors down to the origin of life (except for Echinoderms, which show deceleration instead of acceleration). The obtained parameter values are compatible with those given in Fig. 14.7, which result from a fit that does not include the ancestors of the lineage.

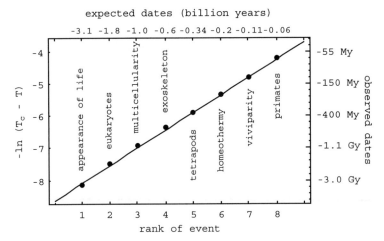

Fig. 14.4: **Log-periodic distribution of major evolutionary leaps (from [102]).** The line corresponds to $T_c = 2.3$ My and $g = 1.73$.

Global tree from origin of life to viviparity (see Figs. 14.2 and 14.4)

{apparition of life: -3500}, {Eukaryote cells: -1750}, {Multicellular stage: -1000}, {Supporting structure: -570}, {Tetrapods: -380}, {Homeothermy (mammals and dinosaurs): -220}, {Viviparity: -120}.

These events exhibit a significant acceleration towards:

$$T_c = -32 \pm 60 \, \text{My}; \quad g = 1.83 \pm 0.03; \quad t_{st} = 36, P < 0.003 \ (N = 7 \, \text{events}).$$

Primates including hominids (see Figs. 14.2 and 14.5)

{prosimian bauplan: -65}, {simian bauplan: -40}, {great apes bauplan: -20}, {Australopithecus bauplan: -5}, {Homo bauplan: -2}, {H. sapiens bauplan: -0.18}

$T_c = 2.1 \pm 1.0 \, \text{Myr}; g = 1.76 \pm 0.01; t_{St} = 110, P < 0.0001 \ (N = 14$ events, including the "global" tree).

In Fig. 14.5, we have included the date of the common Homo-Pan-Gorilla ancestor estimated from genetic distances ($\sim -8 \, \text{Myr}$, [102, 373]). Reversely, this date can be predicted (as a missing point) from the log-periodic accelerating law: we find a compatible date for this event of $-10 \, \text{Myr}$ in the past.

It has been suggested (see [281]) that other events (actually minor ones) should also be taken into account for this lineage, leading to the following

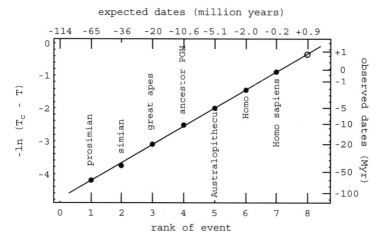

Fig. 14.5: **Log-periodic acceleration of primate evolutionary leaps (from [102]).** The line corresponds to $T_c = 2.3$ My and $g = 1.73$. The next expected evolutionary leap is plotted as a white point.

dates (in Myrs):

$$\{-65, -53, -40, -35, -25, -20, -17, -10, -7, -5, -3.5, -2, -0.18\}.$$

The statistical analysis gives:

$$T_c = 5.8 \pm 4.0 \, \text{My}; \; g = 1.23 \pm 0.01; \; t_{st} = 57, P < 0.001 \; (N = 13 \, \text{events}).$$

The result is still significant, and, moreover, the critical date agrees within error bars (to less than 1σ) with our previous determination. This confirms that T_c is characteristic of the lineage beyond the choice of the events. On the contrary the value of g, which depends on the density of dates, is not conserved, as expected.

Note that, even though the critical date $T_c = 2.1 \pm 1.0$ Myr is significantly different from 0, its difference from present is nevertheless of the order of size of the time difference between the last major leaps (1.8 Myr). Since one expects threshold effects to be added to the theoretical law, in which, unrealistically, the number of events toward the critical time is infinite, it seems more likely that this particular lineage has reached the end of its evolutionary capacity (concerning the characters considered in the changes of bauplans, see [373]).

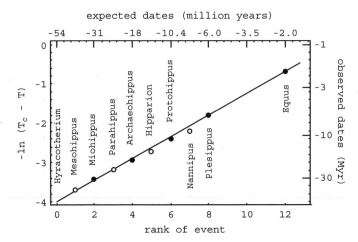

Fig. 14.6: **Log-periodic distribution of Equid evolutionary leaps (from [102]).** The line corresponds to $T_c = 0$ My and $g = 1.32$ (like-colored dots corresponds to a squared scale ratio $g = 1.73$).

Fossil North American equids (see Figs. 14.2 and 14.6)

{Hyracotherium: -54}, {Mesohippus: -38}, {Miohippus: -31}, {Parahippus: -24}, {Archeohippus: -19}, {Hipparion: -15}, {Protohipus: -11}, {Nannipus: -9}, {Plesippus: -6}, {Equus: -2}.

$T_c = -1.0 \pm 2.0$ My; $g = 1.32 \pm 0.01$; $t_{st} = 99$, $P < 0.001$ ($N = 16$ events, including the "global" tree, excluding Equus).

It is remarkable that, for this lineage, the extinction date (about 8000 years ago) is consistent, within uncertainties, with the critical date $T_c \approx 0$.

Rodents (see Figs. 14.2, 14.7 and 14.8)

In the case of rodents, the analysis is different from the other lineages, since it is made on their whole arborescence, according to the data of Hartenberger [223]. We have plotted in Fig. 14.7 the histogram of the distribution of the 61 dates of appearance of rodent families. Well-defined peaks can be identified in this distribution. It is on these peaks that we perform our analysis (Fig. 14.8). However, some uncertainty remains, in particular concerning the large peak after the date of first appartition of the lineage. Three different interpretations are considered.

In Figs. 14.2 and 14.8 we have used the mean value (-50 My) of the first peak. This yields a critical date $T_c = 12 \pm 6$ My in the future. One can

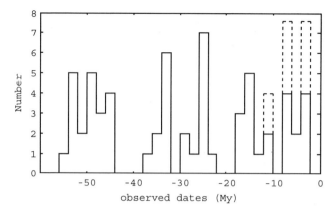

Fig. 14.7: **Distribution of rodents appearance dates (from [380]).** The data is taken from [223]. These data include only a subfraction of the events after -12 My, so that the amplitude of the last peaks is underestimated and has been extrapolated (dotted line).

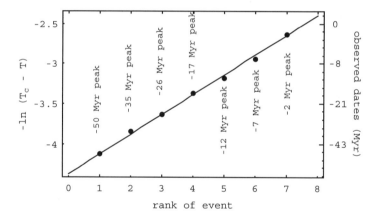

Fig. 14.8: **Acceleration of density peaks of rodent evolutionary leaps (from [102]).** The line corresponds to $T_c = +12$ My and $g = 1.32$.

also singularize the latest date, yielding:

$$\{-56\}, \{-45\}, \{-34\}, \{-26\}, \{-18\}, \{-12\}, \{-7\}, \{-2\}. \text{ One obtains:}$$

$T_c = +7 \pm 3$ My; $g = 1.32 \pm 0.01$; $t_{st} = 78$, $P < 0.001$ ($N = 15$ events, including ancestors in the "global" tree).

But a closer scrutiny of the data suggests that the spurt of branching (that correspond to the sub-peaks inside the main first peak in Fig. 14.7) that followed the group's first appearance actually decelerates. This would be in agreement with the interpretation of these structures in terms of critical phenomena. We find that the deceleration is issued from a critical point at $T_c = -62 \pm 5\,\text{Ma}$, which agrees with the date estimated for the group's first appearance.

Once this initial deceleration is allowed for, the following dates $\{-34, -26, -18, -12, -7, -2)$ exhibit highly significant acceleration towards $T_c = 27 \pm 10\,\text{Ma}$ $(t_{st} = 98, P < 10^{-4})$. It is therefore remarkable that the rodents lineages generally exhibit critical dates, which are further in the future than others lineages. This result supports Gould's statement [204] according to which the species, which show the largest biodiversity, also have the highest potentiality of evolution.

Sauropod dinosaurs (see Figs. 14.2 and 14.9)

Wilson and Sereno [541] have identified five well-defined major events in the evolution of their legs: {Sauropoda: -230}, {Eusauropoda: -204}, {Neosauropoda: -182}, {Titanosauriforms: -167}, {Titanosauria: -156}. These events exhibit a marked log-periodic acceleration toward:

$T_c = -128 \pm 10\,\text{My}; \; g = 1.41 \pm 0.01; \; t_{st} = 122, P < 0.001$ ($N = 10$ events, including ancestors from the "global" tree).

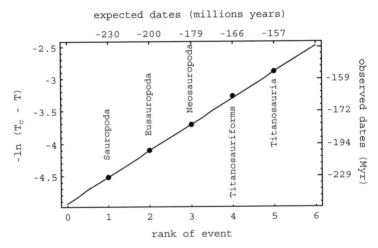

Fig. 14.9: **Major evolutionary leaps of sauropods (from [102]).** The line corresponds to $T_c = -139\,\text{My}$ and $g = 1.50$.

Theropod dinosaurs (see Figs. 14.2)

One can identify from the data of Sereno [488] the following main dates in the evolution of theropods (once again, several events having the same date within uncertainties are counted as one): {Neotetanurae: -227}, {Coelurosauria: -187}, {Maniraptora: -160}, {Aves: -150}, {Euornithes: -145}.

There is a significant acceleration towards:

$T_c = -139 \pm 4 \,\mathrm{My}$; $g = 2.02 \pm 0.02$; $t_{st} = 69$, $P < 0.001$ ($N = 10$ events, including ancestors down to the origin of life).

This supports the existence of a log-periodic acceleration for the whole group of *Saurischia* (Sauropods and Theropods). However, the analysis of the other large dinosaur group, *Ornithischia*, has given no statistically significant structure. This could indicate either that the log-periodicity is not universal and characterizes only some particular lineages or that the data are uncomplete for this group.

Anyway, it is remarkable that for these two groups, the critical date around $-130 \,\mathrm{Myr}$ lies more than $\sim 60 \,\mathrm{Myr}$ before their extinction. This result leads to interpreting this critical date, not as an extinction date, but as the end of the capacity of evolution of the considered lineage [102].

Echinoderms (see Figs. 14.2 and 14.10)

The critical phenomena approach to evolutionary process leads to expecting not only acceleration toward a crisis date, but also deceleration from it. The echinoderm group supports this view. The major events that punctuate their evolution happen at the following dates, according to David and Mooi [127]:

{apparition: -570}, {$-526, -520$}, {-490}, {-430}, {-355}, {-180}.

Processing of this data shows that this group *decelerates* from a critical date $T_c = -575 \pm 25 \,\mathrm{My}$ (see Fig. 14.10). This epoch identifies, within error bars, with the first appearance datum around $-570 \,\mathrm{My}$. We find:

$T_c = -575 \pm 25 \,\mathrm{My}$; $g = 1.67 \pm 0.02$; $t_{st} = 58$, $P < 0.003$ ($N = 5$ events).

14.2.1.4. *Discussion*

Let us first discuss possible biases and uncertainties in our analysis [380]. There is a "perspective" bias, linked to observational data being fossil records observed at the present epoch only. This bias can manifest itself in two ways.

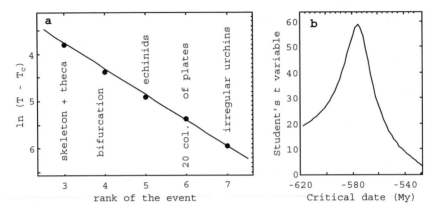

Fig. 14.10: **Log-periodic deceleration of echinoderm evolution leaps.** Comparison of the main dates of the evolution of echinoderms with a log-periodic decelerating law of critical date $t_c = -575\,\mathrm{My}$ and scale ratio $g = 1.67$ (Fig. a). Figure b shows the estimation of the critical date through the optimisation of the Student's t variable (from [380]).

First, the uncertainty on the dates increases with the date itself, so that we expect that $\delta T/T$ be about constant, which could lead to a log-periodic behavior. We have discussed this bias in [102] and we have shown that it cannot account for the observed structure. The additional information given here and in [373] that one observes also decelerations reinforce this conclusion. A second possible form of this bias could be an increasing number of missing events in fossil records for increasing dates in the past. Against such an interpretation, one can recall that the quality of the fossil records, concerning in particular their completeness, has been recently reaffirmed by Kidwell and Flessa [249]. Moreover, the number of missing links needed to compensate for the acceleration seems to be unreasonably large (the interval between major events goes from two billion years at the beginning of life to about two million years now).

In addition, the bias about the choice of dates, in particular in defining which characters are considered to be major ones, has been analyzed here. The solution to this problem lies in the observation that if the acceleration (or deceleration) is real and intrinsic to the lineage under study, its occurrence and the date of convergence T_c ought not to depend (within errors) on the limit applied as to the choice of which events count as important ones. However, there is nothing intrinsic about the scale factor g between intervals, as it decreases as the number of events allowed for increases. We have been able to test this stability of the critical date with the data for which we considered several possible choices (rodents, sauropods) as well as with choices suggested by other workers (primates). We conclude that this uncertainty cannot explain the observed law, which therefore seems to be a genuine one.

However, while log-periodic accelerations or decelerations have been detected in the majority of lineages so far investigated, the question of whether this behavior is systematic or not remains an open one (cf. the general tree for dinosaurs published by Sereno [488]).

Analysis of the values of the critical date for the various lineages leads us to interpret it, in the case of an acceleration, as a limit of the evolutionary capacity of the corresponding group. When a deceleration has been detected, it starts from the appearance date of the lineage.

Let us also stress the fact that the existence of such a law does not mean that the role of chance in evolution is reduced, but instead that randomness may occur within a framework, which may itself be structured (in a partly statistical way). Such structures may find their origin in critical phenomena [373], or, in an equivalent way, in the geometry of intermittency [457].

Indeed, the observed dates follow a log-periodic law only in the mean and show a dispersion around this mean (see [373], p. 320). In other words, this is a statistical acceleration or deceleration, so that the most plausible interpretation is that the discrete T_n values are nothing but the dates of peaks in a continuous probability distribution of the events.

Moreover, it must also be emphasized that this result does not put the average constancy of the mutation rate in question. This is demonstrated by a study of the Cytochrome c tree of branching, which is based on genetic distances instead of geological chronology, and which nevertheless yields a log-periodic acceleration of most lineages [408, 413], and a quasi periodicity, which corresponds to a critical time tending to infinity in some cases. The average mutation rate remains around one mutation for 20 Myr for the last one billion years, so that one cannot escape the conclusion that the number of mutations needed to obtain a major evolutionary leap decreases with time among many lineages and increases for some of them. In other words, this means that, for accelerating lineages, less and less information is needed for each new major leap. This may be understood by a "memory effect", i.e. from the fact that each new evolutionary leap takes place in a bauplan, which is determined by all the previous ones.

14.2.2. *Embryogenesis and human development*

Considering the relationships between phylogeny and ontogeny, it appeared interesting to verify whether the log-periodic law describing the chronology

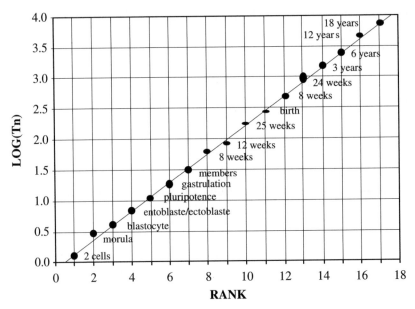

Fig. 14.11: **Log-periodic embryogenesis.** Decimal logarithm $\log(T_n)$ of the dates (in days) of the steps of human development, counted starting from the conception date, plotted in terms of their rank n, showing a log-periodic deceleration with a scale ratio $g = 1.71 \pm 0.01$ [86]. The vertical size of the points gives the confidence interval for the dates (an error bar of ± 12 percent on the first three dates has been assumed).

of several lineages of species evolution may also be applied to the various stages in human embryological development. The result, shown in Fig. 14.11, is that a statistically significant log-periodic deceleration is indeed observed, starting from a critical date which identifies with the conception date [86].

14.2.3. *Evolution of societies*

Many observers have commented on the way historical events accelerate. Grou [212] has shown that the economic evolution since the neolithic can be described in terms of various dominating poles, which are submitted to an accelerating crisis-nocrisis pattern, which has subsequently been quantitatively analysed using log-periodic laws.

Western civilizations

The median dates of the main periods of economic crisis in the history of Western civilization (as listed in [212]) are as follows (we give the dominating pole and the date, in years/JC):

{Neolithic: −6500}, {Egypt: −3000}, {Egypt: −900}, {Grece: −100}, {Rome: +400}, {Byzance: +800}, {Arab expansion: +1100}, {Southern Europ: +1400}, {Nederland: +1650}, {Great-Britain: +1775}, {Great-Britain: +1830}, {Great-Britain: +1880}, {Great-Britain: +1935}, {United States: +1975}.

A log-periodic acceleration with a scale factor $g = 1.32 \pm 0.018$ occurs towards $T_c = 2080 \pm 30$ (see Fig. 14.12). Agreement between the data and the log-periodic law is statistically highly significant ($t_{st} = 145$, $P \ll 10^{-4}$).

Pre-Columbian America

The historical evolution of pre-Columbian America provides an interesting opportunity to test the universality of the law proposed. The median dates of the economic crises of these civilizations are as follows (see [373]):

{Olmeques: −600}, {Classic: 500}, {Mayas: 1000}, {Tolteques: 1350}, {Azteques: 1550}.

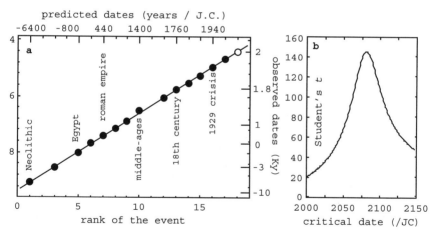

Fig. 14.12: **Log-periodic evolution of economic crises.** Comparison of the median dates of the main economic "crises" of western civilization with a log-periodic accelerating law of critical date $T_c = 2080$ and scale ratio $g = 1.32$ (Fig. a, see detailed nature of the events in text). The last white point corresponds to the predicted next crisis (1997–2000) at the date of the study (1996), as has been later supported in particular by the 1998 and 2000 market crashes, while the next crises are now expected (following this pattern) for (2015–2020), then (2030–2035). Figure b shows the estimation of the critical date through the optimisation of the Student's t variable. This result is statistically significant, since the probability to obtain such a high peak by chance is $P < 10^{-4}$.

A good agreement is obtained between these dates and a log-periodic law of factor 1.76 ± 0.02 and critical point $T_c = 1800 \pm 80$ ($t_{st} = 58$, $P < 5 \times 10^{-3}$).

Discussion: The result obtained for the Western civilization, $T_c = 2080 \pm 30$, has been later confirmed by Johansen and Sornette [241] by an independent study on various market indices, domestic indices and research and development indices, on a time scale of about 200 years, completed by demography on a time scale of about 2000 years. They find critical dates for these various indices in the range 2050–2070, which agrees within error bars with the longer time scale result.

One of the intriguing features of all these results is the frequent occurence of values of the scale ratio g close to $g = 1.73$ and its square root 1.32 (recall that one passes from a value of g to its square root by simply doubling the number of events). This suggests once again a discretization of the values of this scale ratio that may be the result of a probability law (in scale space) showing quantized probability peaks. We have considered the possibility that $g = 1.73 \approx \sqrt{3}$ could be linked to a most probable branching ratio of three [102, 373], while Queiros-Condé [457] has proposed a "fractal skin" model for understanding this value.

14.2.4. *History and geography*

The application of the various tools and methods of the scale relativity theory to history and geography has been proposed by Martin and Forriez [181, 310, 182, 183]. Forriez has shown that the chronology of some historical events (various steps of evolution of a given site) recovered from archeological and historical studies can be fitted by a log-periodic deceleration law with again $g \approx 1.7$ and a retroprediction of the foundation date of the site from the critical date [181, 182]. Moreover, the various differential equation tools developed in the scale relativity approach both in scale and position space, including especially the nonlinear cases of variable fractal dimensions, have been found to be particularly well adapted to the treatment of geographical problems [310, 412].

14.2.5. *Geosciences*

Let us give some examples of applications of critical laws (power laws in $|T - T_c|^\gamma$ and their log-periodic generalizations) to problems encountered in Earth sciences, specifically, earthquakes (South California and Sichuan) and decline of Arctic sea ice.

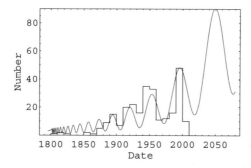

Fig. 14.13: **Log-periodic deceleration of earthquake rate.** The data are taken from the US Geological Survey EarthQuake Data Center (years 1932–2006) and EarthQuake Data Base (Historical earthquakes, years 1500–1932). This rate is well fitted by a power law subjected to a log-periodic fluctuation decelerating since a critical date $T_c = 1796$ (continuous line). The model predicts a future probability peak around the years 2050 [414].

14.2.5.1. *Southern California earthquakes*

The study of earthquakes has been one of the first domains of application of critical and log-periodic laws [495, 9]. The rate of California earthquakes is found to show a very marked log-periodic deceleration [414, 401]. We show indeed in Fig. 14.13 the observed rate of Southern California earthquakes of magnitude larger than five, compared with a log-periodic deceleration law. A clear log-periodic oscillation of probability peaks can be seen in the data, accompanying a power-law increase of their amplitude, starting from a critical date of about 1796. This model allows us to expect future peaks of probability around the years 2050 then 2115.

14.2.5.2. *Sichuan 2008 earthquake*

The May 2008 Sichuan earthquake and its replicas also yields a good example of log-periodic deceleration, but on a much smaller time scale (see Fig 14.14) of some days. We have performed a combined study of the distribution with time of the magnitudes of the aftershocks and of their rate. Both show a continuous decrease to which are added discrete sharp peaks. The peaks, which are common to both diagrams show a clear deceleration according to a log-periodic law starting from the main earthquake of May 12.27, 2008, which allows one to predict the next strongest replicas with a good precision [411]. For example, the peak of aftershocks of 25 May 2008 could be predicted with a precision of 1.5 day from the previous peaks before its occurrence.

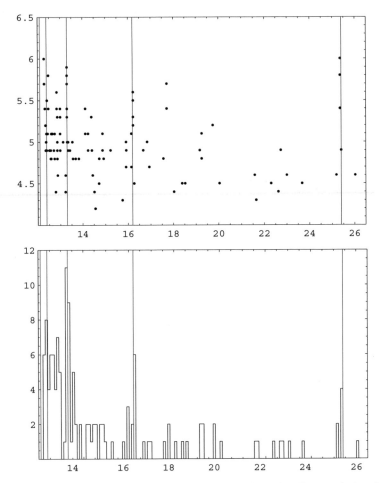

Fig. 14.14: **Aftershocks of May 2008 Sichuan earthquake.** Time evolution during 14 days of the replicas of the May 12, 2008 Sichuan earthquake (data taken from the seismic data bank EduSeis Explorer, http://aster.unice.fr/EduSeisExplorer/form-sis.asp). The top figure gives the magnitudes of the replicas and the bottom figure their rate. The peaks, which are common to both diagrams, show a clear deceleration according to a log-periodic law (vertical lines).

More recently, we have performed a power spectrum analysis of the distribution of $\ln(T_n - T_c)$, where the T_n ($n = 1$ to 167) are the individual dates of the aftershocks. The best fit has been obtained for $T_c = 12.2655$ (day in May 2008), which differs only by 6 minutes from the observed main earthquake date of 12.2694. For this value, a peak of power $p = 15.01$ has been obtained in the power spectrum for a period 0.619 (half the period

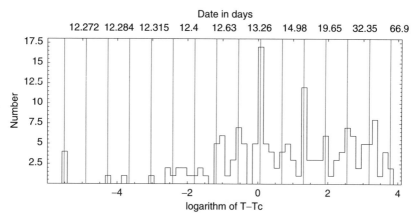

Fig. 14.15: **Log-periodic deceleration of Sichuan aftershocks.** The rate of aftershocks, which have followed the main Sichuan earthquake of May 12.27, 2008 (magnitude 7.9) is shown in function of $\ln(T - T_c)$, for $T_c = 12.2665$ May 2008. The vertical lines indicate the peaks expected according to the best fit period, 0.619 (corresponding to a scale ratio $g = 1.86$).

initially identified in Fig. 14.14), which corresponds to a scale deceleration factor $g = 1.86$. The probability to obtain by chance such a power is only 6×10^{-6}. The histogram of the distribution of the aftershocks values $\ln(T_n - T_c)$ is shown in Fig. 14.15. There are many clear peaks of probability in this distribution, whose dates are highly consistent with a log-periodic deceleration.

14.2.5.3. *Arctic sea ice extent*

It is now well-known that the decrease of arctic sea-ice extent has shown a strong acceleration in 2007 and 2008 with respect to the current models assuming a constant rate ($\approx 8\%$ by decade, line in Fig. 14.16). These models predicted, at the beginning of the years 2000, a total disappearance of the ice at minimum (15 September) for the end of the century. From the view point of these models, the 2007 and now 2008 values (see Fig. 14.16) were totally unexpected.

However, we had proposed, before the knowledge of the 2007 minimum, to fit the data with a critical law of the kind $y = y_0 - a|T - T_c|^\gamma$ [401]. Such an accelerating law has the advantage to include in its structure the fact that one expects the ice to fully disappear after some date, while the constant rate law formally pushed the date of total disappearance to infinity. The fit of the data up to 2006 with the critical law was already far better

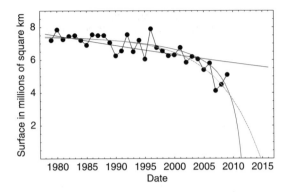

Fig. 14.16: **Evolution of the minimum arctic sea ice extent.** Observed evolution of the minimum arctic sea ice extent, according to the data of the U.S. National Snow and Ice Data Center (NSIDC, http://nsidc.org/), from 1979 to 2009. The minimum occurs around 15 September of each year. This evolution is compared to: (i) the standard linear fit corresponding to an assumed constant rate of extent decrease (line); (ii) a fit by a critical law $y = y_0 - a|T - T_c|^\gamma$ accelerating toward a critical date $T_c = 2013$; (iii) a fit by an exponential law $y = y_0 - a \exp[b(T - T_0)]$ yielding a total melting for 2014–2015. Polynomial fits (quadratic or cubic, not shown) yield a date of 2026 for the disappearance of the arctic sea ice.

than with the constant rate law, and it actually allowed us to predict a full disappearance epoch far closer than previously expected and a low 2007 point [401]. The 2008 point [411] has confirmed the validity of the model (Fig. 14.16). We obtain by the χ^2 method a best fit for the minimum ice surface (in square kilometers), $y = 8 - 12.3 \times |T - 2013|^{-0.86}$. The 2009 point, although higher, is the third lowest value ever registered. With these data, one finds with this model a critical time $T_c = 2013$. The low 2010 point (around $4.6 \, \text{Mkm}^2$) supports this result.

Such an acceleration of the melting may be understood precisely by the fractal nature of the sea ice fracture. A simple model of fractal fracture of the sea ice has been performed by Chmel *et al.* [104]. Their analysis of satellite images confirmed the uniform character of fracture at various scale levels and the fractal geometry of the crack-and-lead pattern. They have identified a time and length invariance in the permanently *fracturing-restoring* structure, which characterizes the Arctic sea-ice cover as a self-organized criticality system. Such a model leads naturally (in two-dimensional) to an exponential increase of the enlightened surface and then of the melting, and to a critical law in three-dimensional.

We have therefore performed a fit of the data up to 2009 by such an exponential model. It first follows very closely the critical power-law model. Though falling less rapidly at the end, it still yields a very near date of full ice melting for about 2014–2015 (see Fig. 14.16). The application of the same method to the mean surface data during August and October months also shows clear accelerations, indicating (in these models) that only one year after the first total melting, the arctic sea can be expected to be free from ice during several months (August to October). The next year values will allow to discriminate between the various models. Anyway, the linear model is yet now clearly excluded, and even a weekly nonlinear quadratic acceleration model leads to a full melting date as near as 2026.

14.2.6. *Predictivity*

Although, at this stage, these studies remain of an empirical nature (it is only a purely chronological analysis, which does not take into account the nature of the events), they nevertheless provide us with a beginning of predictivity. Indeed, the fitting law is a two parameter function (T_c and g) that is applied to time intervals, so that only three events are needed to define these parameters. Therefore the subsequent dates are predicted after the third one, in a statistical way. As already remarked, the predicted dates should be interpreted as the dates of the peaks of probability for an event to happen.

Examples of such a predictivity (or retropredictivity) are:

(i) the retroprediction that the common Homo-Pan-Gorilla ancestor (expected, e.g. from genetic distances and phylogenetic studies), has a more probable date of appearance at ≈ -10 millions years [102]; its fossil has not yet been discovered (this is one of the few remaining "missing links");

(ii) a critical date for the long term evolution of human societies around the years 2050–2080 [373, 241, 380, 213];

(iii) the finding that the critical dates of rodents may reach $+60$ Myrs in the future, showing their large capacity of evolution, in agreement with their known high biodiversity;

(iv) the finding that the critical dates of dinosaurs are about -150 Myrs in the past, indicating that they had reached the end of their capacity of evolution (at least for the specific morphological characters studied) well before their extinction at -65 Myrs;

(v) the finding that the critical dates of North american Equids is, within uncertainties, consistent with the date of their extinction, which may mean that, contrary to the dinosaur case, the end of their capacity of evolution has occured during a phase of environmental change that they have not been able to deal with by the mutation-selection process;

(vi) the finding that the critical date of echinoderms, whose dates of evolutionary leaps decelerate instead of accelerating, is, within uncertainties, the same as that of their apparition during the Precambrian-Cambrian radiation, this supporting the view of the subsequent events as a kind of "scale wave" expanding from this first "shock";

(vii) the log-periodic deceleration of the rate of California earthquakes and the general log-periodic deceleration found in the aftershock structures of many earthquakes [414, 401, 411];

(ix) the studies performed by Brissaud on various evolutive systems (history of the development of particle accelerators [70], of religious orders [73], of Jazz [71], of the discovery of chemical elements [72], of the economic evolution of China and USSR [72]), showing in a systematic way a log-periodic behavior of the chronology of events. In many cases he finds both an acceleration toward a critical date (precursors) then a deceleration from this date (replicas). Moreover, the scale factors, as for species evolution and economic evolution, seem to show systematically peaks of probability for values near 1.3, 1.7 and 2.3 (possibly corresponding to $3^{k/4}$, with $k = 1$ to 3).

It should finally be remarked that, at this level of the analysis, the log-periodic law of species evolution (and other evolutive systems) is only chronological. Concerning the nature of the events themselves, some statements may be made about them, at least regarding the first evolutionary events, in the context of a more sophisticated version of the scale relativity theory, as we shall now see it.

14.3. Applications of Scale Relativity to Biology

One may consider several applications to biology of the various tools and methods of the scale relativity theory, namely, generalized scale laws, macroscopic quantum-type theory and Schrödinger equation in position space then in scale space and emergence of gauge-type fields and their associated charges from fractal geometry [353, 373, 388, 33, 402]. One knows that biology is founded on biochemistry, which is itself based on thermodynamics, to which we contemplate the future possibility to apply the macroquantization tools described in Chapters 5 and 10. Another example of future possible applications is to the description of the growth of polymer chains, which could have consequences for our understanding of the nature of DNA and RNA molecules.

Let us give some explicit examples of such applications.

14.3.1. *Confinement*

The solutions of nonlinear scale equations, such as that involving a harmonic oscillator-like scale force of Sec. 4.3.6 may be meaningful for biological systems. Indeed, its main feature is its capacity to describe a system in which has emerged a clear separation between an inner and an outer region, which is one of the properties of the first prokaryotic cell. We have seen that the effect of a scale harmonic oscillator force results in a confinement of the large scale material in such a way that the small scales may remain unaffected.

Another interpretation of this scale behavior amounts to identifying the zone where the fractal dimension diverges, which corresponds to an increased "thickness" of the material, as the description of a membrane. It is indeed the very nature of biological systems to have not only a well-defined size and a well-defined separation between interior and exterior, but also systematically an interface between them, such as membranes or walls. This is already true of the simplest prokaryote living cells. Therefore this result suggests the possibility that there could exist a connection between the existence of a scale field, e.g. a global pulsation of the system, etc., both with the confinement of the cellular material and with the appearance of a limiting membrane or wall [388]. This is reminiscent of eukaryotic cellular division, which involves both a dissolution of the nucleus membrane and a deconfinement of the nucleus material, transforming, before the division, an eukaryote into a prokaryote-like cell. This could be a key towards a better understanding of the first major evolutionary leap after the appearance of cells, namely the emergence of eukaryotes.

14.3.2. *Morphogenesis*

The Schrödinger equation can be viewed as a fundamental equation of morphogenesis. It has not been yet considered as such, because its unique domain of application was, up to now, the microscopic (molecular, atomic, nuclear and elementary particle) domain, in which the available information was mainly about energy and momentum. Such a situation is now changing thanks to field effect microscopy and atom laser trapping, which begin to allow the observation of quantum-induced geometric shapes at small scales.

However, scale relativity extends the potential domain of application of Schrödinger-like equations to every systems in which the three conditions (infinite or very large number of trajectories, fractal dimension of individual trajectories, local irreversibility) are fulfilled. Macroscopic Schrödinger equations can be constructed, which are not based on Planck's constant $\hbar$, but on constants that are specific to each system and may emerge from their self-organization. In addition, systems, which can be described by hydrodynamics equations including a quantum-like potential also come under the generalized macroscopic Schrödinger approach.

The three conditions above seem to be particularly well adapted to the description of living systems. Let us give a simple example of such an application.

In living systems morphologies are acquired through growth processes. One can attempt to describe such a growth in terms of an infinite family of virtual, fractal and locally irreversible, fluid-like trajectories. Their equation can therefore be written under the form of a fractal geodesic equation, then it can be integrated as a Schrödinger equation or, equivalently, in terms of hydrodynamics-type energy and continuity equations including a quantum-like potential. This last description therefore shares some common points with recent very encouraging works in embryogenesis, which describe the embryo growth by visco-elastic fluid mechanics equations [185, 288]. The addition of a quantum potential to these equations would give them a Schrödinger form, and therefore would allow the emergence of quantized solutions. This could be an interesting advantage for taking into account the organization of living systems in terms of well defined bauplans [135] and the punctuated evolution of species whose evolutive leaps go from one organization plan to another [203].

Let us take a more detailed example of morphogenesis. If one looks for solutions describing a growth from a center, one finds that this problem is formally identical to the problem of the formation of planetary nebulae (see Chapter 13, and, from the quantum point of view, to the problem of particle scattering, e.g. on an atom). The solutions correspond to the case of the outgoing spherical probability wave.

Depending on the potential, on the boundary conditions and on the symmetry conditions, a large family of solutions can be obtained. Considering here only the simplest ones, i.e. central potential and spherical symmetry, the probability density distribution of the various possible values of the angles are given in this case by the spherical harmonics,

$$P(\theta, \varphi) = |Y_{lm}(\theta, \varphi)|^2. \tag{14.2}$$

These functions show peaks of probability for some angles, depending on the quantized values of the square of angular momentum L^2 (measured by the quantum number l) and of its projection L_z on axis z (measured by the quantum number m).

Finally a more probable morphology is obtained by "sending" matter along angles of maximal probability. The biological constraints lead one to skip to cylindrical symmetry. This yields in the simplest case a periodic quantization of the angle θ (measured by an additional quantum number k), that gives rise to a separation of discretized "petals". Moreover, there is a discrete symmetry breaking along the z axis linked to orientation (separation of "up" and "down" due to gravity, growth from a stem). The solutions obtained in this way show floral "tulip-like" shapes (see Fig. 14.17 and [377, 388, 402]).

Coming back to the foundation of the theory, it is remarkable that these shapes are solutions of a geodesic, strongly covariant equation $\hat{d}\mathcal{V}/dt = 0$, which has the form of the Galilean motion equation in vacuum in the absence of external force. Even more profoundly, this equation does not describe the motion of a particle, but purely geometric virtual possible paths; this gives rise to a description in terms of a probability density, which plays the role of a potential for the real particle (if any, since, in the application to elementary particles, we identify the "particles" with the geodesics themselves, i.e. they become pure relative geometric entities devoid of any proper existence). We suggest that this is a striking illustration of the power of the relativity-emptiness principle, as it has been expressed more than two thousand years ago in Buddhist science and philosophy (see e.g. [327]): "form is emptiness, emptiness is form".

Fig. 14.17: **Morphogenesis of a "flower-like" structure, solution of a Schrödinger equation that describes a growth process from a center** ($l = 5, m = 0$). The "petals", "sepals" and "stamen" are traced along angles of maximal probability density. A constant force of "tension" has been added, involving an additional curvature of "petals", and a quantization of the angle θ that gives an integer number of "petals" (here, $k = 5$).

14.3.3. *Origin of life: a new approach?*

The problems of origin are in general more complex than the problems of evolution. Strictly, there is no "origin" and both problems could appear to be similar, since the scientific and causal view amounts to considering that any given system finds its origin in an evolution process. However, systems are in general said to evolve if they keep their nature, while the question is posed in terms of origin when a given system appears from another system of a completely different nature, and moreover, often on time scales, which are very short with respect to the evolution time. An example in astrophysics is the origin of stars and planetary systems from the interstellar medium, and in biology the probable origin of life from a prebiotic medium.

A fundamentally new feature of the scale relativity approach concerning such problems is that the Schrödinger form taken by the geodesic equation can be interpreted as a general tendency for systems to which it applies to make structures, i.e. to lead to self-organization. In the framework of a classical deterministic approach, the question of the formation of a system is always posed in terms of initial conditions. In the new framework, the general existence of stationary solutions allows structures to be formed whatever the initial conditions, in correspondence with the field, the symmetries and the boundary conditions, which become the environmental conditions in biology, and in function of the values of the various conservative quantities that characterize the system.

Such an approach could allow one to ask the question of the origin of life in a renewed way. This problem is analog to the "vacuum" (lowest energy) solutions in a quantum-type description, i.e. of the passage from a non-structured medium to the simplest, fundamental level structures. In astrophysics and cosmology, the problem amounts to understanding the apparition, from the action of gravitation alone, of structures from a highly homogeneous and non-structured medium. In the standard approach to this problem a large quantity of postulated and unobserved dark matter is needed to form structures, and even with this help the result is dissatisfying (see Chapter 13). In the scale relativity framework, we have suggested that an underlying fractal geometry of space involves a Schrödinger form for the equation of motion, leading both to a natural tendency to form structures and to the emergence of an additional potential energy, identified with the "missing mass(-energy)".

The problem of the origin of life, although clearly far more difficult and complex, shows common features with this question of structure formation in cosmology. In both cases one needs to understand the apparition of new structures, functions, properties, etc. from a medium, which does not yet show such structures and functions. In other words, one needs a theory of emergence. We hope that scale relativity is a good candidate for such a theory, since it owns the two required properties: (i) for problems of origin, it gives the conditions under which a weakly structuring or destructuring (e.g. diffusive) classical system may become quantum-like and, therefore, structured; (ii) for problems of evolution, it makes use of the self-organizing property of the quantum-like theory.

We therefore tentatively suggest a new way to tackle the question of the origin of life (and in parallel, of the present functioning of the intracellular medium) [388]. The prebiotic medium on the primordial Earth is expected to have become chaotic. As a consequence, on time scales long with respect to the chaos time (horizon of predictability), the conditions, which underlie the transformation of the motion equation into a Schrödinger-type equation, become fulfilled (complete information loss on angles, position and time leading to a fractal dimension 2 behavior on a range of scales reaching a ratio of at least 10^4–10^5, see Sec. 10.3.2). Since the chemical structures of the prebiotic medium have their lowest scales at the atomic size, this means that, under such a scenario, one expects the first organized units to have appeared at a scale of about $10\,\mu$m, which is indeed a typical scale for the first observed prokaryotic cells (see Fig. 14.1). The spontaneous transformation of a classical, possibly diffusive mechanics, into a quantum-like mechanics, with the diffusion coefficient becoming the quantum self-organization parameter $\mathcal{D}$ would have immediate dramatic consequences: quantization of energy and energy exchanges and therefore of information, apparition of shapes and quantization of these shapes (the cells can be considered as the "quanta" of life), spontaneous duplication and branching properties (see herebelow), etc. Moreover, due to the existence of a vacuum energy in quantum mechanics (i.e. of a non vanishing minimum energy for a given system), we expect the primordial structures to appear at a given non-zero energy, without any intermediate step.

Such a possibility is supported by the symplectic formal structure of thermodynamics [443], in which the state equations are analogous to Hamilton–Jacobi equations. One can therefore contemplate the possibility of a future "quantization" of thermodynamics and then of the chemistry of solutions, leading to a new form of macroscopic quantum (bio-)chemistry, which would hold both for the prebiotic medium at the origin of life and for today's intracellular medium.

In such a framework, the fundamental equation would be the equation of molecular fractal geodesics, which could be transformed into a Schrödinger equation for wave functions ψ. This equation describes a universal tendency to make structures in terms of a probability density P for chemical products (constructed from the distribution of geodesics), given by the squared modulus of the wave function $\psi = \sqrt{P} \times e^{i\theta}$. Each of the molecules being subjected to this probability, which therefore plays the role of a potentiality, it is proportional to the concentration c for a large number of molecules, $P \propto c$ but it also constrains the motion of individual molecules when they are in small number (this is similar to a particle-by-particle Young two-slit experiment).

Finally, the Schrödinger equation may in its turn be transformed into a continuity and Euler hydrodynamic-like system (for the velocity $V = (v_+ + v_-)/2$ and the probability P) with a quantum potential depending on the concentration when $P \propto c$,

$$Q = -2\mathcal{D}^2 \frac{\Delta \sqrt{c}}{\sqrt{c}}. \tag{14.3}$$

This hydrodynamics-like system also implicitly contains as a sub-part a standard diffusion Fokker–Planck equation with diffusion coefficient $\mathcal{D}$ for the velocity v_+ (see Chapters 5 and 10). It is therefore possible to generalize the standard classical approach of biochemistry, which often makes use of fluid equations, with or without diffusion terms (see, e.g. [334, 494]).

Under the point of view of this third representation, the spontaneous transformation of a classical system into a quantum-like system through the action of fractality and irreversibility on small time scales manifests itself by the appearance of a quantum-type potential energy in addition to the standard classical energy balance. We therefore suggest to search whether biological systems are characterized by such an additional potential energy [402]. This would be quite similar to the missing energy of cosmology, usually attributed to a never found "dark matter". This missing energy would be given by the above relation (14.3) in terms of concentrations and could be identified by performing a complete energy balance of biological systems, then by comparing it to the classically expected one.

But we have also shown that the opposite of a quantum potential is a diffusion potential. Therefore, in case of simple reversal of the sign of this potential energy, the self-organization properties of this quantum-like behavior would be immediately turned, not only into a weakly organized classical system, but even into an increasing entropy diffusing and desorganized system. We have tentatively suggested [415, 402] that such a view may provide a renewed way of approach to the understanding of tumors, which are characterized, among many other features, by both energy affinity and morphological desorganization.

14.3.4. Duplication

The passage from the fundamental "vacuum" level to the first excited level now provides one with a rough model of duplication (see Figs. 14.18 and 14.19). Once again, the quantization implies that, in case of energy increase, the system will not increase its size, but will instead be led to jump from a single structure to a binary structure, with no stable intermediate step between the two stationary solutions $n = 0$ and $n = 1$. Moreover, if one comes back to the level of description of individual trajectories, one finds that from each point of the initial one body-structure there exist trajectories that go to the two final structures. In this framework, duplication is expected to be linked to a discretized and precisely fixed jump in energy.

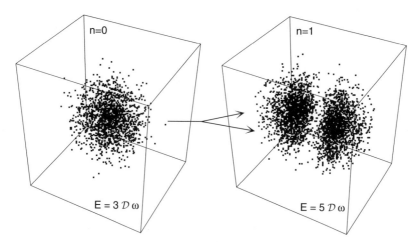

Fig. 14.18: **Model of duplication.** The stationary solutions of the Schrödinger equation in a three-dimensional harmonic oscillator potential can only take discretized morphologies in correspondence with the quantized value of the energy. Provided the energy increases from the one-structure case ($E_0 = 3\mathcal{D}\omega$), no stable solution can exist before it reaches the second quantized level at $E_1 = 5\mathcal{D}\omega$. The solutions of the time-dependent equation show that the system jumps from the one structure to the two-structure morphology.

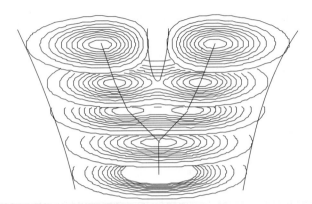

Fig. 14.19: **Model of branching.** Successive solutions of the time-dependent two-dimensional Schrödinger equation in an harmonic oscillator potential are plotted as isodensities. The energy varies from the fundamental level ($n = 0$) to the first excited level ($n = 1$) and as a consequence the system jumps from a one-structure to a two-structure morphology.

It is clear that, at this stage, such a model is extremely far from describing the complexity of a true cellular division, which it did not intend to do. Its interest is to be a generic and general model for a spontaneous duplication process of quantized structures, linked to energy jumps. Indeed, the jump from one to two probability peaks when going from the fundamental level to the first excited level is found in many different situations of which the harmonic oscillator case is only an example. Moreover, this duplication property is expected to be conserved under more elaborated versions of the description provided the asymptotic small scale behavior remains of constant fractal dimension $D_F \approx 2$, such as, e.g. in cell wall-like models based on a locally increasing effective fractal dimension.

14.3.5. *Bifurcation, branching process*

Such a model can also be applied to the first rough description of a branching process (Fig. 14.19), e.g. in the case of a tree growth when the previous structure remains instead of disappearing as in cell duplication.

Note finally that, although such a model is still clearly too rough to claim that it describes biological systems, it may already be improved by combining with it various other functional and morphological elements, which have been obtained. One may apply the duplication or branching process to a system whose underlying scale laws (which condition the derivation of the generalized Schrödinger equation, see Sec. 5.14 include: (i) the model of membrane through a fractal dimension that becomes variable with the distance to a center; (ii) the model of multiple hierarchical levels of organization depending on "complexergy" (see herebelow).

14.3.6. *Nature of first evolutionary leaps*

We have also suggested applications to biology of the new quantum-like mechanics in scale space [388].

In the fractal model of the *tree of life* described hereabove [102], we have voluntarily limited ourselves to an analysis of only the chronology of events (see Fig. 14.2), independently of the nature of the major evolutionary leaps. The suggestion of a quantum-type mechanics in scale space and of the new concept of complexergy (Chapter 8) allows one to reconsider the question.

One may indeed suggest that life evolution proceeds in terms of increasing quantized complexergy. This would account for the existence of punctuated evolution [203] and for the log-periodic behavior of the leap dates, which can be interpreted in terms of probability density of the events, $P = |\psi|^2 \propto \sin^2[\omega \ln(T - T_c)]$. Moreover, one may contemplate the possibility of an understanding of the nature of the events, even though in a rough way as the first step.

Indeed, one can expect the formation of a structure at the fundamental level (lowest complexergy) characterized by one length-scale (see Fig. 8.1 in Chapter 8). Moreover, the most probable value for this scale of formation is predicted to be the "middle" of the scale-space, since the problem is similar to that of a quantum particle in a box, with the logarithms of the minimum scale λ_m and maximum scale λ_M playing the roles of the walls of the box, so that the fundamental level solution has a peak at a scale $\sqrt{\lambda_m \times \lambda_M}$.

The universal boundary conditions are the Planck-length l_P in the microscopic domain and the cosmic scale $\mathbb{L} = \Lambda^{-1/2}$ given by the cosmological constant Λ in the macroscopic domain (see Chapter 12). From the predicted and now observed value of the cosmological constant, one finds $\mathbb{L}/l_\mathbb{P} = 5.3 \times 10^{60}$, so that the mid scale is at $2.3 \times 10^{30} l_\mathbb{P} \approx 40\,\mu$m. A quite similar result is obtained from the scale boundaries of living systems (0.5 Angströms–30 m). This scale of $40\,\mu$m is indeed a typical scale of living cells. Moreover, the first "prokaryot" cells appeared about three Gyrs ago had only one hierarchy level (no nucleus).

In this framework, a further increase of complexergy can occur only in a quantized way. The second level describes a system with two levels of organization, in agreement with the second step of evolution leading to eukaryots about 1.7 Gyrs ago (second event in Fig. 14.2). One expects (in this very simplified model), that the scale of nuclei be smaller than the scale of prokaryots, itself smaller than the scale of eucaryots: this is indeed what is observed.

The following expected major evolutionary leap is a three organization level system, in agreement with the apparition of multicellular forms (animals, plants and fungi) about 1 Gyr ago (third event in Fig. 14.2).[a] It is also expected that the multicellular stage can be built only from eukaryotes, in agreement with the fact that the cells of multicellulars do most of the time have nuclei. More generally, evolved organisms keep in their internal structure the organization levels of the preceeding stages.

The following major leaps correspond to more complicated structures then more complex functions (supporting structures, such as exoskeletons, tetrapody, homeothermy, viviparity), but they are still characterized by fundamental changes in the number of organization levels. Moreover, the first steps in the above model are based on spherical symmetry, but this symmetry is naturally broken at scales larger than $40\,\mu$m, since this is also the scale beyond which the gravitational force becomes larger than the van der Waals force. One therefore expects the evolutionary leaps that follow the apparition of multicellular systems to lead to more complicated structures, such as those of the Precambrian-Cambrian radiation, than can no longer be described by a single scale variable.

14.3.7. *Origin of the genetic code: possible new hints*

Pushing the above reflection further, we intend, in future works, to extend the model to more general symmetries, boundary conditions and constraints. We also emphasize once again that such an approach does not dismiss the role and the importance of the genetic code in biology. On the contrary, we hope that it may help understanding its origin and its evolution.

Indeed, one may suggest that the various biological morphologies and functions are solutions of macroscopic Schrödinger-type equations, whose

---

[a]The possible recent discovery of 2.1 billion years old multicellulars [El Albany *et al.* *Nature* **466**, 100 (2010)] does not contradict this analysis, since in the scale relativity view the events and their date are not absolute but instead depend on causes and conditions. Indeed, these forms can be attributed to accelerated evolution due to significant but temporary increase in oxygen concentration.

solutions are quantized according to integer numbers that represent the various conservative quantities of the system. Among these quantities, one expects to recover the basic physical ones, such as energy, momentum, electric charge, etc. But one may also contemplate the possibility of the existence of prime integrals (conscrvative quantities), which would be specific of biology (or particularly relevant to biology), among which we have suggested the new concept of complexergy, but also new scale "charges" finding their origin in the internal scale symmetries of the biological systems.

The quantization of these various quantities means that any such system would be described by a set of integer numbers, so that one may tentatively suggest that only these numbers, instead of a full continuous and detailed information, would have to be included in the genetic code, which would be by far simpler and would allow to store much more information. In this case the process of genetic code reading, protein synthesis, etc. would be a kind of "analogic solutioner" of Schrödinger equation, leading to the final morphologies and functions. Such a view also offers a new line of research towards understanding the appearance of the genetic code, namely, the transformation of what was a purely chemical process into a support of information and of its implementation (reading and protein synthesis). One of the mysteries of its appearance is indeed the question of why and how an information about a given system may be spontaneously stored into a structure which is different from it.

But the quantization of the energy and of other conservative quantities internal to the system implies a similar quantization of their exchange with the exterior environment, and therefore of the corresponding energies and conservative quantities, which characterize the exterior systems (macromolecules) with which the exchange occurs. A similar information may then be stored in two different systems by this quantized exchange process, and one of these systems becomes an information support for the other. Finally, by endocytosis (similar to the process by which the mitochondria have been later — about two billion years ago — incorporated in cells) the external molecules may become internal, while the exchange process becomes an information reading-like process.

14.3.8. *Systems biology and multiscale integration*

We intend to develop this approach in future works, in particular by including the scale relativity tools and methods in a systems biology framework allowing multiscale integration [33, 402, 408], in agreement with

Noble's "biological relativity" [335] according to which there is no privileged level in living systems.

The intrinsically integrative character of scale relativity should indeed prove extremely powerful by resolving in a very fundamental way the problems of multi-scale integration through identification of fractal-non fractal transitions, of variable fractal dimensions, of the "scale forces" responsible for such variations, and of discrete scale invariance generating a log-periodic behavior for space structure and time evolution. In cases when conditions for the quantum-type regime are fulfilled in biological systems, it could impact biological structures and their evolution through manifestations of a quantum-like potential, such as the number of organization levels related to the conservative quantity complexergy, and help predicting means by which the number and types of these organization levels may be modulated experimentally.

Let us finally illustrate with a simple explicit example the ability of the scale relativity method to provide multi-scale solutions in a spontaneously integrated way. The first step consists in defining the elements of description, which represents the smallest scale considered at the studied level (for example, intracellular "organelles"). Then one writes for these elementary objects an equation of dynamics, which account for the fractality and irreversibility of their motion. Such a motion equation written in fractal space can be integrated under the form of a macroscopic Schrödinger-type equation (no longer based on the microscopic Planck constant, but on a macroscopic constant specific of the system under consideration). Its solutions are wave functions whose modulus squared gives the probability density of distribution of the initial "points".

Actually, the solutions of such a Schrödinger-like equation are naturally multiscaled. It describes, in terms of peaks of probability density, a structuring of the "elementary" objects from which we started (e.g. organelle-like objects structuring at the larger scale cell-like level). Now, while the vacuum state (lowest energy) usually describes one object (a single "cell"), excited states describe multiple objects ("tissue-like" level), each of which being often separated by zones of null density — therefore corresponding to infinite quantum potentials — which may represent "walls" (Fig. 14.20).

A simple two-complementary-fluid model (describing, e.g. hydrophile/ hydrophobe behavior) can easily be obtained, one of them showing probability peaks in the "cells" (and zero probability in the "wall") while the other fluid peaks in the "walls" and have vanishing probability in the "cells". This three-level multi-scale structure results from a general theorem

Fig. 14.20: **Multiscale integration in scale relativity.** Elementary objects at a given level of description (left figure) are organized in terms of a finite structure described by a probability density distribution (second figure from the left). By increasing the energy, this structure spontaneously duplicates (third figure). New increases of energy lead to new duplications (fourth figure), then to a tissue-like organization (fifth figure — the scale of the figures is not conserved).

of quantum mechanics, which remains true for the macroscopic Schrödinger regime considered here, according to which, for a one-dimensional discrete spectrum, the wave function corresponding to the $(n + 1)$th eigenvalue is zero n times [267, 319]. A relevant feature for biological applications is also that these multi-scale structures, described in terms of stationary solutions of a Schrödinger-like equation, depend, not on initial conditions like in a classical deterministic approach, but on environment conditions (potential and boundary conditions).

Moreover, this scale relativity model involves not only the resulting structures themselves but also the way the system may jump from a two-level to a three-level organization. Indeed, the solution of the time-dependent Schrödinger equation describes a spontaneous duplication when the energy of the system jumps from its fundamental state to the first excited state (see Sec. 14.3.4 and Figs. 14.18 and 14.20).

One may even obtain solutions of the same equation organized on more than three levels, since it is known that fractal solutions of the Schrödinger equation do exist (see Sec. 5.8) [47, 222, 403]. An example of such a fractal solution for the Schrödinger equation in a two-dimensional box is given in Fig. 14.21.

Note that the resulting structures are not only qualitative, but also quantitative, since the relative sizes of the various embedded levels can be derived from the theoretical description. Finally, such a "tissue" of individual "cells" can be inserted in a growth equation, which will itself take a Schrödinger form. Its solutions yield a new, larger level of organization, such as the flower-like structure of Fig. 14.17. Then the matching conditions between the small scale and large scale solutions (wave functions) allow to connect the constants of these two equations, and therefore the quantitative

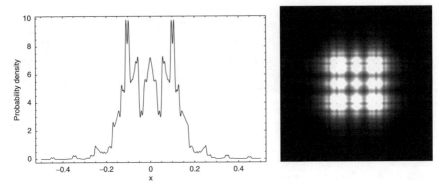

Fig. 14.21:   **Fractal multiscale solutions of the Schrödinger equation.** Left figure:
one-dimensional solution in a box, in terms of position x for a given value of the time
t. This solution reads $\psi(x, t) = (1/\pi) \sum_{n=-N}^{n=N} (-1)^n (n + 1/2)^{-1} \exp\{i\pi[2x(n + 1/2) - t(n + 1/2)^2]\}$, with $N \to \infty$ [47]. Finite resolution approximations of this solution can
be constructed by taking finite values of N. Here the probability density $|\psi|^2$ is drawn
for $N = 100$ and $t = 0.551$. Right figure: fractal multiscale solution in a two-dimensional
box. It is constructed as a product $\psi(x)\psi(y)$ of the left one-dimensional solution.

scales of their solutions. This may be even performed in a global way by
writing a Schrödinger equation in the scale space, whose solutions yield
(still in terms of peaks of probability of presence), the relative values
of the most probable scales. These values depend on universal boundary
conditions in the scale space. For example, these boundaries are, in biology,
the size of atoms at small scale (about 1 Angström) and the maximal size
of individual living systems at large scale (about 10 meters). With such
boundary conditions, the fundamental level is a single probability peak
which lies at a scale given by the mean of their logarithms (about 30
micrometers, a typical cell size). Therefore, such an approach allows one
to suggest an explanation for the universality and invariance of the size of
cells.

Chapter 15

GENERAL CONCLUSION

To conclude, the scale relativity theory amounts to making proposals for the fundamental principles to be adopted in order to set the foundations of the laws that underlie the various physical, biological and other phenomenons with the insistance on the fact that the most fundamental of these principles is the principle of relativity itself. The extraordinary success due to the application of this principle to position, orientation and motion for four centuries is well known.

But, during the last three decades, the various sciences have been faced with an ever-increasing number of new unsolved problems, of which many are linked to questions of scales. It, therefore, seemed natural, in order to deal with these problems at a fundamental and first principle level, to extend theories of relativity by including the scale in the very definition of the coordinate system, then to account for these scale transformations in a relativistic way.

We have attempted to give an extended discussion of the various developments of the theory of scale relativity and of its applications. The aim of this theory is to describe space-time as a continuous manifold without making the hypothesis of differentiability, i.e. by considering the two differentiable and nondifferentiable cases and to physically constrain its possible geometry by the principle of relativity, of both motion and scale.

This is effectively made by using the physical principles (the covariance, equivalence and geodesic principles) that directly derive from the principle of relativity, being actually a re-expression of it. These principles lead in their turn to the construction of covariant derivatives and finally to the writing, in terms of these covariant derivatives, of the motion equations

711

under the form of free-like geodesic equations. Such an attempt is therefore a natural extension of general relativity, since the two-times differentiable continuous manifolds of Einstein's theory that are constrained by the principle of relativity of motion, are particular sub-cases of the new geometry to be built.

Giving up the differentiability hypothesis, i.e. generalizing the geometric description to general continuous manifolds, differentiable or not, involves an extremely large number of new possible structures to be investigated and described. In view of the immensity of the task, we have chosen to proceed by steps, using presently-known physics as a guide. Such an approach is rendered possible by the result according to which the small scale structures, which manifest the nondifferentiability, are smoothed out beyond some relative transitions toward the large scales. One therefore recovers the standard classical differentiable theory as a large scale approximation of this generalized approach. But one also obtains a new geometric theory, which allows one to understand quantum mechanics as a manifestation of an underlying nondifferentiable and fractal geometry and finally to suggest generalizations of it and new domains of application for these generalizations.

Now the difficulty that also makes their interest with theories of relativity is that they are meta-theories rather than theories of some particular systems. Hence, after the construction of special relativity of motion at the beginning of the 20th century, the whole of physics needed to be rendered "relativistic" (from the viewpoint of motion), a task that is not yet fully achieved.

The same is true regarding the program of constructing a fully scale-relativistic science. Whatever the already-obtained successes, the task remains huge, in particular when one realizes that it is no longer only physics that is concerned, but also many other sciences. Its ability to go beyond the frontiers between sciences may be one of the main interests of the scale relativity theory, opening the hope of a refoundation on mathematical principles and on predictive differential equations of a "philosophy of nature" in which physics would no longer be separated from other sciences.

BIBLIOGRAPHY

1. Abbott, L. F. and Wise, M. B. (1981). *Am. J. Phys.* **49**, 37.
2. Adelman-McCarthy, J., Agueros, M. A., Allam, S. S. *et al.* (2006). *Astrophys. J. Suppl.* **162**, 38.
3. Agnese, A. G., Festa, R. (1997). *Phys. Lett.* **A 227**, 165.
4. Aharonian, F. A. *et al.* (1999). *Astron. Astrophys.* **349**, 11.
5. Ahmad, I. *et al.* (APEX Collaboration) (1995). *Phys. Rev. Lett.* **75**, 2658.
6. Ahmed, E. and Mousa, M. H. (1997). *Chaos, Solitons & Fractals* **8**, 805.
7. Aitchison, I. J. R. (1982). *An Informal Introduction to Gauge Field Theories*, Cambridge University Press, Cambridge.
8. Albeverio, S., Blanchard, Ph. and Hoegh-Krohn, R. (1983). *Exp. Math.* **4**, 365.
9. Allègre, C., Le Mouel, J. and Provost, A. (1982). *Nature* **297**, 47.
10. Allen, A. D. (1983). *Speculations in Science and Technology* **6**, 165.
11. Al-Rashid, S. (2006). *Some Applications of Scale Relativity Theory in Quantum Physics*, Ph. D. Thesis, Anbar University.
12. Al-Rashid, S., Habeeb, M. and Ahmad, K. (2007). Preprint.
13. Altaisky, M. V. (1996). *Differential Equations and Dynamical Systems* **4**, 267.
14. Alunni, C. (2001). *Revue de Synthèse* **122**, 147.
15. Alunni, C. (2006). Continental genealogies. Mathematical confrontations in Albert Lautman and Gaston Bachelard in *Virtual Mathematics. The Logic of Difference*, Simon Duffy (Ed.), Clinamen Press, Manchester, p. 65.
16. Alunni, C. (2008). *Proceedings of 7th International Colloquium on Clifford Algebra*, Toulouse, France, 19–29 Mai 2005, *Advances in Applied Clifford Algebra*, **18**.
17. Alunni, C. and Nottale, L. (2011). in Mélanges Gilles Châtelet, *Revue de synthèse*, Springer-Verlog, Berlin, in press.
18. Ambjorn, J., Jurkiewicz, J. and Watabiki, Y. (1995). *Nucl. Phys.* **B 454**, 313 (arXiv:hep-lat/9507014v1).
19. Ambjorn, J., Jurkiewicz, J. and Loll, R. (2005). *Phys. Rev. Lett.* **95**, 171301.
20. Amelino-Camelia, G. (2001). *Phys. Lett.* **B 510**, 255.
21. Amelino-Camelia, G. (2002). *Int. J. Mod. Phys.* **D 11**, 1643.
22. Anderson, J. D. *et al.* (1998). *Phys. Rev. Lett.* **81**, 2858.

23. Anderson, J. D. *et al.* (2002). *Phys. Rev.* **D 65**, 082004.
24. Anglès, P. (2008). *Conformal Groups in Geometry and Spin Structures*, Birkhauser, Boston.
25. Anglès, P. (Ed.) (2008). *Proceedings of 7th International Colloquium on Clifford Algebra*, Toulouse, France, 19–29 Mai 2005, *Advances in Applied Clifford Algebra*, **18**.
26. Ansoldi, S., Aurilia, A. and Spalluci, E. (1997). *Phys. Rev.* **D 56**, 2352.
27. Antoine, C. (2000). *Probation Report*, Ecole Centrale Paris.
28. Anz Meador, P. (2000). *Orbital debris news*, Oct 2000, JSC Houston.
29. Arp, H. (1986). *Astron. Astrophys.* **156**, 207.
30. Ashtekar, A. (1986). *Phys. Rev. Lett.* **57**, 2244.
31. Ashtekar, A. (1987). *Phys. Rev.* **D 36**, 1587.
32. Aspect, A., Dalibart, J. and Roger, G. (1982). *Phys. Rev. Lett.* **49**, 1805.
33. Auffray, C. and Nottale, L. (2008). *Progr. Biophys. Mol. Bio.* **97**, 79.
34. Bacry, H. and Levy-Leblond, J. M. (1968). *J. Math. Phys.* **9**, 1605.
35. Balbinot, R., Bergamini, R. and Comastri, A. *Phys. Rev.* **D 38**, 2415.
36. Balian, R. (1989). *Am. J. Phys.* **57**, 1019.
37. Barbour, J. (1974). *Nature* **249**, 328.
38. Barbour, J. and Bertotti, B. (1982). *Proc. R. Soc. Lond.* **A 382**, 295.
39. Barbour, J. and Pfister, H. (Eds.) (1995). *Mach Principle: From Newton's Bucket to Quantum Gravity*, Birkhäuser, Boston.
40. Beatty, J. K. and Chaikin, A. (Eds.) (1990). *The New Solar System*, Cambridge University Press, Cambridge.
41. Bell, J. S. (1965). *Physics* (N.Y.) **1**, 195.
42. Ben Adda, F. and Cresson, J. (2000). *C. R. Acad. Sci. Paris* **330**, 261.
43. Ben Adda, F. and Cresson, J. (2004). *Chaos, Solitons & Fractals* **19**, 1323.
44. Ben Adda, F. and Cresson, J. (2005). *Applied Mathematics and Computation* **161**, 323.
45. Ben Adda, F. (2007). *Int. J. Pure and Applied Mathematics* **38**, 159.
46. Benedetti, D. (2009). *Phys. Rev. Lett.* **102**, 111303.
47. Berkeley, G. (1734). *The Analyst*, Edited by David R. Wilkins. http://www.maths.tcd.ie/pub/HistMath/People/Berkeley/Analyst/Analyst.html
48. Berry, M. V. (1996). *J. Phys. A: Math. Gen.* **29**, 6617.
49. Bethe, H. and Salpeter, E. (1957). *Quantum Mechanics of One and Two Electron Atoms*, New York Academic Press, New York, p. 122.
50. Biesecker, D. (2001). http://sungrazer.nascom.nasa.gov/
51. Bird, D. J. *et al.* (1994). *Astrophys. J.* **424**, 491.
52. Bird, D. J. *et al.* (1995). *Astrophys. J.* **441**, 144.
53. Bjorken, J. D. and Drell, S. D. (1965). *Relativistic Quantum Mechanics*, McGraw-Hill, New York.
54. Blanchard, Ph. (1984). *Acta Physica Austriaca Suppl.* **XXVI**, 185.
55. Bluhm, R., Kostelecky, V. A. and Tudose, B. (1995). *Phys. Rev.* **A 52**, 2234.
56. Bluhm, R., Kostelecky, V. A. and Porter, B. (1996). *Am. J. Phys.* **64**, 944.
57. Bogoliubov, N. N. and Shirkov, D. V. (1980) *Introduction to the Theory of Quantized Fields*, John Wiley & Sons, New York.

58. Bohm, D. (1952). *Phys. Rev.* **85**, 166.
59. Bohm, D. (1952). *Phys. Rev.* **85**, 180.
60. Bohm, D. (1954). *Quantum Theory*, Constable and Company Ltd., London.
61. Bohm, D. and Vigier, J. P. (1954). *Phys. Rev.* **96**, 208.
62. Bonazzola, S. and Peter, P. (1997). *Astropart. Phys.* **7**, 161.
63. Born, M. (1926). *Z. Physik* **37**, 863.
64. Born, M. (1927). *Nature* **119**, 354.
65. Brahic, A. (Ed.) (1982). *Formation of Planetary Systems*, CNES/Cepadues Editions, Toulouse.
66. Brancewicz, H. K. and Dworak, T. Z. (1980). *Acta Astronomica* **30**, 501.
67. Bransden, B. H. and Joachain, C. J. (2000). *Quantum Mechanics*, Pearson Education Ltd., Harlow, England.
68. Brenig, L. (2007). *J. Phys. A: Math. Gen.* **40**, 4567.
69. Brillouin, L. (1953). *J. of Applied Physics* **24** (9), 1152.
70. Brissaud, I. and Baron, E. (2007). *Cybergeo*, Revue Européenne de Géographie.
71. Brissaud, I. (2007). *Math. Sci. Hum., Mathematics and Social Sciences* **178**, 41.
72. Brissaud, I. (2009). *Des fleurs pour Schrödinger. La relativité d'échelle et ses applications*, Nottale L., Chaline J. and Grou P, Ellipses, Paris, pp. 270–285.
73. Brissaud, I. (2010). *Math. Sci. Hum., Mathematics and Social Sciences*, 2011, in press.
74. Broadhurst, T. J., Ellis, R. S., Koo, D. C. and Szalay, A. S. (1990). *Nature* **343**, 726.
75. Brouwer, D. and Clemence, G. M. (1961). *Methods of Celestial Mechanics*, Academic, New York.
76. Brown, L. S. (1973). *Am. J. Phys.* **41**, 525.
77. Burbidge, G. R. and Burbidge, E. M. (1967). *Astrophys. J. Lett.* **148**, L107.
78. Burkhardt, H., Legerlehner, F., Penso, G. and Vergegnassi, C. (1989). *Z. Phys. C-Particles and Fields* **43**, 497.
79. Callan, C. G. (1970). *Phys. Rev.* **D 2**, 1541.
80. Campagne, J. E. and Nottale, L. (2003). Unpublished preprint.
81. Campesino-Romeo, E., D'Olivo, J. C. and Socolovsky, M. (1982). *Phys. Lett.* **89A**, 321.
82. Campos, A. *et al.* (1995). *Clustering in the Universe*, 30th Rencontres de Moriond (Astrophysics meeting), (Eds.) Maurogordato S., Balkowski C., Tao C. and Tran Thanh Van J. (Frontières), p. 403.
83. Carroll, S. M. and Press, W. H. (1992). *Ann. Rev. Astron. Astrophys.* **30**, 499.
84. Cartan, E. (1953). *Oeuvres Complètes*, Gauthier-Villars, Paris.
85. Carter, B. and Henriksen, R. N. (1991). *J. Math. Phys.* **32**, 2580.
86. Cash, R., Chaline, J., Nottale, L. and Grou, P. (2002). *C.R. Biologies* **325**, 585.
87. Castro, C. (1997). *Found. Phys. Lett.* **10**, 273.
88. Castro, C. and Granik, A. (2000). *Chaos, Solitons & Fractals* **11**, 2167.

89. Castro, C. and Pavsic, M. (2002). *Phys. Lett.* **B 539**, 133.
90. Ceccolini, D. and Lehner, T. and Nottale, L. (2011). *Scale-relativistic formation of structures in an expanding universe*, submitted.
91. Célérier, M. N. and Nottale, L. (2001). arXiv: hep-th/0112213.
92. Célérier, M. N. and Nottale, L. (2003). *Electromagnetic Phenomena*, **T. 3**, N.1 (9), 83 (arXiv: hep-th/0210027).
93. Célérier, M. N. and Nottale, L. (2004). *J. Phys. A: Math. Gen.* **37**, 931 (arXiv: quant-ph/0609161).
94. Célérier, M. N. and Nottale, L. (2004). *Scientific Highlights 2004, Proceedings of the Journées de la SF2A*, Paris 14–18 June 2004, F. Combes, D. Barret, T. Contini, F. Meynadier and L. Pagani (Eds.), EDP Sciences, p. 649 (arXiv:gr-qc/0505012).
95. Célérier, M. N. (2005). *Proceedings of the 22nd Texas Symposium on Relativistic Astrophysics*, Stanford University (2004). Chen, P., Bloom, E., Madejski, G. and Petrosian, V. (Eds.), eConfC041213, 1403.
96. Célérier, M. N. and Nottale, L. (2006). *J. Phys. A: Math. Gen.* **39**, 12565 (arXiv: quant-ph/0609107).
97. Célérier, M. N. (2009). *J. Math. Phys.* **50**, 123101.
98. Célérier, M. N. and Nottale, L. (2010). *Int. J. Mod. Phys. A* **25**, 4239 (arXiv:1009.2934).
99. Chaline, J. and Mein, P. (1979). *Les rongeurs et l'évolution*, Doin, Paris, p. 235.
100. Chaline, J., Laurin, B., Brunet-Lecomte, P. and Viriot, L. (1993). *Modes and Tempo of Evolution in the Quaternary*, Chaline, J. and Werdelin, L. (Eds.), *Quaternary International* **19**, 27.
101. Chaline, J. (1998). *C. R. Acad. Sci. Paris* **326**(IIa), 307.
102. Chaline, J., Nottale, L. and Grou, P. (1999). *C.R. Acad. Sci. Paris* **328**, 717.
103. Chavanis, P. H. (2002). *Astron. Astrophys.* **381**, 340.
104. Chmel, A., Smirnov, V. N. and Astakhov, M. P. (2005). *Physica* **A 357**, 556.
105. Cocke, W. J. (1985). *Astrophys. J.* **288**, 22.
106. Cohen-Tannoudji, C., Diu, B. and Laloë, F. (1977). *Mécanique Quantique*, Hermann, Paris.
107. Coleman, P. H. and Pietronero, L. *Physics Reports* **213**, 311.
108. Coles, P. and Spencer, K. (2003). *MNRAS* **342**, 176.
109. Connes, A. (1994). *Noncommutative Geometry*, Academic Press, New York.
110. Connes, A., Douglas, M. R. and Schwarz, A. *J. High Energy Phys.* **02**, 003 (arXiv:hep-th/9711162).
111. Conway, A. W. (1937). *Proc. Roy. Soc.* **A 162**, 145.
112. Conway, A. W. (1945). *Proc. Roy. Irish Acad.* **A 50**, 98.
113. Corradi, R. L. M. (1999). *Astron. Astrophys.* **354**, 1071.
114. Cortis, A. (2007). *Phys. Rev.* **E 76**, 030102 (1–4).
115. Cresson, J. (2001). *Mémoire d'habilitation à diriger des recherches*, Université de Franche-Comté, Besançon.
116. Cresson, J. (2002). *Chaos, Solitons & Fractals* **14**, 553.

117. Cresson, J. (2003). *J. Math. Phys.* **44**, 4907.

118. Cresson, J. (2006). *International Journal of Geometric Methods in Modern Physics*, **3** (7).

119. Cresson, J. (2007). *J. Math. Phys.* **48**, 033504.

120. Croasdale, M. R. (1989). *Astrophys. J.* **345**, 72.

121. Currie, T. and Hansen, B. (2007). *Astrophys. J.* **666**, 1232.

122. Dale Favro, L. (1960). *Phys. Rev.* **119**, 53.

123. da Rocha, D. and Nottale, L. (2003). *Chaos, Solitons & Fractals* **16**, 565.

124. da Rocha, D. and Nottale, L. (2003). *On the morphogenesis of stellar flows. Application to planetary nebulae*, arXiv: astro-ph/0310031.

125. da Rocha, D. and Nottale, L. (2003). *Winds, Bubbles, and Explosions: a conference to honor John Dyson*, International Colloquium, Pátzcuaro, Michoacán, México, September 9–13, 2002, S. J. Arthur and W. J. Henney (Eds.), *Revista Mexicana de Astronomía y Astrofísica* (Serie de Conferencias) Vol. 15, p. 69.

126. da Rocha, D. (2005). *Gravitational Structuring in Scale Relativity*, Ph.D. thesis, Paris Observatory, http://tel.archives-ouvertes.fr/tel-00010204.

127. David, B. and Mooi, R. (1999). *Bull. Soc. Géol. Fr.* **170** (1), 91.

128. Davis, M. and Peebles, P. J. E. (1983). *Astrophys. J.* **267**, 465.

129. de Bernardis, P. *et al.* (2000). *Nature* **404**, 955.

130. de Broglie, L. (1923). *Comptes Rendus Acad. Sci. Paris* **177**, 507, 548.

131. de Broglie, L. (1925). *Annales de Physique* 10è série, Tome III (Ph.D. Thesis, *Recherches sur la théorie des quanta*, Masson).

132. de Broglie, L. (1926). *Compt. Rend.* **183**, 447.

133. De Gouveia Dal Pino, E. M. *et al.* (1995). *Astrophys. J. Lett.* **442**, L45.

134. de Oliveira Neto, M., Maia, L. A., Carneiro, S. (2004). *Chaos, Solitons & Fractals* **21**, 21.

135. Devillers, C. and Chaline, J. (1993). *Evolution. An Evolving Theory*, Springer Verlag, New York, p. 251.

136. DeWitt, B. (1967). *Phys. Rev.* **160**, 1113.

137. Dezael, F. X. (2003). *Probation Report*, ENS (4th year) and Paris Observatory.

138. Dicke, R. H. (1961). *Nature* **192**, 440.

139. Dirac, P. A. M. (1938). *Proc. Roy. Soc. Lond.* **A 165**, 199.

140. Dirac, P. A. M. (1973). *Proc. Roy. Soc. Lond.* **A 333**, 403.

141. Dittus, H. (2005). *A mission to explore the Pioneer anomaly*, arXiv:gr-qc/0506139.

142. Dohrn, D. and Guerra, F. (1977). *Stochastic behaviour in classical and quantum Hamiltonian systems*, Proceedings of Conference, Como, 20–24 June, (Eds.) G. Casati and J. Ford.

143. Dohrn, D. and Guerra, F. (1985). *Phys. Rev.* **D 31**, 2521.

144. Dubois, D. (2000). *Proceedings of CASYS'1999*, 3rd International Conference on Computing Anticipatory Systems, Liège, Belgium, *Am. Institute of Physics Conference Proceedings* **517**, 417.

145. Dubrulle, B. (1994). *Phys. Rev. Lett.* **308**, 151.

146. Dubrulle, B. and Graner, F. (1996). *J. Phys. (Fr.)* **6**, 797.

147. Dubrulle, B., Graner, F. and Sornette, D. (Eds.) (1998). *Scale invariance and beyond*, Proceedings of Les Houches school, (EDP Sciences, Les Ullis/Springer-Verlag, Berlin, New York).
148. Dunkley, J. *et al.* (2009). *Astrophys. J. Suppl.* **180**, 306.
149. Dwarkadas, V. and Chevalier, R. (1995). *Astrophys. J.* **457**, 773.
150. Dyer, C. C. (1976). *M.N.R.A.S.* **198**, 1033.
151. Eckart, A. and Genzel, R. (1996). *Nature* **383**, 415.
152. Eddington, A. S. (1936). *Theory of Protons and Electrons*, Cambridge University Press.
153. Einasto, J. *et al.* (1997). *Nature* **385**, 139.
154. Einstein, A. (1905). *Annalen der Physik* **17**, 891.
155. Einstein, A. (1905). *Annalen der Physik* **17**, 132.
156. Einstein, A. (1905). *Annalen der Physik* **17**, 549.
157. Einstein, A. (1916). *Annalen der Physik* **49**, 769, translated in *The Principle of Relativity*, Dover, 1923, 1952, p. 111.
158. Einstein, A. (1917). *Sitzungsberichte der Preussichen Akad. d. Wissenschaften*, translated in *The Principle of Relativity*, Dover, p. 177.
159. Einstein, A., Podolsky, B. and Rosen, N. (1936). *Phys. Rev.* **47**, 777.
160. Einstein, A. (1936). *Journal of the Franklin Institute* **221**, 313–347.
161. Einstein, A. (1936). *Science* **84**, 506.
162. Einstein, A. (1948). Letter to Pauli, in *Albert Einstein, Oeuvres choisies*, I, Quanta, Seuil/CNRS, p. 249.
163. Eisenstaedt, J. (1977). *Phys. Rev.* **D 16**, 927.
164. Elbert, J. W. and Sommers, P. (1995). *Astrophys. J.* **441**, 151.
165. El Naschie, M. S. (1992). *Chaos, Solitons & Fractals* **2**, 211.
166. El Naschie, M. S., Rössler, O. and Prigogine, I. (Eds.) (1995). *Quantum Mechanics, Diffusion and Chaotic Fractals*, Pergamon, New York.
167. El Naschie, M. S. (2000). *Chaos, Solitons & Fractals* **11**, 2391.
168. Ehrenfest, P. (1927). *Z. Physik* **45**, 455.
169. Encrenaz, T., Bibring, J. P. and Blanc, M. (1991). *The Solar System*, Springer-Verlag.
170. ESA-ESO Working Group on *Fundamental Cosmology* (2006). arXiv: astro-ph/0610906.
171. Evans, N. W. and Tabachnik, S. (1999). *Nature* **399**, 41.
172. Fenyes, I. (1952). *Z. Physik* **132**, 81.
173. Fetter, A. L. (1965). *Phys. Rev.* **A 138**, 429.
174. Feynman, R. P. (1955). *Progress in Low Temperature Physics*, Gorter, C. J. (Ed.), North Holland Pub. Comp., Amsterdam.
175. Feynman, R. P., Leighton, R. and Sands, M. (1964). *The Feynman Lectures on Physics*, Addison-Wesley.
176. Feynman, R. P. and Hibbs, A. R. (1965). *Quantum Mechanics and Path Integrals*, MacGraw-Hill, New York.
177. Feynman, R. P. (1986). *The Reason for Antiparticles: The 1986 Dirac Memorial Lecture*, Slone J E, Cambridge.
178. Finkel, R. W. (1987). *Phys. Rev.* **A 35**, 1486.
179. Finkelstein, D. R. (1999). *Emptiness and relativity*, preprint.

180. Fletcher, N. H. (2004). *American Journal of Physics* **72**, 5.
181. Fleury, V. (2009). *Eur. Phys. J. Appl. Phys.* **45**, 30101.
182. Forriez, M. (2005). *Etude de la Motte de Boves*, Geography and History Master I report, Artois University.
183. Forriez, M. and Martin, P. (2006). *Geopoint Colloquium: Demain la Géographie*, p. 305.
184. Forriez, M. (2007). *Construction d'un espace géographique fractal. Pour une géographie mathématique et recherche d'une théorie de la forme*, Avignon, Université d'Avignon et des Pays du Vaucluse — Master 2 report, p. 202.
185. Frederick, C. (1976). *Phys. Rev.* **D 13**, 3183.
186. Freund, P. G. O. (1986). *Introduction to Supersymmetry*, Cambridge University Press, Cambridge.
187. Frisch, U. (1995). *Turbulence: The Legacy of A.N. Kolmogorov*, Cambridge University Press, Cambridge.
188. Gaeta, Z. D. and Stroud, C. R. (1990). *Phys. Rev.* **A 42**, 6308.
189. Galileo G. (1623). *Il Saggiatore*, a cura di L. Sosio, Milano, 1965.
190. Galileo G. (1630). *Dialogo sopra i massimi sistemi*, Torino, 1975; *Dialogue sur les deux grands systèmes du monde*, Ed. du Seuil.
191. Galopeau, P., Nottale, L., Ceccolini, D., Da Rocha, D., Schumacher, G. and Tran-Minh, N. (2004). *Scientific Highlights 2004, Proceedings of the Journées de la SF2A*, Paris 14–18 June 2004, F. Combes, D. Barret, T. Contini, F. Meynadier and L. Pagani (Eds.), EDP Sciences, p. 75.
192. Garbaczewski, P. (2008). *Phys. Rev.* **E 78**, 031101.
193. Garcia-Segura, A. *et al.* (1999). *Astrophys. J.* **517**, 767.
194. Garnavich, P. M., Jha, S., Challis, P. *et al.* (1998). *Astrophys. J.* **50**, 74.
195. Gautreau, R. (1983). *Phys. Rev.* **D 27**, 764.
196. Gautreau, R. (1984). *Phys. Rev.* **D 29**, 186.
197. Gaveau, B., Jacobson, T., Kac, M. and Schulman, L. S. (1984). *Phys. Rev. Lett.* **53**, 419.
198. Georgi, H. and Glashow, S. L. (1974). *Phys Rev. Lett.* **32**, 438.
199. Georgi, H., Quinn, H. R. and Weinberg, S. (1974). *Phys Rev. Lett.* **33**, 451.
200. Georgi, H. (1999). *Lie Algebras in Particle Physics*, Perseus books, Reading, Massachusetts.
201. Glashow, S. L. (1961). *Nucl. Phys.* **22**, 579.
202. Gould, S. J. and Eldredge, N. (1977). *Paleobiology* **3**(2), 115.
203. Gould, S. J. (1997). *L'éventail du vivant*, Le Seuil, Paris, p. 304.
204. Gozdziewski, K., Konacki, M. and Wolszczan, A. (2005). *Astrophys. J.* **619**, 1084.
205. Grabert, H., Hänggi, P. and Talkner, P. (1979). *Phys. Rev.* **A 19**, 2440.
206. Graner, F. and Dubrulle, B. (1994). *Astron. Astrophys.* **282**, 262.
207. Grassberger, P. and Procaccia, I. (1983). *Phys. Rev. Lett.* **50**, 346.
208. Green, M. B., Schwarz, J. H. and Witten, E. (1987). *Superstring Theory* (2 Vol.), Cambridge University Press, Cambridge.
209. Greenberger, D. M. (1983). *Found. Phys.* **13**, 903.
210. Greiner, W. and Müller, B. (2000). *Gauge Theory of Weak Interactions*, Springer, Berlin.

211. Gross, E. P. (1961). *Nuevo Cimento* **20**, 454.

212. Grou, P. (1987). *L'aventure économique*, L'Harmattan, Paris.

213. Grou, P., Nottale, L. and Chaline, J. (2004). *Zona Arqueologica, Miscelanea en homenaje a Emiliano Aguirre*, IV Arqueologia, 230, Museo Arquelogico Regional, Madrid.

214. Grundlach, J. H. and Merkowitz, S. M. (2000). *Phys. Rev. Lett.* **85**, 2869.

215. Gunn, J. E. and Tinsley, B. M. (1975). *Nature* **257**, 454.

216. Guthrie, B. N. G. and Napier, W. M. (1991). *M.N.R.A.S.* **253**, 533.

217. Guthrie, B. N. G. and Napier, W. M. (1996). *Astron. Astrophys.* **310**, 353.

218. Guyon, E., Hulin, J. P., Petit, L. (1991). *Hydrodynamique physique*, InterEditions/Editions du CNRS.

219. Hall, M. J. W. (2004). *J. Phys. A: Math. Gen.* **37**, 9549.

220. Halzen, F. and Martin, A. D. (1984). *Quarks and Leptons: An Introductory Course in Modern Particle Physics*, John Wiley and Sons, New York.

221. Hamilton, W. R. (1848). *Trans. Roy. Irish Acad.* **21**, 199.

222. Hammad, F. (2007). *Phys. Lett.* **A 370**, 374.

223. Hartenberger, J. L. (1998). *C.R. Acad. Sci. Paris* **316**(II), 439.

224. Hawking, S. W. (1984). *Phys. Lett.* **B 134**, 403.

225. Heck, A. and Perdang, J. M. (Eds.) (1991). *Applying Fractals in Astronomy*, Springer-Verlag, Berlin.

226. Hermann, R. (1997). *J. Phys. A: Math. Gen.* **30**, 3967.

227. Hermann, R., Schumacher, G. and Guyard, R. (1998). *Astron. Astrophys.* **335**, 281.

228. Herrmann, F. (1997). *A scaling medium representation, a discussion on well-logs, fractals and waves*, Ph.D. Thesis, Delft University of Technology.

229. Hestenes, D. (1966). *Space-Time Algebra*, Gordon and Breach, New York.

230. Heydari-Malayeri, M. (1999). http://www.obspm.fr/actual/nouvelle/jul99.fr.shtml.

231. Hickson, P. (1994). *Atlas of Compact Groups of Galaxies*, Gordon and Breach.

232. Hickson, P. (2003). *Compact Groups of Galaxies*, http://www.astro.ubc.ca/people/hickson/hcg/index.html.

233. Hinshaw, G. *et al.* (2009). *Astrophys. J. Suppl.* **180**, 225 (arXiv:0803.0732 [astro-ph]).

234. Hodapp, K. W., MacQueen, R. M. and Hall, D. (1992). *Nature* **335**, 707.

235. Hogg, D. W. *et al.* (2004). *arXiv: astro-ph/0411197*.

236. Horava, P. (2009). *Phys. Rev. Lett.* **102**, 161301.

237. Issartel, J. P. (2004). *Atmos. Chem. Phys. Discuss.* **4**, 2615.

238. Itzykson, C. and Zuber, J. B. (1980). *Quantum Field Theory*, McGraw-Hill, New York.

239. Jacob, C. (1998). *Probation Report*, Paris Observatory.

240. Jehle, H. (1938). *Zeitschrift fur Astrophysics* **15**, 182.

241. Johansen, A. and Sornette, D. (2001). *Physica* **A 294**, 465.

242. Jumarie, G. (2001). *Int. J. Mod. Phys.* **A 16**, 5061.

243. Jumarie, G. (2006). *Computer and Mathematics* **51**, 1367.

244. Jumarie, G. (2006). *Chaos, Solitons & Fractals* **28**, 1285.

245. Jumarie, G. (2007). *Phys. Lett.* **A 363**, 5.
246. Kalupahana, D. J. (trad.) (1986). *Nagarjuna: The Philosophy of the Middle Way*, State University Press of New York, New York.
247. Karlsson, K. G. *Astron. Astrophys.* **58**, 237.
248. Karoji, H. and Nottale, L. (1976). *Nature* **259**, 31.
249. Kawatra, M. P. and Pathria, R. K. (1966). *Phys. Rev.* **151**, 132.
250. Kidwell, S. M. and Flessa, K. W. (1996). *Annu. Rev. Earth Planet Sci.* **24**, 433.
251. Koch, H. von (1904). *Archiv für Mathematik, Astronomi och Physik* **1**, 681.
252. Kolwankar, K. M. and Gangal, A. D. (1998). *Phys. Rev. Lett.* **80**, 214.
253. Komatsu, E. *et al.* (2009). *Astrophys. J. Suppl.* **180**, 330.
254. Konacki, M. and Wolszczan, A. (2003). *Ap. J. Lett.* **591**, L147.
255. Koroliouk, V. (Ed.) (1983). *Aide mémoire de probabilité et de statistique mathématique*, Mir, Moscow.
256. Koutchmy, S. (1972). *Astron. Astrophys.* **16**, 103.
257. Kraemmer, A. B., Nielson, H. B. and Tze, H. C. (1974). *Nucl. Phys.* **B 81**, 145.
258. Kröger, H. (1997). *Phys. Rev.* **A 55**, 951.
259. Kröger, H. (2000). *Phys. Rep.* **323**, 81.
260. Kuipers, J. B. (1999). *Quaternions and Rotation Sequences*, Princeton University Press.
261. Kundrát, P. and Lokajícek, M. (2003). *Phys. Rev.* **A 67**, 012104.
262. Lambert, D. (1983). *Collins Guide to Dinosaures*, Collins, Hong Kong, p. 256.
263. Lamy, P. (1974). *Astron. Astrophys.* **33**, 191.
264. Lamy, P., Kuhn, J. R., Koutchmy, S. and Smartt, R. N. (1992). *Science* **257**, 1377.
265. Lanczos, C. (1929). *Z. Physik* **57**, 447.
266. Landau, L. and Lifchitz, E. (1967). *Quantum Mechanics*, Mir, Moscow.
267. Landau, L. and Lifchitz, E. (1970). *Mechanics*, Mir, Moscow.
268. Landau, L. and Lifchitz, E. (1970). *Field Theory*, Mir, Moscow.
269. Landau, L. and Lifchitz, E. (1972). *Relativistic Quantum Theory*, Mir, Moscow.
270. Landau, L. and Lifchitz, E. (1972). *Fluid Mechanics*, Mir, Moscow.
271. Landy, S. D. *et al.* (1996). *Astrophys. J. Lett.* **456**, L1.
272. Lang, K. R. (1980). *Astrophysical Formulae*, Springer-Verlag.
273. Lang, K. R. (1992). *Astrophysical Data. Planets and Stars*, Springer-Verlag.
274. Laperashvili, L. V. and Ryzhikh, D. A. (2001). arXiv: hep-th/0110127, Institute for Theoretical and Experimental Physics, Moscow.
275. Lapidus, M. and van Frankenhuysen, M. (2000). *Fractal Geometry and Number Theory: Complex Dimensions of Fractal Strings and Zeros of Zeta Functions*, Birkhäuser.
276. Lapidus, M. L. (2008). *In search of the Riemann Zeros, Strings, Fractal Membranes and Noncommutative Spacetimes*, American Mathematical Society, Providence.
277. Laskar, J. and Robutel, P. (1993). *Nature* **361**, 608.

278. Laskar, J., Joutel, F. and Robutel, P. (1993). *Nature* **361**, 615.
279. Laskar, J. (2000). *Phys. Rev. Lett.* **84**, 3240.
280. Lastovicka, T. (2002). arXiv: hep-ph/0203260.
281. Lecointre, G. and Le Guyader, H. (2001). *Classification phylogénétique du vivant*, Belin, Paris.
282. Lefranc, S. (2000). *Probation Report*, Ecole Nationale Supérieure des Techniques Avancées, Paris.
283. Lefèvre, E. T. (2000). Unpublished preprint.
284. Lehner, T., Nottale, L., di Menza, L., da Rocha, D. and Ceccolini, D. (2004). *Scientific Highlights 2004, Proceedings of the Journées de la SF2A*, Paris 14–18 June 2004, F. Combes, D. Barret, T. Contini, F. Meynadier and L. Pagani (Eds.), EDP Sciences, p. 687.
285. Lehner, T., Nottale, L. and di Menza, L. (2011). *Structure formation by the Schrödinger-Poisson equation*, submitted.
286. Lehner, T. and Nottale, L. (2011). In preparation.
287. Lemaitre, G. (1934). *Proc. Nat. Acad. Sci.* **20**, 12.
288. Lena, P., Viala, Y. *et al.* (1974). *Astron. Astrophys.* **37**, 81.
289. le Noble, F., Fleury, V., Pries, A., Corvol, P., Eichmann, A. and Reneman, R. S. (2005). *Cardiovascular Research* **65**, 619.
290. Levy-Leblond, J. M. (1976). *Am. J. Phys.* **44**, 271.
291. Lichnerowicz, A. (1955). *Théories relativistes de la gravitation et de l'électromagnétisme*, Masson, Paris.
292. Lichtenberg, A. J. and Lieberman, M. A. (1983). *Regular and Stochastic motion*, Springer-Verlag.
293. Liebowitz, B. (1944). *Phys. Rev.* **66**, 343.
294. Lifchitz, E. and Pitayevski, L. (1980). *Statistical Physics*, Part 2, Pergamon Press, Oxford.
295. Lorentz, H. A. (1904). *Proceedings of the Academy of Sciences of Amsterdam* 6, translated in *The Principle of Relativity*, Dover, 1923, 1952, p. 11.
296. Losa, G., Merlini, D., Nonnenmacher, T. and Weibel, E. (Eds.). *Fractals in Biology and Medicine*, Vol. III, Proceedings of Fractal 2000 Third International Symposium, Birkhäuser Verlag.
297. Luminet, J. P., Weeks, J., Riazuelo, A., Lehoucq, R. and Uzan, J. P. (2003). *Nature* **425**, 593.
298. Mach, E. (1893). *Science of Mechanics*, Open Court Publishing Company.
299. MacQueen, R. M. (1968). *Astrophys. J.* **154**, 1059.
300. Madelung, E. (1927). *Zeit. F. Phys.* **40**, 322.
301. Maeda, H., Gurian, J. H. and Gallagher, T. F. (2009). *Phys. Rev. Lett.* **102**, 103001 (1–4).
302. Magueijo, J. and Smolin, L. (2002). *Phys. Rev. Lett.* **88**, 190403.
303. Malhotra, R., Black, D., Eck, A. and Jackson, A. (1992). *Nature* **356**, 583.
304. Malhotra, R. (1993). *Ap. J.* **407**, 266.
305. Mandelbrot, B. and Van Ness, J. W. (1968). *SIAM Review* **10**, 422.
306. Mandelbrot, B. (1975). *Les Objets Fractals*, Flammarion, Paris.
307. Mandelbrot, B. (1982). *The Fractal Geometry of Nature*, Freeman, San Francisco.

308. Mandelbrot, B. (1999). *Multifractals and 1/f Noise*, Springer.
309. Mandelbrot, B. (2007). *Chronicle of books on fractal geometry*, http://www.math.yale.edu/mandelbrot/web_docs/chronicleFractals.doc.
310. Markowich, P. A., Rein, G. and Wolansky, G. (2001). arXiv: math-ph/0101020.
311. Martin, P. and Forricz, M. (2006). *Geopoint Colloquium: Demain la Géographie.*
312. Martin, P. and Nottale, L. (2010). *Karstologia*, submitted for publication.
313. Martinez-Gonzales, E., Sanz, J. L. and Silk, J. (1990). *Ap. J. Lett.* **355**, L5.
314. Mateo, M. *Ann. Rev. A. & A.* **36**, 435.
315. Mayor, M. and Queloz, D. (1995). *Nature* **378**, 355.
316. McClendon, M. and Rabitz, H. (1988). *Phys. Rev.* **A 37**, 3479.
317. McKeon, D. G. C. and Ord, G. N. (1992). *Phys. Rev. Lett.* **69**, 3.
318. McNamara, P. A. and Wu, S. L. (2002). *Rep. Prog. Phys.* **65**, 465.
319. Mellema, G. and Frank, A. (1994). *Astrophys. J.* **430**, 800.
320. Messiah, A. (1959). *Mécanique Quantique*, Dunod, Paris.
321. Milgrom, M. (1983). *Astrophys. J.* **270**, 365.
322. Minkowski, H. (1908). *The Principle of Relativity*, Dover, 1923, 1952, p. 75.
323. Minor Planet Center. (2007). *List of Centaurs and Scattered-Disk Objects*, http://www.cfa.harvard.edu/iau/lists/Centaurs.html.
324. Misner, C. W., Thorne, K. S. and Wheeler, J. A. (1973). *Gravitation*, Freeman, San Francisco.
325. Mizutami, K., Maihara, T., Hiromoto, N. and Takami, H. (1984). *Nature* **312**, 134.
326. Mohr, J. J., Geller, M. J., Fabricant, D. G., Wegner, G., Thorstensen, J. and Richstone, D. O. (1996). *Astrophys. J.* **470**, 724.
327. Mohr, P. J. and Taylor, B. N. (2005). *Rev. Mod. Phys.* **77**, 1 (CODATA 2002).
328. Moulin, T. (2002). *AIP Conference Proceedings* **627**, 253.
329. NASA (2009). http://lambda.gsfc.nasa.gov
330. Nauenberg, M. (1989). *Phys. Rev.* **A 40**, 1133.
331. Nelson, E. (1966). *Phys. Rev.* **150**, 1079.
332. Nelson, E. (1985). *Quantum Fluctuations*, Princeton University Press.
333. Nilson, P. (1973). *Uppsala General Catalogue of Galaxies*, Acta Universitatis Upsaliensis.
334. Noble, D. (2002). *BioEssays*, **24**, 1155.
335. Noble, D. (2006). *The music of Life: Biology Beyond the Genome*, Oxford University Press, Oxford.
336. Noble, D. (2008). *Experimental Physiology*, **93**, 16.
337. Nore, C., Brachet, M. E., Cerda, E. and Tirapegui, E. (1994). *Phys. Rev. Lett.* **72**, 2593.
338. Nore, C. (1995). *Etude de comportements hydrodynamiques par simulation numérique de l'équation de Schrödinger non-linéaire*, Ph.D. Thesis, Paris VII University.
339. Nottale, L. (1980). *Perturbation of the Hubble relation by clusters of galaxies*, Ph.D. Thesis (Doctorat d'Etat), Paris VI University.

340. Nottale, L. (1982). *Astron. Astrophys.* **110**, 9.
341. Nottale, L. (1982). *Astron. Astrophys.* **114**, 261.
342. Nottale, L. (1982). *L'Univers*, Ecole de Goutelas, E. Schatzman (Ed.), *Clusters of galaxies and cosmological tests.*
343. Nottale, L. (1983). *Astron. Astrophys.* **118**, 85.
344. Nottale, L. and Schneider, J. (1984). *J. Math. Phys.* **25**, 1296.
345. Nottale, L. and Hammer, F. (1984). *Astron. Astrophys.* **141**, 144.
346. Nottale, L. (1984). *M.N.R.A.S.* **175**, 429.
347. Nottale, L. (1987). *Astrophysics Lecture*, Ecole Centrale de Paris.
348. Nottale, L. (1988). *Annales de Physique* **13**, 223.
349. Nottale, L. (1989). *Int. J. Mod. Phys. A* **4**, 5047.
350. Nottale, L. (1991). *Progress in Electromagnetics Research Symposium*, Proceedings I.E.E.E., Cambridge, USA, p. 754.
351. Nottale, L. (1991). *Applying Fractals in Astronomy*, A. Heck and J. M. Perdang (Eds.), Springer-Verlag, Berlin, p. 181.
352. Nottale, L. (1992). *Int. J. Mod. Phys. A* **7**, 4899.
353. Nottale, L. (1993). *Fractal Space-Time and Microphysics: Towards a Theory of Scale Relativity*, World Scientific, Singapore.
354. Nottale, L. (1993). *Cellular Automata: Prospects in Astrophysical Applications*, Université de Liège, A. Lejeune and J. Perdang (Eds.), World Scientific, p. 268.
355. Nottale, L. (1994). *Chaos, Solitons and Fractals* **4**, 361. Reprinted in *Quantum Mechanics, Diffusion and Chaotic Fractals*, M.S. El Naschie, O.E. Rössler and I. Prigogine (Eds.), Pergamon (1995), p. 51.
356. Nottale, L. (1994). *Relativity in General*, (Spanish Relativity Meeting 1993), J. Diaz Alonso and M. Lorente Paramo (Eds.), Editions Frontières, Paris, p. 121.
357. Nottale, L. (1994). *Chaos and Diffusion in Hamiltonian Systems*, Proceedings of the fourth workshop in Astronomy and Astrophysics of Chamonix (France), 7–12 February 1994, D. Benest and C. Froeschlé (Eds.), Editions Frontières, p. 173.
358. Nottale, L. (1995). *Chaos, Solitons & Fractals* **6**, 399.
359. Nottale, L. (1995). *Clustering in the Universe*, 30th Rencontres de Moriond (Astrophysics meeting), S. Maurogordato, C. Balkowski, C. Tao and J. Tran Thanh Van (Eds.), Frontières, p. 523.
360. Nottale, L. (1996). *Chaos, Solitons & Fractals* **7**, 877.
361. Nottale, L. (1996). *Astron. Astrophys. Lett.* **315**, L9.
362. Nottale, L. (1997). *Astron. Astrophys.* **327**, 867.
363. Nottale, L. (1997). *Scale invariance and beyond*, Proceedings of Les Houches school, B. Dubrulle, F. Graner and D. Sornette (Eds.), EDP Sciences, Les Ullis/Springer-Verlag, Berlin, New York, p. 249.
364. Nottale, L., Schumacher, G. and Gay, J. (1997). *Astron. Astrophys.* **322**, 1018.
365. Nottale, L. (1998). *Chaos, Solitons & Fractals* **9**, 1035.
366. Nottale, L. (1998). *Chaos, Solitons & Fractals* **9**, 1043.
367. Nottale, L. (1998). *Chaos, Solitons & Fractals* **9**, 1051.

368. Nottale, L. (1998). *La relativité dans tous ses états*, Hachette, Paris.

369. Nottale, L. and Schumacher, G. (1998). *Fractals and beyond: complexities in the sciences*, M. M. Novak (Ed.), World Scientific, Singapore, p. 149.

370. Nottale, L. (1998). *Ciel et Terre*, Bulletin de la Société royale belge d'Astronomie, **114** (2), 63.

371. Nottale, L. (1999). *Chaos, Solitons & Fractals* **10**, 459.

372. Nottale, L. (2000). *Science of the Interface*, Proceedings of International Symposium in honor of Otto Rössler, ZKM Karlsruhe, 18–21 May 2000, H. Diebner, T. Druckney and P. Weibel (Eds.), Genista Verlag, Tübingen, p. 38.

373. Nottale, L., Chaline, J. and Grou, P. (2000). *Les arbres de l'évolution: Univers, Vie, Sociétés*, Hachette, Paris, 379 pp.

374. Nottale, L., Schumacher, G. and Lefèvre, E. T. (2000). *Astron. Astrophys.* **361**, 379.

375. Nottale, L. (2000). Proceedings of 4th Forum Halieumétrique, Juin 1999, *Les espaces de l'halieutique*, D. Gascuel, P. Chavance, N. Bez and A. Biseau (Eds.), IRD Editions, p. 41.

376. Nottale, L. (2001). *Frontiers of Fundamental Physics*, Proceedings of Birla Science Center Fourth International Symposium, 11–13 December 2000, B. G. Sidharth and M. V. Altaisky (Eds.), Kluwer Academic, Kluwer Academic/Plenum Publishers, New York, p. 65.

377. Nottale, L. (2001). *Revue de Synthèse*, **T. 122**, 4e S., No 1, January-March 2001, p. 93.

378. Nottale, L. (2001). *Chaos, Solitons & Fractals* **12**, 1577.

379. Nottale, L. and Tran Minh, N. (2002). *Scientific News*, Paris Observatory, http://www.obspm.fr/actual/nouvelle/nottale/nouv.fr.shtml.

380. Nottale, L., Chaline, J. and Grou, P. (2002). *Fractals in Biology and Medicine, Vol. III*, Proceedings of Fractal 2000 Third International Symposium, G. Losa, D. Merlini, T. Nonnenmacher and E. Weibel (Eds.), Birkhäuser Verlag, Basel, p. 247.

381. Nottale, L. (2002). *Lois d'Echelle, Fractales et Ondelettes*, Traité IC2, P. Abry, P. Gonçalvès et J. Levy Vehel (Eds.), (Hermès Lavoisier), Vol. **2**, Chap. 7, p. 233.

382. Nottale, L. (2002). *Penser les limites. Ecrits en l'honneur d'André Green*, Delachiaux et Niestlé, p. 157.

383. Nottale, L. (2003). *Chaos, Solitons & Fractals* **16**, 539.

384. Nottale, L. (2003). *Electromagnetic Phenomena* **T. 3**, N.1 (9), 24.

385. Nottale, L., Célérier, M. N. and Lehner, T. (2003). *The Pioneer anomalous acceleration: a measurement of the cosmological constant at the scale of the solar system*, arXiv: hep-th/0307093.

386. Nottale, L. (2003). *The Pioneer anomalous acceleration: a measurement of the cosmological constant at the scale of the solar system*, arXiv: gr-qc/0307042.

387. Nottale, L. (2004). *Proceedings of Symposia in Pure Mathematics* **72**, Part 1, p. 57 (American Mathematical Society, Providence).

388. Nottale, L. (2004). *Computing Anticipatory Systems. CASYS'03 - Sixth International Conference* (Liège, Belgique, 11–16 August 2003), Daniel M. Dubois Editor, *American Institute of Physics Conference Proceedings* **718**, 68.

389. Nottale, L. (2004). *Scientific Highlights 2004, Proceedings of the Journées de la SF2A*, Paris 14–18 June 2004, F. Combes, D. Barret, T. Contini, F. Meynadier and L. Pagani (Eds.), EDP Sciences, p. 699.

390. Nottale, L. (2004). *Scientific Highlights 2004, Proceedings of the Journées de la SF2A*, Paris 14–18 June 2004, F. Combes, D. Barret, T. Contini, F. Meynadier and L. Pagani (Eds.), EDP Sciences, p. 373.

391. Nottale, L., Galopeau, P., Ceccolini, D., Da Rocha, D., Schumacher, G. and Tran-Minh, N. (2004). *Extrasolar Planets: Today and Tomorrow*, XIXth IAP Colloquium (Paris, France), 30 June-4 July 2003, J.P. Beaulieu, A. Lecavelier des Etangs and C. Terquem (Eds.), *Astronomical Society of the Pacific Conference Series* **321**, p. 355.

392. Nottale, L. (2005). *Progress in Physics* **1**, 12.

393. Nottale, L. (2005). *Chaos, Solitons & Fractals*, **25**, 797. Reprinted in *Space Time Physics and Fractality*, P. Weibel, G. Ord and O. Rössler, (Eds.), Springer-Verlag, Wien, p. 41.

394. Nottale, L., Célérier, M. N. and Lehner, T. (2006). *J. Math. Phys.* **47**, 032303 (arXiv: hep-th/0605280).

395. Nottale, L. and Lehner, T. (2006). *Numerical simulation of a macroscopic quantum-like experiment: oscillating wave packet*, arXiv: quant-ph/0610201.

396. Nottale, L. (2006). Proceedings of the Sixth International Symposium *Frontiers of Fundamental Physics*, Udine, Italy, 26–29 September 2004, Sidharth, B. G., Honsell, F. and de Angelis, A. (Eds.), Springer, Dordrecht, p. 107.

397. Nottale, L. (2006). *Workshop Pulsars, Theory and Observations*, 16 and 17 January 2006, Institut d'Astrophysique de Paris, http://lpce.cnrs-orleans.fr/~pulsar/PSRworkshop/Talks/L.Nottale.ppt.

398. Nottale, L. and Timar, P. (2006). *Psychanalyse et Psychose* **6**, 195.

399. Nottale, L. and Célérier, M. N. (2007). *J. Phys. A: Math. Theor.* **40**, 14471 (arXiv:0711.2418 [quant-ph]).

400. Nottale, L. (2007). Special Issue (July 2007) on "Physics of Emergence and Organization", *Electronic Journal of Theoretical Physics* **4** No. 16 (II), 187.

401. Nottale, L. (2007). *Les grands défis technologiques et scientifiques du XXIè siècle*, P. Bourgeois et P. Grou (Eds.), Ellipse, Chap. 9, p. 121.

402. Nottale, L. and Auffray, C. (2008). *Progr. Biophys. Mol. Bio.* **97**, 115.

403. Nottale, L. (2008). *Proceedings of 7th International Colloquium on Clifford Algebra and their applications*, 19–29 May 2005, Toulouse, *Advances in Applied Clifford Algebra*, **18**, 917.

404. Nottale, L. and Timar, P. (2008). *Simultaneity: Temporal Structures and Observer Perspectives*, Susie Vrobel, Otto Rössler, Terry Marks-Tarlow, (Eds.), World Scientific, Singapore), Chap. 14, p. 229.

405. Nottale, L. (2009). Proceedings of International Colloquium (Lecce, Italy, 2005), *Albert Einstein and Hermann Weyl: 1955–2005, Open Epistemologic Questions*, Eds. Charles Alunni, Mario Castellana, Demetrio Ria and Arcangelo Rossi, Collection *Pense des sciences* (Europa edizioni, Maglie and Editions Rue d'Ulm, Paris), pp. 141–173.

406. Nottale, L. (2009). *J. Phys. A: Math. Theor.* **42**, 275306 (arXiv: 0812.0941).

407. Nottale, L. (2009). *Quantum-like gravity waves and vortices in a classical fluid*, arXiv: 0901.1270.

408. Nottale, L., Chaline, J. and Grou, P. (2009). *Des fleurs pour Schrödinger: la relativité d'échelle et ses applications*, Ellipses, Paris.

409. Nottale, L. (2009). *Encyclopedia of Complexity and Systems Science* (R.A. Meyers Ed.), Springer, Berlin, pp. 3858–3878.

410. Nottale, L. (2009). *Scaling, Fractals and Wavelets* (Digital Signal and Image processing Series), Edited by Patrice Abry, Paulo Gonalves and Jacques Lvy Vhel (ISTE-Wiley, 2009), Chapter 14, p. 464–498.

411. Nottale, L. (2010). Proceedings of First International Conference on the Evolution and Development of the Universe, 8–9 October 2008, ENS, Paris, France, *Foundations of Science* **15**, 101 (arXiv: 0812.3857 [quant-ph]).

412. Nottale, L., Martin, P. and Forriez, M. (2010). *Revue de Géomatique*, submitted for publication.

413. Nottale, L. (2011). In preparation.

414. Nottale, L., Héliodore, F. and Dubois, J. (2011). *Log-periodic evolution of earthquake rate*, in preparation.

415. Nottale, L., Pocard, M., Rouleau, E. and Grou, P. (2011). In preparation.

416. Novak, M. (Ed.) (1998). *Fractals and Beyond: Complexities in the Sciences, Proceedings of the Fractal 98 conference*, World Scientific.

417. Oldershaw, R. L. (2007). *Astrophys. Space Sci.* **311**, 431.

418. Onorato, M., Osborne, A. R., Serio, M. and Bertone, S. (2001). *Phys. Rev. Lett.* **86**, 5831.

419. Onsager, L. (1949). *Nuevo Cimento* **6** (Suppl. 2), 249.

420. Ord, G. N. (1983). *J. Phys. A: Math. Gen.* **16**, 1869.

421. Ord, G. N. (1992). *Int. J. Theor. Phys.* **31**, 1177.

422. Ord, G. N. and McKeon, D. G. C. (1993). *Ann. Phys.* **272**, 244.

423. Ord, G. N. (1996). *Ann. Phys.* **250**, 51.

424. Ord, G. N. and Galtieri, J. A. (2002). *Phys. Rev. Lett.* **1989**, 250403.

425. Pais, A. (1982). *Subtle is the Lord: the Science and Life of Albert Einstein*, Oxford University Press, New York.

426. Paquet, E. and Chin, S. L. (1991). *J. Phys. B: At. Mol. Opt. Phys.* **24** L579.

427. Parisi, G. and Frisch, U. (1985). *A multifractal model of intermittency*, in M. Ghil *et al.* (Eds.), *Turbulence and Predictability in Geophysical Fluid Dynamics and Climate Dynamics*, North-Holland, Amsterdam, p. 84.

428. (Particle Data Group) Ikasa K. I. *et al.* (1992). *Phys. Rev.* **D 45**, S1.

429. (Particle Data Group) Montanet, L. *et al.* (1994). *Phys. Rev.* **D 50**, 1173.

430. (Particle Data Group) Barnett, R. M. *et al.* (1996). *Phys. Rev.* **D 54**, 1.

431. (Particle Data Group) Caso, C. *et al.* (1998). *The European Physical Journal* **C 3**, 1.

432. (Particle Data Group) Groom, D. E. *et al.* (2000). *The European Physical Journal* **C 15**, 1.

433. (Particle Data Group) Hagiwara, K. *et al.* (2002). *Phys. Rev.* **D 66**, 010001.

434. (Particle Data Group) Eidelman, S. *et al.* (2004). *Phys. Lett.* **B 592**, 1.

435. (Particle Data Group) Yao, W.-M. *et al.* (2006). *J. Phys.* **G 33**, 1.

436. (Particle Data Group) Amsler, C. *et al.* (2008). *Phys. Lett.* **B 667**, 1.

437. Pecker, J. C. and Schatzman, E. (1959). *Astrophysique générale*, Masson, Paris.

438. Peebles, J. (1980). *The Large-Scale Structure of the Universe*, Princeton University Press, Princeton.

439. Perdang, J. M. and Lejeune, A. (Eds.) (1993). *Cellular Automata: Prospects in Astrophysical Applications*, World Scientific.

440. Perlmutter, S., Aldering, G., Della Valle, M. *et al.* (1998). *Nature* **391**, 51.

441. Persic, M. and Salucci, P. (1995). *Astrophys. J. Suppl. Ser.* **99**, 501.

442. Peterson, A. W. (1967). *Astrophys. J.* **148**, L 37.

443. Peterson, M. A. (1979). *Am. J. Phys.* **47**, 488.

444. Petit, J. M. and Hénon, M. (1986). *Icarus* **66**, 536.

445. Pietz, J., Kerp, J., Kalberla, P., Mebold, U., Burton, W. and Hartmann, D. (1996). *Astron. Astrophys.* **308**, L37.

446. Pissondes, J. C. (1999). *J. Phys. A: Math. Gen.* **32**, 2871.

447. Pissondes, J. C. (1999). *Chaos, Solitons and Fractals* **10**, 513.

448. Pitaevskii, L. P. (1961). *Soviet Phys. — JETP* **13**, 451.

449. Pocheau, A. (1994). *Phys. Rev. E* **49**, 1109.

450. Poincaré, H. (1905). *C. R. Acad. Sci. Paris* **140**, 1504.

451. Poincaré, H. (1906). *Rendiconti del Circolo Matematico di Palermo* **XXI**, 17.

452. Polchinski, J. (1998). *String Theories*, Cambridge University Press, Cambridge.

453. Postnikov, M. *Leçons de Géométrie. Groupes et algèbres de Lie*, Leçon 14, Mir, Moscow.

454. Press, W. H., Flannery, B. P., Teukolsky, S. A. and Vetterling, W. T. (1984). *Numerical Recipes*, Cambridge University Press, Cambridge.

455. Preston, H. G. and Potter, F. (2003). arXiv: gr-qc/0303112.

456. Queiros-Condé, D. (2000). *C. R. Acad. Sci. Paris* **330**, 445.

457. Quinn, T. J., Speake, C. C., Richman, S. J., Davis, R. S. and Picard, A. (2001). *Phys. Rev. Lett.* **87**, 111101.

458. Randall, L. and Sundrum, R. (1999). *Phys. Rev. Lett.* **83**, 4690.

459. Rasio, F. A., Nicholson, P. D., Shapiro, S. L. and Teukolsky, S. A. (1992). *Nature* **355**, 325.

460. Rasio, F. A., Nicholson, P. D., Shapiro, S. L. and Teukolsky, S. A. (1993). *Planets around pulsars, ASP Conference Series* **36**, 107.

461. Rastall, P. (1964). *Rev. Mod. Phys.* **36**, 820.

462. Reinisch, G. (1998). *Astron. Astrophys.* **337**, 299.

463. Riess, A. G., Filippenko, A. V., Challis, P. *et al.* (1998). *Astron. J.* **116**, 1009.

464. van Ritbergen, T., Vermaseren, J. A. M. and Larin, S. A. (1997). *Phys. Lett.* **B400**, 379. aXiv: hep-ph/9701390.

465. Rosales, J. L. and Sanchez-Gomez, J. L. (1999). *The Pioneer effect as a manifestation of the cosmic expansion in the solar system*, arXiv: gr-qc/9810085.

466. Rovelli, C. and Smolin, L. (1988). *Phys. Rev. Lett.* **61**, 1155.

467. Rovelli, C. and Smolin, L. (1995). *Phys. Rev.* **D 52**, 5743.

468. Rovelli, C. (2004). *Quantum Gravity*, Cambridge Universty Press, Cambridge.

469. Saar, E. *et al.* (2002). *Astron. Astrophys.* **393**, 1.

470. Saar, S. H. and Brandenburg, A. (1999). *Astrophys. J.* **524**, 295.

471. Salam, A. (1968). in *Elementary Particle Physics*, Ed. N. Svartholm, Almquist, and Wiksells, Stockholm.

472. Schatzman, E. and Praderie, F. (1993). *The Stars*, Springer-Verlag, Berlin.

473. Scheffers, G. (1893). *C. R. Acad. Sc. Berlin* **116**, 1114.

474. Schiff, L. I. (1968). *Quantum Mechanics*, McGraw-Hill, New York.

475. Schneider, S. E. and Salpeter, E. E. (1993). *Astrophys. J.* **385**, 32.

476. Schneider, J. (2008). *The Extrasolar Planets Encyclopaedia*, http://exoplanet.eu/index.php.

477. Schrödinger, E. (1926). *Naturwiss.* **14**, 664.

478. Schrödinger, E. (1935). *Proceedings of the Cambridge Philosophical Society* **31**, 555.

479. Schrödinger, E. (1943). *What is Life — the Physical Aspect of the Living Cell*, Cambridge University Press, Cambridge.

480. Schumacher, G. and Gay, J. (2001). *Astron. Astrophys.* **368**, 1108.

481. Schumacher, G. and Nottale, L. (2002). Unpublished preprint.

482. Schweber, S. S. (1986). *Reviews of Modern Physics* **58**, 449.

483. Sciama, D. W. (1953). *MNRAS* **113**, 34.

484. Scully, M. O., Englert, B. G. and Walther, H. (1991). *Nature* **351**, 111.

485. Seiberg, N. and Witten, E. (1999). *J. High Energy Phys.* **09**, 032 (hep-th/9908142).

486. Semelin, B., Sanchez, N., de Vega, H. J. and Combes, F. (1999). *Phys. Rev.* **D 59**, 125021.

487. Semelin, B., Sanchez, N. and de Vega, H. J. (2001). *Phys. Rev.* **D 63**, 084005.

488. Sereno, P. C. (1999). *Science* **284**, 2137.

489. Serva, M. (1988). *Ann. Inst. Henri Poincaré — Physique théorique* **49**, 415.

490. Sher, M. (1989). *Phys. Rep.* **179**, 273.

491. Sidharth, B. G. (2001). *The Chaotic Universe*, Nova Science Publishers, New York.

492. Sidharth, B. G. (2001). *Chaos, Solitons & Fractals* **12**, 1101.

493. Silk, J. (2001). *Europhysics News* **32/6**, 210.

494. Smith, N. P, Nickerson, D. P., Crampin, E. J. and Hunter, P. J. (2004). *Acta Numerica*, p. 371, Cambridge University Press, Cambridge.

495. Sornette, D. and Sammis, C. G. (1995). *J. Phys. I France* **5**, 607.
496. Sornette, D., Johansen, A. and Bouchaud, J. P. (1996). *J. Phys. I France* **6**, 167.
497. Sornette, D. (1998). *Phys. Rep.* **297**, 239.
498. Sornette, D. (2003). *Why Stock Market Crash*, Princeton University Press, Princeton.
499. Soucail, G., Fort, B., Mellier, Y. and Picat, J. P. (1987). *Astron. Astrophys.* **172**, L14.
500. Spergel, D. N. *et al.* (2003). *Astrophys. J. Suppl.* **148**, 175.
501. Spergel, D. N. *et al.* (2006). *Astrophys. J. Suppl.* **170**, 377. (arXiv: astro-ph/0603449).
502. Stancu, F. (1996). *Group Theory in Subnuclear Physics*, Calendon Press, Oxford.
503. Symanzik, K. (1970). *Comm. Math. Phys.* **18**, 227.
504. Synge, J. L. (1972). *Quaternions, Lorentz transformations, and the Conway-Dirac-Eddington matrices*, Comm. Dublin Inst. Advanced Studies **A 21**.
505. Szapudi, I. and Kaiser, N. (2003). *Ap.J. Lett.* **583**, L1.
506. Takabayasi, T. (1952). *Progr. Theor. Phys.* **8**, 143.
507. Takabayasi, T. (1953). *Progr. Theor. Phys.* **9**, 187.
508. Tegmark, M. *et al.* (2006). *Phys. Rev.* **D 74**, 123507 (arXiv: astro-ph/0308632).
509. Tifft, W. G. (1976). *Astrophys. J.* **206**, 38.
510. Tifft, W. G. (1977). *Astrophys. J.* **211**, 31.
511. Tifft, W. G. and Cocke, W. J., (1984). *Astrophys. J.* **287**, 492.
512. Timar, P. (2002). http://www.spp.asso.fr/main/PsychanalyseCulture/SciencesDeLaComplexite/Items/3.htm.
513. Titschmarch, E. C. (1958). *Proc. Roy. Soc.* (London) **A 245**, 147.
514. Toledo, C. (2006). *Complex Systems*, submitted.
515. Tricot, C. (1994). *Courbes et dimensions fractales*, Springer, Paris.
516. Tricottet, M. (1997). *DEA Probation Report*, ParisVII University.
517. Tricottet, M. and Nottale, L. (2003). Unpublished preprint.
518. Vader, J. P. and Sandage, A. (1991). *Astrophys. J. Lett.* **379**, L1.
519. Valtonen, M. J. *et al.* (1993). *Astron. J.* **105**, 886.
520. Van Waerbeke, L. *et al.* (2001). *Astron. Astrophys.* **374**, 757.
521. Vaucouleurs, G. de (1959). *Astrophys. J.* **130**, 718.
522. Vaucouleurs, G. de (1971). *Pub. A.S.P.* **83**, 113.
523. Vermaseren, J. A. M., Larin, S. A. and van Ritbergen, T. (1997). *Phys. Lett.* **B 405**, 327.
524. von Neumann, J. (1955). *Mathematical Foundations of Quantum Mechanics*, Princeton University Press, Princeton.
525. Vrobel, S., Rössler, O. E. and Marks-Tarlow, T. (Eds.) (2008). *Simultaneity. Temporal Structures and Observer Perspectives*, World Scientific, Singapore.
526. Wang, M. S. and Liang, W. K. (1993). *Phys. Rev.* **D 48**, 1875.
527. Weinberg, S. (1967). *Phys. Rev. Lett.* **19**, 1264.

528. Weinberg, S. (1972). *Gravitation and Cosmology*, John Wiley & Sons, New York.

529. Weinberg, S. (1989). *Rev. Mod. Phys.* **61**, 1.

530. Weisskopf, V. (1939). *Phys. Rev.* **56**, 72.

531. Weisskopf, V. (1989). *La révolution des quanta*, Hachette, Paris.

532. Weizel, W. (1953). *Z. Physik* **134**, 264.

533. Welsh, D. J. A. (1970). *Mathematics Applied to Physics*, E. Roubine (Ed.), Springer-Verlag, Berlin, Chap. IX, p. 464.

534. Wertz, E., Nottale, L., Lefranc, S. and da Rocha, D. (2005). *Catalog of Nilson Galaxy Pairs*, in *Gravitational structuring in Scale Relativity*, D. da Rocha Ph.D. thesis, Paris Observatory, http://tel.archives-ouvertes.fr/tel-00010204, p. 242.

535. Weyl, H. (1918). *Sitz. Preus. Akad. d. Wiss.* English translation in "*The Principle of Relativity*", Dover publications, p. 201.

536. Wheeler, J. A. (1968). *Batelle Rencontres*, C. DeWitt and J. A. Wheeler (Eds.), Benjamin, New York.

537. Wilson, K. G. (1969). *Phys. Rev.* **179**, 1499.

538. Wilson, K. G. (1975). *Rev. Mod. Phys.* **47**, 774.

539. Wilson, K. G. (1979). *Sci. Am.* **241**, 140.

540. Wilson, K. G. (1983). *Rev. Mod. Phys.* **55**, 583.

541. Wilson, J. A. and Sereno, P. C. (1998). *Journal of Vertebrate Paleontology* **2** (18), 1.

542. Woerden, H. V. *et al.* (1999). *Nature* **400**, 138.

543. Wolszczan, A. and Frail, D. A. (1992). *Nature* **355**, 145.

544. Wolszczan, A. (1994). *Science* **264**, 538.

545. Wolszczan, A. (2000). Preprint.

546. Yeazell, J. A. and Stroud, C. R. (1988). *Phys. Rev. Lett.* **60**, 1494.

547. York, M. (2000). *AIP Conf. Proc.* **545**, 104.

548. Zak, M. (2005). *Chaos Solitons & Fractals* **26**, 1019.

549. Zak, M. (2006). *Chaos Solitons & Fractals* **28**, 616.

550. Zastawniak, T. (1990). *Europhys. Lett.* **13**, 13.

551. Zeilinger, A. and Svozil, K. (1985). *Phys. Rev. Lett.* **54**, 2553.

552. Zeldovich Ya., B. (1967). *JETP Lett.* **6**, 316.

553. Zeldovich Ya., B. Ruzmaikin, A. A. and Sokoloff, D. D. (1983). *Magnetic Fields in Astrophysics*, Gordon and Breach, New York.

554. Zwicky, F. (1933). *Helv. Phys. Acta* **6**, 110.

555. Zwicky, F. (1957). *Morphological Astronomy*, Springer-Verlag, Berlin.

INDEX